Lecture Notes in Computer Science 1548

Edited by G. Goos, J. Hartmanis and J. van Leeuwen

Springer

Berlin
Heidelberg
New York
Barcelona
Hong Kong
London
Milan
Paris
Singapore
Tokyo

Armando Martín Haeberer (Ed.)

Algebraic Methodology and Software Technology

7th International Conference, AMAST'98
Amazonia, Brazil, January 4-8, 1999
Proceedings

Volume Editor

Armando M. Haeberer
Universidade Católica do Rio de Janeiro
Departamento de Informática, Pontifícia
Laboratório de Métodos Formais
Rua Marquês De São Vicente, 225
22453-900 Rio de Janeiro, RJ, Brazil
E-mail: armando@inf.puc-rio.br

Cataloging-in-Publication data applied for

Die Deutsche Bibliothek - CIP-Einheitsaufnahme

Algebraic methodology and software technology : 7th international conference ; proceedings / AMAST '98, Amazonia, Brazil, January 4 - 8, 1999. Armando Martín Haeberer (ed.). - Berlin ; Heidelberg ; New York ; Barcelona ; Hong Kong ; London ; Milan ; Paris ; Singapore ; Tokyo : Springer, 1999
(Lecture notes in computer science ; Vol. 1548)
ISBN 3-540-65462-3

CR Subject Classification (1998): F.3-4, D.2, C.3, D.1.6, I.2.3, H.2.1-4, I.1.3

ISSN 0302-9743
ISBN 3-540-65462-3 Springer-Verlag Berlin Heidelberg New York

Typesetting: Camera-ready by author
SPIN 10693122 06/3142 – 5 4 3 2 1 0 Printed on acid-free paper

Preface

AMAST's goal is to advance awareness of algebraic and logical methodology as part of the fundamental basis of software technology. Ten years and seven conferences after the start of the AMAST movement, I believe we are attaining this. The movement has propagated throughout the world, assembling many enthusiastic specialists who have participated not only in the conferences, which are now annual, but also in the innumerable other activities that AMAST promotes and supports.

We are now facing the Seventh International Conference on Algebraic Methodology and Software Technology (AMAST'98). The previous meetings were held in Iowa City, USA (1989 and 1991), in Enschede, The Netherlands (1993), in Montreal, Canada (1995), in Munich, Germany (1996), and in Sydney, Australia (1997). This time it is Brazil's turn, in a very special part of this colorful country – Amazonia.

Thus, "if we have done more it is by standing on the shoulders of giants." The effort started by Teodor Rus, Arthur Fleck, and William A. Kirk at AMAST'89 was consolidated in AMAST'91 by Teodor Rus, Maurice Nivat, Charles Rattray, and Giuseppe Scollo. Then came modular construction of the building, wonderfully carried out by Giuseppe Scollo, Vangalur Alagar, Martin Wirsing, and Michael Johnson, as Program Chairs of the AMAST conferences held between 1993 and 1997. Beside the conferences, a number of AMAST Workshops have been held around the world; four on real-time systems, organized by Teodor Rus, by Didier Begay, by Aurel Cornel, and by Miguel Bertran; one on topology completion in semantics organized by Maurice Nivat, Paul Gastin, and Jan Rutten; and one on algebraic processing of programming languages, organized by Anton Nijholt, Maurice Nivat, and Giuseppe Scollo. The fifth workshop on real-time systems will be held in 1999, organized by Joost-Pieter Katoen at the University of Erlangen, Germany. The mortar for this huge structure was provided by the inspiring tutelage of our General Chair, Maurice Nivat, and the vision of the AMAST Steering Committee constituted by Egidio Astesiano, Robert Berwick, Zohar Manna, Michael Mislove, Anton Nijholt, Maurice Nivat, Jacques Printz, Charles Rattray, Teodor Rus, Giuseppe Scollo, John Staples, Jeannette Wing, and Martin Wirsing.

For AMAST'98, in response to the call for papers, 80 papers and three system demonstrations were submitted. After a strict selection process, 29 papers were chosen for presentation. These, along with the extended abstracts of six invited papers and succinct descriptions of the demonstrations, make up these proceedings.

The number of papers accepted by no means indicates the full scope of papers with the necessary quality for presentation and publication. It reflects only the time restrictions imposed by a four-day conference.

The publication of a special number of Theoretical Computer Science (TCS) with revised versions of the best papers is a tradition of the AMAST conferences. As Martin Wirsing stated in the preface to the AMAST'96 proceedings, "An eventual goal is to establish algebraic and logical methodology as a practically viable and attractive alternative to the prevailing ad hoc approaches to software engineering." In AMAST'97 this trend was obvious, and we had a handful of very good papers on software engineering applications. Thus, it has been decided that this time two special issues will be published, one in TCS with the best theoretically oriented papers, and another in Software Architecture (the journal heir to Science of Computing

Programming) with the best application oriented papers. The full versions of the invited papers will also appear in these special issues.

The six distinguished invited speakers of AMAST'98 are Zhou Chaochen, Thomas Maibaum, Zohar Manna, Don Pigozzi, Glynn Winskel, and Pamela Zave.

As in the previous years, the first day of AMAST'98 will be an Education Day, this time coordinated by Carlos Camarão de Figueiredo. Formal methods training experiences will be presented and discussed.

On behalf of the Program Committee, my thanks to all who submitted papers and system demonstrations, and to the invited speakers for their contributions. We also extend very special thanks to the referees for sharing with us the difficult task of reviewing the set of very good papers received, and giving us assistance in choosing those that should be accepted.

Our gratitude goes to Springer-Verlag for their invaluable cooperation, as always, regarding the edition and publication of these proceedings.

A very special acknowledgement must be made to Edjard de Souza Motta and the *Departamento de Ciência da Computação* of *Universidade do Amazonas*, whose enthusiastic collaboration greatly facilitated the innumerable local organization tasks.

The support of CNPq – *Conselho Nacional de Desenvolvimento Científico e Tecnológico*, of FINEP – *Financiadora de Estudos e Projetos*, and of UNU/IIST – International Institute for Software Technology of the United Nations University, is gratefully acknowledged.

I give personal thanks to the Program Committee for their intense labor during the organization and reviewing process, to José Meseguer for allowing us the use of SRI International facilities for the Program Committee Meeting, and to José Fiadeiro for splendidly organizing the meeting. I would also like to thank Marcelo Frias for his work during the first stretch of the organization job.

I have a very special debt of gratitude, to add to an endless list that I shall never be able to pay off, to my collaborators at the Laboratory of Formal Methods: first, to our Márcia Ferreira, formally Chair of AMAST'98 Organizing Committee, but actually organize-and-do-everything-that's-needed; then to Christiano Braga, Daniela Cardoso, André Carregal, Cassio Gondim, Fernanda Mesquita, Cláudio Terra Prates, and all the others who pitched in when necessary.

November 1998 Armando Haeberer

Program Chair: Armando Haeberer

Program Committee
Vangalur Alagar (Canada)
Egidio Astesiano (Italy)
Gabriel Baum (Argentina)
Chris Brink (South Africa)
Walter Carnielli (Brazil)
Rocco De Nicola (Italy)
José Fiadeiro (Portugal)
Marcelo Frias (Brazil)
Kokichi Futatsugi (Japan)
Dov Gabbay (United Kingdom)
E. Hermann Haeusler (Brazil)
Paola Inverardi (Italy)
Michael Johnson (Australia)
Richard Jullig (United States)
Rafael Lins (Brazil)
Michael Lowry (United States)
Carlos José Pereira de Lucena (Brazil)
Roger Maddux (United States)
Thomas Maibaum (United Kingdom)
Michael Mislove (United States)
Ugo Montanari (Italy)
Peter Mosses (Denmark)
Ístvan Németi (Hungary)
Anton Nijholt (Netherlands)
Daltro Nunes (Brazil)
Fernando Orejas (Spain)
Don Pigozzi (United States)
Ruy de Queiroz (Brazil)
Charles Rattray (United Kingdom)
Teodor Rus (United States)
Giuseppe Scollo (Netherlands)
Michel Sintzoff (Belgium)
Douglas Smith (United States)
Andrzej Tarlecki (Poland)
Paulo Veloso (Brazil)
Martin Wirsing (Germany)
Zhou Chaochen (Macao)

Education Day Chair
Carlos Camarão de Figueiredo

General Chair: Maurice Nivat

Organizing Committee
Chair: Márcia Ferreira

Finances & Local Arrangements:
Daniela Cardoso
Edjard de Souza Motta

Tools and Demos:
Claudio Terra
André Carregal

Proceedings Editing Assistance
Cássio Gondim

Referees

In addition to the members of the Program Committee, external referees reviewed all submitted papers, helping in their selection amongst the many excellent submissions received.

Mario R. F. Benevides
Marco Bernardo
Philippe Besnard
Marcin Bialasik
Michel Bidoit
Michele Boreale
Antonio Brogi
Isabel Cafezeiro
Carlos Caleiro
Maria Victoria Cengarle
Maura Cerioli
Christine Choppy
Cristina Cornes
Andrea Corradini
Marcelo da Silva Correa
Mads Dam
Pedro R. D'Argenio
Luca De Alfaro
Razvan Diaconescu
Theodosis Dimitrakos
Carlos H. Cabral Duarte
Bruno Dutertre
Juan Echagüe
A. Fantechi
Maribel Fernandez
Gianluigi Ferrari
Mario Florido
Burkhard Freitag
Thom Fruehwirth
Fabio Gadducci
Alex Vasconcelos Garcia
Stefania Gnesi
Roberto Gorrieri
Reiko Heckel
Rolf Hennicker
Furio Honsell
Delia Kesner
Alexander Knapp
Mikolaj Konarski
Beata Koninowska
Piotr Kosiuczenko
Francois Lamarche
Kevin Lano
D. Latella
Antonia Lopes
Zhaohui Luo
Maurizio Martelli
Mitihiro Matsumoto
Stephan Merz
Bernd Meyer
Eugenio Moggi
Carlo Montangero
Till Mossakowski
Junbo Ogata
Narciso Marti Oliet
Le Gall Pascale
Alfonso Pierantonio
Christian Prehofer
Francesco Parisi Presicce
Corrado Priami
Rosario Pugliese
Paola Quaglia
Leila Ribeiro Korff
G. Rosolini
Don Sanella
David Schmidt
Hugo Scolnik
Laura Semini
Luis Sierra
Marian Srebrny
Andrzej Szalas
Audrey Tan
Alvaro Tasistro
Enrico Tronci
Steve Vickers
Björn Victor
T.C. Wang
Michel Wermelinger

Table of Contents

In these Proceedings, the invited talks are grouped into a separate corpus. Papers are loosely grouped by theme, as were the presentation sessions. Because of these weak groupings, to give names to the sessions would have been misleading.

Invited Talks (Extended Abstracts)

Presentations

Session 1

Session 2

Duration Calculus, a Logical Approach to Real-Time Systems

Zhou Chaochen

United Nations University
International Institute for Software Technology (UNU/IIST), Macau
zcc@iist.unu.edu, http://www.iist.unu.edu

Abstract. The Duration Calculus (DC) represents a logical approach to formal design of real-time systems. DC is based on interval logic, and uses real numbers to model time, and Boolean-valued (i.e. 0,1-valued) functions over time to model states of real-time systems. The duration of a state in a time interval is the accumulated presence time of the state in the interval. DC extends interval logic with a calculus to specify and reason about properties of state durations. The first paper of DC was published in 1991, and dozens of papers of DC have been published since then, which cover developments of logical calculi, their applications and mechanical support tools. This paper will give a brief introduction to DC and also an overview of the research of DC.

Extended Abstract

1 Real-Time Systems

Let us consider the following two cases of real-time systems.

Deadline Driven Scheduler

The scheduler serves a finite number of processes, say $p_1, p_2, ..., p_m$, which share a single processor. Each process periodically requests a constant amount of processor time. It is assumed that in every T_i time units p_i raises a request to occupy the processor for C_i time units, where $(C_i < T_i)$.

A requirement for the scheduler is to fulfil all real-time requests of the processes. The deadline driven scheduling algorithm is proposed in [2]. It satisfies the requirement under assumptions that the scheduler overhead is negligible and

$$\sum_{i=1}^{m} \frac{C_i}{T_i} \leq 1.$$

In this algorithm, the expiration time of a request is called the deadline of the request. The algorithm *dynamically* assigns the highest priority to processes with the nearest deadline. At any instant, only one of the processes with the highest priority and yet unfulfilled request will be selected to *preempt* the processor.

A.M. Haeberer (Ed.): AMAST'98, LNCS 1548, pp. 1–7, 1998.

Gas Burner

This example is first investigated by [6]. Usually, no gas is *leaking* in a gas burner. However, when a flame failure appears, gas will be leaking. A design of a safe gas burner must assure that the leaking time does not get too long.

Let us assume that the ventilation requested for normal combustion would prevent dangerous accumulation of gas, provided that the proportion of leak time is not more than one twentieth of the elapsed time for any time interval being at least one minute long. This is a real-time requirement for a safe gas burner.

Turning next to the task of design, certain decisions must be taken about how the real-time requirement is to be met. For example, for any period where the requirement is guaranteed, any leak in this period should be detectable and stoppable within one second; and to prevent frequent leaks it is acceptable that after any leak in this period the gas burner rejects switching on gas for thirty seconds. The conjunction of these two decisions implies the original requirement, a fact which should be proved before implementation proceeds.

Both the deadline driven scheduler and the gas burner are real-time systems, although the first one is a software system, and the second is a hybrid system.

Duration Calculus (abbreviated DC) is a logical approach to designing real-time systems. DC uses real numbers to model time, and functions from time to Boolean values (or real values) to model behaviour of real-time systems. Based on interval logic, DC establishes a formal notation to specify properties of real-time systems and a calculus to formally prove those properties.

1.1 State Models

DC starts with *Boolean states* to model behaviour of real-time systems. A *Boolean state model* of a real-time system is a set of Boolean-valued (i.e. $\{0,1\}$-valued) functions over time: $\mathrm{Time} \rightarrow \{0,1\}$, where Time is the set of real numbers. Each Boolean-valued function is a *characteristic* function of a specific aspect of the system behaviour.

In order to prove the correctness of the deadline driven scheduler, we introduce $2m$ states to model the behaviour of the scheduler [8]. They are

$$\begin{array}{ll} \mathrm{Run}_i : \mathrm{Time} \rightarrow \{0,1\}, & i = 1, 2, ..., m \\ \mathrm{Std}_i : \mathrm{Time} \rightarrow \{0,1\}, & i = 1, 2, ..., m \end{array}$$

$\mathrm{Run}_i(t) = 1$ iff p_i is running in the processor at time t. $\mathrm{Std}_i(t) = 1$ iff at time t the current request of p_i is still standing. Namely, the current request of p_i is yet to be fulfilled at time t.

To verify the design decisions of the gas burner against the requirement, one may start with a single Boolean state to model the critical aspect of the system:

$$\mathrm{Leak} : \mathrm{Time} \rightarrow \{0,1\}$$

where Leak(t) $= 1$ iff gas is leaking at time t. However at a later stage of the design one may have to introduce more primitive Boolean states such as Gas and Flame to characterize flowing and burning of gas [5]. Then Leak can be pointwise defined as Boolean expression of Gas and Flame: for any t ($\in$ Time)

$$\text{Leak}(t) \;\widehat{=}\; \text{Gas}(t) \wedge \neg\text{Flame}(t)$$

Boolean operators (e.g. $\neg$ and $\wedge$) for states are therefore included in DC.

1.2 State Durations

DC uses state *durations* as essential measurements of behaviour of real-time systems. A duration of a Boolean state over a time interval is the accumulated presence time of the state in the interval. Let S be a Boolean state (i.e. S : Time $\rightarrow \{0,1\}$), and $[b,e]$ an interval. The duration of state S over $[b,e]$ equals

$$\int_b^e S(t)dt.$$

The real-time requirement of the deadline driven scheduler can be expressed in terms of durations of Run_i. Let us assume that all the processes raise their first request at time 0. Thus, the nth request of p_i is raised at time $(n-1)T_i$ and is expired at time nT_i. Therefore the scheduler fulfils the nth request of p_i, iff the accumulated run time of p_i in the interval $[(n-1)T_i, nT_i]$ equals to the requested time C_i. Namely,

$$\int_{(n-1)T_i}^{nT_i} \text{Run}_i(t)dt = C_i$$

The real-time requirement of the gas burner can be expressed in terms of the durations of Leak: for any interval $[b,e]$

$$(e-b) \geq 60 \; sec. \;\Rightarrow\; 20\int_b^e \text{Leak}(t)dt \leq (e-b)$$

A mathematical formulation of these two requirements can hardly leave out state durations. Since the processor may be preempted *dynamically*, the duration of Run_i has to be used to extract the accumulated running time of p_i, and since gas leaks due to *random* flame failures, the duration of leak has to be used to extract the accumulated leak time of gas.

Distance between states is another important measurement of real-time systems. However state distances can be expressed in terms of state durations. With state durations one can first express a lasting period of a state. Let us assume that a presence of state S lasts for a period of $[c,d]$ $(d > c)$. It can be expressed as that the duration of S in $[c,d]$ is equal to the length of $[c,d]$:

$$\int_c^d S(t) = (d-c) > 0, \qquad \text{(abbreviated } S[c,d]\text{)}$$

if we do not care of instant absence of S. Thus, constraints on lasting periods of states can be expressed in terms of state durations. Consider the first design decision of the gas burner. Let $[b,e]$ be the guarantee period of the gas burner. The first design decision that any leak in $[b,e]$ should not last for longer than one second can be expressed as

$$\forall c,d : b \le c < d \le e.(\text{Leak}[c,d] \;\Rightarrow\; (d-c) \le 1 \; sec.)$$

Constraints on state distances can be expressed in terms of state durations similarly. Consider the second design decision of the gas burner. The second design decision can be stated as that the distance between any two consecutive leaks in the guaranteed period $[b,e]$ must be at least thirty seconds long:

$$\begin{array}{l}\forall c,d : b \le c < d < f < g \le e. \\ (\text{Leak}[c,d] \;\wedge \text{NonLeak}[d,f] \wedge \text{Leak}[f,g]) \;\Rightarrow\; (f-d) \ge 30 \; sec.\end{array}$$

where NonLeak $\widehat{=}$ $\neg$Leak(t).

State durations as integrals of Boolean-valued functions are functions from time intervals to real numbers. DC axiomatizes state durations based on the interval logic proposed in [7], which is a logic for functions of time intervals.

2 Interval Logic

2.1 Interval Variables

In the interval logic, functions of intervals are called *interval variables.* Let

$$\textbf{Intv} \;\widehat{=}\; \{\; [b,e] \mid (b,e \in \textbf{Time}) \wedge (b \le e) \;\},$$

and let $\mathbf{R}$ be the set of real numbers and v_i $(i = 1,2,3,4)$ be interval variables:

$$v_i \;:\; \textbf{Intv} \;\rightarrow\; \mathbf{R} \qquad\qquad \text{for } i = 1,2,3,4.$$

A formula of v_i such as $v_1 \le (v_2 + v_3 \cdot v_4)$ is interpreted in the interval logic as a function from **Intv** to the truth values {tt,ff}: **Intv** $\rightarrow$ {tt,ff}.

Therefore the interval logic provides a *functional* calculus, which specify and reason about properties of functions of intervals in a way such that the arguments of the functions (i.e. the intervals) are not referred explicitly.

Interval *length* is a special interval variable denoted ℓ.

$$\ell \;:\; \textbf{Intv} \;\rightarrow\; \mathbf{R}$$

The duration of a state S can be regarded as another example of interval variable, which is written as $\int S$

$$\textstyle\int S \;:\; \textbf{Intv} \;\rightarrow\; \mathbf{R}$$

For arbitrarily given interval $[b,e]$, the value of the interval variable $\int S$ is the duration of S in $[b,e]$, namely the value of

$$\textstyle\int_b^e S(t)dt.$$

Hence state durations can be introduced into the interval logic as interval variables with specific meaning. The requirement of the gas burner can be expressed in an interval logic of state duration $\int$Leak as

Req

$$\ell \geq 60 \;\Rightarrow\; 20\int \mathrm{Leak} \leq \ell.$$

2.2 Interval Modalities

Interval logic uses *modalities* to define structures (called *reachabilities*) among intervals, such as one interval is a subinterval of another interval, or an interval is made of two adjacent subintervals.

Subinterval Modality $\Diamond$:

For any formula ϕ, $\Diamond\phi$ is a new formula which holds for an interval iff ϕ holds for a *subinterval* of it. Thus, with $\Diamond$, from an interval one can reach its subintervals.
The dual of $\Diamond$ is $\Box$: $\Box\phi \;\widehat{=}\; \neg\Diamond\neg\phi$. Hence, $[b,e]$ satisfies $\Box\phi$, iff any subinterval of $[b,e]$ satisfies ϕ.
With $\Box$ one can formulate in an interval logic of state durations the first design decision of the gas burner. First the mathematical definition of $S[c,d]$ can be transformed into a formula without explicit intervals:

$$(\int S = \ell) \;\wedge\; (\ell > 0) \qquad\qquad \text{(abbreviated } \lceil S \rceil\text{)}$$

Then the following is a formulation of the first design decision

Des-1

$$\Box(\lceil \mathrm{Leak} \rceil \;\Rightarrow\; (\ell \leq 1)).$$

Chop Modality $\frown$:

For formulas ϕ and ψ, an interval satisfies the new formula $\phi\frown\psi$ iff the interval can be chopped into two adjacent subintervals such that the first subinterval satisfies ϕ and the second one satisfies ψ.
With $\frown$ and $\Box$, one can formulate the second design decisions of the gas burner. A simplified formulation of the second design decision is

Des-2

$$\Box((\lceil \mathrm{Leak} \rceil \frown \lceil \neg\mathrm{Leak} \rceil \frown \lceil \mathrm{Leak} \rceil) \;\Rightarrow\; (\ell \geq 30))$$

To prove the correctness of the two design decisions is therefore to prove the validity of the formula

$$(\textbf{Des-1} \wedge \textbf{Des-2}) \;\Rightarrow\; \textbf{Req}$$

In fact the subinterval modality $\Diamond$ can be derived from the chop modality $\frown$, since

$$\Diamond\phi \;\Leftrightarrow\; \mathrm{true}\frown\phi\frown\mathrm{true}$$

A modality is called *contracting*, if the modality only provides access to *inside* parts of a given interval. $\Diamond$ and $\frown$ are two examples of contracting modalities. With the contracting modalities, we have formulated *safety* properties for real-time systems. However contracting modalities cannot specify *unbounded liveness* and *fairness* properties of computing systems. Thus, modalities which provide accesses to outside of a given interval are also necessary. They are called *expanding* modalities. In the literature, there are twelve unary modalities and three binary modalities to define various contracting and expanding interval reachabilities. In [10], it is proved that all fifteen modalities of interval logic can be derived from two simple modalities (*left* and *right neighbourhoods*) in a first order logic with interval length ℓ.

3 Duration Calculus

The research on DC was initiated by the case study [6] in connection with the ProCoS project (ESPRIT BRA 3104 and 7071). Two main observations of this case study were that the notion of time interval was useful and that the notion of state duration was necessary. This led to the first publication of DC [11] in 1991. In [11] a relatively *complete* proof system for state durations was established, which included the following axioms:

DCA1 $\int 0 = 0$

DCA2 $\int 1 = \ell$

DCA3 $\int S \geq 0$

DCA4 $\int S_1 + \int S_2 = \int (S_1 \vee S_2) + \int (S_1 \wedge S_2)$

DCA5 $((\int S = x)\frown(\int S = y)) \Rightarrow (\int S = x + y)$

DCA6 $\int S_1 = \int S_2$, provided $S_1 \Leftrightarrow S_2$ holds in propositional logic

Dozens of papers of DC have been published since 1991. They include

1. the formal calculi for state durations in different models of real-time systems, covering Boolean state, event [12, 9], real state [13], dependability [3], finite divergence [1], super dense computation [9, 4], etc.
2. applications of DC, such as case studies of auto pilot, railway crossing, water level monitor, gas burner, air traffic controller, production cell, motor-load control, inverted pendulum, etc. and formal specification and verification of real-time programs and circuits, and
3. mechanical support tools for DC, including results on completeness, decidability of subclasses and model checking algorithms and implementation of proof assistant and model checker

Acknowledgements

I would thank many colleagues of mine who have contributed to the development and application of DC. My special thanks go to Michael Hansen. The structure of this overview has been discussed among us for many times. Due to the space limit I cannot provide in this extended abstract a complete bibliography about DC. Please access
`http://www.iist.unu.edu/home/Unuiist/newrh/II/2/2/page.html`, and
`http://www.comlab.ox.ac.uk/archive/formal-methods.html#DC`
for more complete references.

References

1. M.R. Hansen, P.K. Pandya and Zhou Chaochen, Finite Divergence, Theoretical Computer Science, vol 138, pp 113-139, 1995
2. C.L. Liu and J.W. Layland, Scheduling Algorithm for Multiprogramming in a Hard Real-Time Environment, Journal of the ACM, vol 20, no 1, pp 46-61 1973
3. Liu Zhiming, A.P. Ravn, E.V. Sørensen and Zhou Chaochen, A Probabilistic Duration Calculus, Dependable Computing and Fault-Tolerant Systems Vol. 7: Responsive Computer Systems, H. Kopetz and Y. Kakuda (Eds.), pp 30-52, Springer-Verlag, 1993
4. Paritosh K. Pandya and Dang Van Hung. Duration Calculus with Weakly Monotonic Time, Proceedings of FTRTFT'98, LNCS 1486, pp 55-64, Springer-Verlag, 1998
5. A.P. Ravn, H. Rischel and K.M. Hansen, Specifying and Verifying Requirements of Real-Time Systems, IEEE Trans. Softw. Eng., 1993
6. E.V. Sørensen, A.P. Ravn and H. Rischel, Control Program for a Gas Burner: Part 1: Informal Requirements, ProCoS Case Study 1, ProCoS Rep. ID/DTH EVS2", 1990
7. J. Halpern, B. Moskowski and Z. Manna, A Hardware Semantics based on Temporal Intervals, ICALP'83, LNCS 154, pp 278-291, 1983
8. Zheng Yuhua and Zhou Chaochen, A Formal Proof of the Deadline Driven Scheduler, Proceedings of FTRTFT'94, H. Langmack, W.-P. de Roever and J. Vytopil (Eds), LNCS 863, pp 756-775, Springer-Verlag, 1994
9. Zhou Chaochen and Michael R. Hansen, Chopping a Point, BCS-FACS 7th Refinement Workshop, Electronic Workshops in Computing, Springer-Verlag, 1996
10. Zhou Chaochen and Michael R. Hansen, An Adequate First Order Logic of Intervals, Technical Report 91, UNU/IIST, 1996
11. Zhou Chaochen and C.A.R. Hoare and A.P. Ravn, A Calculus of Durations, Information Processing Letters, vol 40, no 5, pp 269-276, 1991
12. Zhou Chaochen and Li Xiaoshan, A Mean Value Calculus of Durations, A Classical Mind: Essays in Honour of C.A.R. Hoare, pp 431-451, Prentice Hall International, 1994
13. Zhou Chaochen, A.P. Ravn and M.R. Hansen, An Extended Duration Calculus for Hybrid Systems, Hybrid Systems, R.L. Grossman, A. Nerode, A.P. Ravn, H. Rischel (Eds.), LNCS 736, pp 36-59, Springer-Verlag, 1993

Abstract Algebraic Logic

Don Pigozzi

Deparment of Mathematics
Iowa State University, Ames IA 50011
dpigozzi@iastate.edu

Abstract. The classical method of associating a class of algebras with a logical system is that of Lindenbaum and Tarski. It can be applied to any system with a biconditional $\leftrightarrow$ that is compositional in the sense that it defines a congruence relation on the absolutely free algebra of formulas. The method has been abstacted to provide effective criteria for the algebraizability of a large class of formal systems. One consequence of this work is a uniform method of providing a formal system with algebraic semantics. The main features of the theory of abstract algebraic logic and one of its potential applications are surveyed.

1 Introduction

Equational logic has played a prominent in the applications of logic in computer science. It has found application in a number of different areas including functional programming, term rewriting, specification theory, and relational models of programming semantics. Algebraic logic deals specifically the representation of different logical systems in equational logic, and for this reason is a promising area to look for new applications of logic in computer science. One of the chief paradigms for algebraic logic is the way that the class of Heyting algebras is constructed from Heyting's formalization of Brouwer's intuitionism by the so-called Lindenbaum-Tarski process.

In its traditional form the Lindenbaum-Tarski process relies on the fact that the underlying logic has a biconditional $\leftrightarrow$ that defines logical equivalence. The set of all sentences or formulas are partitioned into logical equivalence classes and then abstracted by the familiar process of forming the quotient. The resulting algebra is called the *Lindenbaum-Tarski algebra* of the original logic. The class of algebras one gets this way by adjoining arbitrary sets of nonlogical axioms becomes the subject of the algebraic study of the original logic. That this process actually leads to an algebra relies on the compositionality of logical equivalence. Nonstandard logical systems, like those that arise in computer science applications, may not have a biconditional with the appropriate properties. And the traditional Lindenbaum-Tarski process is not directly applicable in these cases The abstraction of the Lindenbaum-Tarski process and the investigation of the consequences of its application to this more general class of logical systems is the domain of abstract algebraic logic.

A.M. Haeberer (Ed.): AMAST'98, LNCS 1548, pp. 8–16, 1998.

2 Deductive Systems and Matrix Semantics

By a *language type* we mean a set $\mathcal{L}$ of connectives or operation symbols, depending on whether we are viewing them from a logical or algebraic perspective. The set of formulas over $\mathcal{L}$, in symbols $\mathrm{Fm}_{\mathcal{L}}$, is formed in the usual way. $\mathbf{Fm}_{\mathcal{L}}$ is the corresponding *algebra of formulas*. For any set X of variables, $\mathrm{Fm}_{\mathcal{L}}(X)$ is the set of formulas in which only variables from X occur, and $\mathbf{Fm}_{\mathcal{L}}(X)$ is the corresponding formula algebra. The operation of simultaneously substituting fixed but arbitrary formulas for variables is identified with the unique endomorphism of $\mathbf{Fm}_{\mathcal{L}}$ it determines.

The basic syntactic unit of a k-deductive system is a k-tuple of $\mathcal{L}$-formulas; these are called *k-formulas.* If $\varphi = \langle \varphi_0, \dots, \varphi_{k-1} \rangle$ is a k-formula and $\sigma : \mathbf{Fm}_{\mathcal{L}} \to \mathbf{Fm}_{\mathcal{L}}$ is a substitution, then the *σ-substitution instance of* φ, $\sigma(\varphi)$, is defined to be $\langle \sigma(\varphi_0), \dots, \sigma(\varphi_{n-1}) \rangle$.

Let k be any positive integer. A k-deductive system over $\mathcal{L}$ is a pair $\mathfrak{S} = \langle \mathrm{Fm}_{\mathcal{L}}, \vdash_{\mathfrak{S}} \rangle$ were $\vdash_{\mathfrak{S}} \subseteq \mathcal{P}(\mathrm{Fm}^k_{\mathcal{L}}) \times \mathrm{Fm}^k_{\mathcal{L}}$ satisfies the following conditions for all $\Gamma, \Delta \subseteq \mathrm{Fm}^k_{\mathcal{L}}$ and $\varphi \in \mathrm{Fm}^k_{\mathcal{L}}$.

$$\Gamma \vdash_{\mathfrak{S}} \varphi \text{ for all } \varphi \in \Gamma;$$
$$\Gamma \vdash_{\mathfrak{S}} \varphi \text{ and } \Delta \vdash_{\mathfrak{S}} \psi \text{ for every } \psi \in \Gamma \text{ imply } \Delta \vdash_{\mathfrak{S}} \varphi;$$
$$\Gamma \vdash_{\mathfrak{S}} \varphi \text{ implies } \Gamma' \vdash_{\mathfrak{S}} \varphi \text{ for some } \Gamma' \subseteq_{\omega} \Gamma;$$
$$\Gamma \vdash_{\mathfrak{S}} \varphi \text{ implies } \sigma(\Gamma) \vdash_{\mathfrak{S}} \sigma(\varphi) \text{ for every substitution } \sigma.$$

By a *k-dimensional rule*, or simply a *k-rule*, we mean a pair $\langle \Gamma, \varphi \rangle$, usually written in the form $\dfrac{\Gamma}{\varphi}$, where $\Gamma \cup \{\varphi\} \subseteq \mathrm{Fm}^k_{\mathcal{L}}$. A k-formula φ is a *theorem* of a k-deductive system $\mathfrak{S}$ if $\vdash_{\mathfrak{S}} \varphi$ (i.e., $\emptyset \vdash_{\mathfrak{S}} \varphi$). The rule $\dfrac{\Gamma}{\varphi}$ is a *derived rule* of $\mathfrak{S}$ if $\Gamma \vdash_{\mathfrak{S}} \varphi$. A $T \subseteq \mathrm{Fm}^k_{\mathcal{L}}$ is a *theory* of $\mathfrak{S}$ if it is closed under all derived inference rules.

A large number of different logical systems are either k-deductive systems or can be reformalized as k-deductive systems. All the familiar sentential logics together with their various fragments and refinements are naturally formalized as finitary 1-deductive systems. For example, the classical and intuitionistic sentential logics, the various modal logics (including Lewis's S_4 and S_5), and the multiple-valued logics of Łukasiewicz and Post.

The most important example of a 2-deductive system is equational logic. A 2-formula $\langle \varphi, \psi \rangle$ is to be interpreted as the equation $\varphi \approx \psi$ or, more precisely, as the congruence $\varphi \equiv \psi$.

Let $\mathcal{L} = \{\, \omega_i : i \in I \,\}$ be any language type. The axioms and rules of inference of the system of free equational logic are:

$$\text{(A1)}\ \langle x, x \rangle; \qquad \text{(R1)}\ \frac{\langle x, y \rangle}{\langle y, x \rangle}; \qquad \text{(R2)}\ \frac{\langle x, y \rangle, \langle y, z \rangle}{\langle x, z \rangle};$$

$$(\mathrm{R3}_{\omega})\ \frac{\langle x_0, y_0 \rangle, \dots, \langle x_{n-1}, y_{n-1} \rangle}{\langle \omega x_0 \dots x_{n-1}, \omega y_0 \dots y_{n-1} \rangle}, \quad \text{for each } \omega \in \mathcal{L},\ n \text{ the rank of } \omega.$$

The theories of free equational logic, in the sense defined above, are exactly the congruences on the formula algebras $\mathbf{Fm}_{\mathcal{L}}$.

The equational logic $\mathfrak{S}^{\mathrm{EQL}}\,\mathsf{Q}$ associated with a particular quasivariety Q over the language $\mathcal{L}$ type is an extension of the free equational logic. The axioms and rules of inference of $\mathfrak{S}^{\mathrm{EQL}}\,\mathsf{Q}$ include (A1), (R1), (R2), and (R3$_\omega$), for $\omega \in \mathcal{L}$. If Q is axiomatized by a set Id of identities and a set Qd of quasi-identities, we adjoin the axioms $\langle \varphi, \psi \rangle$, for every identity $\varphi \approx \psi \in \mathrm{Id}$, and the rules

$$\frac{\langle \xi_0, \eta_0 \rangle, \dots, \langle \xi_{n-1}, \eta_{n-1} \rangle}{\langle \varphi, \psi \rangle},$$

for every quasi-identity $(\bigwedge_{i<n} \xi_i \approx \eta_i) \rightarrow \varphi \approx \psi \in \mathrm{Qd}$.

Conversely, any extension of the free equational logic over $\mathcal{L}$ by new axioms and rules is of the form $\mathfrak{S}^{\mathrm{EQL}}\,\mathsf{Q}$ for some quasivariety Q.

The logic of partially ordered algebras can also be formalized as a 2-deductive system with 2-formulas representing the partial ordering, i.e., the 2-formula $\langle \varphi, \psi \rangle$ is now to be interpreted as the inequality (or more precisely the quasi-inequality) $\varphi \leq \psi$. The axioms and rules of inference are the same as for the free equational logic except that the symmetry rule (R1) is omitted. The theories of free quasi-ordered logic are the quasi-orderings on the formula algebra with the property that each fundamental operation is monotone in each argument.

Sequent calculi, with a fusion connective that allows the set of formulas on the left hand side of a sequent to be combined in a single formula, can be reformulated as 2-deductive systems. Here $\langle \varphi, \psi \rangle$ stands for the sequent $\varphi \vdash \psi$. Roughly speaking, a sequent calculus with this property constitutes a generalization of quasi-ordered logic in which the fundamental operations may be anti-monotone with respect to the ordering in some arguments. Linear logic (with the tensor connective $*$ as the fusion connective) is of this kind.

Hyper-equational logic is an applied equational logic that in a certain sense encompasses in a single formalism all the applied equational logics, over all languages types $\mathcal{L}$. It takes the form $\mathfrak{S}^{\mathrm{EQL}}\,\mathsf{V}$ where V is the variety of abstract algebras of clones. Hyper-equational logic constitutes an adequate formalization of all of that part of applied equational logic that deals only with axiomatic extensions of free equational logic (no nonlogical inference rules), and it allows one to reduce the study of all such applied equational logics to the study of the theories of a single applied equational logic.

First-order predicate logic can be transformed into a k-deductive system in much the same way, and this transformation is intimately connected with the classical process of algebraizing first-order logic. In the case of predicate logic we get a 1-deductive system, but an application of the Lindenbaum-Tarski process to the resulting 1-deductive system gives an applied equational logic as in the hyper-equational case.

By focusing on k-deductive systems here we are in effect restricting ourselves to strict universal Horn theories with a single predicate symbol, and in particular without equality. There is nothing essential about the restriction to a single predicate or the elimination of equality. Most of the main results of abstract algebraic logic extend with only minor modifications to arbitrary strict Horn theories, with or without equality, and to an even wider class of logical systems. This includes, for example, the sequent calculi in their usual formulation, and an appropriately chosen subclass of the class of institutions ([11]), π-institutions ([10]), and general logics ([18]).

A pseudo algebraic semantics can be provided for an arbitrary k-deductive system in a uniform way by means of the k-matrix models. The algebraic study of the matrix semantics for deductive systems originated with Łos and Suszko [13] and was systematically developed in [5, 6].

By a *k-matrix* (or simply a *matrix* in contexts in which k is fixed) we mean an ordered pair $\mathfrak{A} = \langle \boldsymbol{A}, F_{\mathfrak{A}} \rangle$ where $\boldsymbol{A}$ is an $\mathcal{L}$-algebra and $F_{\mathfrak{A}}$ is any subset of k-tuples of the universe A of $\boldsymbol{A}$.

Let $\mathfrak{A} = \langle \boldsymbol{A}, F_{\mathfrak{A}} \rangle$ and $\mathfrak{B} = \langle \boldsymbol{B}, F_{\mathfrak{B}} \rangle$ be k-matrices. $\mathfrak{A}$ is a *submatrix* of $\mathfrak{B}$ if $\boldsymbol{A}$ is a subalgebra of $\boldsymbol{B}$ and $F_{\mathfrak{A}} = F_{\mathfrak{B}} \cap A^k$. A *matrix homomorphism* h from $\mathfrak{A}$ to $\mathfrak{B}$ is a homomorphism of the underlying algebras $\boldsymbol{A}$ and $\boldsymbol{B}$ such that $h(F_{\mathfrak{A}}) \subseteq F_{\mathfrak{B}}$ or, equivalently, $F_{\mathfrak{A}} \subseteq h^{-1}(F_{\mathfrak{B}})$. By the *direct product* of a system $\langle \mathfrak{A}_i : i \in I \rangle$ of k-matrices we mean the k-matrix $\langle \prod_{i \in I} \boldsymbol{A}_i, \prod^k_{i \in I} F_{\mathfrak{A}_i} \rangle$, where $\prod^k_{i \in I} F_{\mathfrak{A}_i}$ is the set of all k-tuples $\langle a_0, \ldots, a_{k-1} \rangle$ of elements of $\prod_{i \in I} A_i$ such that $\langle a_0(i), a_1(i), \ldots, a_{k-1}(i) \rangle \in F_{\mathfrak{A}_i}$ for each $i \in I$. A submatrix of a direct product of k-matrices is a *subdirect product* if each projection is surjective.

Let $\mathfrak{A}$ be an arbitrary k-matrix over the language $\mathcal{L}$. $\mathfrak{A}$ defines for each set X of variables a semantic consequence relation $\models_{\mathfrak{A}(X)} \subseteq \mathcal{P}(\mathrm{Fm}^k(X)) \times \mathrm{Fm}^k(X)$ over $\mathcal{L}$. Let $\Gamma \cup \{\varphi\} \subseteq \mathrm{Fm}^k(X)$ and $\varphi \in \mathrm{Fm}^k(X)$. Then $\Gamma \models_{\mathfrak{A}(X)} \varphi$ is the relation that holds between Γ and φ if, for every interpretation $\bar{a}$ of the variables in X in A,

$$\psi^{\boldsymbol{A}}(\bar{a}) \in F_{\mathfrak{A}} \text{ for every } \psi \in \Gamma \quad \text{implies} \quad \varphi^{\boldsymbol{A}}(\bar{a}) \in F_{\mathfrak{A}}.$$

For any class K of k-matrices, $\Gamma \models_{\mathsf{K}(X)} \varphi$ iff $\Gamma \models_{\mathfrak{A}(X)} \varphi$ for every $\mathfrak{A} \in \mathsf{K}$. It is clear that if X and Y are sets of variables such that $X \subseteq Y$ and if $\Gamma \cup \{\varphi\} \subseteq \mathrm{Fm}^k(X)$, then $\Gamma \models_{\mathfrak{A}(X)} \varphi$ iff $\Gamma \models_{\mathfrak{A}(Y)} \varphi$. Thus in the sequel we omit explicit reference to the set of variables and write simply $\Gamma \models_{\mathfrak{A}} \varphi$.

Let $\mathfrak{S}$ be a k-dimensional deductive system. A k-matrix $\mathfrak{A}$ is called a *matrix model* of $\mathfrak{S}$ if $\Gamma \vdash_{\mathfrak{S}} \varphi$ implies $\Gamma \models_{\mathfrak{A}} \varphi$, for every set X of variables every $\Gamma \cup \{\varphi\} \subseteq \mathrm{Fm}^k(X)$. A set $F \subseteq A^k$ is an *$\mathfrak{S}$-filter* on $\boldsymbol{A}$ if $\langle \boldsymbol{A}, F \rangle$ is a model of $\mathfrak{S}$. The set of all $\mathfrak{S}$-filters on $\boldsymbol{A}$ is denoted by $\mathrm{Fi}^{\boldsymbol{A}}\, \mathfrak{S}$ The class of matrix models of $\mathfrak{S}$ is denoted by $\mathsf{Mod}\,\mathfrak{S}$. The $\mathfrak{S}$-matrices are exactly the models (in the first-order sense) of the universal Horn theory associated with $\mathfrak{S}$.

3 The Abstract Lindenbaum-Tarski Process

The abstract Lindenbaum-Tarski process can be applied to obtain a more algebra-like semantics for k-deductive systems. This entails abstracting the notion of logical equivalence. We adapt Leibniz's well-known definition of equality for this purpose. Let $\mathfrak{A} = \langle \boldsymbol{A}, F \rangle$ be an arbitrary matrix. Define the binary relation $\boldsymbol{\Omega}^{\boldsymbol{A}}\, F$ on A by the condition that $\langle a, b \rangle \in \boldsymbol{\Omega}^{\boldsymbol{A}}\, F$ if, for every formula

$\varphi(x,\bar{u}) \in \mathrm{Fm}_{\mathcal{L}}$ with a single variable x and an arbitrary number of parameters $\bar{u} = \langle u_0,\dots,u_{n-1}\rangle$,

$$\varphi^{\boldsymbol{A}}(a,\bar{c}) \in F \quad \text{iff} \quad \varphi^{\boldsymbol{A}}(b,\bar{c}), \qquad \text{for every } \bar{c} = \langle c_0,\dots,c_{n-1}\rangle \in A^n.$$

$\boldsymbol{\Omega}^{\boldsymbol{A}} F$ is a congruence relation on $\boldsymbol{A}$, abstracting Frege's principle of compositionality for logical equivalence. $\boldsymbol{\Omega}^{\boldsymbol{A}} F$ is called the *Leibniz congruence* of F on $\boldsymbol{A}$. It is the largest congruence on $\boldsymbol{A}$ *compatible* with F in the sense that $a \equiv b \pmod{\boldsymbol{\Omega}^{\boldsymbol{A}} F}$ and $a \in F$ implies $b \in F$ The quotient matrix $\mathfrak{A}/\boldsymbol{\Omega}^{\boldsymbol{A}} F = \langle \boldsymbol{A}/\boldsymbol{\Omega}^{\boldsymbol{A}}, F/\boldsymbol{\Omega}^{\boldsymbol{A}} F\rangle$ is called the *reduction* of $\mathfrak{A}$ and is denoted by $\mathfrak{A}^*$. $\mathfrak{A}$ is *reduced* if $\mathfrak{A} \cong \mathfrak{A}^*$. If $\mathfrak{S}$ is a deductive system we denote the class of all reduced models of $\mathfrak{S}$ by $\mathsf{Mod}^*\,\mathfrak{S}$, and $\mathsf{Alg\,Mod}^*\,\mathfrak{S}$ is the class of underlying algebras of reduced models.

In abstract algebraic logic the k-deductive systems are classified into the *algebraic hierarchy.* It reflects the degree to which the reduced matrix semantics of a system behaves like a real algebraic semantics. At the top of the hierarchy are the so-called *algebraizable* deductive systems introduced in [2], and the various weaker notions considered in [7, 12]. The paradigms here are the 1-deductive systems of classical and intuitionistic logic. In this case $\mathsf{Alg\,Mod}^*\,\mathfrak{S}$ is respectively the variety of Boolean and the variety of Heyting algebras. In addition to the classical and intuitionistic propositional logics, most modal logics, the multiple-valued logics of Łukasiewicz and Post, and the first-order predicate logic are algebraizable. The applied equational logics are the primary examples of algebraizable 2-deductive systems.

Lower down in the hierarchy come the *protoalgebraic* and *equivalential* systems ([1, 3, 4, 6]). Here the class of reduced matrix models $\mathsf{Mod}^*\,\mathfrak{S}$ still exhibits strong algebra-like characteristics, but cannot be replaced by a $\mathsf{Alg\,Mod}^*\,\mathfrak{S}$, as is the case for finitely algebraizable systems. Almost all deductive systems that have appeared in the literature are protoalgebraic. The equivalential systems are more specialized but still constitute a much broader class then the algebraizable ones, especially among 2-deductive systems. For simplicity the following survey of the algebraic hierarchy is formulated only for 1-deductive systems.

Let $\Delta(x,y) = \{\, \delta_i(x,y) : i \in I \,\}$ be a finite set of 1-formulas in two variables. $\Delta(x,y)$ is called a *protoequivalence system* for a 1-deductive system $\mathfrak{S}$ if the following are theorems and a derived rule of $\mathfrak{S}$.

$$\Delta(x,x), \qquad \frac{x,\ \Delta(x,y)}{y}.$$

Theorem 1. *Let $\mathfrak{S}$ be a deductive system. The following conditions are equivalent.*

(i) *$\mathfrak{S}$ has a protoequivalence system.*

(ii) *The Leibniz congruence operator is monotonic on $\mathfrak{S}$-filters, that is, for every $\mathcal{L}$-algebra $\boldsymbol{A}$ and all $F, G \in \mathrm{Fi}^{\boldsymbol{A}}\,\mathfrak{S}$, $F \subseteq G$ implies $\boldsymbol{\Omega}^{\boldsymbol{A}} F \subseteq \boldsymbol{\Omega}^{\boldsymbol{A}} G$.*

(iii) $\mathsf{Mod}^*\,\mathfrak{S}$ *is closed under subdirect products.*

A deductive system $\mathfrak{S}$ is *protoalgebraic* if any one, and hence all, of the above conditions hold. Almost all the deductive systems considered in the literature are protoalgebraic. One which is not is the 1-deductive system defined by the class of all matrices of the form $\langle \boldsymbol{A}, F\rangle$ where $\boldsymbol{A}$ is a distributive lattice and F is a lattice filter. Protoalgebraicity seems to be the minimal property required in order for $\mathsf{Mod}^* \mathfrak{S}$ to behave like a reasonable algebraic semantics.

$E(x,y)$ is an equivalence system for a 1-deductive system $\mathfrak{S}$ if it is a protoequivalence system satisfying the additional conditions

$$\frac{E(x,y)}{E(y,x)}, \qquad \frac{E(x,y), E(y,z)}{E(x,z)},$$

$$\frac{E(x_0,y_0),\dots,E(x_{n-1},y_{n-1})}{E(\lambda x_0,\dots,x_{n-1},\ \lambda y_0,\dots,y_{n-1})} \quad \text{for all } \lambda \in \mathcal{L} \ (n \text{ is the rank of } \lambda).$$

Theorem 2. *The following conditions are equivalent for every deductive system* $\mathfrak{S}$

(i) $\mathfrak{S}$ *has an finite equivalence system.*
(ii) *The Leibniz congruence operator has the following properties.*
  (a) *monotonic on* $\mathfrak{S}$*-filters;*
  (b) *commutes with inverse matrix homomorphisms in the sense that, for any* $h\colon \boldsymbol{B} \to \boldsymbol{A}$ *and* $F \in \mathrm{Fi}^{\boldsymbol{A}}$, $\boldsymbol{\Omega}^{\boldsymbol{B}}\, h^{-1}(F) = h^{-1}(\boldsymbol{\Omega}^{\boldsymbol{A}}\, F)$.
  (c) *continuous in the sense that* $\boldsymbol{\Omega}^{\boldsymbol{A}}(\bigcup i \in I F_i) = \bigcup_{i\in I} \boldsymbol{\Omega}^{\boldsymbol{A}}\, F_i$ *for any set of* $\mathfrak{S}$*-filters that is upper directed under inclusion.*
(iii) $\mathsf{Mod}^* \mathfrak{S}$ *is closed under submatrices, direct products, and ultraproducts* (*i.e., forms a quasivariety in the sense of Mal'cev;* [17]).

A deductive system is *finite equivalential* if it satisfies any of the above conditions. The paradigms here are the applied quasi-ordered logics and the special sequent systems consider above, in particular the 2-deductive fragment of linear logic mentioned above.

In the sequel we use $K \approx L$ as an abbreviation for a set of equations $\{\,\kappa_i \approx \lambda_i : i \in I\,\}$. Let $\mathfrak{S}$ be a finitely equivalential deductive system and $E(x,y)$ a finite equivalence system for $\mathfrak{S}$. Then the equational consequence relation of the quasivariety $\mathsf{Alg\,Mod}^* \mathfrak{S}$ can be faithfully interpreted in the consequence relation of $\mathfrak{S}$ in the sense that, for all $K \approx L \cup \{\varphi \approx \psi\} \subseteq \mathrm{Fm}^2_{\mathcal{L}}$,

$$K \approx L \models_{\mathsf{Alg\,Mod}^* \mathfrak{S}} \varphi \approx \psi \quad \text{iff} \quad \{\,E(\kappa,\lambda) : \kappa \approx \lambda \in K \approx L\,\} \vdash_{\mathfrak{S}} E(\varphi,\psi). \qquad (1)$$

The equivalence system E is said to be *invertible* if there is a finite set of equations in two variables that defines a faithful interpretation of the consequence relation of $\mathfrak{S}$ in the equational consequence relation of $\mathsf{Alg\,Mod}^* \mathfrak{S}$ that is the inverse of (1).

Theorem 3. *The following are equivalent for any deductive system $\mathfrak{S}$.*

(i) *$\mathfrak{S}$ has an invertible finite equivalence system.*
(ii) *The Leibniz congruence operator on $\mathfrak{S}$-filters is monotone, continuous, and injective.*
(iii) $\mathsf{Mod}^*\,\mathfrak{S}$ *is a quasivariety in the Mal'cev sense and each matrix in* $\mathsf{Mod}^*\,\mathfrak{S}$ *is uniquely determined by its underlying algebra.*

$\mathfrak{S}$ is *finitely algebraizable* if it satisfies any of the conditions of this theorem. The quasivariety $\mathsf{Alg}\,\mathsf{Mod}^*\,\mathfrak{S}$ is called the *equivalent algebraic* semantics for $\mathfrak{S}$.

4 Interpolation and Modularity

Certain properties of specification languages concerning the notion of modularity of data type specifications can be conveniently formalized in terms of the interpolation properties. Some references to the computer science literature on this subject are [9, 15, 16, 19, 20, 21].

We survey here the connection between two basic interpolation properties and two semantical amalgamation properties, in the context of abstract algebraic logic. Details can be found in [8]

$\mathfrak{S}$ is a k-deductive system over an arbitrary language $\mathcal{L}$. Recall that For each set of (sentential) variables W, $\mathrm{Fm}^k_{\mathcal{L}}(W)$ denotes the set of k-formulas over W in $\mathcal{L}$. For $\varphi \in \mathrm{Fm}^k_{\mathcal{L}}(W)$ and $\Gamma \subseteq \mathrm{Fm}^k_{\mathcal{L}}(W)$, $\mathrm{var}\,\varphi$ and $\mathrm{var}\,\Gamma$ denote respectively the set of variables occurring in φ and in at least one $\psi \in \Gamma$. For any $\Gamma, \Delta \subseteq \mathrm{Fm}^k_{\mathcal{L}}(W)$, $\Gamma \vdash_{\mathfrak{S}} \Delta$ will mean $\Gamma \vdash_{\mathfrak{S}} \varphi$ for all $\varphi \in \Delta$. $\Gamma \vdash_{\mathfrak{S}} \varphi$ means that $\Gamma \vdash_{\mathfrak{S}(W)} \varphi$ for any W such that $\Gamma \cup \{\varphi\} \subseteq \mathrm{Fm}^k(W)$. In the following definition the set W of variables is arbitrary but fixed.

Definition 1. *Let $\mathfrak{S}$ be a k-deductive system and W a set of variables. Let $\Gamma, \Delta \subseteq \mathrm{Fm}^k_{\mathcal{L}}(W)$ and $\varphi, \psi \in \mathrm{Fm}^k_{\mathcal{L}}(W)$.*

(i) *$\mathfrak{S}$ has the* Craig interpolation property *if, whenever* $(\mathrm{var}\,\Gamma \cap \mathrm{var}\,\varphi) \neq \emptyset$, *we have*

$$\Gamma \vdash_{\mathfrak{S}} \varphi \Longrightarrow \exists_{\Gamma' \subseteq \mathrm{Fm}^k_{\mathcal{L}}(\mathrm{var}\,\Gamma \cap \mathrm{var}\,\varphi)} (\Gamma \vdash_{\mathfrak{S}} \Gamma' \;\wedge\; \Gamma' \vdash_{\mathfrak{S}} \varphi).$$

(ii) *$\mathfrak{S}$ has the* Maehara interpolation property *if, whenever* $\mathrm{var}\,\Gamma \cap \mathrm{var}(\Delta \cup \{\varphi\}) \neq \emptyset$, *we have*

$$\Gamma, \Delta \vdash_{\mathfrak{S}} \varphi \Longrightarrow \exists_{\Gamma' \subseteq \mathrm{Fm}^k_{\mathcal{L}}(\mathrm{var}\,\Gamma \cap \mathrm{var}(\Delta \cup \{\varphi\}))} (\Gamma \vdash_{\mathfrak{S}} \Gamma' \;\wedge\; \Gamma', \Delta \vdash_{\mathfrak{S}} \varphi).$$

The first explicit appearance of the Maehara interpolation property in the literature appears to be in [14] where the property is established for the intuitionistic propositional logic. The term "Maehara interpolation property" now seems to be commonly used for this interpolation property in the literature of sentential logic. In the computer science literature it has been called both the "strong" interpolation property and the "modularization" property ([9, 19, 21]).

Definition 2. *Let* L *be any category of* k*-matrices or* $\mathcal{L}$*-algebras.*

(i) *Assume* L*-free extensions exist.* L *has the* categorical flat amalgamation property *if injections are transferable over free injections in* L*, i.e., for all* $\mathfrak{A}, \mathfrak{B}, \mathfrak{C} \in \mathsf{L}$ *and every injective matrix homomorphism* $f: \mathfrak{C} \rightarrowtail \mathfrak{A}$ *and free injective matrix homomorphism* $g: \mathfrak{C} \circ\!\!\rightarrow \mathfrak{B}$*, there exists a* $\mathfrak{D} \in \mathsf{L}$*, a free injection* $h: \mathfrak{A} \circ\!\!\rightarrow \mathfrak{D}$ *and an injection* $k: \mathfrak{B} \rightarrowtail \mathfrak{D}$ *such that* $h \circ f = k \circ g$*.*

(ii) L *has the* categorical modular amalgamation property *if injections are transferable over all matrix homomorphisms in* L*, i.e., for all* $\mathfrak{A}, \mathfrak{B}, \mathfrak{C} \in \mathsf{L}$*, every injective matrix homomorphism* $f: \mathfrak{C} \rightarrowtail \mathfrak{A}$ *and every matrix homomorphism* $g: \mathfrak{C} \to \mathfrak{B}$*, there exists* $\mathfrak{D} \in \mathsf{L}$*, a* $h: \mathfrak{A} \to \mathfrak{D}$*, and an injective* $k: \mathfrak{B} \rightarrowtail \mathfrak{D}$ *such that* $h \circ f = k \circ g$*.*

A clean connection between the interpolation and amalgamation properties seem to require that we go at least up to the level of equivalential systems in the algebraic hierarchy.

Theorem 4. *Let* $\mathfrak{S}$ *be an equivalential deductive system. The following are equivalent.*

(i) $\mathfrak{S}$ *has the Craig interpolation property.*
(ii) $\mathsf{Mod}\,\mathfrak{S}$ *has the categorical flat amalgamation property.*
(iii) $\mathsf{Mod}^*\,\mathfrak{S}$ *has the categorical flat amalgamation property.*

Assume now that $\mathfrak{S}$ *is finitely algebraizable, and let* $\mathsf{Q} = \mathsf{Alg}\,\mathsf{Mod}^*\,\mathfrak{S}$ *be its equivalent algebraic semantics. Then all three of the above conditions are equivalent to*

(iv) Q *has the categorical flat amalgamation property.*

Theorem 5. *Let* $\mathfrak{S}$ *be an equivalential deductive system. The following are equivalent.*

(i) $\mathfrak{S}$ *has the Maehara interpolation property.*
(iii) $\mathsf{Mod}^*\,\mathfrak{S}$ *has the categorical modular amalgamation property.*

Assume now that $\mathfrak{S}$ *is finitely algebraizable, and let* $\mathsf{Q} = \mathsf{Alg}\,\mathsf{Mod}^*\,\mathfrak{S}$ *be its equivalent algebraic semantics. Then both of the above conditions are equivalent to*

(iv) Q *has the modular amalgamation property.*

References

[1] W. J. Blok and D. Pigozzi. Protoalgebraic logics. *Studia Logica*, 45:337–369, 1986.
[2] W. J. Blok and D. Pigozzi. *Algebraizable logics*, volume 396 of *Mem. Amer. Math. Soc.* American Mathematical Society, Providence, January 1989.

[3] W. J. Blok and D. Pigozzi. Algebraic semantics for universal Horn logic without equality. In A. Romanowska and J. D. H. Smith, editors, *Universal Algebra and Quasigroup Theory*, pages 1–56. Heldermann, Berlin, 1992.

[4] J. Czelakowski. *Protoalgebraic Logics*. Kluwer, Amsterdam. to appear.

[5] J. Czelakowski. *Model-theoretic methods in methodology of propositional calculus*. The Institute of Philosophy and Sociology of the Polish Academy of Sciences, Warsaw, 1980.

[6] J. Czelakowski. Equivalential logics I and II. *Studia Logica*, 40:227–236 and 355–372, 1981.

[7] J. Czelakowski and R. Jansana. Weakly algebraizable logics. Preprint 365, Centre de Recerca Matemàtica, Bellaterra, Spain, July 1997.

[8] J. Czelakowski and P. Pigozzi. Amalgamation and interpolation in abstract algebraic logic. Preprint 343, Centre de Recerca Matemàtica, Bellaterra, Spain, September 1996.

[9] R. Diaconescu, J. Goguen, and P. Stefaneas. Logical support for modularization. In G. Huet and G. Plotkin, editors, *Proceedings of a Workshop on Logical Frameworks (Edinburgh, May 1991)*.

[10] J. Fiadeiro and A. Sernadas. Structuring theories on consequence. In D. Sanella and A. Tarlecki, editors, *Recent Trends in Data Type Specification*, volume 332 of *Lecture Notes in Computer Science*, pages 44–72. Springer-Verlag, 1988.

[11] J. Goguen and R. Burstall. Institutions: abstract model theory for specification and programming. *J. Assoc. of Computing Machinery*, 39:95–146, 1992.

[12] B. Herrmann. Equivalential and algebraizable logics. *Studia Logica*, 57:419–436, 1996.

[13] J. Łoś and R. Suszko. Remarks on sentential logics. *Proc. Kon. Nederl. Akad. Van Wetenschappen, Series A*, pages 177–183, 1958.

[14] S. Maehara. Craig's interpolation theorem. *Sŭgaku*, pages 235–237, 1961. Japanese.

[15] T. Maibaum, J. Fiadeiro, and M. Sadler. Stepwise program development in π-institutions. Research report, Imperial College of Science, Technology and Medicine, Department of Computing, 1990.

[16] T. Maibaum and M. Sadler. Axiomatising specification theory. In H.-J. Kreowski, editor, *Recent Trends in Data Type Specification*, volume 116 of *Informatik-Facberichte*, pages 171–177. Springer-Verlag, 1985.

[17] A. I. Mal'cev. *Algebraic Systems*, volume 192 of *Die Grundlehren der mathematischen Wissenschaften in Einzeldarstellungen*. Springer-Verlag, New York, 1973.

[18] J. Meseguer. General logics. In H.-D. Ebbinghaus et al., editor, *Proceedings, Logic Colloquium, 1987*, pages 275–329. North-Holland Publishing Co., Amsterdam, 1989.

[19] P. A. S. Veloso. A new, simpler proof of the modularization theorem for logical specifications. *Bull. of the IGPL*, 1:3–12, 1993.

[20] P. A. S. Veloso. On some logical properties related to modularization and interpolation. Publicações Técnicas ES-285/93, Universidade Federal do Rio de Janeiro, 1993.

[21] P. A. S. Veloso and T. S. E. Maibaum. On the modularisation theorem for logical specifications. Research Report DoC 92/35, Imperial College of Science, Technology and Medicine, Department of Computing, 1992.

Systematising Reactive System Design

Thomas S.E. Maibaum
with
Pauline Kan and Kevin Lano

Department of Computing
Imperial College
180 Queen's Gate
London SW7 2BZ UK
tsem@doc.ic.ac.uk

Abstract. Reactive systems are a very important class of systems for engineering organisations. The role of software is now dominant and its share of system construction costs is ever increasing. However, in spite of much research devoted to reactive systems development, proper engineering languages, methods and tools, as construed by the conventional engineering community, are not available to support the technology. We examine the state of play and suggest a way ahead for putting into place appropriate technology for industry.

1 Introduction

There is a growing interest in providing methods and tools to support the development of (real-time) reactive systems. Reactive systems are a very important class of systems for engineering organisations. The role of software is now dominant and its share of system construction costs is ever increasing. However, in spite of much research devoted to reactive system development, proper engineering languages, methods and tools are not available to support the technology ([18]).

According to [16], the day to day activities of engineers consist of *normal design*, as comprising "the improvement of the accepted tradition or its application under 'new or more stringent conditions'". He goes on to say: "The engineer engaged in such design knows at the outset how the device in question works, what are its customary features, and that, if properly designed along such lines, it has good likelihood of accomplishing the desired task."

[12] discusses this concept of 'normal design', although he does not use this phrase himself. "An engineering handbook is not a compendium of fundamental principles; but it does contain a corpus of rules and procedures by which it has been found that these principles can be most easily and effectively applied to the particular design tasks established in the field. The outline design is already given, determined by the established needs and products." ... "The methods of value are micro-methods,

A.M. Haeberer (Ed.): AMAST'98, LNCS 1548, pp. 17-22, 1998.

closely tailored to the tasks of developing particular well-understood parts of particular well-understood products."

An implied but not explicitly stated view of engineering design is that engineers normally design devices as opposed to systems. A *device*, in this sense, is an entity whose design principles are well defined, well structured and subject to normal design principles. A *system*, in this sense, is an entity that lacks some important characteristics making normal design possible. "Systems are assemblies of devices brought together for a collective purpose." Examples of the former given by [16] are airplanes, electric generators, turret lathes; examples of the latter are airlines, electric-power systems and automobile factories. The software engineering equivalent of devices may include compilers, relational databases, PABXs, etc. Software engineering examples of systems may include air traffic control systems, internet banking systems, ...

It would appear that systems become devices when their design attains the status of being normal, i.e., the level of creativity required in their design becomes one of *systematic choice*, based on *well defined analysis*, in the context of *standard definitions* and *criteria* developed and agreed by the relevant engineers. The implications of this conception of design are clear: the design methods of engineers are specific and heavily sytematised. The implication is that the design environment and supported methods are highly specific and heavily domain dependent. (An obvious conclusion may be that environments designed to be very general in their application are likely to be less than effective in any particular domain, not being able to deal directly with the concepts, notations and methods of that domain.)

The software engineering equivalent of such "normal methods" for the design of devices are best exemplified by systems such as PLANWARE [19] which automatically generates a scheduling program from two inputs by a user: a classification of the kind of problem being solved (amongst four classes organised in a simple hierarchy) and a spreadsheet which is used to record attributes of the specific problem. A lot of theory and past system development experience has gone into the construction of this automated tool, which may be said to resemble a *wizard* as seen in many modern applications. PLANWARE addresses a class of problems for which the required engineering design knowledge can be encapsulated in a one step interaction. Reactive systems as a class are not so simple and require more sophisticated multi-step design methods. Nevertheless, guided by the concept of normal design, we want to look at reactive systems and apply the same principles of *design by classification*.

2 Design by Classification

In PLANWARE, what is actually being classified is the kind of scheduling problem being addressed. This then derives choices amongst potential algorithms to be used and data structure implementations to be generated. These may obviously be regarded as choices amongst particular *patterns* (algorithmic or data oriented, see [13]) that may be relevant to the class of problems. Reactive systems are not algorithms, as such, so what is the equivalent concept of pattern to be applied to classifying reactive systems? The concept of *software architecture* ([11, 17]) provides an appropriate starting point.

Loosely, a software architecture is a definition of the structure of a software system in terms of (functional) components and so-called connectors. The latter are themselves software components, but ones which are meant to standardise the form of interaction between the components being connected. The 'pipes and filters' based scripting facilities of some operating systems may be seen as primitive examples of these sorts of structuring ideas. The problem frame proposals of [17] may also be seen in this light, at least as far as defining structure for requirements are concerned.

In order to achieve normality for reactive system design, we must provide *systematised choice*, based on *well-defined analysis*, in the context of *standard definitions* and *criteria* developed and agreed by the relevant engineers. If we are to use software architecture to systematise choice, then we must find the right architectures for conceivable classes of reactive systems and then organise them in some way, presumably using some notion of hierarchy. These concepts have been explored in the setting of a formalisation of object oriented concepts using theories in temporal or modal logic and universal constructions of category theory for structuring systems from components ([20]). By using such an approach based on object oriented concepts, we can use notions such as inheritance as relations underpinning a concept of architectural hierarchy.

We are not going to attempt to organise the whole subject area of reactive systems in this manner, but we can try the approach within more tightly constrained domains, eventually learning through this process how to organise larger parts of the area. The domains about which we have specific knowledge and which would appear to conform to our objectives include the design of PABXs and of process control systems. Below we describe our experience in this domain, highlighting the concepts of systematised choice, well-defined analysis, and standard definitions and criteria.

3 Process Control Applications

The B language [1, 2] is used for formal specification and design of reactive systems, in particular safety critical systems. It is based on first order predicate logic and set theory. Its operations are defined using an extension, GSL, of Dijkstra's guarded command language. It is a modular language; the modules encapsulate states and operations on these states, and similar in style to that of the language Ada. Procedural Control Theory (PCT) [3, 4] is used to synthesise algorithms for control devices at a high level (abstract designing of behavioural specifications), termed Procedural Controllers. It is built upon standard control engineering methods, and introduces the concept of ordered actions in response to system events. It provides a systematic means of transforming the control logic of a system to finite state machines (FSM), requiring no timed or parameterised transitions. As these FSMs are mathematically guaranteed to model processing behaviour required to meet system specifications, it can be used as a provably correct specification in B. Several case studies that have been developed using the techniques will be described and used as examples.

A set of detailed guidelines has been formed to specify in B controllers for reactive systems. Six steps have been identified [5, 6], these being:

1 **Finite State Machines** are produced for each component in the model, representing individual component behaviour. Two types of transitions exist: controllable transitions, which are output control signals from the controller to the component, and uncontrollable transitions, which are component input events sent to the controller.

2 **Formalisation of Properties** Three types of properties need to be formalised: *Required Reactions to Events*: are actions required in response to all possibly occurring system events; *Safety Properties*: are conditions and actions required for the prevention of hazardous situations; *Temporal Properties*: are representations of liveness and timing requirements. The language of Object Calculus [7] can be used for this representation.

3 **Data and Control Flow Diagrams (DCFDs)** show the communication between controllers and system hardware (as signals from sensor devices to the controller). The controller reacts to the input signals by sending appropriate commands to actuators (such as valves and pumps). The decomposition structure of the controllers is clearly discernible in the DCFDs. There are two types of signals: from sensors to controllers and controller commands to the actuators and successor controllers.

4 **Specification of Controllers in B** Each system event that can occur must have a corresponding operation, describing required actions to be taken. A controller is an 'overseer' in the system. It is possible to have a controller governing each mode of system operation, or each system component.

5 **Implementation of Controllers in B** No ordering of actions is required at the specification level (step 4). The concern is with the set of actions needed. Ordering of actions is introduced during the implementation stage only.

6 **Specification and Implementation of 'Outer Controller' Component** The Outer Controller module interfaces the controlling software with the external world. Detection of input events from system components is by, for example, polling or buffered event transmission techniques.

The specification of the controlling algorithm can be modularised and decomposed into conceptually coherent and verifiable sub-components. Three ways of decomposing the controlling algorithm have been identified. *Horizontal Decomposition of Controllers*: is based on a physical decomposition of the actual system. Events from the external world are copied to two separate sub-control algorithms, say, S1 and S2. Both S1 and S2 will compute their reactions independently of each other. However, this is (i) only feasible where S1 and S2 require the use and control of disjoint sets of actuators; (ii) only practical where few sensors are shared; (iii) responses cannot be time-critical, relative to each other.

Vertical Decomposition of Controllers: is also known as hierarchical decomposition. Events e are handled by the main controller S, which is responsible for certain interactions between components, for example, to maintain system safety invariants. The events e are then forwarded to subordinate controllers S1....Sn, which are responsible for the management of individual subcomponent behaviour. This structure is appropriate in systems where control can be separated and managed at both an aggregate and individual component level, such as in the Chain of Responsibility design pattern, where the responsibility for various control aspects of a system can be delegated to different levels. Again this is usually based upon a physical decomposition of the actual system.

Decomposition by Control Mode: splits the system in terms of modes of operation, or phases. A separate controller is specified for each mode, which will handle the responses for each event occurrence in that mode [8].

The detection of faults by the controlling software can be either inherent in the control algorithm, or delegated to different levels of control (*Chain of Responsibility* pattern). In the flexible production cell, a schedule is created for each blank arriving into the system, giving arrival times and duration at each system location. Schedules can then be derived for each crane, processing unit and deposit belt. The combined schedules form a system timetable, and can be used to detect component failures, by comparing the actual and scheduled progress of each component. Should a difference occur, fault tolerant software can then be initiated. A similar usage is made in train signaling systems. The design pattern of Chain of Responsibility has also been applied to the Steam Boiler example [14] and the flexible cell example, amongst others. It appears that variations of it, which can be organised hierarchically as software architectures, have universal applicability in this process control domain. Detection of invalid data transmission is applied by the outermost controlling level, separate from the detection of component failures (applied by the topmost controlling software). The calculations of responses by individual components are then performed by a lower level controller.

Conclusions

As we learn more about process control applications, we find that a domain specific set of software architectures and control algorithms and a set of design heuristics serve to systematise software design. We believe that this can be taken to the point that a 'design wizard' can be built which would assist conventional control engineers in the construction of highly reliable software based process control systems without the need for intervention by engineers skilled in esoteric skills like model checking or interactive proof. We further believe that this is the way forward in other domains of reactive systems.

References

[1]J. R. Abrial, *The B Book : Assigning Programs to Meanings*, Cambridge University Press, 1996

[2] K. Lano, *The B Language and Method :A Guide to Practical Formal Development*, FACIT series, Springer-Verlag, 1996

[3] G. E. Rotstein, A.Sanchez, S.Macchietto, Procedural Control of Discrete Event Systems. Submitted to *Intl. J. of Control*, 1998

[4] A. Sanchez, Formal Specification and Synthesis of Procedural Controllers for Process Systems, *Springer-Verlag Lecture Notes in Control and Information Sciences*, vol. 212, 1996

[5] P. Kan, Specification and Implementation of Reactive Systems with B, MSc thesis, Imperial College, 1997

[6] K. Lano, P. Kan, J. Bicarregui, Combining Scheduling Theory and Formal Methods in the Development of a Flexible Manufacturing System. Submitted to *Theory and Formal Methods 98*,

[7] J. Fiadeiro, T. Maibaum, Temporal Theories as Modularisation Units for Concurrent System Specification, *Formal Aspects of Computing* 4(3), pp. 239-272, 1992

[8] International Society for Measurement and Control, Control Models and Terminology ISAS88.01-1995, 1995

[9] J. R. Abrial, E. Borger, H. Langmaack, Formal Methods for Industrial Applications: Specifying and Programming the Steam Boiler Control *LNCS 1165*, Springer-Verlag, 1997

[10] M. Ali, Specification of Steam Boiler, MSc thesis, Imperial College, 1998

[11] R. Allen, D. Garlan, Formalising Architectural Connection, *First Int. Workshop on Architectures for Software Systems*, 1995

[12] M. Jackson, Formal Methods and Traditional Engineering, *Journal of Systems and Software: special issue on Formal Methods Technology Transfer*, 40(3), pp. 191-194, 1998

[13] E. Gamma, R. Helm, R. Johnson, J. Vlissides, Design Patterns: Elements of Reusable Object-oriented Software, Addison-Wesley, 1994

[14] PRESTO P4 Project, Integrated Design of Control and Automation Systems, PRESTO Document 200197A11, Centre for Process Systems Engineering, Imperial College, 1997

[15] K. Lano, A. Sanchez, Design of Reactive Control Systems for Event-Driven Operations, *FME 97: Industrial Applications and Strengthened Foundations of Formal Methods*. (J. Fitzgerald, C. B. Jones, P. Lucas Eds.) LNCS 1313, 1997

[16] W. G. Vincenti,, *What Engineers Know and How They Know It*, The Johns Hopkins University Press, 1990

[17] P. Zave, M. Jackson, Four Dark Corners of Requirements Engineering, *ACM TOSEM*, 6(1), pp. 1-30, 1997

[18] A. Haeberer, T. Maibaum, The Very idea of Software Development Environments: A Conceptual Architecture for the ARTS Environment Paradigm, *Proc. of ASE 1998*, IEEE Computer Society Press, pp. 260-269, 1998

[19] L. Blaine, L. Gilham, J. Liu, D. Smith, S. Westfold, PLANWARE – Domain-Specific Synthesis of High Performance Schedulers, *Proc. of ASE 1998*, IEEE Computer Society Press, pp. 270-279, 1998

[20] J. Fiadeiro, T. Maibaum, A Mathematical Toolbox for the Software Architect, *Proc. of the 8th International Workshop on Software Specification and Design*, IEEE Computer Society Press, 1996

Systematic Design of Call-Coverage Features
(Extended Abstract)

Pamela Zave

AT&T Laboratories—Research
Shannon Laboratory
180 Park Avenue, Room D205
Florham Park, New Jersey 07932, USA
pamela@research.att.com
+1 973 360 8676

Abstract. Feature interaction is a severe practical problem in the design and maintenance of telecommunication software. The Distributed Feature Composition (DFC) virtual architecture provides a new formal foundation for large-scale, modular description of telecommunication features. It also provides a semantic structure within which feature interactions can be diagnosed, and undesirable ones can be prevented or cured. This approach is illustrated by a systematic treatment of the interactions among the class of "call coverage" features. Language semantics and analysis techniques link this domain-specific reasoning to formal verification of system properties.

Keywords: *telecommunications, feature interaction, formal methods, modularity, software architecture, analysis, verification.*

1 A Challenge to Formal Methods: The Feature-Interaction Problem in Telecommunications

Many descriptions of systems are organized into modules called *features*. More specifically, the system description is a composition of a base description and some feature descriptions. What distinguishes features from other types of module is that each feature is optional—usually any subset of a valid feature set is also a valid feature set. Feature modularity is popular, particularly for informal descriptions, because it makes system descriptions easier to understand, modify, and extend.

Features can modify or influence each other in describing the overall system behavior. Although the mechanisms through which features interact depend on the language for specifying features and the rule for composing them, feature interactions must be possible in any useful description technique, because many feature interactions are necessary or desirable.

Feature-interaction problems arise when the result of composing acceptable feature specifications is unacceptable in some way. Depending on language or

A.M. Haeberer (Ed.): AMAST'98, LNCS 1548, pp. 23–27, 1998.

composition semantics and engineering choices, a composition might be considered unacceptable because it is inconsistent, incomplete, not associative, nondeterministic, unimplementable, or incorrect with respect to some global constraint on system behavior.

Feature-interactions problems are particularly severe in the domain of telecommunications. Among other reasons, telecommunication systems grow over time to have hundreds or even thousands of features. Desirable feature interactions are common (many features are explicit exceptions to other features), so it is not surprising that undesirable feature interactions are also common. Many new features are technology-driven, which causes problems because new technologies tend to undermine assumptions and invariants upon which older features depend. The scope and severity of these problems are amply illustrated by the proceedings of three workshops on the subject [1,3,4].

Despite much research activity on the problem of feature interaction in telecommunications, there is little sense of progress [8]. The fundamental difficulty is that existing formal description techniques are insufficiently modular. It is too likely that introducing a new feature to a system description will introduce undesirable feature interactions, thus demanding changes to previous features (this is referred to as "non-monotonicity" by Velthuijsen). To manage the complexity of telecommunications successfully, a formal description technique must make it easy to describe features that are independent or that interact with others only in desirable ways. At the same time, there must be no loss of generality.

2 Foundation: The Distributed Feature Composition Virtual Architecture

Distributed Feature Composition (DFC) is a new architecture for the description of telecommunication services, developed by Michael Jackson and myself (full details and feature examples can be found elsewhere [5]). It was designed to capture the full range of behavior of these systems while abstracting away from most of the implementation detail. It was also designed for feature modularity. As it achieves these goals to a significant degree, it provides a good foundation for new work on feature interaction.

In the pipe-and-filter architectural style [7], a filter is an independently scheduled process and a pipe is a buffered communication channel following a uniform protocol. Filters have no communication with other filters except through pipes, and a filter does not know what is on the other ends of its pipes (the configuration of pipes and filters is usually assumed to be static). The advantage of this architectural style is that it is modular with respect to filters.

DFC can be thought of as an adaptation of the pipe-and-filter style to the telecommunication domain, with featureless, internal, two-party voice calls as pipes and with *feature boxes* as filters. Assemblies of pipes and filters, known as *usages,* are created dynamically (through a routing procedure) in response to external service requests. Since the goal is feature modularity, and some features are realized by more than one feature box, different boxes implementing the same feature are allowed to share state.

3 A DFC-Based Method for Preventing Feature-Interaction Problems

A method for preventing feature-interaction problems in DFC-based descriptions has the following steps:

1. Engineers write initial feature descriptions, thinking of the features (as usual) in isolation.
2. An algorithm checks that constraints on descriptions of individual features are satisfied (backtrack if the check fails). The constraints are intended to impose predictability without actually limiting the features that can be described. In particular, they enable us to design algorithms for detecting possible feature interactions.
3. An analysis algorithm generates a list of possible interactions among the features. Some of these interactions will be desirable, and some will not. Both kinds are difficult for people to predice unaided.
4. Engineers decide on the desired behavior, encode it in the feature descriptions, and repeat the method from Step 3 onward until the only remaining feature interactions are desirable ones.

Although these steps are not defined in terms of aspects of DFC, the use of DFC is important for the success of this method, for several reasons:

- Its generality ensures that all desired features can be specified.
- Its abstraction and structural constraints make it possible to detect a broad range of feature interactions algorithmically from descriptions of individual features. Where abstraction and structure are lacking, the only feature interactions that can be detected automatically are violations of explicitly stated external correctness properties. These properties must be formulated manually, and the task of doing so has proven to be extremely difficult [8].
- Its modularity means that the desired behavior can usually be achieved just by describing individual features wisely. It is seldom necessary to increase description complexity or compromise feature modularity by "programming" deliberate cooperation among features.

4 Details of the Method for Call-Coverage Features

The class of *call-coverage* features has a technical definition within the DFC architecture. The class gets its name because the usual purpose of a call-coverage feature is to cover for a callee who cannot or will not answer an incoming call. Some of the best-known features, such as Unconditional Call Forwarding, Call Forwarding on Busy, Call Forwarding on No Answer, Call Screening, and Return Call, are call-coverage features.

Some constraints imposed by the method on feature descriptions ensure that a feature box is transparent (unobservable by other boxes in the usage) except when the specific function of the feature is involved. Other constraints ensure that a feature box notifies its environment about some of its internal activities, such as use of voice channels. This information is necessary for other feature boxes to coordinate their efforts.

Many different types of interaction are detectable. For example, a feature can cancel another under certain circumstances, or take priority over it in responding to a particular condition. One feature can delay another, suspend another, or cause it to receive status signals from another person's telephone as if they came from the feature's subscriber's telephone. Features contend with each other for use of a voice path to communicate with a user by means of announcements and touch-tones.

Despite this variety, most of the time the desired interaction behavior can be achieved simply by adjusting the feature precedences, which determine the order in which feature boxes appear in a usage.

5 Application of the Method to a Useful Set of Call-Coverage Features

Consider a telephone with Caller Identification and rich capabilities for conferencing and time-multiplexing individual calls. With such an interface, a user could easily manipulate multiple incoming, unanswered calls.

Added to this powerful interface, four call-coverage features would provide the user with a great deal of convenience and flexibility:

OM With Outbound Messaging, a callee can leave a message to be played if a particular person calls. Alternatively, the callee can request that a yet-unaswered caller hear a prerecorded message such as, "I will be able to answer you very shortly, please stay on the line."

IM Inbound Messaging gives the caller the ability to leave a voice message for the callee.

SCF With Selective Call Forwarding, the callee can specify ahead of time that calls from a particular person should be forwarded to another destination. Alternatively, the callee can forward a current incoming call instead of answering it.

BUCP Blocking with Urgent Calling Privilege enables the callee to block all incoming calls or, alternatively, to receive only urgent calls from a particular set of privileged callers.

These features are not simple, and their set of possible interactions is large—large enough to contain surprises for even the most experienced feature designer. The method is applicable, however, and the best possible interaction behavior can be specified without any loss of feature modularity.

6 Issues of Language, Analysis, and Verification

The reasoning outlined in Section 4 is domain-specific. It is easily justified, because the possibility of an interaction can be established convincingly by an example. The necessary analysis is defined on feature descriptions written in Promela and Z, as explained elsewhere [9].

One obvious flaw in this treatment is that there is no guarantee against further, undetected feature interactions. It would be valuable to be able to verify general global properties of feature sets. These global properties could include non-interference properties and rules of telephone etiquette.

General-purpose verification of DFC descriptions is likely to be challenging because of the wide-spectrum nature of the descriptions, not to mention the inherent behavioral complexity of telecommunication systems. This goal may be best reached by describing feature sets using one or more different formal notations, for example algebraic specifications of stream processing [2,6].

References

1. L. G. Bouma and H. Velthuijsen, editors. *Feature Interactions in Telecommunications Systems.* IOS Press, 1994.
2. Manfred Broy. Towards a logical basis for systems engineering. In *Proceedings of the Nineteenth International NATO Summer School,* to appear, 1999.
3. K. E. Cheng and T. Ohta, editors. *Feature Interactions in Telecommunications Systems III.* IOS Press, 1995.
4. P. Dini, R. Boutaba, and L. Logrippo, editors. *Feature Interactions in Telecommunication Networks IV.* IOS Press, 1997.
5. Michael Jackson and Pamela Zave. Distributed feature composition: A virtual architecture for telecommunications services. *IEEE Transactions on Software Engineering,* to appear, October 1998.
6. Bernhard Moller. Algebraic structures for program calculation. In *Proceedings of the Nineteenth International NATO Summer School,* to appear, 1999.
7. Mary Shaw and David Garlan. *Software Architecture: Perspectives on an Emerging Discipline.* Prentice-Hall, 1996.
8. Hugo Velthuijsen. Issues of non-monotonicity in feature-interaction detection. In K. E. Cheng and T. Ohta, eds., *Feature Interactions in Telecommunications Systems III,* pages 31-42. IOS Press, Amsterdam, 1995.
9. Pamela Zave. Formal description of telecommunication services in Promela and Z. In *Proceedings of the Nineteenth International NATO Summer School,* to appear, 1999.

Visual Abstractions for Temporal Verification *

Zohar Manna, Anca Browne, Henny B. Sipma, and Tomás E. Uribe

Computer Science Department
Stanford University
Stanford, CA. 94305-9045
manna@cs.stanford.edu

Abstract. *Generalized Verification Diagrams* combine deductive and algorithmic verification to establish general temporal properties of finite- and infinite-state reactive systems. The diagram serves as an abstraction of the system. This abstraction is deductively justified and algorithmically model checked. We present a new simple class of verification diagrams, using Müller acceptance conditions, and show how they can be used to verify general temporal properties of reactive systems.

1 Introduction

Reactive systems maintain an ongoing interaction with their environment, and include discrete, real-time and hybrid systems. *Deductive verification* is based on verification rules, which reduce the system validity of a temporal property to the general validity of first-order *verification conditions*. This methodology is complete relative to the underlying first-order reasoning, and allows the verification of a wide range of infinite-state systems. However, the proofs can be difficult to construct and understand, particularly as the complexity of the system increases.

Verification Diagrams provide a graphical representation of a deductive proof, summarizing the necessary verification conditions, and are therefore easier to construct and understand. *Generalized Verification Diagrams* extend them to be applicable to arbitrary temporal properties, replacing the well-formedness check on the diagram by a finite-state model checking step.

Diagrams can also be seen as an *abstraction* of the system, where properties of the diagram are guaranteed to hold for the system as well. The diagram represents the aspects of the system relevant to the property to be proved, and serves as an intermediary between the system and the property. To prove that a system $\mathcal{S}$ satisfies φ, we can find a diagram Ψ such that

$$\mathcal{L}(\mathcal{S}) \subseteq \mathcal{L}(\Psi) \subseteq \mathcal{L}(\varphi) \ ,$$

* This research was supported in part by the National Science Foundation under grants CCR-95-27927 and CCR-9804100, the Defense Advanced Research Projects Agency under NASA grant NAG2-892, ARO under grants DAAH04-95-1-0317, DAAH04-96-1-0122 and DAAG55-98-1-0471, ARO under MURI grant DAAH04-96-1-0341, and by Army contract DABT63-96-C-0096 (DARPA).

A.M. Haeberer (Ed.): AMAST'98, LNCS 1548, pp. 28–41, 1998.

where $\mathcal{L}(\mathcal{S})$ is the set of computations of $\mathcal{S}$, $\mathcal{L}(\Psi)$ is the set of computations of the diagram, and $\mathcal{L}(\varphi)$ is the set of models of φ.

The inclusion $\mathcal{L}(\mathcal{S}) \subseteq \mathcal{L}(\Psi)$ is proved deductively, by establishing verification conditions, and is equivalent to proving the correctness of an abstraction of $\mathcal{S}$. On the other hand, $\mathcal{L}(\Psi) \subseteq \mathcal{L}(\varphi)$ can be proved algorithmically, viewing the diagram as a finite ω-automaton. As ω-automata, diagrams can capture a class of properties that is strictly more general than those expressible in linear-time temporal logic.

Verification diagrams are thus a way of combining model checking and deductive verification. As abstractions, diagrams can be re-used in proofs of different properties, and provide visual documentation of the behaviors of the system.

In this paper we present a new, simple version of Generalized Verification Diagrams, based on Müller acceptance conditions. The Streett acceptance conditions of [BMS95] are equally expressive and more concise, but the new presentation allows a simpler, alternative definition. We show how these diagrams can be viewed as system abstractions, and used to verify general temporal properties of reactive systems.

Outline: Section 2 presents the basic background material. Section 3 presents the new class of Generalized Verification Diagrams. We discuss some practical issues in Section 4, and present related work in Section 5. In Section 6 we conclude by briefly describing our implementation of these methods in the Stanford Temporal Prover, STeP.

2 Background

2.1 Specifying Systems and Properties

We represent reactive systems as *fair transition systems* [MP95]. A fair transition system $\langle \Sigma, \Theta, \mathcal{T} \rangle$ is given by a set of states Σ, an *initial condition* Θ, which is a subset of Σ, and a set of *transitions* $\mathcal{T}$, each of which is a binary relation over Σ, describing how the system can move from one state to the next.

In our framework, we assume an *assertion language* based on first-order logic. The set Σ of possible system states is defined by a finite set of *system variables* $\mathcal{V}$; each transition τ is described by its *transition relation* $\rho_\tau(\mathcal{V}, \mathcal{V}')$, an assertion over the system variables $\mathcal{V}$ and a set of *primed variables* $\mathcal{V}'$ indicating their next-state values. Similarly, Θ can be expressed as an assertion over the system variables. We assume that $\mathcal{T}$ includes an *idling transition*, whose transition relation is $\mathcal{V} = \mathcal{V}'$. We use the standard triple notation for verification conditions,

$$\{\varphi\}\, \tau\, \{\psi\} \stackrel{\text{def}}{=} (\varphi(\mathcal{V}) \wedge \rho_\tau(\mathcal{V}, \mathcal{V}')) \rightarrow \psi(\mathcal{V}') \ .$$

A *run* of $\mathcal{S}$ is an infinite sequence of states $s_0, s_1, \ldots$, where s_0 satisfies Θ and for every s_i there is a transition $\tau \in \mathcal{T}$ such that (s_i, s_{i+1}) satisfy ρ_τ. In this case we say that τ is *taken* at s_i. The *enabled* predicate characterizes the set of states at which a transition τ can be taken:

$$\textit{enabled}(\tau) \stackrel{\text{def}}{=} \exists \mathcal{V}'.\rho_\tau(\mathcal{V}, \mathcal{V}') \ .$$

$$\textbf{local } x, y : \textbf{integer where } x = 0 \land y = 0$$

$$\left[\begin{array}{l}\textbf{loop forever do}\\ \left[\begin{array}{l}\ell_0\text{: } \textbf{awaitc } x = 0\\ \ell_1\text{: } y := y + 1\end{array}\right]\end{array}\right] \quad \| \quad \left[\begin{array}{l}\textbf{loop forever do}\\ \left[\begin{array}{l}m_0\text{: } \textbf{while } x > 0 \textbf{ do}\\ \quad [m_1\text{: } x := x - 1]\\ m_2\text{: } x := y\end{array}\right]\end{array}\right]$$

–P1– –P2–

Fig. 1. Program LOOPS

Transitions can be marked as *just* or *compassionate*. Just (or *weakly fair*) transitions cannot be continuously enabled without ever being taken. Compassionate (or *strongly fair*) transitions cannot be enabled infinitely often without being taken. Every compassionate transition is also just. A *computation* is a run that satisfies these fairness requirements.

Properties of systems are expressed as formulas in *linear-time temporal logic* (LTL). *Assertions*, or state-formulas, are first-order formulas with no temporal operators, and can include quantifiers. *Temporal formulas* are constructed from assertions, boolean connectives, and the usual *future* ($\Box, \Diamond, \bigcirc, \mathcal{U}, \mathcal{W}$) and *past* ($\boxminus, \diamondminus, \ominus, \mathcal{B}, \mathcal{S}$) temporal operators [MP95]. A *model* of a temporal property φ is an infinite sequence of states $s_1, s_2, \ldots$ that satisfies φ. For a system $\mathcal{S}$, we say that φ is $\mathcal{S}$*-valid* if all the computations of $\mathcal{S}$ are models of φ.

Example 1. Figure 1 shows program LOOPS, written in the Simple Programming Language (SPL) of [MP95]. SPL programs can be naturally translated into corresponding fair transition systems, following the semantics of each of the SPL constructs. To each process corresponds a *control variable*. For LOOPS, the control variables for processes P1 and P2 range over locations $\{\ell_0, \ell_1\}$ and $\{m_0, m_1, m_2\}$. The assertions m_i and ℓ_j are used to indicate the control location for each process, so the initial condition is:

$$\Theta : \ell_0 \land m_0 \land x = 0 \land y = 0 .$$

The **awaitc** transition ℓ_0 is assumed to be a compassionate variant of the just **await** statement from [MP95]. That is, if control resides at ℓ_0 and $x = 0$ infinitely often, then the transition must eventually be taken. All other transitions are assumed to be just. Program LOOPS is an infinite-state system, since the variables x and y can grow beyond any bound. We will show that

$$\varphi : \Box(y > 0 \rightarrow \Diamond\Box(y \geq M))$$

is valid over this program, for any $M > 1$. That is, if $y > 0$ then eventually y will always be greater than the arbitrary constant M. □

We distinguish between *safety* and *progress* properties [MP95]. Intuitively, a safety property means that a particular class of "bad states" will never be reached. For example, *invariance formulas*, of the form $\Box p$ for an assertion p, and *wait-for formulas*, of the form $\Box(p \rightarrow q_1 \mathcal{W} \ldots \mathcal{W} q_n)$, express safety properties. Progress properties, on the other hand, state that certain states will eventually be reached. While safety properties do not depend on the fairness constraints of the system, progress properties do require fairness.

A binary relation $\succ$ is *well-founded* over a domain $\mathcal{D}$ if there are no infinite sequences of elements $e_1, e_2, \ldots$ in $\mathcal{D}$ such that $e_1 \succ e_2 \succ \ldots$ We write $x \succeq y$ iff $x \succ y$ or $x = y$. A *ranking function* δ is a mapping from system states into a *well-founded domain* $(\mathcal{D}, \succ)$.

2.2 Deductive Verification

Verification rules reduce the validity of a given temporal property over a given system $\mathcal{S}$ to the general validity of a set of first-order *verification conditions*. For example, the *general invariance rule*, which proves a property of the form $\Box p$ for an assertion p, requires finding an assertion φ such that the following verification conditions are valid: (1) $\varphi \rightarrow p$ (that is, φ strengthens p), (2) $\Theta \rightarrow \varphi$ (that is, φ holds initially), and (3) $\{\varphi\}\, \tau\, \{\varphi\}$ for each transition $\tau \in \mathcal{T}$ (that is, φ is preserved by all transitions). Other verification rules are available for proving different classes of temporal properties [MP95].

3 Temporal Verification Diagrams

Verification diagrams were introduced by Manna and Pnueli [MP94] as a graphical representation of the verification conditions needed for a deductive proof. As with verification rules, different classes of diagrams are used to prove different classes of temporal properties.

Verification diagrams are generalized in [BMS95] to be applicable to arbitrary temporal properties, and shown to be a *complete* proof method for general (state-quantified) temporal formulas, relative to the reasoning required to establish verification conditions. The Generalized Verification Diagrams we describe below differ in presentation from those in [BMS95, BMS96], but the underlying notions remain the same.

A *Generalized Verification Diagram* (GVD) for a system $\mathcal{S}$ is a directed graph whose nodes are labeled by assertions, where a subset of the nodes is marked as *initial*. The assertion labeling a node n is indicated by $\mu(n)$. For a set of nodes $S : \{n_1, \ldots, n_k\}$, we define

$$\mu(S) \stackrel{\text{def}}{=} \mu(n_1) \vee \ldots \vee \mu(n_k) \ ,$$

where $\mu(\{\}) = \mathit{false}$. For a node n, the set of successor nodes of n is $\mathit{succ}(n)$.

A *run* of a diagram is a sequence $\sigma : s_0, s_1, \ldots$ of states of $\mathcal{S}$ such that there is an associated path $\pi : n_0, n_1, \ldots$ through the diagram, where n_0 is an initial node and for each $i \geq 0$, the state s_i satisfies $\mu(n_i)$.

We use an *acceptance condition* to define the limit behavior of the diagram. The theory of automata on infinite words (ω-automata) provides several types of acceptance conditions [Tho90]. For simplicity, here we choose *Müller* acceptance conditions. These are equally expressive as the more concise *Streett* acceptance conditions used in [BMS95], but allow a more intuitive presentation.

For an infinite path π through a GVD, let $\mathit{inf}(\pi)$ be the set of nodes that appear infinitely often in π. A *Müller acceptance condition* $\mathcal{F}$ is a set *of sets* of nodes. A path π is *accepting* if $\mathit{inf}(\pi) \in \mathcal{F}$. Note that an infinite path must eventually remain in a *strongly connected subgraph* (SCS), so an acceptance condition can always be expressed as a set of diagram SCS's. A *computation* of a diagram Ψ is a run of Ψ that has an associated accepting path. The set of all computations of Ψ is $\mathcal{L}(\Psi)$.

3.1 $(\mathcal{S}, \Psi)$ Verification Conditions

Associated with a GVD Ψ and a system $\mathcal{S}$ are verification conditions that, when valid, ensure that $\mathcal{L}(\mathcal{S}) \subseteq \mathcal{L}(\Psi)$. In this case, we say that Ψ is *$\mathcal{S}$-valid.*

- **Initiation:** Every initial state of $\mathcal{S}$ should be covered by some initial diagram node, that is,
$$\Theta \rightarrow \mu(I) \ ,$$
where I is the set of initial diagram nodes. This implies that every run of $\mathcal{S}$ can start at some initial node of Ψ.
- **Consecution:** For every node n and every transition τ, there is a successor node that can be reached by taking τ (if τ can be taken at all), that is,
$$\mu(n) \wedge \rho_\tau \rightarrow \mu'(\mathit{succ}(n)) \ .$$
Here, $\mu'(\mathit{succ}(n))$ is the result of replacing each system variable x in $\mu(\mathit{succ}(n))$ by x'.

Together, these two conditions imply that every run of $\mathcal{S}$ can remain within Ψ:

Proposition 1. *If a diagram Ψ satisfies the initiation and consecution requirement for a system $\mathcal{S}$, then the runs of $\mathcal{S}$ are a subset of the runs of Ψ.*

Thus, once the above verification conditions are proved, we can conclude that any safety property of Ψ also holds for $\mathcal{S}$.

To preserve progress properties, a second set of verification conditions ensures that every computation of the system can follow an accepting path in the diagram, that is, can always eventually remain in an accepting SCS. Thus, if an SCS S is not accepting, we must show that computations can always leave S, or cannot stay in S forever.

For an SCS S, a *tail*(S)*-computation* is a system computation that has a corresponding path π in the diagram such that $\mathit{inf}(\pi) = S$. An SCS is called *transient* if every *tail*(S)-computation can leave S (so it is also a *tail*(S')-computation, for an SCS $S' \neq S$).

We want to show that every non-accepting SCS is transient. An SCS can be shown to be transient in one of the following three ways:

- **Just exit:** An SCS S has a *just exit*, if there is a just transition τ such that the following verification conditions are valid: for *every* node $m \in S$,

$$\mu(m) \rightarrow enabled(\tau) \quad \text{and} \quad \mu(m) \wedge \rho_\tau \rightarrow \mu'(succ(m) - S) \ .$$

  This means that τ is enabled and can leave the SCS at all nodes. We say that τ is the *just exit transition* for S.
- **Compassionate exit:** An SCS S has a *compassionate exit*, if there is a compassionate transition τ such that the following verification conditions are valid: for *every* node $m \in S$,

$$\mu(m) \rightarrow \neg enabled(\tau) \quad \text{or} \quad \mu(m) \wedge \rho_\tau \rightarrow \mu'(succ(m) - S) \ ,$$

  and for *some* node $n \in S$, τ is enabled at n:

$$\mu(n) \rightarrow enabled(\tau) \ .$$

  This means that for every node in S, either τ is disabled or τ can lead out of S, and there is at least one node n where τ can indeed leave S. We say that n is the *exit node* and τ is the *compassionate exit transition* for S.
- **Well-founded SCS:** An SCS $S : \{n_1, \ldots, n_k\}$ is *well-founded* if there exist ranking functions $\{\delta_1, \ldots, \delta_k\}$, where each δ_i maps the system states into elements of a well-founded domain $(\mathcal{D}, \succ)$, such that the following verification conditions are valid: there is a *cut-set*[1] E of edges in S such that for all edges $(n_1, n_2) \in E$ and every transition τ,

$$\mu(n_1) \wedge \rho_\tau \wedge \mu'(n_2) \rightarrow \delta_1(\mathcal{V}) \succ \delta'_2(\mathcal{V}) \ ,$$

  and for all other edges $(n_1, n_2) \notin E$ in S and for all transitions τ,

$$\mu(n_1) \wedge \rho_\tau \wedge \mu'(n_2) \rightarrow \delta_1(\mathcal{V}) \succeq \delta'_2(\mathcal{V}) \ .$$

  This means that there is no $tail(S)$-computation: it would have to traverse at least one of the edges in E infinitely often, which contradicts the well-foundedness of the ranking functions.

We say that S has a *fair exit* if it has a just or a compassionate exit. Combined with consecution, the fair exit verification conditions ensure that a $tail(S)$-computation can always follow a path that leaves S. Any run of the system that is *forced* to stay within an SCS with a fair exit must be unfair. If S is well-founded, there can be no $tail(S)$-computations. We can now claim:

Proposition 2. *If a GVD Ψ for a system $\mathcal{S}$ satisfies the initiation and consecution requirements, and all non-accepting SCS's have a fair exit or are well-founded, then $\mathcal{L}(\mathcal{S}) \subseteq \mathcal{L}(\Psi)$, that is, Ψ is $\mathcal{S}$-valid.*

---

[1] A cut-set of an SCS S is a set of edges E such that every loop in S contains some edge in E (that is, the removal of E disconnects S).

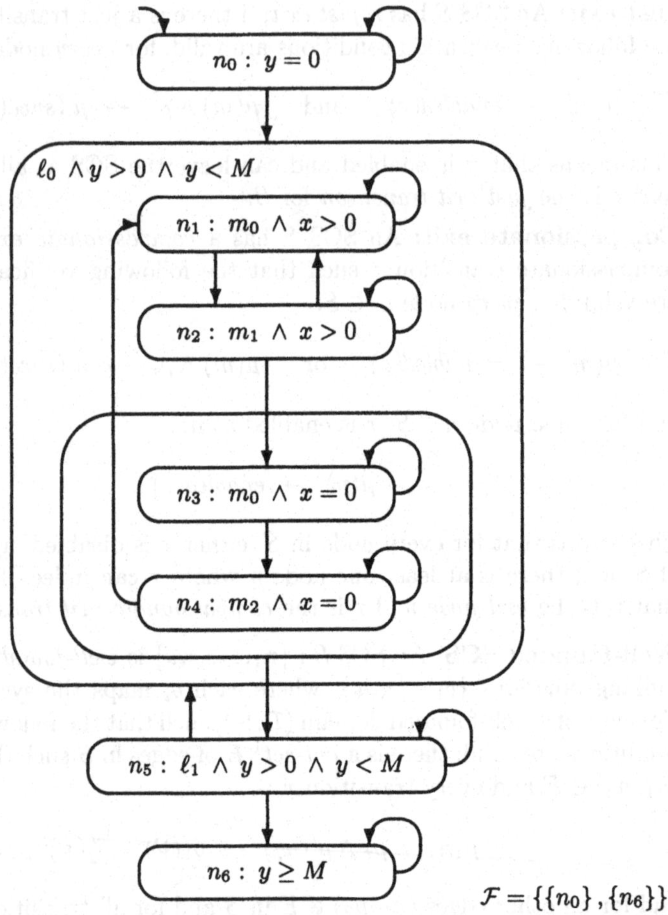

Fig. 2. A GVD Ψ for program LOOPS and property $\varphi : \Box(y > 0 \rightarrow \Diamond\Box(y \geq M))$

Thus, to show that a given diagram Ψ is $\mathcal{S}$-valid, the user must prove initiation and consecution, and specify, for each non-accepting SCS, one of the following:

1. a just exit transition τ;
2. a compassionate exit transition τ and its exit node n; or
3. well-founded ranking functions δ_i and a cut-set E that prove that the SCS is well-founded.

Example 2. Figure 2 shows a GVD for the program LOOPS of Figure 1. The only initial node is n_0. M is a constant, where we assume $M > 1$. *Encapsulation conventions*, based on those of Statecharts [Har87], are used to make diagrams more succinct. Nodes n_3 and n_4 are part of a *compound node* which, together with n_1 and n_2, is part of a larger compound node. An assertion that labels a compound node is added, as a conjunct, to its subnodes. Edges leaving (entering) a compound node are interpreted as leaving (entering) all of its subnodes.

The runs of the diagram include all runs of program LOOPS. Initiation holds, since Θ implies $y = 0$. The consecution conditions can also easily be proved: for instance, the only transitions enabled at node n_1 are the idling transition and transition m_0, which when taken always leads to n_2. However, the diagram has runs that, for example, stay in node n_2 forever. These are not computations of the program, because they are not fair.

The diagram acceptance condition allows the transfer of progress properties from the diagram to the system, if the diagram is shown to be $\mathcal{S}$-valid. The weakest acceptance condition that lets us prove

$$\varphi : \Box(y > 0 \rightarrow \Diamond\Box(y \geq M))$$

is $\mathcal{F} : \{\{n_0\}, \{n_6\}\}$. The presence of $\{n_0\}$ means that computations of the diagram can stay at n_0 with $y = 0$ and never reach $y > 0$. The inclusion of $\{n_6\}$ means that once a diagram computation leaves the safe haven of n_0, it must be able to reach n_6 and stay there.

Thus, to prove that the diagram is $\mathcal{S}$-valid, we must show that all other SCS's are transient:

- It is straightforward to show that $\{n_1\}$, $\{n_2\}$, $\{n_3\}$, $\{n_4\}$ and $\{n_5\}$ have just exit transitions m_0, m_1, m_0, m_2 and ℓ_1, respectively.
- The SCS $\{n_1, n_2, n_3, n_4\}$ has compassionate exit transition ℓ_0, with exit node n_3 or n_4. Transition ℓ_0 is always enabled at n_3 and n_4, leading out of the SCS, and is disabled at n_1 and n_2.
- The SCS $\{n_1, n_2\}$ can be shown to be well-founded with ranking function $\delta_i : x$ at both nodes. The value of x, always positive in this SCS, decreases along the edge $\langle n_2, n_1\rangle$, which is a cut-set for the SCS, and does not change along the other edges.
- The remaining SCS's are of the form $\{\langle n_1, n_2, n_3, n_4\rangle, n_5\}$, where $\langle n_1, \ldots, n_4\rangle$ stands for any nonempty subset of $\{n_1, \ldots, n_4\}$. They can be shown to be well-founded using the ranking function $\delta_i : M - y$ at all nodes. This is well-founded, since $M > y$ within these SCS's. For each SCS, the set of edges that leave n_5 is the cut-set. These edges can only be traversed by transition ℓ_1, which increases y and thus decreases the well-founded order. The transitions on all other edges preserve the value of y. □

3.2 (Ψ, φ) Property Satisfaction

Section 3.1 describes verification conditions that prove that $\mathcal{L}(\mathcal{S}) \subseteq \mathcal{L}(\Psi)$, that is, that the diagram defines a correct abstraction of the system. To prove the $\mathcal{S}$-validity of a property φ, it remains to show that all the computations of the diagram are models of φ, that is,

$$\mathcal{L}(\Psi) \subseteq \mathcal{L}(\varphi) \ .$$

This check can be performed using standard ω-automata model checking, if we can relate the nodes of the diagram with the *atomic assertions* in φ (the ones with no subformulas other than themselves). As in [BMS95], we do this using a *propositional labeling*, where a diagram node n can be labeled with a boolean combination b of atomic assertions of φ, if

$$\mu(n) \rightarrow b$$

is valid. Given a propositional labeling, a GVD Ψ can be seen as a finite-state ω-automaton $\Psi^{\mathcal{A}}$, where each node is a state of the automaton, labeled with the given propositions. The property itself can be seen as an abstract property $\varphi^{\mathcal{A}}$ over its propositions. For instance, the atomic assertions in $\varphi : \Box(y > 0 \rightarrow \Diamond\Box(y \geq M))$ are $p : y > 0$ and $q : y \geq M$, which are now regarded as propositions. The abstract property is then $\varphi^{\mathcal{A}} : \Box(p \rightarrow \Diamond\Box q)$.

In most cases, the node label justification is trivial, since the diagram is drawn with φ in mind and the atomic assertions of φ are usually already present in the node assertions (see Section 4).

Example 3. Consider again the GVD for program LOOPS in Figure 2. Let $p : y > 0$ and $q : y \geq M$. Then we can label the nodes as follows:

n_0 :	$\neg p$	(proving $y = 0 \rightarrow \neg(y > 0)$)
n_1, n_2, n_3, n_4, n_5 :	*true*	(no label is needed for these nodes)
n_6 :	q	

The resulting ω-automaton $\Psi^{\mathcal{A}}$ is shown in Figure 3. The abstract property is $\varphi^{\mathcal{A}} : \Box(p \rightarrow \Diamond\Box q)$. We can now use finite-state model checking to automatically establish that $\Psi^{\mathcal{A}}$ satisfies $\varphi^{\mathcal{A}}$ [Kur94, BCM+92]. Thus, we have proved that $\varphi : \Box(y > 0 \rightarrow \Diamond\Box(y \geq M))$ is valid for program LOOPS. □

From an abstraction point of view, the propositional labeling ensures that the abstract properties model checked for $\Psi^{\mathcal{A}}$ imply the desired properties of the

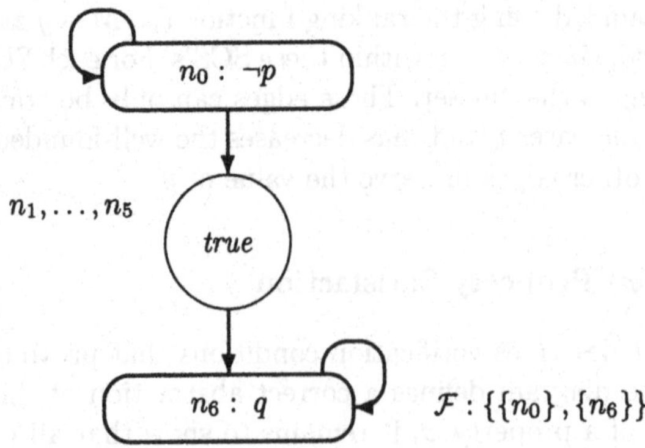

Fig. 3. Automata $\Psi^{\mathcal{A}}$, a propositional labeling of diagram Ψ of Figure 2

original system $\mathcal{S}$. The diagram can then be seen as a *weakly property-preserving abstraction* of the system $\mathcal{S}$ [Uri98].

Once a diagram Ψ is proved to be a correct abstraction, i.e. $\mathcal{L}(\mathcal{S}) \subseteq \mathcal{L}(\Psi)$, it can be used to transfer any property that can be model checked for the diagram over to the system $\mathcal{S}$.

Example 4. The diagram from Figure 2 also lets us prove the $\mathcal{S}$-validity of

$$\Box((\ell_0 \wedge y > 0 \wedge y < M) \rightarrow \Diamond x = 0) \ .$$

In this case, we let $p : \ell_0$, $q : y > 0$, $r : y < M$ and $s : x = 0$, and label the nodes as follows:

$n_0 : \quad \neg p \qquad$ (proving $y = 0 \rightarrow \neg(y > 0)$)
$n_1, n_2 : p \wedge q \wedge r$
$n_3, n_4 : s$
$n_5 : \quad \neg p \qquad$ (proving $\ell_1 \rightarrow \neg\ell_0$)
$n_6 : \quad \neg r \qquad$ (proving $y \geq M \rightarrow \neg(y < M)$)

We can now, as before, model check $\varphi^{\mathcal{A}} : \Box((p \wedge q \wedge r) \rightarrow \Diamond s)$ over the resulting ω-automaton $\Psi^{\mathcal{A}}$. □

4 GVD Templates

As mentioned earlier, [MP94] provide different types of diagrams, depending on the type of temporal property being proved. In particular, INVARIANCE diagrams prove properties of the form $\Box p$, for an assertion p; WAIT-FOR diagrams prove properties of the form $\Box(p \rightarrow q_m \mathcal{W} \ldots \mathcal{W} q_0)$; CHAIN prove response properties, of the form $\Box(p \rightarrow \Diamond q)$, that do not require the use of well-founded orders; and RANK diagrams prove response properties that do. Each type of verification diagram has different well-formedness constraints to ensure that the diagram satisfies the property.

We can define GVD templates for the more common properties, similarly to special-purpose diagrams. For example, to prove an invariance $\Box p$, the acceptance condition includes all SCS's, so no progress verification conditions need to be proved, and each node must be shown to satisfy p. The result is equivalent to the INVARIANCE diagrams of [MP94].

Example 5. Figure 4 provides a GVD template to prove formulas of the form $\Box(p \rightarrow \Diamond \Box q)$. The acceptance condition is

$$\mathcal{F} : \{\{n_0\}, \{n_2\}\} \ .$$

Thus all SCSs appearing in this circle must be shown to be transient: they should have a fair exit or be well-founded. The diagram of Figure 2 is an instance of this template. □

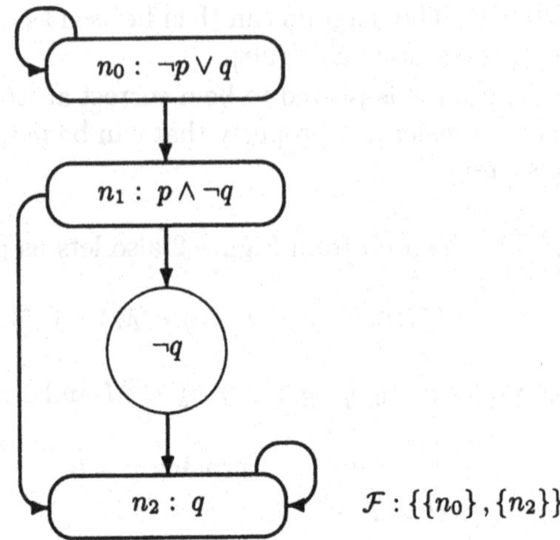

Fig. 4. GVD template to prove properties of the form $\square(p \rightarrow \diamond \square q)$

In general, the starting template for a GVD can be obtained from the *temporal tableau* of φ, a finite-state ω-automaton that represents all the models of φ (see, e.g. [MP95]).

In this case, the property satisfaction check is guaranteed to succeed, and the propositional labeling can be the identity. The user will have to prove consecution over this diagram, and show that all non-accepting SCS's have a fair exit or are well-founded. The structure of the diagram can then be filled in, adding details according to the user's understanding of the system. Verification diagrams can be constructed and checked incrementally. Since the verification conditions are local to the diagram, portions of the diagram can be formally verified, while others are edited until they can be proved correct or an error is found in the system being verified (see Section 6).

5 Related Work

While verification diagrams give a direct proof that all computations of a system satisfy the property φ, *deductive model checking* (DMC) [SUM98] shows that no computation can satisfy its negation. This is done by transforming a *falsification diagram*, which represents the product of a system abstraction and the tableau for $\neg\varphi$. The system abstraction is *refined*, as necessary, until the property is proved or a counterexample is found. At any given point, the falsification diagram includes all the computations of the system that may possibly violate φ.

While a GVD shows that every computation of S *can* follow an accepting path through the diagram, DMC shows that every computation of S *must* end in an unsuitable SCS in the product graph.

Property-preserving abstractions for reactive systems are discussed in, for instance, [CGL94, LGS+95, Dam96]. For a more extensive discussion of deductive-algorithmic verification and abstraction, see [Uri98]. See [Sip98] for more on deductive model checking, diagrams, and their application to the verification of real-time and hybrid systems.

6 Implementation: The STeP System

The Stanford Temporal Prover (STeP) is a tool for the temporal verification of reactive systems [BBC+96, MBB+98]. STeP parses SPL programs into fair transition systems, or can take transition systems directly as input. STeP includes verification rules and diagrams, automatic invariant generation, and symbolic and explicit-state LTL model checking.

The latest version of STeP, 2.0, features a Java graphical user interface that facilitates the construction and verification of diagrams, including GVD's (currently, using Streett acceptance conditions). The system automatically generates the required verification conditions and performs the required ω-automata model checking step. The diagram editor is closely integrated with the proof editor, so that the user can visualize the portions of the diagram that are proved and unproved as the diagram is incrementally constructed.

STeP provides integrated first-order automated reasoning and decision procedures [Bjø98] to facilitate the proof of verification conditions. STeP also uses these procedures for the generation of finite-state abstractions [CU98]. STeP can verify real-time and hybrid systems [BMSU97, MS98], and is being extended to include modular specification and verification [FMS98].

STeP is freely available for research and educational use. For more information, see the STeP web pages, at:

`http://www-step.stanford.edu`

References

[BBC+96] N.S. Bjørner, A. Browne, E.S. Chang, M. Colón, A. Kapur, Z. Manna, H.B. Sipma, and T.E. Uribe. STeP: Deductive-algorithmic verification of reactive and real-time systems. In R. Alur and T.A. Henzinger, editors, *Proc. 8th Intl. Conference on Computer Aided Verification*, vol. 1102 of *LNCS*, pages 415–418. Springer-Verlag, July 1996.

[BCM+92] J.R. Burch, E.M. Clarke, K.L. McMillan, D.L. Dill, and J. Hwang. Symbolic model checking: 10^{20} states and beyond. *Information and Computation*, 98(2):142–170, 1992.

[Bjø98] N.S. Bjørner. *Integrating Decision Procedures for Temporal Verification.* PhD thesis, Computer Science Department, Stanford University, November 1998.

[BMS95] A. Browne, Z. Manna, and H.B. Sipma. Generalized temporal verification diagrams. In *15th Conference on the Foundations of Software Technology and Theoretical Computer Science*, vol. 1026 of *LNCS*, pages 484–498. Springer-Verlag, 1995.

[BMS96] A. Browne, Z. Manna, and H.B. Sipma. Hierarchical verification using verification diagrams. In 2nd *Asian Computing Science Conf.*, vol. 1179 of *LNCS*, pages 276–286. Springer-Verlag, December 1996.

[BMSU97] N.S. Bjørner, Z. Manna, H.B. Sipma, and T.E. Uribe. Deductive verification of real-time systems using STeP. In *4th Intl. AMAST Workshop on Real-Time Systems*, vol. 1231 of *LNCS*, pages 22–43. Springer-Verlag, May 1997.

[CGL94] E.M. Clarke, O. Grumberg, and D.E. Long. Model checking and abstraction. *ACM Trans. on Programming Languages and Systems*, 16(5):1512–1542, September 1994.

[CU98] M.A. Colón and T.E. Uribe. Generating finite-state abstractions of reactive systems using decision procedures. In A.J. Hu and M.Y. Vardi, editors, *Proc.* 10th *Intl. Conference on Computer Aided Verification*, vol. 1427 of *LNCS*, pages 293–304. Springer-Verlag, July 1998.

[Dam96] D.R. Dams. *Abstract Interpretation and Partition Refinement for Model Checking.* PhD thesis, Eindhoven University of Technology, July 1996.

[FMS98] B. Finkbeiner, Z. Manna, and H.B. Sipma. Deductive verification of modular systems. In W.P. de Roever, H. Langmaack, and A. Pnueli, editors, *Compositionality: The Significant Difference, COMPOS'97*, vol. 1536 of *LNCS*, pages 239–275. Springer-Verlag, 1998.

[Har87] D. Harel. Statecharts: A visual formalism for complex systems. *Sci. Comp. Prog.*, 8:231–274, 1987.

[Kur94] R.P. Kurshan. *Computer-Aided Verification of Coordinating Processes: The Automata-Theoretic Approach.* Princeton University Press, 1994.

[LGS+95] C. Loiseaux, S. Graf, J. Sifakis, A. Bouajjani, and S. Bensalem. Property preserving abstractions for the verification of concurrent systems. *Formal Methods in System Design*, 6:1–35, 1995.

[MBB+98] Z. Manna, N.S. Bjørner, A. Browne, M. Colón, B. Finkbeiner, M. Pichora, H.B. Sipma, and T.E. Uribe. An update on STeP: Deductive-algorithmic verification of reactive systems. In *Tool Support for System Specification, Development and Verification*, pages 87–91. Christian-Albrechts-Universitat, Kiel, June 1998. Full version to appear in LNCS.

[MP94] Z. Manna and A. Pnueli. Temporal verification diagrams. In M. Hagiya and J.C. Mitchell, editors, *Proc. International Symposium on Theoretical Aspects of Computer Software*, vol. 789 of *LNCS*, pages 726–765. Springer-Verlag, 1994.

[MP95] Z. Manna and A. Pnueli. *Temporal Verification of Reactive Systems: Safety.* Springer-Verlag, New York, 1995.

[MS98] Z. Manna and H.B. Sipma. Deductive verification of hybrid systems using STeP. In T. Henzinger and S. Sastry, editors, *Hybrid Systems: Computation and Control*, vol. 1386 of *LNCS*, pages 305–318. Springer-Verlag, 1998.

[Sip98] H.B. Sipma. *Diagram-based Verification of Discrete, Real-time and Hybrid Systems.* PhD thesis, Computer Science Department, Stanford University, December 1998.

[SUM98] H.B. Sipma, T.E. Uribe, and Z. Manna. Deductive model checking. To appear in *Formal Methods in System Design*, 1998. Preliminary version appeared in *Proc.* 8th *Intl. Conference on Computer Aided Verification*, vol. 1102 of *LNCS*, Springer-Verlag, pp. 208–219, 1996.

[Tho90] W. Thomas. Automata on infinite objects. In J. van Leeuwen, editor, *Handbook of Theoretical Computer Science*, vol. B, pages 133–191. Elsevier Science Publishers (North-Holland), 1990.

[Uri98] T.E. Uribe. *Abstraction-based Deductive-Algorithmic Verification of Reactive Systems.* PhD thesis, Computer Science Department, Stanford University, December 1998.

A Linear Metalanguage for Concurrency

Glynn Winskel

BRICS*, University of Aarhus, Denmark

Abstract. A metalanguage for concurrent process languages is introduced. Within it a range of process languages can be defined, including higher-order process languages where processes are passed and received as arguments. The metalanguage is provided with two interpretations both of which can be understood as categorical models of a variant of linear logic. One interpretation is in a simple category of nondeterministic domains; here a process will denote its set of traces. The other interpretation, obtained by direct analogy with the nondeterministic domains, is in a category of presheaf categories; the nondeterministic branching behaviour of a process is captured in its denotation as a presheaf. Every presheaf category possesses a notion of (open-map) bisimulation, preserved by terms of the metalanguage. The conclusion summarises open problems and lines of future work.

1 Introduction

Over the last few years, Gian Luca Cattani and I have worked on presheaf models for interacting processes, culminating in Cattani's forthcoming PhD thesis [2]. The work started from the general definition of bisimulation via open maps in [11] which suggested examining a broad class of models for concurrency—presheaf categories. Later we realised that presheaf models can themselves be usefully assembled together in a category in which the maps are colimit-preserving functors. There are two main benefits: one is a general result stating that colimit-preserving functors between presheaf categories preserve open maps and bisimulation [5]; the other that the category of the presheaf models is a form of domain theory for concurrency, with a compositional account of bisimulation, though at the cost that domains are categories rather than special partial orders [17,3].

We originally concentrated on the category of presheaf categories with colimit-preserving functors (or equivalently, the bicategory of profunctors). We've come to realise that by shifting category, to presheaf categories with *connected*-colimit preserving functors, a lot of our work can be done more systematically. (A connected colimit is a colimit of a nonempty connected diagram.) In particular the new category supports a metalanguage in which many of our constructions can be defined once and for all. This is not the only way the metalanguage saves work. Its terms will automatically preserve connected colimits. The metalanguage supports recursive definitions because ω-colimits are examples of connected colimits.

* Basic Research in CS, Centre of the Danish National Research Foundation.

A.M. Haeberer (Ed.): AMAST'98, LNCS 1548, pp. 42–58, 1998.

Connected-colimit preserving functors preserve open-map bisimulation. Consequently terms of the metalanguage preserve open-map bisimulation; if two terms which are open-map bisimilar are substituted for the same variable in a term of the metalanguage then the resulting terms will be open-map bisimilar.

The metalanguage can be interpreted in a wide range of categories. To spare some of the overhead of working with presheaf categories the metalanguage will first be interpreted in a simple category of nondeterministic domains. Equality of terms in this model will coincide with trace equivalence. However the nondeterministic domains are mathematically close to presheaf categories. With a switch of viewpoint, essentially the same constructions lead to an interpretation of the metalanguage in presheaf categories with connected-colimit preserving functors, for which open-map bisimulation is an appropriate equivalence.

2 Presheaf Models Sketched

Let $\mathbb{P}$ be a small category. The category of *presheaves over* $\mathbb{P}$, written $\widehat{\mathbb{P}}$, is the category $[\mathbb{P}^{op}, \mathbf{Set}]$ with objects the functors from $\mathbb{P}^{op}$ (the opposite category) to **Set** (the category of sets and functions) and maps the natural transformations between them.

In our applications, the category $\mathbb{P}$ is thought of as consisting of abstract paths (or computation-path shapes) where a map $e : p \to p'$ expresses how the path p is extended to the path p'. In this paper the categories over which we take presheaves will be (the category presentation of) partial orders; the way a path p extends to a path p' will be unique and a map from a path p to a path p' simply a witness to $p \leq p'$ in the partial order.

A presheaf $X : \mathbb{P}^{op} \to \mathbf{Set}$ specifies for a typical path p the set $X(p)$ of computation paths of shape p. The presheaf X acts on $e : p \to p'$ in $\mathbb{P}$ to give a function $X(e)$ saying how p'-paths in X restrict to p-paths in X. In this way a presheaf can model the nondeterministic branching of a process.

Bisimulation on presheaves is derived from notion of open map between presheaves [10,11]. Open maps are a generalisation of functional bisimulations, or zig-zag morphisms, known from transition systems [11]. Presheaves in $\widehat{\mathbb{P}}$ are *bisimilar* iff there is a span of surjective (i.e., epi) open maps between them.

Because the category of presheaves $\widehat{\mathbb{P}}$ is characterised abstractly as the free colimit completion of $\mathbb{P}$ we expect that colimit-preserving functors between presheaf categories to be useful. They are, but not all operations associated with process languages preserve arbitrary colimits. Prefixing operations only preserve connected colimits while parallel compositions usually only preserve connected colimits in each argument separately. However, the preservation of connected colimits is all we need of a functor between presheaf categories for it to preserve bisimulation.

Proposition 1. *[6] Let $G : \widehat{\mathbb{P}} \to \widehat{\mathbb{Q}}$ be any connected-colimit preserving functor between presheaf categories. Then G preserves surjective open maps and open-map bisimulation.*

Define $\mathcal{C}on$ to be the category consisting of objects partial orders $\mathbb{P}, \mathbb{Q}, \cdots$, with maps $g : \mathbb{P} \to \mathbb{Q}$ the connected-colimit preserving functors $g : \widehat{\mathbb{P}} \to \widehat{\mathbb{Q}}$ between the associated presheaf categories, and composition the usual composition of functors. Define $\mathcal{C}ol$ to be the subcategory of colimit-preserving functors.

3 Categories of Nondeterministic Domains

We obtain nondeterministic domains by imitating the definitions on presheaves but replacing **Set** by the much simpler partial order category **2** with two elements $\mathbf{0}, \mathbf{1}$ ordered by $\mathbf{0} \leq \mathbf{1}$.

Instead of presheaves $\widehat{\mathbb{P}} = [\mathbb{P}^{op}, \mathbf{Set}]$ we now obtain $\widehat{\mathbb{P}} = [\mathbb{P}^{op}, \mathbf{2}]$, functors, and so monotonic functions from $\mathbb{P}^{op}$ to **2**. It's not hard to see that an object x of $\widehat{\mathbb{P}}$ corresponds to a downwards-closed set given by $\{p \in \mathbb{P} \mid x(p) = \mathbf{1}\}$, and that a natural transformation from x to y in $\widehat{\mathbb{P}}$ corresponds to the inclusion of $\{p \in \mathbb{P} \mid x(p) = \mathbf{1}\}$ in $\{p \in \mathbb{P} \mid y(p) = \mathbf{1}\}$. So we can identify $\widehat{\mathbb{P}}$ with the partial order of downwards-closed subsets of $\mathbb{P}$, ordered by inclusion. Thought of in this way it is sensible to think of $\widehat{\mathbb{P}}$ as a nondeterministic domain in the sense of [8]; the order $\widehat{\mathbb{P}}$ has joins got simply via unions so it is certainly a cpo, with least element $\emptyset$, and we can think of the union operation as being a form of nondeterministic sum. It's worth remarking that the domains obtained in this way are precisely the infinitely-distributive algebraic lattices (see e.g. [16]) and that these are just the same as the prime algebraic lattices of [15], and free join completions of partial orders.

There are several choices about what to take as maps between nondeterministic domains. If we eschew "fairness", the most generous we seem to have call for is that of all Scott-continuous functions between the domains. We are interested in maps which are just broad enough to include those operations we associate with interacting processes, operations such prefixing of actions, nondeterministic sum and parallel composition, so we look for a narrower class of maps than continuous functions.

3.1 The Category $\mathcal{D}om_s$

On mathematical grounds it is natural to consider taking maps between nondeterministic domains which preserve their join structure, to choose functions f from $\widehat{\mathbb{P}}$ to $\widehat{\mathbb{Q}}$ which preserve all joins, i.e. so $f(\bigcup X) = \bigcup_{x \in X} f(x)$. Such functions (known often as *additive* functions) compose as usual, have identities and give rise to a category rich in structure. Call this category $\mathcal{D}om_s$ and write $f : \mathbb{P} \to_s \mathbb{Q}$ for a map in $\mathcal{D}om_s$, standing for an additive function from $\widehat{\mathbb{P}}$ to $\widehat{\mathbb{Q}}$. Notice that such maps can be presented in several different ways. Because such maps preserve joins they are determined by their results on just the "complete primes", elements $p\!\downarrow\, \in \widehat{\mathbb{P}}$, for $p \in \mathbb{P}$, such that

$$p\!\downarrow\!(p') = \mathbf{1} \text{ if } p' \leq p, \text{ and } \mathbf{0} \text{ otherwise.}$$

Let $f : \mathbb{P} \to_s \mathbb{Q}$, so $f : \widehat{\mathbb{P}} \to \widehat{\mathbb{Q}}$, and write $f^o : \mathbb{P} \to \widehat{\mathbb{Q}}$ for its restriction such that $f^o(p) = f(p\downarrow)$, for $p \in \mathbb{P}$. As every element x of $\widehat{\mathbb{P}}$ is the join $\bigcup_{p \in x} p\downarrow$ we see that

$$f(x) = \bigcup_{p \in x} f^o(p) .$$

In this way maps $f : \mathbb{P} \to_s \mathbb{Q}$ correspond to monotonic functions $f^o : \mathbb{P} \to \widehat{\mathbb{Q}}$. But monotonic functions $g : \mathbb{P} \to \widehat{\mathbb{Q}}$ are just the same as monotonic functions $g : \mathbb{P} \to [\mathbb{Q}^{op}, \mathbf{2}]$ and, uncurrying, these correspond to monotonic functions $h : \mathbb{P} \times \mathbb{Q}^{op} \to \mathbf{2}$ and so to elements of $\widehat{\mathbb{P}^{op} \times \mathbb{Q}} = [(\mathbb{P}^{op} \times \mathbb{Q})^{op}, \mathbf{2}]$. This suggests that $\mathbb{P}^{op} \times \mathbb{Q}$ is a function space, as indeed is so.

The category $\mathcal{D}\mathbf{om_s}$ is monoidal-closed and in fact carries enough structure to be a categorical model of classical linear logic, the involution of linear logic, $\mathbb{P}^{\perp}$, being given as $\mathbb{P}^{op}$. The tensor of $\mathbb{P}$ and $\mathbb{Q}$ is given by the product of partial orders $\mathbb{P} \times \mathbb{Q}$ and the function space from $\mathbb{P}$ to $\mathbb{Q}$ by $\mathbb{P}^{op} \times \mathbb{Q}$. Its products and coproducts are both given by disjoint unions on objects; for example the usual product of domains $\widehat{\mathbb{P}} \times \widehat{\mathbb{Q}}$ is easily seen to be isomorphic to $\widehat{\mathbb{P} + \mathbb{Q}}$, the nondeterministic domain of the disjoint union $\mathbb{P} + \mathbb{Q}$.

3.2 Lifting

One important construction on domains, that of lifting, is missing. Lifting a domain places a new bottom element below a domain. We can achieve this by adjoining a new element $\perp$ below a copy of $\mathbb{P}$ to obtain $\mathbb{P}_\perp$; a way to realise this is by taking $\perp$ to be the empty set $\emptyset$ and the copy of $\mathbb{P}$ to be $\{p\downarrow \mid p \in \mathbb{P}\}$ so that the order of $\mathbb{P}_\perp$ is given simply by restricting the order of $\widehat{\mathbb{P}}$. Operations on processes, notably prefixing and parallel composition, make essential use of an operation associated with lifting. The operation is the function

$$\lfloor - \rfloor : \widehat{\mathbb{P}} \to \widehat{\mathbb{P}_\perp}$$

such that $\lfloor x \rfloor(\perp) = \mathbf{1}$ and $\lfloor x \rfloor(p\downarrow) = x(p)$ for $x \in \widehat{\mathbb{P}}$. But the function $\lfloor - \rfloor$ is not a map from $\mathbb{P}$ to $\mathbb{P}_\perp$ in $\mathcal{D}\mathbf{om_s}$ as it does not preserve all joins; the problem occurs with the join of the empty set $\bigcup \emptyset$, the least element of $\widehat{\mathbb{P}}$, which is not sent to the least element of $\widehat{\mathbb{P}_\perp}$.

3.3 The Category $\mathcal{D}\mathbf{om}$

To accommodate the function $\lfloor - \rfloor$ we are forced to move to a slightly broader category, though fortunately one that inherits a good many properties from $\mathcal{D}\mathbf{om_s}$. The category $\mathcal{D}\mathbf{om}$ has the same objects, partial orders, but its morphisms from $\mathbb{P}$ to $\mathbb{Q}$, written $f : \mathbb{P} \to \mathbb{Q}$, are functions $f : \widehat{\mathbb{P}} \to \widehat{\mathbb{Q}}$ which need only preserve *nonempty* joins, or more accurately, joins of non-empty sets.

Maps $\mathbb{P} \to \mathbb{Q}$ in $\mathcal{D}\mathbf{om}$ are determined by their action on $\emptyset$ and $p\downarrow$, for $p \in \mathbb{P}$. This is because any $x \in \widehat{\mathbb{P}}$ is trivially the nonempty join with the least element $\emptyset \cup \bigcup_{p \in x} p\downarrow$. Given the way to represent $\mathbb{P}_\perp$ as consisting precisely of the elements $\emptyset$ and $p\downarrow$, for $p \in \mathbb{P}$, there is an embedding $j : \mathbb{P}_\perp \to \widehat{\mathbb{P}}$. So any map $f : \mathbb{P} \to \mathbb{Q}$ is determined by its restriction $f \circ j : \mathbb{P}_\perp \to \widehat{\mathbb{Q}}$. The restriction $f \circ j$ is clearly monotonic. Moreover any monotonic function $g : \mathbb{P}_\perp \to \widehat{\mathbb{Q}}$ has an extension[1] $g^\dagger : \mathbb{P} \to \mathbb{Q}$ in $\mathcal{D}\mathbf{om}$ given by $g^\dagger(x) = \bigcup_{p \in \lfloor x \rfloor} g(p\downarrow)$ for $x \in \widehat{\mathbb{P}}$. The two operations $(-) \circ j$ and $(-)^\dagger$ are mutually inverse. Consequently maps $\mathbb{P} \to \mathbb{Q}$ in $\mathcal{D}\mathbf{om}$ correspond bijectively to maps $\mathbb{P}_\perp \to_s \mathbb{Q}$ in $\mathbf{Dom_s}$,[2] and so to elements in $\widehat{(\mathbb{P}_\perp)^{op} \times \mathbb{Q}}$.

3.4 Fixed Points

The set of maps in $\mathcal{D}\mathbf{om}$ from a path order $\mathbb{P}$ to one $\mathbb{Q}$ inherits an order from elements of the function space $\widehat{(\mathbb{P}_\perp)^{op} \times \mathbb{Q}}$. Operations of the category $\mathcal{D}\mathbf{om}$ will come to preserve nonempty joins of such maps and, in particular, joins of ω-chains. Hence operations F of $\mathcal{D}\mathbf{om}$ taking maps $\mathbb{P} \to \mathbb{Q}$ to maps $\mathbb{P} \to \mathbb{Q}$ will have least fixed points $\mathit{fix}\ F : \mathbb{P} \to \mathbb{Q}$.

3.5 Intuition

How is one to think of the category $\mathcal{D}\mathbf{om}$? The interpretation we'll give and the way in which we define denotational semantics to process languages will have some novelty, though similar uses of categories of nondeterministic domains have been made (see for instance [8]). An object $\mathbb{P}$ is to be thought of as consisting of finite computation paths (each one a "trace" in the sense of [9]), for example the finite string of actions that a CCS or CSP process might perform. The partial order $p \le p'$ on $\mathbb{P}$ is thought of as saying that the computation path p can be extended to the computation path p'. With this intuition in mind we shall call the objects of $\mathcal{D}\mathbf{om}$ *path orders*. An element of $\widehat{\mathbb{P}}$ is a trace set as in [9] and stands for the set of computation paths a nondeterministic process can perform.

A map $f : \mathbb{P} \to \mathbb{Q}$ takes a nondeterministic process with computation paths in $\mathbb{P}$ as input and yields a nonderministic process with computation paths in $\mathbb{Q}$ as output. How is one to understand that a map preserves joins of nonempty sets? Because the map need only preserve nonempty joins it is at liberty to ignore the input process in giving nontrivial output. Because the map preserves all nonempty joins the interaction with the input process has to be conducted in a linear way; the input process cannot be copied to explore its different nondeterministic possibilities, so once started it can only follow a single course of computation, during which it may be interacted with intermittently. It's helpful to think of a map in $\mathcal{D}\mathbf{om}$ as a context which surrounds an input process interacting with the input process occasionally and sometimes interacting with its own environment; whichever computation path the output process (the context surrounding the input process) follows it can only involve the input process following a single computation path.

---

[1] In fact, the left Kan extension along j.

[2] The correspondence is natural in $\mathbb{P}$ and $\mathbb{Q}$ making $\mathcal{D}\mathbf{om}$ the coKliesli category associated to the comonad $(-)_\perp$ on $\mathbf{Dom_s}$ and a reflective subcategory of $\mathbf{Dom_s}$.

4 Constructions on Path Orders

4.1 Tensor

The tensor of path orders $\mathbb{P} \otimes \mathbb{Q}$ is given by the set $(\mathbb{P}_\perp \times \mathbb{Q}_\perp) \setminus \{(\perp, \perp)\}$, ordered coordinatewise, in other words, as the product of $\mathbb{P}_\perp$ and $\mathbb{Q}_\perp$ as partial orders but with the bottom element $(\perp, \perp)$ removed.

Let $f : \mathbb{P} \to \mathbb{P}'$ and $g : \mathbb{Q} \to \mathbb{Q}'$. We define $f \otimes g : \mathbb{P} \otimes \mathbb{Q} \to \mathbb{P}' \otimes \mathbb{Q}'$ as the extension (*cf.* Section 3.3) $h^\dagger$ of a monotonic function

$$h : (\mathbb{P} \otimes \mathbb{Q})_\perp \to \widehat{\mathbb{P}' \otimes \mathbb{Q}'} \ .$$

Notice that $(\mathbb{P} \otimes \mathbb{Q})_\perp$ is isomorphic to the product as partial orders of $\mathbb{P}_\perp \times \mathbb{Q}_\perp$ in which the bottom element is then $(\perp, \perp)$. With this realisation of $(\mathbb{P} \otimes \mathbb{Q})_\perp$ we can define $h : \mathbb{P}_\perp \times \mathbb{Q}_\perp \to \widehat{\mathbb{P}' \otimes \mathbb{Q}'}$ by taking

$$(h(p,q))(p',q') = \lfloor f(p) \rfloor (p') \times \lfloor g(q) \rfloor (q')$$

for $p \in \mathbb{P}_\perp$, $q \in \mathbb{Q}_\perp$ and $(p', q') \in \mathbb{P}' \otimes \mathbb{Q}'$—on the right we use the product, or meet, of $\mathbf{2}$, so $\mathbf{0} \times \mathbf{0} = \mathbf{0} \times \mathbf{1} = \mathbf{1} \times \mathbf{0} = \mathbf{0}$ and $\mathbf{1} \times \mathbf{1} = \mathbf{1}$.

The unit for tensor is the empty path order $\mathbb{O}$.

Elements $x \in \widehat{\mathbb{P}}$ correspond to maps $\tilde{x} : \mathbb{O} \to \mathbb{P}$ sending the empty element to x. Given $x \in \widehat{\mathbb{P}}$ and $y \in \widehat{\mathbb{Q}}$ we define $x \otimes y \in \widehat{\mathbb{P} \otimes \mathbb{Q}}$ to be the element pointed to by $\tilde{x} \otimes \tilde{y} : \mathbb{O} \to \mathbb{P} \otimes \mathbb{Q}$.

4.2 Function Space

The function space of path orders $\mathbb{P} \multimap \mathbb{Q}$ is given by the product of partial orders $(\mathbb{P}_\perp)^{op} \times \mathbb{Q}$. Thus the elements of $\mathbb{P} \multimap \mathbb{Q}$ are pairs, which we write suggestively as $(p \mapsto q)$, with $p \in \mathbb{P}_\perp, q \in \mathbb{Q}$, ordered by

$$(p' \mapsto q') \leq (p \mapsto q) \iff p \leq p' \ \&\ q' \leq q$$

—note the switch in order on the left.

We have the following chain of isomorphisms between partial orders:

$$\mathbb{P} \otimes \mathbb{Q} \multimap \mathbb{R} = (\mathbb{P} \otimes \mathbb{Q})_\perp \times \mathbb{R} \cong \mathbb{P}_\perp \times \mathbb{Q}_\perp \times \mathbb{R} \cong \mathbb{P} \multimap (\mathbb{Q} \multimap \mathbb{R}) \ .$$

This gives isomorphism between the elements $\widehat{\mathbb{P} \otimes \mathbb{Q} \multimap \mathbb{R}}$ and $\mathbb{P} \multimap \widehat{(\mathbb{Q} \multimap \mathbb{R})}$. Thus there is a 1-1 correspondence *curry* from maps $\mathbb{P} \otimes \mathbb{Q} \to \mathbb{R}$ to maps $\mathbb{P} \to (\mathbb{Q} \multimap \mathbb{R})$ in $\mathcal{D}\mathbf{om}$; its inverse is called *uncurry*. We obtain *linear application*, *app* : $(\mathbb{P} \multimap \mathbb{Q}) \otimes \mathbb{P} \to \mathbb{Q}$, as $uncurry(1_{\mathbb{P} \multimap \mathbb{Q}})$.

4.3 Products

The product of path orders $\mathbb{P}\&\mathbb{Q}$ is given by the disjoint union of $\mathbb{P}$ and $\mathbb{Q}$. An element of $\widehat{\mathbb{P}\&\mathbb{Q}}$ can be identified with a pair (x, y), with $x \in \widehat{\mathbb{P}}$ and $y \in \widehat{\mathbb{Q}}$, which provides the projections $\pi_1 : \mathbb{P}\&\mathbb{Q} \to \mathbb{P}$ and $\pi_2 : \mathbb{P}\&\mathbb{Q} \to \mathbb{Q}$. More general, not just binary, products $\&_{i\in I} \mathbb{P}_i$ with projections π_j, for $j \in I$, are defined similarly. From the universal property of products, a collection of maps $f_i : \mathbb{P} \to \mathbb{P}_i$, for $i \in I$, can be tupled together to form a unique map $\langle f_i\rangle_{i\in I} : \mathbb{P} \to \&_{i\in I} \mathbb{P}_i$ with the property that $\pi_j \circ \langle f_i\rangle_{i\in I} = f_j$ for all $j \in I$. The empty product is given by $\mathbb{O}$ and as the terminal object is associated with unique maps $\mathbb{P} \to \mathbb{O}$, constantly $\emptyset$, for any path order $\mathbb{P}$. Finite products are most often written as $\mathbb{P}_1 \& \cdots \& \mathbb{P}_k$.

Each object $\mathbb{P}$ is associated with (nondeterministic) sum operations, a map $\Sigma : \&_{i\in I} \mathbb{P} \to \mathbb{P}$ in $\mathcal{D}\mathbf{om}$ taking an element of the domain, viewed as a tuple $\{x_i \mid i \in I\}$, to its union $\bigcup_{i\in I} x_i$ in $\widehat{\mathbb{P}}$. The empty sum yields $\emptyset \in \mathbb{P}$. Finite sums are typically written as $x_1 + \cdots + x_k$.

Because there are empty elements we can define maps in $\mathcal{D}\mathbf{om_s}$ from products to tensors of path orders. For instance, in the binary case, $\sigma : \mathbb{P}\&\mathbb{Q} \to_s \mathbb{P} \otimes \mathbb{Q}$ in $\mathcal{D}\mathbf{om_s}$ is specified by

$$(x, y) \mapsto (x \otimes \emptyset) + (\emptyset \otimes y) .$$

The composition of such a map with the diagonal map, *viz.*

$$\mathbb{P} \xrightarrow{diag} \mathbb{P}\&\mathbb{P} \xrightarrow{\sigma} \mathbb{P} \otimes \mathbb{P}$$

will play a role later in the semantics of the metalanguage, allowing us to duplicate arguments to maps of a certain kind.

4.4 Lifted Sums

The category $\mathcal{D}\mathbf{om}$ does not have coproducts. However, we can build a useful sum in $\mathcal{D}\mathbf{om}$ with the help of the coproduct of $\mathcal{D}\mathbf{om_s}$ and lifting. Let $\mathbb{P}_i$, for $i \in I$, be a family of path orders. As their lifted sum we take the disjoint union of the path orders $\Sigma_{i\in I}\mathbb{P}_{i\bot}$, over the underlying set $\bigcup_{i\in I}\{i\} \times (\mathbb{P}_i)_\bot$; the latter path order forms a coproduct in $\mathcal{D}\mathbf{om_s}$ with the obvious injections $in_j : \mathbb{P}_{j\bot} \to_s \Sigma_{i\in I}\mathbb{P}_{i\bot}$, for $j \in I$. The *injections* $In_j : \mathbb{P}_j \to \Sigma_{i\in I}\mathbb{P}_{i\bot}$ in $\mathcal{D}\mathbf{om}$, for $j \in I$, are defined to be the composition $In_j(-) = in_j(\lfloor - \rfloor)$. This construction is not a coproduct in $\mathcal{D}\mathbf{om}$. However, it does satisfy a weaker property analogous to the universal property of a coproduct. Suppose $f_i : \mathbb{P}_i \to \mathbb{Q}$ are maps in $\mathcal{D}\mathbf{om}$ for all $i \in I$. Then, there is a unique mediating map

$$f : \Sigma_{i\in I}\mathbb{P}_{i\bot} \to_s \mathbb{Q}$$

in $\mathcal{D}\mathbf{om_s}$ (note the subscript) such that

$$f \circ In_i = f_i$$

for all $i \in I$.

Suppose that the family of maps $f_i : \mathbb{P}_i \to \mathbb{Q}$, with $i \in I$, has the property that each f_i is constantly $\emptyset$ whenever $i \in I$ is different from j and that f_j is $h : \mathbb{P}_j \to \mathbb{Q}$. Write $[h]_j : \sum_{i\in I} \mathbb{P}_{i\perp} \to \mathbb{Q}$ for the unique mediating map obtained for this choice. Then

$$[h]_j(In_j(z)) = h(z) \ , [h]_j(In_i(z)) = \emptyset \text{ if } i \neq j \ , \text{ and } [h]_j(\emptyset) = \emptyset \ .$$

For a general family $f_i : \mathbb{P}_i \to \mathbb{Q}$, with $i \in I$, we can describe the action of the mediating morphism on $x \in \widehat{\Sigma_{i\in I}\mathbb{P}_{i\perp}}$ as $f(x) = \Sigma_{i\in I}[f_i]_i(x)$.

Because lifted sum is not a coproduct we do not have that tensor distributes over lifted sum to within isomorphism. However there is a map in $\mathcal{D}\mathbf{om}_s$

$$dist : \mathbb{Q} \otimes \Sigma_{i\in I}\mathbb{P}_{i\perp} \to_s \Sigma_{i\in I}(\mathbb{Q} \otimes \mathbb{P}_i)_\perp \ ,$$

expressing a form of distributivity, given as the extension $h^\dagger$ of the function

$$h : \mathbb{Q}_\perp \times (\Sigma_{i\in I}\mathbb{P}_{i\perp})_\perp \to \Sigma_{i\in I}(\mathbb{Q}\otimes\mathbb{P}_i)_\perp; \quad h(q,(i,p)) = (i,(q,p))\downarrow, \ h(q,\perp) = \emptyset \ .$$

Unary lifted sums in $\mathcal{D}\mathbf{om}$, when I is a singleton, are an important special case as they amount to lifting.

4.5 Recursive Definitions

Suppose that we wish to model a process language rather like CCS but where processes are passed instead of discrete values, subject to the linearity constraint that when a process is received it can be run at most once. Assume the synchronised communication occurs along channels forming the set A. The path orders can be expected to satisfy the following equations:

$$\mathbb{P} = \mathbb{P}_\perp + \Sigma_{a\in A}\mathbb{C}_\perp + \Sigma_{a\in A}\mathbb{F}_\perp \ , \qquad \mathbb{C} = \mathbb{P} \otimes \mathbb{P} \ , \qquad \mathbb{F} = (\mathbb{P} \multimap \mathbb{P}) \ .$$

The three components of process paths $\mathbb{P}$ represent paths beginning with a silent (τ) action, an output on a channel (a!), resuming as a concretion path (in $\mathbb{C}$), and an input from a channel (a?), resuming as an abstraction path (in $\mathbb{F}$). It is our choice of path for abstractions which narrows us to a *linear* process-passing language, one where the input process can be run at most once to yield a single (computation) path.

Fortunately the simple technique for solving recursive domain equations via information systems in [12] suffices to solve such equations. A path order $\mathbb{P}$ can be regarded as an information system in which every finite subset of $\mathbb{P}$ is consistent and in which the entailment relation is given by the partial order $\leq$ of $\mathbb{P}$, so $\{p'\} \vdash p$ iff $p \leq p'$. Path orders under the order

$$\mathbb{P} \trianglelefteq \mathbb{Q} \iff \mathbb{P} \subseteq \mathbb{Q} \ \& \ (\forall p, p' \in \mathbb{P}.\ p \leq_{\mathbb{P}} p' \iff p \leq_{\mathbb{Q}} p')$$

form a (large) cpo with respect to which all the constructions on path orders we have just seen are continuous (their continuity is verified just as in information systems by showing them monotonic w.r.t. $\trianglelefteq$ and "continuous on token sets"). Solutions to equations like those above are then obtained as (simultaneous) least fixed points.

5 A Metalanguage

Assume that path orders are presented using the constructions with the following syntax:

$$\mathbb{T} ::= \mathbb{O} \mid \mathbb{T}_1 \otimes \mathbb{T}_2 \mid \mathbb{T}_1 \multimap \mathbb{T}_2 \mid \Sigma_{i\in I}\mathbb{T}_{i\perp} \mid \mathbb{T}_1 \& \mathbb{T}_2$$
$$\mid P \mid \mu_j P_1, \cdots, P_k.(\mathbb{T}_1, \cdots, \mathbb{T}_k)$$

All the construction names have been met earlier with the exception of the notation for recursively defined path orders. Above P is drawn from a set of variables used in the recursive definition of path orders; $\mu_j P_1, \cdots, P_k.(\mathbb{T}_1, \cdots, \mathbb{T}_k)$ stands for the j-component (so $1 \le j \le k$) of the $\trianglelefteq$-least solution to the defining equations

$$P_1 = \mathbb{T}_1, \ \cdots, \ P_k = \mathbb{T}_k \ ,$$

in which the expressions $\mathbb{T}_1, \cdots, \mathbb{T}_k$ may contain $P_1, \cdots, P_k$. We shall write $\mu P_1, \cdots, P_k.(\mathbb{T}_1, \cdots, \mathbb{T}_k)$ as an abbreviation for

$$(\mu_1 P_1, \cdots, P_k.(\mathbb{T}_1, \cdots, \mathbb{T}_k), \cdots, \mu_k P_1, \cdots, P_k.(\mathbb{T}_1, \cdots, \mathbb{T}_k)) \ .$$

In future we will often use vector notation and, for example, write $\mu\vec{P}.\vec{\mathbb{T}}$ for the expression above, and confuse a closed expression for a path order with the path order itself.

The operations of Sections 3 and 4 form the basis of a "raw" syntax of terms which will be subject to typing and linearity constraints later:

$$\begin{array}{lll}
t, u, v, \cdots ::= & x, y, z, \cdots & \text{(Variables)} \\
& \emptyset \mid \Sigma_{i\in I} t_i \mid & \text{(Sums)} \\
& rec\ x.t \mid & \text{(Recursive definitions)} \\
& \lambda x.t \mid u \cdot v \mid & \text{(Abstraction and application)} \\
& In_j(t) \mid [t > In_j(x) \Rightarrow u] \mid & \text{(Injections and tests for lifted sums)} \\
& (t, u) \mid [t > (x, -) \Rightarrow u] \mid & \\
& \qquad [t > (-, x) \Rightarrow u] \mid & \text{(Pairing and tests for products)} \\
& t \otimes u \mid [t > x \otimes y \Rightarrow u] & \text{(Tensor operation and tests)}
\end{array}$$

The language is similar to that in [1], being based on a form of pattern matching. In particular $[t > In_j(x) \Rightarrow u]$ "tests" or matches t denoting an element of a lifted sum against the pattern $In_j(x)$ and passes the results of successful matches for x on to u; how the possibly multiple results of successful matches are combined to a final result varies according to the category in which language is interpreted. Accordingly, variables like x in such patterns are binding occurrences and bind later occurrences of the variable in the body, u in this case. We shall take for granted an understanding of free and bound variables, and substitution on raw terms. In examples we'll allow ourselves to use $+$ both in writing sums of terms and lifted sums of path orders.

Let $\mathbb{P}_1, \cdots, \mathbb{P}_k$ be closed expressions for path orders and assume that the variables $x_1, \cdots, x_k$ are distinct. A syntactic judgement

$$x_1 : \mathbb{P}_1, \cdots, x_k : \mathbb{P}_k \vdash t : \mathbb{Q}$$

stands for a map

$$[\![x_1 : \mathbb{P}_1, \cdots, x_k : \mathbb{P}_k \vdash t : \mathbb{Q}]\!] : \mathbb{P}_1 \otimes \cdots \otimes \mathbb{P}_k \to \mathbb{Q}$$

in $\mathcal{D}\mathbf{om}$. We shall typically write Γ, or Δ, for an environment list $x_1 : \mathbb{P}_1, \cdots, x_k : \mathbb{P}_k$. We shall most often abbreviate the denotation map to

$$\mathbb{P}_1 \otimes \cdots \otimes \mathbb{P}_k \xrightarrow{t} \mathbb{Q} \text{ , or even } \Gamma \xrightarrow{t} \mathbb{Q} .$$

Here k may be 0 so the list in the syntactic judgement is empty and the corresponding tensor product the empty path order $\mathbb{O}$.

A linear language will restrict copying and so substitutions of a common term into distinct variables. The counterpart in the models is the absence of a suitable diagonal map from objects $\mathbb{P}$ to $\mathbb{P} \otimes \mathbb{P}$. For example the function $x \mapsto x \otimes x$ from $\widehat{\mathbb{P}}$ to $\widehat{\mathbb{P} \otimes \mathbb{P}}$ is not in general a map in $\mathcal{D}\mathbf{om}$. To see this assume that $\mathbb{P}$ is the discrete order on the set $\{a, b\}$. Then the nonempty join $x = a\!\downarrow \cup\, b\!\downarrow$ is not sent to

$$(a\!\downarrow \otimes\, a\!\downarrow) \cup (b\!\downarrow \otimes\, b\!\downarrow) = \{(a,a), (b,b), (a,\bot), (\bot,b)\}$$

as would be needed to preserve non-empty joins, but instead to

$$x \otimes x = \{(a,a), (b,b), (a,b), (a,\bot), (\bot,b)\}$$

with the extra "cross term" (a, b). Consider a term $t(x, y)$, with its free variables x and y shown explicitly, for which

$$x : \mathbb{P}, y : \mathbb{P} \vdash t(x,y) : \mathbb{Q} \text{ ,}$$

corresponding to a map $\mathbb{P} \otimes \mathbb{P} \xrightarrow{t(x,y)} \mathbb{Q}$ in $\mathcal{D}\mathbf{om}$. This does not generally entail that

$$x : \mathbb{P} \vdash t(x,x) : \mathbb{Q}$$

—there may not be a corresponding map in $\mathcal{D}\mathbf{om}$, for example if $t(x, y) = x \otimes y$. There is however a condition on how the variables x and y occur in t which ensures that the judgement $x : \mathbb{P} \vdash t(x, x) : \mathbb{Q}$ holds and that it denotes the map in $\mathcal{D}\mathbf{om}$ obtained as the composition

$$\mathbb{P} \xrightarrow{diag} \mathbb{P} \& \mathbb{P} \xrightarrow{\sigma} \mathbb{P} \otimes \mathbb{P} \xrightarrow{t(x,y)} \mathbb{Q}$$

—using the maps seen earlier in Section 4.3. Semantically, the map $\mathbb{P} \otimes \mathbb{P} \xrightarrow{t(x,y)} \mathbb{Q}$ has to be essentially a map $\mathbb{P} \& \mathbb{P} \to \mathbb{Q}$, more precisely the left Kan extension of such a map along σ. Syntactically, this is assured if the variables x and y are *not crossed* in t according to the following definition:

Definition 2. Let t be a raw term. Say a set of variables V is *crossed* in t iff there are subterms of t of the form

a tensor $s \otimes u$, an application $s \cdot u$, or a test $[z > u \Rightarrow s]$

for which t has free occurrences of variables from V appearing in both s and u.

For example, variables x and y are crossed in $x \otimes y$, but variables x and y are not crossed in $(x+y) \otimes z$. Note that a set of variables V is crossed in a term t if V contains variables x, y, not necessarily distinct, so that $\{x, y\}$ is crossed in t. We are mainly interested in when sets of variables are *not crossed* in a term.

The term-formation rules are listed below alongside their interpretations as a constructors on morphisms, taking the morphisms denoted by the premises to that denoted by the conclusion. We assume that the variables in any enviroment list which appears are distinct.

Structural rules:

$$\frac{}{x:\mathbb{P} \vdash x:\mathbb{P}}\,,\text{ interpreted as } \frac{}{\mathbb{P} \xrightarrow{1_{\mathbb{P}}} \mathbb{P}}\,.$$

$$\frac{\Delta \vdash t:\mathbb{P}}{\Gamma, \Delta \vdash t:\mathbb{P}}\,,\text{ interpreted as } \frac{\Delta \xrightarrow{t} \mathbb{P}}{\Gamma \otimes \Delta \xrightarrow{\emptyset \otimes 1_{\Gamma}} \mathbb{O} \otimes \Delta \cong \Delta \xrightarrow{t} \mathbb{P}}\,.$$

$$\frac{\Gamma, x:\mathbb{P}, y:\mathbb{Q}, \Delta \vdash t:\mathbb{R}}{\Gamma, y:\mathbb{Q}, x:\mathbb{P}, \Delta \vdash t:\mathbb{R}}\,,\text{ interpreted via } s:\mathbb{Q}\otimes\mathbb{P} \cong \mathbb{P}\otimes\mathbb{Q} \text{ as}$$

$$\frac{\Gamma \otimes \mathbb{P} \otimes \mathbb{Q} \otimes \Delta \xrightarrow{t} \mathbb{R}}{\Gamma \otimes \mathbb{Q} \otimes \mathbb{P} \otimes \Delta \xrightarrow{1_{\Gamma} \otimes s \otimes 1_{\Delta}} \Gamma \otimes \mathbb{P} \otimes \mathbb{Q} \otimes \Delta \xrightarrow{t} \mathbb{R}}\,.$$

Recursive path orders:

$$\frac{\Gamma \vdash t : \mathbb{T}_j[\mu \vec{P}.\vec{\mathbb{T}}/\vec{P}]}{\Gamma \vdash t : \mu_j \vec{P}.\vec{\mathbb{T}}}\,, \qquad \frac{\Gamma \vdash t : \mu_j \vec{P}.\vec{\mathbb{T}}}{\Gamma \vdash t : \mathbb{T}_j[\mu \vec{P}.\vec{\mathbb{T}}/\vec{P}]}\,.$$

where the premise and conclusion of each rule are interpreted as the same map because $\mu_j \vec{P}.\vec{\mathbb{T}}$ and $\mathbb{T}_j[\mu\vec{P}.\vec{\mathbb{T}}/\vec{P}]$ denote equal path orders.

Sums of terms:

$$\frac{}{\Gamma \vdash \emptyset : \mathbb{P}}\,,\text{ interpreted as } \frac{}{\Gamma \xrightarrow{\emptyset} \mathbb{P}}\,,\text{ the constantly } \emptyset \text{ map.}$$

$$\frac{\Gamma \vdash t_i : \mathbb{P} \text{ for all } i \in I}{\Gamma \vdash \Sigma_{i\in I} t_i : \mathbb{P}}\,,\text{ interpreted as } \frac{\Gamma \xrightarrow{t_i} \mathbb{P} \text{ for all } i \in I}{\Gamma \xrightarrow{\langle t_i \rangle_{i\in I}} \&_{i\in I}\mathbb{P} \xrightarrow{\Sigma} \mathbb{P}}\,.$$

Recursive definitions:

$$\frac{\Gamma, x:\mathbb{P} \vdash t:\mathbb{P} \quad \{y, x\} \text{ not crossed for all } y \text{ in } \Gamma}{\Gamma \vdash rec\ x.t : \mathbb{P}}\,,\text{ interpreted as } \frac{\Gamma \otimes \mathbb{P} \xrightarrow{t} \mathbb{P}}{\Gamma \xrightarrow{fix\ F} \mathbb{P}}$$

—see Section 3.4, where for $\Gamma \xrightarrow{g} \mathbb{P}$ the map $F(g)$ is the composition

$$\Gamma \xrightarrow{diag} \Gamma\&\Gamma \xrightarrow{\sigma} \Gamma \otimes \Gamma \xrightarrow{1_\Gamma \otimes g} \Gamma \otimes \mathbb{P} \xrightarrow{t} \mathbb{P}\,.$$

Abstraction:

$$\frac{\Gamma, x:\mathbb{P} \vdash t:\mathbb{Q}}{\Gamma \vdash \lambda x.t : \mathbb{P} \multimap \mathbb{Q}}\,,\text{ interpreted as } \frac{\Gamma \otimes \mathbb{P} \xrightarrow{t} \mathbb{Q}}{\Gamma \xrightarrow{curry\ t} (\mathbb{P} \multimap \mathbb{Q})}\,.$$

Application:

$$\frac{\Gamma \vdash u : \mathbb{P} \multimap \mathbb{Q} \quad \Delta \vdash v : \mathbb{P}}{\Gamma, \Delta \vdash u \cdot v : \mathbb{Q}} \text{,interpreted as } \frac{\Gamma \xrightarrow{u} (\mathbb{P} \multimap \mathbb{Q}) \quad \Delta \xrightarrow{v} \mathbb{P}}{\Gamma \otimes \Delta \xrightarrow{u \otimes v} (\mathbb{P} \multimap \mathbb{Q}) \otimes \mathbb{P} \xrightarrow{app} \mathbb{Q}} .$$

Injections and test for lifted sums:

$$\frac{\Gamma \vdash t : \mathbb{P}_j \text{, where } j \in I}{\Gamma \vdash In_j(t) : \Sigma_{i \in I} \mathbb{P}_{i\perp}} \text{, interpreted as } \frac{\Gamma \xrightarrow{t} \mathbb{P}_j \text{, where } j \in I}{\Gamma \xrightarrow{t} \mathbb{P}_j \xrightarrow{In_j} \Sigma_{i \in I} \mathbb{P}_{i\perp}} .$$

$$\frac{\Gamma, x : \mathbb{P}_j \vdash u : \mathbb{Q} \text{, where } j \in I. \quad \Delta \vdash t : \sum_{i \in I} \mathbb{P}_{i\perp}}{\Gamma, \Delta \vdash [t > In_j(x) \Rightarrow u] : \mathbb{Q}} \text{, interpreted as}$$

$$\frac{\Gamma \otimes \mathbb{P}_j \xrightarrow{u} \mathbb{Q} \text{, where } j \in I. \quad \Delta \xrightarrow{t} \sum_{i \in I} \mathbb{P}_{i\perp}}{\Gamma \otimes \Delta \xrightarrow{1_\Gamma \otimes t} \Gamma \otimes \sum_{i \in I} \mathbb{P}_{i\perp} \xrightarrow{dist} \sum_{i \in I} (\Gamma \otimes \mathbb{P}_i)_\perp \xrightarrow{[-]_j} \mathbb{Q}} .$$

Pairing and tests for products:

$$\frac{\Gamma \vdash t : \mathbb{P} \quad \Gamma \vdash u : \mathbb{Q}}{\Gamma \vdash (t, u) : \mathbb{P} \& \mathbb{Q}} \text{, interpreted as } \frac{\Gamma \xrightarrow{t} \mathbb{P} \quad \Gamma \xrightarrow{u} \mathbb{Q}}{\Gamma \xrightarrow{\langle t, u \rangle} \mathbb{P} \& \mathbb{Q}} .$$

$$\frac{\Gamma, x : \mathbb{P} \vdash u : \mathbb{R} \quad \Delta \vdash t : \mathbb{P} \& \mathbb{Q}}{\Gamma, \Delta \vdash [t > (x, -) \Rightarrow u] : \mathbb{R}} \text{, interpreted as}$$

$$\frac{\Gamma \otimes \mathbb{P} \xrightarrow{u} \mathbb{R} \quad \Delta \xrightarrow{t} \mathbb{P} \& \mathbb{Q}}{\Gamma \otimes \Delta \xrightarrow{1_\Gamma \otimes (\pi_1 \circ t)} \Gamma \otimes \mathbb{P} \xrightarrow{u} \mathbb{R}} .$$

$$\frac{\Gamma, x : \mathbb{Q} \vdash u : \mathbb{R} \quad \Delta \vdash t : \mathbb{P} \& \mathbb{Q}}{\Gamma, \Delta \vdash [t > (-, x) \Rightarrow u] : \mathbb{R}} \text{, interpreted as}$$

$$\frac{\Gamma \otimes \mathbb{Q} \xrightarrow{u} \mathbb{R} \quad \Delta \xrightarrow{t} \mathbb{P} \& \mathbb{Q}}{\Gamma \otimes \Delta \xrightarrow{1_\Gamma \otimes (\pi_2 \circ t)} \Gamma \otimes \mathbb{Q} \xrightarrow{u} \mathbb{R}} .$$

Tensor operation and test for tensor:

$$\frac{\Gamma \vdash t : \mathbb{P} \quad \Delta \vdash u : \mathbb{Q}}{\Gamma, \Delta \vdash t \otimes u : \mathbb{P} \otimes \mathbb{Q}} \text{, interpreted as } \frac{\Gamma \xrightarrow{t} \mathbb{P} \quad \Delta \xrightarrow{u} \mathbb{Q}}{\Gamma \otimes \Delta \xrightarrow{t \otimes u} \mathbb{P} \otimes \mathbb{Q}} .$$

$$\frac{\Gamma, x : \mathbb{P}, y : \mathbb{Q} \vdash u : \mathbb{R} \quad \Delta \vdash t : \mathbb{P} \otimes \mathbb{Q}}{\Gamma, \Delta \vdash [t > x \otimes y \Rightarrow u] : \mathbb{R}} \text{, interpreted as}$$

$$\frac{\Gamma \otimes \mathbb{P} \otimes \mathbb{Q} \xrightarrow{u} \mathbb{R} \quad \Delta \xrightarrow{t} \mathbb{P} \otimes \mathbb{Q}}{\Gamma \otimes \Delta \xrightarrow{1_\Gamma \otimes t} \Gamma \otimes \mathbb{P} \otimes \mathbb{Q} \xrightarrow{u} \mathbb{R}} .$$

Proposition 3. *Suppose $\Gamma, x : \mathbb{P} \vdash t : \mathbb{Q}$. The set $\{x\}$ is not crossed in t.*

Lemma 4. *(Well-formed substitutions) Suppose*

$$\Gamma, x_1 : \mathbb{P}, \cdots, x_k : \mathbb{P} \vdash t : \mathbb{Q}$$

and that the set of variables $\{x_1, \cdots, x_k\}$ *is not crossed in* t. *Suppose* $\Delta \vdash u : \mathbb{P}$ *where the variables of* Γ *and* Δ *are disjoint. Then,*

$$\Gamma, \Delta \vdash t[u/x_1, \cdots, u/x_k] : \mathbb{Q} \ .$$

In particular, as singleton sets of variables are not crossed in well-formed terms we immediately deduce:

Corollary 5. *If* $\Gamma, x : \mathbb{P} \vdash t : \mathbb{Q}$ *and* $\Delta \vdash u : \mathbb{P}$, *where the variables of* Γ *and* Δ *are disjoint, then* $\Gamma, \Delta \vdash t[u/x] : \mathbb{Q}$.

Exploiting the naturality of the various operations used in the semantic definitions, we can show:

Lemma 6. *(Substitution Lemma) Suppose* $\Gamma, x : \mathbb{P} \vdash t : \mathbb{Q}$ *and* $\Delta \vdash u : \mathbb{P}$ *where* Γ *and* Δ *have disjoint variables. Then,*

$$[\![\Gamma, \Delta \vdash t[u/x] : \mathbb{Q}]\!] = [\![\Gamma, x : \mathbb{P} \vdash t : \mathbb{Q}]\!] \circ (1_\Gamma \otimes [\![\Delta \vdash u : \mathbb{P}]\!]) \ .$$

In particular, linear application amounts to substitution:

Lemma 7. *Suppose* $\Gamma \vdash (\lambda x.t) \cdot u : \mathbb{Q}$. *Then,* $\Gamma \vdash t[u/x] : \mathbb{Q}$ *and*

$$[\![\Gamma \vdash (\lambda x.t) \cdot u : \mathbb{Q}]\!] = [\![\Gamma \vdash t[u/x] : \mathbb{Q}]\!] \ .$$

5.1 Extending the Metalanguage

General patterns are well-formed terms built up according to

$$p ::= x \mid \emptyset \mid In_j(p) \mid p \otimes q \mid (p, -) \mid (-, p) \mid p \mapsto p' \ .$$

A test on a pattern $[u > p \Rightarrow t]$ binds the free variables of the pattern p to the resumptions after following the path specified by the pattern in u; because the term t may contain these variables freely the resumptions may influence the computation of t. Such a test is understood inductively as an abbreviation for a term in the metalanguage:

$$\begin{array}{ll}
[u > x \Rightarrow t] \equiv (\lambda x.t) \cdot u \ , \qquad [u > \emptyset \Rightarrow t] \equiv t \ , & \\
{[u > In_j(p) \Rightarrow t] \equiv [u > In_j(x) \Rightarrow [x > p \Rightarrow t]]} & \text{for a fresh variable } x, \\
{[u > (p, -) \Rightarrow t] \equiv [u > (x, -) \Rightarrow [x > p \Rightarrow t]]} & \text{for a fresh variable } x, \\
{[u > (-, p) \Rightarrow t] \equiv [u > (-, x) \Rightarrow [x > p \Rightarrow t]]} & \text{for a fresh variable } x, \\
{[u > p \otimes q \Rightarrow t] \equiv [u > x \otimes y \Rightarrow [p > x \Rightarrow [q > y \Rightarrow t]]]} & \text{for fresh variables } x, y, \\
{[u > (p \mapsto q) \Rightarrow t] \equiv [u > f \Rightarrow [f \cdot p > q \Rightarrow t]]} & \text{for a fresh variable } f.
\end{array}$$

Let $\lambda x \otimes y.t$ stand for $\lambda w.[w > x \otimes y \Rightarrow t]$, where w is a fresh variable, and write $[u_1 > p_1, \cdots, u_k > p_k \Rightarrow t]$ to abbreviate $[u_1 > p_1 \Rightarrow [\cdots [u_k > p_k \Rightarrow t] \cdots]$.

5.2 Interpretation in $\mathcal{C}$on

We can interpret the metalanguage in the category of presheaf models $\mathcal{C}$**on** with essentially the same constructions and operations as those in $\mathcal{D}$**om**, once we replace **2** by **Set** and understand (nonempty) joins as (connected) colimits; the category of presheaf models $\mathcal{C}$**ol** will play the role of $\mathcal{D}\mathbf{om_s}$. Because now domains $[\mathbb{P}^{op}, \mathbf{2}]$ are replaced by presheaf categories $[\mathbb{P}^{op}, \mathbf{Set}]$ we shall often have to make do with isomorphism rather than straight equality.

In fact, to mimic the mathematics behind the interpretation of the metalanguage in $\mathcal{D}$**om**, all that's required of a category $\mathcal{V}$, in place of **2**, is that it has all *colimits* and all *finite products*. Now $\widehat{\mathbb{P}}$, taken to be $[\mathbb{P}^{op}, \mathcal{V}]$, will have all colimits, in particular coproducts to interpret (nondeterministic) sums, and will also support left Kan extensions to play the role of $(-)^\dagger$. We can understand the embedding $(-)\!\downarrow : \mathbb{P} \to \widehat{\mathbb{P}}$ through the initial and terminal objects **0** and **1** of $\mathcal{V}$. The lifting map $\lfloor - \rfloor : \mathbb{P} \to \widehat{\mathbb{P}}$, again defined so $\lfloor x \rfloor(\bot) = \mathbf{1}$ and $\lfloor x \rfloor(p\!\downarrow) = x(p)$, will preserve connected colimits. Using the product of $\mathcal{V}$, instead of that of **2**, we can copy the definition of the functor $\otimes$.

The advantage of this generality is that objects in the category $\mathcal{V}$ don't just have to say whether a path is present in a process but can provide a "measure" of how. If $\mathcal{V}$ is **Set** a process will denote a presheaf X which identifies the set of different ways $X(p)$ in which a path p is realised.

6 Examples

6.1 CCS

As in CCS, assume a set of labels A, a complementation operation producing $\bar{a}$ from a label a, with $\bar{\bar{a}} = a$, and a distinct label τ. In the metalanguage we can specify the path order $\mathbb{P}$ as the $\trianglelefteq$-least solution

$$\mathbb{P} = \mathbb{P}_\bot + \Sigma_{a\in A}\mathbb{P}_\bot + \Sigma_{a\in A}\mathbb{P}_\bot \,.$$

Write the injections from $\mathbb{P}$ into its expression as a lifted sum as $\tau.t$, $a.t$ and $\bar{a}.t$ for $a \in A$ and term t of type $\mathbb{P}$. The curried CCS parallel composition can be defined as the following term of type $\mathbb{P} \multimap (\mathbb{P} \multimap \mathbb{P})$ in the metalanguage:

$$\begin{aligned} Par = rec\ P.\ \lambda x \lambda y.\ & \Sigma_{\alpha\in A\cup\{\tau\}}[x > \alpha.x' \Rightarrow \alpha.(P\cdot x'\cdot y)] + \\ & \Sigma_{\alpha\in A\cup\{\tau\}}[y > \alpha.y' \Rightarrow \alpha.(P\cdot x\cdot y')] + \\ & \Sigma_{a\in A}[x > a.x',\ y > \bar{a}.y' \Rightarrow \tau.(P\cdot x'\cdot y')] \,. \end{aligned}$$

The other CCS operations are easy to encode. Interpreted in $\mathcal{D}$**om** two CCS terms will have the same denotation iff they have same traces (or execution sequences). By virtue of having been written down in the metalanguage the operation of parallel composition will preserve open-map bisimulation when interpreted in $\mathcal{C}$**on**; for this specific $\mathbb{P}$, open-map bisimulation coincides with strong bisimulation. In $\mathcal{C}$**on** we can recover the expansion law for general reasons: the

Substitution Lemmas 6,7 hold in $\mathcal{C}\mathbf{on}$, though with isomorphism replacing equality; the mediating morphism associated with lifted sums are now in $\mathcal{C}\mathbf{ol}$ (the analogue of $\mathcal{D}\mathbf{om_s}$) so that tests for lifted sums distribute over nondeterministic sums. In more detail, write $X|Y$ for $Par \cdot X \cdot Y$, where X and Y are terms of type $\mathbb{P}$. Suppose

$$X = \Sigma_{\alpha \in A \cup \{\tau\}} \Sigma_{i \in I(\alpha)} \alpha.X_i \,, \qquad Y = \Sigma_{\alpha \in A \cup \{\tau\}} \Sigma_{j \in J(\alpha)} \alpha.Y_j \,.$$

Using Lemma 7, and then that the tests distribute over nondeterministic sums,

$$\begin{aligned} X|Y \cong & \Sigma_{\alpha \in A \cup \{\tau\}} [X > \alpha.x' \Rightarrow \alpha.(x'|Y)] + \Sigma_{\alpha \in A \cup \{\tau\}} [Y > \alpha.y' \Rightarrow \alpha.(X|y')] \\ & + \Sigma_{a \in A} [X > a.x',\ Y > \bar{a}.y' \Rightarrow \tau.(x'|y')] \\ \cong & \Sigma_{\alpha \in A \cup \{\tau\}} \Sigma_{i \in I(\alpha)} \alpha.(X_i|Y) + \Sigma_{\alpha \in A \cup \{\tau\}} \Sigma_{j \in J(\alpha)} \alpha.(X|Y_j) \\ & + \Sigma_{a \in A} \Sigma_{i \in I(a), j \in J(\bar{a})} \tau.(X_i|Y_j) \,. \end{aligned}$$

The equation for the path order for CCS with *early value-passing* would be very similar to that above. An equation suitable for *late value-passing* is

$$\mathbb{P} = \mathbb{P}_\perp + \Sigma_{a \in A, v \in V} \mathbb{P}_\perp + \Sigma_{a \in A} (\Sigma_{v \in V} \mathbb{P}_\perp)_\perp \,,$$

though this is not the same equation as in [17] which has $\Sigma_{a \in A}(\Sigma_{v \in V} \mathbb{P})_\perp$ as the final component—perhaps the metalanguage should be broadened to allow this.

6.2 A Linear Higher-Order Process Language

Recall the path orders for processes, concretions and abstractions for a higher-order language in Section 4.5. We are chiefly interested in the parallel composition of processes, $Par_{\mathbb{P},\mathbb{P}}$ of type $\mathbb{P} \otimes \mathbb{P} \multimap \mathbb{P}$. But parallel composition is really a family of mutually dependent operations also including components such as $Par_{\mathbb{F},\mathbb{C}}$ of type $\mathbb{F} \otimes \mathbb{C} \multimap \mathbb{F}$ to say how abstractions compose in parallel with concretions *etc.* All these components can be tupled together in a product using &, and parallel composition defined as a simultaneous recursive definition whose component at $\mathbb{P} \otimes \mathbb{P} \multimap \mathbb{P}$ satisfies

$$\begin{aligned} P|Q = & \Sigma_\alpha [P > \alpha.P' \Rightarrow \alpha(P'|Q)] + \\ & \Sigma_\alpha [Q > \alpha.Q' \Rightarrow \alpha(P|Q')] + \\ & \Sigma_a [P > a?F,\ Q > a!S \otimes R \Rightarrow \tau.(F \cdot S|R)] + \\ & \Sigma_a [P > a!S \otimes R,\ Q > a?F \Rightarrow \tau.(R|F \cdot S)] \,, \end{aligned}$$

where we have chosen suggestive names for the injections and, for instance, $P|Q$ abbreviates $Par_{\mathbb{P},\mathbb{P}} \cdot (P \otimes Q)$. In the summations $a \in A$ and α ranges over $a!, a?, \tau$ for $a \in A$.

7 Problems

The interpretation of the metalanguage in $\mathcal{C}\mathbf{on}$ provides a base from which to examine its equational theory and operational semantics. We should update the treatment of bisimulation in [3] to take better account of $\mathcal{C}$**on** and the metalanguage.

The range of interpretations for the metalanguage indicated in Section 5.2 is restrictive, for example, in requiring $\mathcal{V}$ to be cocomplete. As remarked in [7] there are sensible choices for $\mathcal{V}$ which are not cocomplete—the countable sets for instance, provided we also restrict the path orders to be countable.

Perhaps instantiating $\mathcal{V}$ to some specific category, can help provide a "presheaf model" of a higher-order Pi-Calculus to accompany [4]. This would be a good basis from which to compare and relate with the project of action structures [14].

The metalanguage here cries out for extensions in two directions, one to cope with name generation as in the Pi-Calculus, the other to go beyond linearity. The exponential ! of [17,4] is appropriate to the latter, but its effects on open-map bisimulation are not understood.

The question of how to approach higher-order independence models remains.

How to turn the framework on weak bisimulation and contextual equivalence is the subject of current work based on a lead by Marcelo Fiore.

Gian Luca Cattani and I are working on how to understand open-map bisimulation at higher-order in operational terms [6].

References

1. S. Abramsky. Computational interpretation of linear logic. Tech. Report 90/20, Dept. of Computing, Imperial College, 1990.
2. G. L. Cattani. Forthcoming PhD thesis, CS Dept., University of Aarhus.
3. G. L. Cattani, M. Fiore, and G. Winskel. A Theory of Recursive Domains with Applications to Concurrency. In *Proc. of LICS '98.*
4. G. L. Cattani, I. Stark, and G. Winskel. Presheaf Models for the π-Calculus. In *Proc. of CTCS '97*, LNCS 1290, 1997.
5. G. L. Cattani and G. Winskel. Presheaf Models for Concurrency. In *Proc. of CSL' 96*, LNCS 1258, 1997.
6. G. L. Cattani and G. Winskel. On bisimulation for higher order processes. Manuscript, 1998.
7. G. L. Cattani, A. J. Power and G. Winskel. A categorical axiomatics for bisimulation. In *Proc. of CONCUR'98*, LNCS 1466, 1998.
8. M. Hennessy and G.D. Plotkin. Full abstraction for a simple parallel programming language. In *Proc. of MFCS'79*, LNCS 74, 1979.
9. C.A.R. Hoare. A model for communicating sequential processes. *Tech. Report PRG-22, University of Oxford Computing Lab.*, 1981.
10. A. Joyal and I. Moerdijk. A completeness theorem for open maps. *Annals of Pure and Applied Logic*, 70:51–86, 1994.
11. A. Joyal, M. Nielsen, and G. Winskel. Bisimulation from open maps. *Information and Computation*, 127:164–185, 1996.
12. G.Winskel and K.Larsen. Using information systems to solve recursive domain equations effectively. LNCS 173, 1984.
13. R. Milner. *Communication and concurrency.* Prentice Hall, 1989.
14. R. Milner. Calculi for Interaction. *Acta Informatica* 33, 1996.
15. M. Nielsen, G.D. Plotkin and G. Winskel. Petri nets, Event structures and Domains, part 1. Theoretical Computer Science, vol. 13, 1981.

16. G. Winskel. An introduction to event structures. In *Proc. of REX summer-school in temporal logic, 'May 88*, LNCS 354, 1988.
17. G. Winskel. A presheaf semantics of value-passing processes. In *Proceedings of CONCUR'96*, LNCS 1119, 1996.

Verification of Bounded Delay Asynchronous Circuits with Timed Traces

Tomohiro Yoneda, Bin Zhou[1], and Bernd–Holger Schlingloff[2]

[1] Tokyo Institute of Technology {yoneda,zhou}@cs.titech.ac.jp
[2] Universität Bremen, TZI-BISS hs@tzi.org

Abstract. In this paper, we extend the verification method based on the failure semantics of process algebra and the resulting trace theory by Dill et al. for bounded delay asynchronous circuits. We define a timed conformance relation between trace structures which allows to express both safety and responsiveness properties. In our approach, bounded delay circuits as well as their real-time properties are modelled by time Petri nets. We give an explicit state-exploration algorithm to determine whether an implementation conforms to a specification. Since for IO-conflict free specifications the conformance relation is transitive, this algorithm can be used for hierarchical verification of large asynchronous circuits. We describe the implementation of our method and give some experimental results which demonstrate its efficiency.

Keywords: Trace Theory, Time Petri Nets, Failure Analysis, Conformance Checking, State Space Exploration, Asynchronous Circuits, Hardware Verification, Delay Analysis, Real Time Systems, Computer Aided Verification

1 Introduction

One of the main problems in the design of wafer-scale integrated circuits is the distribution of the global clock signal. Difficulties which arise in the design of large synchronous circuits are clock skews, clock delay estimation in layout design, etc. Therefore, *asynchronous processors* without a global clock are of increasing interest. However, asynchronous circuits are difficult to construct since the timing analysis often is very complex. Because of this reason, asynchronous circuits are usually modelled with a *speed independent model*, where the gate delays are unbounded, or are bounded by an unknown constant. Most of the research on design, synthesis, and verification of asynchronous circuits has been done under this model. Although the speed independent model is quite powerful, the possibility of unbounded delay can force the designer to add additional complexity to the circuit. For example, Muller's C element [MB 59], defined by the truth table in Fig. 1(a), is implemented by the circuit of Fig. 1(b).

This implementation, however, is not correct under the speed independent model. Assuming that each gate can have an unbounded delay, there exists a

A.M. Haeberer (Ed.): AMAST'98, LNCS 1548, pp. 59–73, 1998.

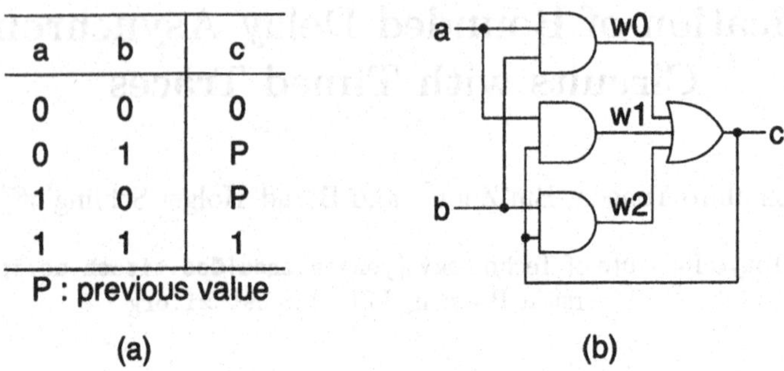

a	b	c
0	0	0
0	1	P
1	0	P
1	1	1

P : previous value

(a)

Fig. 1. Muller's C element: truth table and gate-level implementation.

signal transition sequence in which the output illegally goes down before both inputs go down (suppose all wires initially have the value 0) :

$$a\uparrow\ b\uparrow\ w_0\uparrow\ c\uparrow\ b\downarrow\ w_0\downarrow\ c\downarrow .$$

The reason for this alleged fault is an extremely large delay of the gate with output w_1. With any well-processed VLSI, such a large delay should be impossible. In actual designs, the given circuit can be safely used to implement a C element. Thus, the speed independent model sometimes is not appropriate. In this paper we use a *bounded delay model* to model asynchronous circuits, where with each gate a lower and upper bound for the delay is associated.

In [Dil 88], an efficient verification method for speed independent circuits was proposed, which is based on *trace theory.* The primary advantage of this method is the possibility of hierarchical verification, which greatly reduces the complexity of the verification procedure. However, this method is only suited for verifying safety properties.

In this paper we adapt Dill's verification method to the bounded delay model. First, we show how trace theory can be extended to handle timed traces as well as certain timing requirements. We then describe time Petri nets as an appropriate model for asynchronous bounded delay circuits. Subsequently, we derive an algorithm to check whether an implementation, consisting of a set of modules, meets its specification. Finally, we give some experimental results and concluding remarks.

2 Timed Trace Theory

Let us briefly describe verification based on trace theory. In this method, the specification of a circuit is given as a *trace automaton*, i.e. a finite automaton over an input alphabet $\mathcal{I}$ and output alphabet $\mathcal{O}$. The implementation, which is supposed to be a set of modules, is given as a set of trace automata, each one representing the behavior of its related module. Then, special *composition* and *hiding* operations on trace automata are defined. The implementation *conforms* to the specification, if they agree on the input and output alphabets, respectively, and the implementation can be safely substituted for the specification in every context. This means, that the implementation causes a failure in an environment only if the specification also causes a failure in that environment.

A *failure* of a module in an environment is an output of the module which is not accepted by the environment, or an output of the environment which is not accepted by the module. By this definition, conformance can be expanded to the following requirements: the implementation should be able to handle every input that the specification can handle, and it never produces an output unless the specification can produce it. This in turn can be checked by considering the mirror of the specification, where all inputs are outputs and vice versa. The implementation conforms to the specification iff the result of hiding all internal signal transitions in the implementation and composing it with the mirror of the specification is failure-free.

The verification approach proposed here is the timed version of this method, where time Petri nets and timed traces are used instead of automata and traces. The extension to real-time makes it also possible to verify certain timing properties.

In the rest of this section, we define timed traces and their related notions, and the conformance relation between specification and implementation.

Let $\mathcal{W}$ be a set of wires, and let $\mathbf{Q}$ denote the set of nonnegative rational numbers. For any $w \in \mathcal{W}$ and $t \in \mathbf{Q}$, the tuple (w, t) is called an *event*. Intuitively, (w, t) represents the change of the value of wire w at time t.

Definition 1. *A* (timed) trace x *over* $\mathcal{W}$ *is a finite or infinite sequence of events* $x \triangleq x_1 x_2 \cdots$, *where* $x_i \triangleq (w_i, t_i)$, *such that the following properties are satisfied:*

- *Monotonicity: for all* $0 < i < |x|$, $t_i \leq t_{i+1}$.
- *Progress: if* x *is infinite, then for every* $t \in \mathbf{Q}$ *there exists an index* i *such that* $t_i > t$.

In this definition, $|x|$ denotes the length of trace x. If $|x| = 0$, then x is the *empty trace* ε. For any finite trace x, trace y, and event e, the result of *appending* e or y to x is denoted by $x \circ e$ or $x \circ y$, respectively. x is a *prefix* of y if $y = x$ or $y = x \circ z$ for some trace z. The *projection* of a trace $x \triangleq x_1 \circ x_2 \circ \cdots$ over $\mathcal{W}$ onto another alphabet $\mathcal{W}'$ can be defined as usual:

$$project(x, \mathcal{W}') \triangleq \begin{cases} \varepsilon, & \text{if } x = \varepsilon \\ x_1 \circ y, & \text{if } x_1 = (w_1, t_1),\ w_1 \in \mathcal{W}' \\ y & \text{else,} \qquad\qquad \text{where } y \triangleq project(x_2 \circ x_3 \circ \cdots, \mathcal{W}') \end{cases}$$

Definition 2. *A* module *or* canonical trace structure *is a tuple* $M \triangleq (\mathcal{I}, \mathcal{O}, \mathcal{T})$, *where* $\mathcal{I}$ *is a set of* input wires, $\mathcal{O}$ *is a set of* output wires $(\mathcal{I} \cap \mathcal{O} = \emptyset)$, *and* $\mathcal{T}$ *is a set of traces over* $\mathcal{W} \triangleq \mathcal{I} \cup \mathcal{O}$.

The traces of a module can be regarded as the set of all maximal execution sequences of some transition system. However, trace structures are insensitive to nondeterminism; they can not distinguish between $a \circ (b+c)$ and $(a \circ b)+(a \circ c)$. In timed systems, usually the set of traces will be an infinite (or even uncountable) set of infinite sequences.

Now we consider the composition of several modules. Assume we are given a set $\mathcal{M} \triangleq \{M_1, \cdots, M_n\}$ of modules, where $M_k \triangleq (\mathcal{I}_k, \mathcal{O}_k, \mathcal{T}_k)$, $\mathcal{W}_k \triangleq \mathcal{I}_k \cup \mathcal{O}_k$, and $\mathcal{O}_j \cap \mathcal{O}_k = \emptyset$. That is, each wire is either an input, output, or both; in the latter case we say the wire is *internal.* Any wire can be an output of at most one module, and input of arbitrary many modules. Intuitively, modules are composed by soldering wires with the same name together. Output wires of one module are connected to input wires of other modules. However, in some cases this connection of wires may cause failures in the composed module.

If $M = (\mathcal{I}, \mathcal{O}, \mathcal{T})$, then M *without* w $(M \backslash w)$ is the module $(\mathcal{I}', \mathcal{O}', \mathcal{T}')$, where $\mathcal{I}' = \mathcal{I} - \{w\}$, $\mathcal{O}' = \mathcal{O} - \{w\}$ and $\mathcal{T}' = \mathit{project}(\mathcal{T}, \mathcal{I}' \cup \mathcal{O}')$. Module M *allows* trace x $(M \models x)$ if there exists some trace y such that x is a prefix of y and $\mathit{project}(y, \mathcal{W}) \in \mathcal{T}$. Furthermore, for $\mathcal{M} \triangleq \{M_1, ..., M_n\}$, we say that $\mathcal{M} \models x$ if $M_k \models x$ for all $k \leq n$.

Definition 3. *A* safety failure *of $\mathcal{M}$ is any nonempty finite trace $x \triangleq y \circ (w, t)$, where $w \in \mathcal{O}_k$ for some $k \leq n$, such that $\mathcal{M} \setminus w \models x$, and $M_k \models x$, but $\mathcal{M} \not\models x$.*

Intuitively, a safety failure occurs if any module M_k tries to send an output, but some other module cannot receive this as internal input. $\mathcal{M}$ is *safety failure free,* if no safety failure can occur, i.e., if every output which may be produced by some module can be accepted by all other modules at the same time. Whenever a module can change the value on one of its output wires, all modules which have this wire connected as internal input must be able to process the signal immediately.

Definition 4. *A* timing failure *of $\mathcal{M}$ is any nonempty finite trace $x \triangleq y \circ (w, t)$, where $w \in \mathcal{I}_k$ for some $k \leq n$, such that $\mathcal{M} \setminus w \models x$, and $M_k \models x$, but there is no $x' \triangleq y \circ (w', t')$, where $w' \in \mathcal{I}_k$, and $\mathcal{M} \models x'$.*

Intuitively, a timing failure occurs if some module M_k expects an internal input from some other module which is not provided in time. $\mathcal{M}$ is timing failure free if whenever a module requests a signal on one of its internal input wires, there exists a module which can produce some signal as output within the required time interval. For any set $\mathcal{M} \triangleq \{M_1, \cdots, M_n\}$ of modules, *failure*$(\mathcal{M})$ is the set of all safety and timing failures of $\mathcal{M}$. $\mathcal{M}$ is *failure-free* if *failure*$(\mathcal{M}) = \emptyset$.

Next, we define a conformance relation between a system consisting of a set of modules and a specification given as a single module. Consider a set $\mathcal{M}_C \triangleq \{M_1, \cdots, M_n\}$ of modules, where $M_k \triangleq (\mathcal{I}_k, \mathcal{O}_k, \mathcal{T}_k)$, and a module $M_S \triangleq (\mathcal{I}_S, \mathcal{O}_S, \mathcal{T}_S)$ such that $\mathcal{I}_S \triangleq \bigcup \mathcal{I}_k - \bigcup \mathcal{O}_k$ and $\mathcal{O}_S \subseteq \bigcup \mathcal{O}_k$. Module M_S can be thought of as an abstract specification of the concrete circuit M_C: all external inputs of the circuit M_C appear as inputs of the specification M_S, and some (but not necessarily all) outputs of the circuit M_C are visible in the specification M_S.

Definition 5. $\mathcal{M}_C$ *conforms to* M_S, *if for any module* $M_E \triangleq (\mathcal{O}_S, \mathcal{I}_S, \mathcal{T}_E)$, *whenever* $\{M_S, M_E\}$ *is failure-free, also* $\mathcal{M}_C \cup \{M_E\}$ *is failure-free.*

In other words, the circuit $\mathcal{M}_C$ may have a failure in the environment M_E only if the specification M_S allows a failure in the same context. This conformance relation is reflexive and transitive, but not symmetric: The circuit may be failure-free even in contexts in which the specification fails.

A module M is called *I/O-conflict free*, if for any trace x, and for all events $e_i \triangleq (w_i, \tau_i)$ and $e_o \triangleq (w_o, \tau_o)$ with $w_i \in \mathcal{I}$ and $w_o \in \mathcal{O}$ it holds that $M \models x \circ e_i$ and $M \models x \circ e_o$ implies $M \models x \circ e_i \circ e_o$ and $M \models x \circ e_o \circ e_i$. Since conflicts between inputs and outputs often indicate hazardous situations, specifications usually do not contain such conflicts. Thus, henceforth we assume that all specifications are I/O-conflict free.

Definition 6. *The* mirror module M^m *of a module* $M \triangleq (\mathcal{I}, \mathcal{O}, \mathcal{T})$ *is the module* $M^m \triangleq (\mathcal{O}, \mathcal{I}, \mathcal{T})$; *that is, each input wire in* M^m *is an output wire of* M *and vice versa.*

For any module M, the set $\{M, M^m\}$ is failure-free. Moreover, the following *hierarchy lemma* holds:

Lemma 1. *Consider three modules* M_1, M_2 *and* M_3 *such that* $\mathcal{I}_1 = \mathcal{I}_2 = \mathcal{I}_3$ *and* $\mathcal{O}_1 \supseteq \mathcal{O}_2 = \mathcal{O}_3$. *If* $\{M_1, M_2^m\}$ *is failure-free and* $\{M_2, M_3^m\}$ *is failure-free, then* $\{M_1, M_3^m\}$ *is failure-free.*

Proof. Assume that both $\{M_1, M_2^m\}$ and $\{M_2, M_3^m\}$ are failure-free, and let $x \triangleq y \circ (w, t)$. We have to show that $\{M_1, M_3^m\}$ is failure-free. The following cases have to be considered.

1. First, assume that $w \in \mathcal{O}_1$, $M_1 \models x$ and $M_3^m \setminus w \models x$, and show that $M_3^m \models x$. If $w \notin \mathcal{W}_2$, then from $\mathcal{W}_3 \subseteq \mathcal{W}_2$, we have $w \notin \mathcal{W}_3$. Thus, from $M_3^m \setminus w \models x$, we have $M_3^m \models x$. If $w \in \mathcal{W}_2$, then w cannot be from $\mathcal{I}_2 = \mathcal{O}_2^m$, since in this case it would be impossible to compose M_1 with M_2^m. Hence w must be in $\mathcal{O}_2$. Assume for contradiction that $M_2 \setminus w \not\models x$. Then there must be some $x' = y' \circ (w', t')$ such that x' is an initial part of x, $M_2 \setminus w' \models y'$ and $M_2 \setminus w \not\models x'$. Since $M_1 \models x$ and $M_3 \setminus w \models x$, it follows that $M_1 \models x'$ and $M_3 \setminus w \models x'$. Since $M_2 \setminus w \not\models x'$, we must have $w' \in \mathcal{W}_2$. If $w' \in \mathcal{O}_2$, then $w' \in \mathcal{O}_1$. Since $\{M_1, M_2^m\}$ is safety failure-free, $M_2 \setminus w \models x'$, which is a contradiction. Similarly, if $w' \in \mathcal{I}_2 = \mathcal{I}_3 = \mathcal{O}_3^m$, then since $\{M_2, M_3^m\}$ is safety failure-free, a contradiction arises. Thus, $M_2 \setminus w \models x$. Since $\{M_1, M_2^m\}$ is safety failure-free, $M_2^m \models x$, thus $M_2 \models x$. Since $\{M_2, M_3^m\}$ is safety failure-free, $w \in \mathcal{O}_2$ and $M_3^m \setminus w \models x$ (hypothesis), we have $M_3^m \models x$.
2. The second case is symmetric to the first case: assume that $w \in \mathcal{O}_3^m = \mathcal{I}_3$, $M_3^m \models x$ and $M_1 \setminus w \models x$, and show that $M_1 \models x$. To be able to compose M_1 with M_3^m, the set $\mathcal{O}_1 \cap \mathcal{O}_3^m = \mathcal{I}_3$ must be empty. If $w \notin \mathcal{I}_1$, then $M_1 = M_1 \setminus w$ and there is nothing to show. If $w \in \mathcal{I}_1$, then $\mathcal{I}_1 \subseteq \mathcal{I}_2$ gives $w \in \mathcal{I}_2 = \mathcal{O}_2^m$. Similar to the previous case, $M_2 \setminus w \models x$. Since $\{M_2, M_3^m\}$ is safety failure-free, we can infer that $M_2 \models x$. Since $\{M_1, M_2^m\}$ is safety failure-free, it follows that $M_1 \models x$.

3. Next, assume that $w \in \mathcal{I}_1$, $M_1 \models x$ and $M_3^m \setminus w \models x$, and show that $\{M_1, M_3^m\} \models y \circ (w', t')$ for some $w' \in \mathcal{I}_1$. Since $\{M_1, M_2^m\}$ does not have timing failures, there is some (w_1, t_1) with $w_1 \in \mathcal{I}_1$ such that $\{M_1, M_2^m\} \models y \circ (w_1, t_1)$. Since $\mathcal{I}_1 \subseteq \mathcal{I}_2$ and $\{M_2, M_3^m\}$ does not have timing failures, there is some (w_2, t_2) with $w_2 \in \mathcal{I}_2$ such that $\{M_2, M_3^m\} \models y \circ (w_2, t_2)$. Since $w_2 \in \mathcal{I}_2 = \mathcal{O}_2^m$ and $\{M_1, M_2^m\}$ does not have safety failures, $M_1 \models y \circ (w_2, t_2)$.
4. Finally, assume that $w \in \mathcal{I}_3^m = \mathcal{O}_3$, $M_3^m \models x$ and $M_1 \setminus w \models x$, and show that $\{M_1, M_3^m\} \models y \circ (w', t')$ for some $w' \in \mathcal{O}_3$. Since $\{M_2, M_3^m\}$ is timing failure-free, there is some (w_1, t_1) such that $w_1 \in \mathcal{O}_3$ and $\{M_2, M_3^m\} \models y \circ (w_1, t_1)$. Since $\mathcal{O}_3 \subseteq \mathcal{O}_2 = \mathcal{I}_2^m$ and $\{M_1, M_2^m\}$ is timing failure-free, there is some (w_2, t_2) such that $w_2 \in \mathcal{O}_2$ and $\{M_1, M_2^m\} \models y \circ (w_2, t_2)$. Since $\{M_2, M_3^m\}$ is safety failure-free and $\mathcal{O}_2 \subseteq \mathcal{O}_3$, we have $\{M_1, M_3^m\} \models y \circ (w_2, t_2)$ as desired. □

This lemma can be extended to deal with sets of modules instead of a single modules. From the hierarchy lemma, the following *mirror theorem* can be obtained. It gives a similar characterization of conformance as in [Dil 88]:

Theorem 1. *$\mathcal{M}_C$ conforms to M_S iff $\mathcal{M}_C \cup \{M_S^m\}$ is failure-free.*

Proof. Assume that $\mathcal{M}_C \cup \{M_S^m\}$ has a failure. Then for the environment $M_E = M_S^m$ we have that $\{M_S, M_E\}$ is failure-free, but $\mathcal{M}_C \cup \{M_E\}$ is not failure-free, i.e., $\mathcal{M}_C$ does not conform to M_S.
In the other direction, we have to show that failure-freeness of $\mathcal{M}_C \cup \{M_S^m\}$ implies that $\mathcal{M}_C$ conforms to M_S. Since M_S is a specification for the circuit $\mathcal{M}_C$, $\mathcal{I}_S = \bigcup_k \mathcal{I}_k - \bigcup_k \mathcal{O}_k$ and $\mathcal{O}_S \subseteq \bigcup_k \mathcal{O}_k$. If $\mathcal{M}_C \cup \{M_S^m\}$ is failure-free, then the hierarchy lemma asserts that for any module M_E such that $\mathcal{W}_E = \mathcal{W}_S$ and $\{M_S, M_E\}$ is failure-free, $\mathcal{M}_C \cup \{M_E\}$ must also be failure-free. Thus, $\mathcal{M}_C$ conforms to M_S. □

To get an intuitive understanding of the conformance relation, consider the case of a single module $M_C \triangleq (\mathcal{I}_C, \mathcal{O}_C, \mathcal{T}_C)$ conforming to $M_S \triangleq (\mathcal{I}_S, \mathcal{O}_S, \mathcal{T}_S)$. This amounts to $\mathcal{I}_S = \mathcal{I}_C$, $\mathcal{O}_S \subseteq \mathcal{O}_C$, and for all traces x such that $\{M_C, M_S\} \models x$, and all events $i \triangleq (w_i, t_i)$, $w_i \in \mathcal{I}_S$, and $o \triangleq (w_o, t_o)$, $w_o \in \mathcal{O}_S$, the following holds:

- If $M_S \models x \circ i$, then $M_C \models x \circ i$,
- if $M_C \models x \circ o$, then $M_S \models x \circ o$,
- if $M_S \models x \circ o$, then there exists an $o' \triangleq (w'_o, t'_i)$, $w'_o \in \mathcal{O}_S$ such that $\{M_S, M_C\} \models x \circ o'$, and
- if $M_C \models x \circ i$, then there exists a $i' \triangleq (w'_i, t'_i)$, $w'_i \in \mathcal{I}_S$ such that $\{M_S, M_C\} \models x \circ i'$.

The first and second condition state that $\{M_C, M_S^M\}$ is safety failure-free: every input allowed by M_S is allowed by M_C, and every output allowed by M_C is allowed by M_S. The third condition reflects the definition of timing-failure: as long as M_S^m expects an input, that is, M_S requires an output, M_C should produce some output in time. The fourth condition is similar. If M_C is constructed as an implementation for the specification M_S, then this can be read as:

- The implementation can handle every input that the specification can handle,
- the implementation never produces an output unless the specification produces it,
- if the specification requires an output, the implementation produces it in time, and
- the implementation never expects an input unless the specification expects the input.

Therefore, our definition of the conformance relation includes not only safety properties, but also a certain timing property. In the case of bounded delay asynchronous circuits the absence of timing failure amounts to in-time-responsiveness, which is an important issue for verification. For example, consider the specification of an *or*-gate, where input a or b lead to output c within a certain time. Suppose that this specification is implemented erroneously by an *and*-gate. Then, after sending a to this circuit, it can not produce the output c. However, since the specification requires such an output, this situation leads to a timing failure.

Note that we do not actually compose the modules constituting the implementation. Therefore, in our approach it is not necessary to eliminate so-called autofailures, which arise from internal communication errors in a composed module. Also we do not have an explicit hiding operation: Failures resulting from the effect of hiding variables are transparent to the specification and will also be detected during the verification procedure. However, if we consider only safety-failures in untimed systems, then our notion of conformance is equivalent to the one in [Dil 88].

3 Analysis of Time Petri Nets

In the general setting of the previous section, there was absolutely no restriction posed on the set of traces of a module. To be able to give concrete algorithms, however, this set should at least be recursive, i.e., generated by some kind of automaton. In this section, we consider trace sets generated by one-safe *time Petri nets* [MF 76]. In contrast to timed Petri nets or stochastical Petri nets, which are used in *simulation* for the optimization of processes [TSS 98], time Petri nets have been applied successfully in the *verification* of hard real-time constraints.

One-safe Petri nets can be seen as a subclass of finite automata, where different parallel activities can be modelled by multiple tokens. Therefore, a Petri net model can be much more succinct than the corresponding automaton. Similarly, time Petri nets can be regarded as a subclass of timed automata [AD 92]. Compared with timed automata, the expressive power of time nets with respect to certain timing properties is restricted. This restriction, however, simplifies the analysis: we can check the conformance relation by a simple state space generation algorithm, traversing every state only once in a depth first search manner.

Definition 7. *A* time Petri net *N is a six-tuple,* $N \triangleq (P, T, F, Eft, Lft, \mu_0)$, *where*

- $P \triangleq \{p_1, p_2, \cdots, p_m\}$ *is a finite nonempty set of* places;
- $T \triangleq \{\tau_1, \tau_2, \cdots, \tau_n\}$ *is a finite set of* transitions $(P \cap T = \emptyset)$;
- $F \subseteq (P \times T) \cup (T \times P)$ *is the* flow relation;
- $Eft : T \to \mathbf{Q}$, $Lft : T \to \mathbf{Q} \cup \{\infty\}$ *are functions for the* earliest *and* latest firing times *of transitions, satisfying* $Eft(\tau) \leq Lft(\tau)$ *for all* $\tau \in T$;
- $\mu_0 \subseteq P$ *is the* initial marking *of the net.*

For any transition τ, $\bullet\tau \triangleq \{p \in P \mid (p, \tau) \in F\}$ and $\tau\bullet \triangleq \{p \in P \mid (\tau, p) \in F\}$ denote the *preset* and the *postset* of τ, respectively.

In the following we will restrict ourselves to one-safe Petri nets, where each place can contain at most one token. Therefore, a *marking* μ of N is defined to be any subset of P. A transition is *enabled* in a marking μ if $\bullet\tau \subseteq \mu$ (all its input places have tokens in μ); otherwise, it is *disabled*. Let $enabled(\mu)$ be the set of transitions enabled in μ.

A *state* σ of a time Petri net is a pair $(\mu, clock)$, where μ is a marking and *clock* is a function $T \to \mathbf{Q}$. The *initial state* σ_0 is $(\mu_0, clock_0)$, where $clock_0(\tau) = 0$ for all $\tau \in T$.

The states of time Petri nets change, if time passes or if a transition fires. In state $\sigma \triangleq (\mu, clock)$, time $t \in \mathbf{Q}$ *can pass*, if for all $\tau \in enabled(\mu)$, $clock(\tau) + t \leq Lft(\tau)$. In this case, state $\sigma' \triangleq (\mu', clock')$ is *obtained by passing* t from σ, if

1. $\mu = \mu'$, and
2. for all $\tau \in T$, $clock'(\tau) = clock(\tau) + t$.

In state $\sigma \triangleq (\mu, clock)$, transition $\tau_f \in T$ *can fire*, if $\tau_f \in enabled(\mu)$, and $clock(\tau_f) \geq Eft(\tau_f)$. In this case, state $\sigma' \triangleq (\mu', clock')$ is *obtained by firing* τ_f from σ, if

1. $\mu' = (\mu - \bullet\tau_f) \cup \tau_f\bullet$, and
2. for all $\tau \in T$, $clock'(\tau) = \begin{cases} 0 & \text{if } \tau \in enabled(\mu'), \tau \notin enabled(\mu - \bullet\tau_f) \\ clock(\tau) & \text{else} \end{cases}$.

Intuitively, this can be interpreted as follows: Passing time t does not change the marking, but advances all clock values. Firing a transition τ_f consumes no time, but updates μ and *clock* such that the clock values associated with newly enabled transitions (i.e. transitions which are enabled in μ' but not in $\mu - \bullet\tau_f$) are reset to 0. Clock values of other transitions (i.e. transitions not affected by τ_f) are left unchanged.

In contrast to untimed Petri nets, not all enabled transitions may be firable in a given state; certain firing sequences which can occur without timing may not be possible in the time Petri net. A *run* $\rho \triangleq \sigma_0 \xrightarrow{\tau_1} \sigma_1 \xrightarrow{\tau_2} \sigma_2 \xrightarrow{\tau_3} \cdots$ of N is a finite or infinite sequence of states and transitions such that σ_0 is the initial state, and σ_{i+1} is obtained from σ_i by passing time and then firing transition τ_{i+1}. We write $\sigma_i(\rho)$ for the i-th state of ρ, and similarly $\mu_i(\rho)$ and $clock_i(\rho)$, and omit the argument (ρ) whenever appropriate. A run is *maximal*, if it is infinite or in its last state there is no enabled transition. The *behavior* $B(N)$ of N is the set of all maximal runs of N.

Given any run ρ and $i \geq 0$, we define $time_i(\rho)$ to be the sum of all times t passed between $\sigma_0(\rho)$ and $\sigma_i(\rho)$; that is, $time_0(\rho) \triangleq 0$ and $time_{i+1}(\rho) \triangleq time_i(\rho) + clock_{i+1}(\tau) - clock_i(\tau)$ for some τ which is not newly enabled in μ_{i+1}. A state σ is *reachable* if there exists a finite run whose last state is σ.

Definition 8. *A time Petri net is* one-safe, *if for every state* $\sigma \triangleq (\mu, clock)$ *obtained by passing time from any reachable state* σ', *and for every transition* τ *which can fire in* σ, $\tau \bullet \cap \mu = \emptyset$.

The restriction to one-safe nets simplifies the verification algorithm.

In order to satisfy the progress condition, we assume that time certainly passes in any cyclic behavior of N. For example, this requirement is satisfied if the sum of earliest firing times of transitions forming any loop in N is positive. In the sequel, a *net* will always be a one-safe time Petri net satisfying the above restriction.

Let *wire* be a function from a set of transitions to a set of wires. Every maximal run $\rho \triangleq \sigma_0 \xrightarrow{\tau_1} \sigma_1 \xrightarrow{\tau_2} \cdots$ of a net N *generates* the timed trace $((wire(\tau_1), time_1(\rho)), (wire(\tau_2), time_2(\rho)), \cdots)$. We also say that a net N *represents* the module consisting of all traces generated by maximal runs of N.

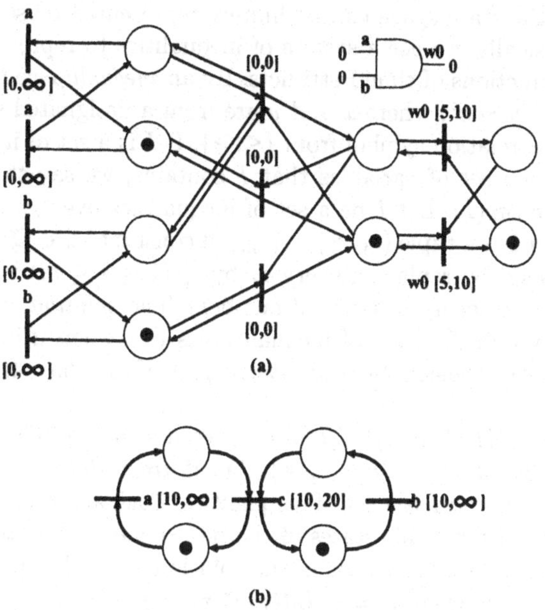

Fig. 2. Nets specifying AND gate and C element

Bounded delay asynchronous circuits can be easily described by nets. For example, an *and*-gate which has inputs a, b and an output w_0 with gate delay [5,10] can be represented by the net shown in Fig. 2(a). In this modelling, we do not distinguish between the change of a wire from 0 to 1 and from 1 to 0. An *or*-gate can be represented similarly. Even though it would be possible to give a more detailled description of gates (e.g., transistor level behavior), for most verification purposes the given net is an adequate representation.

The composition of several gates in a circuit can be described by simply putting together all nets representing single gates. Assuming that all wires in the circuit have unique names, for each transition the corresponding wire can be assigned. Then, the disjoint union of all these nets represents the complete circuit. Thus, the implementation of Muller's C element shown in Fig. 1(b) can be represented by a collection of nets which are similar to the one in Fig. 2(a).

This implementation works correctly under the following assumptions:

1. if an input changes, then the same input never changes again before an output changes, and
2. no input changes before some constant time passes after the change of the output.

The net shown in Fig. 2(b) specifies the behavior of a C-element with these assumptions. Verification consists in showing that the gate-level representation conforms to this specification. This is done by exploring the reachable states of the composed net.

We now describe an algorithm to generate these reachable states of time Petri nets. Since for time Petri nets the time domain consists of rational (not real) numbers, the state space can be finitely represented by sets of systems of inequalities. Basically, we use a system of inequalities to represent a number of different clock functions of time Petri nets. By an *inequality* we mean any string of the form "$x - y \sim c$", where x and y are from a designated set of variables, $c \in \mathbf{Q}$ and $\sim$ is a relation symbol from $\{\leq, \geq\}$. If I is a set of inequalities, then $var(I)$ denotes the set of variables that I contains; we say that I is a set of inequalities *over* $var(I)$. Let I be a set of inequalities over $\{x_1, x_2, \cdots, x_m\}$. A *feasible vector* for I is a tuple $(c_1, c_2, \cdots, c_m)$ of constants $c_i \in \mathbf{Q}$, such that every inequality obtained by replacing every x_i by c_i $(1 \leq i \leq m)$ in any inequality from I holds in the theory of rational numbers. The *solution set* of I is the set of feasible vectors for I. A set of inequalities is *consistent* if its solution set is nonempty. Two sets of inequalities are *isomorphic*, if they have the same solution set.

If the net $N \triangleq (P, T, F, Eft, Lft, \mu^0)$ represents the module $M \triangleq (\mathcal{I}, \mathcal{O}, \mathcal{T})$, we denote this by $M = (\mathcal{I}, \mathcal{O}, N, wire)$. An *abstract state* of the net is a pair (μ, I), where $\mu \subseteq P$ and I is a set of inequalities. Each abstract state denotes an equivalence class of reachable states of the net, namely all states for which the clock values form a feasible vector in the solution set of I. The initial abstract state of N is (μ^0, I_0), where $I_0 \triangleq \{\text{"}Eft(\tau) \leq \underline{\tau} - v \leq Lft(\tau)\text{"} \mid \tau \in enabled(\mu^0)\}$. Here, $\underline{\tau}$ in I_0 is a variable to represent the next firing time of the transition τ. The variable v indicates the initial time point.

The next step is to compute the set of abstract successor states σ' of an abstract state σ of N. To this end we need the notion of *deletion* of a set U of variables from a set I of inequalities. For every such I and U there exists an (up to isomorphism) unique set $I' \triangleq delete(I, U)$ of inequalities over $var(I) - U$, such that the solution set of I' is equal to the solution set of I, projected on $var(I) - U$. For example, if $I \triangleq \{$"$y - x \geq 2$", "$y - x \leq 7$", "$y - z < 3$", "$z - y \leq 11$"$\}$, then $delete(I, \{y\}) = \{$"$x - z < 1$", "$z - x \leq 18$"$\}$. I' can be computed incrementally by a shortest path algorithm in time $O(|var(I)|^2)$[Rok 93, Shi 94].

Let $\sigma \triangleq (\mu, I)$ be an abstract state of N, and $\tau_f \in enabled(\mu)$. Then, $first(\mu, \tau_f) \triangleq \{$"$\underline{\tau} - \underline{\tau}_f \geq 0$"$\mid \tau \in enabled(\mu)\}$ is a set of inequalities describing that τ_f is the first transition which fires in μ. $firable(\sigma) \triangleq \{\tau_f \mid \tau_f \in enabled(\mu), I \cup first(\mu, \tau_f)$ is consistent$\}$ is the set of transitions that can fire earlier than all other transitions in the given marking.

- τ_f is a transition in $firable(\sigma)$.
- μ' is the marking of N obtained by firing transition τ_f.
  That is, $\mu' \triangleq (\mu - \bullet\tau_f) \cup \tau_f \bullet$.
- R is a set of newly enabled transitions obtained by the firing of τ_f. That is, $R \triangleq enabled(\mu') - enabled(\mu - \bullet\tau_f)$.
- $J \triangleq \{$"$\underline{\tau} - \underline{\tau}_{out} \geq Eft(\tau)$"$\mid \tau \in R\} \cup \{$"$\underline{\tau} - \underline{\tau}_{out} \leq Lft(\tau)$"$\mid \tau \in R\}$.
- $J' \triangleq I \cup first(\mu, \tau_{out}) \cup J$.
- $D \triangleq \{\underline{\tau} \mid \tau$ made some transition τ' enabled, and τ' is still enabled in $\mu'\}$.
- $I' \triangleq delete(J_3, \{\underline{\tau} \mid \tau \notin enabled(\hat{\mu}')\} - D)$

Intuitively, J, J', D and I' can be read as follows: J relates the variables of newly enabled transitions to the variable of the fired transition τ_{out}. J' is the union of I, J, and a set of inequalities representing that τ_{out} fires earlier than others. Transitions related to variables in D are currently parents of enabled transitions in μ', and these variables are necessary to check the coverability between the firing domains of transitions. Finally, in I' the variables of disabled transitions except for those in D are deleted. We write $\sigma \xrightarrow{\tau_f} \sigma'$ if $\sigma' \triangleq (\mu', I')$ is a successor of the abstract state $\sigma \triangleq (\mu, I)$ with respect to τ_f.

We now describe how conformance can be checked, using this successor relation between abstract states. We consider a set $\{M_0, M_1, \cdots, M_n\}$ of modules, where $M_i = (\mathcal{I}_i, \mathcal{O}_i, N_i, wire_i)$, $N_i \triangleq (P_i, T_i, Eft_i, Lft_i, \mu_i^0)$, and assume that for $i \neq j$, $P_i \cap P_j = T_i \cap T_j = P_i \cap T_j = \emptyset$. Some module in the set is a mirror of a specification, and input transitions and output transitions must not be in conflict in the module. If there is no confusion, we use the notation *wire* instead of $wire_i$, and $\tau \in M_i$, when $\tau \in T_i$. Let $m(\tau)$ be the module number of τ, i.e., $m(\tau) = i$, if $\tau \in M_i$. Transition τ is called an *output transition* if $wire_{m(\tau)}(\tau) \in \mathcal{O}_{m(\tau)}$, and an *input transition* if $wire_{m(\tau)}(\tau) \in \mathcal{I}_{m(\tau)}$. If $\sigma_i \triangleq (\mu_i, I_i)$, $i \leq n$, are abstract states of the nets N_i, and K is a set of inequalities, we say that $s \triangleq (\sigma_0, \cdots, \sigma_n, K)$ is an abstract state of the module set $\{M_0, M_1, \cdots, M_n\}$.

The initial abstract state is $s_0 \triangleq (\sigma_0^0, \cdots, \sigma_n^0, \emptyset)$. We extend the definitions of *enabled*(μ) and *firable*(σ) with respect to $s \triangleq (\sigma_0, \cdots, \sigma_n, K)$ as follows.

$$enabled(s) \triangleq \{\tau \mid \tau \in enabled(\mu_{m(\tau)})\}, \text{ and}$$

$$globally\_firable(s) \triangleq \{\tau \mid \tau \in enabled(s),\ \ first(s,\tau) \cup \bigcup_{i=0}^{n} I_i \cup K \text{ is consistent}\},$$

where $first(s,\tau) \triangleq \{\text{"}\tau - \tau' \leq 0\text{"} \mid \tau' \in enabled(s)\}$. Furthermore, for an output transition τ_O such that $\tau_O \in globally\_firable(s)$,

$$sync\_trans(\tau_O, s) \triangleq \{\tau \mid wire(\tau) = wire(\tau_O), \tau \in globally\_firable(s)\}.$$

When $\{M_0, M_1, \cdots, M_n\}$ is at $s \triangleq (\sigma_0, \cdots, \sigma_n, K)$, it moves to $s' \triangleq (\sigma_0', \cdots, \sigma_n', K')$ with respect to $\tau_O \in globally\_firable(s)$ by firing all transitions in $sync\_trans(\tau_O, s)$.

- for $1 \leq i \leq n$
  - if $\tau \in sync\_trans(\tau_O, s) \cap T_i$, then $\sigma_i \xrightarrow{\tau} \sigma_i'$, and
  - if $sync\_trans(\tau_O, s) \cap T_i = \emptyset$, then $\sigma_i' = \sigma_i$.
- $K' \triangleq K \cup \{\text{"}\underline{\tau} = \underline{\tau}'\text{"} \mid \tau, \tau' \in sync\_trans(\tau_O, s)\}$.

Let $s \xrightarrow{\tau_O} s'$ denote this state transition relation of the module set. For any transition τ and abstract state σ, the variable $parent(\tau, \sigma)$ indicates which transition enabled τ. Formally, if $\sigma \triangleq (\mu, I)$, $\sigma' \triangleq (\mu', I')$, $\sigma \xrightarrow{\tau} \sigma'$, and $\tau' \in enabled(\sigma')$, then

$$parent(\tau', \sigma') \triangleq \begin{cases} \underline{\tau}, & \text{if } \tau' \in enabled(\mu') - enabled(\mu - \bullet\tau) \\ parent(\tau', \sigma), & \text{otherwise.} \end{cases}$$

For a set I of inequalities, let $earlier(x, y, I)$ be the predicate expressing that $solution(\{\text{"}x > y\text{"}\} \cup I) = \emptyset$, i.e., $earlier(x, y, I)$ holds iff $x \leq y$ for every solution vector of I. We write $earlier(x, y, \sigma_i)$ for $earlier(x, y, I_i)$, where $\sigma_i \triangleq (\mu_i, I_i)$, and $earlier(x, y, s)$ for $earlier(x, y, \bigcup_{i=0}^{n} I_i \cup K)$, where $s \triangleq (\sigma_0, \cdots, \sigma_n, K)$. Let $\tau \in M_i$, $\sigma_i \triangleq (\mu_i, I_i)$, and $\tau \in enabled(s)$.

- $earliest\_firing\_time(s, \tau) \triangleq parent(\tau, \sigma_i) + Eft(\tau)$, and
- $latest\_firing\_time(s, \tau) \triangleq parent(\tau, \sigma_i) + Lft(\tau)$.

A state $s \triangleq (\sigma_0, \cdots, \sigma_n, K)$ is called *safe*, if for every output transition τ_O such that $\tau_O \in globally\_firable(s)$, and for every module $M_j (0 \leq j \leq n)$ such that $wire(\tau_O) \in \mathcal{I}_j$, there exists an input transition τ_I such that $wire(\tau_I) = wire(\tau_O)$, $\tau_I \in enabled(s)$, $earlier(earliest\_firing\_time(s, \tau_I), \tau_O, s)$ holds, and either

1. $earlier(\tau_O, latest\_firing\_time(s, \tau_I), s)$, or
2. for some output transition τ such that $\tau \in enabled(s)$, $earlier(\tau, latest\_firing\_time(s, \tau_I), s)$.

A state $s \triangleq (\sigma_0, \cdots, \sigma_n, K)$ is called *live*, if for every input transition τ_I such that $\tau_I \in globally\_firable(s)$, there exists an output transition τ (of an arbitrary module) such that $\tau \in globally\_firable(s)$.

Let modules $M_1, \cdots, M_n$ be represented by nets $N_1, \cdots, N_n$. A safety failure corresponds to a non-safe state in the reachable state space, and a timing failure occurs if a state can be reached which is not live. In other words, $failure(M_1, M_2, \cdots, M_n)$ is empty, iff every state which is reachable from the initial state of $\langle N_1, \cdots, N_n \rangle$ is both safe and live. Therefore, the verification of conformance between modules can be done by traversing the state space of $\langle N_1, \cdots, N_n \rangle$ and checking if non-safe or non-live states are reachable.

Furthermore, it is possible to replace an abstract description of a module by a more concrete implementation. If $\{M_1, \cdots, M_{k-1}, M_k, M_{k+1}, \cdots, M_n\}$ conforms to M_S, $\{M_{k_1}, ..., M_{k_m}\}$ conforms to M_k, and $(\bigcup_{j=1}^{m} \mathcal{W}_{k_j} - \mathcal{W}_k) \cap \bigcup_{j=1}^{n} \mathcal{W}_j = \emptyset$, then $\{M_1, \cdots, M_{k-1}, M_{k_1}, ..., M_{k_m}, M_{k+1}, \cdots, M_n\}$ conforms to M_S. The set of wires in a specification usually is much smaller than the set of wires in the implementation. Thus, the total computation cost to determine whether $\{M_1, \cdots, M_{k-1}, M_k, M_{k+1}, \cdots, M_n\}$ conforms to M_S and $\{M_{k_1}, \cdots, M_{k_m}\}$ conforms to M_k is significantly smaller than the computation of whether $\{M_1, \cdots, M_{k-1}, M_{k_1}, \cdots, M_{k_m}, M_{k+1}, \cdots, M_n\}$ conforms to M_S. This is the primary advantage of hierarchical verification.

4 Experimental Results

We have implemented the algorithm shown in the previous section on a UNIX workstation in C++. In this section, we present some experimental verification results.

First, our verifier shows that the implementation in Fig. 1(b) is correct with respect to the specification in Fig. 2(b) after traversing 51 states, which takes about one second on a 17 MIPS workstation.

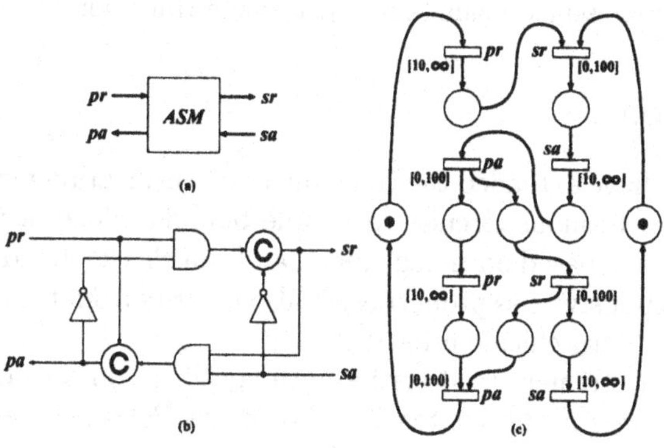

Fig. 3. An automatic sweeping module, gate level implementation, and specification

The second example is a control circuit of the request-acknowledgement handshake mechanism for asynchronous circuits. This circuit called an automatic sweeping module (ASM, for short) has two inputs (a primary request *pr*, a secondary acknowledgement *sa*) and two outputs (a primary acknowledgement *pa*, a secondary request *sr*) (Fig. 3(a)). It has the following functionality:

1. When the primary request goes high with the secondary acknowledgement low, ASM sets the secondary request.
2. When the secondary acknowledgement becomes high, ASM resets the secondary request with setting the primary acknowledgement.
3. When the primary request becomes low, ASM resets the primary acknowledgement.

This functionality with almost the same assumptions as for the C-element is specified with a net as shown in Fig. 3(c). On the other hand, Fig. 3(b) was proposed as the gate level implementation of ASM. We assume that each gate has a delay [5,10].

Our verifier shows that this implementation is correct with respect to the specification in Fig. 3(c). In Table 1, the column *flat* shows the size of the nets, the number of states, and CPU times needed for this verification when C elements are expanded by using their gate level implementations shown in Fig. 1. The column *hierarchical* shows the results of the hierarchical verification. That is, the specification net shown in Fig. 2(b) is used for the verification of ASM. In this case, the total verification time is the sum of the verification times for both ASM and C-element. These results show the advantage of the hierarchical verification as well.

Table 1. Results of verification

	flat			*hierarchical*		
	size[†]	states	CPU time (s)	size[†]	states	CPU time (s)
C-element	–	–	–	p:30, t:34	51	1.3
ASM	p:78, t:90	391	81.8	p:34, t:34	58	1.2
Total	p:78, t:90	391	81.8	p:64, t:68	109	2.5

† : "p:" and "t:" represent the numbers of places and transitions, respectively.

5 Conclusion

In this paper, we have extended the trace theoretic verification method for speed-independent asynchronous circuits to handle bounded delay asynchronous circuits. Our method is based on timed traces, and can check (timed) safety properties as well as responsiveness properties. It also inherits from the original method the possibility of hierarchical verification.

We use time Petri nets to describe both specification and implementation. Time Petri nets are a natural extension of ordinary Petri nets, which are widely used in conventional verification methods. In this formalism, both (timed) properties and bounded delay asynchronous circuits can be described. We have developed a decision algorithm to check whether an implementation is correct with respect to its specification. It is based on state space traversal of a set of time Petri nets, and checking if any failure states are reachable.

First experimental results show that hierarchical verification works extremely well. Nevertheless, the increase of the number of modules can have a bad influence on the verification time. In the future we want to apply partial order analysis techniques [YY 96, YS 97, BM 98] to our method. This could help to further reduce the average complexity of the verification.

Acknowledgments
We would like to thank D. Dill for giving us the program code of his verifier, and Ichiki Honma for his help in implementing a first version of our algorithm.

References

[AD 92] R. Alur and D. Dill: *Automata for Modelling Real-time Systems*; LNCS 600: Real time: Theory in Practice, pp.45–73, (1992).

[BM 98] W. Belluomini and C. Myers: *Verification of Timed Systems Using POSETS* Proc. Int. Conf. on Computer Aided Verification (CAV), Univ. of British Columbia, Vancouver, Canada (1998)

[BD 91] B. Berthomieu and M. Diaz: *Modeling and Verification of Time Dependent Systems using Time Petri Nets*; IEEE Trans. on Software Eng., 17(3):259–273, (1991).

[Dil 88] D. Dill: *Trace Theory for Automatic Hierarchical Verification of Speed-Independent Circuits*; PH.D. Thesis, MIT press, (1988).

[GTM 96] J.D. Garside, S. Temple, and R. Mehra: *The AMULET2e Cache System*; Proc. of the Second International Symposium of ASYNC, pp. 208—217 (1996).

[JM 87] F. Jahanian and A. Mok: *A Graph-Theoretic Approach for Timing Analysis and its Implementation*; IEEE Trans. Comput. C-36(8):961–975 (1987).

[MB 59] D. Muller and W. Bartke: *A Theory of Asynchronous Circuits*; Theory of Switching, Harvard University Press, Masachusetts (1959).

[MF 76] P. Merlin and D. Faber: *Recoverability of Communication Protocols*; IEEE Trans. on Communication, COM-24(9), (1976).

[Rok 93] T. Rokicki: *Representing and Modeling Digital Circuits*; Ph.D Dissertation, Stanford University, (1993).

[Shi 94] W. Shigetaka: *On the Acceleration of Formal Verification Methods*; Thesis, Tokyo Institute of Technology (1994).

[TSS 98] L. Twele, B.-H. Schlingloff, H. Szczerbicka: *Performability Analysis of an Avionics-Interface*; IEEE Symp. on Man, Machine and Cybernetics; San Diego, (1998)

[YS 97] T. Yoneda, B-H. Schlingloff: *Efficient Verification of Parallel Real-Time Systems*; Journal of Formal Methods in System Design 11-2, pp. 187-215, (1997).

[YY 96] T. Yoneda, T. Yoshikawa: *Using Partial Orders for Trace Theoretic Verification of Asynchronous Circuits*; Proc. of the Second International Symposium of ASYNC (1996).

Verification of Temporal Properties of Processes in a Setting with Data

Jan Friso Groote[1,2] and Radu Mateescu[3]*

[1] Cwi, P.O. Box 94079, NL-1090 GB Amsterdam, The Netherlands
[2] Eindhoven University of Technology, P.O. Box 513
NL-5600 MB Eindhoven, The Netherlands
JanFriso.Groote@cwi.nl
[3] INRIA Rhône-Alpes / VASY group, 655, avenue de l'Europe
F-38330 Montbonnot Saint Martin, France
Radu.Mateescu@inria.fr

Abstract. We define a value-based modal μ-calculus, built from first-order formulas, modalities, and fixed point operators parameterized by data variables, which allows to express temporal properties involving data. We interpret this logic over μCRL terms defined by linear process equations. The satisfaction of a temporal formula by a μCRL term is translated to the satisfaction of a first-order formula containing parameterized fixed point operators. We provide proof rules for these fixed point operators and show their applicability on various examples.

1 Introduction

In recent years we have applied process algebra in numerous settings [4,8,12]. The first lesson we learned is that process algebra *pur sang* is not very handy, and we need an extension with data. This led to the language μCRL (*micro Common Representation Language*) [13]. The next observation was that it is very convenient to eliminate the parallel operator from a process description and reduce it to a very restricted form, which we call a *linear process equation* or *linear process operator* [3]. Such an elimination can be done automatically [5,9] and generally yields a compact result, of the same size as the original system description. For proving equations of the form *specification=implementation*, a proof methodology has been developed [14] and has been applied to numerous examples (see e.g. [4,8,11,20]) that all have infinite or unbounded state spaces.

An obvious question that has not been addressed thus far is whether the linear process format can also be employed in proving temporal logic formulas. In this paper we provide a way of doing so that roughly goes as follows. First, we extend the modal μ-calculus [16] to express properties about data, meaning that we include boolean expressions on data variables, parameterization of actions contained in the modalities, quantification over data, and parameterization

* This work has been funded by the grant no. 97-09 of the ERCIM fellowship programme for collaboration between INRIA and CWI.

A.M. Haeberer (Ed.): AMAST'98, LNCS 1548, pp. 74–90, 1998.

of minimal and maximal fixed point operators. A typical example of temporal property expressed in this logic is

$$\big(\nu Y(n{:}\mathrm{N}).\exists m{:}\mathrm{N}.\,\langle a(m+n)\rangle\, Y(m+n)\big)(2)$$

describing the states from which an infinite sequence of actions $a(i_0)a(i_1)a(i_2)\cdots$ can be performed, where $2 \leq i_0 \leq i_1 \leq i_2 \leq \cdots$. Another example of formula is

$$\forall i{:}\mathrm{N}.[a(i)](i > n)$$

stating that whenever an $a(i)$ action can be performed, i must be larger than n.

The second step is to prove that a given linear process satisfies such a temporal formula. To achieve this, we first transform both the process and the temporal formula into a first-order fixed point formula. This approach is similar to the model-checking algorithms in [2,22,1], where a formula of standard μ-calculus (i.e., without data) and a finite state automaton are combined to form a set of fixed point boolean equations, which can be solved in linear time, provided the formula is alternation-free. In our setting, this transformation applies to the full logic (formulas of arbitrary alternation depth), is purely syntactical, and in many cases can be carried out by hand, as both the linear process and the temporal formula are generally quite small.

In order to solve the first-order fixed point formulas obtained in this way, we use the standard proof rules for connectives and quantifiers, and we introduce a set of proof rules for fixed point operators allowing to approximate (towards either satisfaction, or refutation) the fixed point (sub)formulas. If the initial state of the process satisfies an approximation of a maximal fixed point formula, we know that it satisfies the maximal fixed point too. The approximation of minimal fixed points captures the fact that the property expressed by a minimal fixed point formula must be reached in a finite number of steps. These rules reflect the proof principles for safety and liveness properties discussed in [17].

We included a simple example and a slightly more elaborate one, in order to show how the proof method that we propose can be used. We have also successfully applied the method to verify a distributed summing protocol [11], but due to space limitations we have not included it in this paper. All these examples are quite promising, as they show that our method leads to straightforward arguments of validity of the temporal formulas.

Other approaches to prove temporal properties involving data that we are aware of [19,7] use tableau-based methods, often directed towards decomposing the property over the system. The approach we adopt here is different, being intended to facilitate manual verification in the natural deduction style (see also [15]). Since the linear processes obtained from μCRL specifications are generally small, we expect a good applicability of our method to various examples.

The paper is organized as follows. Section 2 defines the linear μCRL processes and their models. Section 3 gives the syntax and semantics of the extended μ-calculus that we propose, together with examples of temporal properties. Section 4 presents the verification method, i.e., the translation into first-order fixed point formulas and the proof rules for extremal fixed points. Finally, Section 5 shows the application of this method on an infinite-state linear μCRL process.

2 Preliminaries

We define below the notions of data expression, linear process, and labeled transition system (LTS), over which the temporal logic formulas will be interpreted.

2.1 Expressions

The set *Exp* of *data expressions* is defined over a set *DVar* of *data variables* and a set *Func* of *functions*. Each data variable $x \in DVar$ has a type D and each function $f \in Func$ has a profile $D_1 \times \cdots \times D_n \rightarrow D$, where $D_1, \ldots, D_n$ are the argument types of f and D is its result type. We write **Val** for the domain containing all the values belonging to the types D. The expressions $e \in Exp$ are defined by the following grammar:

$$e ::= x \mid f(e_1, \ldots, e_n)$$

The set of variables occurring in an expression e is noted $var(e)$.

We define the domain $\mathbf{DEnv} = DVar \rightarrow \mathbf{Val}$ of *data environments*. A data environment $\varepsilon \in \mathbf{DEnv}$ is a partial function mapping data variables into values of their corresponding types. The *support* of an environment ε, noted $supp(\varepsilon)$, denotes the set of variables that are assigned a value in **Val** by ε. An environment mapping the variables $x_1, \ldots, x_n$ respectively to the values $v_1, \ldots, v_n$ is noted $[v_1/x_1, \ldots, v_n/x_n]$. The environment having an empty support is noted $[\,]$. The *overriding* of ε by $[v_1/x_1, \ldots, v_n/x_n]$ is the data environment defined as follows: $(\varepsilon[v_1/x_1, \ldots, v_n/x_n])(x) = \mathit{if}\ \exists i \in [1, n].x = x_i\ \mathit{then}\ v_i\ \mathit{else}\ \varepsilon(x)$.

The semantics of data expressions is given by the interpretation function $[\![.]\!] : Exp \rightarrow \mathbf{DEnv} \rightarrow \mathbf{Val}$, defined inductively below. For an expression e and a data environment ε such that $var(e) \subseteq supp(\varepsilon)$, $[\![e]\!]\,\varepsilon$ denotes the value of e in the context of ε:

$$[\![x]\!]\,\varepsilon \stackrel{\text{def}}{=} \varepsilon(x)$$

$$[\![f(e_1, \ldots, e_n)]\!]\,\varepsilon \stackrel{\text{def}}{=} f([\![e_1]\!]\,\varepsilon, \ldots, [\![e_n]\!]\,\varepsilon)$$

We assume that the domain $\mathbf{Bool} = \{\mathbf{tt}, \mathbf{ff}\}$ of boolean values is predefined, together with the usual operations $\wedge$, $\vee$, $\neg$, and $\rightarrow$. Boolean expressions are denoted by the symbol b.

2.2 Linear Processes

Linear processes share with LTSs the advantage of being a simple, straightforward notation, suitable for further analysis of processes in either automatic or manual form. But they do not share the most important disadvantage, namely the exponential blow-up caused by the parallel operator (see [5]). As we are interested in devising analysis methods for realistic distributed systems, it is clear that LTSs are not satisfactory. Therefore, we use the linear processes, of which we give a definition below.

Let Act be a set of actions, which may be parameterized by data values.

Definition 1. *Let Act* $\subseteq$ Act $\cup \{\tau\}$ *be a finite set of actions and* D, D_a, E_a *be data types. A* linear process *over Act and D is defined by an equation of the following form:*

$$X(x{:}D) = \sum_{a \in Act} \sum_{x_a : E_a} a(e_a) \cdot X(e'_a) \lhd b_a \rhd \delta$$

where x *is a parameter of type* D*, and for each action* $a \in Act$*,* x_a *is a variable of type* E_a*,* e_a *and* e'_a *are expressions of type* D_a *and* D*, respectively, and* b_a *is an expression of type* **Bool***, such that* $var(e_a) \cup var(e'_a) \cup var(b_a) \subseteq \{x, x_a\}$*. The constant* δ*, called* deadlock*, cannot perform any action. The* initial state *of process* X *may be specified by giving an initial value* $v_0 \in D$ *for* x*.*

A linear process expression must be read as follows. If a process is in state x, then it can perform actions $a(e_a)$ provided a value of x_a in E_a can be found such that b_a holds. In such a case, the process ends up in a state e'_a.

For simplicity, we allow at most one data parameter for any action $a \in Act$ (we assume that τ has a dummy parameter) and for each linear process X. Using pairing and projection, the formalization can be straightforwardly used with multiple parameters.

2.3 Transition Systems

We consider a linear μCRL process X as in Definition 1. According to the operational semantics of μCRL [13], the transition system modeling a linear process is defined as follows.

Definition 2. *The* transition system *of a linear process is a quadruple* $M = (S, L, \rightarrow, s_0)$*, where:*

- $S \stackrel{\text{def}}{=} \{X(v) \mid v \in D\}$ *is the set of* states*;*
- $L \stackrel{\text{def}}{=} \{a(v_a) \mid a \in Act \wedge v_a \in D_a\}$ *is the set of* labels*;*
- $\rightarrow \stackrel{\text{def}}{=} \{X(v) \xrightarrow{a(v'_a)} X(v') \mid a \in Act \wedge \exists v_a \in E_a.([\![b_a]\!]\,[v/x, v_a/x_a] \wedge v'_a = [\![e_a]\!]\,[v/x, v_a/x_a] \wedge v' = [\![e'_a]\!]\,[v/x, v_a/x_a])\}$ *is the* transition relation*;*
- $s_0 \stackrel{\text{def}}{=} X(v_0) \in S$ *is the* initial state.

The definition of the initial state of the process is not mandatory, unless there are properties of X that must be explicitly verified on $X(v_0)$.

3 Temporal Logic

The logic we consider is based upon an extension of the modal μ-calculus [16] with data variables, quantifiers, and parameterization, in order to express properties involving data. Other similar value-based formalisms extending the modal μ-calculus have been used in the framework of symbolic transition systems [19] and of the polyadic π-calculus [7].

The logic we propose here contains a set *AForm* of *action formulas* and a set *SForm* of *state formulas*, whose syntax and semantics are defined below. To simplify the notations, we implicitly consider throughout this section a transition system $M = (S, L, \rightarrow, s_0)$, over which the formulas are interpreted.

The action formulas $\alpha \in AForm$ are defined by the following grammar:

$$\alpha ::= a(e) \mid tt \mid \neg\alpha \mid \alpha_1 \wedge \alpha_2 \mid \exists y{:}D.\alpha$$

where $a \in Act$, $e \in Exp$, and $y \in DVar$ is a data variable of type D. The usual derived operators are also allowed: $ff = \neg tt$, $\alpha_1 \vee \alpha_2 = \neg(\neg\alpha_1 \wedge \neg\alpha_2)$, $\alpha_1 \rightarrow \alpha_2 = \neg\alpha_1 \vee \alpha_2$, $\forall y{:}D.\alpha = \neg\exists y{:}D.\neg\alpha$. Data variables are bound by quantifiers in the usual way. The set of free data variables of a formula α is noted $fdv(\alpha)$.

The semantics of action formulas is given by the interpretation function $[\![.]\!] : AForm \rightarrow \mathbf{DEnv} \rightarrow 2^L$, defined inductively below. Given an action formula α and a data environment ε such that $fdv(\alpha) \subseteq supp(\varepsilon)$, $[\![\alpha]\!]\,\varepsilon$ denotes the set of labels satisfying α in the context of ε:

$$\begin{aligned}
[\![a(e)]\!]\,\varepsilon &\stackrel{\text{def}}{=} \{a([\![e]\!]\,\varepsilon)\}\\
[\![tt]\!]\,\varepsilon &\stackrel{\text{def}}{=} L\\
[\![\neg\alpha]\!]\,\varepsilon &\stackrel{\text{def}}{=} L \setminus [\![\alpha]\!]\,\varepsilon\\
[\![\alpha_1 \wedge \alpha_2]\!]\,\varepsilon &\stackrel{\text{def}}{=} [\![\alpha_1]\!]\,\varepsilon \cap [\![\alpha_2]\!]\,\varepsilon\\
[\![\exists y{:}D.\alpha]\!]\,\varepsilon &\stackrel{\text{def}}{=} \textstyle\bigcup_{v \in D} [\![\alpha]\!]\,\varepsilon[v/y].
\end{aligned}$$

The state formulas $\varphi \in SForm$, built over the set *AForm* and over a set *PVar* of *propositional variables*, are defined by the following grammar:

$$\varphi ::= b \mid Y(e) \mid \neg\varphi \mid \varphi_1 \wedge \varphi_2 \mid \langle\alpha\rangle\,\varphi \mid \exists y{:}D.\varphi \mid (\mu Y(y{:}D).\varphi)(e)$$

where $b \in Exp$ is a boolean expression, $Y \in PVar$ is a (parameterized) propositional variable, $\alpha \in AForm$ is an action formula and $y \in DVar$ is a data variable of type D. Besides the usual derived connectives, we also define the box modal operator $[\alpha]\,\varphi = \neg\,\langle\alpha\rangle\,\neg\varphi$ and the maximal fixed point operator $(\nu Y(y{:}D).\varphi)(e) = \neg(\mu Y(y{:}D).\neg\varphi[\neg Y/Y])(e)$, where $\varphi[\neg Y/Y]$ denotes the syntactic substitution of Y by $\neg Y$ in φ. In the sequel, we let σ range over $\{\mu, \nu\}$.

Data variables are bound by quantifiers and by parameterization, and propositional variables are bound by fixed point operators, in the usual way. The sets of free data variables and free propositional variables of φ are noted $fdv(\varphi)$ and $fpv(\varphi)$, respectively. A formula φ is said *closed* if $fdv(\varphi) = \emptyset$ and $fpv(\varphi) = \emptyset$.

We assume that state formulas are *syntactically monotonic*, i.e., for each formula $(\sigma Y(y{:}D).\varphi)(e)$, every free occurrence of Y in φ falls under an even number of negations. This enables to convert any formula φ in *Positive Normal Form* (PNF for short) by pushing the negations downwards to its atomic subformulas and (if necessary) by α-converting it such that there is no variable Y having both free and bound occurrences in φ. In the sequel, we consider only closed state formulas in PNF.

We define the domain $\mathbf{PEnv} = PVar \rightarrow (\mathbf{Val} \rightarrow 2^S)$ of *propositional environments*. A propositional environment $\rho \in \mathbf{PEnv}$ is a partial function mapping propositional variables to functions from the domains of their parameters to sets of transition system states. The support, bracketed notation, and overriding of propositional environments are defined in the same way as for data environments.

The semantics of state formulas is given by the interpretation function $[\![.]\!]$: $SForm \rightarrow \mathbf{PEnv} \rightarrow \mathbf{DEnv} \rightarrow 2^S$, defined inductively below. For a state formula φ, a propositional environment ρ, and a data environment ε such that $fpv(\varphi) \subseteq supp(\rho)$ and $fdv(\varphi) \subseteq supp(\varepsilon)$, $[\![\varphi]\!]\,\rho\varepsilon$ denotes the set of states satisfying φ in the context of ρ and ε:

$$\begin{aligned}
[\![b]\!]\,\rho\varepsilon &\stackrel{\text{def}}{=} \mathit{if}\ \ [\![b]\!]\,\varepsilon\ \mathit{then}\ S\ \mathit{else}\ \emptyset \\
[\![Y(e)]\!]\,\rho\varepsilon &\stackrel{\text{def}}{=} (\rho(Y))([\![e]\!]\,\varepsilon) \\
[\![\varphi_1 \wedge \varphi_2]\!]\,\rho\varepsilon &\stackrel{\text{def}}{=} [\![\varphi_1]\!]\,\rho\varepsilon \cap [\![\varphi_2]\!]\,\rho\varepsilon \\
[\![\langle\alpha\rangle\,\varphi]\!]\,\rho\varepsilon &\stackrel{\text{def}}{=} \{X(v) \in S \mid \exists v' \in D.\exists a \in Act.\exists v_a \in D_a. \\
&\qquad\qquad X(v) \stackrel{a(v_a)}{\longrightarrow} X(v') \wedge a(v_a) \in [\![\alpha]\!]\,\varepsilon \wedge X(v') \in [\![\varphi]\!]\,\rho\varepsilon\} \\
[\![\exists y{:}D.\varphi]\!]\,\rho\varepsilon &\stackrel{\text{def}}{=} \{X(v) \in S \mid \exists v' \in D.X(v) \in [\![\varphi]\!]\,\rho(\varepsilon[v'/y])\} \\
[\![(\mu Y(y{:}D).\varphi)(e)]\!]\,\rho\varepsilon &\stackrel{\text{def}}{=} (\mu\Phi_{\rho\varepsilon})([\![e]\!]\,\varepsilon)
\end{aligned}$$

where the functional $\Phi_{\rho\varepsilon} : (D \rightarrow 2^S) \rightarrow (D \rightarrow 2^S)$, associated to the formula $\mu Y(y{:}D).\varphi$, is defined as $\Phi_{\rho\varepsilon} = \lambda F{:}D \rightarrow 2^S.\lambda v{:}D.\,[\![\varphi]\!]\,(\rho[F/Y])(\varepsilon[v/y])$.

It is straightforward to check that, for state formulas in PNF, every functional $\Phi_{\rho\varepsilon}$ associated to a fixed point (sub)formula is monotonic over $D \rightarrow 2^S$. Since the underlying lattices $D \rightarrow 2^S$ are complete, it follows from Tarski's theorem [21] that every $\Phi_{\rho\varepsilon}$ functional has a unique minimal fixed point $\mu\Phi_{\rho\varepsilon}$ and a unique maximal fixed point $\nu\Phi_{\rho\varepsilon}$.

3.1 Example

We describe a simple infinite state process, together with some temporal properties, in order to illustrate the techniques presented in here. In Section 4.3 we will translate the temporal formulas and in Section 4.5 we will prove the validity of the first-order fixed point formulas that we have obtained this way. The example is given by the following linear process equation, describing a slot machine:

$$\begin{aligned}
X(v{:}\mathrm{N}, b{:}\mathbf{Bool}) = \quad & s \cdot X(v+1, \neg b) \ \triangleleft \ \neg b \ \triangleright \delta\ + \\
& \textstyle\sum_{m:\mathrm{N}} w(m) \cdot X(v-m, \neg b) \triangleleft b \wedge m \leq v \triangleright \delta
\end{aligned}$$

The parameters v and b denote the current amount of money and the current state of the machine, respectively. When b equals *ff*, a user can activate the machine by inserting a coin (action s); afterwards, b becomes *tt* and the machine will deliver the money m won by the user (action $w(m)$). The initial state of the system is $X(v_0, \mathit{ff})$, for some fixed $v_0 \geq 0$. (Actually, the linear process above allows a user to collect any amount of money he wants, but for the sake of the example we do not complicate the slot machine in order to avoid this.)

We are interested in the temporal properties below.

1. A basic liveness property is that, for any amount of money $l \in \mathrm{N}$, the machine can potentially deliver it to a user:
$$\varphi_1 \stackrel{\text{def}}{=} \mu Y.\, \langle w(l)\rangle\, tt \vee \langle tt\rangle\, Y$$
2. A stronger liveness property would be that, for any amount of money $l \in \mathrm{N}$, the machine must eventually deliver it:
$$\varphi_2 \stackrel{\text{def}}{=} \mu Y.\, \langle tt\rangle\, tt \wedge [\neg w(l)]\, Y$$
3. A basic safety property is that every $l \in \mathrm{N}$ won in a $w(l)$ action cannot exceed the initial amount of money v_0 of the machine, updated with the p and r money that have been inserted and won by users since the initial state of the system, respectively:
$$\varphi_3 \stackrel{\text{def}}{=} (\nu Y(p,r{:}\mathrm{N}).\forall l{:}\mathrm{N}.\, [w(l)]\, (l \leq v_0+p-r \wedge Y(p,r+l)) \wedge [s]\, Y(p+1,r))(0,0).$$

Clearly, φ_1 and φ_3 are valid for X, but φ_2 does not hold.

4 Verification

The verification problem consists to check whether a transition system M (given by a linear μCRL process) satisfies a given temporal formula φ. Two different cases are usually distinguished: *global* verification, consisting to decide if all the states of M satisfy φ, and *local* verification, consisting to decide if one particular state (e.g., the initial state s_0) of M satisfies φ. Both instances of the problem can be reduced to the satisfaction of a first-order fixed point formula. First we define the language of first-order fixed point formulas, next we describe the translation of a model M and a state formula φ into a first-order fixed point formula, and finally we provide sound proof rules for reasoning about fixed point operators.

4.1 First-Order Fixed Point Formulas

We define the syntax and semantics of the set *BForm* of *first-order fixed point formulas*, which will be used as an intermediate formalism for verification purposes. The formulas $\psi \in \mathit{BForm}$, built over a set *BVar* of *boolean variables*, are defined by the following grammar:

$$\psi ::= b \mid Z(e) \mid \neg\psi \mid \psi_1 \wedge \psi_2 \mid \exists z{:}D.\psi_1 \mid (\mu Z(z{:}D).\psi_1)(e)$$

where $b \in \mathit{Exp}$ is a boolean expression and $Z \in \mathit{BVar}$ is a (parameterized) boolean variable. The derived boolean, first-order, modal, and fixed point operators are defined as usual. The data and boolean variables are bound in a manner similar to the state formulas φ. The sets of free data variables and free boolean variables of ψ are noted $fdv(\psi)$ and $fbv(\psi)$, respectively. For simplicity, we use only one data parameter in first-order fixed point formulas; the formalization could be easily extended to allow multiple parameters. In the same way as for state formulas, we consider here only closed first-order fixed point formulas that have been translated in PNF.

We introduce the domain $\mathbf{BEnv} = BVar \to (\mathbf{Val} \to \mathbf{Bool})$ of *boolean environments.* A boolean environment $\eta \in \mathbf{BEnv}$ is a partial function mapping boolean variables to predicates over the domains of the data parameters. The support, bracketed notation, and overriding of boolean environments are defined in the same way as for propositional environments.

The semantics of first-order fixed point formulas is given by the interpretation function $[\![.]\!] : BForm \to \mathbf{BEnv} \to \mathbf{DEnv} \to \mathbf{Bool}$, defined inductively below. For a formula ψ, a boolean environment η, and a data environment ε such that $fbv(\psi) \subseteq supp(\eta)$ and $fdv(\psi) \subseteq supp(\varepsilon)$, $[\![\psi]\!]\,\eta\varepsilon$ denotes the truth value of ψ in the context of η and ε:

$$\begin{aligned}
[\![b]\!]\,\eta\varepsilon &\stackrel{\text{def}}{=} [\![b]\!]\,\varepsilon \\
[\![Z(e)]\!]\,\eta\varepsilon &\stackrel{\text{def}}{=} (\eta(Z))([\![e]\!]\,\varepsilon) \\
[\![\psi_1 \wedge \psi_2]\!]\,\eta\varepsilon &\stackrel{\text{def}}{=} [\![\psi_1]\!]\,\eta\varepsilon \wedge [\![\psi_2]\!]\,\eta\varepsilon \\
[\![\exists z{:}D.\psi]\!]\,\eta\varepsilon &\stackrel{\text{def}}{=} \exists v \in D.\, [\![\psi]\!]\,\eta(\varepsilon[v/z]) \\
[\![(\mu Z(z{:}D).\psi)(e)]\!]\,\eta\varepsilon &\stackrel{\text{def}}{=} (\mu\Psi_{\eta\varepsilon})([\![e]\!]\,\varepsilon)
\end{aligned}$$

where the functional $\Psi_{\eta\varepsilon} : (D \to \mathbf{Bool}) \to (D \to \mathbf{Bool})$, associated to the formula $\mu Z(z{:}D).\psi$, is defined as $\Psi_{\eta\varepsilon} = \lambda G{:}D \to \mathbf{Bool}.\lambda v{:}D.\, [\![\psi]\!]\,(\eta[G/Z])(\varepsilon[v/z])$.

The functionals $\Psi_{\eta\varepsilon}$ associated to the first-order fixed point formulas being monotonic, and the underlying lattices $D \to \mathbf{Bool}$ being complete, it follows from Tarski's theorem that each functional $\Psi_{\eta\varepsilon}$ has a unique minimal fixed point $\mu\Psi_{\eta\varepsilon}$ and a unique maximal fixed point $\nu\Psi_{\eta\varepsilon}$.

4.2 Transformation of the Verification Problem

Consider the following linear μCRL process:

$$X(x{:}D) = \sum_{a \in Act} \sum_{x_a : E_a} a(e_a) \cdot X(e'_a) \triangleleft b_a \triangleright \delta$$

As we precised in Section 2.3, the states of the corresponding transition system are identified with terms $X(v)$, where $v \in D$. We assume that the data variables used in the temporal formulas are disjoint from those used in the linear process.

According to the interpretation of state formulas, a state $X(v)$ satisfies a formula φ in the context of a propositional environment ρ and of a data environment ε if and only if $X(v) \in [\![\varphi]\!]\,\rho\varepsilon$. As we will show, this is equivalent to the fact that a first-order fixed point formula $\mathrm{TR}(\varphi)$ is true in the context of a boolean environment $\text{TR}(\rho)$ and of $\varepsilon[v/x]$, where the translations $\text{TR}(\varphi)$ and $\text{TR}(\rho)$, which take the process X as an implicit parameter, are defined below.

Given $\rho \in \mathbf{PEnv}$, the boolean environment $\text{TR}(\rho)$, whose support is given by $supp(\text{TR}(\rho)) = \{Z_Y(x{:}D, y{:}D') \mid Y(y{:}D') \in supp(\rho)\}$, is defined as follows:

$$(\text{TR}(\rho))(Z_Y) \stackrel{\text{def}}{=} \lambda v{:}D, v'{:}D'.(X(v) \in (\rho(Y))(v'))$$

for each $Z_Y \in supp(\text{TR}(\rho))$.

Given $\varphi \in SForm$, the translation $\text{TR}(\varphi)$ is defined inductively below:

$$\begin{aligned}
\text{TR}(b) &\stackrel{\text{def}}{=} b \\
\text{TR}(Y(e)) &\stackrel{\text{def}}{=} Z_Y(x, e) \\
\text{TR}(\varphi_1 \wedge \varphi_2) &\stackrel{\text{def}}{=} \text{TR}(\varphi_1) \wedge \text{TR}(\varphi_2) \\
\text{TR}(\langle\alpha\rangle\, \varphi) &\stackrel{\text{def}}{=} \textstyle\bigvee_{a \in Act} \exists x_a{:}E_a.(b_a \wedge (a(e_a) \models \alpha) \wedge \text{TR}(\varphi)[e'_a/x]) \\
\text{TR}(\exists y{:}D'.\varphi) &\stackrel{\text{def}}{=} \exists y{:}D'.\text{TR}(\varphi) \\
\text{TR}((\mu Y(y{:}D').\varphi)(e)) &\stackrel{\text{def}}{=} (\mu Z_Y(x_Y{:}D, y{:}D').\text{TR}(\varphi)[x_Y/x])(x, e)
\end{aligned}$$

where the predicate $a(e_a) \models \alpha$, expressing that an action $a(e_a)$ satisfies an action formula $\alpha \in AForm$, is defined inductively as follows:

$$\begin{aligned}
a(e_a) \models a'(e') &\stackrel{\text{def}}{=} a = a' \wedge e_a = e' \\
a(e_a) \models tt &\stackrel{\text{def}}{=} tt \\
a(e_a) \models \neg\alpha &\stackrel{\text{def}}{=} \neg(a(e_a) \models \alpha) \\
a(e_a) \models \alpha_1 \wedge \alpha_2 &\stackrel{\text{def}}{=} (a(e_a) \models \alpha_1) \wedge (a(e_a) \models \alpha_2) \\
a(e_a) \models \exists y{:}D.\alpha &\stackrel{\text{def}}{=} \exists y{:}D.(a(e_a) \models \alpha).
\end{aligned}$$

The following lemma states some auxiliary technical properties necessary for showing the correctness of the $\text{TR}(\varphi)$ translation.

Lemma 1. *The following properties hold:*

1. *For all $a \in Act$, $e_a \in Exp$, $\alpha \in AForm$, and $\varepsilon \in$ **DEnv** such that $var(e_a) \cup fdv(\alpha) \subseteq supp(\varepsilon)$:*

$$[\![a(e_a) \models \alpha]\!]\,\varepsilon = (a([\![e_a]\!]\,\varepsilon) \in [\![\alpha]\!]\,\varepsilon).$$

2. *For all $a \in Act$ and $\varphi \in SForm$:*

$$fdv(\text{TR}(\varphi)) \subseteq (fdv(\varphi) \cup \{x\}) \setminus \{x_a\}.$$

3. *For all $\psi \in BForm$, $e \in Exp$, $x \in DVar$, $\eta \in$ **BEnv**, and $\varepsilon \in$ **DEnv** such that $var(e) \cup fdv(\psi) \subseteq supp(\varepsilon)$:*

$$[\![\psi[e/x]]\!]\,\eta\varepsilon = [\![\psi]\!]\,\eta(\varepsilon[[\![e]\!]\,\varepsilon/x]).$$

Proof. Straightforward, by structural induction on α (property 1), on φ (property 2), and on ψ (property 3).

The following proposition expresses the relation between a linear process X, a state formula φ, and the corresponding first-order fixed point formula $\textsc{Tr}(\varphi)$ obtained after translation.

Proposition 1. *Let $X(x{:}D)$ be a linear process as defined above and let φ be a state formula. Then, for any $\rho \in \mathbf{PEnv}$ and $\varepsilon \in \mathbf{DEnv}$ such that $fpv(\varphi) \subseteq supp(\rho)$ and $fdv(\varphi) \subseteq supp(\varepsilon)$:*

$$[\![\varphi]\!]\,\rho\varepsilon = \{X(v) \in S \mid [\![\textsc{Tr}(\varphi)]\!]\,\textsc{Tr}(\rho)(\varepsilon[v/x])\}.$$

Proof. By structural induction on φ, using Lemma 1.

Using the result above, we can now restate the verification problem of a closed state formula φ by a linear process X in terms of the satisfaction of a first-order fixed point formula $\textsc{Tr}(\varphi)$. The global model-checking problem, consisting to verify that the formula is satisfied by every state of the process, becomes:

$\forall v{:}D.(X(v) \in [\![\varphi]\!]\,[\,][\,]) \leftrightarrow$	by Proposition 1
$\forall v{:}D.\,[\![\textsc{Tr}(\varphi)]\!]\,\textsc{Tr}([\,])([\,][v/x]) \leftrightarrow$	by definition of $\textsc{Tr}(\rho)$
$\forall v{:}D.\,[\![\textsc{Tr}(\varphi)]\!]\,[\,][v/x] \leftrightarrow$	by definition of $[\![.]\!]\,\eta\varepsilon$
$[\![\forall x{:}D.\textsc{Tr}(\varphi)]\!]\,[\,][\,]$.	

(Note that we can use empty environments whenever the formulas are closed w.r.t. the corresponding variables.) The local model-checking problem, consisting to verify that the formula is satisfied by the initial state of the process, becomes:

$X(v_0) \in [\![\varphi]\!]\,[\,][\,] \leftrightarrow$	by Proposition 1
$[\![\textsc{Tr}(\varphi)]\!]\,\textsc{Tr}([\,])([\,][v_0/x]) \leftrightarrow$	by definition of $\textsc{Tr}(\rho)$
$[\![\textsc{Tr}(\varphi)]\!]\,[\,][v_0/x] \leftrightarrow$	by definition of $[\![.]\!]\,\eta\varepsilon$
$[\![\forall x{:}D.(x = v_0) \to \textsc{Tr}(\varphi)]\!]\,[\,][\,]$.	

Using the standard proof rules for first-order logic, together with the rules for minimal and maximal fixed point operators that will be given in Section 4.4, we have the basic tools available for proving the first-order fixed point formulas above.

4.3 Example (Continued)

We continue the example from Section 3.1 by giving the translations of the formulas φ_1, φ_2, and φ_3. So, to establish the validity of these formulas we must prove, respectively:

1. $(\mu Z(v{:}\mathrm{N}, b{:}\mathbf{Bool}).(b \wedge l \leq v) \vee (\neg b \wedge Z(v+1, \neg b)) \vee \exists m{:}\mathrm{N}.(b \wedge m \leq v \wedge Z(v-m, \neg b)))(v, b)$;

2. $(\mu Y(v{:}\mathrm{N}, b{:}\mathbf{Bool}).(\neg b \rightarrow Z(v+1, \neg b)) \wedge \forall m{:}\mathrm{N}.((b \wedge m \leq v \wedge m \neq l) \rightarrow Z(v-m, \neg b)))(v, b)$;
3. $(\nu Z(p, r, v{:}\mathrm{N}, b{:}\mathbf{Bool}).\forall l, m{:}\mathrm{N}.((b \wedge m \leq v \wedge m = l) \rightarrow (l \leq v_0 + p - r \wedge Z(p, r+l, v-m, \neg b))) \wedge (\neg b \rightarrow Z(p+1, r, v+1, \neg b)))(0, 0, v, b)$.

4.4 Proof Rules

As shown in Section 4.2, the verification of a data-based temporal logic formula on a linear μCRL process can be reduced to the satisfaction of a first-order formula containing fixed point operators. We provide here proof rules associated to the minimal and maximal fixed point operators. These rules can be naturally used in conjunction with some proof system for first-order logic (e.g., Gentzen's natural deduction system [6]) in order to prove the validity of first-order fixed point formulas.

We first define some auxiliary notations. Consider a fixed point formula $\sigma Z(z{:}D).\psi_1$ representing a predicate over D, and let $\psi_2 \in \mathit{BForm}$ such that $\mathit{fbv}(\psi_2) \subseteq \mathit{fbv}(\psi_1)$ and $\mathit{fdv}(\psi_2) \subseteq \mathit{fdv}(\psi_1)$. The *application* of ψ_1 on ψ_2 is defined as follows:

$$\psi_1[\psi_2] \stackrel{\text{def}}{=} \psi_1[\psi_2[e/z]/Z(e)]$$

Intuitively, $\psi_1[\psi_2]$ is obtained by substituting all the occurrences of $Z(e)$ in ψ_1 by ψ_2, in which all occurrences of z have been replaced with the actual parameter e. The conditions on the variables of ψ_2 ensure that no free variables of ψ_2 become bound in $\psi_1[\psi_2]$. For simplicity, whenever $\mathit{fdv}(\psi_2) = \{z\}$, we will write $\psi_2(e)$ for $\psi_2[e/z]$. We also assume that the domain N of natural numbers is predefined. For every $k \in \mathrm{N}$, the application k times of ψ_1 on ψ_2, noted $\psi_1^k[\psi_2]$, is defined as follows:

$$\psi_1^0[\psi_2] \stackrel{\text{def}}{=} \psi_2,\ \psi_1^{k+1}[\psi_2] \stackrel{\text{def}}{=} \psi_1[\psi_1^k[\psi_2]]$$

Using these notations, the proof rules for minimal and maximal fixed point operators are given below:

$$\frac{\forall k \geq 0.(\psi_2(k) \rightarrow \psi_1^k[\mathit{ff}])}{(\exists k \geq 0.\psi_2(k)) \rightarrow (\mu Z(z{:}D).\psi_2)(z)}\ \textsc{LfpUp} \qquad \frac{\psi_1[\psi_2] \rightarrow \psi_2}{(\mu Z(z{:}D).\psi_1)(z) \rightarrow \psi_2}\ \textsc{LfpDn}$$

$$\frac{\forall k \geq 0.(\psi_1^k[\mathit{tt}] \rightarrow \psi_2(k))}{(\nu Z(z{:}D).\psi_1)(z) \rightarrow (\forall k \geq 0.\psi_2(k))}\ \textsc{GfpDn} \qquad \frac{\psi_2 \rightarrow \psi_1[\psi_2]}{\psi_2 \rightarrow (\nu Z(z{:}D).\psi_1)(z)}\ \textsc{GfpUp}$$

where $\psi_2(k)$ means that the variable k, denoting a natural number, occurs free in ψ_2. Intuitively, the rules LFPUP, GFPUP and LFPDN, GFPDN allow to approximate the extremal fixed points towards satisfaction and towards refutation, respectively. The following proposition states the soundness of these rules.

Proposition 2. *The rules* LFPUP, LFPDN, GFPUP, *and* GFPDN *defined above are sound w.r.t. the semantics of the first-order fixed point formulas* $\psi \in \mathit{BForm}$.

Proof. Given in [10].

4.5 Example (Continued)

We show the use of the rules given above by proving the formulas given in Section 4.3. We consider the three formulas separately. We give the proof of these formulas in extreme detail, such that every reasoning step can be understood.

1. For the first case we let $\psi_1 \stackrel{\text{def}}{=} (b \wedge l \leq v) \vee (\neg b \wedge Z(v+1, \neg b)) \vee \exists m{:}\mathrm{N}.(b \wedge m \leq v \wedge Z(v-m, \neg b))$. In order to apply the rule LFPUP we must find some $\psi_2(k)$. We propose $\psi_2(k) \stackrel{\text{def}}{=} k > \mathit{if}(l \leq v, |\neg b|, 2(l-v) - |\neg b|)$. Here, $\mathit{if}(b, x, y)$ equals x if b holds and y otherwise; $|b|$ equals 1 if b holds and 0 otherwise. (Intuitively, k denotes the minimal number of steps necessary to reach a $w(l)$ action, starting from any state of the system.) Note that the left hand side in the conclusion of LFPUP becomes $\exists k \geq 0.(k > \mathit{if}(l \leq v, |\neg b|, 2(l-v) - |\neg b|))$, which is a tautology. So, if we can prove the premises of LFPUP we have shown that the temporal formula φ_1 is valid in all states of $X(v, b)$.
   The premise of LFPUP has become $\forall k \geq 0.(k > \mathit{if}(l \leq v, |\neg b|, 2(l-v) - |\neg b|) \rightarrow \psi_1^k[\mathit{ff}])$. We prove this premise by induction on k. For $k = 0$ this holds vacuously, because the left hand side of the implication equals falsum. For $k = k' + 1$, we must prove: $k' \geq \mathit{if}(l \leq v, |\neg b|, 2(l-v) - |\neg b|) \rightarrow (b \wedge l \leq v) \vee (\neg b \wedge \psi_1^{k'}[\mathit{ff}](v+1, \neg b)) \vee \exists m{:}\mathrm{N}.(b \wedge m \leq v \wedge \psi_1^{k'}[\mathit{ff}](v-m, \neg b))$. This is done by making a few case distinctions:
   - Suppose b holds and $l \leq v$. Clearly, the statement above is true, as the first disjunct of the right hand side trivially holds.
   - Now, suppose b holds and $l > v$. We want to show that the third disjunct holds. As b holds by assumption, it suffices to show that $\exists m{:}\mathrm{N}.(m \leq v \wedge \psi_1^{k'}[\mathit{ff}](v-m, \neg b))$. Take $m = 0$. The proof obligation reduces to $\psi_1^{k'}[\mathit{ff}](v, \neg b)$. This is implied by the induction hypothesis, because $(\psi_2(k'))(v, \neg b) = k' > 2(l-v) - 1$, which is equivalent in this case to the left hand side $k' \geq 2(l-v)$ of the implication.
   - We still must consider the case where $\neg b$. We show that the second disjunct holds in this case. We must prove that $\psi_1^{k'}[\mathit{ff}](v+1, \neg b)$. The left hand side of the implication becomes $k' \geq \mathit{if}(l \leq v, 1, 2(l-v) - 1)$, which is easily seen (by distinguishing between the cases $l \leq v$, $l = v+1$, and $l > v+1$) to imply $(\psi_2(k'))(v+1, \neg b) = k' > \mathit{if}(l \leq v+1, 0, 2(l-v) - 2)$. So, the proof obligation follows from the inductive hypothesis.

   This finishes the proof of the first temporal formula.
2. We show that this formula does not hold in any state of X. Let ψ_1 be the body of the μZ formula. We apply LFPDN, taking $\psi_2 \stackrel{\text{def}}{=} \mathit{ff}$. The left hand side $\psi_1[\psi_2]$ of the premise looks like $(\neg b \rightarrow \mathit{ff}) \wedge \forall m{:}\mathrm{N}.((b \wedge m \leq v \wedge m \neq l) \rightarrow \mathit{ff})$, which is equivalent to ff. Thus, the fixed point formula is false for all $v \in \mathrm{N}$ and $b \in \mathbf{Bool}$.
3. We show that this formula is satisfied by the initial state of the system. Let ψ_1 be the body of the νZ formula. We must prove that $(v = v_0 \wedge b = \mathit{ff}) \rightarrow (\nu Z(p, r, v{:}\mathrm{N}, b{:}\mathbf{Bool}).\psi_1)(0, 0, v, b)$ for all $v \in \mathrm{N}$ and $b \in \mathbf{Bool}$. We solve this by showing a slightly stronger property, namely that $(v = v_0 + p - r) \rightarrow (\nu Z(p, r, v{:}\mathrm{N}, b{:}\mathbf{Bool}).\psi_1)(p, r, v, b)$, which implies the above boolean

property by instantiating v, b, p, and r with v_0, ff, 0, and 0, respectively. We apply GFPUP, taking $\psi_2 \stackrel{\text{def}}{=} (v = v_0+p-r)$. The premise of GFPUP reduces to $(v = v_0+p-r) \rightarrow (\forall l, m{:}\mathrm{N}.((b \wedge m \leq v \wedge m = l) \rightarrow (l \leq v_0+p-r \wedge v-m = v_0+p-r-l)) \wedge (\neg b \rightarrow v+1 = v_0+p+1-r))$, which is easily seen to be a tautology. Hence, the initial state $X(v_0, ff)$ satisfies φ_3.

5 Application

We present here a more involved verification example using the methodology described in Section 4. Consider the following linear process $Q(q)$ describing a queue q:

$$Q(q) = \textstyle\sum_{d:D} r(d) \cdot Q(in(d,q)) + s(toe(q)) \cdot Q(untoe(q)) \lhd |q| > 0 \rhd \delta$$

Data elements $d \in D$ are inserted in Q via $r(d)$ actions and are delivered by Q via $s(d)$ actions. The $|.|$ operator returns the number of elements in a queue. The *in* function inserts an element into a queue, the *untoe* function eliminates the element which was inserted first into a queue, and the *toe* function returns that element. We assume that the domain D has at least one element. The concatenation of two queues q_1 and q_2 is described by the linear process below:

$$\begin{aligned} Q(q_1,q_2) = \textstyle\sum_{d:D} r(d) \cdot Q(in(d,q_1),q_2) \quad & \lhd \ tt \ \rhd \delta \ + \\ \tau \cdot Q(untoe(q_1), in(toe(q_1),q_2)) \ & \lhd |q_1| > 0 \rhd \delta \ + \\ s(toe(q_2)) \cdot Q(q_1, untoe(q_2)) \quad & \lhd |q_2| > 0 \rhd \delta \end{aligned}$$

The initial state of this process is $Q(nil, nil)$, where *nil* is a function returning an empty queue. In the following paragraphs we present the description and verification of several safety and liveness properties of the process Q.

Property 1. The essential safety property of the system is that every sequence of elements inserted in Q will be delivered in the same order. This can be neatly expressed using a fixed point operator parameterized by a queue q storing all the elements that have been inserted in Q but not yet delivered:

$$\begin{aligned} \varphi_1 \stackrel{\text{def}}{=} (\nu Y(q).\forall d_0{:}D.\, & [r(d_0)]\, Y(in(d_0,q)) \wedge \\ & [s(d_0)]\, (|q| > 0 \wedge toe(q) = d_0 \wedge Y(untoe(q))) \wedge \\ & [\neg \exists d_1{:}D.(s(d_1) \vee r(d_1))]\, Y(q) \\ )(nil) & \end{aligned}$$

This formula captures exactly the desired behaviour of the system: the two concatenated queues must behave as a single queue. (Note the presence of the quantifier in the action formula of the last box modality, in order to express that an action is different from any $s(\ldots)$ or $r(\ldots)$ action.) We verify φ_1 in the initial state $Q(nil, nil)$ of the system. This translates as follows:

$$
\begin{array}{l}
\forall q_1, q_2.(q_1 = nil \wedge q_2 = nil) \rightarrow \\
\quad (\nu Z(q_1, q_2, q).\forall d_0{:}D.\forall d{:}D.(d_0 = d \rightarrow Z(in(d, q_1), q_2, in(d_0, q))) \wedge \\
\qquad\qquad ((|q_2| > 0 \wedge d_0 = toe(q_2)) \rightarrow (|q| > 0 \wedge d_0 = toe(q) \wedge \\
\qquad\qquad\qquad\qquad Z(q_1, untoe(q_2), untoe(q)))) \wedge \\
\qquad\qquad (|q_1| > 0 \rightarrow Z(untoe(q_1), in(toe(q_1), q_2), q)) \\
\quad)(q_1, q_2, nil)
\end{array}
$$

Let ψ_1 be the body of the νZ formula. To show the first-order fixed point formula above, we prove a slightly stronger property, namely that $(q_1 + q_2 = q) \rightarrow (\nu Z(q_1, q_2, q).\psi_1)(q_1, q_2, q)$ for all q_1, q_2, and q, where $q_1 + q_2$ denotes the concatenation of q_1 and q_2. We use the rule GFPUP, taking $\psi_2 \stackrel{\text{def}}{=} (q_1 + q_2 = q)$. The premise $\psi_2 \rightarrow \psi_1[\psi_2]$ of GFPUP reduces to the following three implications:

1. $\forall d_0, d{:}D.(q_1 + q_2 = q \wedge d_0 = d) \rightarrow (in(d, q_1) + q_2 = in(d_0, q))$;
2. $\forall d_0{:}D.(q_1 + q_2 = q \wedge |q_2| > 0 \wedge d_0 = toe(q_2)) \rightarrow (|q| > 0 \wedge d_0 = toe(q) \wedge q_1 + untoe(q_2) = untoe(q))$;
3. $\forall d_0{:}D.(q_1 + q_2 = q \wedge |q_1| > 0) \rightarrow (untoe(q_1) + in(toe(q_1), q_2) = q)$.

These properties can be easily shown using an appropriate axiomatization of the queue operators. Now, by instantiating q to *nil*, and since $(q_1 = nil \wedge q_2 = nil) \rightarrow (q_1 + q_2 = nil)$, this implies that $(q_1 = nil \wedge q_2 = nil) \rightarrow (\nu Z(q_1, q_2, q).\psi_1)(q_1, q_2, nil)$ for all q_1 and q_2. Hence, $Q(nil, nil)$ satisfies φ_1.

Property 2. A simple liveness property (which also implies deadlock freedom) is that every datum $d_0 \in D$ can be potentially inserted in Q by an action $r(d_0)$:

$$\varphi_2 \stackrel{\text{def}}{=} \mu Y.\, \langle r(d_0)\rangle\, tt \vee \langle tt\rangle\, Y$$

The verification of φ_2 in all the states of Q translates as follows:

$$
\begin{array}{l}
\forall q_1, q_2.(\mu Z(q_1, q_2).\exists d{:}D.(d = d_0) \vee \exists d{:}D.Z(in(d, q_1), q_2) \vee \\
\qquad\qquad (|q_1| > 0 \wedge Z(untoe(q_1), in(toe(q_1), q_2))) \vee \\
\qquad\qquad (|q_2| > 0 \wedge Z(q_1, untoe(q_2))) \\
\quad)(q_1, q_2)
\end{array}
$$

We write ψ_1 for the body of the μZ formula. Since the disjunct $\exists d{:}D.(d = d_0)$ is trivially true, ψ_1 reduces to *tt* and, by applying the rule LFPUP with $\psi_2(k) = tt$, it follows that $(\mu Z(q_1, q_2).\psi_1)(q_1, q_2)$ is valid for all values of q_1 and q_2. Hence, φ_2 holds in all states of Q.

Property 3. A more involved liveness property is that every datum d_0 which is inserted in Q by an action $r(d_0)$ will be eventually delivered by an action $s(d_0)$:

$$\varphi_3 \stackrel{\text{def}}{=} [r(d_0)]\, \mu Y.\, \langle tt\rangle\, tt \wedge [\neg s(d_0)]\, Y$$

The verification of φ_3 in all the states of Q translates as follows:

$$\begin{array}{l}\forall q_1, q_2.\forall d{:}D.d = d_0 \rightarrow \\ \quad (\mu Z(q_1, q_2).\forall d{:}D.Z(in(d, q_1), q_2) \wedge \\ \qquad\qquad (|q_1| > 0 \rightarrow Z(untoe(q_1), in(toe(q_1), q_2))) \wedge \\ \qquad\qquad ((|q_2| > 0 \wedge toe(q_2) \neq d_0) \rightarrow Z(q_1, untoe(q_2))) \\ \quad)(in(d, q_1), q_2)\end{array}$$

Let ψ_1 be the body of the μZ formula. Observing that $\psi_1[\mathit{ff}] = \mathit{ff}$, the rule LFPDN leads to $(\mu Z(q_1, q_2).\psi_1)(q_1, q_2) \rightarrow \mathit{ff}$ for every q_1 and q_2. Then, the whole first-order fixed point formula reduces to $\forall d{:}D.d \neq d_0$, which is obviously false. Hence, φ_3 does not hold in any state of Q. This happens because one can always insert data elements into Q (see formula φ_2 above) and, under an unfair scheduling of actions (but see next paragraph), the process may never deliver an element, letting q_1 and q_2 grow unboundedly.

Property 4. We may express the formula φ_3 by taking into account only the execution paths that are *fair* w.r.t. the action $s(d_0)$, i.e., those paths which cannot infinitely often enable $s(d_0)$ without infinitely often executing it:

$$\varphi_4 \stackrel{\text{def}}{=} [r(d_0)]\, \nu Y_1.\, [\neg s(d_0)]\, Y_1 \wedge \mu Y_2.\, \langle s(d_0)\rangle\, tt \vee \langle tt\rangle\, Y_2$$

The formula φ_4 specifies that after d_0 has been inserted in Q, as long as it has not yet been delivered, it is still possible to deliver it. This is an action-based instance of the fairness operator proposed in [18], where it was shown that it expresses the reachability on fair paths.

The verification of φ_4 in all the states of Q translates as follows:

$$\begin{array}{l}\forall q_1, q_2\,.\,(\nu Z_1(q_1, q_2).\forall d{:}D.Z_1(in(d, q_1), q_2) \wedge \\ \qquad\qquad (|q_1| > 0 \rightarrow Z_1(untoe(q_1), in(toe(q_1), q_2))) \wedge \\ \qquad\qquad ((|q_2| > 0 \wedge toe(q_2) \neq d_0) \rightarrow Z_1(q_1, untoe(q_2))) \wedge \\ \qquad\qquad (\mu Z_2(q_1, q_2).(|q_2| > 0 \wedge toe(q_2) = d_0) \vee \exists d{:}D.Z_2(in(d, q_1), q_2) \vee \\ \qquad\qquad\qquad (|q_1| > 0 \wedge Z_2(untoe(q_1), in(toe(q_1), q_2))) \vee \\ \qquad\qquad\qquad (|q_2| > 0 \wedge Z_2(q_1, untoe(q_2))) \\ \qquad\quad)(q_1, q_2) \\ \quad)(in(d_0, q_1), q_2)\end{array}$$

Let ψ_1 be the body of the νZ_1 formula. We show the first-order fixed point formula above by proving a slightly stronger property, namely that $d_0 \in q_1 + q_2 \rightarrow (\nu Z_1(q_1, q_2).\psi_1)(q_1, q_2)$ for all q_1 and q_2, where $\in$ denotes the membership of an element in a queue. (Having shown this, the validity of the first-order fixed point formula above follows by instantiating q_1 with $in(d_0, q_1)$, since $d_0 \in in(d_0, q_1) + q_2$ is trivially true.) We apply the rule GFPUP on ψ_1, taking $\psi_1' \stackrel{\text{def}}{=} d_0 \in q_1 + q_2$. The premise $\psi_1[\psi_1']$ reduces to the following four implications:

1. $(d_0 \in q_1 + q_2) \rightarrow (\forall d{:}D.d_0 \in in(d, q_1) + q_2)$;
2. $(d_0 \in q_1 + q_2 \wedge |q_1| > 0) \rightarrow (d_0 \in untoe(q_1) + in(toe(q_1), q_2))$;
3. $(d_0 \in q_1 + q_2 \wedge |q_2| > 0 \wedge toe(q_2) \neq d_0) \rightarrow (d_0 \in q_1 + untoe(q_2))$;
4. $(d_0 \in q_1 + q_2) \rightarrow (\mu Z_2(q_1, q_2).\psi_2)(q_1, q_2)$

where ψ_2 is the body of the μZ_2 subformula. The first three properties follow easily from an axiomatization of the queue type. We show the last property using the rule LFPUP, by taking $\psi_2'(k) \stackrel{\text{def}}{=} d_0 \in q_1 + q_2 \wedge 2|q_1| + |q_2| \leq k$ (intuitively, k denotes the minimal number of steps in which an element d_0 already present in Q can be delivered). Note that the left hand side in the conclusion of LFPUP becomes $\exists k \geq 0.(d_0 \in q_1 + q_2 \wedge 2|q_1| + |q_2| \leq k)$, which is trivially equivalent to $d_0 \in q_1 + q_2$.

We show the premise $\forall k \geq 0.(\psi_2'(k) \rightarrow \psi_2^k[\mathit{ff}])$ of LFPUP by induction on k. For $k = 0$ this holds vacuously, because $\psi_2'(0)$ is false. For $k = k' + 1$, we must prove that $(d_0 \in q_1 + q_2 \wedge 2|q_1| + |q_2| \leq k' + 1) \rightarrow \psi_2^{k'+1}[\mathit{ff}]$. We distinguish two cases:

- $|q_1| > 0$. We show that the left hand side of the implication above implies the disjunct $|q_1| > 0 \wedge \psi_2^{k'}[\mathit{ff}](untoe(q_1), in(toe(q_1), q_2))$ of $\psi_2^{k'+1}[\mathit{ff}]$. The first conjunct is true by assumption. The second conjunct is implied by the inductive hypothesis, because: (a) $d_0 \in q_1 + q_2 \rightarrow d_0 \in untoe(q_1) + in(toe(q_1), q_2)$, and (b) $2|untoe(q_1)| + |in(toe(q_1), q_2)| = 2|q_1| + |q_2| - 1 \leq k'$.
- $|q_1| = 0$. This implies that $|q_2| > 0$, because $d_0 \in q_1 + q_2$ by hypothesis. If $toe(q_2) = d_0$, then the disjunct $|q_2| > 0 \wedge toe(q_2) = d_0$ of $\psi_2^{k'+1}[\mathit{ff}]$ is true. If $toe(q_2) \neq d_0$, the disjunct $|q_2| > 0 \wedge \psi_2^{k'}[\mathit{ff}](q_1, untoe(q_2))$ of $\psi_2^{k'+1}[\mathit{ff}]$ follows from the inductive hypothesis, because: (a) $d_0 \in q_1 + untoe(q_2)$, and (b) $2|q_1| + |untoe(q_2)| = 2|q_1| + |q_2| - 1 \leq k'$.

This concludes the proof that all the states of Q satisfy φ_4.

Acknowledgements

We are grateful to Yaroslav Usenko, Anubhav Gupta, and to the anonymous referees for their careful reading and judicious comments on this paper.

References

1. Andersen, H.R.: Model Checking and Boolean Graphs. *Theoretical Computer Science*, 126(1):3–30, 1994.
2. Arnold, A., Crubillé, P.: A Linear Algorithm to Solve Fixed-Point Equations on Transition Systems. *Information Processing Letters*, 29:57–66, 1988.
3. Bezem, M.A., Groote, J.F.: Invariants in Process Algebra with Data. In: Jonsson, B., Parrow, J. (eds.): *Proceedings of CONCUR'94 (Uppsala, Sweden)*, LNCS 836, pp. 401–416, Springer Verlag, 1994.
4. Bezem, M.A., Groote, J.F.: A Correctness Proof of a One Bit Sliding Window Protocol in μCRL. *The Computer Journal*, 37(4):289–307, 1994.

5. Bosscher, D., Ponse, A.: Translating a Process Algebra with Symbolic Data Values to Linear Format. In: Engberg, U.H., Larsen, K.G., Skou, A. (eds.), *Proceedings of TACAS'95 (Aarhus, Denmark)*, BRICS Notes Series, pp. 119–130, University of Aarhus, 1995.
6. van Dalen, D.: Logic and Structure. Springer Verlag, 1994.
7. Dam, M.: Model Checking Mobile Processes. *Information and Computation*, 129:35–51, 1996.
8. Fredlund, L.-Å., Groote, J.F., Korver, H.: Formal Verification of a Leader Election Protocol in Process Algebra. *Theoretical Computer Science*, 177:459–486, 1997.
9. Groote, J.F.: A Note on n Similar Parallel Processes. In: Gnesi, S., Latella, D. (eds.), *Proceedings of the 2nd ERCIM Int. Workshop on Formal Methods for Industrial Critical Systems (Cesena, Italy)*, pp. 65–75, 1997. (See also Report CS-R9626, CWI, Amsterdam, 1996).
10. Groote, J.F., Mateescu, R.: Verification of Temporal Properties of Processes in a Setting with Data. Technical Report SEN-R9804, CWI, Amsterdam, 1998.
11. Groote, J.F., Monin, F., Springintveld, J.: A Computer Checked Algebraic Verification of a Distributed Summing Protocol. Computer Science Report 97/14, Dept. of Math. and Comp. Sci., Eindhoven University of Technology, 1997.
12. Groote, J.F., van de Pol, J.C.: A Bounded Retransmission Protocol for Large Data Packets. A Case Study in Computer Checked Verification. In: Wirsing, M., Nivat, M. (eds.), *Proceedings of AMAST'96 (Munich, Germany)*, LNCS 1101, pp. 536–550, Springer Verlag, 1996.
13. Groote, J.F., Ponse, A.: The Syntax and Semantics of μCRL. In: Ponse, A., Verhoef, C., van Vlijmen, S.F.M. (eds.), *Algebra of Communicating Processes*, Workshops in Computing, pp. 26–62, 1994.
14. Groote, J.F., Springintveld, J.: Focus Points and Convergent Process Operators. A Proof Strategy for Protocol Verification. Technical Report 142, Logic Group Preprint Series, Utrecht University, 1995. (See also Technical Report CS-R9566, CWI, Amsterdam, 1995).
15. Kindler, A., Reisig, W., Völzer, H., Walter, R.: Petri Net Based Verification of Distributed Algorithms: an Example. *Formal Aspects of Computing*, 9:409–424, 1997.
16. Kozen, D.: Results on the Propositional μ-calculus. *Theoretical Computer Science* 27, pp. 333–354, 1983.
17. Manna, Z., Pnueli, A.: Adequate Proof Principles for Invariance and Liveness Properties of Concurrent Programs. *Science of Computer Programming* 32:257–289, 1984.
18. Queille, J-P., Sifakis, J.: Fairness and Related Properties in Transition Systems — a Temporal Logic to Deal with Fairness. *Acta Informatica*, 19:195–220, 1983.
19. Rathke, J., Hennessy, M.: Local Model Checking for a Value-Based Modal μ-calculus. Technical Report 5/96, School of Cognitive and Computing Sciences, University of Sussex, 1996.
20. Shankland, C.: The Tree Identify Protocol of IEEE 1394. In: Groote, J.F., Luttik, B., van Wamel, J. (eds.), *Proceedings of the 3rd ERCIM Int. Workshop on Formal Methods for Industrial Critical Systems (Amsterdam, The Netherlands)*, pp. 299–319, 1998.
21. Tarski, A.: A Lattice-Theoretical Fixpoint Theorem and its Applications. *Pacific Journal of Mathematics* 5, pp. 285–309, 1955.
22. Vergauwen, B., Lewi, J.: A Linear Algorithm for Solving Fixed-Point Equations on Transition Systems. *Proceedings of CAAP'92 (Rennes, France)*, LNCS 581, pp. 322–341, Springer Verlag, 1992.

A Logic for Real–Time Systems Specification, Its Algebraic Semantics, and Equational Calculus

Gabriel A. Baum[1], Marcelo F. Frias[2], and Thomas S.E. Maibaum*[3]

[1] Universidad Nacional de La Plata, LIFIA, Departamento de Informática.
gbaum@sol.info.unlp.edu.ar

[2] Universidad de Buenos Aires, Departamento de Computación, and Universidad Nacional de La Plata, LIFIA, Departamento de Informática.
mfrias@sol.info.unlp.edu.ar

[3] Imperial College, 180 Queen's Gate, London SW7 2BZ, U.K.
tsem@doc.ic.ac.uk

Abstract. We present a logic for real time systems specification which is an extension of first order dynamic logic by adding (a) arbitrary atomic actions rather than only assignments, (b) variables over actions which allow to specify systems partially, and (c) explicit time. The logic is algebraized using closure fork algebras and a representation theorem for this class is presented. This allows to define an equational (but infinitary) proof system for the algebraization.

1 Introduction

The motivation for this work is the need to describe *industrial processes* as part of a project for a Brazilian telecommunications company. We want to be able to give formal descriptions of such processes so as to be able to analyze such descriptions. For example, we want to be able to calculate critical paths for tasks in processes, throughput times of processes, etc. We also want to demonstrate correctness of process descriptions in relation to their specifications (where this is appropriate), derive implementations of process specifications in terms of the available concrete apparatus in the factory, validate (using formal techniques) an implementation against its abstract description, and so on. Available languages for describing processes are unsuitable for various reasons, most having to do with the nature of the formalization of such processes being used in the project.

The method used in the project for describing processes (*the method*) is based on the ideas presented in [11]. This method sees the world as being modeled in terms of two (and only two) kinds of entities: products and processes. A *product* is a description of an entity in the real world (*a referent*) in terms of measurable attributes. (Here, we use *measure* and *measurable* in the traditional sense of science and engineering. See [2][11][14].) A *product instance* is characterized by the

* The third author would like to thank the EPSRC(UK), CNPq(Brasil), Imperial College, LMF-DI/PUC-RJ and The Royal Academy of Engineering for their financial support during the conduct of this research.

A.M. Haeberer (Ed.): AMAST'98, LNCS 1548, pp. 91–105, 1998.

values (measures) associated with its attributes (and, implicitly, by the theory of the product, i.e., the defined relationships between the potential measured values of its attributes). Hence, such a product instance may be seen as a *model*, in the sense of logic, of the product. We may see products as being characterized by data types in first order logic, for example.

The distinguishing characteristic of products is that they exist 'independently' at an instant in time, where *time* is used here in its normal scientific sense. (Independence here means that a product is defined without recourse to any other referent or only in terms of other (sub)products. Products of the former kind are called *atomic* products.) In fact, all products have a time attribute whose value in a product instance indicates the time instant at which the values of the attributes were (co)determined, presumably by some appropriate measurement procedures. On the other hand, processes are distinguished entities which do not exist at a time instant, but which have time duration. Further, processes are not independently definable, but are defined in terms of their input and output products.

Processes also model entities of the real world and again are defined in terms of attributes. The method imposes a very restrictive notion of process, namely one in which *all processes have a single input and a single output.* (The reasons for this restriction need not detain us here, except to say that they are methodologically very well motivated. The restriction clearly will have a profound influence on the nature of the language we define below.) Distinguished attributes of a process include the transfer function(s) 'computed' by the process (i.e., how the input product is transformed into the output product), upper and lower bounds on the time taken for the process to execute, a flag indicating whether the process is 'enabled', and so on. The transfer function may be described in terms of an underlying state machine used to organize phases of the process being defined and to 'sense' important external state information required to control the execution of the process. Like products, processes may be defined in terms of 'sub-processes' and we now turn to this language of processes, as its formalization is the subject matter of the paper.

We will use an analogy with conventional sequential programming languages to motivate the nature of the formalization. Consider such a programming language. The programs in the language are constructed from basic commands (usually just assignment) and various control structures. The programming language data types are used to model the inputs and outputs of programs. Let us focus on a program that exhibits simple input/output behavior. We realize this behavior by executing the program on some machine (a real machine for a low level language and an 'abstract' machine for a high level language). Hence, we can see an analogy between inputs/outputs and products and between programs and processes. Both programs and processes are intended to model entities that define families of executions on the machine used to execute the program/process. This is exactly how we want to understand processes, i.e., as defining a class of potential executions over some (abstract) machine.

The following questions must be answered, amongst others, in order to make the analogy more exact: What is the nature of the abstract machine over which processes are defined? What 'data types' are allowed as inputs and outputs of processes? What 'control structures' may be used? What do we mean by 'execution' of a process? (Obvious further questions include: How do we specify required processes? What do we mean by 'refinement' and how do we derive refinements or prove their correctness?)

As to the first of these questions, we do not envisage a single abstract machine which will underpin all potential processes. Rather, we assume that our abstract machine is provided by an object, in the sense of object oriented programming. Such an object has

- a set of internal states;
- a set of methods, with appropriate input and output parameters, that it can execute and which change the internal state;
- a set of potential behaviors that it can exhibit, with the behavior being exhibited 'chosen' by the program being executed.

So, the purpose of a process, like a program, is to 'choose' a particular behavior allowed by the object (our abstract machine). Of course, the object itself is used, in our case, to model the basic capabilities of the organization whose industrial processes are being modeled or prescribed. These basic capabilities may be those of machines (computers, presses, conveyor belts, etc.), or people (programmers, hardware engineers, salesmen, managers, etc.), or even (sub)organizations. The aim of the exercise is to choose from all the potential behaviors (jointly) exhibitable by this abstract machine those which have the appropriate characteristics (i.e., manufacturing a product with appropriate quality and other characteristics and in a dependable manner).

We should add here two important comments about our underlying 'object'. Firstly, such a complex underlying object may itself be built in a structured manner from sub-objects by using standard object oriented structuring methods. See [3][4] for a formal account of this. Secondly, there are lacuna in object oriented programming methods to do with exactly what we are attempting here, i.e., defining a particular subclass of behaviors from those potentially exhibitable by the object. In object oriented programming, this problem of defining *threads of computation* over objects is usually overcome by defining a 'system object' that drives the choice of desirable behaviors. Different applications over the same object base then require different 'system objects' to choose the different behaviors.

The methods of this abstract machine represent the atomic 'machine executable' processes from which our industrial processes will have to be built. Consider the example below, presented using the diagrammatic notation of the method.

The two entities we have discussed, products and processes, are denoted, respectively, by arrows and boxes. (We use lower case letters for process names and upper case letters for products.) Each product is represented in terms of its attributes and each execution of the process will assign to each of these attributes

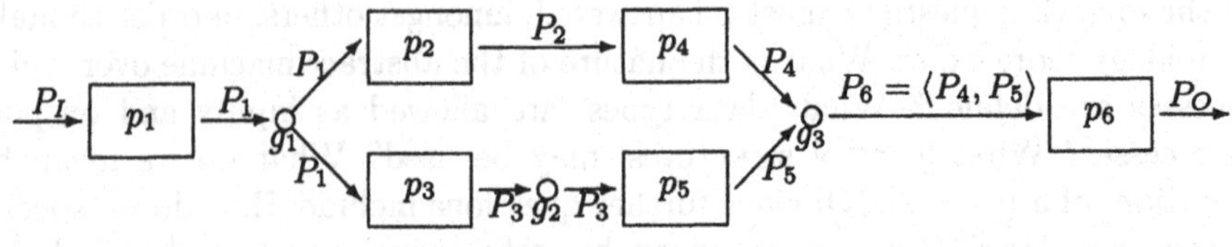

Fig. 1. A complex process example.

specific values from the appropriate domains. Each process is an atomic method from the underlying object. There is a third kind of entity in the diagram that we have not yet discussed. This is the *gate*, which is an artifact of the method used basically for two purposes: i) helping to enforce the single input, single output regime of processes, and ii) acting as guards on processes so as to control the computation. For example, g_1 copies the output product P_1 to create single inputs for each of p_2 and p_3. The gate g_2, on the other hand, is a guard which is intended to stop progress of P_3 until, for example, p_2 has terminated (or even forcing p_5 to wait for an external event, like turning on the machine). Gate g_3 is used to create the single input to process p_6 from the outputs P_4 and P_5 of processes p_4 and p_5, respectively. P_6 is simply a tuple of products synchronized in time. (The single input mechanism here is used to enforce a unique time for the process to be initiated with the required input. The method uses the unique initiation and termination times of processes to attain a specific notion of well definedness and the single input and single output regime is an aid to accomplishing this.) As we see, we need the following constructs in our language:

- sequential composition of processes;
- parallel composition of processes (with parallelism being interpreted as 'don't care' parallelism, in the sense that it is potential parallelism of which an implementer may take advantage);
- nondeterministic choice among processes;
- guards for processes (which may be combined with copying and 'restructuring' of products);
- a loop construct to allow us to define iterative processes with a guard to control the number of iterations.

The language is formalized by extending first order dynamic logic with a parallel combinator and the ability to express real time constraints. (The only construct above not used in dynamic logic is the parallel combinator.) The semantics of dynamic logic uses a notion of transition system that is used to represent the underlying abstract machine capable of executing the atomic processes. The logic is extended with variables over processes so that we can specify abstractly the processes we are interested in building. There is a notion of refinement associated with such specifications, allowing us to demonstrate that a process satisfies its specification. Finally, we demonstrate, using techniques developed in [7][8], how to algebraize this logic and thus obtain an equational proof system for our process formalism. In order to algebraize the logic we will use *omega closure fork algebras* (ω-CFA). These algebras are extensions of relation algebras [13] with three new operators, a pairing operator called *fork* [6][5], a *choice* operator [12] and the Kleene star. A consequence preserving function mapping formulas of the logic to equations in the language of ω-CFA will be defined. We will also present a representation theorem which, together with the mapping, will allow to reason equationally about properties of the logic.

The paper is organized as follows: in Section 2 we will present a first order formalization of objects. In Section 3 will be presented the logic we propose for specifying and reasoning about the properties of processes. In Section 4 we introduce the class of omega closure fork algebras. In Section 5 we present the algebraization. Finally, in Section 6 we present our conclusions about this work.

2 Objects

The first problem we confront when trying to formalize these concepts is that of characterizing the 'abstract machine' over which our processes will be defined. These processes are meant to use the underlying capabilities of the organization, as represented by the behaviors displayed by individual components within the organization. (Such individual components may be people, groups, manufacturing machines, etc.) These behaviors are organized (at least in some abstract sense) into a joint behavior which IS our 'abstract machine'. In the last decade, we have learned to organize such behaviors in terms of concepts used in object oriented programming. Objects are characterized by the data structures that are maintained by the object (seen in terms of the different states of the object) and the methods (which we call actions below) that may be executed by the object and which may change its state.

Hence, we will assume as given some object (which may be very complex and built as a system from less complex components [3]), which represents the potential behaviors of the organization as an abstract machine. This object will represent the actions/methods, state variables, external events/actions and some prescription of allowed behaviors from which individual processes must be built. (We note again the analogy between computers and programs, on the one hand, and the object/abstract machine and processes, on the other. Our processes will be used to define specific classes of behaviors in which we are interested, our required processes, from the very large class allowed by the object.) The definitions below give a somewhat non standard account of objects in terms of the underlying transition system defining the object's allowed behaviors. However, the standard parts of such descriptions (i.e., methods, state variables, etc) are easily distinguishable.

Definition 1. An *object signature* is a pair $\langle A, \Sigma \rangle$ in which $\Sigma = \langle S, F, P \rangle$ is a many-sorted first-order signature with set of sorts S, set of function symbols F and set of predicate symbols P. Among the sorts, we will single out one sort called the *time sort*, denoted by T. A is a set of *action symbols*. To each $a \in A$ is associated a pair $\langle s_1, s_2 \rangle \in (S^*)^2$ called its *arity*. We will denote the input arity of a by $ia(a)$ and the output arity of a by $oa(a)$.

Definition 2. Given an object signature $\mathcal{S} = \langle A, \langle S, F, P \rangle \rangle$, an *object structure for* $\mathcal{S}$ is a structure $\mathcal{A} = \langle \mathbf{S}, \mathbf{A}, \mathbf{F}, \mathbf{P} \rangle$ in which $\mathbf{S}$ is an S-indexed family of nonempty sets, where the set T is the T-th element in $\mathbf{S}$. In general, the set corresponding to sort s will be denoted by s. $\mathbf{A}$ is an A-indexed family of binary relations satisfying the typing constraints of symbols from A, i.e., if $ia(a) = s_1 \ldots s_m \in S^*$ and $oa(a) = s'_1 \ldots s'_n \in S^*$, then $a^{\mathcal{A}}$ (as we will denote the a-th element from $\mathbf{A}$) is contained in $(\mathsf{s}_1 \times \cdots \times \mathsf{s}_m) \times (\mathsf{s}'_1 \times \cdots \times \mathsf{s}'_n)$. To each $f : s_1 \ldots s_k \to s$ in F is associated a function $f^{\mathcal{A}} : \mathsf{s}_1 \times \cdots \times \mathsf{s}_k \to \mathsf{s} \in \mathbf{F}$. To each p of arity $s_1 \ldots s_k$ in P is associated a relation $p^{\mathcal{A}} \subseteq \mathsf{s}_1 \times \cdots \times \mathsf{s}_k \in \mathbf{P}$.

Regarding the domain T associated to the time sort T, we will not deepen on the different possibilities for modeling time, but will rather choose some adequate (with respect to the application we have in mind) representation, as for instance the fields of rational or real numbers, extended with a maximum element ∞. We will distinguish some constants, as 0, ϵ, etc.

3 The Logic, the Relational Variables, and the Time

We will extend a standard notation for specifying and reasoning about programs, namely dynamic logic. What we want to do is define processes/programs over our objects which reflect the intuitive model outlined in the introduction. Dynamic logic starts with basic actions and constructs programs by using certain combinators. The usual basic action is assignment, but we will replace this with the actions of the underlying object. The basic actions will be represented by binary relations. The input and output domains of such relations will be tuples of state variables or a choice of a set of state variables, thus reflecting the single input, single output idea of processes. The combinators are also extended with one to allow us to express (potential) parallelism of processes (defined via the intersection operator for binary relations).

Another important aspect of processes, as we wish to define them, is the real–time aspect. In defining processes, we often want to reason about time: throughput time, critical paths, optimization of processes. This requires that we are able to deal with reasoning about time within the formalism. We adapt a real–time logic developed in [1] which presents an extension of the logic presented in [3]. Each basic action is supplemented with a specification of lower and upper time bounds for occurrences of that action. These bounds may have various interpretations, amongst which we have the following: the lower bound is interpreted as the minimum time that must pass before which the action's effects are committed to happen and the upper bound gives a maximum time by which the action's effects are committed to happen. Specifications of processes will also have associated lower and upper bounds, and refinements will be expected to provably meet these bounds.

In this section we will present the *Product/Process Modeling Logic* (P/PML). Consider the formula $\varphi(x) := [x\,Ax]\beta(x)$ where A is an action term (a binary or n-ary relation) and the notation $[x\,Ax]\beta$ means that "all executions of action A establish the property β". According to our previous discussion about processes and products, we read φ as stating that β is a truth of the system A, then proving the truth of φ can be seen as the *verification* of the property β in the system described by A. Opposed to the previous view, is the notion of an *implicit specification* of a system, in which A is not a ground term, but rather may contain some relational variables that represent subsystems not yet fully determined. In what follows we will denote by *RelVar* the set of relational variables $\{R, S, T, \dots\}$.

Definition 3. Given an object signature $\mathcal{S} = \langle A, \langle S, F, P \rangle \rangle$, the sets of *relational terms* and *formulas* on $\mathcal{S}$ are the smallest sets $RT(\mathcal{S})$ and $For(\mathcal{S})$ such that

1. $a \in RT(\mathcal{S})$ for all $a \in A \cup RelVar \cup \{ 1'_t : t \in S^* \}$.
2. If $r \in RT(\mathcal{S})$ and $ia(r) = oa(r)$, then $r^* \in RT(\mathcal{S})$. We define $ia(r^*) = oa(r^*) = ia(r)$.
3. If $r, s \in RT(\mathcal{S})$, $ia(r) = ia(s)$ and $oa(r) = oa(s)$, then $r+s \in RT(\mathcal{S})$ and $r\cdot s \in RT(\mathcal{S})$. We define $ia(r+s) = ia(r\cdot s) = ia(r)$ and $oa(r+s) = oa(r\cdot s) = oa(r)$.
4. If $r, s \in RT(\mathcal{S})$ and $oa(r) = ia(s)$, then $r;s \in RT(\mathcal{S})$. We define $ia(r;s) = ia(r)$ and $oa(r;s) = oa(s)$.
5. If $\alpha \in For(\mathcal{S})$ is quantifier free and has free variables $x_1, \dots, x_n$ with x_i of sort s_i, then $\alpha? \in RT(\mathcal{S})$ and $ia(\alpha?) = oa(\alpha?) = s_1 \dots s_n$.
6. The set of first-order atomic formulas on the signature Σ is contained in $For(\mathcal{S})$.
7. If $\alpha, \beta \in For(\mathcal{S})$, then $\neg\alpha \in For(\mathcal{S})$ and $\alpha \vee \beta \in For(\mathcal{S})$..
8. If $\alpha \in For(\mathcal{S})$ and x is an individual variable of sort s, then $(\exists x : s)\alpha \in For(\mathcal{S})$.
9. If $\alpha \in For(\mathcal{S})$, $t \in RT(\mathcal{S})$ with $ia(t) = s_1 \dots s_m$ and $oa(t) = s'_1 \dots s'_n$, $\vec{x} = x_1, \dots, x_m$ with x_i of sort s_i, $\vec{y} = y_1, \dots, y_n$ with y_i of sort s'_i and l, u are variables of sort T, then $\left\langle \vec{x}\ {}_l t^u\ \vec{y} \right\rangle \alpha \in For(\mathcal{S})$.

Definition 4. Let $R \in RT(\mathcal{S})$ with $ia(R) = s_1 \dots s_m$ and $oa(R) = s'_1 \dots s'_n$, $\vec{x} = x_1, \dots, x_m$ with x_i of sort s_i, $\vec{y} = y_1, \dots, y_n$ with y_i of sort s'_i, and l, u variables of sort T. An expression of the form $\vec{x}\ {}_l R^u\ \vec{y}$ is called a *timed action term*.

We will assume that a lower and an upper bound are assigned to atomic actions, namely $l_a \in \mathsf{T}$ and $u_a \in \mathsf{T}$ for each action $a \in A$. From the bounds of the atomic actions it is possible to define bounds for complex actions in a quite natural way.

Definition 5. Let $\mathcal{S}$ be an object signature. The functions l and u from $RT(\mathcal{S}) \cup For(\mathcal{S})$ to T are defined as follows[1]:

1. If $a \in A$, then $\mathsf{l}(a) = l_a$ and $\mathsf{u}(a) = u_a$.
2. If $R = X \in RelVar$, then $\mathsf{l}(X) = 0$ and $\mathsf{u}(X) = \infty$.
3. If $R = 1'_t$, with $t \in S^*$, then $\mathsf{l}(R) = 0$ and $\mathsf{u}(R) = \epsilon$ (ϵ being a constant of sort T).
4. If $R = S^*$, then $\mathsf{l}(R) = 0$ and $\mathsf{u}(R) = \infty$.
5. If $R = S+T$, then $\mathsf{l}(R) = \min\{\mathsf{l}(S), \mathsf{l}(T)\}$ and $\mathsf{u}(R) = \max\{\mathsf{u}(S), \mathsf{u}(T)\}$.
6. If $R = S \cdot T$, then $\mathsf{l}(R) = \max\{\mathsf{l}(S), \mathsf{l}(T)\}$ and $\mathsf{u}(R) = \max\{\mathsf{u}(S), \mathsf{u}(T)\}$.
7. If $R = S;T$, then $\mathsf{l}(R) = \mathsf{l}(S)$ and $\mathsf{u}(R) = \mathsf{u}(S) + \mathsf{u}(T)$.
8. If $R = \alpha?$ with $\alpha \in For(\mathcal{S})$ quantifier free and with free variables $\vec{x}$, $\mathsf{l}(R) = \mathsf{l}(\alpha)$ and $\mathsf{u}(R) = \mathsf{u}(\alpha)$.
9. If $\alpha = p(t_1, \ldots, t_k)$, then $\mathsf{l}(\alpha) = l_p \in \mathsf{T}$ and $\mathsf{u}(\alpha) = u_p \in \mathsf{T}$, with $l_p \leq u_p$.
10. If $\alpha = \neg\beta$, then $\mathsf{l}(\alpha) = \mathsf{l}(\beta)$ and $\mathsf{u}(\alpha) = \mathsf{u}(\beta)$.
11. If $\alpha = \beta \mathbf{op} \gamma$ with $\mathbf{op} \in \{\vee, \wedge, \rightarrow\}$, then $\mathsf{l}(\alpha) = \min\{\mathsf{l}(\beta), \mathsf{l}(\gamma)\}$ and $\mathsf{u}(\alpha) = \max\{\mathsf{u}(\beta), \mathsf{u}(\gamma)\}$.
12. If $\alpha = \left\langle \vec{x}\ {}_l R^u\ \vec{y} \right\rangle \beta$, then $\mathsf{l}(\alpha) = \mathsf{l}(R)$ and $\mathsf{u}(\alpha) = \mathsf{u}(R) + \mathsf{u}(\beta)$.

Given a set of sorts $S = \{s_1, \ldots, s_k\}$ and domains $\mathsf{S} = \{\mathsf{s}_1, \ldots, \mathsf{s}_k\}$ for these sorts, by a *valuation of the individual variables of sort s_i* we refer to a function $\nu : IndVar_{s_i} \rightarrow \mathsf{s}_i$. A valuation of the relational variables is a function $\mu : RelVar \rightarrow \mathcal{P}(\mathsf{S}^* \times \mathsf{S}^*)$.

Definition 6. Given a valuation of the individual variables ν and an array of variables $\vec{x} = x_1, \ldots, x_n$, by $\nu(\vec{x})$ we denote the tuple $\langle \nu(x_1), \ldots, \nu(x_n) \rangle$.

Let $\mathcal{A}$ be an object structure and μ a valuation of the relational variables. Given valuations of the individual variables ν and ν' and a timed action term $\vec{x}\ {}_l R^u\ \vec{y}$, by $\nu \left(\vec{x}\ {}_l R^u\ \vec{y} \right) \nu'$ we denote the fact that: $\left\langle \nu(\vec{x}), \nu'(\vec{y}) \right\rangle \in R^{\mathcal{A}}_{\mu}$ (the denotation of the relational term R, formally defined in Def. 7), for every variable z not occurring in $\vec{y}$, $\nu'(z) = \nu(z)$, and, $\nu(l) \leq \mathsf{l}(R)$ and $\nu(u) \geq \mathsf{u}(R)$.

The semantics of formulas is now defined relative to valuations of individual variables and relational variables. In the following definition, the notation $\mathcal{A} \models_{P/PML} \alpha[\nu][\mu]$, is to be read "*The formula α is satisfied in the object structure $\mathcal{A}$ by the valuations ν and μ*".

Definition 7. Let us have an object signature $\mathcal{S} = \langle A, \langle S, F, P \rangle \rangle$ and an object structure $\mathcal{A} = \langle \mathbf{S}, \mathbf{A}, \mathbf{F}, \mathbf{P} \rangle$. Let ν be a valuation of individual variables and μ a valuation of relational variables. Then:

1. If $a \in A$ then $a^{\mathcal{A}}_{\mu}$ is the element with index a in $\mathbf{A}$.
2. If $R \in RelVar$, then $R^{\mathcal{A}}_{\mu} = \mu(R)$.
3. If $R = 1'_t$ with $t = s_1 \ldots s_k$, $R^{\mathcal{A}}_{\mu} = \{ \langle \langle a_1, \ldots, a_k \rangle, \langle a_1, \ldots, a_k \rangle \rangle : a_i \in \mathsf{s}_i \}$.

[1] We will only consider quantifier-free formulas, since these are the ones used for building actions of the form $\alpha?$.

4. If $R = S^*$, with $S \in RT(\mathcal{S})$, then $R_\mu^\mathcal{A}$ is the reflexive-transitive closure of the binary relation $S_\mu^\mathcal{A}$.
5. If $R = S+T$, with $S, T \in RT(\mathcal{S})$, then $R_\mu^\mathcal{A} = S_\mu^\mathcal{A} \cup T_\mu^\mathcal{A}$.
6. If $R = S \cdot T$, with $S, T \in RT(\mathcal{S})$, then $R_\mu^\mathcal{A} = S_\mu^\mathcal{A} \cap T_\mu^\mathcal{A}$.
7. If $R = S;T$, with $S, T \in RT(\mathcal{S})$, then $R_\mu^\mathcal{A}$ is the composition of the binary relations $S_\mu^\mathcal{A}$ and $T_\mu^\mathcal{A}$.
8. If $R = \alpha?$ with $\alpha \in For(\mathcal{S})$ quantifier free and with free variables $\vec{x} = x_1, \ldots, x_n$, then $R_\mu^\mathcal{A} = \left\{ \left\langle \nu(\vec{x}), \nu(\vec{x}) \right\rangle : \mathcal{A} \models_{P/PML} \alpha[\nu][\mu] \right\}$.
9. If $\varphi = p(t_1, \ldots, t_n)$ with $p \in P$, $\mathcal{A} \models_{P/PML} \varphi[\nu][\mu]$ if $\langle t_{1\nu}^\mathcal{A}, \ldots, t_{n\nu}^\mathcal{A} \rangle \in p^\mathcal{A}$.
10. If $\varphi = \neg\alpha$, then $\mathcal{A} \models_{P/PML} \varphi[\nu][\mu]$ if $\mathcal{A} \not\models_{P/PML} \alpha[\nu][\mu]$.
11. If $\varphi = \alpha \vee \beta$, $\mathcal{A} \models_{P/PML} \varphi[\nu][\mu]$ if $\mathcal{A} \models_{P/PML} \alpha[\nu][\mu]$ or $\mathcal{A} \models_{P/PML} \beta[\nu][\mu]$.
12. If $\varphi = (\exists x : s)\alpha$, then $\mathcal{A} \models_{P/PML} \varphi[\nu][\mu]$ if there exists $a \in \mathbf{s}$ such that $\mathcal{A} \models_{P/PML} \alpha[\nu_x^a][\mu]$ (ν_x^a, as usual, denotes the valuation that agrees with ν in all variables but x, and satisfies $\nu_x^a(x) = a$).
13. If $\varphi = \left\langle \vec{x}\ {}_l R^u\ \vec{y} \right\rangle \alpha$, then $\mathcal{A} \models_{P/PML} \varphi[\nu][\mu]$ if there exists a valuation ν' such that $\nu \left(\vec{x}\ {}_l R^u\ \vec{y} \right) \nu'$ and $\mathcal{A} \models_{P/PML} \alpha[\nu'][\mu]$.

Example 1. The example shows how the real-time features of the specification language (P/PML) play a decisive role in the election of the implementations of processes. A manufacturer of candy vending machines wants to manufacture machines with the following characteristics. If the machine has candy, then, after money has been deposited, at most a time K_1 passes before candy is delivered. If the machine is empty, then at most a time K_2 can pass before the transaction is finished. If the machine can be fully replenished in time to meet the K_2 upper bound, then it should be replenished, otherwise, the money should be given back.

Let us model the part of the behavior of the machine after money has been introduced and until candy has been delivered or the money was given back[2].

$$\forall m, l, u, x\, (\$\_in?(m) = \mathsf{t} \wedge x = \#candy(m) > 0 \wedge u \geq K_1 \wedge l = 0$$

$$\Rightarrow [m\ {}_l VM^u m]\, (\#candy(m) = x - 1 \wedge delivered?(m) = \mathsf{t}))$$

$$\forall m, l, u\, (\$\_in?(m) = \mathsf{t} \wedge \#candy(m) = 0 \wedge u \geq K_2 \wedge l = 0$$

$$\Rightarrow [m\ {}_l VM^u m]\, (\#candy(m) = max\_candy - 1 \wedge delivered?(m) = \mathsf{t}))$$

$$\forall m, l, u\, (\$\_in?(m) = \mathsf{t} \wedge \#candy(m) = 0 \wedge u < K_2 \wedge l = 0$$

$$\Rightarrow [m\ {}_l VM^u m]\, (delivered?(m) = \mathsf{f} \wedge money\_back?(m) = \mathsf{t}))$$

If the manufacturer believes that a consumer can wait for candy 3 minutes without loosing his patience, then K_2 can be set to 3 minutes in the specification.

---

[2] Given an object m of the class "*vending machine*", the method *$_in?* tests if money has been deposited. Method *#candy* retrieves the amount of candy left in the machine. The method *delivered?* tests if candy has been delivered, and *money_back?* tests if the money has been returned to the customer. The constant *max_candy* stands for the maximum amount of candy the machine can contain. A formal specification of the class is not given by lack of space.

Let us assume that as a constraint, this part of the machine must be built using some of the following processes:

- *RETURN PRODUCT* (that returns candy provided the machine is not empty. Its lower time bound is 0 and the upper time bound is 3 seconds).
- *REPLENISH* (that fully replenishes the machine. Its lower time bound is 0 and the upper time bound will be discussed later).
- *RETURN MONEY* (that gives the customer its money back. Its lower time bound is 0 and the upper time bound is 3 seconds)

If the machine is to be placed inside a convenience store, then as soon as the machine is emptied a clerk will replenish it, and therefore, a reasonable upper time bound for the replenishing action might be 2 minutes. Then, the following action shows a feasible implementation:

$$(\#candy(m) > 0)?; RETURN\ PRODUCT$$
$$+ (\#candy(m) = 0)?; REPLENISH; RETURN\ PRODUCT\ .$$

If the machine is to be placed in a subway station, then it may be expected that it will not be replenished more than twice a day. Then, the upper time bound for the replenishing action might for instance be 12 hours. In this case, the previously described process does not satisfy the specification, but the following one does:

$$(\#candy(m) > 0)?; RETURN\ PRODUCT$$
$$+ (\#candy(m) = 0)?; RETURN\ MONEY\ .$$

4 Omega Closure Fork Algebras

Equational reasoning based on substitution of equals for equals is the kind of manipulation that is performed in many information processing systems. The role of equational logics in development of formal methods for computer science applications is increasingly recognized and various tools have been developed for modeling user's systems and carrying through designs within the equational framework (Gries and Schneider [10], Gries [9]).

In this section we present the *calculus for closure fork algebras* (CCFA), an extension of the *calculus of relations* (CR) and of the *calculus of relations with fork* [5]. Because of the non enumerability of the theory of dynamic logic, the CCFA cannot provide an adequate algebraization. In order to overcome this restriction we will define the calculus ω-CCFA by adding an infinitary equational inference rule. From the calculus we define the class ω-CFA of the *omega closure fork algebras* and a representation theorem is presented, showing that the Kleene star as axiomatized, indeed characterizes reflexive-transitive closure.

In the following paragraphs we will introduce the *Calculus for Closure Fork Algebras* (CCFA).

Definition 8. Given a set of relation symbols R, the set of CCFA terms on R is the smallest set $TCCFA(R)$ satisfying: $R \cup RelVar \cup \{0, 1, 1'\} \subseteq TCCFA(R)$. If $x \in TCCFA(R)$,then $\{\breve{x}, x^*, x^\diamond\} \subseteq TCCFA(R)$. If $x, y \in TCCFA(R)$,then $\{x{+}y, x{\cdot}y, x;y, x \nabla y\} \subseteq TCCFA(R)$.

The symbol $^\diamond$ denotes a *choice* function (see [12, §3], which is necessary in order to prove Thm. 1.

Definition 9. Given a set of relation symbols R, the set of CCFA formulas on R is the set of identities $t_1 = t_2$, with $t_1, t_2 \in$ TCCFA(R).

Definition 10. Given terms $x, y, z, w \in$ TCCFA(R), the identities defined by the following conditions are axioms:
Identities axiomatizing the relational calculus [13],
The following three axioms for the fork operator:

$$x \nabla y = (x;(1' \nabla 1)) \cdot (y;(1 \nabla 1')),$$
$$(x \nabla y);(z \nabla w)^{\smile} = (x;\breve{z}) \cdot (y;\breve{w}),$$
$$(1' \nabla 1)^{\smile} \nabla (1 \nabla 1')^{\smile} \leq 1'.$$

The following three axioms for the choice operator, taken from [12, p. 324]:

$$x^\diamond;1;\breve{x}^\diamond \leq 1', \qquad \breve{x}^\diamond;1;x^\diamond \leq 1',$$
$$1;(x \cdot x^\diamond);1 = 1;x;1.$$

The following two axioms for the Kleene star:

$$x^* = 1' + x;x^*, \qquad x^*;y \leq y + x^*;(\overline{y} \cdot x;y).$$

Let us denote by $1'_U$ the partial identity $Ran\left(\overline{1 \nabla 1}\right)$. Then, the axiom $1;1'_U;1 = 1$ (which states the existence of a nonempty set of non-splitting elements) is added.

The rules of inference for the calculus CCFA are those of equational logic. Note that x^* is the smallest reflexive and transitive relation that includes x.

Definition 11. We define the calculus ω-CCFA as the extension of the CCFA obtained by adding the following inference rule[3]:

$$\frac{\vdash 1' \leq y \qquad x^i \leq y \vdash x^{i+1} \leq y}{\vdash x^* \leq y}$$

Definition 12. We define the class of the *omega closure fork algebras* (ω-CFA) as the models of the identities provable in ω-CCFA.

The standard models of the ω-CCFA are the *Proper Closure Fork Algebras* (PCFA for short). In order to define the class PCFA, we will first define the class •PCFA.

[3] Given $i > 0$, by x^i we denote the relation inductively defined as follows: $x^1 = x$, and $x^{i+1} = x;x^i$.

Definition 13. Let E be a binary relation on a set U, and let R be a set of binary relations. A •PCFA is a two sorted structure with domains R and U $\langle R, U, \cup, \cap, ^{-}, \emptyset, E, ;, Id, ^{\smile}, \nabla, ^{\diamond}, ^{*}, \star \rangle$ such that

1. $\bigcup R \subseteq E$,
2. $\star : U \times U \rightarrow U$ is an injective function when its domain is restricted to the set E,
3. If we denote by Id the identity relation on the set U, then $\emptyset$, E and Id belong to R,
4. R is closed under set choice operator defined by the condition:

$$x^{\diamond} \subseteq x \qquad \text{and} \qquad |x| = 1 \iff x \neq \emptyset.$$

5. R is closed under set union ($\cup$), intersection ($\cap$), complement relative to E ($^{-}$), composition of binary relations ($;$), converse ($^{\smile}$), reflexive-transitive closure (*) and fork (∇), defined by $S \nabla T = \{ \langle x, \star(y, z) \rangle : x S y \text{and} x T z \}$.

Note that $x^{\diamond}$ denotes an arbitrary pair in x, then $x^{\diamond}$ is called a *choice* operator.

Definition 14. We define the class PCFA as **Rd**•PCFA where **Rd** takes reducts to structures of the form $\langle R, \cup, \cap, ^{-}, \emptyset, E, ;, Id, ^{\smile}, \nabla, ^{\diamond}, ^{*} \rangle$.

Note that given $\mathbf{A} \in$ PCFA, the terms $(1' \nabla 1)^{\smile}$ and $(1 \nabla 1')^{\smile}$ denote respectively the binary relations $\{ \langle a \star b, a \rangle : a, b \in A \}$ and $\{ \langle a \star b, b \rangle : a, b \in A \}$. Thus, they behave as projections with respect to the injection $\star$. We will denote these terms by π and ρ, respectively.

From the operator fork we define $x \otimes y = (\pi ; x) \nabla (\rho ; y)$. The operator $\otimes$ (*cross*), when interpreted in an proper closure fork algebra behaves as a parallel product: $x \otimes y = \{ \langle a \star b, c \star d \rangle : \langle a, c \rangle \in x \wedge \langle b, d \rangle \in y \}$.

A relation R is *constant* if satisfies: $\breve{R} ; R \leq 1'$, $1 ; R = R$, and $R ; 1 = 1$. Constant relations are alike constant functions, i.e., they relate every element from the domain to a single object[4]. We will denote the constant whose image is the value a by C_a.

Definition 15. We denote by FullPCFA the subclass of PCFA in which the relation E equals $U \times U$ for some set U and R is the set of all binary relations contained in E.

Similarly to the relation algebraic case, where every proper relation algebra (PRA) **A** belongs to[5] **ISP**FullPRA, it is easy to show that every PCFA belongs to **ISP**FullPCFA. We finally present the representation theorem for ω-CFA.

Theorem 1. *Given* $\mathbf{A} \in \omega$-CFA, *there exists* $\mathbf{B} \in$ PCFA *such that* **A** *is isomorphic to* **B**.

[4] This comment is in general a little strong and applies to *simple* algebras, but is nevertheless useful as an intuitive aid for the non specialist.

[5] By **I**, **S** and **P** we denote the closure of an algebraic class under isomorphic copies, subalgebras and direct products, respectively.

5 Interpretability of P/PML in ω-CCFA

In this section we will show how theories on P/PML can be interpreted as equational theories in ω-CCFA. This is very useful because allows to reason equationally in a logic with variables over two different sorts (individuals and relations).

Definition 16. Let S, F and P be sets consisting of sort, function and relation symbols, respectively. By ω-CCFA$^+(S,A,F,P)$ we denote the extension of ω-CCFA obtained by adding the following equations as axioms.

1. For each $s, s' \in S$ $(s \neq s')$, the equations $1'_s + 1'_U = 1'_U$ and $1'_s \cdot 1'_{s'} = 0$ (elements from types do not split, and different types are disjoint).
2. For each $a \in A$ with $ia(a) = s_1 \dots s_k$ and $oa(a) = s'_1 \dots s'_n$, the equation $(1'_{s_1} \otimes \cdots \otimes 1'_{s_k})\,;a;\,(1'_{s'_1} \otimes \cdots \otimes 1'_{s'_n}) = a$.
3. For each $f : s_1 \dots s_k \to s \in F$, $\breve{f};f + 1'_s = 1'_s$ and $(1'_{s_1} \otimes \cdots \otimes 1'_{s_k})\,;f = f$, stating that f is a functional relation of the right sorts.
4. For each p of arity $s_1 \dots s_k$ in P, the equation $(1'_{s_1} \otimes \cdots \otimes 1'_{s_k})\,;p;1 = p$, stating that p is a right-ideal relation expecting inputs of the right sorts.

Definition 17. A *model* for the calculus ω-CCFA$^+(S,A,F,P)$ is a structure $\mathcal{A} = \langle \langle \mathbf{A}, S^{\mathcal{A}}, A^{\mathcal{A}}, F^{\mathcal{A}}, P^{\mathcal{A}} \rangle, m \rangle$ where: $\mathbf{A} \in \omega$-CFA. $S^{\mathcal{A}}$ is a set of disjoint partial identities, one for each sort symbol in S. $A^{\mathcal{A}}$ is a set of binary relations, one for each action symbol $a \in A$. Besides, if $ia(a) = s_1 \dots s_k$ and $oa(a) = s'_1 \dots s'_n$, then $a^{\mathcal{A}}$ satisfies the condition in item 2 of Def. 16. $F^{\mathcal{A}}$ is a set of functional relations, one for each function symbol in F. Besides, if $f : s_1 \dots s_k \to s$, then $f^{\mathcal{A}}$ satisfies the conditions in item 3 of Def. 16. $P^{\mathcal{A}}$ is a set of right ideal relations, one for each predicate symbol $p \in P$. Besides, if p has arity $s_1 \dots s_k$, then $p^{\mathcal{A}}$ satisfies the conditions in item 4 of Def. 16. $m : \mathit{RelVar} \to \mathbf{A}$.

Noce that the mapping m in a ω-CCFA$^+(S,A,F,P)$ model extends homomorphically to arbitrary relational terms. For the sake of simplicity, we will use the same name for both.

In the following paragraphs we will define a function mapping formulas from $P/PML(S,A,F,P)$ to ω-CCFA$^+(S,A,F,P)$ formulas. In the next definitions, σ is a sequence of numbers increasingly ordered. Intuitively, the sequence σ contains indices of those individual variables that appear free in the formula (or term) being translated. By $\mathit{Ord}(n,\sigma)$ we will denote the position of the index n in the sequence σ, by $[\sigma \oplus n]$ we denote the extension of the sequence σ with the index n, and by $\sigma(k)$ we denote the element in the k-th position of σ. In what follows, $t^{;n}$ is an abbreviation for $t; \cdots ;t$ (n times). For the sake of completeness, $t^{;0}$ is defined as 1'. We will denote by $\mathit{IndTerm}(F)$ the set of terms from P/PML built from the set of constant and function symbols F. By $\mathit{RelDes}(K)$ we denote the set of terms from ω-CCFA that are built from the set of relation constants K.

Definition 18. The function $\delta_\sigma : \mathit{IndTerm}(F) \to \mathit{RelDes}(F)$, mapping individual terms into relation designations, is defined inductively by the conditions:

1. $\delta_\sigma(v_i) = \begin{cases} \rho;^{Ord(i,\sigma)-1};\pi & \text{if } i \text{ is not the last index in } \sigma, \\ \rho;^{Length(\sigma)-1} & \text{if } i \text{ is the last index in } \sigma. \end{cases}$
2. $\delta_\sigma(f(t_1,\ldots,t_m)) = (\delta_\sigma(t_1)\nabla\cdots\nabla\delta_\sigma(t_m));f$ for each $f \in F$.

Given a sequence σ such that $Length(\sigma) = l$ and an index n ($n < \omega$) such that v_n has sort s, we define the term $\Delta_{\sigma,n}$ ($n < \omega$) by the condition[6]

$$\Delta_{\sigma,n} = \begin{cases} \delta_\sigma(v_{\sigma(1)})\nabla\cdots\nabla\delta_\sigma(v_{\sigma(k-1)})\nabla 1_s\nabla\delta_\sigma(v_{\sigma(k+1)})\nabla\cdots\nabla\delta_\sigma(v_{\sigma(l)}) \\ \qquad\qquad\qquad\qquad\qquad\qquad \text{if } k = Ord(n,[\sigma\oplus n]) < l, \\ \delta_\sigma(v_{\sigma(1)})\nabla\cdots\nabla\delta_\sigma(v_{\sigma(l-1)})\nabla 1_s \qquad \text{if } Ord(n,[\sigma\oplus n]) = l. \end{cases}$$

Notation 1 Let σ be a sequence of indices of individual variables of length n. Let $\vec{x}= \langle x_1,\ldots,x_k\rangle$ be a vector of variables whose indices occur in σ. We will denote by $\Pi_{\sigma,\vec{x}}$ the relation that given a tuple of values for the variables whose indices appear in σ, projects the values corresponding to the variables appearing in $\vec{x}$. For example, given $\sigma = \langle 2,5,7,9\rangle$ and $\vec{x}= \langle v_2, v_7\rangle$, $\Pi_{\sigma,\vec{x}} = \{ \langle a_1 \star a_2 \star a_3 \star a_4, a_1 \star a_3\rangle : a_1,a_2,a_3,a_4 \in A \}$. Similarly, $Arrange_{\sigma,\vec{x}}$ denotes the relation that, given two tuples of values (one for the variables with indices in σ and the other for the variables in $\vec{x}$), produces a new tuple of values for the variables with indices in σ updating the old values with the values in the second tuple. For the previously defined σ and $\vec{x}$, we have $Arrange_{\sigma,\vec{x}} = \{ \langle (a_1 \star a_2 \star a_3 \star a_4) \star (b_1 \star b_2), b_1 \star a_2 \star b_2 \star a_4\rangle : a_1,a_2,a_3,a_4,b_1,b_2 \in A \}$. Note that these two relations can be easily defined using the projections π and ρ previously defined.

Definition 19. The mappings $M : RT(S) \to RelDes(A)$ and $T_\sigma : For(S) \to RelDes(A \cup F \cup P)$ are mutually defined by

$M(a) = a$ for each $a \in A \cup RelVar$, $\quad M(1'_{s_1\ldots s_k}) = 1'_{s_1} \otimes \cdots \otimes 1'_{s_k}$,
$M(R^*) = M(R)^*$, $\quad M(R{+}S) = M(R){+}M(S)$,
$M(R{\cdot}S) = M(R){\cdot}M(S)$, $\quad M(R;S) = M(R);M(S)$,
$M(\alpha?) = T_{\sigma_\alpha}(\alpha)\cdot 1'$,
$T_\sigma(p(t_1,\ldots,t_k)) = (\delta_\sigma(t_1)\nabla\cdots\nabla\delta_\sigma(t_k))\,;p$, $\quad T_\sigma(\neg\alpha) = \overline{T_\sigma(\alpha)}$,
$T_\sigma((\exists v_n : s)\,\alpha) = \Delta_{\sigma,n};T_{[\sigma\oplus n]}(\alpha)$, $\quad T_\sigma(\alpha \vee \beta) = T_\sigma(\alpha){+}T_\sigma(\beta)$,

$$T_\sigma\left(\left\langle \vec{x}\;{}_l R^u\; \vec{y}\right\rangle \alpha\right) = \left(\begin{array}{c} 1' \\ \nabla \\ \Pi_{\sigma,\vec{x}};M(R) \end{array}\right);Arrange_{\sigma,\vec{y}};T_\sigma(\alpha)\cdot\left((v_l;\leq)\cdot C_{l(R)}\right);1\cdot\left((v_u;\geq)\cdot C_{u(R)}\right);1.$$

We will denote by $\vdash_{\omega\text{-CCFA}}$ the provability relation in the calculus ω-CCFA. The next theorem states the interpretability of theories from P/PML as equational theories in ω-CCFA.

Theorem 2. *Let $\Gamma \cup \{\varphi\}$ be a set of P/PML formulas without free individual variables. Then,* $\Gamma \models_{P/PML} \varphi \iff \{ T_{\langle\rangle}(\gamma) = 1 : \gamma \in \Gamma \} \vdash_{\omega\text{-CCFA}} T_{\langle\rangle}(\varphi) = 1$.

[6] By 1_s we denote the relation $1;1'_s$.

6 Conclusions

We have presented a logic (P/PML) for formal real–time systems specification and construction. This logic is an extension of dynamic logic by considering arbitrary atomic actions, an operator for putting processes in parallel, and explicit time. We have also presented an equational calculus in which theories of P/PML can be interpreted, thus enabling the use of equational inference tools in the process of systems construction.

References

1. Carvalho, S.E.R., Fiadeiro, J.L. and Haeusler, E.H., *A Formal Approach to Real–Time Object Oriented Software*. In Proceedings of the Workshop on Real–Time Programming, pp. 91–96, sept/1997, Lyon, France, IFAP/IFIP.
2. Fenton, N. E., *Software Metrics. A Rigorous Approach*, International Thomson Computer Press, 1995.
3. Fiadeiro, J. L. L. and Maibaum, T. S. E., *Temporal Theories as Modularisation Units for Concurrent System Specifications*, Formal Aspects of Computing, Vol. 4, No. 3, (1992), 239–272.
4. Fiadeiro, J. L. L. and Maibaum, T. S. E., *A Mathematical Toolbox for the Software Architect*, in Proc. 8th International Workshop on Software Specification and Design, J. Kramer and A. Wolf, eds., (1995) (IEEE Press), 46–55.
5. Frias M. F., Baum G. A. and Haeberer A. M., *Fork Algebras in Algebra, Logic and Computer Science*, Fundamenta Informaticae Vol. 32 (1997), pp. 1–25.
6. Frias, M. F., Haeberer, A. M. and Veloso, P. A. S., *A Finite Axiomatization for Fork Algebras*, Logic Journal of the IGPL, Vol. 5, No. 3, 311–319, 1997.
7. Frias, M. F. and Orlowska, E., *A Proof System for Fork Algebras and its Applications to Reasoning in Logics Based on Intuitionism*, Logique et Analyze, vol. 150–151–152, pp. 239–284, 1995.
8. Frias, M. F. and Orlowska, E., *Equational Reasoning in Non–Classical Logics*, Journal of Applied Non Classical Logic, Vol. 8, No. 1–2, 1998.
9. Gries, D., *Equational logic as a tool*, LNCS 936, Springer–Verlag, 1995, pp. 1-17.
10. Gries, D. and Schneider, F. B., *A Logical Approach to Discrete Math.*, Springer–Verlag, 1993.
11. Kaposi, A. and Myers, M., *Systems, Models and Measures*, Springer–Verlag London, Formal Approaches to Computing and Information Technology, 1994.
12. Maddux, R.D., *Finitary Algebraic Logic*, Zeitschr. f. math. Logik und Grundlagen d. Math. vol. 35, pp. 321–332, 1989.
13. Maddux, R.D., *Relation Algebras*, Chapter 2 of Relational Methods in Computer Science, Springer Wien New York, 1997.
14. Roberts, F. S., *Measurement Theory, with Applications to Decision–Making, Utility and the Social Sciences*, Addison–Wesley, 1979.

Effective Recognizability and Model Checking of Reactive Fiffo Automata

Gregoire Sutre[1], Alain Finkel[1], Olivier Roux[2], and Franck Cassez[2]

[1] LSV, ENS Cachan & CNRS URA 2236, France,
{sutre, finkel}@lsv.ens-cachan.fr
[2] IRCyN, EC Nantes & CNRS UMR 6597, France,
{roux, cassez}@lan.ec-nantes.fr

Abstract. Our work intends to verify reactive systems with event memorization specified with the reactive language Electre. For this, we define a particular behavioral model for Electre programs, Reactive Fiffo Automata (RFAs), which is close to Fifo Automata. Intuitively, a RFA is the model of a reactive system which may store event occurrences that must not be immediately taken into account. We show that, contrarily to lossy systems where the reachability set is recognizable but not effectively computable, (1) the reachability set of a RFA is recognizable, and (2) it is effectively computable. Moreover, we also study the relationships between RFAs and Finite Automata and we prove that (3) from a trace language point of view, inclusions between RFAs and Finite Automata are undecidable and (4) the linear temporal logic LTL on states without the temporal operator next is decidable for RFAs, while LTL on transitions is undecidable.

1 Introduction

Objectives. The aim of this work is to verify reactive systems [MP92] with event memorization specified with the reactive language Electre [CR95]. A reactive program is supposed to react instantaneously to occurrences of events. A particular feature of the Electre language is that it is possible to store occurrences of events in order to process them later. The number of stored occurrences is unbounded. Consequently the behavioral model for an Electre program has an unbounded number of states and verification with standard model-checking techniques cannot be used on this model. Roux & Cassez have verified Electre programs by bounding the number of stored occurrences [CR97]. This paper deals with analysis of transition systems produced by compilation of Electre programs, *without any assumption on the boundedness of the number of stored occurrences.*

Related work. The behavioral model for Electre programs [CR95] is close to *Communicating Finite State Machines (CFSMs)* or *Fifo Automata.* However, this class has the power of Turing Machines since it's possible to simulate any Turing Machine by a system of two CFSMs [BZ83, FM97]. The reachability problem is decidable for systems with the recognizable channel property [Pac87], but

A.M. Haeberer (Ed.): AMAST'98, LNCS 1548, pp. 106–123, 1998.

this result cannot be easily used in general because this property is undecidable. Decidability results have been established for particular classes of Fifo Automata. The reachability problem is decidable for linear Fifo Automata, which can be simulated by colored Petri Nets [FC87, JJ93]. The reachability problem is decidable for lossy systems and the reachability set of a lossy system is recognizable [AJ93], but it is not effectively computable [CFP96]. Half-duplex systems and quasi-stable systems have a recognizable reachability set and it is effectively computable [CF97]. Semi-algorithms computing a symbolic representation for the reachability set of a Fifo Automaton have also been established [BG96, BGWW97, BH97, Que96, ABJ98].

Our contribution. Our work intends to establish similar results for the new class of *Reactive Fiffo Automata (RFAs)* [CR97] that models Electre programs. The three main results of the paper are:

1. the reachability set of a Reactive Fiffo Automaton is recognizable (section 4),
2. the reachability set of a Reactive Fiffo Automaton is effectively computable (section 4),
3. the linear temporal logic LTL without the temporal operator next (LTL\X) is decidable for Reactive Fiffo Automata (section 5). This result especially allows to check liveness and safety properties.

We also analyse the relationships between Reactive Fiffo Automata and Finite Automata and we prove that from a trace language point of view, inclusions between RFAs and Finite Automata are undecidable. Semantic models of other reactive languages are finite automata: in this sense, the expressiveness of Electre is strictly greater as the semantic model is a RFA.

Outline of the paper. Section 2 recalls several definitions we use throughout the paper. In section 3 we introduce the behavioral model for Electre programs which is a Reactive Fiffo Automaton. Section 4 is devoted to the proving that the reachability set of a RFA is recognizable and effectively computable. In section 5 we examine the relationships between Reactive Fiffo Automata and Finite Automata. Eventually we give in section 6 directions for future work.

Several proofs are not included in this paper, but they can be found in a longer version [1].

2 Preliminaries

Here are some basics on words and transition systems. Let Σ be an alphabet (a finite, non empty set). We write Σ^* for the set of all finite *words* $x_1x_2\cdots x_k$ with $x_i \in \Sigma$, and ε is the empty word. For two words $x, y \in \Sigma^*$, $x \sqcup\!\sqcup y$ is their shuffle: $x \sqcup\!\sqcup y = \{x_1y_1x_2y_2\cdots x_ny_n \;/\; x = x_1x_2\cdots x_n$ and $y = y_1y_2\cdots y_n$ with

[1] Available from the authors.

$x_i, y_i \in \Sigma^*\}$. If $x \in \Sigma^*$ is a word and $e \in \Sigma$ is a letter, we write $|x|_e$ for the number of occurrences of e in x. For two words $x, y \in \Sigma^*$, x is a *subword* of y iff $y \in x \sqcup\!\sqcup \Sigma^*$.

A *transition system* is a structure $TS = (S, s_0, A, \rightarrow)$ where S is a set of *states*, s_0 is the *initial state*, A is a finite set of *actions* and $\rightarrow\, \subseteq S \times A \times S$ is a set of *transitions*. We note $\xrightarrow{*}$ for the reflexive transitive closure of $\rightarrow$. An *execution* is a finite or infinite sequence of transitions $\left(s_i \xrightarrow{\alpha_i} s'_i\right)_{i \geq 1}$ such that for all $i \geq 1$, $s_{i+1} = s'_i$. Furthermore, we write $s_1 \xrightarrow{\alpha_1\alpha_2\cdots\alpha_n} s_{n+1}$ whenever we have $s_1 \xrightarrow{\alpha_1} s_2 \xrightarrow{\alpha_2} s_3 \cdots s_n \xrightarrow{\alpha_n} s_{n+1}$. A state s is said to be *reachable* in TS iff there exists an execution from the initial state $s_0 \xrightarrow{*} s$. The *reachability set* of TS, noted RS(TS), is the set of all reachable states in TS.

Let us also recall some decision problems for transition systems. The *Reachability Problem* is, given a transition system TS and a state s of TS, to determine whether s is reachable in TS. The *Reccurent Reachability Problem* is, given a transition system TS and a state s of TS, to determine whether there exists an execution in TS in which s appears infinitely often. The *Finite Reachability Set Problem* is, given a transition system TS, to determine whether the reachability set of TS is finite. The *Inclusion of Reachability Sets Problem* is, given two transition systems TS_1 and TS_2, to determine whether the reachability set of TS_1 contains the reachability set of TS_2. The *Termination Problem* is, given a transition system TS, to determine whether all executions in TS are finite.

3 RFA: A Model for Reactive Systems with Event Memorization

Electre is a reactive language aimed at specifying and programming real-time applications. Due to the types of these applications, we need to cope with events of different nature, for instance: a relevant classification concerns their memorization properties. To this extent, Electre provides for two sorts of events: fleeting or memorized.

These features are essential in the programming of some automated applications: real industrial experiments have been carried out in the field of embedded systems in cars and in the avionics (namely the SNECMA company [2] and the CERT/ONERA laboratory [BBnRR98]) but of course they are too big to be reported in this paper.

In order to understand the need for memorization of events, consider a conveyor which brings items to be manufactured at a rate which may differ from the rate of the manufacturing machine. To process all the incoming items, we have to memorize the pending items. Thus, the ability of the language to express memorization of events can ease the task of specification. This particular feature becomes crucial in the case there are many events the memorizations of which can be interleaved.

[2] Work partially supported under a three years grant number 765 358 L.

To deal with the memorization, we start with a finite model of Electre programs (Control Automaton, Definition 1) which does not take into account the memorization. Then, a list of stored occurrences of events is added to this finite model in order to deal with the ordered and multiple memorizations of the events occurrences. Thus, we obtain a Reactive Fiffo Automaton (RFA, Definition 2): a stored occurrence of an event is processed as soon as possible and priority is given to the oldest stored occurrence, hence the name *First In First Fireable Out (fiffo)*.

This memorization issue is completely defined in the semantics of the Electre language: this accounts for the semantic model of programs (RFA) which is the subject of this section.

The RFA model can be used for simulations or real executions (tools have been developed namely SILEX for simulations, EXILE for executions). A specific real-time executive based on the RFA model is run in EXILE and provides an efficient execution.

In this section, we first give a brief description of the language. Then, we define a behavioral model for Electre programs which is a Reactive Fiffo Automaton (RFA).

3.1 The Electre Reactive Language

Overview of the Language. An Electre program describing the behavior of a process is made of three types of components:

modules: which are tasks of the process without blocking points: each instance of a module is a piece of executable code which can be either active, preempted or idle,

events: which can be software or hardware originated: each occurrence of an event is a signal which can be either memorized or not,

operators: combining the two previous components (for instance parallelism, sequence, preemption or launching (of a module by an event), repetition, and so on).

The term reactive means that the system controlling the process is to react *instantaneously* to any event occurring in the environment. As a running example, we will focus on an Electre program for describing the well-known readers/writers problem.

The Readers/Writers Problem. The *readers/writers* problem was originally stated and solved in [CHP71]. There are several variations on this problem, all involving priorities. We specify our readers/writers problem here, with the following requirements:

- several readers can read the book simultaneously,
- when a writer writes the book, no other process (reader or writer) can access the book.

To specify the problem in the Electre language with two readers and two writers, we proceed as follows:

- the processes readers and writers are what we called *modules*
  - $READ_1$ (respectively $READ_2$) refers to the module for reader 1 (respectively reader 2) to read the book,
  - $WRITE$ refers to the module for both writer 1 and writer 2 to write the book,
- a request to read or write the book is an *event*
  - r_1 (respectively r_2) is a request for reading the book made by reader 1 (respectively reader 2),
  - w_1 (respectively w_2) is a request for writing the book made by writer 1 (respectively writer 2).

An Electre program that specifies the behavior of the system is presented in Figure 1.

```
PROGRAM Readers&Writers ;
     loop
        await
               { r₁ : READ₁ || r₂ : READ₂ }
           or
               #w₁  : WRITE
           or
               #w₂  : WRITE
     end loop ;
END Readers&Writers ;
```

Fig. 1. Readers/writers with no multiple memorization for reading requests

We shall not go into details about the syntax of the Electre language; the meaning of the above written program can be summed up as follows:

1. a request for reading (r_1 or r_2) is a *standard* event (no qualifier before them); this means that
   - if the request can not be taken into account at the time the event occurs then the request is:
     - ignored if it has already been stored;
     - stored otherwise;

     (it means that a standard event is memorized at most once).
   - on the contrary if the request can be taken into account, the corresponding module ($READ_1$ for request r_1) is launched (this is the meaning of the symbol ":"),
2. the activities of readers 1 and 2 may be run simultaneously (symbol "||") and when $READ_1$ is being run request r_2 can be taken into account (the converse when $READ_2$ is being run holds),

3. the writing activity $WRITE$ and the parallel activity $READ_1$ and $READ_2$ are in mutual exclusion (symbol "`or`"),
4. the requests for writing w_1 and w_2 can be satisfied one at a time (symbol "`or`") between the events w_1 and w_2;
5. the program consists in a cycle (structure "`loop` — `end loop`") of waiting until one of the events r_1, r_2, w_1 or w_2 occurs (structure "`await`"),
6. w_1 and w_2 are *multiple storage* events (prefixed by "`#`", Σ_M in Definition 1): occurrences of these events may be memorized an unbounded number of times.
   r_1 and r_2 are *single storage* events (not prefixed, Σ_S in Definition 1): only one occurrence of these events may be memorized at one time.
   Memorizable events are either multiple or single storage events. Other events are fleeting events.

3.2 From Electre Programs to RFAs

The first step towards the behavioral model for Electre programs is a *Control Automaton.* Each transition of this automaton indicates what is to be done upon the occurrence of event. It is built according to the semantics of the language [CR95].

On the example of the readers and writers, we obtain the automaton depicted in Figure 2.

It must be interpreted as follows:

- each X module completion (written end_X) is a fleeting event,
- **immediate processing:** whenever the occurrence of an event x can be taken into account, the transition labeled x is triggered (e.g. $q_0 \xrightarrow{r_1} q_1$),

1.(*a*) **memorization/sending:** whenever the occurrence of a memorizable event x cannot be taken into account, it is stored, and the transition is labeled !x (e.g. $q_2 \xrightarrow{!w_1} q_2$); moreover, there is no state change in the Control Automaton (Definition 1, 1.(*a*)),

1.(*b*) **batch processing/reception:** whenever a stored occurrence of a memorizable event x is processed, the transition labeled ?x is triggered (e.g. $q_0 \xrightarrow{?w_2} q_3$). Batch processing an event has the same effect as the immediate processing of the same event (Definition 1, 1.(*b*)),

2. the automaton is complete w.r.t. memorizable events (Definition 1, 2.).

This does not state when the transitions are triggered. We will define the operational semantics of the Control Automaton in Definition 2.

In the sequel, we focus on the memorizable event: consequently, immediate processing transitions are abstracted in τ-transition (e.g. $q_0 \xrightarrow{r_1} q_1$ becomes $q_0 \xrightarrow{\tau} q_1$).

Now, we can give a formal definition of a Control Automaton:

Definition 1 (Control Automaton). *A* Control Automaton *is a finite transition system $C = (Q, q_0, A, \rightarrow_C)$, where:*

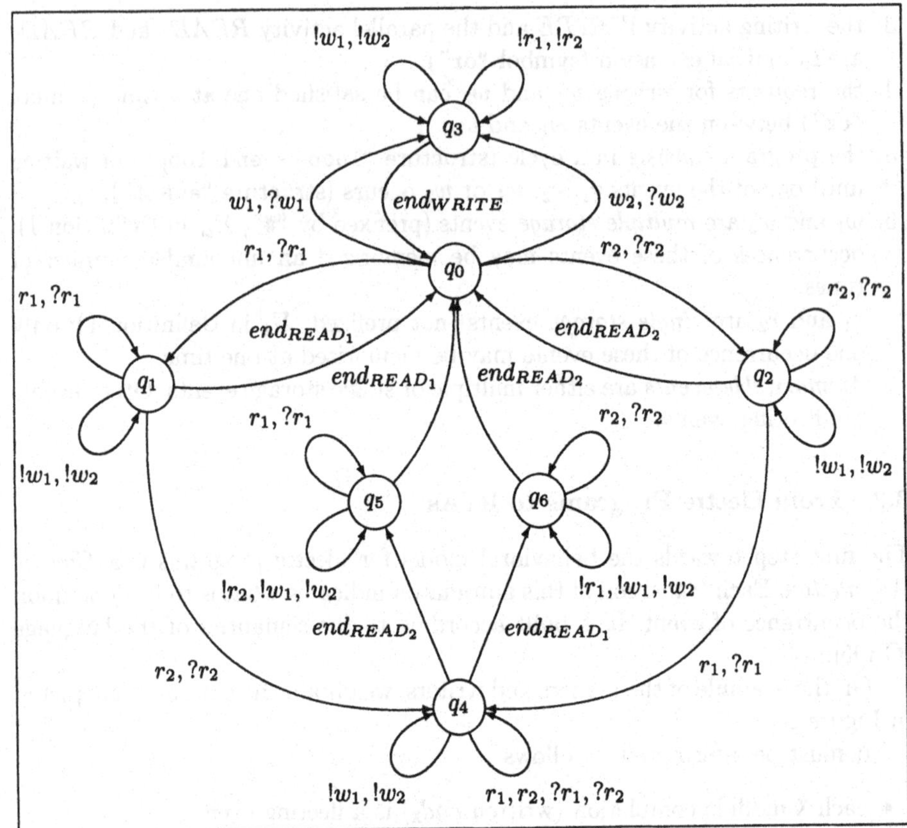

Fig. 2. The Reactive Fiffo Automaton for the readers/writers

- Q *is a finite set of* control *states, and,*
- q_0 *is the initial state, and,*
- $A = (\{!, ?\} \times \Sigma) \cup \{\tau\}$ *is a finite set of actions such that* Σ *is an alphabet and* Σ_M, Σ_S *are subsets of* Σ *verifying:* $\Sigma_M \cap \Sigma_S = \emptyset$ *and* $\Sigma_M \cup \Sigma_S = \Sigma$, *and,*
- $\rightarrow_C$ *is any finite set of transitions, verifying the two following properties:*
  1. *for all* $q, q' \in Q$ *and* $e \in \Sigma$*:*
     (a) *if* $q \xrightarrow{!e}_C q'$, *then* $q = q'$,
     (b) *if* $q \xrightarrow{?e}_C q'$, *then* $q \xrightarrow{\tau}_C q'$,
  2. *for all* $q \in Q$ *and* $e \in \Sigma$, *we have either* $q \xrightarrow{!e}_C q$ *or there exists a state* q' *such that* $q \xrightarrow{?e}_C q'$.

  For every control state $q \in Q$, *we write* $\Sigma_q = \{e \in \Sigma \mid q \xrightarrow{!e}_C q\}$.

Remark 1. Every reachable state of a Control Automaton is reachable by an execution containing only τ-transitions.

A Control Automaton is built for every Electre program. A *fiffo queue* is then added to the Control Automaton to take the memorisation and batch processing of events into account. This is formally defined by the Reactive Fiffo Automaton (RFA).

Definition 2 (Reactive Fiffo Automaton). *The* Reactive Fiffo Automaton *R associated with a Control Automaton $C = (Q, q_0, (\{!, ?\} \times \Sigma) \cup \{\tau\}, \rightarrow_C)$ is the potentially infinite transition system $R = (S, s_0, A, \rightarrow_R)$ defined as follows:*

- *$S = Q \times \Sigma^*$ is the set of states, and,*
- *$s_0 = (q_0, \varepsilon)$ is the initial state, and,*
- *$A = (\{!, ?\} \times \Sigma) \cup \{\tau\}$ is the set of actions, and,*
- *the set of transitions $\rightarrow_R$ is the smallest subset of $S \times A \times S$ verifying:*
  1. *if $q \xrightarrow{\tau}_C q'$ then forall $w \in \Sigma_q{}^*$, $(q, w) \xrightarrow{\tau}_R (q', w)$, and*
  2. *if $q \xrightarrow{!e}_C q'$ then forall $w \in \Sigma_q{}^*$, $(q, w) \xrightarrow{!e}_R (q', w')$, where w' is defined by:*
     (a) *$w' = w$ if $e \in \Sigma_S$ and $|w|_e \geq 1$,*
     (b) *$w' = we$ otherwise, and*
  3. *if $q \xrightarrow{?e}_C q'$ then forall $w_1 \in \Sigma_q{}^*, w_2 \in \Sigma^*$ $(q, w_1 e w_2) \xrightarrow{?e}_R (q', w_1 w_2)$,*

Definition 3 (Stability). *A state (q, w) of a Reactive Fiffo Automaton is* stable *iff $w \in \Sigma_q{}^*$. Otherwise, it is* unstable.

The definition of a Reactive Fiffo Automaton corresponds to the informal semantics of the fiffo queue given in the beginning of this section:

- conditions 1 and 2 give priority to batch processings: stored occurrences of events are processed as soon as possible,
- condition 3 corresponds to the fiffo order: in a batch processing, priority is given to the oldest stored occurrence.

Example 1. Keeping our readers/writers example, let us consider the RFA R associated with the Control Automaton C described in Figure 2. An execution of R is for example:

$$(q_0, \varepsilon) \xrightarrow{w_1} (q_3, \varepsilon) \xrightarrow{!r_1} (q_3, r_1) \xrightarrow{!w_2} (q_3, r_1 w_2) \xrightarrow{!r_2} (q_3, r_1 w_2 r_2) \xrightarrow{endWRITE} (q_0, r_1 w_2 r_2)$$

$$\downarrow ?r_1$$

$$\ldots (q_3, \varepsilon) \xleftarrow{?w_2} (q_0, w_2) \xleftarrow{endREAD_2} (q_6, w_2) \xleftarrow{endREAD_1} (q_4, w_2) \xleftarrow{?r_2} (q_1, w_2 r_2)$$

Three relevant observations can be done on this example :

- $(q_0, r_1 w_2 r_2)$ is an unstable state: priority is given to the processing of the first memorized occurrence (r_1),
- hence, even though $q_3 \xrightarrow{endWRITE} q_0 \xrightarrow{w_1} q_3 \ldots$ is an execution of the Control Automaton C, $(q_3, r_1 w_2 r_2) \xrightarrow{endWRITE} (q_0, r_1 w_2 r_2) \xrightarrow{w_1} (q_3, r_1 w_2 r_2) \ldots$ is not an execution of the RFA R, because the last transition is not a transition of R,

- in the transition $(q_1, w_2r_2) \xrightarrow{?r_2} (q_1, w_2)$ above, it can be noticed that the processing of the memorized occurrences are done in the *First In First Fireable Out* order (which is not strictly the *fifo* order).

4 Computation of the Recognizable Reachability Set of a RFA

We prove, in this section, that the reachability set of a Reactive Fiffo Automaton is recognizable and that it is effectively computable. This result especially allows us to decide the Reachability Problem, the Finite Reachability Set Problem and the Inclusion of Reachability Sets Problem for RFAs.

In the following, we consider a Reactive Fiffo Automaton R associated with a Control Automaton $C = (Q, q_0, (\{!, ?\} \times \Sigma) \cup \{\tau\}, \rightarrow)$.

Our first result states that the reachability set of a Reactive Fiffo Automaton is recognizable. This property comes essentially from condition 1.(a) of Definition 1. When an event e may be memorized, it is possible to memorize e^n for any $n \geq 0$.

Intuitively speaking, a RFA cannot count, but it takes the fiffo order (which is very close to the *fifo* order) into account. The fact that it cannot count allows the recognizability of its rechability set. Petri nets are orthogonal: they allow to count but not to retain the fifo ordering. So these two partially analysable models are based on different assumptions.

Hence, the fiffo queue of a RFA behaves like a fiffo queue capable of both lossiness and duplication errors [CFP96]. It follows that the reachability set of a RFA is recognizable.

Theorem 1. *The reachability set of R is recognizable.*

Proof. The proof is similar to the proof that the reachability set of a lossy system is recognizable [AJ93]. Let $\preceq$ be the well ordering over $Q \times \Sigma^*$ defined by $(q, w) \preceq (q', w')$ iff $q = q'$ and w is a subword of w'. Assume the reachability set RS(R) of R is downward closed. Then Compl(RS(R)) is upward closed. Since $\preceq$ is well ordering, Compl(RS(R)) has a finite set M of minimal elements, which gives a recognizable description of Compl(RS(R)):

$$\text{Compl}(\text{RS}(R)) = \bigcup_{(q,w)\in M} (q, w \sqcup\!\sqcup \Sigma^*)$$

As Compl(RS(R)) is recognizable, we obtain that RS(R) is recognizable. It remains to prove that RS(R) is downward closed.

Let $(q, w_1ew_2) \in \text{RS}(R)$, with $w_1, w_2 \in \Sigma^*$ and $e \in \Sigma$. We show that $(q, w_1w_2) \in \text{RS}(R)$. Since $(q, w_1ew_2) \in \text{RS}(R)$, there exists an execution $\pi = (q_0, \varepsilon) \xrightarrow{*} (q, w_1ew_2)$ in R, which may be decomposed as follows:

$$\pi = (q_0, \varepsilon) \xrightarrow{\sigma_1} (q, x) \xrightarrow{!e} (q, xe) \xrightarrow{\sigma_2} (q, w_1ew_2)$$

where the event occurrence **e** memorized by the transition $t = (q, x) \xrightarrow{!\mathbf{e}} (q, x\mathbf{e})$ is the event occurrence **e** in $w_1\mathbf{e}w_2$. Hence $|\sigma_2|_{?e} \leq |x|_e$ and π' defined below is still an execution of R, as it is possible to remove the transition t from π [3]:

$$\pi' = (q_0, \varepsilon) \xrightarrow{\sigma_1} (q, x) \xrightarrow{\sigma_2} (q, w_1 w_2)$$ □

However, this does not prove that the reachability set of a Reactive Fiffo Automaton is *effectively* recognizable. For instance in the case of lossy systems, it has been shown that the reachability set is recognizable, *but not computable* [CFP96].

The following of this section is devoted to the proving that the recognizable reachability set of a Reactive Fiffo Automaton is *computable*.

Definition 4. *For every $q \in Q$, we call* language of the fiffo queue in the control state q, *written $\mathcal{L}_R(q)$, the set:*

$$\mathcal{L}_R(q) = \{w \in \Sigma^* \ / \ (q_0, \varepsilon) \xrightarrow{*} (q, w)\}$$

It is clear, according to the definition of a Reactive Fiffo Automaton that if a control state $q \in Q$ is not reachable in the Control Automaton C then $\mathcal{L}_R(q) = \emptyset$. We will in the following deal with the control states $q \in \mathrm{RS}(C)$ reachable in C.

Notation. For every $F \subseteq \Sigma$, we write $\mathcal{S}(F^*)$ for the set of words over F containing at most one occurrence of each single storage event, $\mathcal{S}(F^*) = \{w \in F^* \ / \ \forall e \in \Sigma_S, |w|_e \leq 1\}$.

Let us notice that for any $F \subseteq \Sigma$, the language $\mathcal{S}(F^*)$ is regular. Moreover, according to the definition of a Reactive Fiffo Automaton, a single storage event can appear at most once in the fiffo queue part of a reachable state. More formally:

Remark 2. For all reachable state (q, w) of R, we have $w \in \mathcal{S}(\Sigma^*)$.

Lemma 1. *Let $q \in \mathrm{RS}(C)$ be a control state of R reachable in C. If $w \in \mathcal{S}(\Sigma_q^*)$ then (q, w) is reachable in R.*

Using Lemma 1 and Remark 2, it is easy to infer that the recognizable set of *stable* reachable states in R is computable, because it may be written as $\bigcup_{q \in \mathrm{RS}(C)} (q, \mathcal{S}(\Sigma_q^*))$.

We will now, in the following lemma, also deal with the *unstable* reachable states of R.

Lemma 2. *Let (q, w) be a stable reachable state of R and $(q, w) \xrightarrow{\tau} (q_1, w) \xrightarrow{?e_1?e_2\cdots?e_k} (q', w')$ be an execution of R. Then we have:*

[3] Intuitively, the memorized event occurrence **e** could have not occured, since memorization does not change the control state

1. $e_1 e_2 \cdots e_k \in \mathcal{S}(\Sigma_q^*)$ *and,*
2. $w' \in \mathcal{S}(X^*)$, *where* $X = \Sigma_q \setminus (\Sigma_S \cap \{e_1, e_2, \cdots, e_k\})$ *and* $\{e_1, e_2, \cdots, e_k\}$ *denotes the set associated with the multiset consisting of the elements* e_1, $e_2, \cdots, e_k$.

We now define, for all $(q, F) \in Q \times 2^{\Sigma}$, the set CoReach$(q, F)$, which will allow us to prove that the reachability set of R is recognizable. Intuitively, we define CoReach(q, F) so that if a control state p is in CoReach(q, F), then the states in $(q, \mathcal{S}(F^*))$ are reachable from the set of stable states $(p, \mathcal{S}(\Sigma_p^*))$ and hence are reachable in R.

Definition 5 (CoReach). *Let $q \in Q$ be a control state of R and $F \subseteq \Sigma$ be a subset of Σ. The set* CoReach$(q, F) \subseteq Q$ *is the set of control states $p \in Q$ such that there exists an execution $p \xrightarrow{\tau} q_1 \xrightarrow{?e_1?e_2\cdots?e_k} q$ in C verifying:*

1. $e_1 e_2 \cdots e_k \in \mathcal{S}(\Sigma_p^*)$ *and,*
2. $F = \Sigma_p \setminus (\Sigma_S \cap \{e_1, e_2, \cdots, e_k\})$.

Let us remark that for all $(q, F) \in Q \times 2^{\Sigma}$, the set $\{F \subseteq \Sigma \ / \ \text{CoReach}(q, F) \neq \emptyset\}$ is finite. The following theorem gives a precise description of the reachability set of a Reactive Fiffo Automaton, which will allow us to prove that the recognizable reachability set of a Reactive Fiffo Automaton is computable.

Theorem 2. *For every control state $q \in \text{RS}(C)$ reachable in C, we have:*

$$\mathcal{L}_R(q) = \mathcal{S}(\Sigma_q^*) \bigcup \left[\bigcup_{F \subseteq \Sigma \ / \ \text{CoReach}(q,F) \neq \emptyset} \mathcal{S}(F^*) \right]$$

Proof. Let us prove the inclusion from left to right. Assume $q \in \text{RS}(C)$ and $w \in \mathcal{L}_R(q)$. Two cases may arise:

- (q, w) is stable: then according to the definition of a stable state, we have $w \in {\Sigma_q}^*$. In this way, Remark 2 leads to $w \in \mathcal{S}(\Sigma_q^*)$.
- (q, w) is unstable: as $w \in \mathcal{L}_R(q)$, there exists an execution $\pi = (q_0, \varepsilon) \xrightarrow{*} (q, w)$ in R. Since (q_0, ε) is stable, π contains a stable state, and we call (q_s, w_s) the *last stable state* of π. As (q, w) is unstable, we come to $(q, w) \neq (q_s, w_s)$. Hence (q_s, w_s) is the source of a transition in π and we get that π may be written as:

$$\pi \quad = \quad (q_0, \varepsilon) \xrightarrow{*} (q_s, w_s) \xrightarrow{\alpha} (q_1, w_1) \xrightarrow{\sigma} (q, w)$$

  As (q_s, w_s) is the last stable state of π, we obtain:
  - on one hand $\alpha \notin \{?\} \times \Sigma$ and on the other hand $\alpha \notin \{!\} \times \Sigma$ because otherwise, (q_1, w_1) would be a stable state. Therefore $\alpha = \tau$ and $w_1 = w_s$.
  - every state in $(q_1, w_1) \xrightarrow{\sigma} (q, w)$ is unstable. Therefore, $\sigma \in (\{?\} \times \Sigma)^*$ and we assume in the following that σ is written $?e_1?e_2\cdots?e_k$.

Let us assume that $F = \Sigma_{q_s} \setminus (\Sigma_S \cap \{e_1, e_2, \cdots, e_k\})$. We can apply Lemma 2 so that:

- $e_1 e_2 \cdots e_k \in \mathcal{S}(\Sigma_{q_s}^*)$. Moreover $q_s \xrightarrow{\tau} q_1 \xrightarrow{?e_1?e_2\cdots?e_k} q$ is an execution of C. Therefore $q_s \in \mathrm{CoReach}(q, F)$ so that $\mathrm{CoReach}(q, F) \neq \emptyset$.
- $w \in \mathcal{S}(F^*)$.

Finally:

$$w \in \mathcal{S}(\Sigma_q^*) \bigcup \left[\bigcup_{F \subseteq \Sigma \ / \ \mathrm{CoReach}(q,F) \neq \emptyset} \mathcal{S}(F^*) \right]$$

Let us prove the inclusion from right to left. Assume $q \in \mathrm{RS}(C)$. We notice that according to Lemma 1, for all $w \in \mathcal{S}(\Sigma_q^*)$, (q, w) is reachable in R. Now assume $\mathrm{CoReach}(q, F) \neq \emptyset$ and $w \in \mathcal{S}(F^*)$. Since $\mathrm{CoReach}(q, F) \neq \emptyset$, there exists a control state p and an execution $p \xrightarrow{\tau} q_1 \xrightarrow{?e_1?e_2\cdots?e_k} q$ in C such that:

- $e_1 e_2 \cdots e_k \in \mathcal{S}(\Sigma_p^*)$ and,
- $F = \Sigma_p \setminus (\Sigma_S \cap \{e_1, e_2, \cdots, e_k\})$.

As $w \in \mathcal{S}(F^*)$, we have for all $e \in \Sigma_S$:

- if $e \in \{e_1, \ldots, e_k\}$ then $e \notin F$, hence $e \notin \mathrm{alph}(w)$. Therefore $|(e_1 \ldots e_k) \cdot w|_e = |e_1 \ldots e_k|_e \leq 1$.
- if $e \notin \{e_1, \ldots, e_k\}$ then $|(e_1 \ldots e_k) \cdot w|_e = |w|_e \leq 1$.

Consequently, in both cases, we have $|(e_1 \ldots e_k) \cdot w|_e \leq 1$. We remark that $(e_1 \ldots e_k) \cdot w \in \Sigma_p{}^*$, hence $(e_1 \ldots e_k) \cdot w \in \mathcal{S}(\Sigma_p^*)$. We can apply Lemma 1 so that $(p, (e_1 \ldots e_k) \cdot w)$ is reachable in R. Because $(p, (e_1 \ldots e_k) \cdot w) \xrightarrow{\tau} (q_1, (e_1 \ldots e_k) \cdot w) \xrightarrow{?e_1 \ldots ?e_k} (q, w)$ is an execution of R, we obtain that (q, w) is reachable in R. Finally:

$$w \in \mathcal{L}_R(q)$$

□

Now let us present the main result of this section which says that a regular expression for the recognizable reachability set of a Reactive Fiffo Automaton is effectively computable.

Theorem 3. *There exists an algorithm computing a regular expression for $\mathcal{L}_R(q)$, for every $q \in Q$.*

From the previous theorem, one may easily deduce the following corollary.

Corollary 1. *The Reachability Problem, the Finite Reachability Set Problem, the Inclusion of Reachability Sets Problem, the Control State Reachability Problem and the Termination Problem are decidable for RFAs.*

While recognizability of the reachability set of a Reactive Fiffo Automaton comes essentially from condition 1.(a) of Definition 1, effectivity crucially depends on condition 2. of Definition 1. As a matter of fact, let us show that if we extend Control Automata in removing condition 2. of Definition 1 then the reachability set is still recognizable but it becomes not effectively computable. Indeed, for every machine capable of both lossiness and duplication errors, one may construct a generalized RFA having the same reachability set. Now it is known that machines capable of both lossiness and duplication errors have a non effective recognizable reachability set [CFP96].

Example 2. Let us consider the RFA R modelling our readers/writers example. The reachability set of R is:

$$\bigcup_{i \in \{0,1,\cdots,6\}} (q_i, \mathcal{S}(F_i^*))$$

with $F_0 = F_3 = \{r_1, r_2, w_1, w_2\}$, $F_1 = F_5 = \{r_2, w_1, w_2\}$, $F_2 = F_6 = \{r_1, w_1, w_2\}$ and $F_4 = \{w_1, w_2\}$. The state $(q5, w_1 r_2 w_2 w_2)$ is reachable while the state $(q5, r_1)$ is not reachable. All the control states of R are reachable. The reachability set of R is infinite and R does not terminate.

Remark 3. We have implemented an optimized algorithm which computes simultaneously for all control state q, a regular expression for $\mathcal{L}_R(q)$. This algorithm has a complexity of $O(K(|\Sigma|) \cdot |Q| \cdot | \rightarrow |)$, where $K(|\Sigma|) = O(2^{|\Sigma|})$ is the complexity of subset operations over Σ. Hence, the various decision problems of Corollary 1 are decidable with the same complexity.

The reachability set of a Reactive Fiffo Automaton with a non empty initial fiffo queue is still recognizable and effectively computable [Sut97]. This result especially allows us to decide the Recurrent Reachability Problem for RFAs.

The previous theorems are very useful when considering practical aspects: simulation and verification. Indeed, the compilation of an Electre program produces a RFA (given by its associated Control Automaton), which is used for simulation and verification purposes. This RFA leads to a C program, which is then compiled to produce an executable file. Clearly, the control states which are not reachable do not need to be included in the C program. This is also the case for the transitions of the Control Automaton which are not quasi-live (a transition t of a Control Automaton is *quasi-live* if there exists an execution containing t in the associated RFA). Fortunately, the Quasi-liveness Problem is decidable for RFAs.

Proposition 1. *The Quasi-liveness Problem is decidable for RFAs.*

Proof. If $t = (q \rightarrow q')$ is a τ-transition or if t is an emission transition of the Control Automaton C, then t is quasi-live if and only if q is a reachable control state of C, which is decidable. A reception transition $q \xrightarrow{?e} q'$ of C is quasi-live if and only if $\mathcal{L}_R(q) \cap {\Sigma_q}^* e \Sigma^* \neq \emptyset$, which is decidable because a regular expression for $\mathcal{L}_R(q)$ is computable. □

5 Relationships between Reactive Fiffo Automata and Finite Automata

In the previous section, we have precisely described the reachability set of a Reactive Fiffo Automaton. We now analyse the set of executions of a RFA, and we establish a comparison between RFAs and Finite Automata. We first study the general case of trace inclusion between a RFA and a Finite Automaton. We then analyse the model checking of LTL.

5.1 Trace Inclusions

One may believe that Reactive Fiffo Automata are essentially equivalent to Finite Automata, because reception transitions are not blocking (as for each reception transition $q \xrightarrow{?e} q'$ of a Control Automaton, there exists a τ-transition $q \xrightarrow{\tau} q'$) and emission transitions can be repeated arbitrarily often. But the following undecidability results prove that this is not the case.

We define the *trace language* of a RFA in the usual way : we introduce a new alphabet of actions $\mathcal{A}$ and every transition of a Control Automaton is labelled by an *action* $a \in \mathcal{A}$ or by the empty word ε. The set of *finite (resp. infinite) traces* $T^*(R)$ (resp. $T^\omega(R)$) of a RFA R is the set of finite (resp. infinite) words on $\mathcal{A}$ corresponding to finite (resp. infinite) executions of R.

It is clear that for every regular language L (resp. ω-regular language L), there exists a RFA R such that $T^*(R) = L$ (resp. $T^\omega(R) = L$). The following theorem shows that from a trace language point of view, inclusions between RFAs and Finite Automata are undecidable.

Theorem 4. *The four following problems are undecidable :*

i) Given a RFA R and a regular language L, is $T^*(R) \subseteq L$?
ii) Given a RFA R and a regular language L, is $T^*(R) \supseteq L$?
iii) Given a RFA R and an ω-regular language L, is $T^\omega(R) \subseteq L$?
iv) Given a RFA R and an ω-regular language L, is $T^\omega(R) \supseteq L$?

As RFAs contain all Finite Automata, we obtain the following corollary.

Corollary 2. *The two following problems are undecidable :*

i) Given two RFAs R_1 and R_2, is $T^*(R_1) \subseteq T^*(R_2)$?
ii) Given two RFAs R_1 and R_2, is $T^\omega(R_1) \subseteq T^\omega(R_2)$?

5.2 Model Checking with LTL

We prove, in this section, that the linear temporal logic LTL [MP92, Eme90] without the temporal operator next, which we denote by LTL\X, is decidable for Reactive Fiffo Automata. This result especially allows to check liveness and safety properties, and also to decide the Reccurent State Problem for RFAs.

Notation. For every transition system $TS = (S, s_0, A, \rightarrow)$, we write $\mathcal{L}^\omega(TS)$ for the set of ω-sequences of states corresponding to infinite executions of TS:

$$\mathcal{L}^\omega(TS) = \{s_0 s_1 \cdots s_n \cdots \; / \; \forall i \in \mathbb{N}, s_i \rightarrow s_{i+1}\}$$

In the following, we consider a finite set AP of atomic propositions on which are based LTL formulas. Unless specified, we assume that the atomic propositions label control states. Two ω-sequences of states are *equivalent modulo stuttering* iff they display the same ω-sequence of states when two repeated consecutive states are seen as one state only.

Theorem 5 ([Lam83]). *Two ω-sequences of states equivalent modulo stuttering satisfy the same* LTL\X *formulas.*

We will now prove that a Control Automaton C and the Reactive Fiffo Automaton R associated with C satisfy the same LTL\X formulas. In this way, Model-checking of LTL\X is decidable for Reactive Fiffo Automata. In order to demonstrate this result, we introduce a Finite Automaton, Restr(C), built from C. We will actually show that C, R and Restr(C) satisfy the same LTL\X formulas.

Definition 6 (Restricted Control Automaton). *The* Restricted Control Automaton Restr(C) *associated with a Control Automaton* $C = (Q, q_0, (\{!, ?\} \times \Sigma) \cup \{\tau\}, \rightarrow)$ *is the finite transition system* Restr(C) $= (S, s_0, A, \rightarrow_{\text{Restr}(C)})$ *defined as follows:*

- $S = Q \cup \overline{Q}$, *with* $\overline{Q} = \{\overline{q} \; / \; q \in Q\}$ *a copy of* Q, *is the set of states, and,*
- $s_0 = q_0$ *is the initial state, and,*
- $A = (\{!\} \times \Sigma) \cup \{\tau\}$ *is the set of actions, and,*
- *the set of transitions* $\rightarrow_{\text{Restr}(C)}$ *is the smallest subset of* $S \times A \times S$ *verifying for all* $q, q' \in Q$ *the two following properties:*
  1. *if* $q \xrightarrow{\tau} q'$ *then* $q \xrightarrow{\tau}_{\text{Restr}(C)} q'$, *and,*
  2. *if* $q \xrightarrow{!e} q$ *then we have* $q \xrightarrow{!e}_{\text{Restr}(C)} \overline{q}$ *and also* $\overline{q} \xrightarrow{!e}_{\text{Restr}(C)} \overline{q}$.

We notice that, according to point 2. of Definition 1, each state of a Control Automaton is the source of a transition. Furthermore, this property holds for Reactive Fiffo Automata and Restricted automata too. In this way, model checking of LTL is well-defined for these transitions system.

In the following, in order to simplify the presentation, *we identify states* q *and* $\overline{q}$ *of a Restricted Control Automaton.* Moreover, if $Q \times \Sigma^*$ is the set of states of a Reactive Fiffo Automaton, we will write proj_Q for the projection on control states: proj_Q is the morphism $proj_Q : (Q \times \Sigma^*)^* \rightarrow Q^*$ defined by $\text{proj}_Q(q, w) = q$.

The two following lemmas express close relations between Control Automata, Reactive Fiffo Automata and Restricted Control Automaton.

Lemma 3. *Let* R *be the Reactive Fiffo Automaton associated with a Control Automaton* $C = (Q, q_0, (\{!, ?\} \times \Sigma) \cup \{\tau\}, \rightarrow)$. *We have:*

$$\mathcal{L}^\omega(\text{Restr}(C)) \subseteq \text{proj}_Q(\mathcal{L}^\omega(R)) \subseteq \mathcal{L}^\omega(C)$$

Lemma 4. *Let $C = (Q, q_0, (\{!, ?\} \times \Sigma) \cup \{\tau\}, \rightarrow)$ be a Control Automaton. For every ω-sequence $\nu \in \mathcal{L}^\omega(C)$, there exists an ω-sequence $\nu' \in \mathcal{L}^\omega(\mathrm{Restr}(C))$ such that ν and ν' are equivalent modulo stuttering.*

We now present the main result of this section, which especially allow to check liveness and safety properties on RFAs.

Theorem 6. *Model-checking of* LTL\X *is decidable for Reactive Fiffo Automata and is* PSPACE*-complete.*

Proof. Let R be a RFA associated to a Control Automaton $C = (Q, q_0, (\{!, ?\} \times \Sigma) \cup \{\tau\}, \rightarrow)$. According to Lemma 3, to Lemma 4 and to Theorem 5, we obtain that C, R and $\mathrm{Restr}(C)$ satisfy the same LTL\X formulas. Now, model-checking of LTL\X is PSPACE-complete for Finite Automata [SC85], which concludes the proof. □

Corollary 3. *The Reccurent Control State Problem is decidable for Reactive Fiffo Automata.*

We finally show that the model-checking of LTL with atomic propositions on transitions is undecidable. Let us remark that this result is stronger than Teorem 4 *iii*).

Theorem 7. *Model-checking of* LTL *with atomic propositions on transitions is undecidable for Reactive Fiffo Automata.*

6 Conclusion

In this work, we were interested in the verification of Reactive Fiffo Automata, an new class of infinite systems that models Electre programs.

We have shown in this paper that Reactive Fiffo Automata have a recognizable reachability set and that it is effectively computable. This result allows to decide several problems involved in verification of RFAs, for instance the Reachability Problem, the Finite Reachability Set Problem and the Inclusion of Reachability Sets Problem.

This work is a first step towards the assesment of response times of the system for taking into account memorized occurrences of events. This will be of a great significance for real-life systems.

We have also analysed the relationships between Reactive Fiffo Automata and Finite Automata. We have proved that from a trace language point of view, inclusions between RFAs and Finite Automata are undecidable. But fortunately, we obtained that the linear temporal logic LTL without the temporal operator next (LTL\X) is decidable for Reactive Fiffo Automata. This result especially allows to check liveness and safety properties.

The decidability of fragments of CTL has already been investigated in [Sut97]. The decidability of LTL (with atomic propositions on states) and the decidability of CTL remain open problems.

References

[ABJ98] P. A. Abdulla, A. Bouajjani, and B. Jonsson. On-the-fly analysis of systems with unbounded, lossy fifo channels. In *Proc. of the* 10th *Conference on Computer-Aided Verification (CAV)*, 1998.

[AJ93] P. Adulla and B. Jonsson. Verifying programs with unreliable channels. In *Proc. of the 8th IEEE Symposium on Logic in Computer Science*, 1993.

[BBnRR98] F. Boniol, A. Burgueño, O. Roux, and V. Rusu. Étude d'un modèle hybride discret - continu pour la spécification de systèmes temps-réel embarqués. Rapport de contrat CERT/ONERA - IRCyN N^0 DERI 3703-33, February 1998.

[BG96] B. Boigelot and P. Godefroid. Symbolic verification of communication protocols with infinite state spaces using qdds. In *Proc. of the* 8th *Conference on Computer-Aided Verification (CAV)*, volume 1102, pages 1–12. LNCS, August 1996.

[BGWW97] B. Boigelot, P. Godefroid, B. Willems, and P. Wolper. The power of qdds. In *Proceedings of SAS'97*, September 1997.

[BH97] A. Bouajjani and P. Habermehl. Symbolic reachability analysis of FIFO-channel systems with nonregular sets of configurations. In *Proc. of the 24th International Colloquium on Automata, Languages, and Programming (ICALP)*, volume 1256, pages 560–570. LNCS, July 1997.

[BZ83] D. Brand and P. Zafiropulo. On communicating finite-state machines. *JACM*, 30(2):323–342, 1983.

[CF97] G. Cécé and A. Finkel. Programs with quasi-stable channels are effectively recognizable. In *Proc. of the* 9th *Conference on Computer-Aided Verification (CAV)*, volume 1254, pages 304–315. LNCS, June 1997.

[CFP96] G. Cécé, A. Finkel, and I. S. Purushothaman. Unreliable channels are easier to verify than perfect channels. *Information and Computation*, 124(1):20–31, 1996.

[CHP71] P. J. Courtois, F. Heymans, and D. L. Parnas. Concurrent control with "readers" and "writers". *Communications of the ACM*, 14(10):667–668, October 1971.

[CR95] F. Cassez and O. Roux. Compilation of the Electre reactive language into finite transition systems. *Theoretical Computer Science*, 146(1–2):109–143, July 1995.

[CR97] F. Cassez and O. Roux. Modelling and verifying reactive systems with event memorisation. Revised version submitted, 1997.

[Eme90] E. A. Emerson. *Handbook of Theoretical Computer Science*, chapter 16, pages 996–1072. Elsevier Science Publishers, 1990.

[FC87] A. Finkel and A. Choquet. Simulation of linear fifo nets by petri nets having a structured set of terminal markings. In *Proc. of the 8th European Workshop on Application and Theory of Petri Nets, Saragoza*, pages 95–112, 1987.

[FM97] A. Finkel and P. McKenzie. Verifying identical communicating processes is undecidable. *Theoretical Computer Science*, 174:217–230, 1997.

[JJ93] T. Jéron and C. Jard. Testing for unboundedness of fifo channels. *Theoretical Computer Science*, 113:93–117, 1993.

[Lam83] L. Lamport. What good is temporal logic ? In *Information Processing'83. Proc. IFIP 9th World Computer Congress*, pages 657–668. North-Holland, September 1983.

[MP92] Z. Manna and A. Pnueli. *The Temporal Logic of Reactive and Concurrent Systems: Specification.* Springer-Verlag, 1992.

[Pac87] J. K. Pachl. Protocol description and analysis based on a state transition model with channel expressions. In *Proc. of Protocol Specification, Testing and Verification, VII*, 1987.

[Que96] Y. M. Quemener. *Vérification de protocoles à espace d'états infini représentable par une grammaire de graphes.* PhD thesis, Université de Rennes 1 (FRANCE), 1996.

[SC85] A. P. Sistla and E. M. Clarke. The complexity of propositional linear temporal logics. *Journal of the ACM*, 32(3):733–749, 1985.

[Sut97] G. Sutre. Vérification de propriétés sur les automates à file réactifs produits par compilation de programmes Electre. Mémoire de DEA, Univ. Paris VII et Ecole Polytechnique, 1997.

Combining Methods for the Livelock Analysis of a Fault-Tolerant System

Bettina Buth[1], Jan Peleska[2], and Hui Shi[3]

[1] BISS, Bremen Institute for Safe Systems, bb@informatik.uni-bremen.de
[2] JP Software-Consulting, Bremen, jp@informatik.uni-bremen.de
[3] University of Bremen, shi@informatik.uni-bremen.de

Abstract. This article presents experiences gained from the verification of communication properties of a large-scale real-world embedded system by means of formal methods. This industrial verification project was performed for a fault-tolerant system designed and implemented by Daimler-Benz Aerospace for the International Space Station ISS and focused essentially on deadlock and livelock analysis. The approach is based on CSP specifications and the model-checking tool FDR. The tasks are split into manageable subtasks by applying abstraction techniques for restricting the specifications to the essential communication behavior, modularization according to the process structure, and a set of generic theories developed for the application.

1 Introduction

The acceptance of Formal Methods in industries essentially depends on their scalability, i.e. their applicability in large scale realistic industrial projects. An important aspect is the availability of suitable tools, but from our experience this is but one aspect. The diverse nature of system components and the techniques used in the different steps of the development process require the use of a combination of methods for the development as well as for the analysis of these components. In this paper we report experiences in using a combination of methods for the analysis of a large software system, namely the fault-tolerant data management system for the International Space Station (ISS). It was necessary to use these methods in order to break down the respective tasks into manageable sub-tasks.

The project started in 1995 when JP Software-Consulting in collaboration with the Bremen Institute for Safe System (BISS) were contracted by Daimler-Benz Aerospace (DASA) to perform an analysis of a fault-tolerant data management system for the ISS. Up to now various aspects of the system correctness have been investigated:

- freedom of deadlock (see for example Buth et al. [3]),
- freedom of livelock,
- correct implementation of voting algorithms,
- correct implementation of the Byzantine Agreement Protocol,

A.M. Haeberer (Ed.): AMAST'98, LNCS 1548, pp. 124–139, 1998.

- performance properties depending on clock rates,
- Hardware-In-The-Loop tests for the overall system.

Different approaches were employed for these analysis tasks. These comprise abstraction of the occam code to CSP pprocesses and reasoning on these abstractions, Hoare-style program verification for sequential code, and the use of stochastic Petri nets for the throughput analysis.

Whereas the results presented in Buth et.al. [3] relate to the deadlock analysis, this article focuses on the techniques used for the livelock analysis. Rather than presenting in-depth theoretical results our goal is to describe how we overcame the essential problems of size and complexity of a realistic large-scale system by employing a combination of methods for the analysis. Due to the confidential nature of the material unfortunately it is not possible in this framework to make available details of the original code or our specifications (these are available only in the internal reports [4,5]).

In the following section a brief overview of the fault tolerant computer is given. Section 3 provides an overview of the methods used for the analysis. In Section 4 our experiences are described and put into relation to the deadlock analysis. The conclusion outlines future work, especially with respect to tool support for similar analysis projects.

2 The Fault Management System

The software to be analyzed is part of a fault tolerant computer to be used in the International Space Station (ISS) to control space station assembly, reboost operations for flight control, and data management for experiments carried out in the space station.

The overall architecture consists of up to four communicating lanes, each providing services for the applications. Each of these lanes is structured into an application services layer (ASS), a fault management layer (FML), and the avionics interface (AVI). The ASS resides on the application layer board and contains table driven services for the application software and the operating system. The AVI is in charge of the MIL Bus protocol handling according to predefined timing slot allocations. These are defined in an input/output table. The function of the FML is twofold: First, it provides the interface between the ASS and AVI of one lane, transferring messages from AVI to ASS and vice versa. Second, it performs the data transfer between lanes thus allowing communication between the fault management layers of all lanes. This communication is the basis for error detection, error correction, lane isolation (in the case of an unrecoverable error), and lane reintegration. In each lane, the application layer plus ASS runs on a customized Matra board using a SPARC CPU. Both FML and AVI reside on separate transputer boards. The lanes communicate only at FML level using the transputer links. Each FML uses up to three links for communication with the other lanes, and one link (link 0) for communication with the AVI. Data transfer with the ASS is performed using a VME interface. See Figure 1 for the architecture of a full four-lane system.

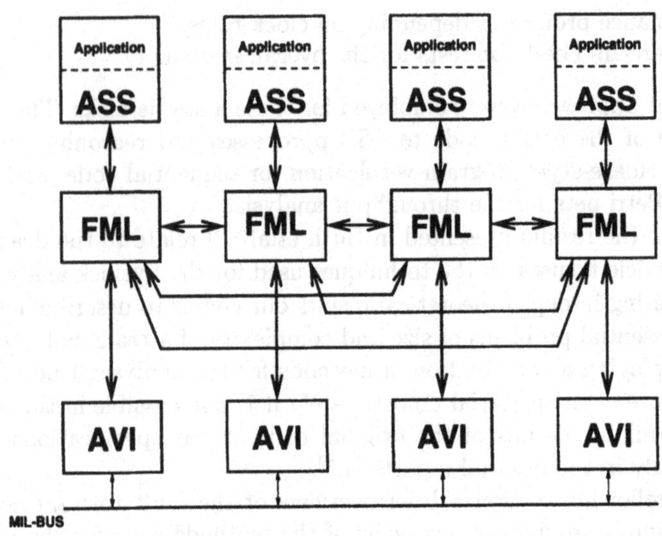

Fig. 1. FTC architecture

3 Livelock Analysis with CSP

The general idea for the analysis of communication properties of occam programs as proposed here is to exploit the fact that occam channel communication can easily be modelled in CSP. For CSP a wide range of theories and suitable tools like theorem provers and model checkers are available. The original idea for the livelock analysis performed in this project is to use model checking and the tool FDR [10]. After manually abstracting the occam programs to CSP processes the systems turned out to be too large for a direct approach using FDR. Thus it is necessary to decompose the task and use other techniques for combining these results to obtain an overall result for the full system. Some of these techniques are tool supported, others currently are only performed unassisted.

Compositionality theories allow to exploit general results from CSP theory to reduce the overall task of checking system properties to checks of properties of components without additional justifications. Similarly the use of *generic theories* in this framework: by proving that CSP processes are refinements of instances of such generic theories certain properties can be derived for the concrete process which are used for the proof of livelock freedom. The additional obligation, namely the proof that the original component is indeed a refinement of the instance of the generic theory, can be discharged using FDR. In cases where neither of these approaches is directly applicable it is necessary to further decompose the system and use property oriented techniques as *liveness induction* or *dependency analysis* to combine the results.

In high level concurrent programming languages, such as occam it is conventional for communication channels between two processes to be concealed from the environment. This can potentially cause a form of divergence known as *livelock*

The objective of our analysis is to investigate occurrence of internal divergence only. A CSP (or occam) system X is called *livelock free* with respect to interface channels $c_1, c_2, \ldots, c_n$, if the system will never engage in an unbounded sequence of (internal) communications without interleaved (visible) communications on the interface $c_1, c_2, \ldots, c_n$. This property is as important as the *absence of deadlocks* since it ensure the interaction of the system with its environment.

For real-time applications, periodic internal events such as time-scheduled interrupts cannot cause divergence as long as their period is long enough to allow for other activities in between. These events may cause cycles of internal communications which are intended and not to be considered as livelock, because their occurrence is guarded by time intervals where other communication events are possible. To reflect this properly in an untimed CSP model, we regard these channels as part of the interface for the purpose of livelock analysis.

The verification of livelock freedom of AVI and FML is discussed in this document. We wish to justify that they are livelock free with respect to their respective external interfaces. In the following we are going to present the techniques applied for this purpose. The first subsections describe methods used for both AVI and FML, whereas Subsection 3.4 and Subsection 3.5 refer to different techniques used in FML and AVI respectively for deriving results for larger sets of processes that can not be obtained using model-checking.

3.1 Abstract Interpretation

For the analysis of livelock freedom it is unnecessary to inspect every detail of the occam code, since only a subset of the programmed statements influences the communication behavior. It is therefore possible to generate a CSP specification which represents an abstract version of the original occam process P showing only the amount of detail which is relevant for communication behavior. Such a CSP specification $\mathcal{A}(P)$ is called a *livelock-valid abstract interpretation* of the corresponding occam process P, if

> *Whenever P runs into a livelock situation this implies that $\mathcal{A}(P)$ may run into a livelock situation, too.*

If a livelock-valid abstract interpretation $\mathcal{A}(P)$ is available, we can analyse $\mathcal{A}(P)$ instead of P: If $\mathcal{A}(P)$ is free of livelocks, the same must hold for P.

The basic approach to construct valid abstract interpretations uses four steps in the translation from occam to CSP:

1. Every sequential algorithm whose results do not influence communication behavior is deleted.
2. Each occam channel protocol is reduced to a simpler one such that its values influence the communication behavior in a different way.
3. Every occam IF-construct `IF condition THEN P ELSE Q` may be replaced by the internal choice operator of CSP yielding $P \sqcap Q$.

4. If valid abstract interpretations $\mathcal{A}(P), \mathcal{A}(Q)$ for two processes P and Q are available and these interpretations use the same protocol on their communication interface I, then $\mathcal{A}(P) \underset{I}{\parallel} \mathcal{A}(Q)$ is a valid abstract interpretation of P and Q operating in parallel. Using this fact, larger abstract interpretations can be built from existing ones.

If a livelock-valid abstraction of an occam process is not livelock free, two cases are possible: Either the occam process contains a livelock or we have introduced too high a degree of nondeterminism in its CSP abstraction, for example, through using the first three techniques. Since we intuitively assume that the occam process should turn out to be livelock free, we try to find another valid abstraction which is more deterministic than the one we have constructed. The correctness of the abstraction methods can be shown by induction on occam processes. A more detailed description of our abstraction methods is given in [3].

3.2 Model-Checking and Refinement Properties

The specification language CSP (Communicating Sequential Processes) is associated with a formal method allowing to verify properties of parallel systems. (See Hoare [11] and the recent book by Roscoe [17] for more details). CSP processes proceed by engaging in communications. Processes may be composed by operators which require synchronization on some communications. This, rather than assignments to shared state variables, is the fundamental means of interaction between agents. The theory of CSP has classically been based on mathematical models remote from the language itself. These models have been based on observable behaviors of processes such as *traces*, *failures* and *divergiences*. The semantic of a CSP process can be given in three models:

- In the *trace model* a process is represented by the set of finite sequences of communications it can perform.
- In the *failures model* a process is represented by its set of traces as above and also by its failures – a set of communications it can refuse after a sequence of communications.
- The *failures-divergence model* extends the failures model with the divergences of a process – the traces during or after which the process can perform an infinite sequence of consecutive internal actions.

Every CSP specification consisting of finite-state processes with finite-value channels can be translated into a finite *transition graph* representation. This graph contains all the semantic information of the original CSP specification. As a consequence, every property of the specification – as, for example, livelock freedom – can be verified by exhaustive analysis of the transition graph. Moreover, such an analysis can be mechanised. The FDR tool provides this mechanisation and has been used for all model checking results about the abstract interpretations described in this document.

Two CSP processes P_1 and P_2 can be compared in the failures-divergence model with regard to the following property:

P_2 can perform only communications P_1 may also perform, but P_2 refuses less and diverges less than P_1.

If this is the case, P_2 is called a *failures-divergence refinement* of P_1. This is denoted by $P_1 \sqsubseteq_{FD} P_2$. Failures-divergence refinement preserves deadlock freedom and livelock freedom and restrictions on possible communication sequences. If P_1 is free of livelocks and observes a safety condition about communication sequences, then every process refining P_1 in the failures-divergence model has these properties, too. If P_1 and P_2 are finite-state CSP specifications the refinement relation $P_1 \sqsubseteq_{FD} P_2$ can be verified by model checking.

A typical application of this refinement relation in our context is the fact that

given a valid abstract interpretation $\mathcal{A}(P)$ of P and a process Q such that $Q \sqsubseteq_{FD} \mathcal{A}(P)$, then Q is also a valid abstract interpretation of P.

The second refinement relation applied to prove livelock freedom is *trace refinement*.

Process P_2 is a trace refinement of P_1 (denoted by $P_1 \sqsubseteq_T P_2$), if every trace of communications that P_2 may perform is also a trace of P_1.

Failures-divergence refinement implies trace refinement, but for certain situations it is easier to deal with the latter, because the investigation of trace refinement properties does not require the analysis of refusal sets which is necessary for failures-divergence refinement checking. Formally speaking, the livelock freedom of a process X with internal communications L and external communications C means that there exists a *bounding function* $f : \mathbb{N}_0 \rightarrow \mathbb{N}_0$, such that

$$\#(s \upharpoonright L) \leq f(\#(s \upharpoonright C))$$

where $\#(s \upharpoonright L)$ is the number of local communications in s and the length of the trace s restricted to communications in C is denoted by $\#(s \upharpoonright C)$. An equivalent way of expressing the property of livelock freedom is to say that $X \backslash L$ (treating L as internal channels of X) must not diverge.

Our main application of trace refinement is given by the following theorem which obviously holds because the refining process only runs through traces which are possible for the abstract process as well:

Theorem 1. *Let P_1 be a process over alphabet $\{|c_1, c_2, \ldots, c_n|\} \cup L$ with bounding function $f : \mathbb{N}_0 \rightarrow \mathbb{N}_0$ such that every trace s of P_1 satisfies*

$$\#(s \upharpoonright L) \leq f(\#(s \upharpoonright \{|c_1, c_2, \ldots, c_n|\}))$$

Then f is also a bounding function for every trace of P_2 which refines process P_1.

□

As a consequence of this theorem, we can apply the following verification strategy to prove that a process P_2 is free of livelocks:

- Construct a "simple" process P_1 for which the existence of a bounding function is obvious.
- Verify by model checking that $P_1 \sqsubseteq_T P_2$ holds. This implies livelock freedom of P_2.

In general it is helpful to take simple processes from generic theories with known properties as the goal of this further abstraction. Examples for such generic theories are pipes and buffers, or the more specialized processes *MUXOUT* below.

We are going to take the process *Input Distributor*, which is one of the modules of FML, as an example to show the application of Theorem 1. The structure[1] of *Input Distributor* is given by Figure 2.

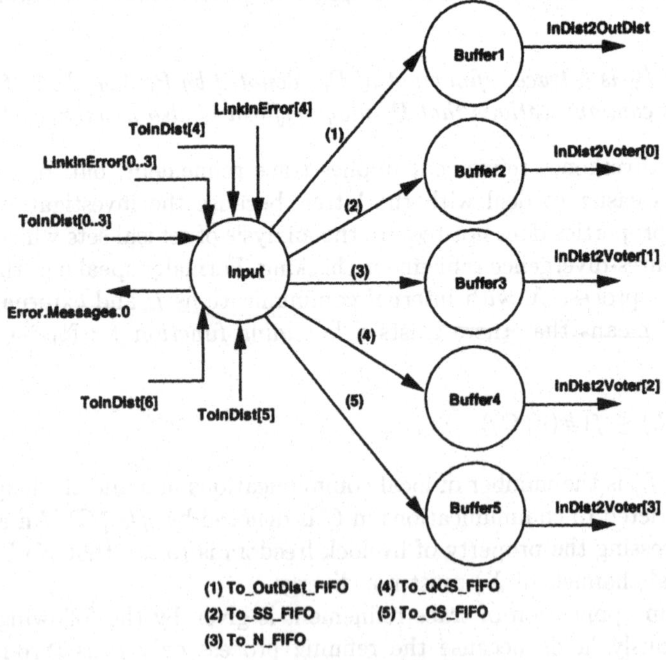

Fig. 2. Input Distributor

At first we give a general CSP specification

MUXOUT(*max*, *max*, *IN*, *OUT*)

[1] For presenting graphs of processes and their communication we use data flow diagrams as used in the original FTC documentation. See for example De Marco [9] for a detailed description.

of processes which will always be ready for inputs on channels IN after at most max outputs on channels OUT.

$$
\begin{aligned}
&MUXOUT(n, max, IN, OUT) = \\
&\qquad if\ (n == 0) \\
&\qquad then\ MUXIN(max, IN, OUT) \\
&\qquad else\ (MUXIN(max, IN, OUT) \\
&\qquad\qquad \sqcap \\
&\qquad\qquad (\ \sqcap\ y : OUT \bullet (y \rightarrow MUXOUT(n-1, max, IN, OUT))))
\end{aligned}
$$

$$
\begin{aligned}
&MUXIN(max, IN, OUT) = \\
&\qquad \sqcap\ x : IN \bullet (x \rightarrow MUXOUT(max, max, IN, OUT))
\end{aligned}
$$

A bounding function of the CSP specification $MUXOUT(max, max, IN, OUT)$ is $f(x) = max * x$. Now we can use the FDR tool for verifying the following refinement relation:

$$MUXOUT(4, 4, IN, OUT) \mid\mid\mid CHAOS(M) \sqsubseteq_T \mathcal{A}(Input\ Distributor)$$

where

$$
\begin{aligned}
IN &= \{\mid ToInDist \mid\} \\
OUT &= \{\mid To\_SS\_FIFO, To\_N\_FIFO, To\_QCS\_FIFO, \\
&\quad\ To\_CS\_FIFO, To\_OutDist\_FIFO \mid\} \\
M &= \{\mid Error\_Messages.0, LinkInError \mid\} \\
CHAOS(A) &= \sqcap\ a : A \bullet (a \rightarrow CHAOS(A)) \sqcap \mathrm{STOP}
\end{aligned}
$$

and M is the set of those channels the communication behavior of which is irrelevant ($CHAOS(M)$). Process $STOP$ refuse to do anything. So, the outputs on channels To_SS_FIFO, To_N_FIFO, To_QCS_FIFO, To_CS_FIFO and $To\_OutDist\_FIFO$ are bounded by the inputs on channels $ToInDist[0..6]$.

3.3 Compositionality and Generic Theories

A real-world system such as FML usually consists of a number of processes which communicate and each of them may be comprised of several subprocesses. An example from FML for such a process is the process *Input Distributor*. How to compose the analysis results for each of these subprocesses and processes for the livelock verification of the overall system is the most difficult activity during the livelock analysis. In the following we will present some of the techniques and theoretical results we applied.

The first result follows from the simple theory of non-divergent pipes described in [11, p. 156]:

Theorem 2. *Suppose* CSP *process P is structured as*

$$P = (X \underset{LEFT}{\|} B \underset{RIGHT}{\|} Y)$$

where X is inputting from a set of channels IN and placing data on channels $LEFT$ into a buffer B and Y is inputting from B on $RIGHT$ and outputting on channels OUT without feedback to IN. Then, if

1. *the buffer B has bounded capacity,*
2. *the number of buffer inputs on LEFT is bounded by the number of X-inputs on IN,*
3. *the buffer outputs on RIGHT are destructive (that is, each entry is removed from the buffer by reading it on a RIGHT-channel),*

the number of B-outputs on RIGHT is bounded by the number of X-inputs on IN. If in addition

4. *the number of Y-outputs on OUT is bounded by the number of entries in the buffer,*

the number of Y-outputs on OUT is also bounded by the X-inputs on IN.
□

Using the first result of the theorem we can show that the outputs on channels *InDist2Voter* and *InDist2OutDist* in the process *Input Distributor* are bounded by the inputs on channels *ToInDist*[0..6], since all buffers involved are bounded buffers.

CSP provides several operators, such as parallel operator ($\|$) and interleaving operator ($|||$), which allow to construct new processes out of existing ones. A crucial property of CSP is the fact that refinement is preserved under compositions involving these operators:

If $P_i \sqsubseteq_{FD} Q_i$ *for* $i : 0..n$ *and* ω *is an* n*-ary operator,*
then $\omega(P_0, \ldots, P_n) \sqsubseteq_{FD} \omega(Q_0, \ldots, Q_n)$ *holds.*

If livelock freedom of $\omega(Q_0, \ldots, Q_n)$ can not be established because of the problem size, find simpler processes $P_0, \ldots, P_n$ such that $P_i \sqsubseteq Q_i$ for each $i \in \{1, \ldots, n\}$ and $\omega(P_0, \ldots, P_n)$ is livelock free. If this property can be established, compositionality combined with refinement imply the livelock freedom of $\omega(Q_0, \ldots, Q_n)$ as well.

3.4 Liveness Induction

Another result which we applied during the FML analysis was developed by Roscoe (detailed in [7]). It allows to derive livelock freedom of a collection of *triple-disjoint* communicating processes:

Definition 1 (Triple-Disjoint Process System). *A system of processes is called triple-disjoint, if no communication requires the participation of more than two processes.*

Theorem 3. *Suppose* $V = \langle P_1, \ldots, P_n \rangle$ *is a triple-disjoint system of non-divergent processes such that for every* P_i *in* V

$$P_i \backslash (\bigcup_{j<i} (\alpha(P_i) \cap \alpha(P_j)))$$

is divergence-free. Then

$$(P_1 \parallel \ldots \parallel P_n)\backslash H \quad \textit{where } H = (\bigcup_{j \neq i}(\alpha(P_i) \cap \alpha(P_j)))$$

is also divergence-free.
□

In the above theorem $\alpha(P_i)$ $(1 \leq i \leq n)$ denotes the alphabet of P_i. Informally speaking, this theorem states that if no process P_i in the system can perform an infinite sequence of communications with P_j (for all $j \leq i$), then the system is livelock free with respect to its incoming and outgoing channels.

Theorem 3 is useful in many cases, but finding an order on the processes of a parallel system may be very ineffective. An FML lane consists of 8 processes which have a complicated communication behavior. Consequently, it is very difficult to arrange them in a proper order for applying the theorem. One could try a large number of combinations (in our application at worst 8!); there are 7 proof obligations for each combination.

For this reason we have developed a new technique, *liveness induction*, which can be used to verify livelock freedom at the channel level. To verify that process P is livelock free with respect to channels of interface I, the induction is based on a set $\mathcal{L}$ of channels, inductively defined by

1. The interface I is contained in $\mathcal{L}$.
2. If $\{\!|c_1, \ldots, c_n|\!\}$ is contained in $\mathcal{L}$ and there exist a channel $d \notin I$ and a bounding function $f : \mathbb{N}_0 \to \mathbb{N}_0$, such that $\#(s \upharpoonright \{\!|d|\!\}) \leq f(\#(s \upharpoonright \{\!|c_1, \ldots, c_n|\!\}))$ then d is contained in $\mathcal{L}$, too.
3. No other channel is in $\mathcal{L}$.

Liveness induction is now applied according to the rules of the following theorem (see Peleska et.al. [14]):

Theorem 4. *Let $P_1, \ldots, P_n$ form a triple-disjoint system of concurrent processes, such that each P_i is livelock free with respect to its local interface $I_i =_{df} \alpha(P_i) - L_i$, where L_i is the set of its internal channels. Then $(P_1 \parallel \ldots \parallel P_n)$ is non-divergent with respect to global interface $I \subseteq \bigcup_{i=1}^{n} I_i$, if $\mathcal{L} =_{df} \bigcup_{i=1}^{n} I_i$ can be inductively constructed according to the rules given above.*
□

This theorem can be proved by induction on the number of processes involved based on the compositionality of bounding functions. If communications on channel c_2 is bounded by communications on c_1 with bounding function f_1, and communications on c_3 is bounded by communications on c_2 with bounding function f_2, then communications on c_3 is bounded by communications on c_1 with $f_1 \circ f_2$ as a bounding function.

Liveness induction was used for the proof of FML in the following way: After showing the livelock freedom of each top-level process of FML we treat all the processes except the process *Link Interface* as a single process, say *REST*. Let $SYS = \langle REST, \mathcal{A}(Link\ Interface)\rangle$, it is a triple-disjoint system. The following assertion can be proved using FDR:

$$\mathcal{A}(Link\ Interface)\backslash(\alpha(\mathcal{A}(Link\ Interface)) \cap \alpha(REST)) \text{ is free of livelock}$$

If we can show that *REST* is livelock free, then the whole system is livelock free, too, using Theorem 3.

The next verification target is to show the livelock freedom of *REST*, which is reached by liveness induction. At beginning, $\mathcal{L}$ contains all channels in the interface of *REST*, such as $ToInDist[0..6]$. Then, all channels bounded by the channels in $\mathcal{L}$ are contained in $\mathcal{L}$. For example, communications on channels $InDist2Voter[0..3]$ and $InDist2OutDist$ are bounded by the communications on $ToInDist$ in *Input Distributor* (proved in 3.2), so $InDist2Voter[0..3]$ and $InDist2OutDist$ are contained in $\mathcal{L}$, too. In this way, we have shown all channels involved in the communications of $REST$ are containted in $\mathcal{L}$.

Note that liveness induction is a forward analysis for livelock freedom, and contrasts with the approach described in the next section.

3.5 Dependency Analysis

Whereas the software structure of FML allowed to use the repeated abstraction by using generic processes $MUXOUT$and livelock induction, the complex communication behavior of AVI required a different approach. It turned out that there are three obstacles:

- even some of the main processes are too complex to be verified using model-checking directly;
- it is not possible to decompose the task into subtasks based on the compositional theory since the communication behavior turns out to be too complex;
- for the same reason no further abstraction is possible.

This combination prevents the applicability of liveness induction as presented in the previous section. As an alternative a special form of dependency analysis was used for investigating cycles in the communication graph of processes which could not be checked using FDR directly due to the size of the state space. This analysis is a backwards analysis that determines all possible chains of communication events that may lead to a specific communication. The analysis technique is derived from tree analysis techniques as the *Fault Tree Analysis* (for the use of Fault Tree Analysis in software applications see for example Lyu [13])

Consider a set of processes $P_1, \ldots, P_n$ which have been proved to be free of livelock individually. Let $SYS = P_1 \parallel \ldots \parallel P_n$ be a net of these processes with internal synchronization communications $\mathcal{L}$. The *communication graph* of SYS displays the channels and the direction of communication between the processes $P_i, i \in \{1, \ldots, n\}$. If this graph for SYS contains cycles it is not necessarily the case that these cycles can actually occur. If this is the case for all cycles SYS is livelock free. For the analysis of AVI there are two reasons that a cycle can not actually occur:

- the CSP processes involved do not produce traces that correspond to the cycle; this is for example the case if one of the communications does not depend on its predecessor in the cycle,

- the cycle is possible in the CSP process, but additional information derived from the occam code allow to exclude the occurrence of this situation.

For the analysis of AVI most of the cycles could be invalidated due to the first cause. The approach is illustrated in the following using a small example from AVI.

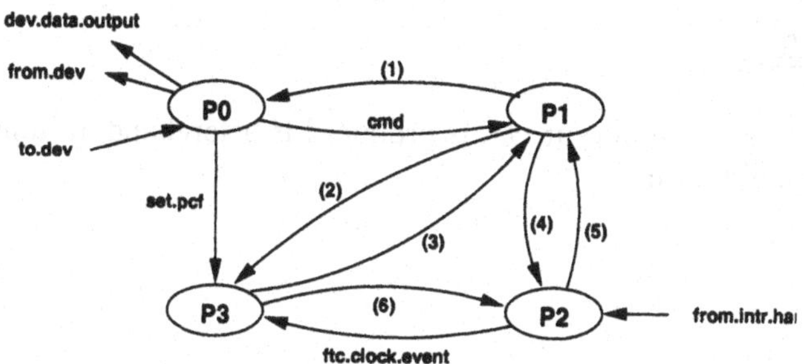

(1) c.stat, real.reset.01, real.reset.02

(2) autosync.on, reset.autoframing

(3) autoframing.intr.error, EOM.ctrl, start.timeout

(4) reset.intr.handler, kill.pending.EOM.intr

(5) int.request.err, spurious.intr, reset.timeout, eos.channel

(6) EOM.intr.req, FTC.clock.intr.req.chan, kill.pending.ftc.clock.intr

Fig. 3. BC communication graph

The structure of module BC (broadcast unit) of AVI is presented in the communication graph of Figure 3. The individual processes Pi, $i \in \{0, \ldots, 3\}$ can easily be checked for livelock freedom using FDR, i.e. it is guaranteed that no non-terminating internal communications can occur within the Pi. Each of these processes is itself a system of several subprocesses, which are denoted $P_{i1}, \ldots, P_{in_i}$. In order to ensure that the whole BC module is free of livelock it suffices to prove that none of the cycles in the graph can actually occur[2].

Regard for example the cycle involving P0 and P1. A livelock situation can only occur if output cmd directly depends on the events c_stat, real.reset.01, or real.reset.02 and vice versa. But this is not the case as code inspection of the CSP abstraction shows: the output of cmd only depends on the external input to_dev. In order to prove this it is necessary to trace the preceding communications leading to the output through the internal subprocesses of P0.

This type of analysis can be performed systematically for all cycles. In general it is only necessary to find one edge in a cycle that does not depend on its

[2] This does not guarantee a fair treatment of the inputs; such a property has to be proved using different means.

predecessors. This breaks a cycle and makes it uncritical with respect to livelock. The analysis itself is documented using techniques gleaned from fault-tree analysis. For each relevant communication it is traced by which other communications, resp. sequence of communications it can be triggered. The following notations are used:

For channels $ch_1, \ldots ch_n$ and process P

$$ch_1 \overset{\boxed{P}}{\longleftarrow} ch_2$$

denotes that in P an output $ch1$ is produced if beforehand an input on channel $ch2$ did occur.

ch_1

$\quad \overset{\boxed{P}}{\longleftarrow} ch_2$

$\quad \ldots$

$\quad \overset{\boxed{P}}{\longleftarrow} ch_n$

means that an output on channel ch_1 is only possible after any of the inputs $ch_i, i \in \{2, \ldots, n\}$.

As an example the dependencies of event `kill_pending_EOM_intr` from `P1` to `P2` are presented here:

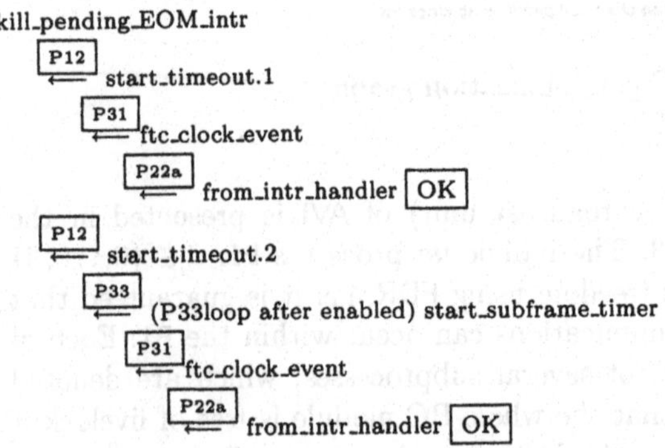

Note that it is necessary to trace the communications through several subprocesses not only of process `P1` and `P2` but also of process `P3`. Furthermore, it is essential that the dependency analysis is complete, i.e. that every sequence of communication leading to the event under consideration are investigated. This means that in the example above `start_timeout.1` and `start_timeout.2` are the inputs to process `P1` which can cause an output of `kill_pending_EOM_intr`.

In the case of BC and other units of the AVI this kind of analysis was used to ensure that most of the cycles in the communication graphs do not occur and to identify critical cycle situations as well. For AVI all the latter situations could be excluded due to further reasoning based on the original code.

4 Livelock Analysis - Some Experiences

Starting point for the livelock analysis for both AVI and FML were the abstractions that had been used in the deadlock analysis (see Buth et.al. [4]). Due to the hierarchical architecture of both components the freedom of livelock for each of the main processes was tried to be established first. During this phase some adjustments in the abstractions were necessary to eliminate non-determinism and formalize the new proof obligations. One example is the explicit introduction of timer processes in order to avoid divergence through timer dependent events. These did not pose any problems for the deadlock analysis.

During this first step it became soon obvious that for the subprocesses of both FML and AVI it was not feasible to use model-checking directly, even in cases where it was possible for the deadlock analysis. On the one hand this is due to the changes in the abstraction which enlarge the state space, on the other hand the problem arises since the states themselves are larger. The reason for this is that livelock analysis uses the *failure-divergence* model of CSP, while the deadlock analysis could be performed within the *failures* model. The internal representation of the states has to contain the additional information about the divergence sets and thus is larger. In some cases it was thus not possible to derive even the results for the main processes of FML and AVI. Additionally it was necessary to find a suitable approach of combining the results from the livelock analysis of the main processes to obtain a result for the two units as a whole.

In general two different approaches are possible in this situation:

- further abstraction and exploitation of the theorems of preservation of the results under refinement,
- further decomposition of components, separate analysis for each basic unit, and derivation of properties for the combined units.

While the first proved to be a suitable way of dealing with the main processes of FML the complex communication behavior of AVI made it necessary to pursue the second approach. In both cases it was necessary to employ suitable means for combining the results for the overall unit.

5 Conclusion

The main obstacle for the analysis of livelock freedom or other properties of realistic industrial systems is their size and complexity. Without suitable approaches for splitting the task into manageable subtasks the attempt of using formal methods and related tools in this area is futile. For our project it proved to be necessary to use a combination of techniques in order to decompose the goal appropriately. These techniques comprise

- abstraction of occam code to CSP,
- use of the compositional proof theory of CSP,

- abstraction to instances of processes from generic theories,
- model checking (for sufficiently small subprocesses),
- liveness induction
- dependency analysis

While there is tool support for model checking, namely the system FDR, the other techniques had to be applied manually. Future work will comprise the search and if necessary development of tools supporting these tasks.

One useful tool would be one that interactively guides the abstraction process, not only for occam programs but for other compatible languages as well. In order to gain more confidence in the abstraction process, it will be necessary to theoretically justify the preservance of specification properties by the abstraction steps. The idea is to classify abstractions with respect to livelock, deadlock, and general safety properties. This will allow to reach a basis for the other methods in a systematic and even tool supported way.

Another aspect of abstraction is the use of generic theories. Here it will be necessary to look for more generic patterns and their specific properties. Some such patterns like buffer, systolic arrays, and multiplexer could be found in the software of our project, others will be found during other case studies. Again it will be useful to classify these theories with respect to their specific properties in order to develop heuristics for their application in the context of abstractions. Furthermore, it will be necessary to prove the properties for the generic theories in a formal way. We would like to employ tools like HOL, Isabelle, or PVS for this task. Some work in this direction can for example be found in Buth et al. [2].

Additional abstraction methods for abstracting from irrelevant details can be used for certain verification obligations that are not covered by refinement relations. During the verification of the Byzantine Agreement Protocol implementation we have developed several such abstraction methods. For example, *data abstraction*, which abstracts a process from concrete data information of communications, is proved to be useful for protocol verifications, where some of the properties of message distributions can be verified without the data information of messages. For different applications different abstraction methods have been or will be developed, which together with compositional theories and generic theories will form a method framework for our future activities in this area.

Tool support is also feasible for the dependency analysis. A rough idea is to use an approach with three phases:

- in the first phase the graph is build from the abstraction, i.e. the CSP processes;
- in the second phase all cycles are marked;
- in the third phase the cycles are investigated with respect to the dependencies of their constituent edges.

This kind of backwards cycle analysis could be complemented by a forward analysis for determining the bounds of the communication load of the cycles.

This could help or replace the livelock induction process. Tools for these tasks might be found with groups working with graph transformation systems or in the area of graph theory in general.

References

1. Brock, N.A., Jackson, D.M.: Formal Verification of a Fault Tolerant Computer. In *Proceedings of 1992 Digital Avionics Systems Conference.* (1992)
2. Buth, B., Cardell-Oliver, R., Peleska, J.: Combining tools for the verification of fault-tolerant systems. In Berghammer, R., Buth, B., Peleska, J. (eds.), *Tools for Software Development and Verification*, volume 1 of *BISS Monographs*. Shaker-Verlag, 1998.
3. Buth, B., Kouvaras, M., Peleska, J., Shi, H.: Deadlock analysis for a fault-tolerant system. In Johnson, M. (ed.), *Algebraic Methodology and Software Technology. Proceedings of the AMAST'97*, number 1349 in LNCS, pages 60–75. Springer, December 1997.
4. Buth, B., Peleska, J.: *Daimler-Benz Aerospace – Project DMS-R, FTC Development – Verification of Avionics Interface AVI.* Technical Report, JP Software-Consulting, (1996).
5. Buth, B., Peleska, J., Shi, H.: *Daimler-Benz Aerospace – Project DMS-R, FTC Development – Fault Management Layer (FML): Verification of Deadlock Freedom.* Technical Report, JP Software-Consulting, (1996).
6. Daimler-Benz Aerospace: *DMS-R FTC Detailed Design Document Volume 3 (FML Software)*
7. Dathi, N.: *Deadlock and Deaklock-Freedom*, Oxford University, D. Phil Thesis,1990.
8. Davies, J.: *Specification and Proof in Real-Time CSP.* Cambridge University Press (1993).
9. De Marco, T.: *Structured Analysis and System Specification* Englewood Cliffs, N.J., Prentice Hall 1979.
10. Formal Systemes: *Failures Divergence Refinement FDR2* Preliminary Manual. Formal Systems (Europe) Lts (1995).
11. Hoare, C.A.R.: *Communicating Sequential Processes.* Prentice-Hall Internationaal (1985).
12. Lamport, L., Shostak, R., Pease, M.: *The Byzantine Generals Problem*, In: ACM Transactions on Programming Languages and Systems, Vol.4, Nr. 3, (1982)
13. Lyu, M. R. (ed.): *Handbook of Software Reliability Engineering*, IEEE Computer Society Press, Computing McGraw-Hill
14. Peleska, J., Shi, H.: *Daimler-Benz Aerospace – Project DMS-R, FTC Development – Fault Management Layer (FML): Verification of Livelock Freedom.* Technical Report, JP Software-Consulting, (1998).
15. Roscoe, A. W.: *Model-Checking CSP.* In: A Classical Mind, Eassys in Honour of C.A.R. Hoare. Prentice-Hall Internationaal (1994).
16. Roscoe, A. W.: *CSP and determinism in security modelling.* In: IEEE Symposium of Security and Privacy, (1995).
17. Roscoe, A. W.: *The Theory and Practice of Concurrency* Prentice-Hall International (1997).

Separating Sets by Modal Formulas

Bernhard Heinemann

Fachbereich Informatik, FernUniversität Hagen
D–58084 Hagen, Germany
phone: ++49-2331-987-2714
fax: ++49-2331-987-319
bernhard.heinemann@fernuni-hagen.de

Abstract. Subsequently, we introduce a reasoning formalism which in particular allows to express that certain sets in a system of subsets of a given set are disjoint. The main purpose of considering such a family of subsets is to be able to investigate how knowledge grows as subsets shrink in the course of time. We actually introduce a *trimodal logic*: we have a system containing operators for knowledge and time, of which the latter corresponds to the effort of measurement and reminds of the nexttime operator of temporal logic; an operator separating sets is added then. So–called *subset tree models* appear as the relevant semantical structures. We present an axiomatization of the set of valid formulas encompassing the three operators and their interaction. Afterwards the completeness of the given axiomatization is proved. We also give arguments showing that the logic is decidable.

1 Introduction

In recent years a distinguished extension of propositional modal logic received some attention by people who are interested in the logical foundations of AI. The main concern of this logic is *reasoning about knowledge*. It was presented in the paper [15] for the first time and was studied in detail in [3]. We briefly descibe its basic ideas.

The logical operators which are present in the formal system quantify over points in a set, X, and over sets in a system of subsets of X, $\mathcal{O}$, respectively. Given a so–called *subset frame* $\mathcal{S} = (X, \mathcal{O})$, the elements of $\mathcal{O}$ can be interpreted as sets of *knowledge* states. The logical language describes the change of these sets, which is caused by certain knowledge gaining procedures like improving the precision of one's measurement or, more generally, by spending computational effort.

On the other hand, the ability to formalize certain aspects of *topological* reasoning constitutes another field of application. In fact, the *shrinking* of a set within a system of sets reminds of that kind of approximation procedures which are modeled mathematically by topological means. So a part of topology is represented by the logic, and it can actually be determined *what* part (see [3]). Because of this relationship with topology the elements of $\mathcal{O}$ are called the *opens*; the logic is called *topological modal logic* conveniently. Quite recently certain systems of

A.M. Haeberer (Ed.): AMAST'98, LNCS 1548, pp. 140–153, 1998.

topological modal logic have also been used to express properties of *hybrid systems* formally [2].

As it was already indicated, the language contains a modal operator $\Box$ quantifying over sets. Actually, $\Box$ models *descent* in $\mathcal{O}$ (w.r.t. set inclusion). This reflects the fact that an increasing of the knowledge of an agent corresponds to a shrinking of the set of states she considers to be alternative. Thus reasoning about knowledge in this way leads to dealing with shrinking procedures as in topology. Consequently, $\Box$ comes out as an *effort operator* in both contexts.

The structure of $(\mathcal{O}, \subseteq)$ may be very complicated in general. So the effect of $\Box$ is hardly to grasp.[1] However, presupposing a *functional* interpretation of $\Box$, i.e., viewing $\Box$ as a *nextstep* operator, not just leads to a *temporal* interpretation of the subset space logic in a stronger sense, but forces $(\mathcal{O}, \subseteq)$ to be a *subset tree* in particular. A corresponding logical system of knowledge and (discrete) time was studied in [11]. More special properties of the tree structure on the set of opens have been examined, too, including finite height [9] and binary ramification [10]. The results obtained so far concern *completeness* of a respective axiomatization of the set of valid formulas, *decidability* of this set, and the determination of the *complexity* of the satisfiability problem.

It is very desirable to extend the framework indicated above in order to make the language more and more appropriate for the description of phenomena occurring in connection with the notion of knowledge or, in other words, for the specification of properties involving knowledge. Moreover, to capture different and more subtle aspects of topological reasoning formally is a challenging task as well. As to the first point of view, one would like to control, for instance, the decomposition of a set of knowledge states at a future time point by means of formulas. Concerning the second, one should be able to deal with topological *separation* adequately. To master *complementation* in topological modal logic is wanted, too, since it appears naturally in spaces where the opens are in fact *clopen* like the frequently used *Cantor space* of all infinite 0–1–sequences (equipped with the prefix topology) (see [10]).

In the present paper we add a modal operator expressing *separation* to the modal logic of subset trees.[2] Our aim is to give a complete axiomatization and a decision procedure of the set of validities in this case. To this end some basic notions and facts from modal logic are required, which are contained in the first sections of [1] and [8], respectively; concerning the logic of knowledge, [4] is used as a standard reference book.

The paper is organized as follows: In Section 2 we present the syntax and the semantics of the underlying logical language of separation, LS, which is based upon the above mentioned system of knowledge and time. Thus we define a trimodal language containing a *knowledge operator*, K, a *nextstep operator*, $\bigcirc$, and

[1] Nevertheless, for some very interesting special classes of subset frames the corresponding modal logic could be determined. The reader should consult [3], [5] and [6] to get to know these logics.

[2] The set–valued complementation operator is touched on in the concluding remarks.

a *separation operator*, S. Afterwards we introduce a logical system **S** which is aimed at producing every semantically valid formula of LS, in Section 3. In Section 4 we outline the proof of the *completeness* of the system **S**. However, only the matters of a corresponding model construction are given and no verifications.[3] Describing how to get *decidability* of our logic we proceed even more sketchy, in Section 5. The final section contains some concluding remarks. In particular, we discuss how complementation may be incorporated into the logic of subset spaces and how it may be related to separation.

2 The Logical Language

We introduce the syntax and semantics of a language, LS, in which, e.g., certain linear time properties of knowledge can be expressed and sets of knowledge states can be separated at future time points.

Let PV be a recursively enumerable set of strings, called *propositional variables* (denoted by upper case Roman letters). Based on PV, the set $\mathcal{F}$ of LS–formulas (denoted by lower case Greek letters) is defined by the following clauses:

- $PV \cup \{\top\} \subseteq \mathcal{F}$;
- $\alpha, \beta \in \mathcal{F} \Longrightarrow \neg\alpha, K\alpha, \bigcirc\alpha, S\alpha, (\alpha \wedge \beta) \in \mathcal{F}$;
- no other strings belong to $\mathcal{F}$.

We use common conventions denoting formulas and, especially, the following abbreviations:

$$L\alpha \text{ for } \neg K \neg \alpha \text{ and } T\alpha \text{ for } \neg S \neg \alpha.$$

As the nextstep operator turns out to be self–dual there need not be a corresponding abbreviation in case of $\bigcirc$.

The semantical structures are triples (X, d, σ) specified by the subsequent definition.

Definition 1. *Let X be a non-empty set, and let $d = (E_j)_{j \in N}$ be a sequence of equivalence relations on X such that every class of E_j is the union of some classes of E_{j+1}, for all $j \in \mathbb{N}$. Then the pair $\mathcal{F} = (X, d)$ is called a subset tree frame.*

Let a mapping $\sigma : PV \times X \longrightarrow \{0, 1\}$ be given additionally. Then σ is called a valuation, and the triple $\mathcal{M} = (X, d, \sigma)$ is called a subset tree model (based on $\mathcal{F}$).

The set of all equivalence classes w.r.t. the relations E_j ($j \in \mathbb{N}$) is called the set of *opens* of $\mathcal{F}$. Note that in every subset tree model the set of opens contained in any equivalence class of the relation E_0 forms in fact a tree w.r.t. (reverse) set inclusion such that no two opens on the same level intersect. Thus, in particular, subset tree models are treelike in the sense of [6]. Moreover, an open need not

[3] More details as well as the missing proofs will be published in the full version of this paper.

nessecarily decompose into at least two proper subsets at the next step; so the desired separation property must vacuously be true in the unramified case. The precise meaning of the operator S is given in the subsequent definition.

Let U_j^x denote the equivalence class of x w.r.t. the relation E_j. A pair x, U_j^x (designated without brackets) is called a *neighbourhood situation* of $\mathcal{F}$. Using this notation we introduce the validity relation for LS–formulas.

Definition 2 (Semantics of LS). *Let a subset tree model $\mathcal{M} = (X, d, \sigma)$ and a neighbourhood situation x, U_j^x of $\mathcal{M}$ be given. Then we define*

$$\begin{array}{lcl} x, U_j^x \models_{\mathcal{M}} A & :\Longleftrightarrow & \sigma(A, x) = 1 \\ x, U_j^x \models_{\mathcal{M}} \neg\alpha & :\Longleftrightarrow & x, U_j^x \not\models_{\mathcal{M}} \alpha \\ x, U_j^x \models_{\mathcal{M}} \alpha \wedge \beta & :\Longleftrightarrow & x, U_j^x \models_{\mathcal{M}} \alpha \text{ and } x, U_j^x \models_{\mathcal{M}} \beta \\ x, U_j^x \models_{\mathcal{M}} K\alpha & :\Longleftrightarrow & y, U_j^y \models_{\mathcal{M}} \alpha \text{ for all } y \in U_j^x \\ x, U_j^x \models_{\mathcal{M}} \bigcirc\alpha & :\Longleftrightarrow & x, U_{j+1}^x \models_{\mathcal{M}} \alpha \\ x, U_j^x \models_{\mathcal{M}} S\alpha & :\Longleftrightarrow & \text{for all } y \in U_j^x : \text{ if } U_{j+1}^y \cap U_{j+1}^x = \emptyset, \\ & & \text{then } y, U_{j+1}^y \models_{\mathcal{M}} \alpha, \end{array}$$

for all $A \in PV$ and $\alpha, \beta \in \mathcal{F}$.

In case $x, U_j^x \models_{\mathcal{M}} \alpha$ is valid we say that α *holds in $\mathcal{M}$ at the neighbourhood situation* x, U_j^x; moreover, the formula $\alpha \in \mathcal{F}$ *holds in* $\mathcal{M}$ (denoted by $\models_{\mathcal{M}} \alpha$), iff it holds in $\mathcal{M}$ at every neighbourhood situation. If there is no ambiguity, we omit the index $\mathcal{M}$ subsequently. Note that the semantics of the operator K is the intended one in contexts where K is interpreted as knowledge since $U_j^x = U_j^y$ if $y \in U_j^x$. Furthermore, according to the definition of the validity of $S\alpha$ the separation happens within the given open at the next time point. For convenience of the reader, we also mention the semantics of the dual T of S:

$$\begin{array}{lcl} x, U_j^x \models_{\mathcal{M}} T\alpha & :\Longleftrightarrow & \text{there is a } y \in U_j^x \text{ such that} \\ & & U_{j+1}^y \cap U_{j+1}^x = \emptyset \text{ and } y, U_{j+1}^y \models_{\mathcal{M}} \alpha. \end{array}$$

This definition shows in particular that the dual operator of S is stronger than the compound operator "$L\bigcirc$".

Finishing this section we present a small example. Let

$$X := \{ f \mid f : \mathbb{N} \longrightarrow \{0, 1\} \},$$

and let U_j^f be defined by

$$U_j^f := \{ g \in X \mid g(i) = f(i) \text{ for } i = 0, 1, \ldots, j \}$$

for all $f \in X$ and $j \in \mathbb{N}$. The resulting subset tree frame $\mathcal{C}$, which is important for many applications, is called the *Cantor space* (by abuse of notation; see the introductory section). Let A represent the proposition "the actual output is 1". Then the formula SA holds at every neighbourhood situation f, U_j^f of $\mathcal{C}$ such that f satisfies $f(j+1) = 0$ (in every model interpreting A properly).

3 The System S

Subsequently we present lists of axioms and rules, respectively, which constitute a logical system **S**. Later on we show that the set of **S**–theorems coincides with the set of LS–validities, i.e., a formula $\alpha \in \mathcal{F}$ is **S**–derivable iff it holds in all subset tree models.

We take the following schemes of formulas as axioms:

(1) All $\mathcal{F}$–instances of propositional tautologies
(2) $K(\alpha \rightarrow \beta) \rightarrow (K\alpha \rightarrow K\beta)$
(3) $K\alpha \rightarrow \alpha$
(4) $K\alpha \rightarrow KK\alpha$
(5) $L\alpha \rightarrow KL\alpha$
(6) $(A \rightarrow \bigcirc A) \wedge (\neg A \rightarrow \bigcirc \neg A)$
(7) $\bigcirc(\alpha \rightarrow \beta) \rightarrow (\bigcirc\alpha \rightarrow \bigcirc\beta)$
(8) $\bigcirc \neg \alpha \leftrightarrow \neg \bigcirc \alpha$
(9) $\bigcirc L \alpha \rightarrow L \bigcirc \alpha$
(10) $S(\alpha \rightarrow \beta) \rightarrow (S\alpha \rightarrow S\beta)$
(11) $\bigcirc\alpha \wedge T\beta \rightarrow L(T\alpha \wedge \bigcirc\beta)$
(12) $\bigcirc K\alpha \wedge S\alpha \rightarrow K \bigcirc \alpha$,

for all $A \in PV$ and $\alpha, \beta \in \mathcal{F}$.

Adding rules, we get a logical system designated **S**. In fact, modus ponens as well as necessitation w.r.t. each modality are present:

$$(1)\ \frac{\alpha, \alpha \rightarrow \beta}{\beta} \quad (2)\ \frac{\alpha}{K\alpha} \quad (3)\ \frac{\alpha}{\bigcirc\alpha} \quad (4)\ \frac{\alpha}{S\alpha},$$

for all $\alpha, \beta \in \mathcal{F}$. — For convenience, we comment on some of the axioms. The schemes (3), (4) and (5) represent the standard axioms of knowledge. They characterize reflexivity, transitivity, and the euclidean property, respectively, of the accessibility relation in Kripke frames, which are the common semantical domains of modal logic. The scheme (8) corresponds in this sense with functionality. (9) relates the *nexttime* operator to the dual of K, saying that $\bigcirc\alpha$ is possible at the actual neighbourhood situation whenever α is possible at the succeeding one. (The converse is not always true.) This axiom determines the interaction between knowledge and time. Schemes of this form are typical of the systems considered in topological modal logic. Axioms (2), (7) and (10) first of all have a proof–theoretical meaning. (11) and (12) relate to each other the way the three modal operators of our language work; note that (12) represents a certain reversal of the dual of (9). Finally, we should say a few words about the scheme (6). It has to be added because we want the valuation to be independent of the time component of a neighbourhood situation. This requirement simplifies the definition of the semantics, but it clearly implies that the system is not closed under substitution.

Using the above definitions the *soundness* of the axioms and the rules w.r.t. the intended structures can easily be established.

Proposition 1. *Axioms (1) to (12) hold in every subset tree model, and rules (1) to (4) preserve the validity of formulas.*

While soundness of the calculus is easy to see we need some preparations in order to prove also its *completeness*. To this end we use the *canonical model* $\widetilde{\mathcal{M}}$ of the system **S** extensively. The canonical model is formed in the usual way ([8], §5); i.e., the accessibility relations on $\widetilde{\mathcal{M}}$ induced by the modal operators K, $\bigcirc$ and S are defined as follows:

$$s \xrightarrow{L} t :\iff \{\alpha \in \mathcal{F} \mid K\alpha \in s\} \subseteq t$$

$$s \xrightarrow{\bigcirc} t :\iff \{\alpha \in \mathcal{F} \mid \bigcirc\alpha \in s\} \subseteq t$$

$$s \xrightarrow{T} t :\iff \{\alpha \in \mathcal{F} \mid S\alpha \in s\} \subseteq t,$$

for all *maximal* **S**–*consistent* sets s, t from the carrier set of $\widetilde{\mathcal{M}}$ (which we designate C).[4] Note that the distinguished valuation of the canonical model is defined by

$$\sigma(A, s) = 1 :\iff A \in s \quad (A \in PV, s \in C).$$

We are going to present some useful properties of the canonical model for which the special axioms of our system are responsible. First of all, however, we remind of the *truth lemma*.

Lemma 1. *Let us denote the usual satisfaction relation of multimodal logic by* $\models$, *and let* $\vdash$ *designate* **S**–*derivability. Then it holds that*

(a) for all $\alpha \in \mathcal{F}$ *and* $s \in C$

$$\widetilde{\mathcal{M}} \models \alpha[s] \text{ iff } \alpha \in s, \text{ and}$$

(b) for all $\alpha \in \mathcal{F}$

$$\widetilde{\mathcal{M}} \models \alpha \text{ iff } \vdash \alpha.$$

The results stated in parts (a) and (b) of the following proposition are well–known; as to (c) compare with [11], Proposition 7.

Proposition 2. *(a) The relation* $\xrightarrow{L}$ *is an equivalence relation on the set* C.

(b) The relation $\xrightarrow{\bigcirc}$ *is a function on* C.

(c) Let $s, t, u \in C$ *be given such that* $s \xrightarrow{\bigcirc} t \xrightarrow{L} u$. *Then there exists a point* $v \in C$ *satisfying* $s \xrightarrow{L} v \xrightarrow{\bigcirc} u$.

Part (a) is forced by axioms (3), (4), and (5), essentially, while axioms (8) and (9) imply the assertion of (b) and (c) respectively.

The impact of the scheme (11) on the canonical model is described in the subsequent proposition (see also Figure 1).

[4] In the context of topological modal logic the notion of (maximal) **S**–consistency is the same as in ordinary modal logic.

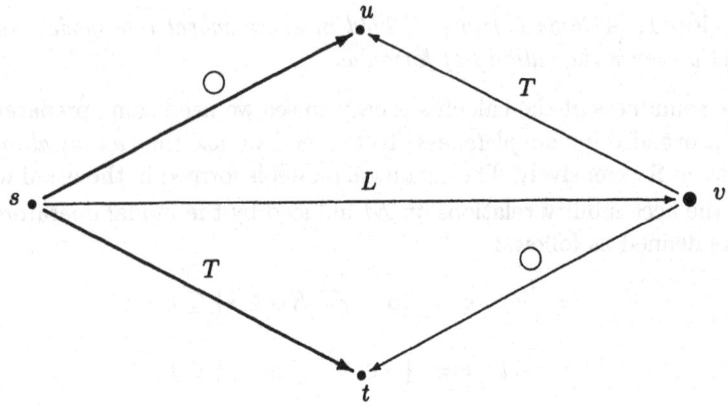

Fig. 1. A diagram illustrating Proposition 3

Proposition 3. *Let $s, t, u \in C$ be given such that $s \xrightarrow{T} t$ and $s \xrightarrow{\bigcirc} u$. Then there exists an element $v \in C$ satisfying $s \xrightarrow{L} v$, $v \xrightarrow{\bigcirc} t$ and $v \xrightarrow{T} u$.*

Finally, axiom (12) has some effect on the interaction of the accessibility relations on the canonical model as well:

Proposition 4. *Let $s, t, u \in C$ be given such that $s \xrightarrow{L} t \xrightarrow{\bigcirc} u$. Then there exists an element $v \in C$ satisfying $s \xrightarrow{\bigcirc} v \xrightarrow{L} u$, or $s \xrightarrow{T} u$ is valid.*

As a consequence we get that the relation $\xrightarrow{L} \circ \xrightarrow{\bigcirc}$ is the union of the relations $\xrightarrow{\bigcirc} \circ \xrightarrow{L}$ and $\xrightarrow{T}$. This follows from the previous proposition, parts (b) and (c) of Proposition 2, and Proposition 3.

Corollary 1. $\xrightarrow{L} \circ \xrightarrow{\bigcirc} = \xrightarrow{\bigcirc} \circ \xrightarrow{L} \cup \xrightarrow{T}$.

4 Completeness

In this section we sketch how completeness of the proposed logical system **S** can be obtained by constructing a subset tree model that falsifies a given non–derivable formula. The main idea is to use a HENKIN–like construction. Such a construction seems to be necessary because it is not clear how to obtain the separation property for sets directly on the canonical model.[5]

Let $\alpha \in \mathcal{F}$ be non–**S**–derivable. Then there exists a maximal **S**–consistent set $t \in C$ containing $\neg \alpha$. Fix t and a distinguished point x_0 from a sufficiently large set of points, Y. We can construct a sequence of triples (X_n, d_n, s_n) inductively such that for every $n \in \mathbb{N}$

[5] The reader should note that step by step constructions are used also in other, more advanced parts of modal logic like the *mosaic method* ([14], Section 5.5.2).

- X_n is a finite set containing x_0;
- $d_n : X_n \times D_n \longrightarrow \mathcal{P}_0(X_n)$ is a partial function, where D_n is an initial segment of $\mathbb{N}$ and $\mathcal{P}_0(X_n)$ is the set of non–empty subsets of X_n; on its domain, d_n satisfies the condition of Definition 1 (required for d there); additionally, $d_n(x_0, 0) = X_n$ holds;
- $s_n : X_n \times D_n \longrightarrow C$ is a partial function such that $s_n(x, m)$ is defined iff $d_n(x, m)$ is defined; moreover, for all $x \in X_n$ and $m \in D_n$ the following assertions are valid:
  - for all $y \in X_n$: if $y \in d_n(x, m)$, then $s_n(x, m) \xrightarrow{L} s_n(y, m)$;
  - if $m + 1 \in D_n$ and $d_n(x, m + 1)$ exists, then $s_n(x, m) \xrightarrow{\bigcirc} s_n(x, m + 1)$;
  - if $m + 1 \in D_n$, $y \in d_n(x, m)$, and $d_n(x, m + 1), d_n(y, m + 1)$ exist such that $d_n(x, m + 1) \cap d_n(y, m + 1) = \emptyset$, then
  $$s_n(x, m) \xrightarrow{T} s_n(y, m + 1) \quad (y \in X_n);$$
  - $s_n(x_0, 0) = t$.

The structures (X_n, d_n, s_n) represent *approximations* to the desired model; especially, d_n turns out to be a finite sequence of equivalence relations.
The construction ensures that for all $n \in \mathbb{N}$ it holds that

- $X_n \subseteq X_{n+1}$ and $D_n \subseteq D_{n+1}$,
- $d_{n+1}(x, m) \supseteq d_n(x, m)$ for all $x \in X_n, m \in D_n$, and
- $s_{n+1} \upharpoonright_{X_n \times D_n} = s_n$.

Furthermore, the following conditions are guaranteed:

- if $L\beta \in s_n(x, m)$, then there is some $k > n$ and some $y \in d_k(x, m)$ such that $\beta \in s_k(y, m)$;
- if $\bigcirc\beta \in s_n(x, m)$, then $\beta \in s_k(x, m + 1)$ for some $k > n$;
- if $T\beta \in s_n(x, m)$, then there are $k > n$ and $y \in d_k(x, m)$ such that
$$d_k(x, m + 1), d_k(y, m + 1) \text{ exist, } d_k(x, m + 1) \cap d_k(y, m + 1) = \emptyset,$$
$$\text{and } \beta \in s_k(y, m + 1).$$

Now suppose that we have carried out the construction meeting these requirements successfully. Let (X, d, s) be the *limit* of the structures (X_n, d_n, s_n), i.e.,

- $X = \bigcup_{n \in \mathbb{N}} X_n$,
- d is defined as the prolongation of all of the mappings d_n, and
- s is given by $s(x, m) := s_n(x, m)$ for the smallest number n such that $s_n(x, m)$ is defined.

Letting
$$\sigma(A, x) = 1 :\Longleftrightarrow A \in s(x, 0)$$
for all $A \in PV$ and $x \in X$, the resulting structure is the one we are looking for, as will be seen below. We state the following easy consequences of the above conditions first.

Proposition 5. *$\mathcal{M} := (X, d, \sigma)$ is a subset tree model such that $d(x_0, 0) = X$; in particular, d and s are total functions. Moreover, the following properties hold for all $m \in \mathbb{N}$, $x \in X$, and $\beta \in \mathcal{F}$:*

- $(\forall y \in X)\ [y \in d(x,m) \Longrightarrow s(x,m) \xrightarrow{L} s(y,m)]$;
- $L\beta \in s(x,m) \Longrightarrow (\exists y \in d(x,m))\ \beta \in s(y,m)$;
- $s(x,m) \xrightarrow{\bigcirc} s(x,m+1)$;
- $\bigcirc\beta \in s(x,m) \Longrightarrow \beta \in s(x,m+1)$;
- *for all $y \in X$:*

$$y \in d(x,m) \wedge d(x,m+1) \cap d(y,m+1) = \emptyset \Longrightarrow s(x,m) \xrightarrow{T} s(y,m+1);$$

- *if $T\beta \in s(x,m)$, then*

$$(\exists y \in d(x,m))\ [d(x,m+1) \cap d(y,m+1) = \emptyset \wedge \beta \in s(y,m+1)];$$

- $s(x_0, 0) = t$.

Applying this proposition, the subsequent lemma can be proved by a structural induction. Note that axiom scheme (6) has to be used in case of a propositional variable.

Lemma 2. *Let $\mathcal{M} = (X, d, \sigma)$ be the subset tree model just constructed, and let s be the function from above. Then for all $\beta \in \mathcal{F}$ and every neighbourhood situation x, U_j^x of $\mathcal{M}$ we have that*

$$x, U_j^x \models_{\mathcal{M}} \beta \iff \beta \in s(x,j).$$

Since $\neg\alpha \in t = s(x_0, 0)$ and $U_0^{x_0} = X$, we get the desired completeness theorem:

Theorem 1. *The non–S–derivable formula α is falsified in the subset tree model $\mathcal{M}$ at the neighbourhood situation x_0, X.*

What remains to be done in order to finish the proof of the theorem is to carry out the inductive definition of the triples (X_n, d_n, s_n) such that the conditions stated before Proposition 5 are fulfilled. This makes up the main technical part of the proof, in which the properties of the accessibility relations on the canonical model proved in Section 3 come into play.

The basic idea is to "realize" every "existential" formula $L\beta$, $\bigcirc\beta$, and $T\beta$, respectively, which is contained in some already attached maximal **S**–consistent set $s_n(x,m)$, in a future step of the construction. For this purpose we must carefully arrange the order of the construction steps. Actually, we have to choose enumerations μ_n, ν_n, κ_n of the sets

$\{(L\beta, x, m) \in \mathcal{F} \times X_n \times D_n \mid s_n(x,m)$ is defined and $L\beta \in s_n(x,m)\}$,
$\{(\bigcirc\beta, x, m) \in \mathcal{F} \times X_n \times D_n \mid s_n(x,m)$ is defined and $\bigcirc\beta \in s_n(x,m)\}$,
and
$\{(T\beta, x, m) \in \mathcal{F} \times X_n \times D_n \mid s_n(x,m)$ is defined and $T\beta \in s_n(x,m)\}$,

respectively, in each step, too. Then the inductive definition of the model proceeds along a "global" enumeration λ, which "schedules" processing correctly to triples $(L\beta, x, m)$, $(\bigcirc\beta, x, m)$, and $(T\beta, x, m)$, such that all of the above sets are exhausted eventually; in particular, every triple $(\bigcirc\gamma, x, m)$ always precedes a triple $(T\beta, x, m)$ w.r.t. the global enumeration. All this can be done in a way which is similar to that carried out in [3], Section 2.2. So we need not regard the enumerations any more in the following.

We omit the simple start of the inductive definition of (X_n, d_n, s_n) and treat only the case "$T\beta$" of the induction step, in which, however, the case "$\bigcirc\beta$" has to be considered as well. So let (X_n, d_n, s_n) be already defined and assume that $T\beta \in s_n(x, m)$ is going to be realized in step $n+1$.

Now we choose $y \in Y \setminus X_n$, and we define $X_{n+1} := X_n \cup \{y\}$. According to the above mentioned property of the global enumeration we may assume that $m+1 \in D_n$. Therefore, we let $D_{n+1} := D_n$. Our next task is to define the mapping d_{n+1}. It is given by the following conditions:

$$d_{n+1}(v,l) := \begin{cases} d_n(v,l) \cup \{y\} & \text{if } v \in d_n(v,l) \text{ and } l \leq m \\ \{y\} & \text{if } v = y \text{ and } l = m+1 \\ d_n(v,l) & \text{otherwise,} \end{cases}$$

for all $v \in X_{n+1}$ and $l \in D_{n+1}$.

The definition of s_{n+1} is somewhat more complicated. First, there is an element $t \in C$ of the canonical model such that $s_n(x,m) \xrightarrow{T} t$ and $\beta \in t$, as it is known from standard modal proof theory. Since there exists a formula γ such that the triple $(\bigcirc\gamma, x, m)$ has already been considered, we get

$$s_n(x,m) \xrightarrow{\bigcirc} u \text{ for some } u \in C$$

because of some former construction step. Now Proposition 3 applies. Thus we obtain an element $v \in C$ satisfying

$$s_n(x,m) \xrightarrow{L} v \text{ and } v \xrightarrow{\bigcirc} t.$$

Finally, let $v_0, \ldots, v_m = v$ be points of C such that

$$s_n(x,j) \xrightarrow{L} v_j \text{ as well as } v_0 \xrightarrow{\bigcirc} v_1 \xrightarrow{\bigcirc} \ldots \xrightarrow{\bigcirc} v_m$$

is valid $(j = 0, \ldots, m)$; the existence of these elements is guaranteed by Proposition 2(c). Then we define s_{n+1} by

$$s_{n+1}(z,l) := \begin{cases} s_n(z,l) & \text{if } z \neq x, y \\ s_n(x,l) & \text{if } z = x \text{ and } l \leq n \\ t & \text{if } z = y \text{ and } l = m+1 \\ v_j & \text{if } z = y \text{ and } l = j \;\; (j = 0, \ldots, m) \end{cases}$$

for all $z \in X_{n+1}$ and $l \in D_{n+1}$.

The validity of almost all requirements on X_{n+1}, d_{n+1} and s_{n+1}, which are formulated at the beginning of this section, is easy to see now. We only concentrate on the subsequent one presently:

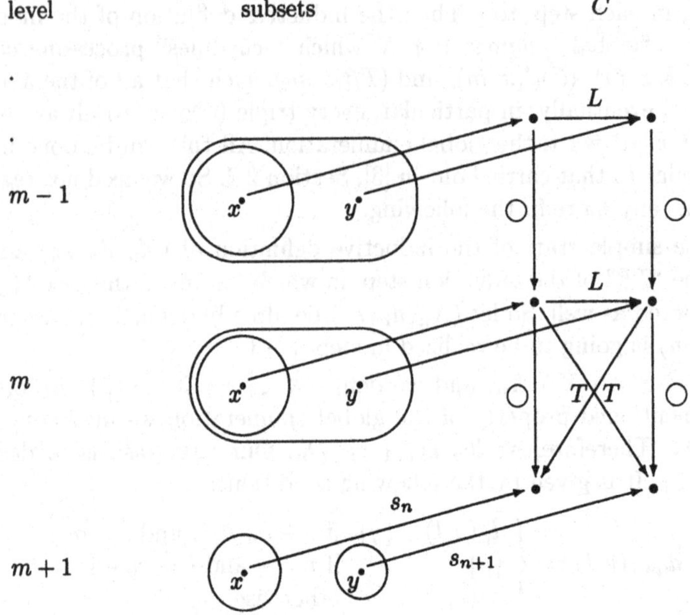

Fig. 2. A single step in the model construction

if $m+1 \in D_{n+1}$, $y \in d_{n+1}(x,m)$, and $d_{n+1}(x,m+1), d_{n+1}(y,m+1)$ exist such that $d_{n+1}(x,m+1) \cap d_{n+1}(y,m+1) = \emptyset$, then

$$(*) \quad s_{n+1}(x,m) \xrightarrow{T} s_{n+1}(y,m+1) \quad (y \in X_{n+1}).$$

At this stage of the construction Proposition 4 plays its part by which it is implied that condition $(*)$ is in fact valid in case

$$s_{n+1}(x,m+1) \xrightarrow{L} s_{n+1}(y,m+1)$$

does never hold if $d_{n+1}(x,m+1) \cap d_{n+1}(y,m+1) = \emptyset$. The latter is true due to the proceeding in the "$\bigcirc\beta$"–case, in which all elements $w \in d_n(x,m)$ such that $s_n(w,m)$ has a $\bigcirc$–sucessor being $\xrightarrow{L}$–equivalent to the $\bigcirc$–sucessor of $s_n(x,m)$ are collected in $s_{n+1}(x,m+1)$.[6] Thus the above requirement is met. — Figure 2 illustrates the step of the model construction just described.

As it was already announced, further details are omitted in this draft; they will be included in the full version of this paper.[7]

One of the main results of the given paper is obtained as an easy consequence of Proposition 1 and Theorem 1 now.

[6] An easy consideration shows that nothing bad can happen in the "$L\beta$"–case then, too.

[7] See also [12].

Corollary 2. *The system* **S** *is sound and complete w.r.t. the class of subset tree models.*

5 Decidability

The completeness proof outlined in the previous section does not give the finite model property of the system **S**. Now we introduce appropriate KRIPKE structures w.r.t. which **S** is sound and complete as well and the finite model property *is* satisfied. This gives decidability of the logic.

Definition 3. *Let* $\mathcal{M} := (W, \{R, Q, P\}, \sigma)$ *be a trimodal model (i.e., W is a non-empty set, R, Q and P are binary relations on W, and σ is a valuation). Then $\mathcal{M}$ is called an LS-model, iff*

- *R is an equivalence relation and*
- *Q is a function on W;*
- *the relation $R \circ Q$ equals the union of the relations $Q \circ R$ and P;*
- *for all $s, t \in W$ such that sQt it holds that*

$$\mathcal{M} \models A[s] \text{ iff } \mathcal{M} \models A[t] \quad (A \in PV).$$

Note that the relation R corresponds with the modality K; accordingly, Q and $\bigcirc$ as well as P and S are related. We get the following theorem.

Theorem 2. *A formula $\alpha \in \mathcal{F}$ is* **S***-derivable, iff it holds in every LS-model.*

In order to prove the finite model property one has to use a suitable *filtration* of the canonical model $\widetilde{\mathcal{M}}$ of the system **S**. Due to the results in Section 3, $\widetilde{\mathcal{M}}$ actually is an *LS*-model, and the filtration can be arranged in a way such that this property is preserved essentially.
Let $\alpha \in \mathcal{F}$ be a formula for which we want to find a finite model. Then the filtration of the canonical model is carried out w.r.t. the set $\widetilde{X}$ of formulas which is defined as follows ($sf(\alpha)$ designates the set of subformulas of α):

$X^{\neg} := sf(\alpha) \cup \{\neg\gamma \mid \gamma \in sf(\alpha)\}$;
$X^{\wedge} := X^{\neg}$ joined with the set of all finite conjunctions of distinct elements of $X^{\neg}$;
$X^{L} := \{L\beta \mid \beta \in X^{\wedge}\}$;
$\widetilde{X} := X^{\wedge} \cup X^{L}$.

The main purpose of the above definition is to guarantee that the crucial properties of the relations $\xrightarrow{L}$, $\xrightarrow{\bigcirc}$, and $\xrightarrow{T}$ are retained by passing to the filtrations. In fact, this can nearly be achieved if one takes the *minimal* filtrations of the respective accessibility relations.

Lemma 3. *Let $\bar{C}$ be the $\widetilde{X}$-filtration of C, and let R, Q and P be the minimal filtration of $\xrightarrow{L}$, $\xrightarrow{\bigcirc}$, and $\xrightarrow{T}$, respectively. Moreover, let $\bar{\sigma}$ be induced by the distinguished valuation of the canonical model. Then, apart from the functionality of Q, $\bar{\mathcal{M}} := (\bar{C}, \{R, Q, P\}, \bar{\sigma})$ is an LS-model.*

The deficiency left over can be rectified by an unwinding procedure followed by an appropriate model surgery. The details of the corresponding constructions remind of the proceeding in usual linear time temporal logic; see [8], §9.

To sum up, we can state:

Theorem 3. *The system* **S** *satisfies the finite model property w.r.t. the class of LS–models.*

As a corollary we obtain decidability of the logic, as desired.

Corollary 3. *The set of all* **S**–*derivable formulas is decidable.*

6 Concluding Remarks

In several applications the set of opens $\mathcal{O}$ of the considered subset frame is closed under complementation, as it is the case for the Cantor space mentioned above, e.g. One would like to treat logically such a richer, hence more complicated structure of $\mathcal{O}$ as well. To this end we regard *complementation as negation* at a preliminary stage, i.e., we let an open and its complement be distinguishable by a formula and its negation respectively. This allows us to retain the basic logical language from Section 2, leading to the following semantics of the complementation operator, which we designate C:

$$\begin{aligned} x, U_j^x \models C\alpha :\iff & \forall y \in U_j^x : \\ & y \in U_{j+1}^x \Longrightarrow y, U_{j+1}^y \models \alpha, \text{ and} \\ & y \notin U_{j+1}^x \Longrightarrow y, U_{j+1}^y \not\models \alpha, \end{aligned}$$

for all neighbourhood situations x, U_j^x of the underlying subset tree model. Actually, C is expressible in the former framework:

Proposition 6. *For all $\alpha \in \mathcal{F}$, the formulas $C\alpha$ and $S\neg\alpha\wedge\bigcirc K\alpha$ are equivalent in every subset tree model.*

In this way we get a weak solution to the problem of integrating complementation in the modal logic of subset spaces. A stronger version should provide an operator acting more directly on the given complement–closed system of sets. But we have to change the formalism then: either the monotonicity requirement which is expressed by axiom scheme (9) has to be given up, or the functionality of the modal operator quantifying over sets. Especially the first way goes beyond what is typical of topological modal logic, whereas for the second the logical language used presently is not fitting. In any case, further research should provide a proper integration of other set–valued operators into the modal logic of subset spaces. The treatment of complementation seems to be particularly interesting since there are connections with logics of different nature [7].

Another topic is the question of *efficiency* of the logic given here. Since ramification occurs in LS–models, the complexity of deciding the validity of a formula is presumably high. In fact, PSPACE–hardness of this problem can be shown utilizing methods due to LADNER [13].

Concluding we summarize the issues of the given paper: An extension of the subset space logic of knowledge and time by a modal operator which separates sets has been obtained. An axiomatization of the set of formulas valid in every subset tree model was given and proved to be sound and semantically complete. Furthermore, decidability of the set of validities was shown.

References

1. Chellas, B. F.: Modal Logic: An Introduction. Cambridge University Press, Cambridge (1980)
2. Artemov, S., Davoren, J., Nerode, A.:. Topological Semantics for Hybrid Systems. In: Adian, S., Nerode, A. (eds.): Logical Foundations of Computer Science, LFCS'97. Lecture Notes in Computer Science, Vol. 1234. Springer–Verlag, Berlin Heidelberg New York (1997) 1–8
3. Dabrowski, A., Moss, L.S., Parikh, R.: Topological Reasoning and The Logic of Knowledge. Annals of Pure and Applied Logic **78** (1996) 73–110
4. Fagin, R., Halpern, J.Y., Moses, Y., Vardi, M.Y.: Reasoning about Knowledge. MIT Press, Cambridge(Mass.) (1995)
5. Georgatos, K.: Knowledge Theoretic Properties of Topological Spaces. In: Masuch, M., Polos, L. (eds.): Knowledge Representation and Uncertainty. Lecture Notes in Computer Science, Vol. 808. Springer–Verlag, Berlin Heidelberg New York (1994) 147–159
6. Georgatos, K.: Reasoning about Knowledge on Computation Trees. In: MacNish, C., Pearce, D., Pereira, L.M. (eds.): Logics in Artificial Intelligence, JELIA'94. Lecture Notes in Computer Science, Vol. 838. Springer–Verlag, Berlin Heidelberg New York (1994) 300–315
7. Georgatos, K.: private communication
8. Goldblatt, R.: Logics of Time and Computation. Center for the Study of Language and Information, Stanford (1987)
9. Heinemann, B.: 'Topological' Modal Logic of Subset Frames with Finite Descent. In: Kautz, H., Selman, B. (eds.): Proceedings of the Fourth International Symposium on Artificial Intelligence and Mathematics, AI/MATH–96. Fort Lauderdale (1996) 83–86
10. Heinemann, B.: On binary computation structures. Mathematical Logic Quaterly **43** (1997) 203–215
11. Heinemann, B.: A topological generalization of propositional linear time temporal logic. In: Prívara, I., Ružička, P. (eds.): Mathematical Foundations of Computer Science (22nd International Symposium, MFCS'97). Lecture Notes in Computer Science, Vol. 1295. Springer–Verlag, Berlin Heidelberg New York (1997) 289–297
12. Heinemann, B.: Including Separation in the Modal Logic of Subset Spaces. Technical Report, No. 231. FernUniversität, Hagen (1998)
13. Ladner, R.E.: The Computational Complexity of Provability in Systems of Modal Propositional Logic. SIAM Journal of Computing **6** (1977) 467–480
14. Marx, M., Venema, Y.: Multi–Dimensional Modal Logic. Kluwer, Dordrecht (1996)
15. Moss, L.S., Parikh R.: Topological Reasoning and The Logic of Knowledge. In: Moses, Y. (ed.): Theoretical Aspects of Reasoning about Knowledge, TARK 1992. Morgan Kaufmann, San Francisco (1992) 95–105

Interpolation in Modal Logic*

Maarten Marx

Institute for Logic, Language and Information, University of Amsterdam, The Netherlands

Abstract. The interpolation property and Robinson's consistency property are important tools for applying logic to software engineering. We provide a uniform technique for proving the Interpolation Property, using the notion of bisimulation. For modal logics, this leads to simple, easy–to–check conditions on the logic which imply interpolation. We apply this result to fibering of modal logics and to modal logics of knowledge and belief.

Keywords: Interpolation, modal logic (of knowledge and belief), fibering

In 1957, W. Craig proved the interpolation theorem for first order logic. Since Craig's paper interpolation has become one of the standard properties that one investigates when designing a logic, though it hasn't received the status of a completeness or a decidability theorem. One of the main reasons why a logic should have interpolation is because of "modular theory building". Assuming compactness, interpolation in modal logic is equivalent to the following property [Marx and Areces1997] (which is the semantical version of Robinson's consistency lemma)

> If two theories T_1, T_2 both have a model, and they don't contradict each other on the common language (i.e., there is no formula θ built up from atoms occurring both in T_1 and in T_2 such that $T_1 \models \theta$ and $T_2 \models \neg\theta$), then $T_1 \cup T_2$ has a model.

The property is not only intuitively valid for scientific reasoning, it also has practical (and computational) consequences. In practice it shows up in the incremental design, specification and development of software, and has received quite some attention in that community (cf., e.g., [Maibaum and Sadler1984, Renardel de Lavalette1989].)

In the real world, we see this phenomenon with mergers of two companies (which might already have ties in the form of shared daughters or stocks). Often it is a highly non-trivial problem to merge their respective databases as well. Clearly Robinson's consistency property can help here: if the logic behind the databases has the property, the two can be combined precisely when they don't contradict each other on their common part. (This seems a necessary precondition for the merger of the two companies as well...)

* The author is supported by the Institute of Logic, Language and Information, Amsterdam, The Netherlands, and by UK EPSRC grant No. GR/K54946.

A.M. Haeberer (Ed.): AMAST'98, LNCS 1548, pp. 154–163, 1998.

1 Interpolation in Modal Logic

In this section, we deal with general modal logics and give a structural description of a large number of them with interpolation. This description is based on the notion of bisimulation. We start with the necessary definitions.

1.1 Preliminaries

Modal logic. A *modal similarity type* S is a pair (O, ρ) with O a set of logical connectives and $\rho : O \longrightarrow \omega$ a function assigning to each symbol in O a finite rank or arity. We call $\mathcal{L}(\mathsf{K}_S)$ a *modal logic* of type $S = (O, \rho)$, if $\mathcal{L}$ is a tuple $(\mathsf{Fml}_S, \mathsf{K}_S, \Vdash_S)$ in which,

- Fml_S is the smallest set containing countably many propositional variables, and which is closed under the Boolean connectives and the connectives in O.
- K_S is a class of frames of the form $(W, R^\diamond)_{\diamond \in O}$, in which W is a non-empty set, and each $R^\diamond$ is a subset of $W^{\rho\diamond+1}$.
- $\Vdash_S$ is the usual truth–relation from modal logic between models over frames in K, worlds and formulas. For the modal connectives it is defined as

$$\mathfrak{M}, x \Vdash \diamond(\varphi_1, \ldots, \varphi_{\rho\diamond}) \stackrel{\text{def}}{\Longleftrightarrow} (\exists x_1 \ldots x_{\rho\diamond}) : R^\diamond x x_1 \ldots x_{\rho\diamond} \ \& \\ \mathfrak{M}, x_1 \Vdash \varphi_1 \ \& \ldots \& \ \mathfrak{M}, x_{\rho\diamond} \Vdash \varphi_{\rho\diamond}$$

Logical consequence is defined globally, i.e., $Th \models_{\mathcal{L}(\mathsf{K})} \varphi$ iff for every K model in which Th is true in every world, φ is also true in every world. If $\mathfrak{F}$ is a frame, we use F to denote its universe. For $\mathfrak{F}$ a frame and $x_1 \ldots x_n$ elements of its domain, $R_{\mathfrak{F}} x_1, \ldots, x_n$ abbreviates $\mathfrak{F} \models R x_1, \ldots, x_n$. Let P be a set of propositional variables. We call a logic $\mathcal{L}(\mathsf{K})$ *canonical*, if, for every choice of P, the canonical frame of $\mathcal{L}(\mathsf{K})$ (cf., [Hughes and Creswell1984]) is a member of K.

Interpolation. In the literature we find the following formulations of the interpolation property for compact logics (cf., e.g., Maksimova [Maksimova1991], Rodenburg [Rodenburg1992]):

AIP A logic $\mathcal{L}$ has the *Arrow Interpolation Property* (AIP) if, whenever $\models_{\mathcal{L}} \varphi \to \psi$, there exists a formula θ such that $\models_{\mathcal{L}} \varphi \to \theta$, $\models_{\mathcal{L}} \theta \to \psi$ and $\Sigma(\theta) \subseteq \Sigma(\varphi) \cap \Sigma(\psi)$.

TIP A logic $\mathcal{L}$ has the *Turnstile Interpolation Property* (TIP) if, whenever $\varphi \models_{\mathcal{L}} \psi$, there exists a formula θ such that $\varphi \models_{\mathcal{L}} \theta$, $\theta \models_{\mathcal{L}} \psi$ and $\Sigma(\theta) \subseteq \Sigma(\varphi) \cap \Sigma(\psi)$.

SIP A logic $\mathcal{L}$ has the *Splitting Interpolation Property* (SIP) if, whenever $\Gamma_0 \cup \Gamma_1 \models_{\mathcal{L}} \psi$, there exists a formula θ such that $\Gamma_0 \models_{\mathcal{L}} \theta$, $\Gamma_1 \cup \{\theta\} \models_{\mathcal{L}} \psi$ and $\Sigma(\theta) \subseteq \Sigma(\Gamma_0) \cap (\Sigma(\Gamma_1) \cup \Sigma(\psi))$.

Remark 1. Recall that the local consequence relation is defined as follows: $\Gamma \models^{loc} \varphi$ iff for every model $\mathfrak{M}$, and every world w in $\mathfrak{M}$ which satisfies Γ, φ is also satisfied in w. If we use the *local* consequence relation in the definitions of SIP, AIP en TIP, the three notions of interpolation collapse into one, because we have $\varphi \models^{loc} \psi \iff \models \varphi \to \psi$.

For canonical modal logics, the three forms of interpolation are related as follows.

Proposition 1. *Let $\mathcal{L}(\mathsf{K})$ be a canonical modal logic.*
(i) *If $\mathcal{L}(\mathsf{K})$ has AIP, then it has TIP.*
(ii) *$\mathcal{L}(\mathsf{K})$ has TIP if and only if it has SIP.*

Proof. We use the facts that any canonical modal logic is compact both in the local and the global sense [Hughes and Creswell1984], and that we can switch from the global to the local perspective by $\varphi \models \psi$ iff $\{\Box^n\varphi \mid n < \omega\} \models^{loc} \psi$ ([van Benthem1983]: Lemma 2.33). We prove the proposition for the uni-modal case. Clearly this extends to any modal similarity type.

(i) Assume $\varphi \models \psi$. This holds iff $\{\Box^n\varphi \mid n < \omega\} \models^{loc} \psi$, iff (by compactness) $\Box^{m*}\varphi \models^{loc} \psi$ for some m, where $\Box^{m*}\varphi = \varphi \wedge \Box\varphi \wedge \Box\Box\varphi \wedge \ldots \wedge \Box^m\varphi$. Iff –by the deduction theorem– $\models \Box^{m*}\varphi \to \psi$. But then, by AIP, there is an interpolant θ such that $\models \Box^{m*}\varphi \to \theta$ and $\models \theta \to \psi$. Whence $\varphi \models \theta$ and $\theta \models \psi$.

(ii) The direction from SIP to TIP is trivial. For the other direction, assume $\Gamma_0 \cup \Gamma_1 \models \psi$. As above we obtain, $\Box^{m*}\varphi_0 \wedge \Box^{k*}\varphi_1 \models^{loc} \psi$, where $\Gamma_0 \models \varphi_0$ and $\Gamma_1 \models \varphi_1$. Then by the deduction theorem, $\Box^{m*}\varphi_0 \models^{loc} \Box^{k*}\varphi \to \psi$. Whence, $\Box^{m*}\varphi_0 \models \Box^{k*}\varphi \to \psi$. By TIP, we find an interpolant θ such that $\Box^{m*}\varphi_0 \models \theta$ and $\theta \models \Box^{k*}\varphi \to \psi$. Whence, $\varphi_0 \models \theta$ and $\varphi_1, \theta \models \psi$, so also $\Gamma_0 \models \theta$ and $\Gamma_1, \theta \models \psi$.

SIP and modularization. There is a vast amount of literature relating SIP to applications in computer science, in particular to the notions of modularization and specification. The importance of modularization for stepwise refinement of formal specifications has been noted by several researchers [Bergstra et al.1990], [Ehrig and Mahr1985], [Maibaum et al.1984], [Andréka et al.1994], [Renardel de Lavalette, 1989]. In [Maibaum and Sadler1984] it is shown that a logic "supports specification" (in their terminology: the consequence relation preserves conservative extensions under implementation) if and only if it has SIP. [Renardel de Lavalette, 1989] contains a similar equivalence theorem.

Bisimulation. Let Fml be a modal language of type $S = (O, \rho)$, and let $\mathfrak{M}{=}(\mathfrak{F}, \mathsf{v}_1)$ and $\mathfrak{N}{=}(\mathfrak{G}, \mathsf{v}_2)$ be two models of that type. We say that $\mathfrak{M}$ and $\mathfrak{N}$ are (Fml, S)–*bisimular* (notation: $\mathfrak{M} \rightleftharpoons_S^{\mathsf{Fml}} \mathfrak{N}$) if there exists a relation $B \subseteq F \times G$ such that, for every propositional variable p in Fml and every pair $(x, y) \in B$ it holds that $x \in \mathsf{v}_1(p) \iff y \in \mathsf{v}_2(p)$, and for every $\Diamond \in O$,

- if $R^{\Diamond}_{\mathfrak{F}} x_0 x_1 \ldots x_{\rho\Diamond}$ and $x_0 B y_0$, then there exists $y_1, \ldots, y_{\rho\Diamond} \in G$ such that $R^{\Diamond}_{\mathfrak{G}} y_0 y_1 \ldots y_{\rho\Diamond}$ and $x_i B y_i$ *(forward condition)*,
- similarly in the other direction *(backward condition)*.

The relation B is called an (Fml, S)*-bisimulation*. If the domain of B equals F, and its range equals G, we call B a *zigzag-connection*. A function $f : F \longrightarrow G$ is called an (Fml, S)*-zigzagmorphism* from $\mathfrak{M}$ onto $\mathfrak{N}$ (notation $\mathfrak{M} \overset{f}{\twoheadrightarrow} \mathfrak{N}$), if f is surjective and f is a (Fml, S)-bisimulation. Note that the forward condition then states that f is a homomorphism. The same definition of bisimulation applies to frames, just forget the clause about the valuations. The important point about bisimulations is that worlds which stand in the bisimulation relation verify the same Fml–formulas. Whenever Fml and S are clear from the context, we drop them as a prefix. [de Rijke1993]: Chapter 6 contains a thoruogh study of bisimulations in modal logic.

Bisimulation–products. Let $\mathfrak{F}$ and $\mathfrak{G}$ be two type-S frames, and let $B \subseteq F \times G$ be an S–zigzag–connection. The frame[1] $(\mathfrak{F} \times \mathfrak{G}) \upharpoonright_B$ is called the *bisimulation–product* of $\mathfrak{F}$ and $\mathfrak{G}$.

Proposition 2. *Every bisimulation–product is a subdirect product, and the projection functions π_0 and π_1 are zigzagmorphisms*

Proof. The product is subdirect because the bisimulation is a zigzag–connection. Hence the projections are surjective homomorphisms. To show the backward condition, we reason as follows. Let $\mathfrak{H}$ be a bisimulation–product of $\mathfrak{F}$ and $\mathfrak{G}$. (We prove the statement for a binary relation and for π_0 only; all other cases are similar.) Suppose $R_{\mathfrak{F}}\pi_0(x)y$. Because $\mathfrak{F}$ and $\mathfrak{G}$ are bisimular and $\pi_0(x)B\pi_1(x)$, we find a y' such that $R_{\mathfrak{G}}\pi_1(x)y'$ and yBy'. But then, $(y, y') \in H$ and $R_{\mathfrak{H}}x(y, y')$, which is what was needed.

Remark 2. A notion which is sometimes easier to work with is that of zigzag–products (cf., [Marx1995]). A frame $\mathfrak{H}$ is a *zigzag–product* of two frames $\mathfrak{F}$ and $\mathfrak{G}$, if $\mathfrak{H}$ is a subdirect product of $\mathfrak{F}$ and $\mathfrak{G}$, and the projections are zigzagmorphisms. By the last proposition, every bisimulation–product is a zigzag–product. The other side also holds: if $\mathfrak{H}$ is a zigzag–product of $\mathfrak{F}$ and $\mathfrak{G}$, then H is a zigzag–connection, whence $\mathfrak{H}$ is a bisimulation–product. (The easy proof is left to the reader.) This equivalence shows that zigzag–products form an elegant way of describing all zigzag–connections between two frames.

1.2 Interpolation via Bisimulation

Theorem 1. *Let $\mathcal{L}(\mathsf{K})$ be a canonical modal logic. If K is closed under bisimulation–products, then $\mathcal{L}(\mathsf{K})$ enjoys all types of interpolation.*

Corollary 1. *Let $\mathcal{L}(\mathsf{K})$ be a canonical modal logic. If K is defined by a universal Horn theory, then $\mathcal{L}(\mathsf{K})$ enjoys all types of interpolation.*

[1] For $B \subseteq F \times G$, $(\mathfrak{F} \times \mathfrak{G}) \upharpoonright_B$ denotes the substructure with universe B of the direct product of the structures $\mathfrak{F}$ and $\mathfrak{G}$. Such a substructure of a direct product is called *subdirect* if the projections are surjective.

Proof. By theorem 1, since every bisimulation–product is a subdirect product, and universal Horn sentences are preserved under the last.

PROOF OF THEOREM 1. Let $\mathcal{L}(\mathsf{K})$ be a canonical modal logic of an arbitrary type S. Assume that K is closed under bisimulation–products. By proposition 1, it is sufficient to show that the logic has AIP. We reason by contraposition. Let φ, ψ be arbitrary $\mathcal{L}(\mathsf{K})$ formulas. Suppose that there is no $\mathcal{L}(\mathsf{K})$ formula which is an interpolant for $\varphi \rightarrow \psi$. We have to show that $\varphi \wedge \neg\psi$ is satisfiable. Let $\mathsf{Fml}_\varphi, \mathsf{Fml}_\psi$ and $\mathsf{Fml}_{\varphi\psi}$ be the $\mathcal{L}(\mathsf{K})$ languages generated by the propositional variables in φ, the ones in ψ, and the common variables, respectively. Let $\mathfrak{M}_\varphi=(\mathfrak{F}_\varphi, \mathsf{v}_\varphi)$ and $\mathfrak{M}_\psi=(\mathfrak{F}_\psi, \mathsf{v}_\psi)$ be the canonical models (cf., Hughes–Creswell [Hughes and Creswell1984]) of the languages Fml_φ and Fml_ψ, respectively.

CLAIM. (i) $B \stackrel{\text{def}}{=} \{(w,v) \in F_\varphi \times F_\psi : \mathsf{Fml}_{\varphi\psi} \cap w = \mathsf{Fml}_{\varphi\psi} \cap v\}$ is an $(\mathsf{Fml}_{\varphi\psi}, S)$–zigzag–connection between $\mathfrak{M}_\varphi$ and $\mathfrak{M}_\psi$.
(ii) There exists a $(w,v) \in B$ such that $\mathfrak{M}_\varphi, w \Vdash \varphi$ and $\mathfrak{M}_\psi, v \Vdash \neg\psi$.

PROOF OF CLAIM. (i). The domain of B equals F_φ, because, for every $w \in F_\varphi$, the set $\mathsf{Fml}_{\varphi\psi} \cap w$ is Fml_ψ–consistent, whence can be extended to a maximal consistent set $v \in F_\psi$. So $(w,v) \in B$. The argument for range and F_ψ is symmetric. The condition about valuations is satisfied by definition. We show the forward condition, the backward condition is shown similarly. For notational convenience, we deal with a binary relation. Suppose $R^\diamond_{\mathfrak{F}_\varphi}xy$ and xBx'. By definition of canonical model, $R^\diamond_{\mathfrak{F}_\varphi}xy \stackrel{\text{def}}{\Longleftrightarrow} (\forall\theta \in y) : \diamond\theta \in x$. Hence, using xBx', we have $(\forall\theta \in (\mathsf{Fml}_{\varphi\psi} \cap y)) : \diamond\theta \in x'$. But then, by general modal-logical considerations, $(\mathsf{Fml}_{\varphi\psi} \cap y)$ can be extended to an element $y' \in F_\psi$ such that $R^\diamond_{\mathfrak{F}_\psi}x'y'$. Clearly yBy'.
To prove (ii), create the set $\{\theta \in \mathsf{Fml}_{\varphi\psi} : \models \varphi \rightarrow \theta\} \cup \{\neg\theta \in \mathsf{Fml}_{\varphi\psi} : \models \theta \rightarrow \psi\}$. Since we assumed there is no interpolant for φ and ψ, we have $\not\models \varphi \rightarrow \bot$ and $\not\models \top \rightarrow \psi$, hence the two sets are satisfiable. Suppose to the contrary that the union is not satisfiable. Then we have some $\theta_1, \theta_2 \in \mathsf{Fml}_{\varphi\psi}$ such that $\models \theta_1 \wedge \neg\theta_2 \rightarrow \bot$, $\models \varphi \rightarrow \theta_1$ and $\models \theta_2 \rightarrow \psi$. But then θ_1 is an interpolant, which contradicts the assumption. So the union is satisfiable, and we can extend it to a maximal set u in $\mathsf{Fml}_{\varphi\psi}$. Clearly also $u \cup \{\varphi\}$ and $u \cup \{\neg\psi\}$ are satisfiable, so we can extend these two sets to $w \in F_\varphi$ and $v \in F_\psi$. By the definitions, $(w,v) \in B$ and $\mathfrak{M}_\varphi, w \Vdash \varphi$ and $\mathfrak{M}_\psi, v \Vdash \neg\psi$. ◀

Define the frame $\mathfrak{F}^* = (\mathfrak{F}_\varphi \times \mathfrak{F}_\psi)\restriction_B$, the valuation v^* for the union of the two languages as $(w,v) \in \mathsf{v}^*(p) \iff w \in \mathsf{v}_\varphi(p)$ *or* $v \in \mathsf{v}_\psi(p)$, and let $\mathfrak{M}^* = (\mathfrak{F}^*, \mathsf{v}^*)$. v^* is well-defined, because for $(w,v) \in B$ and p in the common language, $w \in \mathsf{v}_\varphi(p)$ iff $v \in \mathsf{v}_\psi(p)$. By canonicity, $\mathfrak{F}_\varphi$ and $\mathfrak{F}_\psi$ are in K. $\mathfrak{F}^*$ is a bisimulation–product of these two, whence, by assumption, $\mathfrak{F}^* \in \mathsf{K}$. By (ii) of the claim, there exists worlds w and v such that wBv, $\mathfrak{M}_\varphi, w \Vdash \varphi$ and $\mathfrak{M}_\psi, v \Vdash \neg\psi$. By proposition 2 and the definition of v^*, the projections π_0 and π_1 are $(\mathsf{Fml}_\varphi, S)$–, and (Fml_ψ, S)–zigzagmorphisms respectively. This gives us $\mathfrak{M}^*, (w,v) \Vdash \varphi \wedge \neg\psi$. Because $\mathfrak{F}^* \in \mathsf{K}$, this means that $\varphi \wedge \neg\psi$ is satisfiable in the logic $\mathcal{L}(\mathsf{K})$. QED

Remark 3. [Németi1985] uses a construction similar to bisimulation–products, in order to show that the algebraic counterpart of weakened first–order logic has the strong amalgamation property. [Marx1995] contains an overview of the connections between interpolation properties of logics and amalgamation properties of their algebraic counterparts.

2 Interpolation in Fibered Modal Logic

[van Benthem1996] defines the following strenghtening of the interpolation property. A logic $\mathcal{L}$ has the strong AIP if, whenever $\models_{\mathcal{L}} \varphi \to \psi$, there exists a formula θ such that $\models_{\mathcal{L}} \varphi \to \theta$, $\models_{\mathcal{L}} \theta \to \psi$ and $\Sigma(\theta) \subseteq \Sigma(\varphi) \cap \Sigma(\psi)$ and all modalities occuring in θ occur both in φ and in ψ.

Strong interpolation only makes sense in multi-modal logics. The following example shows that it is really stronger than AIP. Let $\mathcal{L}$ be the modal logic of two unary modalities $\Diamond_1$ and $\Diamond_2$ with the only condition on the frames that $\forall xy(R^{\Diamond_1}xy \to R^{\Diamond_2}xy)$. It is easy to see that $\mathcal{L}$ is a canonical modal logic which can be axiomatized by the K-axioms plus $\Diamond_1 p \to \Diamond_2 p$. By Corollary 1, $\mathcal{L}$ has AIP. But strong AIP fails as is easy to see using the $\mathcal{L}$-valid formula $\Diamond_1\top \to \Diamond_2\top$.

So axioms stating an interaction between the modalities can destroy strong AIP. The next theorem shows that without interaction axioms, strong AIP is fairly common. For multi-S5 the theorem is proved in [van Benthem1996]. Here we use a different argument, which is the obvious adaptation of the proof of Theorem 1.

Theorem 2 (van Benthem–Marx). *Let $\mathcal{L}(\mathsf{K})$ be a canonical modal logic. If K is defined by a universal Horn theory without interaction axioms, then $\mathcal{L}(\mathsf{K})$ enjoys strong AIP.*

Proof. The argument is very similar to that of Theorem 1. Let $\mathcal{L}(\mathsf{K})$ be a modal logic of an arbitrary type S as in the Theorem. We reason by contraposition. Let φ, ψ be arbitrary $\mathcal{L}(\mathsf{K})$ formulas. Suppose that there is no $\mathcal{L}(\mathsf{K})$ formula which is a strong interpolant for $\varphi \to \psi$. We have to show that $\varphi \wedge \neg\psi$ is satisfiable. Let Fml_φ and Fml_ψ be the $\mathcal{L}(\mathsf{K})$ languages generated by the propositional variables in φ, and the ones in ψ, respectively. Let S' be the type of the modalities occuring in φ and ψ. $\mathsf{Fml}'_{\varphi\psi}$ denotes the S' language generated from the variables occuring in both φ and ψ. We use the same (notation for the) models as in the proof of Theorem 1. Because we only assumed there is no strong interpolant, we have to define a weaker zigzag-connection B:

Claim. (i) $B \stackrel{\text{def}}{=} \{(w,v) \in F_\varphi \times F_\psi : \mathsf{Fml}'_{\varphi\psi} \cap w = \mathsf{Fml}'_{\varphi\psi} \cap v\}$ is an $(\mathsf{Fml}'_{\varphi\psi}, S')$–zigzag–connection between $\mathfrak{M}_\varphi$ and $\mathfrak{M}_\psi$.
(ii) There exists a $(w,v) \in B$ such that $\mathfrak{M}_\varphi, w \Vdash \varphi$ and $\mathfrak{M}_\psi, v \Vdash \neg\psi$.

Proof of Claim. As before. ◀

Now we define the common model $\mathfrak{M}^* = (\mathfrak{F}^*, \mathsf{v}^*)$. $\mathfrak{F}^*$ is defined as follows:

- its domain is B,
- for all modalities $\diamond$ in S', $R^{\diamond}_{\mathfrak{F}^*}(x_1, x'_1)\dots(x_n, x'_n)$ only if $R^{\diamond}_{\mathfrak{F}_\varphi} x_1 \dots x_n$ and $R^{\diamond}_{\mathfrak{F}_\psi} x'_1 \dots x'_n$,
- for modalities $\diamond$ occurring in φ but not in ψ, $R^{\diamond}_{\mathfrak{F}^*}(x_1, x'_1)\dots(x_n, x'_n)$ only if $R^{\diamond}_{\mathfrak{F}_\varphi} x_1 \dots x_n$,
- for modalities $\diamond$ occurring in ψ but not in φ, $R^{\diamond}_{\mathfrak{F}^*}(x_1, x'_1)\dots(x_n, x'_n)$ only if $R^{\diamond}_{\mathfrak{F}_\psi} x'_1 \dots x'_n$, and
- for all other modalities $\diamond$, $R^{\diamond}_{\mathfrak{F}^*} = \emptyset$.

v^* is defined as before: $(x, x') \in \mathsf{v}^*(p)$ iff $x \in \mathsf{v}_\varphi(p)$ or $x' \in \mathsf{v}_\psi(p)$.

It is easy to see that the projections π_0 and π_1 are $(\mathsf{Fml}_\varphi, S_\varphi)$–, and $(\mathsf{Fml}_\psi, S_\psi)$–zigzagmorphisms respectively (here S_φ denotes the modal type of all modalities occuring in φ and similar for S_ψ). This gives us $\mathfrak{M}^*, (w, v) \Vdash \varphi \wedge \neg\psi$, by (ii) of the Claim.

So we are done if $\mathfrak{F}^*$ is a frame of the logic. $\mathfrak{F}^*$ will satisfy the conditions for the common modalities for the same reason as before. Moreover, any universal Horn sentence concerning the non-common modalities is valid in $\mathfrak{M}^*$ if it is valid in the respective models. Conditions for modalities not occuring in φ or ψ are trivially satisfied. So, since there are no interaction axioms, we covered all conditions, whence $\mathfrak{F}^* \in \mathsf{K}$.

Multi-modal logics without interaction arise with combining logics using the *dovetailing* approach (cf. [Gabbayforthcoming]). E.g., by Theorem 2.10 in op. cit., it follows that

Theorem 3 (Gabbay). *Assume $\mathcal{L}(\mathsf{K}_i)$, $i \in I$, are all canonical modal logics of type S_i with K_i defined by a set of first-order conditions Σ_i. Then $\mathcal{L}^D_I$ (the dovetailing of $\mathcal{L}(\mathsf{K}_i)$) is canonical, of type $\bigcup_{i \in I} S_i$, and the class of frames with which $\mathcal{L}^I_D$ is complete is the class of all $\bigcup_{i \in I} S_i$-frames satisfying $\bigcup_{i \in I} \Sigma_i$.*

So Theorem 2 can be used to prove the strong AIP for dovetailed modal logics.

3 Modal Logics for Knowledge and Belief

Interpolation in the Standard System

In this section, we apply theorem 1 to modal logics of knowledge and belief. In the literature of philosophical logic, systems for knowledge and belief were studied in the 1960's [Hintikka1962]. In the 1980's, these notions became one of the central themes in the field of AI [Halpern and Moses1985], and are gaining their place in the field of computer science [Meyer et al.1991]. It now seems conventional to take the system S5 for knowledge and weak S5 (or KD45) for belief (cf., [Halpern and Moses1985], [Hintikka1962], [Meyer et al.1991]). This logic has all types of interpolation by theorem 1. To be precise, the logic KB of knowledge and belief is a modal logic with two unary box–type modalities K (for knowledge) and B (for belief). It is usually presented axiomatically as follows: on top of the basic **K** axioms for K and B one assumes that,

- one does not believe false assertions ($\neg\mathsf{B}\bot$);
- believers have positive ($\mathsf{B}\varphi \to \mathsf{BB}\varphi$), as well as negative ($\neg\mathsf{B}\varphi \to \mathsf{B}\neg\mathsf{B}\varphi$) introspection;
- knowledge should moreover also be veridical ($\mathsf{K}\varphi \to \varphi$);
- for interaction between the two notions, the following axioms are proposed in the literature:
  - knowledge implies belief ($\mathsf{K}\varphi \to \mathsf{B}\varphi$), and
  - "one is conscious of one's beliefs" ($\mathsf{B}\varphi \to \mathsf{KB}\varphi$).

Theorem 4. *Let KB^* be a logic of knowledge and belief which is defined axiomatically by adding any subset of the above principles to the basic* **K** *derivation system for this type. Then KB^* has all types of interpolation.*

Note that strong interpolation fails because of the axiom $\mathsf{K}\varphi \to \mathsf{B}\varphi$.

Proof. All the principles are well–known Sahlqvist forms [Sahlqvist1975]. Hence, such a logic is canonical and complete with respect to the class of frames defined by the Sahlqvist correspondents. By theorem 4.3 in [Hoek1993], all the principles, except $\neg\mathsf{B}\bot$, correspond to universal Horn sentences. The formula $\neg\mathsf{B}\bot$ corresponds to $\forall x \exists y R^{\mathsf{B}}xy$. All these sentences are preserved under bisimulation–products. So, by theorem 1, the logic has interpolation.

Interpolation in Other Systems

The logic KB is just one possibility; for some applications one might want to introduce appropriate modalities to handle e.g. multiple agents or common knowledge. [Hoek1993] presents a range of principles from which one can design its own logic of knowledge and belief. In this article, the following types of formulas are defined. Let $\mathsf{X}, \mathsf{Y}, \mathsf{Z}$ be arbitrary box–type epistemic operators. Then formulas of the form:

(a) $\mathsf{X}\varphi \to \mathsf{YZ}\varphi$ are called *positive introspection formulas*
(b) $\neg\mathsf{X}\varphi \to \mathsf{Y}\neg\mathsf{Z}\varphi$ are called *negative introspection formulas*
(c) $\mathsf{XY}\varphi \to \mathsf{Z}\varphi$ are called *positive extraspection formulas*
(d) $\mathsf{X}\neg\mathsf{Y}\varphi \to \neg\mathsf{Z}\varphi$ are called *negative extraspection formulas*
(e) $\mathsf{X}(\mathsf{Y}\varphi \to \varphi)$ are called *trust formulas*.

We will show what these principles mean for the interpolation property. Theorem 4.3 in [Hoek1993] implies that all these principles are canonical, and correspond to the following frame conditions:

$$(a)\ \forall xyz(R^{\mathsf{Y}}xy \;\&\; R^{\mathsf{Z}}yz \Rightarrow R^{\mathsf{X}}xz)$$
$$(b)\ \forall xyz(R^{\mathsf{Y}}xy \;\&\; R^{\mathsf{Z}}xz \Rightarrow R^{\mathsf{X}}yz)$$
$$(c)\ \forall xy(R^{\mathsf{Z}}xy \Rightarrow \exists z(R^{\mathsf{X}}xz \;\&\; R^{\mathsf{Y}}zy)$$
$$(d)\ \forall x\exists y(R^{\mathsf{Y}}xy \;\&\; \forall z(R^{\mathsf{Z}}yz \Rightarrow R^{\mathsf{X}}xz))$$
$$(e)\ \forall xy(R^{\mathsf{X}}xy \Rightarrow R^{\mathsf{Y}}yy)$$

(a), (b) and (e) are preserved under bisimulation–products, because they are universal Horn. It is not difficult to show preservation of (d) as well. Theorem 5.6.6

in [Marx1995] shows that the density principle (c) might result in the loss of even the weakest form of interpolation. But if R^{X} or R^{Y} is reflexive, (c) becomes trivially true, thereby not endangering the interpolation property.

So we can conclude that interpolation is often available in modal logics of knowledge and belief.

References

[Andréka et al.1994] Andréka, H., Németi, I., and Sain, I. (1994). Craig property of a logic and decomposability of theories. In Dekker, P. and Stokhof, M., editors, *Proceedings of the 9th Amsterdam Colloquium*, pages 87–92, Universiteit van Amsterdam. Institute for Logic, Language and Computation, University of Amsterdam.

[Bergstra et al.1990] Bergstra, J., Heering, J., and Klint, P. (1990). Module algebra. *Journal of the Association for Computing Machinery*, 37:335–372.

[de Rijke1993] de Rijke, M. (1993). *Extending Modal Logic*. PhD thesis, Institute for Logic, Language and Computation, University of Amsterdam, Universiteit van Amsterdam. ILLC Dissertation Series 1993–4.

[Ehrig and Mahr1985] Ehrig, H. and Mahr, B. (1985). *Fundamentals of Algebraic Specifications, 1:Equations and Initial Semantics*. Springer Verlag, Berlin.

[Gabbayforthcoming] Gabbay, D. (forthcoming). *Fibring Logics*. Oxford University Press.

[Halpern and Moses1985] Halpern, J. and Moses, Y. (1985). A guide to the modal logic of knowledge and belief. In *Proceedings IJCAI–85*, pages 480–490, Los Angeles, CA.

[Hintikka1962] Hintikka, J. (1962). *Knowledge and Belief*. Cornell University Press, Ithaca, NY.

[Hoek1993] Hoek, W. (1993). Systems for knowledge and belief. *J. Logic Computat.*, 3(2):173–195.

[Hughes and Creswell1984] Hughes, G. and Creswell, M. (1984). *A Companion to Modal Logic*. Methuen.

[Maibaum and Sadler1984] Maibaum, T. and Sadler, M. (1984). Axiomatising specification theory. In Kreowski, H., editor, *3rd Abstract Data Type Workshop*, Fachbereich Informatik 25, pages 171–177. Springer Verlag.

[Maibaum et al.1984] Maibaum, T., Veloso, P., and Sadler, M. (1984). Logical specification and implementation. In Joseph, M. and Shyamasundar, R., editors, *Foundations of Software Technology and Theoretical Computer Science*, LCNS 186, pages 13–30. Spinger Verlag.

[Maksimova1991] Maksimova, L. (1991). Amalgamation and interpolation in normal modal logics. *Studia Logica*, L(3/4):457–471.

[Marx1995] Marx, M. (1995). *Algebraic Relativization and Arrow Logic*. PhD thesis, Institute for Logic, Language and Computation, University of Amsterdam. ILLC Dissertation Series 1995–3.

[Marx and Areces1997] Marx, M. and Areces, C. (1997). Failure of interpolation in combined modal logics. Technical Report CS–RR–326, University of Warwick, Department of Computer Science. submitted for publication.

[Meyer et al.1991] Meyer, J.-J., Hoek, W., and Vreeswijk, C. (1991). Epistemic logic for computer science: A tutorial, Parts I and II. *EATCS Bulletin*, 44:242–270. (Part II is in Vol. 45, 256–287).

[Németi1985] Németi, I. (1985). Cylindric–relativised set algebras have strong amalgamation. *Journal of Symbolic Logic*, 50(3):689–700.

[Renardel de Lavalette1989] Renardel de Lavalette, G. (1989). Modularisation, parametrisation and interpolation. *J. Inf. Process. Cybern. EIK*, 25(5/6):283–292.

[Rodenburg1992] Rodenburg, P. (1992). Interpolation in equational logic. P9201, Programming Research Group, University of Amsterdam.

[Sahlqvist1975] Sahlqvist, H. (1975). Completeness and correspondence in the first and second order semantics for modal logic. In Kanger, S., editor, *Proc. of the Third Scandinavian Logic Symposium Uppsala 1973*, Amsterdam. North–Holland.

[van Benthem1983] van Benthem, J. (1983). *Modal Logic and Classical Logic*. Bibliopolis, Naples.

[van Benthem1996] van Benthem, J. (1996). Modal foundations for predicate logic. In Orlowska, E., editor, *Memorial Volume for Elena Rasiowa*, Studia Logica Library. Kluwer Academic Publishers, Dordrecht.

Building Models of Linear Logic
(Extended Abstract)

Valeria de Paiva[1] and Andrea Schalk[2]

[1] School of Computer Science, University of Birmingham
[2] Dept of Pure Maths and Math Stats, Cambridge University

Abstract. A generic method for constructing categorical models of Linear Logic is provided and instantiated to yield traditional models such as coherence spaces, hypercoherences, phase spaces, relations, etc. The generic construction is modular, as expected. Hence we discuss multiplicative connectives, modalities and additive connectives in turn. Modelling the multiplicative connectives of Linear Logic is a generalisation of previous work, requiring a few non-standard concepts. More challenging is the modelling of the modalities '!' (and, respectively '?'), which is achieved in the surprisingly general setting of this construction by considering !-candidates and showing that they exist and constitute a modality, under appropriate conditions.

1 Introduction

This paper recasts some well-known models of Linear Logic into a more general framework, that allows us to explicate some of their similarities and differences. It is pleasing (and surprising) to find that coherence spaces (Girard's original domain-theoretic model of Linear Logic), hypercoherences (Ehrhard's categorical explanation of sequentiality), phase spaces and even (the category of) sets and relations **Rel** can all be seen as specific instances of our generic construction. This generic construction can cope both with the intuitionistic and the classical flavours of Linear Logic and it allows us to model fragments of Linear Logic, in a modular fashion, as one would expect to be able to.

The motivation for this generic construction arose from comparing Chu's construction [Bar79] with dialectica categories [dP89]. We wanted to discover how far could we get when modelling Linear Logic, simply by mapping into some (linear) algebraic structure, forgetting all about any built-in duality. However, this work can be understood independently from the original motivation. It gives an account of several well-known models of Linear Logic in a unified framework, making them all instances of our particular categorical construction.

This paper is organised as follows. We first review coherence spaces and hypercoherences, our motivating examples. Then we describe our generic construction and show that the categories we obtain have a multiplicative structure. In the next section we describe the modality or exponential "!", which is the hard part when modelling Linear Logic. Then we describe the additives and discuss

A.M. Haeberer (Ed.): AMAST'98, LNCS 1548, pp. 164–177, 1998.

some other examples of models that can be seen as instances of our generic construction. This extended abstract provides all the definitions and a few hints, but full proofs can be found in the long version of the paper [dPS98].

2 Motivating Examples

We first present a different perspective on two well-known models of Linear Logic, namely coherence spaces and hypercoherences.

Recall that a coherence space X is given by a set $|X|$ (its 'web'), and a reflexive binary relation $\supset$ on $|X|$. We use $\frown$ to denote the relation resulting from removing the diagonal from $\supset$. Viewing this model from a different angle, we encode this structure via a function α_X from $|X| \times |X|$ to the three element ordered set $\mathbf{3} := \{\smile < 1 < \frown\}$, where $\langle x, x' \rangle$ in $|X| \times |X|$ is mapped to 1 iff $x = x'$, and to $\frown$ iff $x \frown x'$. We hope overloading the symbol $\frown$ with two totally different meanings does not cause confusion.

A morphism of coherence spaces is given by a relation $R \colon |X| \longrightarrow\!\!\!\!\!+\ |Y|$ such that $x\ R\ y$, $x'\ R\ y'$ and $x \supset x'$ imply $y \supset y'$. In the representation of coherence spaces as maps from $|X| \times |X|$ to $\mathbf{3}$, this is equivalent to the condition:

$$\langle x, x' \rangle\ R \times R\ \langle y, y' \rangle \text{implies}\ \alpha_X \langle x, x' \rangle \leq \alpha_Y \langle y, y' \rangle.$$

The ordered set $\mathbf{3}$ can easily be seen to carry a symmetric monoidal closed structure. If one considers the operations $\bullet$ and $\triangleright$ as follows

$\bullet$	$\frown$	1	$\smile$
$\frown$	$\frown$	$\frown$	$\smile$
1	$\frown$	1	$\smile$
$\smile$	$\smile$	$\smile$	$\smile$

$\triangleright$	$\frown$	1	$\smile$
$\frown$	$\frown$	$\smile$	$\smile$
1	$\frown$	1	$\smile$
$\smile$	$\frown$	$\frown$	$\frown$

then $\mathbf{3}$ is a symmetric monoidal closed poset, where $\bullet$ stands for tensor product and $\triangleright$ for linear implication.

The tensor product $X \otimes Y$ of two coherence spaces X and Y has as underlying set $|X| \times |Y|$ and its tensor product structure is defined by $\langle x, y \rangle \supset \langle x', y' \rangle$ iff $x \supset x'$ and $y \supset y'$. In terms of a function $(|X| \times |Y|) \times (|X| \times |Y|) \longrightarrow \mathbf{3}$, this amounts to a mapping sending $\langle \langle x, y \rangle, \langle x', y' \rangle \rangle$ to $\alpha_X \langle x, x' \rangle \bullet \alpha_Y \langle y, y' \rangle$. Similarly, the linear hom $X \multimap Y$ of X and Y, has $|X| \times |Y|$ as its underlying set. In this case $\langle x, y \rangle \supset \langle x', y' \rangle$ iff ($x \supset x'$ implies $y \supset y'$) and ($x \frown x'$ implies $y \frown y'$). Again, this can easily be expressed as a function $(|X| \times |Y|) \times (|X| \times |Y|) \longrightarrow \mathbf{3}$, namely the one sending the pair $\langle \langle x, y \rangle, \langle x', y' \rangle \rangle$ to $\alpha_X \langle x, x' \rangle \triangleright \alpha_Y \langle y, y' \rangle$. Thus the tensor of coherence spaces is obtained using the tensor product of $\mathbf{3}$ and the linear hom of coherence spaces is similarly obtained from the linear hom in $\mathbf{3}$. We say that the symmetric monoidal closed structure on $\mathbf{3}$ induces that of the category of coherence spaces.

Now consider a second example of a model of linear logic, namely that of hypercoherences [Ehr93]. A hypercoherence X is given by a set $|X|$ (also called the 'web'), and a subset $\Gamma(X)$ of the set of finite non-empty subsets of $|X|$ containing all singletons. This can be encoded as a function $\alpha_X \colon \mathcal{P}_{\mathsf{fne}}|X| \longrightarrow \mathbf{3}$, where $\mathcal{P}_{\text{fne}}$ denotes the (finite, non-empty) powerset functor: α_X maps a finite subset a of $|X|$ to 1 iff a is a singleton, and to $\frown$ iff it is an element of $\Gamma(X)$. Again, this representation does capture the tensor product and the linear hom of hypercoherences in terms of the same operations on **3**.

Morphisms of hypercoherences also fit well into this representation. A morphism of hypercoherences is a relation $R\colon |X| \longrightarrow\!\!\!\!\!\!+\;\, |Y|$ such that for every finite subset E of R, the following conditions are satisfied:

(i) If $\pi_1(E) \in \Gamma(X)$ then $\pi_2(E) \in \Gamma(Y)$.
(ii) If $\pi_2(E)$ is a singleton then $\pi_1(E)$ is a singleton.

The finite powerset functor on the category of sets and relations **Rel** is defined on morphisms by $a\ \mathcal{P}_{\text{fne}}R\ b$, where $a \in \Gamma(X)$, $b \in \Gamma(Y)$, iff there is a (finite) subset E of R with $\pi_1(E) = a$ and $\pi_2(E) = b$. Then R being a morphism of hypercoherences is equivalent to the condition:

$$a\ \mathcal{P}_{\text{fne}}R\ b \text{ implies } \alpha_X(a) \leq \alpha_Y(b)$$

These observations suggest a pattern. As objects of an unifying model we consider functions $\alpha\colon FA \longrightarrow L$, where F is a functor on the category of sets and relations, **Rel**, A is a set and L is a poset which, as a category, is a symmetric monoidal closed category (smcc). We write (A, α) for these objects and call α the structure on A. A morphism from (A, α) to (B, β) is a relation $R\colon A \longrightarrow\!\!\!\!\!\!+\;\, B$ such that for a in FA, b in FB, a (FR) b implies $\alpha(a) \leq \beta(b)$. These stipulations do define a general category. To obtain a tensor-product on this category, we need a natural transformation with components $\sigma_{A,B}\colon F(A \times B) \longrightarrow FA \times FB$, satisfying suitable properties to make $A \times B$ (with a natural structure map) the tensor product of (A, α) and (B, β). Similarly, we obtain a linear function space, and we demonstrate that the other connectives of Linear Logic can also be handled in this setting. The construction sketched turns out to be flexible enough to model both Classical and Intuitionistic Linear Logic, depending on which conditions we choose to impose on the poset L. For a model of Full Intuitionistic Linear Logic we use what has been called a *lineale* in previous work [dP90].

3 The Basic Construction

The category of sets and relations **Rel** was one of the very first models of Linear Logic, but **Rel** is a very collapsed model, as different logical connectives are modelled by the same categorical structure. Both products and co-products in **Rel** have the disjoint union as underlying set, whereas tensor and 'par' are both modelled via the cartesian product, which also serves as the set underlying the linear function space. Moreover the truthvalues 1 and $\perp$, the units for tensor and par coincide. Our aim is to build models for linear logic which suffer less collapse than the category **Rel** we start with. The ingredients we use are the following: An algebraic model L for the multiplicative-additive fragment of the logic we are aiming for, an endo-functor on **Rel** (satisfying certain properties which we will state later) which allows for different flavours of the model and the categorical structure of **Rel**. Note that we do *not* require modalities in L.

Since some of the arrows we will consider are functions whereas others are relations, it is advisable to have different notations for the two. We denote an arrow in **Rel** from A to B by $A \longrightarrow\!\!\!\!\!\!\!+\;\; B$. Furthermore, given $R: A \longrightarrow\!\!\!\!\!\!\!+\;\; B$ we will use R^{op} to name the morphism $B \longrightarrow\!\!\!\!\!\!\!+\;\; A$ defined by $y\ R^{op}\ x$ iff $x\ R\ y$. This will not stop us from using the usual arrow symbol for relations which we know to be functions, even though we consider them as arrows in **Rel**. All symbols referring to categorical constructs such as $\times$, $+$, $\otimes$ and $\Box$ (the symbol we use for Girard's 'par' connective) are meant to be interpreted in terms of **Rel** rather than **Sets**. We view the structure on L to consist of morphisms in **Rel**, such as $\bullet: L \otimes L \longrightarrow L$ or $\triangleright: L \multimap L \longrightarrow L$ (notice the use of plain arrows here).

There are a number of ways of obtaining linear logic modalities on **Rel**. Basically in addition to the constructs mentioned above we need a monoidal comonad such that all free co-algebras of this comonad are commutative comonoids, and morphisms between them have to respect the comonoid structure. However, in **Rel**, all morphisms can easily be 'turned around' (passing from R to R^{op}), which means that instead of comonads and comonoids we might as well consider monads and monoids which are more familiar. In order to make **Rel** a linear category, take any monad whose Eilenberg-Moore algebras carry a commutative monoid structure. This could be, for example, the free commutative monoid, or the finite powerset monad. In order to obtain the desired comonad, just turn around all the morphisms. For our generic construction it is of no importance which way of modelling ! is chosen—they will all make the resulting category a linear one.

3.1 L_F-sets

We are now ready to introduce formally the category we are interested in.

Definition 1. *Let F be an endo-functor on* **Rel** *and let L be a closed poset. An L_F-set is a pair (A, α), where α is a function $FA \longrightarrow L$. Given two L_F-sets (A, α) and (B, β), we define a morphism of L_F-sets to be a relation $R: A \longrightarrow\!\!\!\!\!\!\!+\;\; B$ such that $x\ (FR)\ y$ implies $\alpha(x) \leq \beta(y)$, for $x \in FA$, $y \in FB$.*

We denote a morphism using our symbol for relations, *ie* $R: (A, \alpha) \longrightarrow\!\!\!\!\!\!\!+\;\; (B, \beta)$. Equivalently, we require a morphism to satisfy the inequality $\alpha \leq \beta \circ FR$, represented by the 'weakly commuting' diagrams drawn as follows:

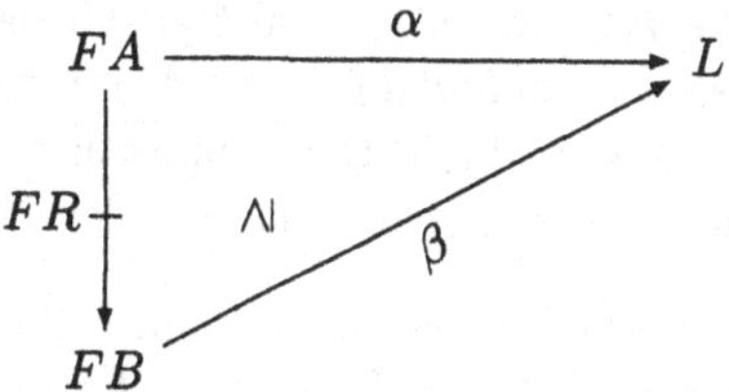

Here $\leq$ is a pre-order defined on hom-sets with co-domain L as follows: For $\alpha, \alpha' \in$ **Rel**(FA, L) we say that $\alpha \leq \alpha'$ iff for all $x \in FA$, $x\ \alpha\ l$ and $x\ \alpha'\ l'$ implies $l \leq l'$.

We shall need a number of useful properties of this pre-order. These are stated in the following lemma which makes statements about the interaction of the order $\leq$ and composition $\circ$, tensor and linear function space.

Lemma 1. *(i) Let $R: B \longrightarrow\!\!\!\!\!+\, A$ and $S, S': A \longrightarrow\!\!\!\!\!+\, L$ be relations. If $S \leq S'$ then $S \circ R \leq S' \circ R$.*
(ii) Let $R, R': A \longrightarrow\!\!\!\!\!+\, L$ and $S, S': B \longrightarrow\!\!\!\!\!+\, L$. If $R \leq R'$ and $S \leq S'$ then $\bullet \circ R \otimes S \leq \bullet \circ R' \otimes S'$.
(iii) Let $R, R': A \longrightarrow\!\!\!\!\!+\, L$ and $S, S': L \longrightarrow\!\!\!\!\!+\, B$. If $R \leq R'$ and $S^{op} \leq S'^{op}$ then $\triangleright \circ S \multimap R \leq \triangleright \circ S' \multimap R'$.
(iv) Let $\alpha, \beta: A \longrightarrow L$. Then $\alpha \leq \beta$ if and only if α is less than or equal to β in the pointwise order for functions.

Identities in L_F-**Sets** are identity relations, and composition is also taken from **Rel**. It is not difficult to check that this defines a category which we call L_F-**Sets**.

4 Multiplicative Structure

We claim that the categories L_F-**Sets** are suitable for modelling the connectives of linear logic, and that in general they suffer less collapse than **Rel**. To define a symmetric monoidal closed structure on L_F-**Sets** we make use of the fact that both, **Rel** and L carry such a structure. Let σ be a natural transformation between the bifunctors $F(\_ \otimes \_)$ and $F\_ \otimes F\_$ such that all its components are functions. Given this, we define a tensor-product on L_F-**Sets**.

Definition 2. *The* tensor product *of two L_F-sets (A, α) and (B, β) has as its underlying set $A \otimes B$ and the structure is given by:*

$$F(A \otimes B) \xrightarrow{\sigma_{A,B}} FA \otimes FB \xrightarrow{\alpha\otimes\beta} L \otimes L \xrightarrow{\bullet} L$$

For morphisms, let the tensor product be the one from **Rel**.

It is not difficult to prove that this defines a symmetric monoidal structure on L_F-sets but the proof demonstrates the need for a natural transformation between bifunctors $F(\_ \otimes \_) \longrightarrow F\_ \otimes F\_$ with properties which are slightly non-standard. These correspond to four commuting diagrams, spelled out in the appendix. A functor F endowed with a natural transformation satisfying these commuting diagrams we call a *comonoidal functor*. A unit for this tensor product is given by (I, ι_1), where $I = \{*\}$ and ι_1 is the composition of the (only) function $FI \longrightarrow I$ and the function $e: I \longrightarrow L$ that picks the identity 1 for $\bullet$ in L.

Proposition 1. *Let L be a closed poset, let F be an endo-functor on* **Rel**, *and let the natural transformations $\sigma: F(\_ \otimes \_) \longrightarrow F\_ \otimes F\_$ and $\sigma_I: FI \longrightarrow I$ make F a* comonoidal *functor, and assume that σ_I as well as all components of σ are functions. Then the tensor product defined above makes L_F-***Sets** *a symmetric monoidal category.*

If we assume that there is a dinatural transformation τ between the mixed variance functors $F(\_\multimap\_)$ and $F\_\multimap F\_$ such that all components of it are functions, then we can define a linear function space on L_F-**Sets**.

Definition 3. *The underlying set of the 'internal hom' of two L_F-sets (A,α) and (B,β) is $A\multimap B$, and the structure is defined by: $F(A\multimap B) \xrightarrow{\tau_{A,B}} FA \otimes FB \xrightarrow{\alpha^{op}\multimap\beta} L\otimes L \xrightarrow{\triangleright} L$.*

Again, on morphisms we use the usual definition on **Rel**. It is not difficult to show that we thus obtain a functor $\multimap$ of mixed variance as desired. In order to provide the desired adjoint for $\otimes$, however, we need to demand a close relationship between F, σ, τ and the units and co-units of the adjunctions between $\otimes$ and $\multimap$ in **Rel**, say η and ε. This gives us another two commuting diagrams. A functor F that satisfies these extra conditions (diagrams in the appendix) could be called *co-exponentiable.* We still need a preliminary lemma which shows that the adjunction in **Rel** and the 'internal' one of L are related:

Lemma 2. *If α is a function $A \longrightarrow L$ then the following diagrams are (weakly) commutative:*

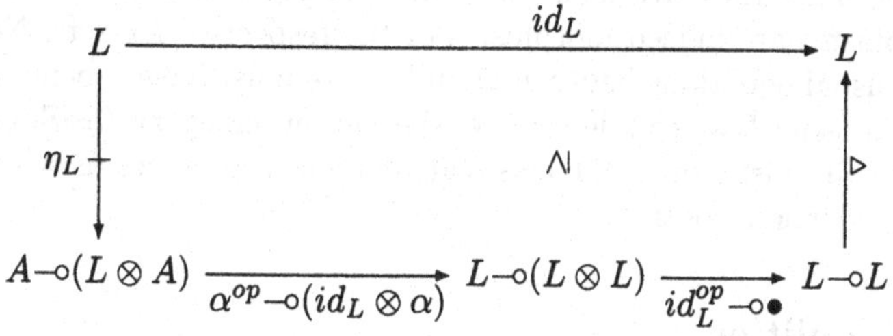

$$\begin{array}{ccccc}
(A\multimap L)\otimes A & \xrightarrow{(\alpha^{op}\multimap id_L)\otimes\alpha} & (L\multimap L)\otimes L & \xrightarrow{\triangleright\otimes id_L} & L\otimes L \\
\varepsilon_L \downarrow & & \geq & & \downarrow \bullet \\
L & & \xleftarrow{\quad id_L \quad} & & L
\end{array}$$

Now we can prove.

Theorem 1. *If F is comonoidal and co-exponentiable then the category L_F-**Sets** is symmetric monoidal closed.*

Since **Rel** is compact closed, it is not a priori necessary for the natural transformations σ and τ to be different, but as we think of generalizing our results to a more categorical setting it seems appropriate to keep them distinct. Note that it is mostly the structure on L that is responsible for L_F-**Sets** not being compact closed: even if τ and σ are the same, L_F-**Sets** will not be compact closed as long as L is not.

4.1 Negation and 'Par'

If the algebraic model L carries a par operation, we can define one for L_F-**Sets** exactly as we defined the tensor product (re-using σ), and so long as par and tensor are interpreted differently on L, these two will not collapse for the category L_F-**Sets**.

Definition 4. *Set* $(A,\alpha)\Box(B,\beta)$ *as* $(A\otimes B, \Box\circ(\alpha\otimes\beta)\circ\sigma_{A,B})$. *The unit for par is given by* $(I,\iota_\bot)$, *where* $\iota_\bot$ *is the composition of* $\sigma_I\colon FI \longrightarrow I$ *and the function* $I \longrightarrow L$ *that picks out the unit* $\bot$ *for par in* L.

We obtain a definition for negation also from the corresponding structure on **Rel** and L, the simplest way of writing it being $(A,\alpha)^\bot := (A,\alpha)\multimap(I,\iota_\bot)$. In other words, the underlying set for $(A,\alpha)^\bot$ is A, and the structure is given by composing α with the negation on L, *ie* $l \longmapsto l\multimap\bot =: l^\bot$.

If L is a model for the multiplicative additive fragment of (classical) Linear Logic then certain equalities hold in L, for example $l^{\bot\bot} = l$, $l^\bot \otimes m^\bot = (l\Box m)^\bot$ etc. These translate into isomorphisms of L_F-sets, *ie* $(A,\alpha)^{\bot\bot} \cong (A,\alpha)$, $(A,\alpha)^\bot \otimes (B,\beta)^\bot = ((A,\alpha)\Box(B,\beta))^\bot$. etc. The underlying sets are obviously isomorphic (actually they are identical), and the equations on L ensure that these isomorphisms are indeed morphisms of the respective L_F-sets. Note that to obtain the usual equations between $\Box$ and $\multimap$ we must have τ being equal to σ. Hence by choosing L we can determine whether our category L_F-**Sets** will be more in the intuitionistic or in the classical vein which provides us with a close and relatively effortless control.

5 The Modalities

To obtain a linear category, we need a monoidal comonad on L_F-**Sets** such that all free co-algebras carry a commutative comonoid structure. We assume that such a structure is already given on **Rel**, namely we assume a monoidal comonad $(!, \epsilon, \delta, m_I, m)$ and natural transformations on the category of free algebras with components $d_A\colon !A \longrightarrow !A\otimes !A$ and $e\colon !A \longrightarrow I$ making $!A$ a commutative comonoid. We will abuse notation and refer to the desired monoidal comonad on L_F-**Sets** again as $(!, \epsilon, \delta, m_I, m)$. This is not problematic since it is obtained by 'lifting' the structure on **Rel**.

The underlying set of $!(A,\alpha)$ is $!A$, but we have to work hard to obtain a suitable structure on that set. Since morphisms of L_F-sets are just relations (satisfying a certain inequality), the definition of ! on morphisms remains the one in **Rel**, but we must ensure that this will indeed satisfy the desired inequality and thus be a morphism of L_F-sets. Similarly, we want to keep the components of the natural transformations ϵ, δ, m, d and e as well as the definition of m_I as the ones in **Rel**. For that we have to choose the structure on $!(A,\alpha)$ in such way that all these become morphisms in L_F-**Sets**. This results in seven inequalities that we want to hold. But three of those inequalities are more central then the remaining four. Hence we have the following definition.

Definition 5. *A* !-candidate for (A, α) *is a function* $t: F!A \longrightarrow L$ *that satisfies the following inequalities:*

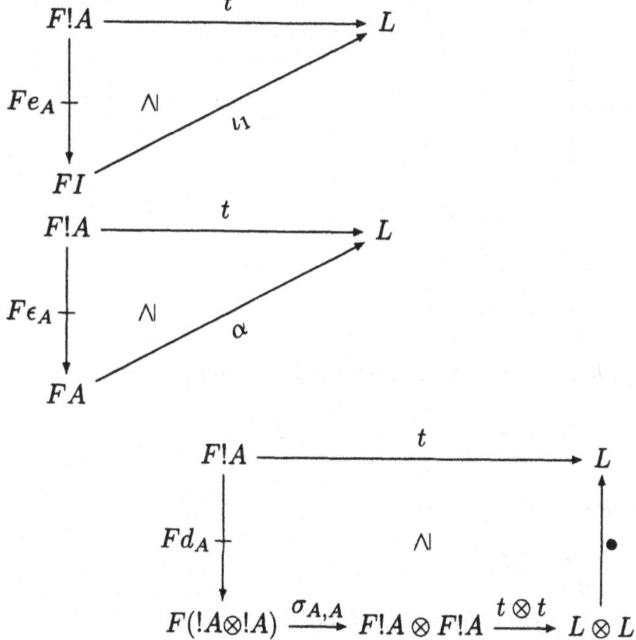

In other words, a !-candidate t for (A, α) makes $(!A, t)$ a co-monoid 'co-generated' by (A, α). And among all the !-candidates there is a canonical one that will yield the desired object, namely the (pointwise) join of all !-candidates—if it exists. To ensure that, we assume from now on that L has all joins, *ie* is a complete lattice.

Proposition 2. *The pointwise join of all !-candidates for an* L_F*-set* (A, α) *is another !-candidate.*

There is a crucial property which all the categories of L_F-sets share which assures that the join of all !-candidates will result in the desired linear structure for the category L_F-**Sets**.

Lemma 3. [Fill-in property] *Let* L *be a complete lattice, let* (A, α) *be an* L_F*-setand let* $R: A \longrightarrow\!\!\!\!\!\!\!+\ \ B$ *be any relation. Then there is* $\beta: FB \longrightarrow L$ *making* (B, β) *an* L_F*-set such that*

(i) The relation R *is a morphism* $(A, \alpha) \longrightarrow (B, \beta)$ *of* L_F*-sets,* ie *the following diagram commutes weakly:*

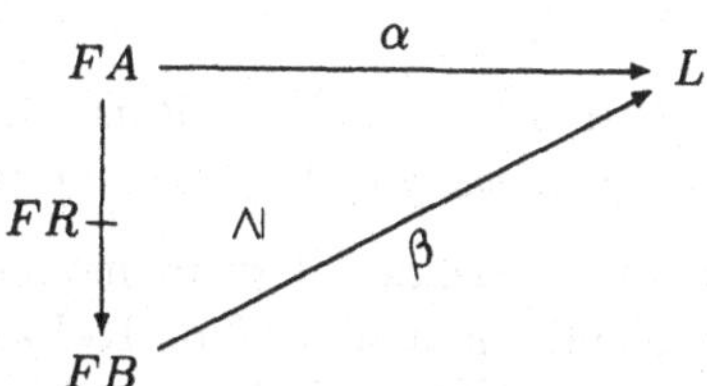

(ii) If $S: B \longrightarrow C$ such that $S \circ R$ is a morphism of L_F-sets $(A, \alpha) \longrightarrow (C, \gamma)$ then S is a morphisms $(B, \beta) \longrightarrow (C, \gamma)$, ie *the weak commutativity of*

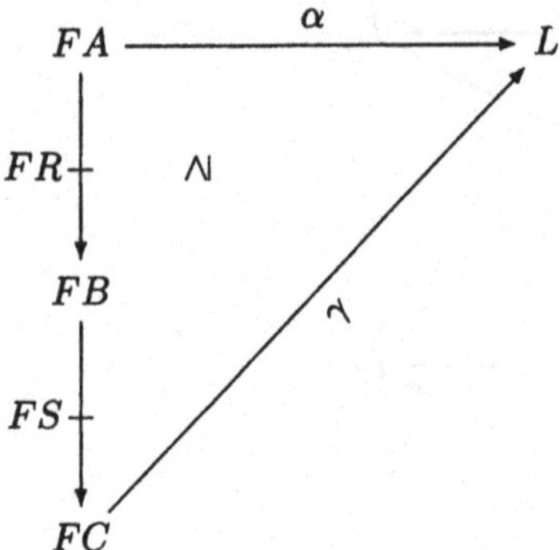

implies that the following diagram is also weakly commutative:

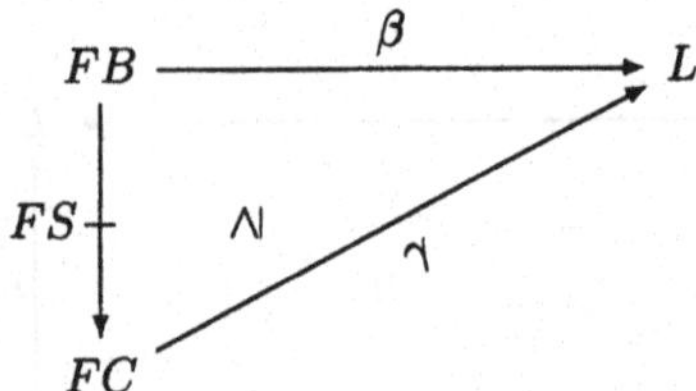

(iii) The function β is uniquely defined by (i) and (ii).

We still need to explain how to obtain !-candidates for the tensor-product of two L_F-sets.

Lemma 4. *If t and u are !-candidates for (A, α) and (B, β) respectively, then the function obtained via the fill-in property (Lemma 3) from the following diagram is a !-candidate for $(A, \alpha) \otimes (B, \beta) = (A \otimes B, \bullet \circ t \otimes u \circ \sigma_{A,B})$:*

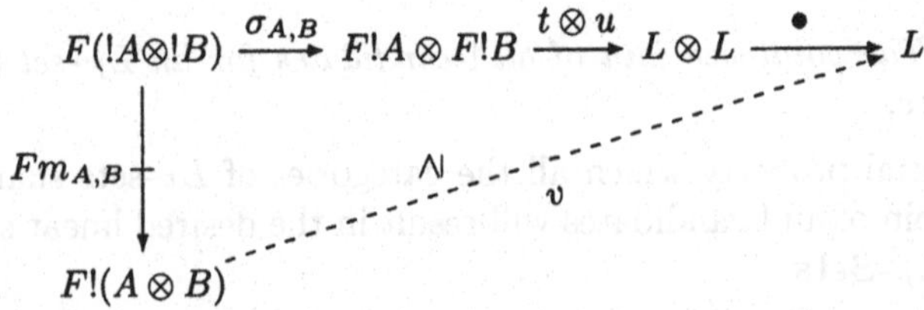

The whole description of $!A$ and especially the fill-in property do not look very categorical, but given the generality of this approach (after all, F is just an arbitrary endo-functor on **Rel**), one cannot expect a constructive definition of the desired structure $F!A \longrightarrow L$. It is rather surprising that we can get away with so general a situation. Using *linear category* in the sense of Bierman [Bie95] we obtain:

Theorem 2. *If L is a complete lattice and closed as a poset, then, under the assumptions of Theorem 1, any category of L_F-sets is a linear category.*

Having defined !, we can use it to define ? if we are modelling (classical) Linear Logic–alternatively, we can take the monad for ? on **Rel** and define a monad on L_F-**Sets** from there. The two approaches will lead to the same result.

6 The Additives

To model the additive connectives we assume—as we did for the modalities—that L is a complete lattice, which results in the fill-in property of Lemma 3. Obviously, the dual of that lemma also holds. We shall need one version for products and the other for co-products.

Looking at co-products we have a candidate for the underlying set of $(A, \alpha) + (B, \beta)$, namely $A + B$, and candidates for the embeddings, namely inl and inr from **Rel**. It remains to determine the structure, *ie* a function $F(A + B) \longrightarrow L$ that will give us the desired universal property. We obtain this function from the fill-in property again.

Consider the following diagrams:

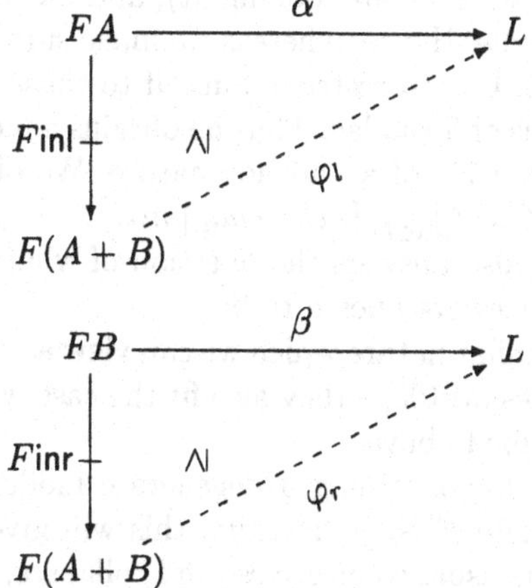

The co-product is given by $(A+B, \varphi_l \vee \varphi_r)$ where the structure is the pointwise join of those two functions. Obviously, products can be obtained via the dual of this process.

Proposition 3. *If L is a complete lattice then any category of L_F-sets has all products and coproducts.*

7 Examples and Properties

7.1 Phase Spaces and Completeness

First of all, if F is a constant functor, mapping everything to a one element set, then the structure map α picks out an element of L. The operations on these L_F-**Sets** are as defined on L, except !, obviously, which is a derived operation, and in that case Theorem 2 tells us how to define a modality for L.

One particular instance of this would be that of phase spaces: Recall that a phase space M consists of a commutative monoid and a subset $\perp$ of M. For subsets X of M, negation is defined via

$$X^{\perp} := \{m \in M \mid \forall n \in X.mn \in \perp\}.$$

In this case, let $L := \{X \subseteq M \mid X = X^{\perp\perp}\}$—in [Gir87], Girard calls these sets 'facts'. They form a complete lattice with respect to $\subseteq$ since facts are closed under arbitrary intersection. The tensor is given via $X \otimes Y := \{mn \mid m \in X, n \in Y\}^{\perp\perp}$. We will not repeat here how the other connectives are defined. It is interesting to compare Girard's exponentials to the ones obtained from our technique. Girard in fact, gives more than one interpretation for !, but we are interested in the one in [Gir95], where $!X = (X \cap I)^{\perp\perp}$ ($I = 1^{\perp\perp}$ is the unit for $\otimes$, where 1 is the unit of the monoid M). We define ! as the largest function $t: L \longrightarrow L$ satisfying $t(X) \leq I$, $t(X) \leq X$ and $t(X) \leq t(X) \otimes t(X)$. (Note that due to the nature of F, it does not matter with which definition of ! on **Rel** we start.) Obviously, Girard's definition satisfies the first two of those inequalities, but not the third. However, if a formula ϕ is provable in Linear Logic, then 1 is an element of its interpretation (no matter which phase space we are looking at), and for facts X containing 1, the desired inequality is true. Hence, whereas we make sure that $!X \leq !X \otimes !X$ is true for all elements of L, Girard restricts himself to those elements which can possibly be interpretations of formulae. Thus he obtains a nice explicit definition for ! which, however, has a bit of an ad hoc nature. We obtain the somewhat less appealing formula $!X := \bigcup_{n \in N}\{m_1 \cdots m_n \mid m_1, \ldots, m_n \in X\}^{\perp\perp}$.

The above discussion also answers the question of whether our semantics is complete, since phase spaces are known to be.

Another model of a similar nature which we can view as a category of L_F-sets are Mitchell's IE-quantales [Mit95]—they also fit the case where F is a constant functor of the kind described above.

The other possibility for obtaining a degenerate model is to have L be a singleton. In the case where F is the identity, this will give us the category of sets and relations with the usual connectives (and collapses, of course), and with whatever definition of ! we start.

7.2 Coherence Spaces and Hypercoherences Revisited

We have seen in Section 2 that coherence spaces can be encoded as L_F-sets, by choosing L to be $\mathbf{3}$, and F to be the diagonal functor Δ for the tensor product on **Rel**. If we denote by G the functor obtained from mapping a coherent space to the corresponding $\mathbf{3}_\Delta$-set, we get:

Proposition 4. *The functor G is full and faithful and preserves the monoidal closed structure on the category of coherence spaces as well as products and coproducts.*

The image of the embedding G consists of all $\mathbf{3}_\Delta$-sets whose structure map takes the value 1 exactly on the diagonal. A few calculations show that the modalities we obtain for $\mathbf{3}_\Delta$-sets if we take ! to be the finite powerset functor on **Rel** correspond to the ones described in [Gir95]. To obtain the ones from [Gir87] we have to adjust our definition of !: this is because the underlying set for $!(A, \alpha)$ is always $!A$ in our construction, *ie* α cannot be used to determine a subset of $!A$ as in the usual version of the modalities for coherence spaces. However, we can provide for that to some degree: Under the assumptions of Theorem 2, let there be a subset $!_\alpha A$ of $!A$ for every (A, α). Further assume that the restrictions and co-restrictions of the linear structure on **Rel** to these subsets, and the equations between those are still valid. If F preserves inclusion of relations, then this results in another linear category. We will not go into the details of the proof here—basically, it consists on showing that the notion of !-candidate can be adapted to those circumstances.

The other example we looked at in Section 2 was that of hypercoherences. To encode those, we chose $L = \mathbf{3}$ and $F = \mathcal{P}_{\text{fne}}$, the finite powerset functor on **Rel**. By defining the resulting functor on morphisms the same way we defined G above, we again obtain a full and faithful embedding that preserves the monoidal closed structure. Its image in the category of $\mathbf{3}_{\mathcal{P}_{\text{fne}}}$-sets is given by those (X, α_X) which take the value 1 if and only if the argument is a singleton. However, the image of that embedding is not closed under products and co-products (the property of the structure taking the value 1 exactly on singletons is not preserved under these constructions). The modalities, however, can be expressed in the category of $\mathbf{3}_{\mathcal{P}_{\text{fne}}}$-sets as described in the section on coherence spaces.

7.3 Lamarche's $Q^n_{\mathcal{A}}$-Coherences

Lamarche's attempt to find a generalization of models for Linear Logic such as hypercoherences and coherence spaces led him to the introduction of what he calls $Q^n_{\mathcal{A}}$-coherences [Lam95], where Q is a quantale. These can be viewed as $Q_{\mathcal{P}_n}$-sets, where $\mathcal{P}_n$ is a powerset functor which only considers sets up to cardinality n. (The additional parameter $\mathcal{A}$ specifies a subset of Q which is used to 'mark' the singleton sets.) Any category of $Q^n_{\mathcal{A}}$-coherences can be embedded into the category of $Q_{\mathcal{P}_n}$-sets, and this embedding preserves all multiplicative and additive connectives, is full and faithful.

8 Conclusions

It is pleasing to see that so many, in principle very different, models of Linear Logic can be cast into the framework of L_F-**Sets**. We remark that the unification happens not only on the level of representing the previous models as L_F-**Sets**, but also in that the relevant categorical structure is preserved by the representation.

It is worth pointing out that there is a constructive version of our construction - provided that the functors (F and !) involved satisfy a simple finitariness condition, we can model Intuitionistic Linear Logic even if L only has finite meets and joins.

Furthermore, there is a relatively simple instantiation of our construction which, unlike hypercoherences and coherence spaces, does not validate the mix rule: Instead of choosing **3** as the underlying poset, take the four element lattice. This will, of course, also ensure that all the constants are different. We believe there should be many applications for these very discriminating kinds of models, but have not had the time to work them out.

Still there are two classes of models that we have not been able to fit into this general framework: dialectica models and games models are not straightforward L_F-**Sets**. This is due to the fact that these models have an 'external' duality - which is achieved by giving two components to every object, and swapping them by negation. We cannot realistically hope to capture this via an internal duality, such as the one on L, which is the one we use to obtain the duality of L_F-**Sets**. We are independently pursuing the work on these two classes of models using other methodologies.

We would like to thank Martin Hyland for numerous discussions on the subject of this paper. Thanks also to Paul Taylor for his diagrams package.

References

[Bar79] Michael Barr. **-Autonomous Categories.* Monographs in Mathematics. Springer-Verlag, 1979.

[Bie95] G. M. Bierman. What is a categorical model of intuitionistic linear logic? In *Proceedings of the Conference Typed Lambda-Calculi and Applications*, volume 902 of *LNCS*, pages 341–356, 1995.

[dP89] Valeria de Paiva. A dialectica model of linear logic. In *Proceedings of the Conference Category Theory in Computer Science*, volume 389 of *LNCS*, pages 341–356, 1989.

[dP90] V. de Paiva. Lineales. Unpublished Manuscript, 1990.

[dPS98] V. de Paiva and A. Schalk. Poset-valued sets or how to build models for linear logic. Manuscript, 1998.

[Ehr93] Thomas Ehrhard. Hypercoherences: a strongly stable model of linear logic. *Math. Struct. in Comp. Science*, 3:365–385, 1993.

[Gir87] Jean-Yves Girard. Linear logic. *Theoretical Computer Science*, 50:1–102, 1987.

[Gir95] J.-Y. Girard. Linear logic: its syntax and semantics. In J.-Y. Girard, Y. Lafont, and L. Regnier, editors, *Advances in linear logic.* London Mathematical Society Lecture Note Series, Cambridge University Press, 1995. Available by anonymous ftp from lmd.univ-mrs.fr as /pub/girard/Synsem.ps.Z.

[Lam95] François Lamarche. Generalizing choerence spaces and hypercoherences. *Electr. Notes in Theor. Comp. Science*, 1, 1995.

[Mit95] W Mitchell. The carcinogenic example. Manuscript available from `http://www.cs.man.ac.uk/ bill/Car.ps`, 1995.

Appendix A: Diagrams for Definitions

We must explain the notions of comonoidal and co-exponentiable functors.

Definition 6. *Let* **C** *be a (symmetric) monoidal category and let F be an endofunctor on* **C**. *We say that the natural transformation of bifunctors* $\sigma: F(\_ \otimes \_) \longrightarrow F\_ \otimes F\_$ *(where all the components are functions) and the map* $\sigma_I: FI \longrightarrow I$ *make F a* (symmetric) co-monoidal *functor iff the following (overleaf) diagrams commute, where α, ρ, λ and γ denote the usual isomorphisms in (symmetric) monoidal categories.*

$$\begin{array}{ccc} F(I\otimes A) & \xrightarrow{\sigma_{I,A}} & FI\otimes FA \\ {\scriptstyle F\lambda_A}\downarrow & & \downarrow{\scriptstyle \sigma_I\otimes id_{FA}} \\ FA & \xrightarrow{\lambda_A^{-1}} & I\otimes FA \end{array} \qquad \begin{array}{ccc} F(A\otimes I) & \xrightarrow{\sigma_{A,I}} & FA\otimes FI \\ {\scriptstyle F\rho_A}\downarrow & & \downarrow{\scriptstyle id_{FA}\otimes\sigma_I} \\ FA & \xleftarrow{\rho_A} & FA\otimes I \end{array}$$

$$\begin{array}{ccccc} F((A\otimes B)\otimes C) & \xrightarrow{\sigma_{A\otimes B,C}} & F(A\otimes B)\otimes FC & \xrightarrow{\sigma_{A,B}\otimes id_{FC}} & (FA\otimes FB)\otimes FC \\ {\scriptstyle F\alpha_{A,B,C}}\downarrow & & & & \downarrow{\scriptstyle \alpha_{FA,FB,FC}} \\ F(A\otimes(B\otimes C)) & \xrightarrow{\sigma_{A,B\otimes C}} & FA\otimes F(B\otimes C) & \xrightarrow{id_{FA}\otimes\sigma_{B,C}} & FA\otimes(FB\otimes FC) \end{array}$$

And, in the case of symmetry:

$$\begin{array}{ccc} F(A\otimes B) & \xrightarrow{\sigma_{A,B}} & FA\otimes FB \\ {\scriptstyle F\gamma_{A,B}}\downarrow & & \downarrow{\scriptstyle \gamma_{FA,FB}} \\ F(B\otimes A) & \xrightarrow{\sigma_{B,A}} & FB\otimes FA \end{array}$$

Definition 7. *Let* **C** *be a (symmetric) monoidal closed category and let* F *be a (symmetric) co-monoidal endo-functor on* **C**. *We say that the natural transformation of bifunctors* $\tau\colon F(\_\multimap\_) \longrightarrow F\_\multimap F\_$ *make* F *a* co-exponentiable *functor iff the following diagrams commute, where* ε *and* η *are the units of the adjunction that makes* **C** *(symmetric) monoidal closed.*

$$\begin{array}{ccc} FA & \xrightarrow{\eta_{FA}^{FB}} & FB\multimap(FA\otimes FB) \\ {\scriptstyle F\eta_A^B}\downarrow & & \uparrow{\scriptstyle id_{FB}{}^{op}\multimap\sigma_{A,B}} \\ F(B\multimap(A\otimes B)) & \xrightarrow{\tau_{B,A\otimes B}} & FB\multimap F(A\otimes B) \end{array}$$

$$\begin{array}{ccc} F((B\multimap C)\otimes B) & \xrightarrow{F\varepsilon_C^B} & FC \\ {\scriptstyle \sigma_{B\multimap C,B}}\downarrow & & \uparrow{\scriptstyle \varepsilon_{FC}} \\ F(B\multimap C)\otimes FB & \xrightarrow{\tau_{B,C}\otimes id_{FB}} & (FB\multimap FC)\otimes FB \end{array}$$

Term Rewriting in a Logic of Special Relations*

W. Marco Schorlemmer

Institut d'Investigació en Intel·ligència Artificial
Consell Superior d'Investigacions Científiques
Departament de Llenguatges i Sistemes Informàtics
Universitat Politècnica de Catalunya
marco@lsi.upc.es

Abstract. Although there exist logics that extend the expressiveness of order-sorted equational logic using additional binary relations besides equality in their logical theories, standard equational rewriting is still the foundation of their operational semantics. But rewriting is not necessarily restricted to the replacement of equals by equals only, and can be generalized to other 'special' binary relations. I show in this paper that by applying rewrite techniques to logical theories considered as instances of a general 'logic of special relations' we can unify and hence simplify the computational analysis within these theories.

1 Introduction

Since the beginning of the algebraic specification discipline for formal program development, there have been various attempts to further increase the limited expressiveness of conditional order-sorted equational logic. Some examples in this direction are classified algebras [21], unified algebras [17], type algebras [12], galactic algebras [15] and membership algebras [16].

From a model-theoretic point of view, these frameworks extend the expressiveness of conditional order-sorted equational logic by means of a semantic treatment of sorts by using additional binary relations besides equality in their sentences. For instance, Manca, Salibra, and Scollo's *equational type logic* "can be viewed as Horn clausal logic with equality and one (binary) predicate, *viz.* type assignment" [12]. I will call these binary relations, that play a central role in the theories of these logics, *special relations*, because they have certain properties I claim can be computationally exploited by term rewriting. I am thinking of properties like reflexivity, symmetry or antisymmetry, monotonicity or antimonotonicity, congruence, transitivity or compositeness with other special relations. For instance, in the above mentioned equational type logic, the following properties or relationships between special relations 'equality' (=) and 'type assignment' (:) hold, for all x, y, z:

$$x = y \wedge y = z \implies x = z$$
$$x : y \wedge y = z \implies x : z$$
$$x = y \wedge y : z \implies x : z$$

* Supported by CICYT-project MODELOGOS (TIC 97-0579-C02-01)

A.M. Haeberer (Ed.): AMAST'98, LNCS 1548, pp. 178–195, 1998.

These relationships are actually specific instances of a general relation-algebra sentence $\alpha;\beta \sqsubseteq \gamma$, where α, β, and γ denote arbitrary binary relations, ';' is composition of relations, and partial order '$\sqsubseteq$' captures implication. Furthermore, in equational type logic every function symbol f is monotonic in all its argument positions i with respect to special relation 'equality': $x = y \Longrightarrow f(\ldots,\overset{i)}{x},\ldots) = f(\ldots,\overset{i)}{y},\ldots)$. In general, a specific function symbol f that is monotone in its i-th argument position with respect to a pair of relations α and β satisfies the following implication: $x \; \alpha \; y \Longrightarrow f(\ldots,\overset{i)}{x},\ldots) \; \beta \; f(\ldots,\overset{i)}{y},\ldots)$.

Although dealing with several binary relations at once, extensions of order-sorted equational logic like equational type logic or membership equational logic still base their respective proof calculi on standard equational rewriting, as are the cases of equational type rewrite systems [12] and membership/rewriting systems [2], respectively. These proof calculi are therefore too restrictive and suffer from a difficult and heterogeneous treatment of their respective deduction mechanisms. But, recently it has been shown that rewriting itself goes beyond equational logic, because rewriting is not necessarily restricted to the replacement of equals by equals only. Levy and Agustí studied mechanisms for automating the deduction in theories involving subset inclusions by means of bi-rewrite systems [11]. Bachmair and Ganzinger based on Levy and Agustí's work their generalization from superposition calculi for full first-order theories with equality to ordered chaining calculi for theories with arbitrary binary relations [1].

It is due to these previous observations that I look at these extensions of order-sorted equational logic as particular instances of a more general *logic of special relations*, that I briefly introduce in Sect.2. In addition I believe that the basic properties of special relations can be captured by term rewriting, and that some interesting computational issues can be naturally studied within a proof calculus relying on a suitable notion of 'term rewriting along binary relations', as defined in Sect.3, which focuses on what I think constitutes the 'bare bones of term rewriting', namely first, the *replacement* of a term by another —applying a given rewrite rule—, second, the successive and meaningful *composition* of several replacements, and third, the possible, but not necessary, application of replacements *within the structure* of a term. In Sect.4 I show that, for instance, in the particular case of membership equational logic [2], such 'kernel' of term rewriting along binary relations uniformly captures, under a unique general notion of local confluence, some important decidability properties of its theories, like sort-decreasingness or descendingness.

2 A Logic of Special Relations

Signatures of the logic of special relations are tuples $\Omega = (S^*, \Sigma)$, where

- $S^* = (S^*, ;, 1, ^{\smile}, \sqsubseteq)$ is a partially ordered free monoid with an anti-involution generated over a set S of special binary relation symbols. The monoid's multiplication ';' and neutral element '*1*' are interpreted as relation composition and identity relation, respectively. The anti-involution '$^{\smile}$' is interpreted as relation conversion. Composition and conversion are order preserving.

- Σ is a ranked alphabet of function symbols, which may be monotonic or antimonotonic in their argument positions with respect to a pair of special relation symbols of S.

I'm going to treat monotonicity and antimonotonicity as inherent features of the signature's function symbols, in the same sense as their arities. For this purpose I use the notion of *polarity*, inspired by Manna and Waldinger's work on special-relation rules [13,14]. For example, let $|x|$ denote the cardinality function applied to the set x. We have that for all x, y, $x \subseteq y \Longrightarrow |x| \leq |y|$, *i.e.* the cardinality function is monotonic in its unique argument position. I will say that its argument position has *positive* polarity (or is positive) with respect to $(\subseteq, \leq)$. In another example, let $x \setminus y$ denote set difference between sets x and y. We have that for all x, y, z, $x \subseteq y \Longrightarrow z \setminus y \subseteq z \setminus x$, *i.e.* the set difference function is antimonotonic in its second argument. I will say that its second argument position has *negative* polarity (or is negative) with respect to $(\subseteq, \subseteq)$.

When I say that an argument position is positive (or negative) I do not exclude the possibility that it has both polarities. In general, when an argument position has some polarity (either positive, negative or both) I will just say that it is *polarized*.

Without loss of generality, in the rest of this paper I will only refer to positive polarities of argument position, since if a position has negative polarity with respect to a pair of relations, I express this polarity as a positive one in the following way: For any argument position i of any function symbol f in Σ, the i-th argument position of f is negative with respect to (α, β) if and only if it is positive with respect to $(\breve{\alpha}, \beta)$ if and only if it is positive with respect to $(\alpha, \breve{\beta})$.

I also extend polarities with respect to composite relations and with respect to the identity relation in the following way: For any argument position i of any function symbol f in Σ, if the i-th argument position of f is positive with respect to both, (α, β) and (α', β'), then it is also positive with respect to $((\alpha; \alpha'), (\beta; \beta'))$. The i-th argument position of f is positive with respect to $(\mathit{1}, \mathit{1})$.

There is a relationship between polarities and the partial order relation $\sqsubseteq$ on relations, which determines a signature to be correctly stated:

Definition 1. *A signature (S^*, Σ) is said to be correct if for any argument position i of any function symbol f in Σ, and any relations α, β, and γ in S we have that*

- *if $\gamma \sqsubseteq \beta$ and the i-th argument position of f is positive with respect to (α, γ), then it is also positive with respect to (α, β),*
- *if $\alpha \sqsubseteq \gamma$ and the i-th argument position of f is positive with respect to (γ, β), then it is also positive with respect to (α, β).*

From now on I only consider correct signatures. As usual, $\mathcal{T}_\Sigma(\mathcal{X})$ denotes the set of first-order Σ-terms over a denumerable set $\mathcal{X}$ of variables. *Sentences* are expressions $s\ \alpha\ t$, where $s, t \in \mathcal{T}_\Sigma(\mathcal{X})$ and $\alpha \in S^*$. A *substitution* $\sigma = \langle x_1 \mapsto t_1, \ldots, x_n \mapsto t_n \rangle$ is a map from a finite subset of variables $\{x_1, \ldots, x_n\} \subseteq \mathcal{X}$ to terms, and can be uniquely extended to a mapping from terms to terms and from sentences to sentences. A *theory presentation* is a pair (Ω, Γ), where Ω is a signature and Γ is a set of Ω-sentences, also called *axioms*.

An interpretation in this logic is a Σ-algebra $\mathcal{A}$ together with an assignment to each $\alpha \in S^*$ of a set $[\![\alpha]\!] \subseteq \mathcal{A} \times \mathcal{A}$, such that, for all $\alpha, \beta \in S^*$, $a, b \in \mathcal{A}$, $f \in \Sigma_n$, and $i \in [1 \ldots n]$

- $(a, b) \in [\![\alpha;\beta]\!]$ iff there exists $c \in \mathcal{A}$ such that $(a, c) \in [\![\alpha]\!]$ and $(c, b) \in [\![\beta]\!]$
- $(a, a) \in [\![1]\!]$
- $(a, b) \in [\![\breve{\alpha}]\!]$ if and only if $(b, a) \in [\![\alpha]\!]$
- $[\![\alpha]\!] \subseteq [\![\beta]\!]$ whenever $\alpha \sqsubseteq \beta$
- $(a, b) \in [\![\alpha]\!]$ implies $([\![f]\!](\ldots, \overset{i)}{a}, \ldots), [\![f]\!](\ldots, \overset{i)}{b}, \ldots)) \in [\![\beta]\!]$ whenever the i-th position of f is positive with respect to (α, β)

We say that such an algebra $\mathcal{A}$ satisfies a sentence $s\ \alpha\ t$ if and only if for each assignment $\rho : \mathcal{X} \to \mathcal{A}$, $([\![s]\!]_\rho, [\![t]\!]_\rho) \in [\![\alpha]\!]$, where $[\![\ ]\!]_\rho$ is the unique Σ-homomorphism extending ρ.

In order to capture a large variety of models of specification, it is necessary to endow the logic with a more general model theory than the one presented in this paper, namely one based on the categorical theory or relations, also known as *allegories* [6], as I have shown in detail in [20].

3 Term Rewriting Along Binary Relations

Given a term t, Let $t|_p$ denote the subterm occurring at position p, represented in Dewey decimal notation. When this occurrence is replaced with term s, we denote that by $t[s]_p$. The polarity of argument positions of functions can be easily extended to subterm positions p within a term t in the following way:

Definition 2.

1. *Position i in a term $f(t_1, \ldots, t_n) \in \mathcal{T}_\Sigma(\mathcal{X})$ is* positive *with respect to a pair of relations (α, β), if and only if the i-th argument position of f is positive with respect to (α, β).*
2. *For every term $u \in \mathcal{T}_\Sigma(\mathcal{X})$, subterm positions p and q, and relations $\alpha, \beta \in S^*$, $p.q$ in u is* positive *with respect to (α, β) if and only if there exists a relation $\gamma \in S^*$ such that the polarity of q in $u|_p$ with respect to (α, γ) and the polarity of p in u with respect to (γ, β) are both positive.*

Obviously this extension captures all the negative polarities through the relation existing between positive and negative polarities, as discussed in the previous section.

3.1 A General Notion of Term Rewriting

The term rewriting approach to theorem proving in equational logic is based on the fact that we can use the equations of a given theory as rewrite rules, by imposing a specific directionality to the equations. In the same sense we may prove theorems of a given theory in our logic with special relations by considering its atomic formulae as rewrite rules, too. We do this either considering an atomic formula $s\ \alpha\ t$ as a rule from left to right, in which case we write $s \xrightarrow[\alpha]{} t$, or else from right to left, in which case we write $s \xleftarrow[\alpha]{} t$. Since sentences $s\ \alpha\ t$ and $t\ \breve{\alpha}\ s$ are equivalent, we may also write $t \xleftarrow[\breve{\alpha}]{} s$ and $t \xrightarrow[\breve{\alpha}]{} s$, respectively. If, given a *theory presentation* $((S^*, \Sigma), \Gamma)$ we interpret the axioms in Γ as rewrite rules in the sense explained above, then we may call Γ also a *term rewriting system*, generalizing in this way the standard notion of term rewriting system (where rewrite rules are actual equations). Consequently I redefine the notion of term rewriting as follows:

Definition 3. *Given a term rewriting system Γ, a rewrite rule $l \xrightarrow[\alpha]{} r$ in Γ, and a term s, I say that s* rewrites along γ to t, *written $s \xrightarrow[\gamma,\Gamma]{} t$, if there exist a relation γ in S^* and a substitution σ such that $\sigma(l) = s|_p$ for a subterm position p that is positive with respect to (α, γ), and $t = s[\sigma(r)]_p$.*

In general I will write $s \xrightarrow[\Gamma]{} t$, if there exists some relation γ in S^* such that $s \xrightarrow[\gamma,\Gamma]{} t$, i.e. $\xrightarrow[\Gamma]{} = \bigcup_{\gamma \in S^*} \xrightarrow[\gamma,\Gamma]{}$. Deviating from its standard definition, I will call $\xrightarrow[\Gamma]{}$ a *rewrite relation.* The standard definition of a rewrite relation is that of a binary relation over terms that is closed both under context application —the 'replacement property'— and under substitutions —the 'fully invariant property'— (see e.g. [4]). My 'redefinition' of rewrite relation differs from the standard one in that, according to Definition 3, $\xrightarrow[\Gamma]{}$ satisfies a weaker 'replacement property', namely that the relation is closed under context application *only on positively polarized positions* with respect to a pair of relations.

Notation: Given the rewrite relation $\xrightarrow[\Gamma]{}$ induced by the term rewriting system Γ, I write $\xrightarrow[\Gamma]{+}$ and $\xrightarrow[\Gamma]{*}$ for its transitive and reflexive-transitive closures, respectively. In particular, I write $s \xrightarrow[\gamma,\Gamma]{+} t$ if there exist terms $s_0, \ldots, s_n \in \mathcal{T}_\Sigma(\mathcal{X})$ and relations $\alpha_1, \ldots, \alpha_n \in S^*$, $n > 0$, such that

$$s = s_0 \xrightarrow[\alpha_1,\Gamma]{} s_1 \xrightarrow[\alpha_2,\Gamma]{} s_2 \xrightarrow[\alpha_3,\Gamma]{} \cdots \xrightarrow[\alpha_n,\Gamma]{} s_n = t \quad \text{and} \quad \alpha_1; \cdots ; \alpha_n \sqsubseteq \gamma\ ,$$

and I write $s \xrightarrow[\gamma,\Gamma]{*} t$ when $n \geq 0$. It is obvious that $\xrightarrow[\Gamma]{+} = \bigcup_{\gamma \in S^*} \xrightarrow[\gamma,\Gamma]{+}$ and $\xrightarrow[\Gamma]{*} = \bigcup_{\gamma \in S^*} \xrightarrow[\gamma,\Gamma]{*}$. Analogously, I will write their respective symmetric closures with $\xleftrightarrow[\gamma,\Gamma]{}$, $\xleftrightarrow[\Gamma]{}$, $\xleftrightarrow[\Gamma]{+}$, and $\xleftrightarrow[\Gamma]{*}$.

In the rest of this paper I will drop the subscript Γ if the term rewriting system is clear from context.

3.2 A Proof Calculus Based on Term Rewriting

I'm interested in using the general notion of term rewriting along relations introduced in the previous section in order to prove if a given atomic formula is or is not a theorem of a given theory in our logic of special relations. We know from standard equational term rewriting that term rewriting systems need to have several properties —the Church-Rosser and termination properties— in order to provide decision algorithms for the equational theory they embed. In this subsection I'm going to analyze how these properties translate to the present general notion of term rewriting system. Of course, due to the additional generality and expressiveness of the logic with special relations, I expect the required properties to be much more subtle than in the equational case.

Definition 4. *A* proof *of the atomic formula $s \ \gamma \ t$ in a term rewriting system Γ is the sequence of rewrites of the form*

$$s = s_0 \underset{\alpha_1}{\longleftrightarrow} s_1 \underset{\alpha_2}{\longleftrightarrow} \cdots \underset{\alpha_n}{\longleftrightarrow} s_n = t \ ,$$

$n \geq 0$, such that $\alpha_1; \cdots ; \alpha_n \sqsubseteq \gamma$, i.e. *$s \overset{*}{\underset{\gamma}{\longleftrightarrow}} t$. Recall that when $n = 0$, $\alpha_1; \cdots ; \alpha_n = 1$.*

Since I'm interested in exploiting rewriting along relations in a computational way, I will look for sufficient conditions a term rewriting system Γ, such that every proof of an atomic formula can also be proved by a rewrite proof:

Definition 5. *A* rewrite proof *of an atomic formula $s \ \gamma \ t$ in a term rewriting system Γ is a proof of the particular form:*

$$s \underset{\alpha_1}{\longrightarrow} \cdots \underset{\alpha_n}{\longrightarrow} u \underset{\beta_m}{\longleftarrow} \cdots \underset{\beta_1}{\longleftarrow} t$$

$n, m \geq 0$, such that $\alpha_1; \cdots ; \alpha_n; \beta_m; \cdots ; \beta_1 \sqsubseteq \gamma$, i.e. *$s \xrightarrow[\alpha_1;\cdots;\alpha_n]{*} u \xleftarrow[\beta_m;\cdots;\beta_1]{*} t$.*

If every proof has a rewrite proof, a decision procedure is straightforward: In order to prove $s \ \gamma \ t$, we compute the sets

$$A = \{(v, \delta) \in \mathcal{T}_\Sigma(\mathcal{X}) \times S^* \mid s \overset{*}{\underset{\delta}{\rightarrow}} v\} \quad B = \{(v, \delta) \in \mathcal{T}_\Sigma(\mathcal{X}) \times S^* \mid v \overset{*}{\underset{\delta}{\leftarrow}} t\}$$

and check if there exist $u \in \mathcal{T}_\Sigma(\mathcal{X})$ and $\alpha, \beta \in S^*$ such that $(u, \alpha) \in A$, $(u, \beta) \in B$, and $\alpha; \beta \sqsubseteq \gamma$. Notice that this decision procedure is not based on normal-form computation as in the case of equational term rewriting, but instead computes the whole rewrite trees starting from s and t. In order to have a decision *algorithm*, sets A and B need to be always finite. We need therefore *termination* and *finite branching*. They are defined in a similar way as with standard rewriting.

3.3 Confluence Properties

If we want that given a term rewriting system Γ, every proof of an atomic formula in Γ can also be proved by a rewrite proof, Γ needs to be not only a finitely branching and terminating term rewriting system, but also a Church-Rosser one. Let us define what the Church-Rosser property and its closely related property of 'local confluence' for term rewriting along relations look like:

Definition 6. *A term rewriting systems Γ is* Church-Rosser *if, for any pair of terms $s, t \in \mathcal{T}_\Sigma(\mathcal{X})$ and relation $\alpha \in S^*$, such that $s \overset{*}{\underset{\alpha}{\longleftrightarrow}} t$, there exists a term $u \in \mathcal{T}_\Sigma(\mathcal{X})$ and relations $\gamma, \delta \in S^*$, such that $s \overset{*}{\underset{\gamma}{\longrightarrow}} u \overset{*}{\underset{\delta}{\longleftarrow}} t$, and $\gamma; \delta \sqsubseteq \alpha$. It is* locally confluent *if, for any three terms $s, t, v \in \mathcal{T}_\Sigma(\mathcal{X})$ and pair of relations $\alpha, \beta \in S^*$, such that $s \underset{\alpha}{\longleftarrow} v \underset{\beta}{\longrightarrow} t$, there exists a term $u \in \mathcal{T}_\Sigma(\mathcal{X})$ and relations $\gamma, \delta \in S^*$, such that $s \overset{*}{\underset{\gamma}{\longrightarrow}} u \overset{*}{\underset{\delta}{\longleftarrow}}$, and $\gamma; \delta \sqsubseteq \alpha; \beta$.*

We will say that a peak having a rewrite proof *converges*. The following propositions are true in our general setting:

Proposition 7. *A terminating term rewriting system Γ is Church-Rosser if and only if it is locally confluent.*

Proof. By standard techniques for proving confluence of abstract reduction relations (see e.g. [9]). The only additional complexity to have in mind is that additional inclusions of compositions of relations ($\gamma; \delta \sqsubseteq \alpha; \beta$) need to be taken into account. □

Recall, that, in standard term rewriting local confluence reduces to convergence of all critical pairs. Critical pairs are formed by non-trivially overlapping left-hand sides of rewrite rules of the term rewriting system at hand[1]. One can then attempt to complete a terminating but non-Church-Rosser term rewriting system into a Church-Rosser one, by adding those critical pairs without a rewrite proof as new rewrite rules to the system. It is therefore very convenient to see how this extends to term rewriting along relations.

3.4 Critical Peaks

Many aspects of the following discussion have already been thoroughly studied within the context of standard rewriting (see e.g. [4]), but it is worth to work through them again within our general framework of rewriting along relations, in order to highlight the subtleties we have to deal with now.

Given a theory presentation $((S^*, \Sigma), \Gamma)$, let s, t, v be terms in $\mathcal{T}_\Sigma(\mathcal{X})$, α, β relations in S^*, and let us consider Γ as a term rewriting system. A peak $s \underset{\alpha}{\longleftarrow} v \underset{\beta}{\longrightarrow} t$ in Γ is the result of rewriting with two (not necessarily distinct) rewrite rules $r_1 \underset{\alpha'}{\longleftarrow} l_1$ and $l_2 \underset{\beta'}{\longrightarrow} r_2$ in Γ on (not necessarily distinct) subterm

[1] A non-trivial overlap means an overlap on non-variable subterm positions.

positions p and q in v, respectively. There are two different cases to consider: Either the rewrite rules do not overlap, but are applied on two disjoint subterm positions in v (disjoint case), or else they do overlap (overlap case).

Subterm positions p and q may have a common prefix, i.e. there exist an r such that $p = r.p'$ and $q = r.q'$, and in the previous two cases it is desirable to neglect the fragment of the term above r since it does not take part in the actual peak, it is just a context $w[\ ]_r$ put around the terms forming it. But to be able to strip off this context we need to proof that context application, when possible, preserves local confluence.

Context Application: Context application preserving local confluence means, that if a peak $s \underset{\alpha}{\leftarrow} v \underset{\beta}{\rightarrow} t$ converges, then by applying a context $w[\ ]_r$ around the terms involved, the resulting peak $w[s]_r \underset{\alpha'}{\longleftarrow} w[v]_r \underset{\beta'}{\longrightarrow} w[t]_r$ converges, too.

Though this is obviously true in standard term rewriting, its validity in the framework of our general notion of term rewriting is not that straightforward. A context $w[\ ]_r$ can be applied around the terms if there exist relations $\alpha', \beta' \in S^*$ such that r in w is positive with respect to (α, α') and with respect to (β, β'). But then the resulting peak converges only if there also exist relations $\gamma', \delta' \in S^*$ such that r in w is positive with respect to (γ, γ') and with respect to (δ, δ'), and, in addition, $\gamma'; \delta' \sqsubseteq \alpha'; \beta'$ (see Fig.1).

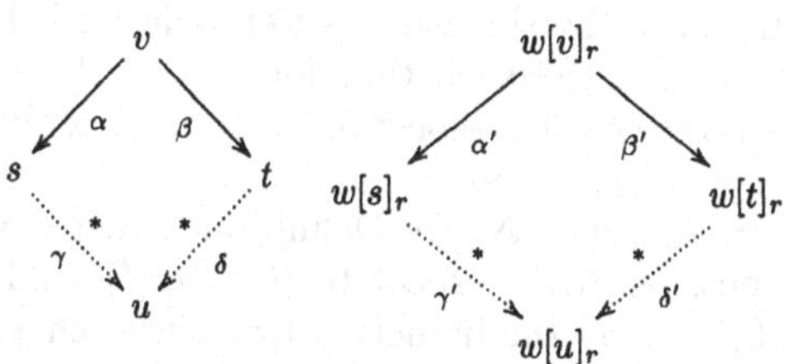

Fig. 1. Context application preserving local confluence

Unfortunately this is not always true in general, and we will need to put additional conditions on the polarity of our signature's function symbols for context application to preserve local confluence. In order to simplify the treatment of these conditions, I've already chosen to define the monotonicity and antimonotonicity properties of function symbols only with respect to special relations in S (see Sect.2), so that the polarities of their argument positions with respect to general binary relations in S^* depend on their polarity with respect to special relations in S and on the partial order in S^* .

Before giving the additional conditions to put on signatures, let me first state the following proposition:

Proposition 8. *Let* $\breve{S} = \{\breve{\alpha} \mid \alpha \in S\}$. *For every relation* $\alpha \in S^*$, *there exist* $\alpha_1, \ldots, \alpha_n \in S \cup \breve{S}$, $n \geq 0$, *such that* $\alpha = \alpha_1; \cdots; \alpha_n$.

Proof. By definition, since S^* is the domain of a free monoid with an anti-involution. □

I now define the notion of *well-polarized* signatures, in order to capture those special cases for which context application is going to preserve local confluence:

Definition 9. *A signature (S^*, Σ) is said to be* well-polarized *if whenever for a function symbols $f \in \Sigma$, argument position i in f, and pair of relations $\alpha, \beta \in S^*$ we have that i in f is positive with respect to (α, β) (where $\alpha = \alpha_1; \cdots; \alpha_n$, and for all $j \in [1 \ldots n]$, $\alpha_j \in S \cup \check{S})^2$, then there exist relations $\beta_1, \ldots, \beta_n \in S^*$ such that for all $j \in [1 \ldots n]$, i in f is positive with respect to (α_j, β_j), and $\beta_1; \cdots; \beta_n \sqsubseteq \beta$.*

Lemma 10. *Let (S^*, Σ) be a well-polarized signature, and let α, β, and γ be three relations in S^*, w a term in $\mathcal{T}_\Sigma(\mathcal{X})$, r a position in w, such that r in w is positive with respect to $((\alpha; \beta), \gamma)$. Then there exist relations α' and β' in S^* such that r in w is positive with respect to (α, α') and with respect to (β, β'), and $\alpha'; \beta' \sqsubseteq \gamma$.*

Proof. By induction on the length of position $r = d_1. \cdots .d_l$, $l \geq 1$, $d_j \in \mathbb{N}$. Let $\alpha = \alpha_1; \cdots; \alpha_n$ and $\beta = \beta_1; \cdots; \beta_m$, $m, n \geq 0$ such that for all $i \in [1 \ldots n]$ and $j \in [1 \ldots m]$, $\alpha_i, \beta_j \in S \cup \check{S}$ (Proposition 8), and suppose f is the top-most function symbol of w, i.e. $w = f(s_1, \ldots, s_p)$ and for all $k \in [1 \ldots p]$, $s_k \in \mathcal{T}_\Sigma(\mathcal{X})$.

1. If $l = 1$ then $r = d$, $d \in \mathbb{N}$. By Definition 2 d in f is positive with respect to $((\alpha; \beta), \gamma)$, and since the signature is well-polarized there exist relations $\alpha'_1, \ldots, \alpha'_n, \beta'_1, \ldots, \beta'_m \in S^*$ such that for all $i \in [1 \ldots n]$ and for all $j \in [1 \ldots m]$, d in f is positive with respect to (α_i, α'_i) and with respect to (β_j, β'_j), and $\alpha'_1; \cdots; \alpha'_n; \beta'_1; \cdots; \beta'_m \sqsubseteq \gamma$.
2. If $l > 1$ then $r = d.r'$, $d \in \mathbb{N}$. By Definition 2 there exists $\gamma' \in S^*$ such that r' in s_d is positive with respect to $((\alpha; \beta), \gamma')$, and d in f is positive with respect to (γ', γ). By the induction hypothesis there exist relations α' and β' in S^* such that r' in s_d is positive with respect to (α, α') and with respect to (β, β'), and $\alpha'; \beta' \sqsubseteq \gamma'$. Therefore d in f is also positive with respect to $((\alpha'; \beta'), \gamma)$. Since the signature is well-polarized and by reasoning analogously to the base case, there exist relations α'' and β'' in S^* such that d in f is positive with respect to (α', α'') and with respect to (β', β''), and $\alpha''; \beta'' \sqsubseteq \gamma$. By Definition 2 r in w is positive with respect (α, α'') and with respect to (β, β'').

□

Indeed, when dealing with well-polarized signatures, local confluence is closed under context application:

Proposition 11. *If the peak $s \underset{\alpha}{\longleftarrow} v \underset{\beta}{\longrightarrow} t$ converges, then, for each term $w \in \mathcal{T}_\Sigma(\mathcal{X})$, position r in w, and relations $\alpha', \beta' \in S^*$ such that r in w is positive with respect to (α, α') and with respect to (β, β'), the peak $w[s]_r \underset{\alpha'}{\longleftarrow} w[v]_r \underset{\beta'}{\longrightarrow} w[t]_r$ converges, too.*

---

[2] Such relations α_j always exist due to Proposition 8.

Proof. Since the peak $s \underset{\alpha}{\leftarrow} v \underset{\beta}{\rightarrow} t$ converges, there exists a term $u \in \mathcal{T}_\Sigma(\mathcal{X})$ and relations $\gamma, \delta \in S^*$ such that $s \overset{*}{\underset{\gamma}{\rightarrow}} u \overset{*}{\underset{\delta}{\leftarrow}} t$ and $\gamma;\delta \sqsubseteq \alpha;\beta$. We have that r in w is positive with respect to $((\alpha;\beta),(\alpha';\beta'))$, therefore r in w is also positive with respect to $((\gamma;\delta),(\alpha';\beta'))$. By Lemma 10 there exist relations $\gamma',\delta' \in S^*$ such that r in w is positive with respect to (γ,γ') and (δ,δ'). Consequently $w[s]_r \overset{*}{\underset{\gamma'}{\rightarrow}} w[u]_r \overset{*}{\underset{\delta'}{\leftarrow}} w[t]_r$. Furthermore, and also by Lemma 10, $\gamma';\delta' \sqsubseteq \alpha';\beta'$. □

Disjoint Case: Knowing that we can strip off the context not involved in the rewritings that form the actual peak, let us now see the case when the peak is due to the application of two rewrite rules on disjoint positions $p = i.p'$ and $q = j.q'$, $i \neq j$. Let us proof first the following lemma:

Lemma 12. *Let $((S^*,\Sigma),\Gamma)$ be a theory presentation. If for all function symbols $f \in \Sigma_n$, $n \in \mathbb{N}$, argument positions $i,j \in [1\ldots n]$, $i \neq j$, and special relations $\alpha,\alpha',\beta,\beta' \in S$ we have that, whenever both, the i-th and j-th argument positions of f are positive with respect to (α',α) and (β',β), respectively, $\alpha;\beta = \beta;\alpha$, then this is also true for all general binary relations $\alpha,\alpha',\beta,\beta' \in S^*$.*

Proof. Let $\alpha = \alpha_1;\cdots;\alpha_n$ and $\beta = \beta_1;\cdots;\beta_m$, $m,n \geq 0$ such that for all $i \in [1\ldots n]$ and $j \in [1\ldots m]$, $\alpha_i,\beta_j \in S \cup \breve{S}$ (Proposition 8). The proof is by induction over n and m. If $n = 0$, $1;\beta = \beta;1$ (analogously for $m = 0$). If $n,m > 0$, let $\bar{\alpha} = \alpha_2;\cdots;\alpha_n$ and $\bar{\beta} = \beta_2;\cdots;\beta_m$. There exist relations $\alpha_1',\beta_1' \in S \cup \breve{S}$ and relations $\bar{\alpha}',\bar{\beta}' \in S^*$, such that

- the i-th argument position in f is positive with respect to (α_1',α_1) and with respect to $(\bar{\alpha}',\bar{\alpha})$
- the j-th argument position in f is positive with respect to (β_1',β_1) and with respect to $(\bar{\beta}',\bar{\beta})$

Therefore $\alpha_1;\beta_1 = \beta_1;\alpha_1$ and by the induction hypothesis $\alpha_1;\bar{\beta} = \bar{\beta};\alpha_1$, $\bar{\alpha};\beta_1 = \beta_1;\bar{\alpha}$, and $\bar{\alpha};\bar{\beta} = \bar{\beta};\bar{\alpha}$. Consequently $\alpha;\beta = \alpha_1;\bar{\alpha};\beta_1;\bar{\beta} = \alpha_1;\beta_1;\bar{\alpha};\bar{\beta} = \beta_1;\alpha_1;\bar{\beta};\bar{\alpha} = \beta_1;\bar{\beta};\alpha_1;\bar{\alpha} = \beta;\alpha$. □

Proposition 13. *Let $((S^*,\Sigma),\Gamma)$ be a theory presentation such that (S^*,Σ) is a well-polarized signature. If for all function symbols $f \in \Sigma_n$, $n \in \mathbb{N}$, argument positions $i,j \in [1\ldots n]$, $i \neq j$, and special relations $\alpha,\alpha',\beta,\beta' \in S$ we have that, whenever both, the i-th and j-th argument positions of f are positive with respect to (α',α) and (β',β), respectively, $\alpha;\beta = \beta;\alpha$, then all peaks due to the application of two rewrite rules in Γ on disjoint subterm positions converge.*

Proof. Let $s \underset{\alpha}{\leftarrow} v \underset{\beta}{\rightarrow} t$ be a peak due to the application of two rewrite rules $r_1 \underset{\alpha'}{\longleftarrow} l_1$ and $l_2 \underset{\beta'}{\longrightarrow} r_2$ in Γ on two disjoint subterm positions $p = i.p'$ and $q = j.q'$ of v, $i \neq j$, and let f be the top-most function symbol of term v. We have that $s = v[\sigma_1(r_1)]_p$ and $t = v[\sigma_2(r_2)]_q$, where σ_1 and σ_2 are the unifiers of

$v|_p$ with l_1, and $v|_q$ with l_2, respectively. It follows that subterm positions p and q in v are positive with respect to (α', α) and (β', β), respectively, and therefore there exist relations $\alpha'', \beta'' \in S^*$, such that the i-th and j-th argument positions of f are positive with respect to (α'', α) and (β'', β), respectively. Furthermore, since positions p and q are disjoint, there exist $u = s[\sigma_2(r_2)]_q = t[\sigma_1(r_1)]_p$, such that $s \underset{\beta}{\longrightarrow} u \underset{\alpha}{\longleftarrow} t$. In addition, by Lemma 12, $\alpha; \beta = \beta; \alpha$, which means that the peak converges. Finally, by Proposition 11, context application preserves local confluence. □

Overlap Case: Unlike in standard term rewriting, in our case we will need to look for overlaps on non-variable *and* variable subterm positions in order to generate critical atoms *and* variable instance atoms, and eventually check if they have a rewrite proof.

Let us define these atoms formally. Let $((S^*, \Sigma), \Gamma)$ be a theory presentation, where Γ is considered a term rewriting system.

Definition 14. *If $l \underset{\alpha}{\longrightarrow} r$ and $s \underset{\beta}{\longrightarrow} t$ are two rewrite rules in Γ and α, β, and γ in S^* are relations such that p is a negative position with respect to (α, γ) of a non-variable subterm of s, and σ is the most general unifier of $s|_p$ and l, then the atomic formula $\sigma(s[r]_p)\ \gamma; \beta\ \sigma(t)$ is a* critical atom.

Definition 15. *If $l \underset{\alpha}{\longrightarrow} r$ and $s \underset{\beta}{\longrightarrow} t$ are two rewrite rules in Γ and α, β, and γ in S^* are relations such that p is a position of a variable subterm x of s, σ is a substitution such that $\sigma(x)$ has l as subterm at position q, but $\sigma(y) = y$ for all $y \neq x$, position $p.q$ in $\sigma(s)$ is a negative position with respect to (α, γ), then the critical atom $\sigma(s)[r]_{p.q}\ \gamma; \beta\ \sigma(t)$ is a* variable instance atom.

Proposition 16. *Let $((S^*, \Sigma), \Gamma)$ be a theory presentation satisfying the same conditions as the one in Proposition 13. The rewrite system Γ is locally confluent if and only if all critical and variable instance atoms have a rewrite proof.*

Proof. By standard techniques of term rewriting (see e.g. [9]). □

3.5 Practical Theorem Proving

Several important differences to standard equational term rewriting appear, which are important for the practicability of the term rewriting approach to deductions with binary relations, when the signature includes functions which are monotonic or antimonotonic with respect to a pair of relations. In general, two rewrite rules may give rise to infinite many critical and variable instance atoms. We can limit the number of critical atoms by putting some additional restrictions on our theories (we are going to see that on an example in the next section), but dealing with variable instance atoms is unfeasible in practice, and nothing is gained compared to a strategy by which monotonicity axioms are treated unspecifically. Therefore we may consider rewriting on subterm positions only when variable instance atoms are generally unnecessary. In particular, one way to do that is the approach followed by Bachmair and Ganzinger in [1], which consists of allowing rewriting on subterm positions only along relations that are symmetric. I suggest another approach that can also be useful to avoid variable instance atoms, namely by exploiting the notion of polarity. Proposition 13 provides a first example were the notion of polarity serves to control term rewriting. In addition, we may allow rewriting along non-symmetric relations only on subterm positions of ground terms, or, more generally, of terms without variables on polarized subterm positions. A detailed case analysis of the feasibility of subterm rewriting in relationship to the notion of polarity of subterm positions remains to be done and would be outside the scope of this paper, though I studied these issues in some depth within the context of rewriting logic [18,19].

4 Decidability Conditions for an Instance of the Logic

From the previous discussion we may conclude that a proof calculus for the logic of special relations based on a general notion of term rewriting along binary relations is unfeasible in practice. In order to overcome tractability problems of the calculus, quite restrictive conditions need to be put on the special relations or on the axioms of our theories. Despite of these severe drawbacks the results of extending rewrite techniques beyond equality unify and hence simplify in an elegant way some aspects of the computational analysis for specification paradigms based on special relations. In this section I show how such general view of term rewriting captures several up to now distinct conditions for the decidability by term rewriting of theories in a particular instance of the logic of special relations —namely membership equational logic— under a single notion of local confluence.

4.1 Membership Equational Logic

In general a signature $\Omega = (\mathcal{K}, \Sigma, \{S_K\}_{K\in\mathcal{K}})$ in membership equational logic consists of a many-kinded signature $(\mathcal{K}, \Sigma)$ and a family $\{S_K\}_{K\in\mathcal{K}}$ of sets of sorts. For the sake of simplicity I will be concerned only with the 'one-kinded' case, *i.e.* when $\mathcal{K}$ is a singleton set, and the family $\{S_K\}_{K\in\mathcal{K}}$ consists of only one set of sorts. Atomic formulae are membership assertions $t : s$ or equations $t = t'$, where t and t' are in $\mathcal{T}_\Sigma(\mathcal{X})$ and s is in S_K. Sentences are expressions $\varphi_1 \wedge \cdots \wedge \varphi_n \Longrightarrow \psi$, $n \geq 0$, where φ_i and ψ are atomic formulae. Sentences of the particular form $x : s_1 \Longrightarrow x : s_2$, where $x \in \mathcal{X}$, are called *subsort sentences*, because they induce a subsort relation $\leqslant$ over the sorts of S_K. I will consider subsort sentences as atomic formulae of the form $s_1 \leqslant s_2$. Theory presentations are pairs (Ω, Γ), where Ω is signature and Γ is a set of sentences. For further details see [2]. For the subsequent discussion, and since theories in the logic of special relations as presented in this paper only have atomic sentences, I will treat the unconditional fragment of membership equational logic, with the exception of subsort sentences.

4.2 Sort-Decreasingness

When considering completion procedures for order-sorted rewrite system, one has to face the problem that order-sorted replacement of equals by equals is not complete in general, and consequently one has to pose the restriction of 'sort-decreasingness' on the rewrite rules of the rewrite system [8]. A rewrite rule is sort-decreasing if the sort to which the right hand side of the rule belongs is sub-sort of the one to which the left hand side belongs. This restriction extends to the completion process, being an new source of failure, in addition to unorientable equations.

Several unsatisfying and complicated ways to solve this problem have been suggested [8,10,7,3], but it turns out that a semantic treatment of sorts provides an elegant solution to the problems posed by the sort-decreasingness requirement. Within the framework of membership equational logic, Bouhoula, Jouannaud, and Meseguer study a Knuth-Bendix-like completion procedure that avoids non-sort-decreasing rewrites by adding semantic preserving membership assertions to the original theory presentation, in a way similar to adding semantic preserving equations when divergent critical pairs among rewrite rules arise (see [2] for further details).

But a theory in membership equational logic is a particular theory in the logic of special relations, involving three different special relations $S = \{\leqslant, :, =\}$ standing for 'subsort', 'membership', and 'equality' respectively, and where elements in S^* are partially ordered by the minimal partial order such that

$$=;= \;\sqsubseteq\; = \qquad \leqslant;\leqslant \;\sqsubseteq\; \leqslant \qquad =;: \;\sqsubseteq\; : \qquad :;\leqslant \;\sqsubseteq\; :$$

Furthermore, we only allow sentences of the form $s\ \alpha\ t$, where α is a special relation in S, not a general binary relation in S^*. Consequently, in order to check for local confluence of the term rewriting system associated to the theory presentation we will have to consider overlaps on the left hand sides of five possible distinct rewrite rules[3]:

$$s \underset{=}{\longrightarrow} t \qquad s \underset{\in}{\longrightarrow} t \qquad s \underset{\ni}{\longrightarrow} t \qquad s \underset{\leqslant}{\longrightarrow} t \qquad s \underset{\geqslant}{\longrightarrow} t$$

But, in the very special case of membership equational logic, suitable restrictions on the ordering on terms '$\succ$' governing the orientation of our rewrite rules allow us to exclude some of the multiple cases of critical atoms we would otherwise have to consider.

1. By requiring all operator symbols for term construction to precede in the term ordering all sort constants we avoid rules of the form $s \underset{\ni}{\longrightarrow} t$.
2. By requiring the term ordering on sort constants to resemble the sort hierarchy, *i.e.* $s_2 \succ s_1$ whenever $s_2 \geqslant s_1$, we avoid rules of the form $s \underset{\leqslant}{\longrightarrow} t$.

[3] I use the symbol '$\in$' instead of ':' as type assignment, because it is a non-symmetric symbol like the special relation it denotes. For the subsequent discussion, and when I am concerned with rewriting, I will follow this convention.

Only three rewrite relations, $\underset{=}{\longrightarrow}$, $\underset{\in}{\longrightarrow}$ and $\underset{\geqslant}{\longrightarrow}$ remain to be considered, and therefore only cases (a) and (b) shown in Fig.2 of locally confluent critical peaks need to be checked. The finiteness of S guarantees always that only a finite amount of critical atoms needs to be considered. Furthermore, since function symbols for term construction are kept separate from sort symbols, which are only allowed to be constants in membership assertions, the unique polarized proper subterm positions are those with respect to $(=,=)$, and because of the symmetry of '=' no variable instance atoms need to be considered (*cf.* the discussion in Sect.3.5). Cases (a) and (b) of Fig.2 correspond to the conventional notion on 'confluence' (rewriting along equality) and a weaker notion of *sort-decreasingness* as the one defined in [2]. I will show the latter observation in more detail.

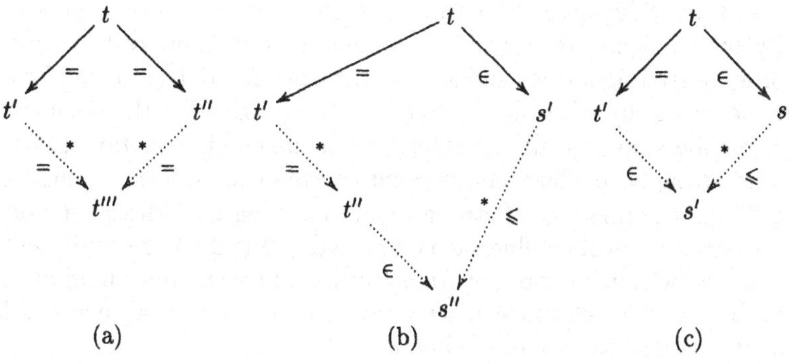

Fig. 2. Cases of local confluence in membership equational logic

In the framework of membership equational logic sort-decreasingness is defined together with the notion of 'critical reduced membership', which I state here adapted to our framework and also to the unconditional case.

Definition 17. *Given rewrite rules* $t \underset{\in}{\longrightarrow} s$ *and* $l \underset{=}{\longrightarrow} r$ *and* $\sigma(t|_p) = \sigma(l)$ *for some non-variable position* p *and most general unifier* σ, *then* $\sigma(t[r]_p) : s$ *is a* critical reduced membership.

It is obvious that a critical reduced membership is an actual critical atom as defined in definition 14 (besides the fact that here we are dealing only with *special* relations).

Definition 18. *A critical reduced membership* $t' : s$ *is* sort-decreasing *if there exists* s' *such that* $s \overset{*}{\underset{\geqslant}{\longrightarrow}} s'$ *and* $t' \underset{\in}{\longrightarrow} s'$.

Analyzing this definition within our framework, it is obvious that sort-decreasingness is actually too strong a condition for decidability of equality and membership statements, because of the unnecessarily required *one-step* rewrite

$t' \underset{\in}{\longrightarrow} s'$ shown in Fig.2(c). We have seen in Sect.3.3 that —given termination— local confluence suffices for decidability of atomic formulae in theories with special relations. In membership equational theories the weaker sort-decreasing condition that actually suffices for decidability of equality, membership and subsort assertions is the one depicted in Fig.2(b), and is closely related to Eker and Meseguer's notion of *descendingness* [5], who, following a different approach than the one presented here, also suggested to weaken the sort-decreasingness requirement in membership equational theories.

4.3 On Regularity

Eker and Meseguer's *descendingness* differs from local confluence as depicted in Fig.2(b) in that it involves the notion of *least sort* and hence assumes strong regularity of the signature. Regularity is another well-know restriction usually put on order-sorted signatures and assures the existence of a least sort for a given term in the hierarchy of sorts. Although not necessary for the decidability of equality, membership and subsort assertions, membership equational logic takes regular and strongly regular signatures into account for efficiency purposes.

Actually, in the presence of strong regularity, Eker and Meseguer's descendingness notion and local confluence as depicted in Fig.2(b) are equivalent. This is easily proved within the present framework, and constitutes an additional argument in favor of the elegance of term rewriting along special relations for the study of such issues, as we can see below.

Definition 19. *A membership equational theory (and by extension the term rewriting system) is* strongly regular *if for each term t there exists a sort s, such that $t \underset{\in}{\longrightarrow} s$ and, whenever there exists another sort s' such that $t \underset{\in}{\longrightarrow} s'$, then $s' \overset{*}{\underset{\geqslant}{\longrightarrow}} s$. We say that s is the* least sort *of t.*

The following is an alternative definition of descendingness to the one given by Eker and Meseguer. It is given with respect to critical reduced memberships as defined in Definition 17 for comparison with Definition 18 of sort-decreasingness.

Definition 20. *A critical reduced membership $t' : s$ from rewrite rules $t \underset{\in}{\longrightarrow} s$ and $l \underset{=}{\longrightarrow} r$, for which s is least sort of t, is* descending, *if there exists a term t'' with least sort s' such that $t' \overset{*}{\underset{=}{\longrightarrow}} t''$ and $s \overset{*}{\underset{\geqslant}{\longrightarrow}} s'$.*

Now we are ready to prove the equivalence between descendingness and local confluence in the presence of strong regularity.

Proposition 21. *Given a strongly regular term rewriting system, all critical atoms between rules $\underset{=}{\longrightarrow}$ and $\underset{\in}{\longrightarrow}$ are locally confluent if and only if all critical reduced memberships are descending.*

Proof. For the if direction, let $t' : s'$ be a critical atom formed from rewrite rules $t \underset{=}{\longrightarrow} t'$ and $t \underset{\in}{\longrightarrow} s'$ and let s be the least sort of t. Then $t' : s$ is a critical reduced membership formed from rewrite rules $t \underset{=}{\longrightarrow} t'$ and $t \underset{\in}{\longrightarrow} s$, and by descendingness there exists a term t'' with least sort s'' such that $t' \overset{*}{\underset{=}{\longrightarrow}} t''$ and $s \overset{*}{\underset{\geqslant}{\longrightarrow}} s''$, and by strong regularity $s' \overset{*}{\underset{\geqslant}{\longrightarrow}} s$. Therefore $t' : s'$ is locally confluent (see Fig.3(a)). For the only if direction, let $t' : s$ be a critical reduced membership formed from rewrite rules $t \underset{=}{\longrightarrow} t'$ and $t \underset{\in}{\longrightarrow} s$, where s is least sort of t'. Then by local confluence there exists a term t'' and a sort s' such that $t' \overset{*}{\longrightarrow} t''$, $t'' \underset{\in}{\longrightarrow} s'$, and $s \overset{*}{\underset{\geqslant}{\longrightarrow}} s'$. By strong regularity t'' has a least sort s'' and therefore $s' \overset{*}{\underset{\geqslant}{\longrightarrow}} s''$, consequently $t' : s$ is descending (see Fig.3(b)). □

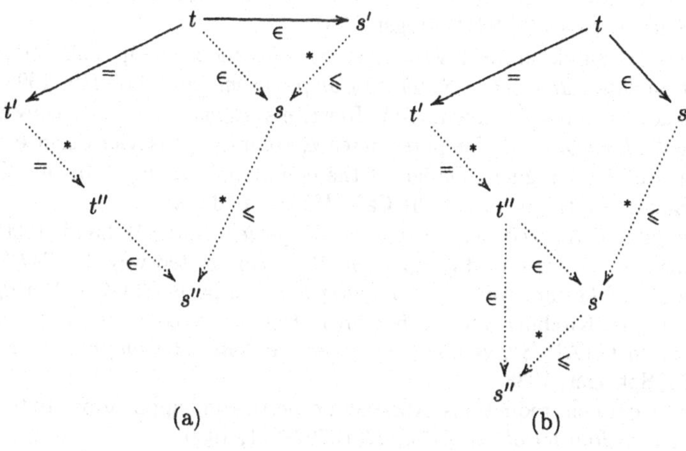

Fig. 3. Descendingness and local confluence are equivalent, provided strong regularity

5 Conclusions

Many logics extending order-sorted equational logic use special relations as basic building blocks for specifications. I have attempted to formally capture this fact by means of a *logic of special relations*, which I have briefly introduced in this paper. Its detailed presentation can be found in [20]. The purpose here was to develop the proof-theoretical aspect of the logic, and to show how several conditions for the decidability of membership equational theories —which are specific theories in the logic of special relations— like sort-decreasingness, descendingness, or regularity, can easily be expressed by a general notion of local confluence when term rewriting is generalized along binary relations. It still remains to extend the present framework to conditional term rewriting, in analogy to conditional equational term rewriting.

Acknowledgments: I am grateful to Jaume Agustí, Jordi Levy and José Meseguer for fruitful discussions concerning the research issues presented in this paper.

References

1. L. Bachmair and H. Ganzinger. Ordered chaining calculi for first-order theories of binary relations. Technical Report MPI-I-95-2-009, Max-Plank-Institut für Informatik, 1995.
2. A. Bouhoula, J.-P. Jouannaud, and J. Meseguer. Specification and proof in membership equational logic. In *TAPSOFT'97*, volume 1214 of *Lecture Notes in Computer Science*, pages 67–92. Springer, 1997.
3. H. Comon. Completion of rewrite systems with membership constraints, Part I: Deduction rules. *Journal of Symbolic Computation*, 25(4):397–419, 1998.
4. N. Dershowitz and J.-P. Jouannaud. Rewrite systems. In J. van Leeuwen, editor, *Handbook of Theoretical Computer Science*, volume B. Elsevier Science, 1990.
5. S. Eker and J. Meseguer. Design of the equational proving tools for Cafe, Mar. 1997. Slides of a talk given at the Cafe Workshop, Japan.
6. P. J. Freyd and A. Scedrov. *Categories, Allegories.* North-Holland, 1990.
7. H. Ganzinger. Order-sorted completion: The many-sorted way. In *TAPSOFT '89*, volume 351 of *Lecture Notes in Computer Science*, pages 244–258. Springer, 1989.
8. I. Gnaedig, C. Kirchner, and H. Kirchner. Equational completion in order-sorted algebras. In *CAAP '88*, volume 299 of *Lecture Notes in Computer Science*, pages 165–184. Springer, 1988.
9. G. Huet. Confluent reductions: Abstract properties and applications to term rewriting systems. *Journal of the ACM*, 27(4):797–821, 1980.
10. C. Kirchner, H. Kirchner, and J. Meseguer. Operational semantics of OBJ-3. In *ICALP '88*, volume 317 of *Lecture Notes in Computer Science.* Springer, 1988.
11. J. Levy and J. Agustí. Bi-rewrite systems. *Journal of Symbolic Computation*, 22(3):279–314, 1996.
12. V. Manca, A. Salibra, and G. Scollo. Equational type logic. *Theoretical Computer Science*, 77:131–159, 1990.
13. Z. Manna and R. Waldinger. Special relations in automated deduction. *Journal of the ACM*, 33(1):1–59, 1986.
14. Z. Manna and R. Waldinger. The special-relation rules are incomplete. In *Automated Deduction - CADE-11*, volume 607 of *Lecture Notes in Artificial Intelligence*, pages 492–506. Springer, 1992.
15. A. Mégrelis. Partial algebra + order-sorted algebra = galactic algebra. In *Logical Foundations of Computer Science*, volume 620 of *Lecture Notes in Computer Science*, pages 314–325. Springer, 1992.
16. J. Meseguer. Membership algebras, July 1996. Lecture at the *Dagstuhl Seminar on Specification and Semantics.*
17. P. Mosses. Unified algebras and institutions. In *Principles of Programming Languages Conference*, pages 304–312. ACM Press, 1989.

18. W. M. Schorlemmer. Bi-rewriting rewriting logic. In *First International Workshop on Rewriting Logic and its Applications*, volume 4 of *Electronic Notes in Theoretical Computer Science*. Elsevier Science, 1996.
19. W. M. Schorlemmer. Un càlcul de demostració per a la lògica de reescriptura. Research Report IIIA 96/28, Institut d'Investigació en Intel·ligèncial Artificial (CSIC), Dec. 1996. In Catalan.
20. W. M. Schorlemmer. Rewriting logic as a logic of special relations. In *Second International Workshop on Rewriting Logic and its Applications*, volume 15 of *Electronic Notes in Theoretical Computer Science*. Elsevier Science, 1998.
21. W. W. Wadge. Classified algebras. Research Report 46, Depatment of Computer Science, University of Warwick, 1982.

Abstraction Barriers in Equational Proof

Jo Erskine Hannay

LFCS, Division of Informatics, University of Edinburgh, Scotland
joh@dcs.ed.ac.uk

Abstract. Module constructs in programming languages have protection mechanisms hindering unauthorised external access to internal operators of data types. In some cases, granting external access to internal operators would result in serious violation of a data type's specified external properties. In order to reason consistently about specifications of such data types, it is necessary in general to incorporate a notion of protective abstraction barrier in proof strategies as well. We show how this can be done in equational calculus by simply restricting the congruence axiom, and see how the motivation for this naturally arises from FI and FRI approaches to specification refinement.

1 Introduction

Many programming languages have encapsulation mechanisms that hide internal detail of data types. Besides providing abstraction from uninteresting detail, these encapsulation mechanisms also provide vital protection of a data type's internal workings, to which direct access might otherwise enable a user to create havoc. Consider for example a data type implementation of sets in SML by sorted non-repeating lists. If granted access to the set constructor, a user might generate things (s)he thinks represent sets but which do not according to the data type. Then applying operators which assume the correct representation might give wrong answers. The power to enforce a suitable *abstraction barrier* between a module and the surrounding program is thus not just an organisational nicety, but also essential for program soundness. We here address these latter aspects of encapsulation, i.e. those pertaining to its *logical* or *protective*, as opposed to *organisational*, necessity.

Algebraic specification is viewed in this paper in a refinement setting as described in e.g. [13, 14] or [10]. In such a setting data types are viewed as algebras, and in several schemes, e.g. [12], [3] specifications and programs are written in a uniform language, so that specifications are abstract multi-modeled descriptions of a data type, while program modules are concrete monomorphic executable descriptions of the same. A refinement process then seeks to develop in a sound methodical way the latter from the former. In this setting, the need for abstraction barriers arises naturally in algebraic specifications as well. The specificational and semantic formalisms of algebraic specification have structural constructs, which if combined in the right order provide protective encapsulation, as for example in the *forget-identify* (FI) and the *forget-restrict-identify* (FRI) approach to refinement [18].

A.M. Haeberer (Ed.): AMAST'98, LNCS 1548, pp. 196–213, 1998.

The broad issue of this paper is that when reasoning about specifications and programs, e.g. when doing refinement proofs, one needs to take into consideration *abstraction barriers in proof methods* as well. This is because information about hidden parts of a data type may have to be used when reasoning about its external properties. In this paper we look specifically at proof obligations arising from the FI and FRI implementation schemes, i.e. implementing a data type by hiding details in, and then quotienting, another data type. Moreover, we wish to show how an abstraction barrier can easily be enforced in equational logic, so we look here at equational specifications. This means we will consider the case when the congruence with which the quotienting is done can be expressed by equations. It would then be proof-technically convenient if these latter equations could be used in an equational calculus directly in conjunction with other equations specifying the data type. Our result is that this is indeed possible, provided one incorporates the appropriate abstraction barrier *in the calculus itself.* It suffices to restrict the congruence (monotonicity) axiom to contexts without designated hidden symbols, i.e. imposing *referential opacity*, see [11, 16] for other uses of referential opacity. Without such an abstraction barrier, the resulting set of equations may be inconsistent since (the axioms for) hidden operators might not respect the intended equality predicate.

Several proof system schemata for structured specifications exist, see [7] for an overview, and the standard way by which quotienting is dealt with is by introducing a predicate symbol and explicitly axiomatising the congruence in terms of that symbol [20]. This also goes for the behavioural equalities viz. congruences dealt with in [7], where the axiomatisations are in general infinitary, although in [9] this problem is taken to higher order logic and finitary axiomatisation is then possible. Our approach is beneficial to mechanised reasoning because it remains finitary, first-order and purely equational. In some cases it also allows one to do behavioural verification more directly because now we can safely do proofs w.r.t. behavioural quotients instead of having to axiomatise behavioural equalities.

We will assume that the specifications to which the hiding and quotienting operators are applied are basic or "flat". It should be noted that this is not such a great restriction. Any first-order specification built from a basic specification by applying the standard specification building operators **sum**, **derive**, **translate** [20] can be algorithmically normalised to a basic specification with a **derive** operator outermost [19, 5, 2]. The other relevant operators are **abstract**, **behaviour** and **quotient**. In a refinement context the two former, it can be argued, should be seen as meta-operators and should only be applied outermost [1]. A similar argument can be made for **quotient**.

In Sect. 2 relevant notions are given as well as motivating examples. In Sect. 3 a calculus is presented which is sound and complete w.r.t. the model class of an equational instance of an FI structure. In Sect. 4 we present a calculus with an ω-rule which is shown sound and complete for the semantics of an equational instance of the FRI approach. The FI case is a special case of the FRI case and the completeness proof of the latter immediately gives a completeness proof for the former. Omitted proofs and more detail may be found in [6].

2 Preliminaries and Motivation

A basic knowledge of notions within universal algebra and algebraic specification is assumed, see [18, 13]. Below we give some notions and simplifying assumptions central to the paper. We will be dealing with many-sorted algebraic specifications whose semantics will be given as classes of total many-sorted algebras with non-empty carriers. Fix a signature $\Sigma = \langle S, \Omega \rangle$. The class of Σ-algebras is denoted by $\Sigma\mathbf{Alg}$. For Σ-algebras A and B the class of Σ-homomorphisms from A to B is denoted by $\Sigma\mathbf{Alg}(A, B)$. We also write $\phi : A \to B$ to indicate that ϕ is a homomorphism from A to B. Throughout this paper, fix X as a U-sorted set of variables, where U includes all sorts involved. The Σ-term algebra, i.e. the free Σ-algebra generated by X^S where the S-sorted X^S is given by $X^S_s = X_s$ for $s \in S$, is denoted by $T_\Sigma(X)$. For a Σ-context $c[\square]$ we write $c \in T_\Sigma(X)$ instead of $c \in T_{\Sigma\cup\{\square\}}(X)$. All signatures $\Sigma = \langle S, \Omega \rangle$ are assumed to be *sensible* w.r.t. to a designated set $I \subseteq S$, i.e. for the I-sorted X^I given by $X^I_s = X_s$ for $s \in I$, we assume that the free Σ-algebra generated by X^I, denoted $T_\Sigma(X^I)$, is non-empty. If $I = \emptyset$ this amounts to assuming that there is at least one constant in Ω of every sort s in S. We write $T_\Sigma(\emptyset)$ as G_Σ.

For signatures $\Sigma = \langle S, \Omega \rangle, \Sigma' = \langle S', \Omega' \rangle$, a *signature morphism* $\sigma : \Sigma \to \Sigma'$ maps the sorts and operator symbols of Σ to those of Σ' such that sorts are preserved. For a Σ'-algebra A the *σ-reduct* $A|_\sigma$ of A is the Σ-algebra with carriers $(A|_\sigma)_s = A_{\sigma(s)}$ for each sort $s \in S$ and $f^{A|_\sigma} = \sigma(f)^A$ for each $f \in \Omega$. For any Σ'-congruence $\sim^A$ on A, $\sim^A|_\sigma$ is defined as $\sim^A|_{\sigma_s} = \sim^A{}_{\sigma(s)}$ for each sort $s \in S$. For any Σ'-homomorphism $\phi : A \to B$, $\phi|_\sigma : A|_\sigma \to B|_\sigma$ is the Σ-homomorphism defined by $\phi|_{\sigma_s} = \phi_{\sigma(s)}$ for each sort $s \in S$. In case $\Sigma \subseteq \Sigma'$ and $\sigma : \Sigma \hookrightarrow \Sigma'$ is the inclusion morphism, we write $A|_\Sigma$, $\sim^A|_\Sigma$, $\phi^A|_\Sigma$, and we might write $a \sim^A a'$ in place of $a \sim^A|_\Sigma a'$, since $\sim^A|_{\Sigma_s} = \sim^A{}_s$ for $s \in S$.

If σ is not surjective, the effect is that of *hiding*, i.e. removing, those carriers and operators of A which are not interpretations of symbols in $\sigma(\Sigma)$.

The class of all Σ-algebras that are models (in the standard sense) of a set of axioms Φ is denoted $Mod_\Sigma(\Phi)$.

2.1 Congruence Induced by a Set of Equations

The following standard notion is central. For a set of Σ-equations $E \subseteq T_\Sigma(X) \times T_\Sigma(X)$, the *congruence* $\sim^A_E$ *induced by* E on any Σ-algebra A is defined as the least Σ-congruence containing $\{\langle \phi(l), \phi(r) \rangle \mid \langle l, r \rangle \in E,\ \phi : T_\Sigma(X) \to A\}$. This definition is equivalent to demanding the least *equivalence relation* containing $\{\langle \phi(c[l]), \phi(c[r]) \rangle \mid \langle l, r \rangle \in E, c \in T_\Sigma(X),\ \phi : T_\Sigma(X) \to A\}$, i.e. the relation inductively defined by

$$\textit{induce} : \frac{}{\phi(c[l]) \sim^A_E \phi(c[r])};\quad \langle l, r \rangle \in E, c \in T_\Sigma(X),\ \phi : T_\Sigma(X) \to A$$

$$\textit{refl} : \frac{}{a \sim^A_E a} \qquad \textit{sym} : \frac{a \sim^A_E a'}{a' \sim^A_E a} \qquad \textit{trans} : \frac{a \sim^A_E a'', a'' \sim^A_E a'}{a \sim^A_E a'}$$

The quotient w.r.t. to $\sim^A_E$ is written A/E. Of course, usually $s \sim^{T_\Sigma(X)}_E t$ is written $E \vdash s = t$.

2.2 Abstraction Barriers by Specificational Structure - FI and FRI

Henceforth, we tacitly assume that every class of algebras presented is closed under isomorphism. A *basic specification* is a pair $\langle \Sigma, \Phi \rangle$. Its *loose* semantics $[\![\langle \Sigma, \Phi \rangle]\!]$ is $Mod_\Sigma(\Phi)$. A number of *specification building operators* exist for constructing *structured* specifications. A common one which will be used in examples is **enrich** SP **by sorts** S' **ops** Ω' **axioms** Φ' with semantics $\{A \in \Sigma'\mathbf{Alg} \mid A|_\Sigma \in [\![SP]\!] \ \wedge\ A \models \Phi'\}$, where $\Sigma' = \langle S \cup S', \Omega \cup \Omega' \rangle$. (This **enrich** operator can be expressed by the **sum** operator.) In this paper we are interested in encapsulation and in particular encapsulation out of logical necessity. Our focus is therefore on the two operators **derive** SP **by** σ whose semantics for a signature morphism σ, is $\{A|_\sigma \mid A \in [\![SP]\!]\}$, and **quotient** SP **by** E whose semantics for a set of equations E, is $\{A/E \mid A \in [\![SP]\!]\}$. The particular structure of interest is

$$\textbf{quotient}\quad (\textbf{derive}\ \langle \Sigma, E \rangle\ \textbf{by}\ incl : \Sigma^e \hookrightarrow \Sigma)\quad \textbf{by}\ E' \tag{1}$$

with semantics $\{(A|_{\Sigma^e})/E' \mid A \in Mod_\Sigma(E)\}$. If $\Sigma^e \subsetneq \Sigma$ then the signature fragment $\Sigma^h = \Sigma \setminus \Sigma^e$ is outside the image of *incl*, so the reduct construct hides the interpretations of operator symbols and sorts in Σ^h. Structure is the essential abstraction barrier here: It is crucial that the hiding **derive** step is done before quotienting, since quotienting in the presence of hidden operators might give inconsistency in the sense illustrated in the following example.

Example 1. Following [13], a specification SP' is a *refinement* of SP, written $SP \rightsquigarrow SP'$ iff $[\![SP']\!] \subseteq [\![SP]\!]$. A nice feature in refinement settings is the provision for using an implementation of one data type to implement another. In the example below from [13], the specification Set is refined by using Bag and specification building operators. This reuses any refinement Bag $\overset{*}{\rightsquigarrow}$ SP'' previously done for Bag. In particular if Bag has been refined to an executable module, then this code is reused when implementing Set. Specifically we have

spec Set **is**
  sorts nat, set
  ops empty : set, add : nat × set → set
    in : nat × set → bool
  axioms $\mathsf{add}(x, \mathsf{add}(x, s)) = \mathsf{add}(x, s)$
    $\mathsf{add}(x, \mathsf{add}(y, s)) = \mathsf{add}(y, \mathsf{add}(x, s))$
    $\mathsf{in}(x, \mathsf{empty}) = \mathsf{false}$
    $\mathsf{in}(x, \mathsf{add}(y, s)) = \mathsf{if}\ x =_{\mathsf{nat}} y\ \mathsf{then\ true\ else\ in}(x, s)$

spec Bag **is**
  sorts nat, bag
  ops empty : bag, add : nat × bag → bag
    count : nat × bag → nat
  axioms $\mathsf{add}(x, \mathsf{add}(y, b)) = \mathsf{add}(y, \mathsf{add}(x, b))$
    $\mathsf{count}(x, \mathsf{empty}) = 0$
    $\mathsf{count}(x, \mathsf{add}(y, b)) = \mathsf{if}\ x =_{\mathsf{nat}} y\ \mathsf{then\ succ(count}(x, b))\ \mathsf{else\ count}(x, b)$

The idea is to put an appropriate interface on bags as specified by Bag, so that they look like sets as specified by Set. This may be done safely by adding in as an interface operator, then hiding its implementation in terms of count and then identifying bags that represent the same set. First in is added:

spec Bag+ **is**
  enrich Bag **by**
    ops in : nat × bag → bool
    axioms in(x, b) = count$(x, b) > 0$

Then we use an instance of the structure (1):

spec SetbyBag **is**
  quotient
    derive Bag+ **by** $\sigma = \iota[\mathsf{set} \mapsto \mathsf{bag}]$
  by $E' : \{\mathsf{add}(x, \mathsf{add}(x, s)) = \mathsf{add}(x, s)\}$

where σ is the signature morphism from the signature of Set to that of Bag+ which is the identity on everything except the sort set which is renamed to bag. (For simplicity (1) was stated using an inclusion morphism. In this example, the signature morphism is not an inclusion proper. However, the renaming from set to bag is trivial.) The morphism is not surjective thus hiding count. This specification is structured so that count is hidden *before* quotienting. If this were not done, the specification would be inconsistent relative to the intended semantics on nat, since any model $B = A/E'$ would then have to satisfy e.g. $2 = 1$, by

$$2 = \mathsf{count}^B(x, \mathsf{add}^B(x, \mathsf{add}^B(x, \mathsf{empty}^B))) = \mathsf{count}^B(x, \mathsf{add}^B(x, \mathsf{empty}^B)) = 1$$

and now it would be too late for hiding count^B. However, the above structure ensures the appropriate abstraction barrier and the desired semantics.

In an executable implementation of SetbyBag, the derive operator might be implemented by an encapsulation mechanism hindering outside access to count, and the quotient operator might be implemented by an equality predicate.

The task is now to prove Set $\rightsquigarrow$ SetbyBag. This paper presents a calculus allowing a direct approach to proofs w.r.t. the general structure (1), and hence particularly w.r.t. SetbyBag for this example. ○

The specification structure (1) in general, and the specification SetbyBag of Example 1 in particular, are instances of the common *forget-identify* (FI) implementation strategy of algebraic specification. Even more common is the strategy of *forget-restrict-identify* (FRI), which involves restricting to the unique reachable sub-algebra after reducting and before quotienting.

Let $\Sigma = \langle S, \Omega \rangle$ and let $S' \subseteq S$. The set $I = S \setminus S'$ might be thought of as designated *input* sorts. A Σ-algebra A is *reachable on* S' if there is no proper Σ-subalgebra whose I-sorted carriers are the same as those of A. Equivalently, let $X^I \subseteq X$ denote the I-sorted variables of X. Then A is reachable on S' iff for every $a \in A$ there is a term $t \in T_\Sigma(X^I)$ such that $\phi(t) = a$ for some homomorphism $\phi : T_\Sigma(X^I) \to A$. Any Σ-algebra has a unique Σ-subalgebra which is reachable on S', denoted $R_{S'}(A)$. The *restriction* $R{\restriction}^{B'}_{A'}$ of a relation $R \subseteq A \times B$ is here taken to be $R \cap A' \times B'$. For any Σ-homomorphism $\phi : A \to B$, the Σ-homomorphism $R_{S'}(\phi) : R_{S'}(A) \to R_{S'}(B)$ (*qua* relation) is defined to be $\phi\restriction^B_{R_{S'}(A)}$ (the range of $\phi\restriction^B_{R_{S'}(A)}$ is $R_{S'}(B)$). For a Σ-congruence $\sim$ on A, the Σ-congruence $R_{S'}(\sim)$ is defined as $\sim\restriction^{R_{S'}(A)}_{R_{S'}(A)}$.

The semantics of the specification **restrict** SP **on** S' is $\{R_{S'}(A) \mid A \in [\![SP]\!]\}$. Specifications with the **restrict** operator are normalisable to the form mentioned prefatorially, but with infinitary axioms. However, in refinement we could again claim **restrict** as a meta-operator to be applied outermost. The FRI approach then, is in our context represented by the specification structure

$$\textbf{quotient}\ \ (\textbf{restrict}\ \ (\textbf{derive}\ \langle\Sigma, E\rangle\ \textbf{by}\ incl : \Sigma^e \hookrightarrow \Sigma)\ \ \textbf{on}\ S')\ \textbf{by}\ E' \qquad (2)$$

where $\Sigma^e = \langle S^e, \Omega^e\rangle$, $S' \subseteq S^e$. Its semantics is $\{R_{S'}(A|_{\Sigma^e})/E' \mid A \in Mod_\Sigma(E)\}$. Note that the input sorts I are now $S^e \setminus S'$. There is a range of model classes according to the choice of S'. The case $S' = \emptyset$ gives $R_\emptyset(A|_{\Sigma^e}) = A|_{\Sigma^e}$, and corresponds to FI. The case $S' = S^e$ is ground term denotability.

Example 2. Consider the specification

```
spec SetEnr is
  enrich Set by
    ops remove : nat × set → set
    axioms in(x, remove(x, s)) = false
```

In this example sets as specified by SetEnr are implemented by lists where equal elements occur consecutively. (One might at lower levels of implementation wish to keep a record of insertions. Also, formulating the example in this way will nicely illustrate the use of referential opacity.) We do this by putting an appropriate interface on basic lists, i.e. starting from

```
spec List is
  sorts nat, list
  ops nil : list, _ :: _ : nat × list → list
```

we add interface operators:

```
spec List+ is
  enrich List by
  ops empty : list, add : nat × list → list,
       remove : nat × list → list, in : nat × list → bool
  axioms empty = nil
         add(x, nil) = x :: nil
         add(x, y :: l) = if x =nat y then x :: y :: l else y :: add(x, l)
         in(x, nil) = false
         in(x, y :: l) = if x =nat y then true else in(x, l)
         remove(x, nil) = nil
         remove(x, y :: nil) = if x =nat y then nil else y :: nil
         remove(x, y :: z :: l) = if x =nat y then
                                     if x =nat z then remove(x, l) else z :: l
                                  else y :: remove(x, z :: l)
```

Notice that remove is optimised by using the fact that we intend to represent sets by lists in which equal elements are stored consecutively. However, this representation has to be guaranteed by imposing a suitable abstraction barrier. We use the FRI construct (2):

spec SetbyCanonicalList **is**
  quotient
   restrict
    derive List+ **by** $\sigma = \iota[\mathsf{set} \mapsto \mathsf{list}]$
   on {set}
  by E' : {add(x, add(y, s)) = add(y, add(x, s)),
      add(x, add(x, s)) = add(x, s)}

where σ is the signature morphism from the signature of Set to that of List+ which is the identity on everything except for the renaming of set to list. It is not surjective thus hiding nil and ::. Semantically, no model of SetbyCanonicalList has interpretations of nil and ::, so the only way of generating lists is by the interpretations of the operator symbols empty and add which in the initial model will generate the canonical lists with which we intend to represent sets. However, we also have to restrict to the least reachable sub-algebra on {set}, because the reduct operator only takes away operators and entire carriers, and leaves all carriers which are interpretations of sorts in $\sigma(\Sigma)$ intact. Without the restrict step, models would not necessarily satisfy in(x, remove(x, s)) = false, since s would then range over all lists hence also non-canonical lists. Note that we must prove [[SetbyCanonicalList]] $\models$ in(x, remove(x, s)) = false to verify that SetbyCanonicalList is a refinement of Set. ○

2.3 Overview of Main Results

This paper presents sound and complete equational calculi for the FI and FRI structured semantics as formulated in schemes (1) and (2). The usefulness of such calculi are apparent in refinement scenarios as those in Examples 1 and 2. The calculi will be generalisations in a certain sense of calculi for the flat basic cases, as explained in the following. For a basic specification $SP = \langle \Sigma, E \rangle$ we have by Birkhoff for the equational calculus $\vdash$

$$[\![SP]\!] \models s = t \;\Leftrightarrow\; T_\Sigma(X)/E \models s = t \;\Leftrightarrow\; E \vdash s = t \qquad *$$

Here $T_\Sigma(X)/E$ is a *classifying* model of $[\![SP]\!]$. Now let K^{FI} be the semantics of the FI structure (1). The first main result will be a calculus $\vdash^{\mathrm{FI}}$ and a classifying model $T_{K^{\mathrm{FI}}}$ such that

$$K^{\mathrm{FI}} \models s = t \;\Leftrightarrow\; T_{K^{\mathrm{FI}}} \models s = t \;\Leftrightarrow\; \vdash^{\mathrm{FI}} s = t$$

Secondly, recall that for the basic specification SP we have for $\vdash^\omega$, i.e. the equational calculus augmented with the ω-rule,

$$\mathit{Reach}([\![SP]\!]) \models s = t \;\Leftrightarrow\; G_\Sigma/E \models s = t \;\Leftrightarrow\; \vdash^\omega s = t \qquad **$$

where $\mathit{Reach}([\![SP]\!])$ is the subclass of $[\![SP]\!]$ consisting of all algebras reachable on the sorts S of Σ, i.e. ground term denotable algebras, or computation structures. Now, in the FRI approach we are interested in classes of reachable sub-algebras, rather than sub-classes of reachable algebras. However in a flat equational setting these two are the same: Let $R_{S'}(\mathit{Mod}_\Sigma(E)) = \{R_{S'}(A) \mid A \in \mathit{Mod}_\Sigma(E)\}$ and $\mathit{Reach}_{S'}(\mathit{Mod}_\Sigma(E)) = \{A \in \mathit{Mod}_\Sigma(E) \mid A \text{ is reachable on } S'\}$.

Fact 1. $R_{S'}(Mod_\Sigma(E)) = Reach_{S'}(Mod_\Sigma(E))$

This correspondence means that we can utilise the ω-rule also when considering $R_{S'}([\![SP]\!])$. In fact it is follows that for arbitrary $S' \subseteq S$,

$$R_{S'}([\![SP]\!]) \models s = t \;\Leftrightarrow\; R_{S'}(T_\Sigma(X)/E) \models s = t \;\Leftrightarrow\; \vdash^\omega_{S'} s = t$$

where $\vdash^\omega_{S'}$ denotes the standard equational calculus augmented by the following parametrised ω-rule.

$$\frac{\forall \tau : T_\Sigma(X) \to R_{S'}(T_\Sigma(X))\ .\ \vdash \tau(s) = \tau(t)}{\vdash s = t}$$

The special case $S' = \emptyset$ is simply $*$. The case $S' = S$ is $**$, in which case $R_{S'}(T_\Sigma(X)/E) \cong G_\Sigma/E$ is the initial object of $Mod_\Sigma(E)$.

Now let $K^{\mathrm{FRI}}_{S'}$ be the semantics the FRI structure (2). By analogy to the basic case we will as a second main result devise a calculus $\vdash^{\mathrm{FRI}}_{S'}$ with a parametrised ω rule and classifying model $T_{K^{\mathrm{FRI}}_{S'}}$ such that

$$K^{\mathrm{FRI}}_{S'} \models s = t \;\Leftrightarrow\; T_{K^{\mathrm{FRI}}_{S'}} \models s = t \;\Leftrightarrow\; \vdash^{\mathrm{FRI}}_{S'} s = t$$

Analogously to the basic case above, the classifying model $T_{K^{\mathrm{FRI}}_{S'}}$ will in the case $S' = S$ be the initial object of K^{FI}.

As a curio, there is an aspect in which the analogy does not hold. In the basic case $R_{S'}(T_\Sigma(X)/E)$ is free on $X_{S \setminus S'}$ in $[\![SP]\!]$, assuming that E does not identify any variables. However, in general $T_{K^{\mathrm{FRI}}_{S'}}$ is not free on $X_{S\setminus S'}$ in or for K^{FI}, except in the case $S' = S$.

3 FI Approach — A Referentially Opaque Calculus

In the following we shall develop a calculus for structured specifications of the form (1), for $\Sigma^e = \langle S^e, \Omega^e\rangle \subseteq \Sigma = \langle S, \Omega\rangle$, E a set of Σ-equations and E' a set of Σ^e-equations. The calculus implements a protective abstraction barrier in the form of referential opacity. The model class of (1) is given by $\{(A|_{\Sigma^e})/E' \mid A \in Mod_\Sigma(E)\}$ and will be denoted by K^{FI} throughout. We will give a calculus that is equationally sound and complete for K^{FI}.

Algebras in K^{FI} are of the form $(A|_{\Sigma^e})/E'$ where A is a model for E. The classifying model is $(T_\Sigma(X)/E|_{\Sigma^e})/E'$ (Theorem 2). Viewing for the moment $T_\Sigma(X)/E|_{\Sigma^e}$ as a "term-algebra" T, we directly get an "abstract" calculus for K^{FI} by considering $\sim^T_{E'}$ on T and the classifying model T/E'. This is a generalisation of the basic case $*$ in Sect. 2.3 where $E \vdash$ is given directly by $\sim^{T_\Sigma(X)}_E$. The abstract calculus thus operates on elements of T, i.e. congruence classes q of $T_\Sigma(X)/E|_{\Sigma^e}$. Notice that each q has the form $[t]_E$ for $t \in T_\Sigma(X)|_{\Sigma^e}$, and recall that in general $T_{\Sigma^e}(X) \not\supseteq T_\Sigma(X)|_{\Sigma^e}$, because for any $s \in S^e$, $T_\Sigma(X)|_{\Sigma^e s} = T_\Sigma(X)_s$.

Of course, instead of this abstract calculus we would rather have a calculus operating on terms. We obtain this by "opening up" the congruence classes q and then building a calculus over E' on $T_\Sigma(X)|_{\Sigma^e}$. Opening up the congruence classes necessitates importing the calculus $E \vdash$.

Definition 1 (Calculus $\vdash^{\mathrm{FI}}$). *For all $u, v \in T_\Sigma(X)|_{\Sigma^e}$,*

$$importE : \frac{E \vdash u = v}{\vdash^{\mathrm{FI}} u = v}$$

$$induceE' : \frac{}{\vdash^{\mathrm{FI}} \phi(c[l]) = \phi(c[r])}; \langle l, r\rangle \in E', c \in T_{\Sigma^e}(X),\ \phi : T_{\Sigma^e}(X) \to T_\Sigma(X)|_{\Sigma^e}$$

$$refl : \frac{}{\vdash^{\mathrm{FI}} u = u} \qquad sym : \frac{\vdash^{\mathrm{FI}} u = v}{\vdash^{\mathrm{FI}} v = u} \qquad trans : \frac{\vdash^{\mathrm{FI}} u = w, \vdash^{\mathrm{FI}} w = v}{\vdash^{\mathrm{FI}} u = v}$$

In rule *induceE'* in Definition 1, the contexts are $T_{\Sigma^e}(X)$-contexts, giving referential opacity w.r.t. $T_\Sigma(X)|_{\Sigma^e}$. This is a direct consequence of the definition of congruence on a Σ^e-algebra ($T_\Sigma(X)|_{\Sigma^e}$) induced by a set of Σ^e-equations (E') (Sect. 2.1). In this way the abstraction barrier provided by the reduct construct in the semantics K^{FI}, gets its counterpart in the calculus in the form of referential opacity. Notice that in fact $\phi(c[\square]) \in T_\Sigma(X)|_{\Sigma^e}$. However, the fact that $c[\square]$ is a $T_{\Sigma^e}(X)$-context, ensures the essential property that all operator symbols in the path from $\square$ to the root of $\phi(c[\square])$ (seen as a tree of sub-terms) are from Ω^e.

Note that the calculus in Definition 1 is given by the Σ^e-congruence on $T_\Sigma(X)|_{\Sigma^e}$ induced by the set of Σ^e-equations

$$E^{\mathrm{FI}} = (\sim_E^{T_\Sigma(X)})\restriction_{T_\Sigma(X)|_{\Sigma^e}} \ \cup\ E' \tag{3}$$

We shall make use of this observation later.

All algebras in K^{FI} are Σ^e-algebras, so satisfaction by K^{FI} only has meaning for Σ^e-equations. However, it is necessary for completeness that the calculus considers substitutions into $T_\Sigma(X)|_{\Sigma^e}$ of Σ^e-equations, since $(T_\Sigma(X)/E|_{\Sigma^e})/E'$ is in K^{FI}. This is just a manifestation of the above discussion, where it was motivated from considering what will turn out to be the classifying model $(T_\Sigma(X)/E|_{\Sigma^e})/E'$ that $\vdash^{\mathrm{FI}}$ must be defined over $T_\Sigma(X)|_{\Sigma^e}$.

Theorem 2 (Soundness and completeness). *Let K^{FI} be the semantics $\{(A|_{\Sigma^e})/E' \mid A \in Mod_\Sigma(E)\}$ of (1). For all $u, v \in T_{\Sigma^e}(X)$,*

$$K^{\mathrm{FI}} \models u = v \ \Leftrightarrow\ T_\Sigma(X)/E|_{\Sigma^e}/E' \models u = v \ \Leftrightarrow\ \vdash^{\mathrm{FI}} u = v$$

Proof: This follows from Theorem 5 by observations 3 and 4. □

Example 3. By Theorem 2, the calculus $\vdash^{\mathrm{FI}}$ can be used in verifying the refinement postulated in Example 1, namely Set $\rightsquigarrow$ SetbyBag. The calculus ensures the safe interaction between the set E of equations associated with Bag+ and the set E' of equations introduced in the quotienting step forming SetbyBag. For instance, although $\vdash^{\mathrm{FI}} \mathsf{add}(x, \mathsf{add}(x, \mathsf{empty})) = \mathsf{add}(x, \mathsf{empty})$, referential opacity prevents us from inferring $\vdash^{\mathrm{FI}} \mathsf{count}(x, \mathsf{add}(x, \mathsf{add}(x, \mathsf{empty}))) = \mathsf{count}(x, \mathsf{add}(x, \mathsf{empty}))$, which would have given $\vdash^{\mathrm{FI}} 2 = 1$. The inference is illegal because count is a hidden operator symbol. Referential opacity ensures soundness and is an appropriate abstraction barrier in the calculus. ○

4 FRI Approach

We now address the FRI approach in which reducts are restricted to reachable subalgebras on certain sorts. We consider the FRI specification structure (2). Again, for $\Sigma^e = \langle S^e, \Omega^e \rangle \subseteq \Sigma = \langle S, \Omega \rangle$, let E be a set of equations over $T_\Sigma(X)$ and let E' be a set of equations over $T_{\Sigma^e}(X)$. Let $S' \subseteq S^e$. The model class of (2) is given by $\{R_{S'}(A|_{\Sigma^e})/E' \mid A \in Mod_\Sigma(E)\}$ and will be denoted by $K_{S'}^{\mathrm{FRI}}$ throughout.

Observation 3. *For $S' = \emptyset$, $K_{S'}^{\mathrm{FRI}} = K^{\mathrm{FI}}$.*

4.1 A Restricted Calculus with ω-rule

Algebras in $K_{S'}^{\mathrm{FRI}}$ are of the form $R_{S'}(A|_{\Sigma^e})/E'$ where A is a model for E. The classifying model is $R_{S'}(T_\Sigma(X)/E|_{\Sigma^e})/E'$ (Theorem 5). As we did for the FI case, viewing for the moment $R_{S'}(T_\Sigma(X)/E|_{\Sigma^e})$ as a "term-algebra" T', we directly get an "abstract" calculus for K^{FRI} by considering $\sim_{E'}^{T'}$ on T' and the classifying model T'/E'. The abstract calculus thus operates on elements of T', i.e. congruence classes q of $R_{S'}(T_\Sigma(X)/E|_{\Sigma^e})$, and each q has the form $[t]_E$ for $t \in R_{S'}(T_\Sigma(X)|_{\Sigma^e})$. Again we obtain a term calculus by opening up the congruence classes and importing $E \vdash$. This calculus is defined over $R_{S'}(T_\Sigma(X)|_{\Sigma^e})$ and is given by the congfuence on $R_{S'}(T_\Sigma(X)|_{\Sigma^e})$ induced by the set of Σ^e-equations

$$E^{\mathrm{FRI}} = (\sim_E^{T_\Sigma(X)}) \upharpoonright_{R_{S'}(T_\Sigma(X)|_{\Sigma^e})} \ \cup \ E'$$

Remember now that as K^{FRI} consists of Σ^e-algebras, we are interested in satisfiability of Σ^e statements, i.e. Σ^e-equations. However, depending on S' it may be the case that $T_{\Sigma^e}(X) \not\subseteq R_{S'}(T_\Sigma(X)|_{\Sigma^e})$, in which case the calculus will not respond to all Σ^e-equations. Hence, we supply an ω-rule dependent on S'.

Definition 2 (Calculus $\vdash_{S'}^{\mathrm{FRI}}$). *The calculus $\vdash_{S'}^{\mathrm{FRI}}$ is given by the following single rule. For all $u, v \in T_{\Sigma^e}(X)$,*

$$\omega_{S'} : \frac{\forall \tau : T_{\Sigma^e}(X) \to R_{S'}(T_\Sigma(X)|_{\Sigma^e}) \ . \ \tau(u) \sim_{E^{\mathrm{FRI}}}^{R_{S'}(T_\Sigma(X)|_{\Sigma^e})} \tau(v)}{\vdash_{S'}^{\mathrm{FRI}} u = v}$$

To spell that out, let $\vdash_{S'}^{\mathrm{FI}}$ be the following calculus. For all $u, v \in R_{S'}(T_\Sigma(X)|_{\Sigma^e})$,

$$importE : \frac{E \vdash u = v}{\vdash_{S'}^{\mathrm{FI}} u = v}$$

$$induceE' : \frac{}{\vdash_{S'}^{\mathrm{FI}} \phi(c[l]) = \phi(c[r])} ; \ \begin{array}{l}\langle l, r\rangle \in E', c \in T_{\Sigma^e}(X), \\ \phi : T_{\Sigma^e}(X) \to R_{S'}(T_\Sigma(X)|_{\Sigma^e})\end{array}$$

$$refl : \frac{}{\vdash_{S'}^{\mathrm{FI}} u = u} \qquad sym : \frac{\vdash_{S'}^{\mathrm{FI}} u = v}{\vdash_{S'}^{\mathrm{FI}} v = u} \qquad trans : \frac{\vdash_{S'}^{\mathrm{FI}} u = w, \vdash_{S'}^{\mathrm{FI}} w = v}{\vdash_{S'}^{\mathrm{FI}} u = v}$$

Now $\vdash_{S'}^{\mathrm{FRI}}$ is given by the following rule. For all $u, v \in T_{\Sigma^e}(X)$,

$$\omega_{S'} : \frac{\forall \tau : T_{\Sigma^e}(X) \to R_{S'}(T_\Sigma(X)|_{\Sigma^e}) \ . \ \vdash_{S'}^{\mathrm{FI}} \tau(u) = \tau(v)}{\vdash_{S'}^{\mathrm{FRI}} u = v}$$

Observation 4. $\vdash^{\mathrm{FRI}}_{S'}$ *subsumes* $\vdash^{\mathrm{FI}}$*: If* $S' = \emptyset$ *then the rule* $\omega_{S'}$ *adds nothing to* $\vdash^{\mathrm{FI}}_{S'}$*, so for* $u, v \in T_{\Sigma^e}(X)$*,* $\vdash^{\mathrm{FRI}}_{\emptyset} u = v \Leftrightarrow \vdash^{\mathrm{FI}}_{\emptyset} u = v \Leftrightarrow \vdash^{\mathrm{FI}} u = v$.

Theorem 5 (Soundness and completeness). *Let* $K^{\mathrm{FRI}}_{S'}$ *be the semantics* $\{R_{S'}(A|_{\Sigma^e})/E' \mid A \in Mod_{\Sigma}(E)\}$ *of (2). For any* $u, v \in T_{\Sigma^e}(X)$*,*

$$K^{\mathrm{FRI}}_{S'} \models u = v \Leftrightarrow R_{S'}(T_{\Sigma}(X)/E|_{\Sigma^e})/E' \models u = v \Leftrightarrow \vdash^{\mathrm{FRI}}_{S'} u = v$$

Proof: The proof is split into Lemmas 6 and 9 below. □

Lemma 6 (Completeness).

$$K^{\mathrm{FRI}}_{S'} \models u = v \Rightarrow R_{S'}(T_{\Sigma}(X)/E|_{\Sigma^e})/E' \models u = v \Rightarrow \vdash^{\mathrm{FRI}}_{S'} u = v$$

Proof: Suppose $K^{\mathrm{FRI}}_{S'} \models u = v$. By definition $R_{S'}(T_{\Sigma}(X)/E|_{\Sigma^e})/E' \models u = v$. Lemma 7 gives $\forall \tau : T_{\Sigma^e}(X) \to R_{S'}(T_{\Sigma}(X)|_{\Sigma^e})$. $[[\tau(u)]_E]_{E'} = [[\tau(v)]_E]_{E'}$. Lemma 8 then gives $\forall \tau : T_{\Sigma^e}(X) \to R_{S'}(T_{\Sigma}(X)|_{\Sigma^e})$. $\vdash^{\mathrm{FI}}_{S'} \tau(u) = \tau(v)$. Finally, the rule $\omega_{S'}$ gives $\vdash^{\mathrm{FRI}}_{S'} u = v$. □

Lemma 7. *For* $u, v \in T_{\Sigma^e}(X)$*,*

$$R_{S'}(T_{\Sigma}(X)/E|_{\Sigma^e})/E' \models u = v$$
$$\Downarrow$$
$$\forall \tau : T_{\Sigma^e}(X) \to R_{S'}(T_{\Sigma}(X)|_{\Sigma^e}) \;.\; [[\tau(u)]_E]_{E'} = [[\tau(v)]_E]_{E'}$$

where $[w]_E$ *denotes the congruence class of* w *in* $R_{S'}(T_{\Sigma}(X)/E|_{\Sigma^e})$*, and* $[q]_{E'}$ *denotes the congruence class of* q *in* $R_{S'}(T_{\Sigma}(X)/E|_{\Sigma^e})/E'$.

Proof: Suppose $R_{S'}(T_{\Sigma}(X)/E|_{\Sigma^e})/E' \models u = v$, i.e.

$$\forall \varphi : T_{\Sigma^e}(X) \to R_{S'}(T_{\Sigma}(X)/E|_{\Sigma^e})/E' \;.\; \varphi(u) = \varphi(v)$$

Define $\psi_E : T_{\Sigma}(X) \to T_{\Sigma}(X)/E$ as $\psi_E(u) = [u]_E$. For any $\tau : T_{\Sigma^e}(X) \to R_{S'}(T_{\Sigma}(X)|_{\Sigma^e})$, we get $R_{S'}(\psi_E|_{\Sigma^e}) \circ \tau : T_{\Sigma^e}(X) \to R_{S'}(T_{\Sigma}(X)/E|_{\Sigma^e})$.

Let $\psi_{E'} : R_{S'}(T_{\Sigma}(X)/E|_{\Sigma^e}) \to R_{S'}(T_{\Sigma}(X)/E|_{\Sigma^e})/E'$ be defined as $\psi_{E'}(q) = [q]_{E'}$. Then $\psi_{E'} \circ (R_{S'}(\psi_E|_{\Sigma^e}) \circ \tau) : T_{\Sigma^e}(X) \to R_{S'}(T_{\Sigma}(X)/E|_{\Sigma^e})/E'$, and so $[[\tau(u)]_E]_{E'} = \psi_{E'} \circ (R_{S'}(\psi_E|_{\Sigma^e}) \circ \tau)(u) = \psi_{E'} \circ (R_{S'}(\psi_E|_{\Sigma^e}) \circ \tau)(v) = [[\tau(v)]_E]_{E'}$.
□

Lemma 8. *Let* $[w]_E$ *denote the congruence class of* w *in* $R_{S'}(T_{\Sigma}(X)/E|_{\Sigma^e})$*, and* $[q]_{E'}$ *denote the congruence class of* q *in* $R_{S'}(T_{\Sigma}(X)/E|_{\Sigma^e})/E'$*. For* $u, v \in R_{S'}(T_{\Sigma}(X)|_{\Sigma^e})$*,*

$$[[u]_E]_{E'} = [[v]_E]_{E'} \Rightarrow \vdash^{\mathrm{FI}}_{S'} u = v$$

Proof: Suppose $[[u]_E]_{E'} = [[v]_E]_{E'}$, that is, $[u]_E \sim_{E'}^{R_{S'}(T_\Sigma(X)/E|_{\Sigma^e})} [v]_E$. Induction on the construction of $\sim_{E'}^{R_{S'}(T_\Sigma(X)/E|_{\Sigma^e})}$.

induce : For some $\langle l, r\rangle \in E'$, $c \in T_{\Sigma^e}(X)$, $\varphi : T_{\Sigma^e}(X) \to R_{S'}(T_\Sigma(X)/E|_{\Sigma^e})$, we have $\varphi(c[l]) \sim_{E'}^{R_{S'}(T_\Sigma(X)/E|_{\Sigma^e})} \varphi(c[r])$ and $u \in \varphi(c[l])$ and $v \in \varphi(c[r])$. It is a fact that $R_{S'}(T_\Sigma(X)/E|_{\Sigma^e}) \cong R_{S'}(T_\Sigma(X)|_{\Sigma^e})/R_{S'}(\sim_E^{T_\Sigma(X)}|_{\Sigma^e})$. So by Fact 11 φ can be factored into $\psi_E \circ \tau$ for some $\tau : T_{\Sigma^e}(X) \to R_{S'}(T_\Sigma(X)|_{\Sigma^e})$, and we have $\varphi(c[l]) = [\tau(c[l])]_E$ and $\varphi(c[r]) = [\tau(c[r])]_E$. So then $E \vdash u = \tau(c[l])$ and $E \vdash \tau(c[r]) = v$, which by *importE* gives $\vdash_{S'}^{\mathrm{FI}} u = \tau(c[l])$ and $\vdash_{S'}^{\mathrm{FI}} \tau(c[r]) = v$. By *induceE'* we have $\vdash_{S'}^{\mathrm{FI}} \tau(c[l]) = \tau(c[r])$, and *trans* then gives $\vdash_{S'}^{\mathrm{FI}} u = v$.

refl : Then $[u]_E = [v]_E$. So $E \vdash u = v$, which by *importE* gives $\vdash_{S'}^{\mathrm{FI}} u = v$.

sym and *trans*: These are dealt with by the i.h. and *sym* and *trans* of $\vdash_{S'}^{\mathrm{FI}}$.
□

Lemma 9 (Soundness). $\vdash_{S'}^{\mathrm{FRI}} u = v \;\Rightarrow\; K_{S'}^{\mathrm{FRI}} \models u = v$

Proof: Fix $A \in Mod_\Sigma(E)$ arbitrarily. Suppose $\vdash_{S'}^{\mathrm{FRI}} u = v$, for $u, v \in T_{\Sigma^e}(X)$. The way this is possible is via the only rule $\omega_{S'}$, and so we must have $\forall \tau : T_{\Sigma^e}(X) \to R_{S'}(T_\Sigma(X)|_{\Sigma^e})$. $\vdash_{S'}^{\mathrm{FI}} \tau(u) = \tau(v)$. By Lemma 10 we then get for any $\tau : T_{\Sigma^e}(X) \to R_{S'}(T_\Sigma(X)|_{\Sigma^e})$ and $\psi : R_{S'}(T_\Sigma(X)|_{\Sigma^e}) \to R_{S'}(A|_{\Sigma^e})$ that

$$\psi(\tau(u)) \sim_{E'}^{R_{S'}(A|_{\Sigma^e})} \psi(\tau(v)). \qquad *$$

Fix $\phi' : T_{\Sigma^e}(X) \to R_{S'}(A|_{\Sigma^e})$ arbitrarily. We now show that there exist $\tau : T_{\Sigma^e}(X) \to R_{S'}(T_\Sigma(X)|_{\Sigma^e})$ and $\psi : R_{S'}(T_\Sigma(X)|_{\Sigma^e}) \to R_{S'}(A|_{\Sigma^e})$ such that $\phi' = \psi \circ \tau$, i.e. making the left-hand part of following diagram commute:

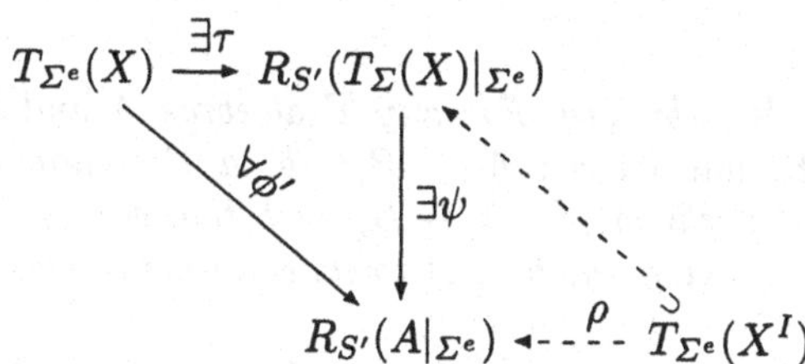

For any $x \in X_s$, $s \in S^e$, let $a = \phi'(x)$. There is some $t_a \in T_{\Sigma^e}(X^I)$ and $\rho : T_{\Sigma^e}(X^I) \to A|_{\Sigma^e}$ such that $\rho(t_a) = a$. We determine τ by defining $\tau(x) = t_a$ ($T_{\Sigma^e}(X^I) \subseteq R_{S'}(T_\Sigma(X)|_{\Sigma^e})$ because $X^I \subseteq R_{S'}(T_\Sigma(X)|_{\Sigma^e})$ and $R_{S'}(T_\Sigma(X)|_{\Sigma^e})$ is a Σ^e-algebra).

Determine ψ as follows. Let $\psi_\rho : T_\Sigma(X) \to A$ be determined by

$$\psi_\rho(x) = \begin{cases} \rho(x), & x \in X_s,\ s \in I = S^e \setminus S' \\ a_\perp, & x \in X_s,\ s \in S \setminus I, \text{ for some choice } a_\perp \end{cases}$$

Define $\psi = R_{S'}(\psi_\rho|_{\Sigma^e})$. Then for any $t \in T_{\Sigma^e}(X^I)$, $\psi(t) = R_{S'}(\psi_\rho|_{\Sigma^e})(t) = \psi_\rho(t) = \rho(t)$.

So for any $x \in X_s$, $s \in S^e$, $\psi(\tau(x)) = \psi(t_a) = \rho(t_a) = a = \phi'(x)$, and so $\phi' = \psi \circ \tau$.

Together with $*$ this gives $\phi'(u) = \psi(\tau(u)) \sim_{E'}^{R_{S'}(A|_{\Sigma^e})} \psi(\tau(v)) = \phi'(v)$. Now, ϕ' was arbitrary so we get $\forall \phi' : T_{\Sigma^e}(X) \to R_{S'}(A|_{\Sigma^e}) \;.\; \phi'(u) \sim_{E'}^{R_{S'}(A|_{\Sigma^e})} \phi'(v)$. Consider any $\phi : T_{\Sigma^e}(X) \to R_{S'}(A|_{\Sigma^e})/E'$. Fact 11 gives a $\phi' : T_{\Sigma^e}(X) \to R_{S'}(A|_{\Sigma^e})$ such that $\phi = \psi_{E'} \circ \phi'$. So for all $\phi : T_{\Sigma^e}(X) \to R_{S'}(A|_{\Sigma^e})/E'$ we have $\phi(u) = [\phi'(u)]_{E'} = [\phi'(v)]_{E'} = \phi(v)$, i.e. $R_{S'}(A|_{\Sigma^e})/E' \models u = v$. □

Lemma 10. *Let $A \in Mod_\Sigma(E)$. For $u, v \in R_{S'}(T_\Sigma(X)|_{\Sigma^e})$,*

$$\vdash_{S'}^{\mathrm{FI}} u = v \;\Rightarrow\; \forall \psi : R_{S'}(T_\Sigma(X)|_{\Sigma^e}) \to R_{S'}(A|_{\Sigma^e}) \;.\; \psi(u) \sim_{E'}^{R_{S'}(A|_{\Sigma^e})} \psi(v)$$

Proof: Suppose $\vdash_{S'}^{\mathrm{FI}} u = v$, for $u, v \in R_{S'}(T_\Sigma(X)|_{\Sigma^e})$. Induction on the construction of $\vdash_{S'}^{\mathrm{FI}}$. Fix $\psi : R_{S'}(T_\Sigma(X)|_{\Sigma^e}) \to R_{S'}(A|_{\Sigma^e})$ arbitrarily.

importE: Since $A \in Mod_\Sigma(E)$ we have for any $\varphi : T_\Sigma(X) \to A$ that $\varphi(u) = \varphi(v)$. Determine $\varphi_\psi : T_\Sigma(X) \to A$ by

$$\varphi_\psi(x) = \begin{cases} \psi(x), & x \text{ a variable in } R_{S'}(T_\Sigma(X)|_{\Sigma^e}) \\ a_\perp, & x \text{ a variable otherwise in } X^S, \text{ for some choice } a_\perp \end{cases}$$

Then for $w \in R_{S'}(T_\Sigma(X)|_{\Sigma^e})$, we have $\varphi_\psi(w) = \psi(w)$, and so $\psi(u) = \varphi_\psi(u) = \varphi_\psi(v) = \psi(v)$. By *refl* of $\sim_{E'}^{R_{S'}(A|_{\Sigma^e})}$, $\psi(u) \sim_{E'}^{R_{S'}(A|_{\Sigma^e})} \psi(v)$.

induceE': Then $u = \tau(c[l])$, $\tau(c[r]) = v$ for some $\langle l, r\rangle \in E'$, $c \in T_{\Sigma^e}(X)$ and $\tau : T_{\Sigma^e}(X) \to R_{S'}(T_\Sigma(X)|_{\Sigma^e})$. By *induce* of $\sim_{E'}^{R_{S'}(A|_{\Sigma^e})}$ we have $\rho(c[l]) \sim_{E'}^{R_{S'}(A|_{\Sigma^e})} \rho(c[r])$ for $\rho : T_{\Sigma^e}(X) \to R_{S'}(A|_{\Sigma^e})$. Now $\psi \circ \tau : T_{\Sigma^e}(X) \to R_{S'}(A|_{\Sigma^e})$, and so $\psi(u) = \psi(\tau(c[l])) \sim_{E'}^{R_{S'}(A|_{\Sigma^e})} \psi(\tau(c[r])) = \psi(v)$.

refl, *sym* and *trans*: These are dealt with by the i.h. and *refl*, *sym* and *trans* of $\sim_{E'}^{R_{S'}(A|_{\Sigma^e})}$. □

Fact 11. *Let Σ be arbitrary. For any Σ-algebras A and B, let $\sim$ be any Σ-congruence on B, and let $\phi : A \to B/\!\sim$ be a Σ-homomorphism. Then there exists a Σ-homomorphism $\phi' : A \to B$, such that $\phi = \psi_\sim \circ \phi'$ where $\psi_\sim$ is the Σ-homomorphism taking any b in B to its equivalence class $[b]_\sim$ in $B/\!\sim$.*

Example 4. The calculus $\vdash_{S'}^{\mathrm{FRI}}$ can thus be used in verifying the refinement postulated in Example 2, namely Set $\rightsquigarrow$ SetbyCanonicalList. Referential opacity ensures the safe and sound interaction between the set E of equations associated with List+ and the set E' of equations introduced in the quotienting step forming SetbyCanonicalList. For instance, although $\vdash_{S'}^{\mathrm{FRI}} \mathsf{add}(x, \mathsf{add}(x, \mathsf{add}(y, \mathsf{empty}))) = \mathsf{add}(y, \mathsf{add}(x, \mathsf{add}(x, \mathsf{empty})))$, referential opacity hinders the inference

$$\vdash_{S'}^{\mathrm{FRI}} \mathsf{in}(y, \mathsf{remove}(y, y :: \mathsf{add}(x, \mathsf{add}(x, \mathsf{add}(y, \mathsf{empty}))))) = \mathsf{in}(y, \mathsf{remove}(y, y :: \mathsf{add}(y, \mathsf{add}(x, \mathsf{add}(x, \mathsf{empty})))))$$

which would have given $\vdash_{S'}^{\mathrm{FRI}}$ true = false. The inference is illegal because :: is a hidden operator symbol. Also, completeness is secured by the $\omega_{S'}$-rule for $S' = \{\mathrm{set}\}$. We have $\vdash_{S'}^{\mathrm{FRI}} \mathsf{in}(x, \mathsf{remove}(x, s)) = \mathsf{false}$ because for this to hold it is only required that $\vdash_{S'}^{\mathrm{FRI}} \mathsf{in}(x, \mathsf{remove}(x, s\tau)) = \mathsf{false}$ holds for all instances $s\tau$ generated by empty and add. ○

4.2 Coincidence of Initial Models

Fact 12. *If E is sufficiently complete w.r.t. Σ^e, i.e. for every ground term $g \in G_\Sigma$ there is a ground term $g^e \in G_{\Sigma^e}$ such that $E \vdash g = g^e$ then the class K^{FI} has an initial object, namely $G_\Sigma/E|_{\Sigma^e}/E'$.*

If E is not sufficiently complete in the sense above, then K^{FI} may or may not have an initial object.

Theorem 13. *Suppose E is sufficiently complete w.r.t. Σ^e, i.e. for every ground term $g \in G_\Sigma$ there is a ground term $g^e \in G_{\Sigma^e}$ such that $E \vdash g = g^e$. By Fact 12 the initial object $G_\Sigma/{\approx}|_{\Sigma^e}$ of K^{FI} exists. Then for $S' = S$,*

$$R_{S'}(T_\Sigma(X)/E|_{\Sigma^e})/E' \cong G_\Sigma/E|_{\Sigma^e}/E'$$

Note for $S' = S$ that $R_{S'}(T_\Sigma(X)/E|_{\Sigma^e})/E'$ is initial in $K^{\mathrm{FRI}}_{S'}$ by virtue of it being the classifying model, and a Σ^e-computation structure.

4.3 Forget-Identify-Restrict (FIR)

The FRI structure (2) is not equivalent to the structure in which the restrict step is done outermost, i.e. FRI is not equivalent to forget-identify-restrict (FIR), for a counter-example see [4]. Let $K^{\mathrm{FIR}}_{S'} = \{R_{S'}(A|_{\Sigma^e}/E') \mid A \in Mod_\Sigma(E)\}$. A sound and complete calculus $\vdash^{\mathrm{FIR}}_{S'}$ for $K^{\mathrm{FIR}}_{S'}$ is given by the following rule. For all $u, v \in T_{\Sigma^e}(X)$,

$$\omega_{S'} : \frac{\forall \tau : T_{\Sigma^e}(X) \to R_{S'}(T_\Sigma(X)|_{\Sigma^e}) \;.\; \vdash^{\mathrm{FI}}_{\emptyset} \tau(u) = \tau(v)}{\vdash^{\mathrm{FIR}}_{S'} u = v}$$

Again, $\vdash^{\mathrm{FI}}$ is subsumed by $\vdash^{\mathrm{FIR}}_{S'}$. Also, for $S' = S$, we have that the classifying model of $K^{\mathrm{FIR}}_{S'}$ is isomorphic to the initial model of K^{FI}.

5 Discussion and Conclusions

We are concerned with the idea of enforcing protective abstraction barriers in proof methods that reflect abstraction barriers in (the semantics of) programs and more generally in program specifications. In this paper we have shown that for equational forms of the FI and FRI approaches to refinement, it suffices to use primitive equational logic with an abstraction barrier in the form of referential opacity. Although we have only discussed the flat case, we claim generality in light of the normal form result for specifications [19, 5, 2] and the argument that **quotient** should only be employed outermost [1].

The calculi devised in this paper could form a basis upon which adaptations of semi-automated proof systems could be done. For instance, the referential opacity present in the calculi suggests altering the rewrite and completion-based method of *proof by consistency* by constraining the generation of critical pairs according to the context in the overlap of rewrite rules. Note that the results concerning the coincidence of initial models is then relevant. In particular, the $\omega_{S'}$ rule of the FRI calculus is relevant for the FI case at initiality. Note also, that the $\omega_{S'}$ rule is particularly interesting when constructors are hidden as in Example 2. Under certain sufficient completeness conditions, the rule states that it is enough to do induction over "abstract constructors", like empty and add, even when functions are defined over "concrete" constructors, as is remove over nil and :: in Example 2.

Behavioural proofs are often simplified greatly by introducing behavioral lemmas, i.e. propositions that are true according to behavioural equality but not true literally. The referentially opaque calculus ensures an appropriate abstraction barrier so that such lemmas may be inserted soundly into the proof environment. The assumption of *stability* introduced in [15] ensures the safe insertion of such lemmas too. Although the notion of stability applies at a different level of the refinement process, there seems to be a relationship worthwhile looking at between the abstraction barriers provided by stability and referential opacity.

In the setting of behavioural refinement, semantically one does not need quotienting, and the restrict operator is also superfluous since one can speak in terms of partial behavioural congruences. However, for proving behavioural refinement steps, the calculi developed here are useful. One way of proving behavioural refinement steps is to consider the *behaviour* of algebras [7]. The behaviour of an algebra is its quotient by a behavioural congruence, and in the case where this congruence is partial, a restrict step has to be done. Hence we regain the FI and FRI situation.

We remarked in Sect. 3 that the calculus in Definition 1 is given by the Σ^e-congruence on $T_\Sigma(X)|_{\Sigma^e}$ induced by the set of Σ^e-equations E^{FI} given by (3). We could therefore have defined K^{FI} as $\{C|_{\Sigma^e}/E^{\mathrm{FI}} \mid C \in \Sigma\mathbf{Alg}\}$, and expressed the classifying model as $T_\Sigma(X)|_{\Sigma^e}/E^{\mathrm{FI}}$. In a sense this flattens the structured view of the semantics we had in the former characterisation of K^{FI}. In fact we can flatten things even more by considering the following relation.

Definition 3 (Referentially Opaque Congruence). *For any set E of Σ-equations and any set E' of Σ^e-equations, define $\approx^A$ on any Σ-algebra A as the least equivalence containing*

$$\{\langle\phi(c[l]),\phi(c[r])\rangle \mid \langle l,r\rangle \in E, c \in T_\Sigma(X),\ \phi : T_\Sigma(X) \to A\} \ \bigcup$$
$$\{\langle\phi(c[l]),\phi(c[r])\rangle \mid \langle l,r\rangle \in E', c \in T_{\Sigma^e}(X),\ \phi : T_{\Sigma^e}(X) \to A|_{\Sigma^e}\}$$

i.e. the relation inductively defined by

$$\mathit{induceE} : \frac{}{\phi(c[l]) \approx^A \phi(c[r])};\quad \langle l,r\rangle \in E, c \in T_\Sigma(X),\ \phi : T_\Sigma(X) \to A$$

$$\mathit{induceE'} : \frac{}{\phi(c[l]) \approx^A \phi(c[r])};\quad \langle l,r\rangle \in E', c \in T_{\Sigma^e}(X),\ \phi : T_{\Sigma^e}(X) \to A|_{\Sigma^e}$$

$$\mathit{refl} : \frac{}{a \approx^A a} \qquad \mathit{sym} : \frac{a \approx^A a'}{a' \approx^A a} \qquad \mathit{trans} : \frac{a \approx^A a'', a'' \approx^A a'}{a \approx^A a'}$$

Note that $\approx^A$ is not (in general) a Σ-congruence on A. However, if we define the reduct $\approx^A|_\sigma$ of $\approx^A$ w.r.t. a signature morphism $\sigma : \Sigma' \to \Sigma$ as for congruences, i.e. $(\approx^A|_\sigma)_s = \approx^A{}_{\sigma(s)}$, then $\approx^A|_{\Sigma^e}$ is a Σ^e-congruence on $A|_{\Sigma^e}$. It is easy to show:

Fact 14. *Let* $\overline{K}^{\mathrm{FI}} = \{A|_{\Sigma^e}/(\approx^A|_{\Sigma^e}) \mid A \in \Sigma\mathbf{Alg}\}$. *Then* $\overline{K}^{\mathrm{FI}} = K^{\mathrm{FI}}$

In a behavioural context, we can in fact use congruences of the form $\approx^A|_{\Sigma^e}$ to give behaviours of algebras. Here we manage to generate the behavioural congruence in the manner of a congruence induced by equations, even in the presence of problematic hidden operators. By Fact 14 the calculus $\vdash^{\mathrm{FI}}$ lets us do proofs accordingly.

Now having considered $A|_{\Sigma^e}/(\approx^A|_{\Sigma^e})$, one might ask what $(A/\approx^A)|_{\Sigma^e}$ is, and in particular what $A/\approx^A$ is. For a Σ-algebra A and a congruence $\sim$ on A, $(A/\sim)|_{\Sigma^e}$ is of course equal to $A|_{\Sigma^e}/(\sim|_{\Sigma^e})$. Since $\approx$ is not in general a congruence on A, considering an obvious naive definition of $A/\approx^A$ wouldn't necessarily give a Σ-algebra. There are however, reasons to consider various other definitions of $A/\approx^A$. Let us give the as yet tentative structure $A/\approx^A$ the name Q_A.

For example, we could define Q_A as the Σ^e-algebra, having as carriers $A_s/\approx^A_s$ for each $s \in S$, and standard interpretations of each $f \in \Omega^e$. Note that Q_A then has carriers for all sorts in S. If we want a direct proof of the soundness and completeness of the calculus $\vdash^{\mathrm{FI}}$ rather than deriving it is a sub-case of the FRI result, this definition of Q_A enables an easy direct proof. This is due to the flat structure, but also to the ability of $Q_{T_\Sigma(X)} = T_\Sigma(X)/\approx^{T_\Sigma(X)}$ to characterise all derivations involved in $\vdash^{\mathrm{FI}}$, also the ones giving the equations (theorems) imported by *importE*.

Hidden operator symbols, i.e. those in $\Omega \setminus \Omega^e$ have no interpretation in the above tentative definition of Q_A. Viewing Q_A as a Σ-structure would demand that hidden operator symbols get an interpretation. These interpretations could be intensional operators, i.e. operators not respecting the equality predicate of the data type. For instance, looking to Example 1, count^{Q_A} would not respect the equality predicate given by the idempotency and associativity axioms for add^{Q_A}. Considering intensional operators in data types is not an alien concept. Hiding and quotienting do not only occur in software development, but abound elsewhere in mathematics too. At the very foundations of real analysis, the reals are defined as a quotient of a set of Cauchy-sequences. The n'th approximant function is then intensional, and from a constructivist point of view, so is every discontinuous function [17]. Indeed in a constructive setting it might be prudent to add intensional operators to a data type [8] (a choice operator for quotient types). Note then that in $(A/\approx^A)|_{\Sigma^e}$ the reduct operator is now applied outermost, in contrast to the FI- and FRI-models we considered earlier. One could speculate if there might be an adjunction between the appropriate reduct functor and some free functor. The free functor would then add intensional operators, and it would seem imperative to find a definition of structures with intensional operators, and in particular a definition of Q_A as a Σ-structure, such that intensional operators are added in a manner in which they bring with them the appropriate abstraction barrier.

Acknowledgments Thanks are due to Don Sannella, Martin Hofmann and the anonymous referees for valuable suggestions and comments on drafts of this paper. This research has been supported by EPSRC grant GR/K63795, and NFR (Norwegian Research Council) grant 110904/41.

References

[1] M. Bidoit, D. Sannella, and A. Tarlecki. Behavioural encapsulation. CoFI Language Design Study Note, 1996. Available at `ftp://ftp.brics.dk/Projects/CoFI/StudyNotes/Lang/MB+DTS+AT-1.ps.Z`.

[2] M.V. Cengarle. *Formal Specification with Higher-Order Parameterization.* PhD thesis, Fakultät für Mathematik, LMU, München, 1994.

[3] O.-J. Dahl and O. Owe. Formal development with ABEL. Forskningsraport 552, Institutt for informatikk, Universitetet i Oslo, 1991.

[4] H. Ehrig, H.-J. Kreowski, B. Mahr, and P. Padawitz. Algebraic implementation of abstract data types. *Theoretical Computer Science*, 20:209–263, 1982.

[5] J. Farrés-Casals. *Verification in ASL and Related Specification Languages, Report CST-92-92.* PhD thesis, Dept. of Computer Science, University of Edinburgh, 1992.

[6] J.E. Hannay. Referential opacity in equational reasoning. Tech. rep. ECS-LFCS-98-398, LFCS, Division of Informatics, Univ. of Edinburgh, 1998.

[7] R. Hennicker. Structured specifications with behavioural operators: Semantics, proof methods and applications. Habilitationsschrift, Inst. für Informatik, LMU, München, 1997.

[8] M. Hofmann. *Extensional Concepts in Intensional Type Theory, Report CST-117-95 and Tech. Report ECS-LFCS-95-327.* PhD thesis, Dept. of Computer Science, University of Edinburgh, 1995.

[9] M. Hofmann and D. Sannella. On behavioural abstraction and behavioural satisfaction in higher-order logic. *Theoretical Computer Science*, 167:3–45, 1996.

[10] C. Morgan. *Programming from Specifications, 2nd ed.* Prentice Hall International Series in Computer Science; C.A.R. Hoare, Series Editor. Prentice-Hall, UK, 1994.

[11] X. Qian and A. Goldberg. Referential opacity in nondeterministic data refinement. *ACM LoPLaS*, 2(1–4):233–241, 1993.

[12] D. Sannella. Formal development in extended ML. In *Proc. 3rd BCS/FACS Workshop on Refinement*, pages 99–130, Hursley Park, 1991. Springer Workshops in Computing.

[13] D. Sannella and A. Tarlecki. Toward formal development of programs from algebraic specifications: implementations revisited. *Acta Inform.*, 25(3):233–281, 1988.

[14] D. Sannella and A. Tarlecki. Essential concepts of algebraic specification and program development. *Formal Aspects of Computing*, 9:229–269, 1997.

[15] O. Schoett. *Data Abstraction and the Correctness of Modular Programming.* PhD thesis, University of Edinburgh, 1986.

[16] H. Søndergaard and P. Sestoft. Referential transparency, definiteness and unfoldability. *Acta Inform.*, 27(6):505–517, 1990.

[17] A.S. Troelstra and D. van Dalen. *Constructivism in Mathematics, An Introduction*, volume 121 of *Studies in Logic and The Foundations of Mathematics.* North Holland, 1988.

[18] M. Wirsing. Algebraic specification. In J. van Leeuwen, editor, *Handbook of Theoretical Computer Science*, chapter 13, pages 675–788. Elsevier, 1990.

[19] M. Wirsing. Structured specifications: Syntax, semantics and proof calculus. In F.L. Bauer, W. Brauer, and H. Schwichtenberg, editors, *Logic and Algebra of Specification, Intl. Summer School Marktoberdorf*, NATO ASI Series F, pages 411–442. Springer, 1993.

[20] M. Wirsing. Algebraic Specification Languages: An Overview. In E. Astesiano, G. Reggio, and A. Tarlecki, editors, *Recent Trends in Data Type Specification*, LNCS, pages 81–115. Springer, 1994.

A Synergy Between Model-Checking and Type Inference for the Verification of Value-Passing Higher-Order Processes*

Mourad Debbabi, Abdelkader Benzakour, and Béchir Ktari

Computer Science Department,
Laval University,
Quebec, Canada
{debabi,benzako,ktari}@@ift.ulaval.ca

Abstract. In this paper, we present a formal verification framework for higher-order value-passing process algebra. This framework stems from an established synergy between type inference and model-checking. The language considered here is based on a sugared version of an implicitly typed λ-calculus extended with higher-order synchronous concurrency primitives. First, we endow such a syntax with a semantic theory made of a static semantics together with a dynamic semantics. The static semantics consists of an annotated type system. The dynamic semantics is operational and comes as a two-layered labeled transition system. The dynamic semantics is abstracted into a transitional semantics so as to make finite some infinite-state processes. We describe the syntax and the semantics of a verification logic that allows one to specify properties. The logic is an extension of the modal μ-calculus for handling higher-order processes, value-passing and return of results.

1 Motivation and Background

Concurrent, functional and imperative programming languages emerged as a multi-paradigmatic alternative appropriate for the development of concurrent and distributed applications. Such languages harmoniously combine syntactic compactness together with higher semantic expressiveness. Furthermore, they support functional abstraction (latent computations) and process abstraction (latent communications). Their expressivity is significantly increased by the higher-order aspect i.e. functions, pointers, channels and processes are first-class computable values (mobile values). Consequently, they cover both data and control aspects.

Concurrent and distributed systems are very often subjected to safety requirements. Accordingly, it is mandatory to have analysis and validation tools whereby one can formally guarantee the correctness of their behaviors with respect to the expected requirements. Model-checking refers to a formal, automatic

* This research has been funded by a grant from FCAR (Fonds pour la Formation de Chercheurs et l'Aide la Recherche), Quebec, Canada.

A.M. Haeberer (Ed.): AMAST'98, LNCS 1548, pp. 214–230, 1998.

and exhaustive verification technique. It consists of the extraction of a model from a formal description of the system to be verified. That model is afterwards checked against a logical or a behavioral specification. Obviously, from the decidability standpoint, the infiniteness of the model is a limiting factor to the feasibility of model-checking. One solution, is the application of abstraction techniques, which aim to abstract an infinite model to a finite one, in such way that if some property holds for the abstracted model, it also holds for the original model.

The main contribution of this paper is a new approach for the verification of higher-order value-passing processes. This approach rests on an established synergy between model-checking and type inference. Such a synergy is achieved thanks to three major results. First, we present an abstraction technique that aims to derive finite models (transition systems) from concurrent and functional programs. The models extracted are rich enough to cope with the verification of data and control aspects of concurrent and distributed applications. Indeed, starting from a concrete dynamic semantics of a core-syntax, we derive an abstract dynamic semantics. The computable values that may be a source of infiniteness are abstracted into finite representations that are types. By doing so, a large class of infinite models will likely be reduced to finite verifiable models. Second, we present a temporal logic that is used to express data and control properties of concurrent and functional programs. Such a logic is defined as an extension of the propositional modal μ-calculus of Kozen [14] to handle communication, value-passing and higher-order objects. The logic is semantically interpreted over the abstract dynamic semantics. Third, we present a verification algorithm based on model-checking techniques. In fact, since the model is finite, the usual algorithms may be easily accommodated to the model-checking of our logic. As an example, we present an accommodation of the Emerson's algorithm.

Here is the way the rest of this paper is organized. Section 2 is devoted to the presentation of the related work. Section 3 is dedicated to the presentation of the language core-syntax considered in this work. In Section 4, we present the static semantics of our core-syntax. In Section 5, we present the dynamic operational semantics. The latter is abstracted in Section 6. The syntax and semantics of the verification logic is given in Section 7. A detailed discussion of the model-checking algorithm is presented in Section 8. Finally, a few concluding remarks and a discussion about further research are ultimately sketched as a conclusion in Section 9.

2 Related Work

The first attempt in the design of concurrent and distributed languages mainly consisted in extending some imperative languages with concurrency and distribution primitives. Accordingly, this gave rise to languages such as Ada, Chill, Modula 2 and Occam. Lately, a great deal of interest has been expressed in concurrent and functional programming. This interest is motivated by the fact that functional programming demonstrated an extensive support of abstraction through the use of abstract data types and the composition of higher-order functions. Accordingly, plenty of languages (Concurrent ML [19, 20], Facile [11], LCS [2], etc.), calculi (CHOCS [23, 24], π-calculus [15, 16]) and semantic theories [3, 4, 8, 10, 13, 17, 18] has been advanced.

Verification techniques could be structured in two major approaches: deductive techniques and semantic-based techniques. Deductive techniques consist of the use of a logic together with the associated theorem prover. Verification is performed by deduction and is usually semi-automatic. Semantics-based verification techniques, also known as model-checking techniques, consist of the automatic extraction of a model from the program to be verified. This model approximates the dynamic behaviors of the program. Afterwards, the model is checked against another model (the specification) or against a logical specification. Logical specifications are usually expressed as formulae in modal temporal logics. In [21, 22], the author addresses the verification by proposing a methodology for generating semantically safe abstract, regular trees for programs that do not possess obvious, finite, state-transition diagram depictions. One primary result of this research is that one can, from infinite data sets, generate finite structures for model-checking. Furthermore, the methodology proposed can deal with various model infinity sources like computable values, infinite process and channel creation. In [5], the author addresses the verification by model-checking of a shared-memory concurrent imperative programming language. The author uses abstract interpretation on a true-concurrent operational semantics based on higher-dimensional transition systems. In [6], the author addresses the verification of CML programs. He presents an operational semantics for CML based on infinite domains of higher-dimensional automata. The author uses dual abstract interpretation to derive finite automata that represent sound but imprecise semantics of programs.

Recently, a surge of interest has been devoted to the verification of higher-order processes in the presence of value-passing. In [1] the authors address the specification and verification problem for process calculi such as CHOCS, CML and Facile where processes or functions are transmissible values. Their work takes place in the context of a static treatment of restriction and of a bisimulation-based semantics. They put the emphasis on (Plain) CHOCS. They show that CHOCS bisimulation can be characterized by an extension of Hennessy-Milner logic including a constructive implication, or function space constructor. Towards a proof system for the verification of process specifications, they present an infinitary sound and complete proof system for the fragment of the calculus not handling restriction. In [7], the author introduces a temporal logic for the polyadic π-calculus based on fixed point extensions of Hennessy-Milner logic. A proof system and a decision procedure are developed based on Stirling and Walker's approach to model-checking the μ-calculus using constants. A proof system and a decision procedure are obtained for arbitrary π-calculus processes with finite control.

3 Language

In this section, we present the Concurrent ML core-syntax considered in this work. We have kept the number of constructs to a bare minimum so as to facilitate a more compact and complete description of our verification framework. The BNF syntax of the core language is presented in Table 1.

Table 1. The core syntax

$\text{Exp} \ni e ::= x \mid v \mid e\, e' \mid \texttt{rec}\ f(x) \texttt{=>}\ e$	(Expressions)
$\mid (e, e') \mid e\ ;\ e' \mid \texttt{let}\ x = e\ \texttt{in}\ e'\ \texttt{end}$	
$\mid \texttt{if}\ e\ \texttt{then}\ e'\ \texttt{else}\ e''\ \texttt{end}$	
$\mid \texttt{spawn}(e) \mid \texttt{sync}(e) \mid \texttt{receive}(e) \mid \texttt{transmit}(e, e')$	
$\mid \texttt{choose}(e, e') \mid \texttt{channel}()$	
$\text{Val} \ni v ::= c \mid \texttt{fn}\ x \texttt{=>}\ e$	(Values)
$\text{Cst} \ni c ::= () \mid \texttt{true} \mid \texttt{false} \mid \texttt{num}\ n$	(Constants)

Along this paper, we will write $m_{x_1,x_2,\ldots}$, the map m excluding the associations of the form $x_i \mapsto \_$. Given two maps m and m', we will write $m \dagger m'$ the overwriting of the map m by the associations of the map m' i.e. the domain of $m \dagger m'$ is $dom(m) \cup dom(m')$ and we have $(m \dagger m')(a) = m'(a)$ if $a \in dom(m')$ and $m(a)$ otherwise.

4 Static Semantics

Our intention here is to endow our core-syntax with a static semantics. The latter is a standard annotated effect type system. We introduce the following static domains:

1. The domain of *regions*: regions are intended to abstract channels. Their domain consists in the disjoint union of a countable set of constants ranged over by r and variables ranged over by ϱ. We will use $\rho, \rho', \ldots$ to represent values drawn from this domain.
2. The domain of *side* and *communication effects* is inductively defined by:

$$\sigma ::= \emptyset \mid \varsigma \mid \sigma \cup \sigma' \mid create(\rho, \tau) \mid in(\rho, \tau) \mid out(\rho, \tau)$$

We use $\emptyset$ to denote an empty effect and ς to denote an effect variable. The communication effect $create(\rho, \tau)$ represents the creation, in the region ρ, of a channel that is a medium for values of type τ. The term $in(\rho, \tau)$ denotes the communication effect resulting from receiving a value of type τ on a channel in the region ρ and $out(\rho, \tau)$ denotes the communication effect resulting from sending a value of type τ on a channel in the region ρ. The effect $\sigma \cup \sigma'$

stands for an effect that represent an upper approximation of σ and σ' (effect cumulation). Actually, only one of the two effects, σ or σ', will emerge at the dynamic evaluation. We write $\sigma \sqsupseteq \sigma' \Leftrightarrow \exists\sigma''.\sigma = \sigma' \cup \sigma''$. Equality on effects is modulo ACUI (Associativity, Commutativity and Idempotence) with $\emptyset$ as the neutral element.

3. The domain of *types* is inductively defined by:

$$\tau ::= unit \mid int \mid bool \mid \alpha \mid \tau \times \tau' \mid chan_\rho(\tau) \mid event_\sigma(\tau) \mid \tau \xrightarrow{\sigma} \tau'$$

The term $chan_\rho(\tau)$ is the type of channels in the region ρ that are intended to be media for values of type τ. The term $\tau \xrightarrow{\sigma} \tau'$ is the type of functions that take parameters of type τ to values of type τ' with a latent effect σ. By latent effect, we refer to the effect generated when the corresponding function expression is evaluated. The type $event_\sigma(\tau)$ denotes inactive processes having potential effect (latent effect σ) that are expected to return a value of type τ once their execution terminated.

Table 2 presents the static semantics of our core language.

Table 2. The typing rules

$$\text{(cte)} \quad \frac{\tau \lhd \mathit{TypeOf}(c)}{\mathcal{E} \vdash c : \tau, \emptyset}$$

$$\text{(var)} \quad \frac{\tau \lhd \mathcal{E}(x)}{\mathcal{E} \vdash x : \tau, \emptyset}$$

$$\text{(abs)} \quad \frac{\mathcal{E}_x \dagger [x \mapsto \tau] \vdash e : \tau', \sigma}{\mathcal{E} \vdash \mathbf{fn}\ x \Rightarrow e : \tau \xrightarrow{\sigma} \tau', \emptyset}$$

$$\text{(app)} \quad \frac{\mathcal{E} \vdash e : \tau \xrightarrow{\sigma} \tau', \sigma' \quad \mathcal{E} \vdash e' : \tau, \sigma''}{\mathcal{E} \vdash (e\ e') : \tau', ((\sigma';\sigma'');\sigma)}$$

$$\text{(let)} \quad \frac{\mathcal{E} \vdash e : \tau, \sigma \quad \mathcal{E}_x \dagger [x \mapsto \mathit{Gen}(\mathcal{E}, \tau, \sigma)] \vdash e' : \tau', \sigma'}{\mathcal{E} \vdash \mathbf{let}\ x = e\ \mathbf{in}\ e'\ \mathbf{end} : \tau', (\sigma;\sigma')}$$

$$\text{(pair)} \quad \frac{\mathcal{E} \vdash e : \tau, \sigma \quad \mathcal{E} \vdash e' : \tau', \sigma'}{\mathcal{E} \vdash (e, e') : \tau \times \tau', (\sigma;\sigma')}$$

$$\text{(seq)} \quad \frac{\mathcal{E} \vdash e : \tau, \sigma \quad \mathcal{E} \vdash e' : \tau', \sigma'}{\mathcal{E} \vdash e\ ;\ e' : \tau', (\sigma;\sigma')}$$

$$\text{(if)} \quad \frac{\mathcal{E} \vdash e : \mathit{bool}, \sigma \quad \mathcal{E} \vdash e' : \tau, \sigma' \quad \mathcal{E} \vdash e'' : \tau, \sigma''}{\mathcal{E} \vdash \mathbf{if}\ e\ \mathbf{then}\ e'\ \mathbf{else}\ e''\ \mathbf{end} : \tau, \sigma \cup \sigma' \cup \sigma''}$$

$$\text{(rec)} \quad \frac{\mathcal{E}_{x,f} \dagger [x \mapsto \tau, f \mapsto \tau \xrightarrow{\sigma} \tau'] \vdash e : \tau', \sigma}{\mathcal{E} \vdash \mathbf{rec}\ f(x) \Rightarrow e : \tau \xrightarrow{\sigma} \tau', \emptyset}$$

$$\text{(obs)} \quad \frac{\mathcal{E} \vdash e : \tau, \sigma \quad \mathit{Observe}(\mathcal{E}, \tau, \sigma) \sqsubseteq \sigma'}{\mathcal{E} \vdash e : \tau, \sigma'}$$

The static semantics manipulates sequents of the form $\mathcal{E} \vdash e : \tau, \sigma$, which state that under some typing environment $\mathcal{E}$ the expression e has type τ and effect σ. We also define type schemes of the form $\forall v_1, \ldots, v_n.\tau$, where v_i can be type, region or effect variable. A type τ' is an instance of $\forall v_1, \ldots, v_n.\tau$, noted $\tau' \lhd \forall v_1, \ldots, v_n.\tau$, if there exists a substitution θ defined over $v_1, \ldots, v_n$ such that $\tau' = \theta\tau$.

Type generalization in this type system states that a variable cannot be generalized if it is free in the type environment $\mathcal{E}$ or if it is present in the inferred effect:

$$Gen(\mathcal{E}, \tau, \sigma) = \texttt{let } v_{1..n} = fv(\tau) \backslash (fv(\mathcal{E}) \cup fv(\sigma)) \texttt{ in } \forall v_{1..n}.\tau \texttt{ end}$$

where $fv(\_)$ denotes the set of free variables. The observation criterion was introduced in order to report only effects that can affect the context of an expression.

$$\begin{aligned} Observe(\mathcal{E}, \tau, \sigma) = & \{\varsigma \in \sigma \mid \varsigma \in fv(\mathcal{E}) \cup fv(\tau)\} \\ & \cup \{create(\rho, \tau') \in \sigma \mid \rho \in fr(\mathcal{E}) \cup fr(\tau) \wedge \tau' \in \text{S\_T}\} \\ & \cup \{in(\rho, \tau') \in \sigma \mid \rho \in fr(\mathcal{E}) \cup fr(\tau) \wedge \tau' \in \text{S\_T}\} \\ & \cup \{out(\rho, \tau') \in \sigma \mid \rho \in fr(\mathcal{E}) \cup fr(\tau) \wedge \tau' \in \text{S\_T}\} \end{aligned}$$

where S_T is the domain of types, $fr(\mathcal{E})$ stands for the set of free channel regions in the static environment $\mathcal{E}$. The function *TypeOf* allows the typing of built-in primitives as defined in the Table 3.

Table 3. The initial static basis

$$\begin{array}{lll} TypeOf = & [\ () & \mapsto unit, \\ & \texttt{true} & \mapsto bool, \\ & \texttt{false} & \mapsto bool, \\ & \texttt{num}\ n & \mapsto int, \\ & \texttt{channel} & \mapsto \forall \alpha, \varrho, \varsigma.unit \xrightarrow{\varsigma \cup create(\varrho,\alpha)} chan_{\varrho}(\alpha), \\ & \texttt{receive} & \mapsto \forall \alpha, \varrho, \varsigma, \varsigma'.chan_{\varrho}(\alpha) \xrightarrow{\varsigma} event_{\varsigma' \cup in(\varrho,\alpha)}(\alpha), \\ & \texttt{transmit} & \mapsto \forall \alpha, \varrho, \varsigma, \varsigma'.chan_{\varrho}(\alpha) \times \alpha \xrightarrow{\varsigma} event_{\varsigma' \cup out(\varrho,\alpha)}(unit), \\ & \texttt{choose} & \mapsto \forall \alpha, \varsigma, \varsigma', \varsigma'', \varsigma'''.event_{\varsigma}(\alpha) \times event_{\varsigma'}(\alpha) \xrightarrow{\varsigma''} event_{\varsigma''' \cup \varsigma \cup \varsigma'}(\alpha), \\ & \texttt{spawn} & \mapsto \forall \varsigma, \varsigma'.(unit \xrightarrow{\varsigma} unit) \xrightarrow{\varsigma'} unit, \\ & \texttt{sync} & \mapsto \forall \alpha, \varsigma, \varsigma'.event_{\varsigma}(\alpha) \xrightarrow{\varsigma'} \alpha \\ &] & \end{array}$$

5 Concrete Dynamic Semantics

In this section, we endow our core-syntax with a dynamic operational semantics. The latter is now standard and will be defined here as a two-layered labeled transition system following [12]. First of all, we need to introduce some semantic domains and to extend the expression syntax to intermediate expressions (expressions that may occur during the dynamic evaluation). As illustrated in Table 4, we introduce six semantic categories.

Table 4. The semantic categories

CVal $\ni cv$	$::= c \mid \mathtt{fn}\ x \Rightarrow i \mid k \mid ev \mid (cv, cv)$	(Computable Values)
Evt $\ni ev$	$::= \langle ec, cv \rangle$	(Events)
IExp $\ni i$	$::= e \mid cv \mid cv\ e \mid \mathtt{let}\ x = cv\ \mathtt{in}\ e\ \mathtt{end}$	
	$\mid\ (cv, e) \mid cv; e$	(Intermediate Expressions)
ECon $\ni ec$	$::= \mathtt{receive} \mid \mathtt{transmit} \mid \mathtt{choose}$	(Event Constructors)
Com $\ni com$	$::= k?cv \mid k!cv$	(Communications)
Act $\ni a$	$::= com \mid \epsilon \mid \lambda(k) \mid \phi(cv)$	(Actions)

The semantic category CVal is ranged over by *cv* and corresponds to the domain of computable values. The semantic category Evt of events is ranged over by *ev*. An event is a pair consisting of the event constructor *ec* and its argument *cv*. An event constructor *ec* is a member of the syntactic domain ECon. The semantic category IExp is ranged over by *i* and corresponds to the domain of intermediate expressions. The semantic category Com is ranged over by *com* and corresponds to the domain of communications. Input communications are of the form $k?cv$ where k is a channel computable value and *cv* is another computable value that will be received on the channel k. Output communications are of the form $k!cv$ where k is a channel computable value and *cv* is another computable value that will be sent along the channel k. The semantic category Act is ranged over by a and corresponds to the domain of actions. The silent action ϵ denotes internal moves. A creation of a channel k is considered as an action and is written $\lambda(k)$. A process spawning of a value *cv* is considered as an action and is written $\phi(cv)$.

The operational semantics is structured in two layers, one for expressions in isolation and one multiset of expressions running in parallel. These two layers involve three transition relations whose definitions are given hereafter.

5.1 Expression Semantics

The first relation, written $\_ \stackrel{\_}{\Longrightarrow} \_ \subseteq \text{Evt} \times \text{Com} \times \text{IExp}$, is a transition relation that is meant to define the communication potential of events. The rules that define this relation are presented in Table 5.

A transition of the form $ev \stackrel{com}{\Longrightarrow} i$ intuitively means that the event *ev* has the potential of performing the communication *com* (when `sync` is applied to the event) and then it will behave as the intermediate expression i.

Table 5. The semantic rules of the relation $\_\overset{-}{\Longrightarrow}\_$

$$(\text{transmit})\ \langle\mathtt{transmit},(k,cv)\rangle \overset{k!cv}{\Longrightarrow} ()\quad(\text{receive})\ \langle\mathtt{receive},k\rangle \overset{k?cv}{\Longrightarrow} cv$$

$$(\text{choose}_1)\ \frac{ev_1 \overset{com}{\Longrightarrow} i}{\langle\mathtt{choose},(ev_1,ev_2)\rangle \overset{com}{\Longrightarrow} i}\quad(\text{choose}_2)\ \frac{ev_2 \overset{com}{\Longrightarrow} i}{\langle\mathtt{choose},(ev_1,ev_2)\rangle \overset{com}{\Longrightarrow} i}$$

The second relation, written $\_\overset{-}{\hookrightarrow}\_ \subseteq \text{IExp}\times\text{Act}\times\text{IExp}$, is the one that defines the operational semantics of processes. A transition of the form $i \overset{a}{\hookrightarrow} i'$ intuitively means that by performing the action a, the intermediate expression i will behave as i'. The rules that define this relation are presented in Table 6.

Table 6. The concrete operational semantics of processes

$$(\text{app}_1)\ \frac{i_1 \overset{a}{\hookrightarrow} i_1'}{i_1\ i_2 \overset{a}{\hookrightarrow} i_1'\ i_2}\qquad(\text{app}_2)\ \frac{i \overset{a}{\hookrightarrow} i'}{cv\ i \overset{a}{\hookrightarrow} cv\ i'}$$

$$(\text{beta}_1)\ (\mathtt{fn}\ x \Rightarrow i)\ cv \overset{\epsilon}{\hookrightarrow} i[cv/x]\qquad(\text{beta}_2)\ ec\ cv \overset{\epsilon}{\hookrightarrow} \langle ec, cv\rangle$$

$$(\text{pair}_1)\ \frac{i_1 \overset{a}{\hookrightarrow} i_1'}{(i_1,i_2) \overset{a}{\hookrightarrow} (i_1',i_2)}\qquad(\text{pair}_2)\ \frac{i \overset{a}{\hookrightarrow} i'}{(cv,i) \overset{a}{\hookrightarrow} (cv,i')}$$

$$(\text{seq}_1)\ \frac{i_1 \overset{a}{\hookrightarrow} i_1'}{i_1;i_2 \overset{a}{\hookrightarrow} i_1';i_2}\qquad(\text{seq}_2)\ \frac{i \overset{a}{\hookrightarrow} i'}{cv;i \overset{a}{\hookrightarrow} cv;i'}$$

$$(\text{chan})\ \mathtt{channel}() \overset{\lambda(k)}{\hookrightarrow} k\qquad(\text{spawn}_1)\ \mathtt{spawn}\ cv \overset{\phi(cv)}{\hookrightarrow} ()$$

$$(\text{rec})\ \mathtt{rec}\ f(x) \Rightarrow i \overset{\epsilon}{\hookrightarrow} \mathtt{fn}\ x \Rightarrow i[(\mathtt{rec}\ f(x) \Rightarrow i)/f]$$

$$(\text{if}_1)\ \frac{i_1 \overset{a}{\hookrightarrow} i_1'}{\mathtt{if}\ i_1\ \mathtt{then}\ i_2\ \mathtt{else}\ i_3\ \mathtt{end} \overset{a}{\hookrightarrow} \mathtt{if}\ i_1'\ \mathtt{then}\ i_2\ \mathtt{else}\ i_3\ \mathtt{end}}$$

$$(\text{if}_2)\ \mathtt{if\ true\ then}\ i_1\ \mathtt{else}\ i_2\ \mathtt{end} \overset{\epsilon}{\hookrightarrow} i_1$$

$$(\text{if}_3)\ \mathtt{if\ false\ then}\ i_1\ \mathtt{else}\ i_2\ \mathtt{end} \overset{\epsilon}{\hookrightarrow} i_2$$

$$(\text{let}_1)\ \frac{i_1 \overset{a}{\hookrightarrow} i_1'}{\mathtt{let}\ x = i_1\ \mathtt{in}\ i_2\ \mathtt{end} \overset{a}{\hookrightarrow} \mathtt{let}\ x = i_1'\ \mathtt{in}\ i_2\ \mathtt{end}}$$

$$(\text{let}_2)\ \mathtt{let}\ x = cv\ \mathtt{in}\ i\ \mathtt{end} \overset{\epsilon}{\hookrightarrow} i[cv/x]$$

$$(\text{sync})\ \frac{ev \overset{com}{\Longrightarrow} i}{\mathtt{sync}\ ev \overset{com}{\hookrightarrow} i}$$

5.2 Program Semantics

Now, in order to define the third transition relation we need to introduce the following semantic functions and domains. We denote by $CV[i]$ the set of channels k occurring in an intermediate expression i.

We view a program as a multiset of intermediate expressions. We let Prog be IExp-MultiSet i.e. the set of program multisets. The set of channels that occur in a multiset P is obtained by including the channels in each intermediate expression. The operational semantics of programs is based on the evolution of the so-called configurations. We define a K-configuration, and we write $K :: P$, to be a pair where the first component K is the set of all channels allocated up to a certain point, and the second component P is a program (a multiset of intermediate expressions). Let Chan be the set of channel computable values. The domain of K-configuration Conf_K is defined as follows:

$$\text{Conf}_K = \{K :: P \mid K \in \text{Chan} \wedge P \in \text{Prog} \wedge CV[P] \subseteq K\}$$

The semantics of K-configurations is given in terms of the labeled transition system $(\text{Conf}_K, \text{Com} \cup \{\epsilon\}, \longrightarrow)$. The transition relation $\longrightarrow$ is defined as the smallest subset of $\text{Conf}_K \times \text{Com} \cup \{\epsilon\} \times \text{Conf}_K$ closed under the rules presented in Table 7.

Table 7. The operational semantics of programs

(action)	$\dfrac{i \overset{a}{\hookrightarrow} i'}{K :: \{\!\!\{i\}\!\!\} \overset{a}{\longrightarrow} K \cup Msg(a) :: \{\!\!\{i'\}\!\!\}}$
(channel)	$\dfrac{i \overset{\lambda(k)}{\hookrightarrow} i'}{K :: \{\!\!\{i\}\!\!\} \overset{\epsilon}{\longrightarrow} K \cup \{k\} :: \{\!\!\{i'\}\!\!\}}\ k \notin K$
(spawn$_2$)	$\dfrac{i \overset{\phi(cv)}{\hookrightarrow} i'}{K :: \{\!\!\{i\}\!\!\} \overset{\epsilon}{\longrightarrow} K :: \{\!\!\{i', cv()\}\!\!\}}$
(communication)	$\dfrac{i_1 \overset{k?cv}{\hookrightarrow} i'_1,\quad i_2 \overset{k!cv}{\hookrightarrow} i'_2}{K :: \{\!\!\{i_1, i_2\}\!\!\} \overset{\epsilon}{\longrightarrow} K :: \{\!\!\{i'_1, i'_2\}\!\!\}}$
(isolation)	$\dfrac{K :: P_1 \overset{a}{\longrightarrow} K' :: P'_1}{K :: P_1 \cup P_2 \overset{a}{\longrightarrow} K' :: P'_1 \cup P_2}$

The function *Msg* extracts the set of channels that are transmitted in a communication.

6 Abstract Dynamic Semantics

In this section, we describe an abstract dynamic semantics derived from the concrete dynamic semantics viewed in the previous section. The motivation is to abstract computable values that could be a source of infiniteness. These values are abstracted into finite representations that are types. By doing so, we ensure that a large class of infinite models will likely be reduced to finite verifiable models.

The abstract semantic categories are illustrated in Table 8. The abstract semantic category AVal is ranged over by *cv* and corresponds to the domain of abstract values. The semantic category AFExp is ranged over by *afe* and

Table 8. The abstract semantic categories

AVal $\ni cv$	$::= ac \mid k_{chan_\rho(\tau)} \mid ev \mid afe \mid (cv, cv)$	(Abstract Values)
AFExp $\ni afe$	$::= \mathtt{fn}\ x \ \mathtt{=>}\ i \mid \tau \xrightarrow{\sigma} \tau'$	(Abstract Functional Exp.)
ACst $\ni ac$	$::= c \mid int \mid bool \mid unit$	(Abstract Constants)
AChan $\ni k_{chan_\rho(\tau)}$	$::= k \mid chan_\rho(\tau)$	(Abstract Channels)
AEvt $\ni ev$	$::= \langle ec, cv\rangle \mid event_\sigma(\tau)$	(Abstract Events)
ECon $\ni ec$	$::= \mathtt{receive} \mid \mathtt{transmit} \mid \mathtt{choose}$	(Event Constructors)
AIExp $\ni i$	$::= e \mid cv \mid cv\ e \mid (cv, e) \mid cv; e$ $\mid \mathtt{let}\ x = cv\ \mathtt{in}\ e\ \mathtt{end}$	(Abstract Intermediate Exp.)
ACom $\ni com$	$::= k_{chan_\rho(\tau)}?\tau \mid k_{chan_\rho(\tau)}!cv$	(Abstract Communications)
AAct $\ni a$	$::= com \mid \epsilon \mid \lambda(k_{chan_\rho(\tau)}) \mid \phi(cv)$	(Abstract Actions)
Loc $\ni l$	$::= n \mid n.l$ where $n \in \mathbb{N}$	(Locations)

corresponds to the abstract functional expressions. The abstract semantic category ACst is ranged over by ac and corresponds to the domain of abstract constants. The abstract semantic category AChan is ranged over by $k_{chan_\rho(\tau)}$ and corresponds to the domain of abstract channels. This category includes channel computable values together with the type of channels. The abstract semantic category AEvt is ranged over by ev and corresponds to the domain of abstract events. This category includes events together with the type of events. The abstract semantic category AIExp is ranged over by i and corresponds to the domain of abstract intermediate expressions. The abstract semantic category ACom is ranged over by com and corresponds to the domain of abstract communications. Abstract input communications are of the form $k_{chan_\rho(\tau)}?\tau$ and stand for the action of receiving values, abstracted by their type, on an abstract channel. Abstract output communications are of the form $k_{chan_\rho(\tau)}!cv$ and stand for the transmission of an abstract value along an abstract channel. The abstract semantic category AAct is ranged over by a and corresponds to the domain of actions. The abstract semantic category Loc is ranged over by l and corresponds to the domain of locations.

6.1 Expression Semantics

The rules that define the relation $\_ \overset{\cdot}{\Longrightarrow} \_ \subseteq$ AEvt $\times$ ACom $\times$ AIExp are presented in Table 9.

The (transmit) and (receive) rules are changed to reflect the use of abstract values. For example, the second one means that the event $\langle \mathtt{receive}, k_{chan_\rho(\tau)}\rangle$ has the potential of performing the communication $k_{chan_\rho(\tau)}?\tau$, and then behaves as the abstract value τ. This means that the possible values received in the channel $k_{chan_\rho(\tau)}$ are abstracted as their type. This abstraction ensures that large class of infinite models will likely be reduced to finite models. The rules

Table 9. The abstract semantic rules of the relation $\_ \overset{-}{\Longrightarrow} \_$

(transmit)	$\langle \mathbf{transmit}, (k_{chan_\rho(\tau)}, cv)\rangle \overset{k_{chan_\rho(\tau)}!cv}{\Longrightarrow} ()$
(receive)	$\langle \mathbf{receive}, k_{chan_\rho(\tau)}\rangle \overset{k_{chan_\rho(\tau)}?\tau}{\Longrightarrow} \tau$
(choose_1)	$\dfrac{ev_1 \overset{com}{\Longrightarrow} i}{\langle \mathbf{choose}, (ev_1, ev_2)\rangle \overset{com}{\Longrightarrow} i}$
(choose_2)	$\dfrac{ev_2 \overset{com}{\Longrightarrow} i}{\langle \mathbf{choose}, (ev_1, ev_2)\rangle \overset{com}{\Longrightarrow} i}$

that define the relation $\_ \overset{-}{\underset{-}{\hookrightarrow}} \_ \subseteq \text{AIExp} \times \text{AAct} \times \text{Loc} \times \text{AIExp}$ are presented in Table 10.

For instance, the rule (beta_3) is defined to evaluate the application of a class of functions that have the same type $(\tau \overset{\sigma}{\longrightarrow} \tau')$ to an abstract value cv.

6.2 Program Abstract Semantics

As the concrete semantics, the program abstract semantics is based on the evolution of K-configurations. Definitions of the domain of K-configuration and the semantic function *Msg* remain the same except that the latter is defined over abstract values. The semantics of K-configuration is given in terms of the labeled transition system $(\text{Conf}_K, \text{ACom} \cup \{\epsilon\}, \longrightarrow)$. The transition relation $\longrightarrow$ is defined as the smallest subset of $\text{Conf}_K \times \text{ACom} \cup \{\epsilon\} \times \text{Conf}_K$ closed under the rules presented in Table 11.
where $i'_1[cv]_l$ stands for the term i'_1 in which the subterm at location l will be replaced by cv.

6.3 Correctness of the Abstraction

The correctness of the abstraction is assured since there is an equivalence between the abstract transition graph and the concrete one. In fact, by unfolding in the abstract graph each transition containing an abstract term by the equivalent set of transitions composed by concrete values, we transform an abstract transition graph into a concrete one.

7 A Modal Logic for Concurrent ML

In this section we introduce a logic that allows one to specify properties of expressions. The logic we consider may be viewed as a variant of the modal μ-calculus [14], or the Hennessey-Milner Logic with recursion. In the proposed logic, modal formulae can also be used to express communication, value-passing and result returns. This logic is semantically interpreted over the abstract dynamic semantics.

Table 10. The abstract operational semantics of processes

(app₁) $\dfrac{i_1 \overset{a}{\underset{l}{\hookrightarrow}} i_1'}{i_1\ i_2 \overset{a}{\underset{1.l}{\hookrightarrow}} i_1'\ i_2}$ (app₂) $\dfrac{i \overset{a}{\underset{l}{\hookrightarrow}} i'}{cv\ i \overset{a}{\underset{2.l}{\hookrightarrow}} cv\ i'}$

(beta₁) $(\mathbf{fn}\ x \Rightarrow i)\ cv \overset{\epsilon}{\underset{0}{\hookrightarrow}} i[cv/x]$ (beta₂) $ec\ cv \overset{\epsilon}{\underset{0}{\hookrightarrow}} < ec, cv >$

(beta₃) $(\tau \overset{\sigma}{\longrightarrow} \tau')(cv) \overset{\epsilon}{\underset{0}{\hookrightarrow}} \tau'$ (chan) $\mathtt{channel}() \overset{\lambda(k_{chan_\rho(\tau)})}{\underset{0}{\hookrightarrow}} k_{chan_\rho(\tau)}$

(pair₁) $\dfrac{i_1 \overset{a}{\underset{l}{\hookrightarrow}} i_1'}{(i_1, i_2) \overset{a}{\underset{1.l}{\hookrightarrow}} (i_1', i_2)}$ (pair₂) $\dfrac{i \overset{a}{\underset{l}{\hookrightarrow}} i'}{(cv, i) \overset{a}{\underset{2.l}{\hookrightarrow}} (cv, i')}$

(seq₁) $\dfrac{i_1 \overset{a}{\underset{l}{\hookrightarrow}} i_1'}{i_1; i_2 \overset{a}{\underset{1.l}{\hookrightarrow}} i_1'; i_2}$ (seq₂) $\dfrac{i \overset{a}{\underset{l}{\hookrightarrow}} i'}{cv; i \overset{a}{\underset{2.l}{\hookrightarrow}} cv; i'}$

(spawn₁) $\mathbf{spawn}\ cv \overset{\phi(cv)}{\underset{0}{\hookrightarrow}} ()$

(rec) $\mathtt{rec}\ f(x) \Rightarrow i \overset{\epsilon}{\underset{0}{\hookrightarrow}} \mathbf{fn}\ x \Rightarrow i[(\mathtt{rec}\ f(x) \Rightarrow i)/f]$

(if₁) $\dfrac{i_1 \overset{a}{\underset{l}{\hookrightarrow}} i_1'}{\mathbf{if}\ i_1\ \mathbf{then}\ i_2\ \mathbf{else}\ i_3\ \mathbf{end} \overset{a}{\underset{1.l}{\hookrightarrow}} \mathbf{if}\ i_1'\ \mathbf{then}\ i_2\ \mathbf{else}\ i_3\ \mathbf{end}}$

(if₂) $\mathbf{if\ true\ then}\ i_1\ \mathbf{else}\ i_2\ \mathbf{end} \overset{\epsilon}{\underset{0}{\hookrightarrow}} i_1$

(if₃) $\mathbf{if\ false\ then}\ i_1\ \mathbf{else}\ i_2\ \mathbf{end} \overset{\epsilon}{\underset{0}{\hookrightarrow}} i_2$

(let₁) $\dfrac{i_1 \overset{a}{\underset{l}{\hookrightarrow}} i_1'}{\mathbf{let}\ x = i_1\ \mathbf{in}\ i_2\ \mathbf{end} \overset{a}{\underset{1.l}{\hookrightarrow}} \mathbf{let}\ x = i_1'\ \mathbf{in}\ i_2\ \mathbf{end}}$

(let₂) $\mathbf{let}\ x = cv\ \mathbf{in}\ i\ \mathbf{end} \overset{\epsilon}{\underset{0}{\hookrightarrow}} i[cv/x]$

(sync) $\dfrac{ev \overset{com}{\Longrightarrow} i}{\mathbf{sync}\ ev \overset{com}{\underset{0}{\hookrightarrow}} i}$

Table 11. The abstract operational semantics of programs

(action) $\dfrac{i \overset{a}{\underset{l}{\hookrightarrow}} i'}{K :: \{\!|i|\!\} \overset{a}{\longrightarrow} K \cup Msg(a) :: \{\!|i'|\!\}}$

(channel) $\dfrac{i \overset{\lambda(k_{chan_\rho(\tau)})}{\underset{l}{\hookrightarrow}} i'}{K :: \{\!|i|\!\} \overset{\epsilon}{\longrightarrow} K \cup \{k_{chan_\rho(\tau)}\} :: \{\!|i'|\!\}}\ k_{chan_\rho(\tau)} \notin K$

(spawn₂) $\dfrac{i \overset{\phi(cv)}{\underset{l}{\hookrightarrow}} i'}{K :: \{\!|i|\!\} \overset{\epsilon}{\longrightarrow} K :: \{\!|i', cv()|\!\}}$

(Communication) $\dfrac{i_1 \overset{k_{chan_\rho(\tau)}?\tau}{\underset{l}{\hookrightarrow}} i_1', \quad i_2 \overset{k_{chan_\rho(\tau)}!cv}{\underset{l'}{\hookrightarrow}} i_2'}{K :: \{\!|i_1, i_2|\!\} \overset{\epsilon}{\longrightarrow} K :: \{\!|i_1'[cv]_l, i_2'|\!\}}$

(Isolation) $\dfrac{K :: P_1 \overset{a}{\longrightarrow} K' :: P_1'}{K :: P_1 \cup P_2 \overset{a}{\longrightarrow} K' :: P_1' \cup P_2}$

7.1 Syntax

The syntax of formulae is presented in Table 12. We refer to this logic as L_μ.

Table 12. The logic

ψ ::= `tt` | `ff` | X | $\neg\psi$ | $\psi \vee \psi'$ | $\psi \wedge \psi'$ |	(Boolean Expressions)
| <*a*> ψ | <*return*(*cv*)>	(Diamond Formulae)
| [*a*] ψ | [*return*(*cv*)]	(Box Formulae)
| $\mu X.\psi$	(Greatest Fixpoint Formulae)
| $\nu X.\psi$	(Least Fixpoint Formulae)
a ::= k *dir* *cv* | ϵ	(Actions)
dir ::= ! | ?	(Directions)

The symbols $\neg$, $\vee$ and $\wedge$ respectively represent negation, disjunction and conjunction. The symbol <*a*> (resp. <*return*(*cv*)>) is a modal operator indexed by a (resp. by *return*(*cv*)) known as the diamond. The meaning of modalized formulae appeal to transition behavior of a program. For instance, a program satisfies the formula <*a*> ψ if it can evolve to some K-configuration obeying ψ by performing an action a. The actions can either be the silent action ϵ or the communication actions $k!cv$ or $k?cv$. Furthermore, a program satisfies the formula <*return*(*cv*)> if it can return the value *cv*. In the same way, the symbol [*a*] (resp. [*return*(*cv*)]) is a modal operator known as box. A program satisfies the formula [*a*] ψ if after every performance of an action a, each result K-configuration satisfies ψ. Furthermore, a program satisfies the formula [*return*(*cv*)] if it returns necessarily the value *cv*. Variables are ranged over by X. The formulae $\mu X.\psi$ (resp. $\nu X.\psi$) is a recursive formula where the least fixpoint operator μ (resp. greatest fixpoint operator ν) binds all free occurrences of X in ψ. An occurrence of X is free if it is not within the scope of a binder μX or νX. Note that like the μ-calculus, all occurrences of X in ψ must appear inside the scope of an even number of negations. This is to ensure the existence of fixpoints.

7.2 Semantics

Formulae are interpreted over models of the form M = <*ST*, *L*>, where $ST = (\mathrm{Conf}_K, \mathrm{ACom} \cup \{\epsilon\}, \longrightarrow)$, and environment of the form $e = [X_i \mapsto P_i]$ which maps variables X_i to sets of K-configurations. Semantically, formulae of the logic correspond to sets of K-configurations for which they are true. The meaning function $[\![.]\!]_e^M : L_\mu \rightarrow 2^C$ is described in Table 13. The set C refers to the set of K-configurations.

Intuitively, all K-configurations satisfy the formula `tt` while there are no K-configurations that satisfy `ff`. The meaning of a variable X is simply the K-configurations that are bound to X in the environment e. Negation, disjunction and conjunction are interpreted in a classical way. The meaning of formulae <*a*>

Table 13. The semantic

$$
\begin{array}{l}
[\![\mathtt{tt}]\!]_e^M = \mathcal{C} \\
[\![\mathtt{ff}]\!]_e^M = \emptyset \\
[\![X]\!]_e^M = e(X) \\
[\![\neg\psi]\!]_e^M = \mathcal{C} \setminus [\![\psi]\!]_e^M \\
[\![\psi_1 \vee \psi_2]\!]_e^M = [\![\psi_1]\!]_e^M \cup [\![\psi_2]\!]_e^M \\
[\![\psi_1 \wedge \psi_2]\!]_e^M = [\![\psi_1]\!]_e^M \cap [\![\psi_2]\!]_e^M \\
[\![\texttt{<}\epsilon\texttt{>}\ \psi]\!]_e^M = \{c \in \mathcal{C} \mid \exists c'.c \xrightarrow{\epsilon} c' \wedge c' \in [\![\psi]\!]_e^M\} \\
[\![\texttt{[}\epsilon\texttt{]}\ \psi]\!]_e^M = \{c \in \mathcal{C} \mid \forall c'.c \xrightarrow{\epsilon} c' \Rightarrow (c' \in [\![\psi]\!]_e^M)\} \\
[\![\texttt{<}k!cv\texttt{>}\ \psi]\!]_e^M = \{c \in \mathcal{C} \mid \exists c', cv'.c \overset{k!cv'}{\rightarrow} c' \wedge c' \in [\![\psi[cv'/cv]]\!]_e^M \wedge cv \preceq cv'\} \\
[\![\texttt{[}k!cv\texttt{]}\ \psi]\!]_e^M = \{c \in \mathcal{C} \mid \forall c', \exists cv'.c \overset{k!cv'}{\rightarrow} c' \Rightarrow (c' \in [\![\psi[cv'/cv]]\!]_e^M \wedge cv \preceq cv')\} \\
[\![\texttt{<}k?cv\texttt{>}\ \psi]\!]_e^M = \{c \in \mathcal{C} \mid \exists c', \tau \text{ where } \mathit{TypeOf}(cv) = \tau.\ c \overset{k?\tau}{\rightarrow} c' \wedge c' \in [\![\psi[\tau/cv]]\!]_e^M\} \\
[\![\texttt{[}k?cv\texttt{]}\ \psi]\!]_e^M = \{c \in \mathcal{C} \mid \forall c', \exists \tau \text{ where } \mathit{TypeOf}(cv) = \tau.\ c \overset{k?\tau}{\rightarrow} c' \Rightarrow c' \in [\![\psi[\tau/cv]]\!]_e^M\} \\
[\![\texttt{<}return(cv)\texttt{>}]\!]_e^M = \{c \in \mathcal{C} \mid \exists n \in \mathbb{N}.c \xrightarrow{a_1} c_1 \xrightarrow{a_2} c_2 \ldots \xrightarrow{a_n} cv\} \\
[\![\texttt{[}return(cv)\texttt{]}]\!]_e^M = \{c \in \mathcal{C} \mid \forall cv', \exists n \in \mathbb{N}.c \xrightarrow{a_1} c_1 \xrightarrow{a_2} c_2 \ldots \xrightarrow{a_n} cv' \Rightarrow (cv = cv')\} \\
[\![\mu X.\psi]\!]_e^M = \bigcap\{C \subseteq \mathcal{C} \mid [\![\psi]\!]_{e[X \mapsto C]}^M \subseteq C\} \\
[\![\nu X.\psi]\!]_e^M = \bigcup\{C \subseteq \mathcal{C} \mid C \subseteq [\![\psi]\!]_{e[X \mapsto C]}^M\}
\end{array}
$$

ψ are K-configurations c that can evolve, by performing an action a, to some K-configuration c' such that c' is part of the meaning of ψ. More accurately, if the action a is an output communication action involving a value cv, then we must ensure the existence of a constant cv' such that $cv \preceq cv'$. The preorder relation $\preceq$ is defined on abstract values as below:

$$cv \preceq cv' \iff \exists\theta.\theta(cv') = cv$$

where θ is a substitution. Moreover, the K-configuration c' must be part of the meaning of the formula ψ in which each occurrence of cv is replaced by cv'. If the action a is an input communication action involving a value cv, then we must ensure the existence of a type τ such that $\mathit{TypeOf}(cv) = \tau$. And the K-configuration c' must be part of the meaning of the formula ψ in which each occurrence of cv is replaced by the type τ.

The meaning of formulae `[`a`]` ψ are K-configurations c such that after every action a, each result K-configuration c' is part of the meaning of ψ. The meaning of formulae <$return(cv)$> are K-configurations c that can evolve through n transitions such that the resulting K-configuration is the value cv. In the same way, the meaning of formula `[`$return(cv)$`]` are K-configurations that when they evolve through n transitions, the resulting K-configurations must be the value cv. The meaning of the fixpoint formulae is the same as defined in the μ-calculus. Hence the greatest fixpoint is given as the union of all post-fixpoints whereas the least fixpoint is the intersection of all pre-fixpoints.

8 A Model-Checking Algorithm

In this section, we present an adaptation of the model checking algorithm proposed by Emerson and Lei [9]. Table 13 contains an algorithm that determines whether or not a structure $M = \langle \mathcal{ST}, L\rangle$ is a model for a formula ψ_0.

Table 14. Symbolic model-checking algorithm

Function $MC(\psi'_0, M)$
  var C', C_i;
begin
  case ψ'_0 **of**
    tt: $C' = \mathcal{C}$;
    ff: $C' = \emptyset$;
    X: $C' = C_i$;
    $\neg\psi$: $C' = \mathcal{C} \setminus MC(\psi, M)$;
    $\psi_1 \vee \psi_2$: $C' = MC(\psi_1, M) \cup MC(\psi_2, M)$;
    $\psi_1 \wedge \psi_2$: $C' = MC(\psi_1, M) \cap MC(\psi_2, M)$;
    $\langle\epsilon\rangle\ \psi$: $C' = \{c \in \mathcal{C} \mid \exists c'.c \overset{\epsilon}{\to} c' \wedge c' \in MC(\psi, M)\}$;
    $[\epsilon]\ \psi$: $C' = \{c \in \mathcal{C} \mid \forall c'.c \overset{\epsilon}{\to} c' \Rightarrow (c' \in MC(\psi, M))\}$;
    $\langle k!cv\rangle\ \psi$: $C' = \{c \in \mathcal{C} \mid \exists c', cv'.c \overset{k!cv'}{\to} c' \wedge c' \in MC(\psi[cv'/cv], M) \wedge cv \preceq cv'\}$;
    $[k!cv]\ \psi$: $C' = \{c \in \mathcal{C} \mid \forall c', \exists cv'.c \overset{k!cv'}{\to} c' \Rightarrow c' \in MC(\psi[cv'/cv], M) \wedge cv \preceq cv'\}$;
    $\langle k?cv\rangle\ \psi$: $C' = \{c \in \mathcal{C} \mid \exists c', \tau$ where $TypeOf(cv) = \tau.$
        $c \overset{k?\tau}{\to} c' \wedge c' \in MC(\psi[\tau/cv], M)\}$;
    $[k?cv]\ \psi$: $C' = \{c \in \mathcal{C} \mid \forall c', \exists\tau$ where $TypeOf(cv) = \tau.$
        $c \overset{k!cv'}{\to} c' \Rightarrow c' \in MC(\psi[\tau/cv], M)\}$;
    $\langle return(cv)\rangle$: $C' = \{c \in \mathcal{C} \mid \exists n \in \mathbb{N}.c \overset{a_1}{\to} c_1 \overset{a_2}{\to} c_2 \ldots \overset{a_n}{\to} cv\}$;
    $[return(cv)]$: $C' = \{c \in \mathcal{C} \mid \forall cv', \exists n \in \mathbb{N}.c \overset{a_1}{\to} c_1 \overset{a_2}{\to} c_2 \ldots \overset{a_n}{\to} cv' \Rightarrow (cv = cv')\}$;
    $\mu X.\psi$: $C_i = \emptyset$; **repeat** $C' = C_i$; $C_i = MC(\psi, M)$; **until** $C' = C_i$;
    $\nu X.\psi$: $C_i = \mathcal{C}$; **repeat** $C' = C_i$; $C_i = MC(\psi, M)$; **until** $C' = C_i$;
  end;
  return(C');
end.

The algorithm follows these three steps:

1. Convert the formula ψ_0 to its equivalent PNF ψ'_0.
2. Compute the set C' of K-configurations in which ψ'_0 holds.
3. if $C' \neq \emptyset$ then M is a model for ψ_0 else it's not a model for ψ_0.

9 Conclusion

In this paper, we have considered the problem of formal and automatic verification of data and control aspects for higher-order value-passing process algebra. Our contribution is a new approach that rests on an established synergy between model-checking and type inference. Such a synergy is achieved thanks to three results: First, starting from a concrete dynamic semantics we derive an abstract dynamic semantics. By doing so, we ensure that infinite models will likely be reduced to finite verifiable models. The source of infiniteness are the computable values. The solution is to abstract these values into finite representation that are types. Second, starting from the propositional modal μ-calculus, we define a logic that handles communication, value-passing , result returns, and higher-order objects. The logic is semantically interpreted over the abstract dynamic semantics. Finally, we propose a verification algorithm based on model-checking techniques. Since the model is finite and thanks to the soundness of abstract dynamic semantics, the usual algorithms may be easily accommodated to the model-checking of our logic. We present an accommodation of the Emerson's algorithm.

As future work, we plan to investigate abstraction techniques for dealing with other model infinity sources such as infinite process and channel creation. To that end, we will take advantage of the pioneering work done by D. Schmidt in [22] on the abstract interpretation of small step semantics. Furthermore, we are interested in tracking infinities that may arise from arithmetic manipulation. For that, we will explore the emerging application of Presburger arithmetic to handle this problem. Finally, as a downstream result of this research, we hope to come up with practical tools that address the verification of higher-order concurrent systems.

References

[1] R. M. Amadio and M. Dam. Reasoning about higher-order processes. *In Proc. of CAAP'95, Aarhus, Lecture Notes in Computer Science*, 915, 1995.

[2] B. Berthomieu. Implementing CCS, the LCS experiment. Technical Report 89425, LAAS CNRS, 1989.

[3] D. Bolignano and M. Debbabi. A coherent type inference system for a concurrent, functional and imperative programming language. In *Proceedings of the AMAST'93 Conference*. Springer Verlag, June 1993.

[4] D. Bolignano and M. Debbabi. A semantic theory for CML. In *Proceedings of the TACS'94 Conference*. Springer Verlag, April 1994.

[5] R. Cridlig. Semantic analysis of shared-memory concurrent languages using abstract model-checking. In *Symposium on Partial Evaluation and Program Manipulation*, 1995.

[6] R. Cridlig. Semantic analysis of concurrent ML by abstract model-checking. In *International Workshop on Verification of Infinite State Systems*, 1996.

[7] Mads Dam. Model checking mobile processes. *Information and Computation*, 129(1):35–51, 25 August 1996.

[8] M. Debbabi. *Intégration des paradigmes de programmation parallèle, fonctionnelle et impérative : fondements sémantiques.* PhD thesis, Université Paris Sud, Centre d'Orsay, July 1994.
[9] E. Allen Emerson and Chin-Laung Lei. Efficient model checking in fragments of the propositional mu-calculus (extended abstract). In *Proceedings, Symposium on Logic in Computer Science*, pages 267–278, Cambridge, Massachusetts, 16–18 June 1986. IEEE Computer Society.
[10] William Ferreira, Matthew Hennessy, and Alan Jeffrey. A theory of weak bisimulation for core CML. *ACM SIGPLAN Notices*, 31(6):201–212, June 1996.
[11] A. Giacalone, P. Mishra, and S. Prasad. Facile: A symmetric integration of concurrent and functional programming. *International Journal of Parallel Programming*, 18(2):121–160, April 1989.
[12] K. Havelund and K. G. Larsen. The fork calculus. In A. Lingas, R. Karlsson, and S. Carlsson, editors, *Proceedings* 20th *ICALP*, volume 700 of *Lecture Notes in Computer Science*. Springer Verlag, 1993.
[13] M. Hennessy and A. Ingólfsdóttir. A theory of communicating processes with value passing. In *Proc. 17th ICALP, LNCS.* Springer Verlag, 1990.
[14] D. Kozen. Results on the propositional mu-calculus. *Theoretical Computer Science*, 23, 1983.
[15] R. Milner. The polyadic π-calculus: A tutorial. Technical report, Laboratory for Foundations of Computer Science, Department of Computer Science, University of Edinburgh, 1991.
[16] R. Milner, J. Parrow, and D. Walker. A calculus of mobile processes. Technical report, Laboratory for Foundations of Computer Science, Department of Computer Science, University of Edinburgh, 1989.
[17] Flemming Nielson and Hanne Riis Nielsen. From CML to process algebra. In E. Best, editor, *Proceedings of CONCUR'93*, LNCS 715, pages 493–508. Springer-Verlag, 1993.
[18] Hanne Riis Nielson and Flemming Nielson. Higher-order concurrent programs with finite communication topology. In *Conference Record of the 21st ACM SIGPLAN-SIGACT Symposium on Principles of Programming Languages (POPL'94)*, pages 84–97, Portland, Oregon, January 17–21, 1994. ACM Press. Extended abstract.
[19] J.H. Reppy. Concurrent programming with events - the Concurrent ML manual. Technical report, Department of Computer Science, Cornell University, November 1990.
[20] J.H. Reppy. CML: A higher-order concurrent language. In *Proceedings of the ACM SIGPLAN '91 PLDI*, pages 294–305. SIGPLAN Notices 26(6), 1991.
[21] D.A. Schmidt. Natural-semantics-based abstract interpretation. In *Proc. 2d Static Analysis Symposium, Glasgow, Sept. 1995*, Lecture Notes in Computer Science 983, pages 1–18. Springer-Verlag, Berlin, 1995.
[22] D.A. Schmidt. Abstract interpretation of small-step semantics. In *Proc. 5th LOMAPS Workshop on Analysis and Verification of Multiple-Agent Languages, Stockholm, June 1996*, Lecture Notes in Computer Science 1192, pages 76–99. Springer-Verlag, Berlin, 1997.
[23] Bent Thomsen. A calculus of higher order communicating systems. In *Conference Record of the Sixteenth Annual ACM Symposium on Principles of Programming Languages*, pages 143–154, Austin, Texas, January 1989.
[24] Bent Thomsen. Plain CHOCS. A second generation calculus for higher order processes. *Acta Informatica*, 30, 1993.

A Trace-Based Refinement Calculus for Shared-Variable Parallel Programs

Jürgen Dingel

School of Computer Science
Carnegie Mellon University
Pittsburgh, PA 15213, USA
jurgend@cs.cmu.edu

Abstract. We present a refinement calculus for shared-variable parallel programs. The calculus allows the stepwise formal derivation of a low-level implementation from a trusted high-level specification. It is based on a trace-theoretic semantic model that supports local variable declaration and fair parallel composition. Compositionality is achieved through assumption-commitment reasoning. The refinement rules are syntax-directed in the sense that each rule corresponds to a specific language construct. The calculus is applicable to terminating and non-terminating programs and supports reasoning about liveness properties like termination and eventual entry. A detailed example is given and related work is reviewed.

1 Introduction

Formal support for the design and verification of parallel programs has been an important research topic for a long time. Some of the approaches, for instance, attempt to generalize Hoare logic to a parallel setting and suggest syntax-directed proof systems [15, 12, 11, 18, 17]. In a different approach, Back and his colleagues generalize sequential programming by grouping independent transitions into actions that are assumed to be executed atomically [2]. The resulting Action systems thus inherit a lot of the theory of sequential programming despite the presence of concurrency. Other approaches require a more radical departure from sequential programming and defy easy classification, eg., [5, 6, 14, 4].

Independent of these efforts, traces have been realized as the adequate tool for modeling concurrent computation [16, 7, 3]. Our point of departure here is Brookes' transition trace semantics. In [3], a combination of transition traces (sequences of pairs of states) together with two straightforward closure conditions (stuttering and mumbling) gives rise to an elegant, fully abstract, denotational semantics $\mathcal{T}$ for a language that includes local variable declarations, synchronization and fair parallelism. The semantics validates several natural laws of concurrent programming, like, for instance, the commutativity and associativity of fair parallel composition

$$C_1 \| C_2 =_{\mathcal{T}} C_2 \| C_1 \qquad [C_1 \| C_2] \| C_3 =_{\mathcal{T}} C_1 \| [C_2 \| C_3]$$

A.M. Haeberer (Ed.): AMAST'98, LNCS 1548, pp. 231–247, 1998.

or the idempotence of **skip**

$$C;\mathbf{skip} =_{\mathcal{T}} \mathbf{skip};C =_{\mathcal{T}} C\|\mathbf{skip} =_{\mathcal{T}} C$$

where $C_1 =_{\mathcal{T}} C_2$ abbreviates $\mathcal{T}[\![C_1]\!] = \mathcal{T}[\![C_2]\!]$.

In the present paper, we argue that Brookes' transition trace semantics provides an ideal formal basis for the study of stepwise refinement for parallel programs. The pleasant meta-theory and robustness of $\mathcal{T}$ allows for the development of a refinement calculus that, in our opinion, constitutes a contribution even in the face of a large body of existing related work. More precisely, we use a notion of context-sensitive approximation for transition traces that was introduced in [8] to define a syntax-directed refinement calculus that supports compositional reasoning, local variables, fairness and reasoning about liveness properties like termination or eventual entry. [1] Preliminary results indicate that our approach also is applicable to distributed notions of concurrency.

The next section reviews some necessary background and is mostly based on [3, 8]. Section 3 introduces our notion of refinement. Section 4 presents a detailed example. Section 5 concludes and discusses related work.

2 Background

2.1 Syntax and Semantics of Programs

Our notion of program is non-standard in the sense that it allows for very abstract descriptions of computations. More precisely, some programs in our setting are either too abstract to be executable or do not have a direct computational interpretation and thus should be viewed as specifications rather than executable code. The most basic program components are *atomic statements* of the form $V{:}[P,Q]$, where V is a finite set of variables and P and Q are assertions [13]. It is meant to describe a single atomic transition, which transforms a state satisfying P into one satisfying Q by just changing the variables in V. A random assignment which may set x to any natural number can thus be described by $\{x\}{:}[tt, x \geq 0]$. An idling, or stuttering, step is expressed as **skip** $\equiv \emptyset{:}[tt,tt]$. To be able to refer to the value a variable held initially, i.e., at the beginning of the transition, we reserve "hooked" variables $\overleftarrow{x}$ in Q. The meaning of the multiple assignment statement $x,y{:=}x+1,0$, for example, is thus captured by $\{x,y\}{:}[tt, x = \overleftarrow{x} + 1 \wedge y = 0]$. If an assertion does not contain hooked variables it is called *unary*. Otherwise it is called *binary*. In a statement $V{:}[P,Q]$, P must be unary, whereas Q may be unary or binary. The semantics of atomic statements is conveniently captured by characteristic formulas.

Definition 1. *Let Var denote the set of all program variables and let ι be a metavariable that ranges over program variables. Given an atomic statement $V{:}[P,Q]$, its* characteristic formula $cf_{V:[P,Q]}$ *is given by the predicate*

[1] A program C is said to have the *eventual entry property* if control always eventually gets past every `await` statement in C.

$$cf_{V:[P,Q]} \equiv \overleftarrow{P} \wedge Q \wedge \forall \iota \in Var - V.\iota = \overleftarrow{\iota}$$

where $\overleftarrow{P}$ *abbreviates the substitution of all free unhooked variables in* P *by their hooked counterpart. We interpret a binary assertion* Q *over pairs of states* (s, s') *where* s *assigns values to hooked variables and* s' *to the unhooked ones. More precisely,* $(s, s') \models Q$ *iff replacing the hooked variables in* Q *by their values in* s *and replacing the unhooked variables in* Q *by their values in* s' *makes* Q *true.*

For instance, the statement $x, y{:=}x + 1, 0$ has the characteristic formula

$$cf_{x,y:=x+1,0} \quad \equiv \quad x = \overleftarrow{x} + 1 \wedge y = 0 \wedge \forall \iota \in Var - \{x, y\}.\iota = \overleftarrow{\iota} \; .$$

More complex programs can be built using sequential and parallel composition, disjunction, iteration, quantification, and hiding. Programs are ranged over by C, D. An important extension to the standard shared-variable parallel language involves labels. Consider the parallel composition $C_1 \| C_2$. In order to be able to distinguish the transitions of C_1 from those of C_2, we allow for C_1 to be enclosed in angle brackets to form $\langle C_1 \rangle \| C_2$. A program that contains exactly one subprogram enclosed in angle brackets is called *labeled.* A program that contains no angle brackets is *unlabeled.* The following grammar generates labeled and unlabeled programs:

$$\begin{aligned} C &::= V{:}[P,Q] \mid C_1 ; C_2 \mid \langle D \rangle \mid C_1 \vee C_2 \mid C_1 \| C_2 \mid C^* \mid C^+ \mid C^\omega \mid \\ &\quad\; \forall x \in I.C \mid \textbf{new } x = v \textbf{ in } C \\ D &::= V{:}[P,Q] \mid D_1 ; D_2 \mid D_1 \vee D_2 \mid D_1 \| D_2 \mid D^* \mid D^+ \mid D^\omega \mid \\ &\quad\; \forall x \in I.D \mid \textbf{new } x = v \textbf{ in } D \end{aligned}$$

where I is some index set over the domain of x, that is, $I \subseteq Dom_x$. *Contexts,* ranged over by E, are unlabeled programs with exactly one hole.

$$\begin{aligned} E &::= [] \mid C ; E \mid E ; C \mid C \vee E \mid E \vee C \mid C \| E \mid E \| C \mid E^* \mid E^+ \mid E^\omega \mid \\ &\quad\; \textbf{new } x = v \textbf{ in } E \end{aligned}$$

A context E gives rise to a program $E[C]$, if the hole in E is replaced by C. Very often, we will consider a labeled statement $\langle C \rangle$ in some context, that is, $E[\langle C \rangle]$ yields the labeled program that is obtained by replacing the hole in E by $\langle C \rangle$. We call a context E *parallel,* if the hole is in the scope of a parallel composition, that is, if there are E_1, E_2 and C such that $E \equiv E_1[E_2 \| C]$. A context is *sequential* if it is not parallel.

Transition Traces Let $s, s', s_i \in \Sigma$ denote states, that is, mappings from the finite set of program variables *Var* to values. Transition traces[2]

$$(s_0, s_0')(s_1, s_1') \ldots (s_i, s_i') \ldots$$

have proven very useful for the definition of compositional models of shared-variable concurrency [16, 7, 3]. One such trace represents a possible "interactive"

[2] Sometimes also called *potential* or *partial computations* or *extended sequences.*

computation of a command in which state changes made by the command (from s_i to s'_i) are interleaved by state changes made by its environment (from s'_i to s_{i+1}). The meaning of a program is given by a set of transition traces. To describe the meaning of a labeled program $\langle C_1 \rangle \| C_2$ we will consider *labeled* transition traces of the form

$$(s_0, l_0, s'_0)(s_1, l_1, s'_1) \ldots (s_i, l_i, s'_i) \ldots$$

where each transition carries a label l from the set $\Lambda \equiv \{p, e\}$. A transition labeled with p was caused by a statement inside the angle brackets, that is, by C_1, and is called a *program transition*. A transition with e is due to C_2 and is called an *environment transition*. By describing a labeled program by means of labeled transition traces we thus regard it as an *open system* while singling out the transitions made by a specific part of the program. In other words, $\langle C_1 \rangle \| C_2$ can be thought of as an open system whose environment is known to at least comprise C_2.

In the semantics, trace sets will be closed under two conditions: stuttering and mumbling. These two conditions were used in [3] to achieve full abstraction. They correspond, respectively, to reflexivity and transitivity of the $\rightarrow^*$ relation in a conventional operational semantics. Given a set T of traces, the closure under stuttering and mumbling $T^\dagger$ is the smallest set which contains T and satisfies:

Stuttering If $\alpha\beta \in T^\dagger$ then $\alpha(s, p, s)\beta \in T^\dagger$ and $\alpha(s, e, s)\beta \in T^\dagger$ and
Mumbling 1. if $\alpha(s, l, s)(s, l', s')\beta \in T^\dagger$ then $\alpha(s, l', s')\beta \in T^\dagger$ and
2. if $\alpha(s, l, s')(s', l', s')\beta \in T^\dagger$ then $\alpha(s, l, s')\beta \in T^\dagger$ and
3. if $\alpha(s, l, s')(s', l, s'')\beta \in T^\dagger$then $\alpha(s, l, s'')\beta \in T^\dagger$.

Before the denotational semantics of programs is presented, we introduce some notation and define a few operations on traces and sets of traces. The concatenation $T_1; T_2$ and the infinite iteration operation T^ω are defined as

$$\begin{aligned} T_1; T_2 &= \{\alpha\beta \mid \alpha \in T_1 \wedge \beta \in T_2\}^\dagger \\ T^\omega &= \{\alpha_0 \ldots \alpha_n \ldots \mid \forall i \geq 0. \alpha_i \in T\}^\dagger. \end{aligned}$$

T^* denotes the smallest set containing T and the empty trace, closed under stuttering, mumbling and concatenation. Fair parallel composition is modeled by fair interleaving of sets of traces

$$T_1 \| T_2 = \bigcup \{\alpha_1 \| \alpha_2 \mid \alpha_1 \in T_1 \wedge \alpha_2 \in T_2\}^\dagger$$

where $\alpha \| \beta$ is the set of all traces built by fairly interleaving α and β. A precise definition of this operation can be found in [3] and is omitted here.

Local Variables $[s|x = v]$ denotes the state that is like s except that the value of x is updated to v. Let $\alpha \equiv (s_0, l_0, s'_0)(s_1, l_1, s'_1) \ldots (s_i, l_i, s'_i) \ldots$ be a transition trace. The trace $\langle x = v \rangle \alpha$ is like α except that x is initialized to v in the first state and that the value of x is retained across points of possible interference. More precisely, $\langle x = v\rangle\alpha$ is

$$([s_0|x = v], l_0, s'_0)([s_1|x = s'_0(x)], l_1, s'_1)\ldots([s_i|x = s'_{i-1}(x)], l_i, s'_i)\ldots$$

The trace $\alpha\backslash x$ on the other hand describes a computation like α except that it never changes the value of x. That is, $\alpha\backslash x$ is

$$(s_0, l_0, [s'_0 \mid x = s_0(x)])(s_1, l_1, [s'_1 \mid x = s_1(x)])\ldots(s_i, l_i, [s'_i \mid x = s_i(x)])\ldots.$$

We are now ready to present the semantics. Given a set T, let T^∞ denote $T^* \cup T^\omega$. Let $\mathcal{P}^\dagger(T)$ denote the set of all subsets of T that are closed under stuttering and mumbling.

Definition 2. *The semantic function $\mathcal{T}$ maps labeled and unlabeled programs and statements to $\mathcal{P}^\dagger((\Sigma \times \Lambda \times \Sigma)^\infty)$ and is defined as $\mathcal{T}_e[\![\_]\!]$ where $\mathcal{T}_l[\![\_]\!]$ for $l \in \Lambda$ is given by*

$$\begin{aligned}
\mathcal{T}_l[\![V{:}[P,Q]]\!] &= \{(s,l,s') \mid (s,s') \models cf_{V:[P,Q]}\}^\dagger\\
\mathcal{T}_e[\![\langle C\rangle]\!] &= \mathcal{T}_p[\![C]\!]\\
\mathcal{T}_l[\![C_1 ; C_2]\!] &= \mathcal{T}_l[\![C_1]\!]; \mathcal{T}_l[\![C_2]\!]\\
\mathcal{T}_l[\![C_1 \vee C_2]\!] &= \mathcal{T}_l[\![C_1]\!] \cup \mathcal{T}_l[\![C_2]\!]\\
\mathcal{T}_l[\![C_1 \| C_2]\!] &= \mathcal{T}_l[\![C_1]\!] \| \; \mathcal{T}_l[\![C_2]\!]\\
\mathcal{T}_l[\![C^*]\!] &= (\mathcal{T}_l[\![C]\!])^*\\
\mathcal{T}_l[\![C^\omega]\!] &= (\mathcal{T}_l[\![C]\!])^\omega\\
\mathcal{T}_l[\![\forall x \in I.C]\!] &= \{\alpha \mid \forall v \in I.\alpha \in \mathcal{T}_l[\![C[v/x]]\!]\}^\dagger\\
\mathcal{T}_l[\![\mathbf{new}\ x = v\ \mathbf{in}\ C]\!] &= \{\alpha\backslash x \mid \langle x = v\rangle\alpha \in \mathcal{T}_l[\![C]\!]\}^\dagger.
\end{aligned}$$

The traces of **new** $x = v$ **in** C do not change the value of x and are obtained by executing C under the assumption that x is set to v initially and that the environment cannot change the value of x.

The standard shared-variable parallel programming language that was used in [15], for instance, is embedded into our setting through the following abbreviations. Note how the **await** statement is implemented using busy waiting. Let e range over arithmetic and boolean expressions and let B be a boolean expression.

$$\begin{aligned}
\{B\} &\equiv \emptyset{:}[B, B]\\
\mathbf{skip} &\equiv \{tt\}\\
x{:=}e &\equiv x{:}[tt, x = \overleftarrow{e}]\\
\mathbf{if}\ B\ \mathbf{then}\ C_1\ \mathbf{else}\ C_2 &\equiv (\{B\}; C_1) \vee (\{\neg B\}; C_2)\\
\mathbf{while}\ B\ \mathbf{do}\ C &\equiv ((\{B\}; C)^*; \{\neg B\}) \vee (\{B\}; C)^\omega\\
\mathbf{await}\ B\ \mathbf{then}\ V{:}[P,Q] &\equiv V{:}[P \wedge B, Q] \vee \{\neg B\}^\omega.
\end{aligned}$$

Additionally, we will use the following abbreviations. Remember that *Var* denotes the set of all program variables. Given a set of variables V, let $\mathcal{P}(V)$ denote the set of all predicates (unary assertions) over V. Let $\Gamma \subseteq \mathcal{P}(Var)$.

$$\begin{aligned}
C^\infty &\equiv C^* \vee C^\omega & pre\ B &\equiv Var{:}[tt, \overleftarrow{B} \Rightarrow B]\\
inv\ e &\equiv Var{:}[tt, e = \overleftarrow{e}] & pre\ \Gamma &\equiv \forall B \in \Gamma.pre\ B\\
inv^\infty\ e &\equiv (inv\ e)^\infty & pre^\infty\ \Gamma &\equiv (pre\ \Gamma)^\infty
\end{aligned}$$

C^∞ denotes finite and infinite iteration over C. *inv* e denotes the most general atomic transition that leaves the value of the expression e *invariant*, that is, unchanged. For boolean expressions B we will need the weaker property that the value of B is unchanged across a transition only if it is true. We say that B is *preserved*.

Definition 3. *A trace* $(s_0, l_0, s_0')(s_1, l_1, s_1')\ldots$ *is* connected *if we have* $s_i' = s_{i+1}$ *for all* $i \geq 0$. *The* executions $\mathcal{E}[\![C]\!]$ *of a program* C *are its connected transition traces. Let* C *be a labeled or unlabeled program. Then,*

$$\mathcal{E}[\![C]\!] \quad = \quad \{\alpha \in \mathcal{T}[\![C]\!] \mid \alpha \textit{ is connected}\}.$$

Let $C_1 \subseteq_\mathcal{T} C_2$ and $C_1 =_\mathcal{T} C_2$ abbreviate $\mathcal{T}[\![C_1]\!] \subseteq \mathcal{T}[\![C_2]\!]$ and $\mathcal{T}[\![C_1]\!] = \mathcal{T}[\![C_2]\!]$ respectively. Similarly for $\mathcal{E}$.

2.2 Approximation

A very natural notion of program approximation arises through transition trace inclusion. inv^∞ x is the most general program that never changes the value of x. For instance, a program C always leaves the value of x invariant in all contexts iff $C \subseteq_\mathcal{T} inv^\infty\ x$.

Lemma 1. *1. If* $C_1 \subseteq_\mathcal{T} C_2$, *then* $C_1 \subseteq_\mathcal{E} C_2$.
2. $V_1{:}[P_1, Q_1] \subseteq_\mathcal{T} V_2{:}[P_2, Q_2]$ *iff* $cf_{V_1:[P_1,Q_1]} \Rightarrow cf_{V_2:[P_2,Q_2]}$.
3. Trace inclusion is a congruence, *that is,* $C_1 \subseteq_\mathcal{T} C_2$ *implies* $E[C_1] \subseteq_\mathcal{T} E[C_2]$ *for all* E.
4. If $x \notin fv(C)$ *then* **new** $x = v$ **in** $C =_\mathcal{T} C$.

Trace inclusion between two programs C_1 and C_2 implies that in *all* possible contexts the executions of C_1 are contained in those of C_2 in the same context. Thus, whenever we want to do refinement in a *specific* context, trace set inclusion may be too strong, because it does not incorporate information about that particular context. In other words, $\subseteq_\mathcal{T}$ is not context-sensitive. We now present a notion of approximation that is context-sensitive and that will form the basis of our stepwise refinement method.

Definition 4. *Let* C_1 *and* C_2 *be unlabeled programs and* E *be a context.* $C_1 \geq_E C_2$ *iff* $E[\langle C_1\rangle] \supseteq_\mathcal{E} E[\langle C_2\rangle]$.

$C_1 =_E C_2$ abbreviates $C_1 \leq_E C_2$ and $C_2 \leq_E C_1$. Intuitively, $C_1 \geq_E C_2$ if E causes C_2 to exhibit only transitions that can be matched by C_1. In other words, E cannot force C_2 to go beyond what C_1 can do.

Example 1. Let $E_1 \equiv y{:}[tt, y > 0]; [[\,] \| z{:=}0]$ and $E_2 \equiv y{:}[tt, y > 0]; [[\,] \| y{:=}0]$. We have $x{:}[tt, x > \overleftarrow{x}] \geq_{E_1} x{:=}x + y$, and $x{:}[tt, x > \overleftarrow{x}] \not\geq_{E_2} x{:=}x + y$. The second approximation fails because in a state with $y \leq 0$, $x{:=}x + y$ has transitions that cannot be matched by $x{:}[tt, x > \overleftarrow{x}]$.

Lemma 2. *$E[\langle C_1\rangle] \supseteq_{\mathcal{E}} E[\langle C_2\rangle]$ implies $E[C_1] \supseteq_{\mathcal{E}} E[C_2]$.*

It seems natural to attempt to distinguish contexts with respect to their "discriminating power" [8, 9]. The context $E_1 \equiv [] \,\|\, Var{:}[tt, tt]^*$, for instance, can do any transition at any time. The context $E_2 \equiv [] \,\|\, inv^* x$, however, can only do those transitions that leave x invariant. Every approximation that holds with respect to E_1 will also hold with respect to E_2, whereas the converse is not true. E_1 is more general and thus has more discriminating power. In [8, 9] we discuss how context-approximation formalizes assumption-commitment reasoning and thus allows for modular proofs of the approximation $C_1 \supseteq_{\mathcal{E}} C_2$.

3 Refinement

Our notion of refinement is based on assumption-commitment reasoning in the spirit of [11, 18] to achieve compositionality. To illustrate the main idea, consider, for instance, the programs $C \equiv x{:=}x{+}1$ and $C' \equiv x{:=}2$. Assuming an initial state that satisfies $x = 1$ and a parallel context that preserves $x = 1$, every transition of C' can be matched by C and thus C can be refined into C' (and vice versa). If, moreover, the parallel context also preserves $x = 2$ we can conclude that x will have value 2 upon termination. Also, C' preserves all predicates P for which $\overleftarrow{P} \wedge cf_{x:=2} \Rightarrow P$. In our calculus this will be expressed by

$$x{:=}x+1 \succ^{x=2,\Delta}_{x=1,\{x=1,x=2\}} x{:=}2$$

where $\Delta \equiv \{P \mid \overleftarrow{P} \wedge cf_{x:=2} \Rightarrow P\}$. Sometimes the refinement requires the introduction of new auxiliary variables that, for instance, store a temporary result or step over an array. Suppose, for example, that we want to split the assignment $C \equiv x{:=}2{\cdot}x{+}y$ into a sequence of simpler ones $C' \equiv t{:=}2{\cdot}x\,;x{:=}t{+}y$. C' introduces the auxiliary variable t. Obviously, not every transition of $t{:=}2\cdot x\,;x{:=}t+y$ can be matched by $x{:=}2\cdot x+y$. However, every transition that does not affect the new, introduced variable t still can be matched. In other words, C can match every transition of C' modulo the changes to t. Formally,

$$x{:=}2\cdot x+y \succ^{x=3\wedge y=1\wedge t=2,\Delta}_{x=1\wedge y=1,\Gamma,\{t\}} t{:=}2\cdot x\,;x{:=}t+y$$

where $\Gamma \equiv \{x=1, y=1, t=2, x=3\}$ and $\Delta \equiv \{P \mid \overleftarrow{P} \wedge cf_{t:=2\cdot x} \Rightarrow P\} \cap \{P \mid \overleftarrow{P} \wedge cf_{x:=t+y} \Rightarrow P\}$. The following definition formalizes this idea and forms the heart of our calculus. To capture the partial correctness behaviour of a program, we will adopt the standard Hoare-triple notation. $\{P\}\ C\ \{Q\}$ expresses that every finite execution of C with an initial state satisfying P will end in a state satisfying Q.

Definition 5. *Let P, Q be predicates and Γ, Δ be sets of predicates, that is, $P, Q \in \mathcal{P}(Var)$ and $\Gamma, \Delta \subseteq \mathcal{P}(Var)$. We will assume that Γ and Δ are closed under logical equivalence, that is, $P \in \Gamma$ and $P \Leftrightarrow P'$ imply $P' \in \Gamma$. Also, let $V \subseteq Var$. We say that C'* refines C with respect to P, Γ, V, Q and Δ,

$$C \succ^{Q,\Delta}_{P,\Gamma,V} C'$$

for short, iff we have that

1. *C' approximates C in the given context, that is, if $V \equiv \{x_1, \ldots, x_n\}$, then for all $1 \leq i \leq n$ and $v_i \in Dom_{x_i}$ such that $P \Rightarrow P[v_1/x_1, \ldots, v_n/x_n]$, we have*
$$C \geq_{\mathbf{new}\ x_1=v_1,\ldots,x_n=v_n\ \mathbf{in}\ \{P\};[[]\|pre^{\infty}\Gamma]} C'$$
and
2. *C' preserves the predicates in Δ, that is,*
$$C' \leq_{\{P\};[[]\|pre^{\infty}\Gamma]} pre^{\infty}\Delta \text{ and}$$
3. *if C' and the parallel context terminate, they do so in a state satisfying Q, that is,*
$$\{P\}\ [C'\|pre^{\infty}\Gamma]\ \{Q\}.$$

Informally, $C \succ^{Q,\Delta}_{P,\Gamma,V} C'$ expresses that assuming an initial state that satisfies P and a parallel context that preserves the predicates in Γ, then every transition of C' can be matched by C modulo the changes to variables in V and will preserve all properties in Δ. Also, if C' and the parallel context terminate, they will do so in a state satisfying Q. Thus, assumption-commitment reasoning is harnessed for a notion of program transformation.

Refinement Rules Refinement is governed by the syntax-directed rules in Figures 1 and 2 which form our refinement calculus. Due to space limitations we do not prove the soundness of the rules here. However, we briefly explain the intuition behind some of them.

ATOM The first premise ensures that for every transition (s_2, s_2') of an atomic statement A_2 there is a transition (s_1, s_1') of A_1 such that s_2 coincides with s_1 modulo the variables in V and s_2' coincides with s_1' modulo the variables in V. The second premise shows that Q holds immediately after termination of A_2. The third premise ensures that P and Q are preserved by the environment.

PAR This is where keeping track of the assumptions Γ and the commitments Δ pays off and allows the formulation of a compositional rule. Guarantees and assumptions have to mutually imply each other. The requirements Γ_1 of C_1 have to be contained in the guarantees of C_2 and vice versa. This rule is similar in spirit to corresponding rules using assumption-commitment reasoning (eg., [11, 18]).

NEW-INTRO Like rule NEW, this rule weakens the assumptions and strengthens the commitments. It is a straight-forward consequence of NEW and Lemma 1.4 below.

WHILE-INTRO This rule allows the replacement of a finite iteration with loop condition B and invariant I by a **while** loop. Moreover, the body of the loop may be refined. To show termination of the resulting **while** loop it recasts the well-known total correctness rule for **while** loops in trace-theoretic terms

$$\text{ATOM}\quad \frac{(\exists V \cup \overleftarrow{V}\,.\,\overleftarrow{P} \wedge cf_{A_2}) \Rightarrow (\exists V \cup \overleftarrow{V}\,.\,\overleftarrow{P} \wedge cf_{A_1}) \quad (\overleftarrow{P} \wedge cf_{A_2}) \Rightarrow Q \quad \{P, Q\} \subseteq \Gamma}{A_1 \succ^{Q,\Delta}_{P,\Gamma,V} A_2}$$

where A_1 and A_2 are atomic statements and $\Delta \subseteq \{P \mid \overleftarrow{P} \wedge cf_{A_2} \Rightarrow P\}$ and $\exists\{x_1, \ldots, x_n\}.P$ stands for $\exists x_1, \ldots, x_n.P$.

$$\text{SEQ}\quad \frac{C_1 \succ^{Q_1,\Delta_1}_{P,\Gamma_1,V_1} C_1' \quad C_2 \succ^{Q,\Delta_2}_{Q_1,\Gamma_2,V_2} C_2'}{C_1\,;C_2 \succ^{Q,\Delta_1 \cap \Delta_2}_{P,\Gamma_1 \cup \Gamma_2, V_1 \cup V_2} C_1'\,;C_2'}$$

$$\text{WHILE}\quad \frac{C \succ^{I,\Delta}_{I \wedge B,\Gamma,V} C' \qquad I \Rightarrow (B \Leftrightarrow B')}{\textbf{while } B \textbf{ do } C \succ^{I \wedge \neg B',\Delta}_{I,\Gamma \cup \{\neg B'\},V} \textbf{while } B' \textbf{ do } C'}$$

$$\text{PAR}\quad \frac{C_1 \succ^{Q_1,\Gamma_2 \cup \Delta}_{P_1,\Gamma_1,V_1} C_1' \qquad C_2 \succ^{Q_2,\Gamma_1 \cup \Delta}_{P_2,\Gamma_2,V_2} C_2'}{C_1 \| C_2 \succ^{Q_1 \wedge Q_2,\Delta}_{P_1 \wedge P_2,\Gamma_1 \cup \Gamma_2,V_1 \cup V_2} C_1' \| C_2'} \qquad \text{STAR}\quad \frac{C \succ^{P,\Delta}_{P,\Gamma,V} C'}{C^* \succ^{P,\Delta}_{P,\Gamma,V} (C')^*}$$

$$\text{NEW}\quad \frac{C \succ^{Q,\Delta}_{P,\Gamma,V \cup \{x\}} C' \qquad P \Rightarrow P[v/x]}{\textbf{new } x = v \textbf{ in } C \succ^{\exists x.Q,\Delta'}_{\exists x.P,\exists x.\Gamma,V} \textbf{new } x = v \textbf{ in } C'}$$

where $\exists x.\Gamma \equiv \{\exists x.P \mid P \in \Gamma\}$ and $\Delta' \equiv \Delta \cup \{P \mid \exists x.P \in \Delta\}$.

$$\text{WEAK}\quad \frac{C_1' \succ^{Q',\Delta'}_{P',\Gamma',V'} C_2'}{C_1 \succ^{Q,\Delta}_{P,\Gamma,V} C_2}$$

where $C_1' \subseteq_{\mathcal{T}} C_1$, $C_2 \subseteq_{\mathcal{T}} C_2'$, $P \Rightarrow P'$, $Q' \Rightarrow Q$, $\Gamma' \subseteq \Gamma$, $\Delta \subseteq \Delta'$, and $V' \subseteq V$.

Fig. 1. Refinement rules

and also transfers it to a concurrent setting. Remember that *Var* denotes all program variables. Given a measure m, the statement a_m decreases m if it is not zero and leaves it unchanged if it is zero. Since each iteration brings m closer to 0 and the environment cannot increase m due to Γ_m, m must eventually be set to 0, which implies $\neg B$ and thus termination of the loop.

AWAIT-INTRO This rule allows the introduction of the synchronization statement **await** with condition B. To show that B eventually becomes and remains true, the parallel context D is shown to decrease m until it is 0 and then leave it unchanged. Note that the correctness of this rule relies on the fairness of parallel composition.

Notation and Properties Let $C \succ^{Q,\Delta}_{P,\Gamma} C'$ stand for $C \succ^{Q,\Delta}_{P,\Gamma,\emptyset} C'$. To motivate the next abbreviation, consider the programs $C_1 \| C_2$ and $C_1\,;C_2$ and suppose that we want to refine C_2. Although C_1 is left unchanged, we still need to deter-

NEW-INTRO $$\frac{C \succ^{Q,\Delta}_{P,\Gamma,V\cup\{x\}} C' \qquad P \Rightarrow P[v/x] \qquad x \notin fv(C)}{C \succ^{\exists x.Q,\Delta'}_{\exists x.P,\exists x.\Gamma,V} \textbf{new } x = v \textbf{ in } C'}$$

where $\exists x.\Gamma \equiv \{\exists x.P \mid P \in \Gamma\}$ and $\Delta' \equiv \Delta \cup \{P \mid \exists x.P \in \Delta\}$.

In the next two rules let $a_m \equiv Var{:}[tt, \overleftarrow{m}= 0 \rightarrow m = 0|m < \overleftarrow{m}]$ where $B \rightarrow P|Q$ abbreviates $(B \Rightarrow P) \wedge (\neg B \Rightarrow Q)$ and $\Gamma_m \equiv \{m \leq n | n \in \mathbb{N}$.

WHILE-INTRO If

$$C \succ^{I,\Delta}_{B\wedge I,\Gamma,V} C' \qquad \text{and} \qquad (\neg B \wedge I) \Rightarrow Q$$

and there exists an arithmetic expression m over the free variables in B and C' such that $m \geq 0$, and $m = 0 \Rightarrow \neg B$, and

$$(inv^* m; a_m; inv^* m)^+ \supseteq_T C'$$

then

$$(\{B \wedge I\}; C)^*; \{Q\} \succ^{Q,\Delta}_{I,\Gamma\cup\{Q\}\cup\Gamma_m,V} \textbf{while } B \textbf{ do } C'.$$

AWAIT-INTRO If

$$\{P_1, \Gamma\} \quad [V{:}[B \wedge P_2, Q_2] \| D] \quad \{Q_1, \Delta\}$$

and there exists an arithmetic expression m over the free variables in B and D such that $m \geq 0$, and $m = 0 \Rightarrow B$, and

$$(inv^* m; a_m)^\omega \vee (inv^* m; a_m; inv^* m)^*; \{m = 0\}; inv^* m \supseteq_T D.$$

then

$$[V{:}[B \wedge P_2, Q_2] \| D] \succ^{Q_1,\Delta}_{P_1,\Gamma\cup\Gamma_m} [\textbf{await } B \textbf{ then } V{:}[P_2, Q_2] \textbf{ end} \| D].$$

Fig. 2. Rules for introducing **while** and **await**

mine which guarantees C_1 can make under which assumptions. The statement $\{P, \Gamma\}\ C_1\ \{Q, \Delta\}$, which abbreviates $C_1 \succ^{Q,\Delta}_{P,\Gamma} C_1$, will be convenient in these situations. Note that in his generalization of Owicki and Gries' Hoare logic [18], Stirling employs a statement with almost precisely the same meaning.

Consider the two sets of predicates $\mathcal{P}(Var)$ and $\mathcal{P}(\emptyset)$. $\mathcal{P}(Var)$ contains all predicates over Var. Thus, an environment that preserves all predicates in $\mathcal{P}(Var)$ cannot change any state in any way. $\mathcal{P}(\emptyset)$ on the other hand, contains only the constant predicates tt and ff (and their equivalents). Since tt and ff are always preserved, $\mathcal{P}(\emptyset)$ places no restrictions and thus allows the environment to change any state arbitrarily.

Lemma 3.1 and 3.2 state that refinement with respect to an environment that preserves *all* predicates, implies execution inclusion (and vice versa). Both follow directly from the definitions. Lemma 3.3 is an instance of the weakening rule.

Lemma 3. *1. If* $C \succ^{Q,\Delta}_{P,\mathcal{P}(Var)} C'$ *then* $\{P\};C \supseteq_{\mathcal{E}} \{P\};C'$ *and* $\{P\}\ C'\ \{Q\}$ *for all* Δ.
2. If $\{P\};C \supseteq_{\mathcal{E}} \{P\};C'$ *then* $C \succ^{tt,\mathcal{P}(\emptyset)}_{P,\mathcal{P}(Var)} C'$.
3. If $C_1 \supseteq_{\mathcal{T}} C_2$ *and* $C_2 \succ^{Q,\Delta}_{P,\Gamma,V} C_3$ *then* $C_1 \succ^{Q,\Delta}_{P,\Gamma,V} C_3$. *Also, if* $C_2 \subseteq_{\mathcal{T}} C_3$ *and* $C_1 \succ^{Q,\Delta}_{P,\Gamma,V} C_2$ *then* $C_1 \succ^{Q,\Delta}_{P,\Gamma,V} C_3$.

Moreover, refinement is transitive.

Lemma 4. *If* $C_1 \succ^{Q,\Delta_1}_{P,\Gamma_1,V_1} C_2$ *and* $C_2 \succ^{Q,\Delta_2}_{P,\Gamma_2,V_2} C_3$ *then* $C_1 \succ^{Q,\Delta_1\cap\Delta_2}_{P,\Gamma_1\cup\Gamma_2,V_1\cup V_2} C_3$.

General Refinement Methodology Suppose we have shown

$$C_i \succ^{Q,\Delta_i}_{P,\Gamma_i} C_{i+1}$$

for $1 \leq i < n$. Using transitivity (Lemma 4) this implies

$$C_1 \succ^{Q,\bigcap_i \Delta_i}_{P,\bigcup_i \Gamma_i} C_n$$

which yields

$$\{P\};C_1 \supseteq_{\mathcal{E}} \{P\};C_n \quad \text{and} \quad \{P\}\ C_n\ \{Q_n\}$$

with weakening and Lemma 3.1. Thus, every execution α of C_n that starts in a state satisfying P also is an execution of C_1 and whenever α is finite, the last state satisfies Q_n. Note that this refinement methodology assumes that all C_i have a non-empty set of executions. We thus have to be careful not to introduce a trivial refinement without executions. Note, however, that all programs that contain the standard programming language constructs only and that are thus part of the language considered by Owicki and Gries [15], do have non-empty sets of executions. Consequently, whenever the most refined program C_n is syntactically well-formed in the sense of [15], the entire refinement is non-trivial.

4 Example

The following example has also been used in [1, 17]. Suppose $n \geq 1$ bank accounts are represented by an array $A[1..n]$. Let a and b with $1 \leq a, b \leq n$ be two distinguished accounts. We want to develop a program which computes the sum s over all entries in A and concurrently also transfers \$20 from account a to account b. We start with a high-level program C_1 that is easily seen to be correct. In the following, let ΣA stand for $\Sigma A \equiv \Sigma_{i=1}^{n} A[i]$. The first two refinement steps are summarized in Figure 3.

$$C_1 \equiv \left[s:=\Sigma A \parallel A[a], A[b]:=A[a]-20, A[b]+20 \right]$$

$$C_2 \equiv \textbf{new } k=1, t=0 \textbf{ in} \left[\begin{array}{l} \{k,t\}:[tt,tt]^*; \\ \{t=\Sigma A\}; \\ s:=\Sigma A \end{array} \middle\| \; A[a], A[b]:=A[a]-20, A[b]+20 \right]$$

$$C_3 \equiv \textbf{new } k=1, t=0 \textbf{ in} \left[\begin{array}{l} \{k,t\}:[tt,tt]^*; \\ \{t=\Sigma A\}; \\ s:=t \end{array} \middle\| \; A[a], A[b]:=A[a]-20, A[b]+20 \right]$$

Fig. 3. Refinements C_1 through C_3

Refining C_1 into C_2 The first refinement step introduces two local variables k and t, and a finite loop that modifies these two variables only and that is required to terminate in a state in which t contains the sum over A.

Let Q_s be the postcondition of the left parallel subprogram and let $P_{a,b}$ and $Q_{a,b}$ be the pre- and post-condition of the right parallel subprogram, that is,

$$Q_s \equiv s = \Sigma A \qquad P_{a,b} \equiv A[a] = v_1 \wedge A[b] = v_2$$
$$Q_{a,b} \equiv A[a] = v_1 - 20 \wedge A[b] = v_2 + 20$$

where v_1, v_2 are integers. Formally, this refinement is based on

$$\begin{aligned} & s:=\Sigma A \\ =_{\mathcal{T}} \; & \textbf{skip}^*; \textbf{skip}; s:=\Sigma A && (1) \\ \succ^{Q_s,\{P_{a,b},Q_{a,b}\}}_{tt,\{Q_s\},\{k,t\}} \; & \{k,t\}:[tt,tt]^*; \{t=\Sigma A\}; s:=\Sigma A. && (2) \end{aligned}$$

Equation (1) follows from the closure condition, which implies that the meaning of a program is invariant under the addition of finite stuttering. Approximation (2) is obtained by ATOM, SEQ, and STAR. Using Lemma 3.3 this implies

$$s:=\Sigma A \succ^{Q_s,\{P_{a,b},Q_{a,b}\}}_{tt,\{Q_s\},\{k,t\}} \{k,t\}:[tt,tt]^*; \{t=\Sigma A\}; s:=\Sigma A. \qquad (3)$$

Another application of ATOM also yields

$$\{P_{a,b}, \{P_{a,b}, Q_{a,b}\}\} \; A[a], A[b]:=A[a]-20, A[b]+20 \; \{Q_{a,b}, \Delta\} \qquad (4)$$

where $\Delta \equiv \{Q_s, t=\Sigma A\} \cup \mathcal{P}(\{k,t,s\})$ and $\mathcal{P}(V)$ denotes the set of all predicates over the variables in V. With an application of PAR to (3) and (4) and then of NEW-INTRO we obtain

$$C_1 \succ^{Q_{a,b} \wedge Q_s, \mathcal{P}(\emptyset)}_{P_{a,b},\{P_{a,b},Q_{a,b},Q_s\}} C_2.$$

Refining C_2 into C_3 If the predicate $t = \Sigma A$ is preserved by the environment, then the abstract assignment $s\!:=\!\Sigma A$ can safely be replaced by $s\!:=\!t$. Formally, we have

$$s\!:=\!\Sigma A \succ^{Q_s,\{P_{a,b},Q_{a,b}\}}_{t=\Sigma A,\{Q_s,t=\Sigma A\}} s\!:=\!t.$$

The derivation of

$$C_2 \succ^{Q_{a,b}\wedge Q_s,\mathcal{P}(\emptyset)}_{P_{a,b},\{P_{a,b},Q_{a,b},Q_s\}} C_3$$

then is determined by the structure of C_2 and C_3 in a syntax-directed fashion and thus omitted. Note that according to (4) the right parallel subprogram preserves the predicates Q_s and $t = \Sigma A$. More precisely, $\{Q_s, t = \Sigma A\} \subseteq \Delta$.

For the remaining three refinement steps consider Figure 4.

$$\begin{array}{l}
C_4 \equiv \textbf{new } k=1, t=0 \textbf{ in} \\
\quad \left[\begin{array}{l} (\{k \le n \wedge I\};\{k,t\}\!:\![tt,tt])^*; \\ \{t=\Sigma A\}; \\ s\!:=\!t \end{array} \middle\| \; A[a],A[b]\!:=\!A[a]-20, A[b]+20 \right] \\
I \equiv k-1 \le n \wedge t = \Sigma_{i=1}^{k-1} A[i]
\end{array}$$

$$\begin{array}{l}
C_5 \equiv \textbf{new } k=1, t=0 \textbf{ in} \\
\quad \left[\begin{array}{l} \textbf{while } k \le n \textbf{ do} \\ \quad t\!:=\!t+A[k];k\!:=\!k+1 \\ \textbf{od}; \\ s\!:=\!t \end{array} \middle\| \; \{A[a],A[b]\}\!:\![P,Q] \right] \\
P \equiv (k < a \wedge k < b) \vee (k > a \wedge k > b) \\
Q \equiv A[a] = \overleftarrow{A[a]} - 20 \wedge A[b] = \overleftarrow{A[b]} + 20.
\end{array}$$

$$\begin{array}{l}
C_6 \equiv \textbf{new } k=1, t=0 \textbf{ in} \\
\quad \left[\begin{array}{l} \textbf{while } k \le n \textbf{ do} \\ \quad t\!:=\!t+A[k];k\!:=\!k+1 \\ \textbf{od}; \\ s\!:=\!t \end{array} \middle\| \begin{array}{l} \textbf{await } P \textbf{ then} \\ \quad A[a],A[b]\!:=\!A[a]-20, A[b]+20 \\ \textbf{end} \end{array} \right]
\end{array}$$

Fig. 4. Refinements C_4 through C_6

Refining C_3 into C_4 We now equip the loop in C_3 with a termination condition $B \equiv k \le n$ and an invariant $I \equiv k - 1 \le n \wedge t = \Sigma_{i=1}^{k-1} A[i]$. Formally, we show

$$\{k,t\}\!:\![tt,tt]^* \supseteq_{\mathcal{T}} (\{k \le n \wedge I\};\{k,t\}\!:\![tt,tt])^*.$$

Using congruence (Lemma 1.3), we get $C_3 \supseteq_{\mathcal{T}} C_4$ which implies

$$C_2 \succ^{Q_{a,b}\wedge Q_s,\mathcal{P}(\emptyset)}_{P_{a,b},\{P_{a,b},Q_{a,b},Q_s\}} C_4$$

by Lemma 3.3 and the previous refinement.

Refining C_4 into C_5 This refinement step modifies the two parallel subprograms in C_4 simultaneously.
Left subprogram: We want to use rule WHILE-INTRO to replace the Kleene star construct by a **while** loop. This rule requires us to refine the loop body first and prove that I is indeed an invariant of the refined loop. More precisely, we show

$$\begin{aligned} & \{k,t\}\text{:}[tt,tt]^* \\ \supseteq_T\ & \{k,t\}\text{:}[tt,tt]\,;\{k,t\}\text{:}[tt,tt] \\ \succ^{I,\{P_{a,b},Q_{a,b}\}}_{k\leq n\wedge I,\{k\leq n,I\}}\ & t\text{:=}t+A[k]\,;k\text{:=}k+1 \end{aligned}$$

using ATOM and SEQ. Next, we show that $t = \Sigma A$ holds upon termination of the loop, that is, $k > n \wedge I \Rightarrow t = \Sigma A$. Moreover, we need to find an arithmetic expression m that allows us to prove termination of the **while** loop. Let $m \equiv cond(n+1 > k, n+1-k, 0)$ where

$$cond(B,e_1,e_2) = \begin{cases} e_1, & \text{if } B \\ e_2, & \text{otherwise.} \end{cases}$$

We check each of the conditions of rule WHILE-INTRO. Clearly, $m \geq 0$ and $m = 0 \Rightarrow k > n$. Using Lemma 1.2 to deduce trace inclusion for atomic statements and the congruence of $\mathcal{T}$ (Lemma 1.3) it can be shown that

$$(inv^* m\,; a_m\,; inv^* m)^+ \supseteq_T inv\ m\,; a_m \supseteq_T t\text{:=}t+A[k]\,;k\text{:=}k+1.$$

Thus,

$$\begin{array}{lcl} (\{k\leq n\wedge I\};\{k,t\}\text{:}[tt,tt])^*\,; & & \textbf{while } k\leq n \textbf{ do} \\ \{t=\Sigma A\}; & \succ^{Q_s,\{P_{a,b},Q_{a,b}\}}_{I,\Gamma\cup\Gamma_m} & \quad t\text{:=}t+A[k]\,;k\text{:=}k+1 \\ s\text{:=}t & & \textbf{od}; \\ & & s\text{:=}t \end{array}$$

where $\Gamma \equiv \{k \leq n, Q_s, I, t = \Sigma A\}$ and $\Gamma_m \equiv \{m \leq n | n \in \mathbb{N}\}$ using WHILE-INTRO and SEQ.
Right subprogram: The above refinement is subject to the constraints Γ and Γ_m. However, in its current form the right subprogram does not meet these constraints. In particular, it does not preserve the invariant I. The transferred money may be counted twice: once on account a and again on b. The solution is to restrict the transition of the interfering component such that it cannot disturb the computation of the other. This is achieved by postulating that the transition which transfers \$20 from account a to account b preserves the value of $\Sigma_{i=1}^{k-1} A[i]$ and thus the predicate $t = \Sigma_{i=1}^{k-1} A[i]$ for all values of k. Let

$$Q \equiv A[a] = \overleftarrow{A}[a] - 20 \wedge A[b] = \overleftarrow{A}[b] + 20$$

$$R \equiv \Sigma_{i=1}^{k-1} A[i] = \Sigma_{i=1}^{\overleftarrow{k}-1} \overleftarrow{A}[i]\,.$$

Then, by ATOM

$$\begin{aligned} & A[a],A[b]\text{:=}A[a]-20,A[b]+20 \\ \succ^{Q_{a,b},\Delta}_{P_{a,b},\{P_{a,b},Q_{a,b}\}}\ & \{A[a],A[b]\}\text{:}[tt,Q\wedge R] \end{aligned} \qquad (5)$$

where $\Delta \equiv \{Q_s, I, t = \Sigma A\} \cup \mathcal{P}(\{k, t, s\})$. We refine this further by restricting the transfer to states in which either $k < a$ and $k < b$, or $k > a$ and $k > b$. Let $P \equiv (k < a \wedge k < b) \vee (k > a \wedge k > b)$. Then, by ATOM

$$\{A[a], A[b]\}{:}[tt, Q \wedge R] \succ^{Q_{a,b},\Delta}_{P_{a,b},\{P_{a,b},Q_{a,b}\}} \{A[a], A[b]\}{:}[P, Q]. \quad (6)$$

Thus, by (5), (6) and transitivity (Lemma 4)

$$\begin{aligned} & A[a], A[b] := A[a] - 20, A[b] + 20 \\ \succ^{Q_{a,b},\Delta}_{P_{a,b},\{P_{a,b},Q_{a,b}\}} & \{A[a], A[b]\}{:}[P, Q]. \end{aligned}$$

This concludes the refinement of the right parallel component.

Note that the refined right subprogram now meets the constraints placed on it by the left subprogram. That is, $\Gamma \cup \Gamma_m \subseteq \Delta$. Thus, by PAR and NEW we get

$$C_4 \succ^{Q_{a,b} \wedge Q_s, \mathcal{P}(\emptyset)}_{P_{a,b},\{P_{a,b},Q_{a,b},Q_s\}} C_5.$$

Refining C_5 into C_6 This refinement step will replace $\{A[a], A[b]\}{:}[P, Q]$ by

$$\textbf{await } P \textbf{ then } A[a], A[b] := A[a] - 20, A[b] + 20$$

using rule AWAIT-INTRO. Let $m \equiv cond(k > max(a, b), 0, max(a, b) - k + 1)$. Clearly, $m \geq 0$ and $m = 0 \Rightarrow (k < a \wedge k < b) \vee (k > a \wedge k > b)$. The third condition can be shown as follows. First, note that $D =_{\mathcal{T}} D_1 \vee D_2$ where

$$\begin{aligned} D \equiv\ & \textbf{while } k \leq n \textbf{ do} \\ & \quad t := t + A[k]; k := k + 1 \\ & \textbf{od}; \\ & s := t \end{aligned}$$

and

$$D_1 \equiv (\{k \leq n\}; t := t + A[k]; k := k + 1)^{\omega}$$

and

$$D_2 \equiv (\{k \leq n\}; t := t + A[k]; k := k + 1)^{*}; \{k > n\}; s := t.$$

Using Lemma 1.2 to deduce trace inclusion between atomic statements and the congruence of $\mathcal{T}$ (Lemma 1.3) we can show that

$$\begin{aligned} & D_1 \subseteq_{\mathcal{T}} (inv^{*}m; a_m; inv^{*}m)^{\omega} \text{ and} \\ & D_2 \subseteq_{\mathcal{T}} (inv^{*}m; a_m; inv^{*}m)^{*}; \{m = 0\}; inv^{*}m. \end{aligned}$$

Note that $k > n$ implies $k > max(a, b)$ and thus $m = 0$. The third condition follows. Thus,

$$C_5 \succ^{Q_{a,b} \wedge Q_s, \mathcal{P}(\emptyset)}_{P_{a,b},\{P_{a,b},Q_{a,b},Q_s\}} C_6$$

by rule AWAIT-INTRO.

Putting It All Together By transitivity we get

$$C_1 \succ^{Q_{a,b} \wedge Q_s, \mathcal{P}(\emptyset)}_{P_{a,b}, \{P_{a,b}, Q_{a,b}, Q_s\}} C_6.$$

With weakening and Lemma 3.1 this implies the desired result

$$\{P_{a,b}\}; C_1 \sqsupseteq_{\mathcal{E}} \{P_{a,b}\}; C_6 \quad \text{and} \quad \{P_{a,b}\}\ C_6\ \{Q_{a,b} \wedge Q_s\}.$$

5 Conclusion

We have presented a syntax-directed refinement calculus for terminating and non-terminating shared-variable parallel programs. It is based on Brookes' transition trace semantics [3] from which it inherits the support for local variables and fairness. The calculus distinguishes specifications and programs neither syntactically nor semantically. Moreover, it allows for reasoning about both safety and liveness properties like termination and eventual entry. The calculus has been used for a completely rigorous verification of a class of n-process mutual exclusion algorithms which includes the tie-breaker, the bakery and the ticket algorithm [10]. Eventual entry was proved using a slight variant of rule AWAIT-INTRO. The calculus has allowed us to verify alternative, somewhat surprising implementations of some of these mutual exclusion algorithms. Our approach also is applicable to a distributed, message-passing setting in which channels are modeled as variables ranging over infinite queues with asynchronous (non-blocking) send and synchronous (blocking) receive. We have formally derived distributed implementations of the prefix sum algorithm and of an all-pair shortest path algorithm for unweighted graphs.

Related Work This paper defines refinement in terms of context-sensitive approximation on transition traces (Definition 5) which was introduced in [8]. Based on [8], tentative steps towards a refinement calculus for UNITY where taken in [9]. However, the refinement relation employed there is based solely on trace inclusion (context-sensitive approximation) and contains neither program guarantees nor partial correctness behaviour. Moreover, the refinement rules are not syntax-directed and rather ad-hoc. The proof systems in [11, 18] both use rely-guarantee (assumption-commitment) reasoning to achieve compositionality. However, whereas Jones employs logical formulas to specify the behaviour of the program and its environment, Stirling uses sets of predicates (invariants) like we do. The work in [17] augments Jones' work with an explicit notion of refinement. Back's refinement calculus for Action systems [2] also models refinement explicitly. However, his calculus is not syntax-directed but rather a more or less arbitrary collection of program transformation rules. All of the above mentioned approaches differ from ours at least in that they lack support for fairness and liveness properties like eventual entry.

References

[1] G.R. Andrews and F.B. Schneider. Concepts for concurrent programming. In J.W. de Bakker, W.-P. de Roever, and G. Rozenberg, editors, *ESPRIT/LPC Advanced School on Current Trends in Concurrency*, volume 224 of *LNCS*. Springer Verlag, 1985.

[2] R.J.R. Back and K. Sere. Stepwise refinement of action systems. In *Mathematics of Program Construction*. Springer Verlag, 1989. LNCS 375.

[3] S.D. Brookes. Full abstraction for a shared-variable parallel language. *Information and Computation*, 127(2):145–163, June 1996.

[4] M. Broy, F. Dederichs, C. Dendorfer, M. Fuchs, T.F. Gritzner, and R. Weber. The design of distributed systems — an introduction to FOCUS. Technical Report TUM-I9202, TU München, Januar 1992.

[5] K.M. Chandy and J. Misra. *Parallel program design: a foundation*. Addison Wesley, 1988.

[6] P. Collette. *Design of Compositional Proof Systems Based on Assumption-Commitment Specifications — Application to UNITY*. PhD thesis, Université Catholique de Louvain, Belgium, June 1994.

[7] F.S. de Boer, J.N. Kok, C. Palamidessi, and J.J.M.M. Rutten. The failure of failures in a paradigm of asynchronous communication. In *CONCUR '91*, pages 111–126. Springer Verlag, 1991.

[8] J. Dingel. Modular verification for shared-variable concurrent programs. In U. Montanari and V. Sassone, editors, *CONCUR '96*, LNCS 1119, pages 703–718. Springer Verlag, 1996.

[9] J. Dingel. Approximating UNITY. In *Second International Conference on Coordination Models and Languages*, LNCS 1282, pages 320–337. Springer Verlag, September 1997.

[10] J. Dingel. The development and verification of n-process mutual exclusion algorithms using invariants and refinement. Draft, 1998.

[11] C.B. Jones. *Development Methods for Computer Programs Including a Notion of Interference*. PhD thesis, Oxford University, 1981.

[12] L. Lamport. The 'Hoare logic' of concurrent programs. *Acta Informatica*, 14:21–37, 1980.

[13] C. Morgan. The specification statement. *ACM Transactions on Programming Languages and Systems*, 10(3), January 1989.

[14] E.-R. Olderog. *Nets, terms and formulas*. Cambridge Tracts in Theoretical Computer Science 23. Cambridge University Press, 1991.

[15] S.S. Owicki and D. Gries. An axiomatic proof technique for parallel programs. *Acta Informatica*, 6:319–340, 1976.

[16] D. Park. On the semantics of fair parallelism. In D. Bjørner, editor, *Abstract Software Specifications*, LNCS 86. Springer Verlag, 1979.

[17] X. Qiwen and H. Jifeng. A theory of state-based parallel programming: Part I. In J. Morris, editor, *4th BCS-FACS Refinement Workshop*, 1991.

[18] C. Stirling. A generalization of Owicki-Gries' Hoare logic for a concurrent while language. *Theoretical Computer Science*, 89:347–359, 1988.

Consistency of Partial Process Specifications

Maarten Steen, John Derrick, Eerke Boiten, and Howard Bowman

Computing Laboratory, University of Kent at Canterbury
Canterbury, Kent CT2 7NF, UK. M.W.A.Steen@ukc.ac.uk

Abstract. The structuring of the specification and development of distributed systems according to *viewpoints*, as advocated by the Reference Model for Open Distributed Processing, raises the question of when such viewpoint specifications may be considered *consistent* with one another. In this paper, we analyse the notion of consistency in the context of formal process specification. It turns out that different notions of correctness give rise to different consistency relations. Each notion of consistency is formally characterised and placed in a spectrum of consistency relations. An example illustrates the use of these relations for consistency checking.

1 Introduction

There is a growing awareness in distributed software engineering that the development of complex distributed systems can no longer be seen as a linear, top-down activity. It is now widely advocated to structure the specification and development of such systems according to, so called, *viewpoints*. Prominent examples of viewpoint oriented development models are the Reference Model for Open Distributed Processing (RM-ODP) [9], the Viewpoint Oriented Software Engineering (VOSE) framework [5], and object oriented analysis and design models, such as [2].

In contrast with the traditional 'waterfall' model of development, where an initial, abstract specification is stepwise refined to a final, concrete specification, viewpoint models allow specifiers to split up the complete specification of a complex system into a number of viewpoint specifications each concentrating on a particular concern or aspect of the system. Individual viewpoint specifications can then be developed further relatively independent of one another. The RM-ODP, for example, defines five viewpoints — enterprise, information, computational, engineering, and technology — from which distributed systems may be described.

One of the main problems in any multiple viewpoint approach to specification is defining and establishing that the various viewpoint specifications are *consistent* with one another. This problem becomes particularly challenging when we consider that different specification techniques may be applicable to different viewpoints. The ODP information viewpoint, for example, can be expressed quite naturally in Z, whereas LOTOS is considered more suitable for the computational viewpoint [16].

A.M. Haeberer (Ed.): AMAST'98, LNCS 1548, pp. 248–262, 1998.

In some viewpoint models consistency is defined as a simple set of syntactic constraints. The Booch method [2] (supported by the Rational Rose[1] tool) for object oriented design, for example, requires that there is a corresponding operation in a Class Diagram for each message in a Sequence Diagram. Here, however, we are concerned with *behavioural*, or *semantic*, consistency.

In this paper, we analyse the consistency problem for a substantial number of process algebraic specification techniques. Process algebra provides a rich theory for the specification of behaviour. Therefore, this work should provide the formal foundations for consistency checking techniques for more 'user-friendly' behavioural specification notations, such as State Charts and Sequence Diagrams. In fact, the consistency relations identified in this paper are directly applicable to all specification formalisms of which the semantics can be expressed using labelled transition systems, traces, refusals or failures, e.g., CSP [8], CCS [15], and Object-Z [6].

2 Process Specification

We introduce a simple process algebraic language similar to CCS and CSP for the description of process behaviour. The syntax is borrowed from LOTOS [1]:

$$\mathcal{P} ::= \textsf{stop} \mid \alpha; \mathcal{P} \mid \mathcal{P} [] \mathcal{P} \mid \mathcal{P} |[\mathsf{A}]| \mathcal{P} \mid \textsf{hide}\ A\ \textsf{in}\ \mathcal{P} \mid X$$

Here it is assumed that a set of action labels $\mathbf{L}$ is given. Then, $\alpha \in \mathbf{L} \cup \{\tau\}$; $\tau \notin \mathbf{L}$ is the unobservable, or internal, action; $A \subseteq \mathbf{L}$; and X is a process name. We will assume that a definition exists for each process name used. Process definitions are written $X := p$, where p is a behaviour expression that can again contain process names, including possibly X itself, thus making the definition recursive.

Semantically, process behaviour can be modelled in many different ways. In the following, we consider labelled transition systems, traces, refusals and some combinations of the latter two.

2.1 Labelled Transition Systems

Definition 1. *A* labelled transition system *is a structure* $(S, L, \longrightarrow, s_0)$, *where* S *is a set of states,* L *is a set of action labels,* $\longrightarrow\ \subseteq S \times (L \cup \{\tau\}) \times S$ *is a transition relation, and* $s_0 \in S$ *is the initial state.*

Each behaviour description is associated, in the usual manner, with a labelled transition system through the axioms and inference rules given in Table 1.

Often labelled transition systems are considered to be too concrete to abstractly specify system behaviour. It is therefore customary to interpret process specifications via, so called, *implementation relations* [13,3]. These are relations between a domain of implementations and a domain of specifications that formalise a particular notion of correctness. They may, for example, abstract from

[1] Rational Rose is a trade mark of the Rational Software Corporation.

Table 1. Inference rules

	$\vdash \alpha; p \xrightarrow{\alpha} p$
$p \xrightarrow{\alpha} p'$	$\vdash p [] q \xrightarrow{\alpha} p'$
$q \xrightarrow{\alpha} q'$	$\vdash p [] q \xrightarrow{\alpha} q'$
$p \xrightarrow{\alpha} p', \alpha \notin A$	$\vdash p \,\lvert[A]\rvert\, q \xrightarrow{\alpha} p' \,\lvert[A]\rvert\, q$
$q \xrightarrow{\alpha} q', \alpha \notin A$	$\vdash p \,\lvert[A]\rvert\, q \xrightarrow{\alpha} p \,\lvert[A]\rvert\, q'$
$p \xrightarrow{\alpha} p', q \xrightarrow{\alpha} q', \alpha \in A$	$\vdash p \,\lvert[A]\rvert\, q \xrightarrow{\alpha} p' \,\lvert[A]\rvert\, q'$
$p \xrightarrow{\alpha} p', \alpha \notin A$	$\vdash \textbf{hide}\ A\ \textbf{in}\ p \xrightarrow{\alpha} \textbf{hide}\ A\ \textbf{in}\ p'$
$p \xrightarrow{\alpha} p', \alpha \in A$	$\vdash \textbf{hide}\ A\ \textbf{in}\ p \xrightarrow{\tau} \textbf{hide}\ A\ \textbf{in}\ p'$
$p \xrightarrow{\alpha} p', X := p$	$\vdash X \xrightarrow{\alpha} p'$

the internal behaviour of an implementation and only verify whether the externally observable behaviour corresponds to the behaviour described in the specification.

2.2 Traces and Refusals

Let $\mathbf{L}^*$ denote the set of all strings over the set of observable actions $\mathbf{L}$. Elements of $\mathbf{L}^*$ are also called *traces*. The empty string, or empty trace, is denoted ϵ and σ is used to range over $\mathbf{L}^*$. Concatenation of traces is represented by juxtaposition.

In Table 2 the notion of transition is generalised to traces. We further define $Tr(p)$, the set of traces of a process p, $Out(p, \sigma)$, the set of possible actions after the trace σ, and $Ref(p, \sigma)$, the sets of actions refused by a process p after the trace σ:

Definition 2.

$$Tr(p) \stackrel{def}{=} \{\sigma \in \mathbf{L}^* \mid p \stackrel{\sigma}{\Longrightarrow}\}$$
$$Out(p, \sigma) \stackrel{def}{=} \{a \in \mathbf{L} \mid \exists p' \bullet p \stackrel{\sigma}{\Longrightarrow} p' \text{ and } p' \stackrel{a}{\Longrightarrow}\}$$
$$Ref(p, \sigma) \stackrel{def}{=} \{X \subseteq \mathbf{L} \mid \exists p' \bullet p \stackrel{\sigma}{\Longrightarrow} p' \text{ and } \forall a \in X \bullet p' \stackrel{a}{\not\Longrightarrow}\}$$

Table 2. Trace relations

Notation	Meaning
$\stackrel{\epsilon}{\Longrightarrow}$	$(\xrightarrow{\tau})^*$, i.e., the reflexive and transitive closure of $\xrightarrow{\tau}$
$p \stackrel{a\sigma}{\Longrightarrow} p'$	$\exists q, q' \bullet p \stackrel{\epsilon}{\Longrightarrow} q \xrightarrow{a} q' \stackrel{\sigma}{\Longrightarrow} p'$
$p \stackrel{\sigma}{\Longrightarrow}$	$\exists p' \bullet p \stackrel{\sigma}{\Longrightarrow} p'$
$p \stackrel{\sigma}{\not\Longrightarrow}$	$\nexists p' \bullet p \stackrel{\sigma}{\Longrightarrow} p'$

2.3 Implementation Relations

A large number of implementation relations has been defined over labelled transition systems [7]; each one capturing a different notion of correctness. In this paper, we consider only the most prominent trace and/or refusal based implementation relations from process algebra. Our selection is largely based on a pioneering study on implementation relations by Brinksma et al. [3].

Definition 3. *Let $p, s \in \mathcal{P}$ be processes, then we define the following relations:*

name	*denotation*	*definition*
trace refinement	$p \leq_{\text{tr}} s$	$Tr(p) \subseteq Tr(s)$
trace equivalence	$p \approx_{\text{tr}} s$	$Tr(p) = Tr(s)$
conformance	$p\,\mathbf{conf}\,s$	$\forall \sigma \in Tr(s) \bullet Ref(p,\sigma) \subseteq Ref(s,\sigma)$
reduction	$p\,\mathbf{red}\,s$	$p \leq_{\text{tr}} s$ *and* $p\,\mathbf{conf}\,s$
extension	$p\,\mathbf{ext}\,s$	$s \leq_{\text{tr}} p$ *and* $p\,\mathbf{conf}\,s$
testing equivalence	$p \approx_{\text{te}} s$	$p\,\mathbf{red}\,s$ *and* $s\,\mathbf{red}\,p$

Perhaps the simplest implementation relation is *trace refinement*. It only verifies that the implementation cannot perform sequences of observable actions (traces) that are not allowed by the specification. This is useful for capturing, so called, *safety properties*. However, we cannot use it to specify that anything *must* happen. *Trace equivalence* is slightly stronger in that it requires that the implementation and specification have the same possible traces. Another notion of validity is captured by the *conformance* relation (**conf**), derived from testing theory. It requires for each trace of the specification, that the implementation can only refuse to do whatever the specification refuses after that trace. The *reduction* relation (**red**), sometimes referred to as testing preorder or failure preorder, is the intersection of trace refinement and conformance. It gives rise to a specification technique with which one can specify both that certain actions must happen and that certain traces are not allowed. The *extension* relation, on the other hand, allows that more traces are added in the implementation, as long as the implementation is still conformant to its specification. The strongest implementation relation considered here is *testing equivalence*. It requires that the observable behaviour of implementation resp. specification cannot be distinguished through external testing.

Process specifications, and in fact any other trace/refusal based specifications, can be interpreted under any of the implementation relations defined above to yield a different specification formalism [10] for system behaviour. In a multiple viewpoint approach to specification potentially all these formalisms may be used simultaneously. Below, we show how different viewpoints may require different implementation relations to adequately capture their intended meaning.

2.4 Example Viewpoint Specifications

Consider the specification of a simple vending machine using the ODP viewpoints. (It is outside the scope of this paper to give definitions for the five ODP viewpoints. The interested reader is referred to [14] or the standard itself [9].)

From the **enterprise viewpoint** one might like to specify the following policies, divided in permissions and obligations:

Permissions The system is permitted to exhibit any of the following traces of behaviour: {ϵ, coin, coin.coffee, coin.tea, coin.coffee.coin, coin.tea.coin, ...}. This could be captured by the following specification, when interpreted under the trace refinement relation ($\leq_{tr}$):

Perm := coin; (coffee; Perm [] tea; Perm)

Obligations The system user is obliged to always first insert a coin into the machine. The following specification captures this. Here we have decided to interpret the specification under the extension relation (**ext**), so the specification does not prohibit any other behaviour.

Obl := coin; **stop**

From the **computational viewpoint** the system is viewed as a computational object providing a computational interface upon which its environment (the user) can invoke one of three operations: coin, coffee and tea.

Comp := τ; coin; (τ; coffee; Comp [] τ; tea; Comp)
[] τ; coffee; Comp
[] τ; tea; Comp

If the coin operation is invoked, the system will respond by offering its environment either coffee or tea. In case one of the other two operations is invoked by the environment, the system will return to its initial state. Non-determinism is used to indicate that not all of these operations need to be present in an implementation. Therefore, any reduction (**red**) is considered a correct implementation.

From the **engineering viewpoint** the system might be viewed as being composed of two components, a money handler (MH) and a drinks dispenser (DD), that communicate via a channel. As the channel is only introduced for internal communication it is hidden from the environment. The following specification of the engineering viewpoint is interpreted under the testing equivalence relation ($\approx_{te}$).

Eng := **hide** channel **in** MH |[channel]| DD
MH := coin; channel; MH
DD := channel; (coffee; DD [] tea; DD)

The obvious question now is whether all these viewpoint specifications are consistent with one another.

3 Consistency

The purpose of this section is to define (necessary and sufficient) conditions for viewpoint specifications to be consistent. For the moment we will concentrate on *binary consistency*, i.e., consistency between two specifications. Informally, we call two specifications consistent if, and only if, they have at least one implementation in common, i.e., if there is an implementation that satisfies both specifications. The definition of consistency is thus parameterised on the notion of correctness that each specification is subjected to. As we have shown above, different viewpoint specifications may be subjected to interpretation under differing implementation relations. Therefore, each combination of implementation relations, $\mathbf{imp}_1, \mathbf{imp}_2$, gives rise to a different consistency relation, denoted $\mathcal{C}_{\mathbf{imp}_1,\mathbf{imp}_2}$.

Definition 4. *Let* $\mathbf{imp}_1, \mathbf{imp}_2$ *be implementation relations, then consistency between specifications subject to* $\mathbf{imp}_1$ *and specifications subject to* $\mathbf{imp}_2$ *is a binary relation* $\mathcal{C}_{\mathbf{imp}_1,\mathbf{imp}_2}$ *such that, for any* $s_1, s_2 \in \mathcal{P}$,

$$s_1\, \mathcal{C}_{\mathbf{imp}_1,\mathbf{imp}_2}\, s_2 \overset{def}{\Longleftrightarrow} \exists p \in \mathcal{P} \bullet p\, \mathbf{imp}_1\, s_1 \wedge p\, \mathbf{imp}_2\, s_2.$$

Considering $\leq_{\mathrm{tr}}$, $\approx_{\mathrm{tr}}$, **conf**, **red**, **ext** and $\approx_{\mathrm{te}}$ as instantiations for $\mathbf{imp}_1$ and $\mathbf{imp}_2$ in the definition of binary consistency, we obtain 36 different notions of consistency. Whenever $\mathbf{imp}_1 = \mathbf{imp}_2$, we speak of *balanced consistency*, denoted $\mathcal{C}^2_{\mathbf{imp}}$. Section 3.1 deals with these (six) cases. The issue of *unbalanced consistency*, the remaining 30 cases, is discussed in section 3.2. Omitted proofs may be found in [17].

It is useful sometimes to use the following alternative characterisation of consistency as the composition of two implementation relations:

Proposition 5. *For any two implementation relations* $\mathbf{imp}_1, \mathbf{imp}_2$,

$$\mathcal{C}_{\mathbf{imp}_1,\mathbf{imp}_2} = \mathbf{imp}_1^{-1} \circ \mathbf{imp}_2\,.$$

3.1 Balanced Consistency

This section largely summarises results from [18], where we considered only the balanced consistency problem.

Since both specifications (in the binary case) are subject to the same implementation relation, binary, balanced consistency is a *symmetric* relation.

Proposition 6. *For any implementation relation* $\mathbf{imp}$, $\mathcal{C}^2_{\mathbf{imp}} = (\mathcal{C}^2_{\mathbf{imp}})^{-1}$.

We consider the six cases of binary, balanced consistency, denoted $\mathcal{C}^2_{\mathbf{imp}}$ for $\mathbf{imp} \in \{\leq_{\mathrm{tr}}, \approx_{\mathrm{tr}}, \mathbf{conf}, \mathbf{red}, \mathbf{ext}, \approx_{\mathrm{te}}\}$. For two of these, $\mathbf{imp}$ is instantiated with an equivalence relation. It is easily established that the consistency relation is equal to the implementation relation in those cases. Of the four remaining balanced consistency relations, three turn out to hold for any two specifications.

Theorem 7.

1. $\mathcal{C}^2_{\approx_{\mathrm{tr}}} = \approx_{\mathrm{tr}}$
2. $\mathcal{C}^2_{\approx_{\mathrm{te}}} = \approx_{\mathrm{te}}$

3. $C^2_{\leq_{tr}} = \mathcal{P} \times \mathcal{P}$
4. $C^2_{\mathbf{conf}} = \mathcal{P} \times \mathcal{P}$
5. $C^2_{\mathbf{ext}} = \mathcal{P} \times \mathcal{P}$

Proof. The first two results follow from the symmetry and transitivity of $\approx_{tr}$ and $\approx_{te}$. The remaining cases are proved by exhibiting a bottom element in the respective refinement lattices. Such a bottom element is presented by a process $\bot$ such that $\forall s \bullet \bot\, \mathbf{imp}\, s$. The existence of such a bottom element implies consistency, since $s_1\, C^2_{\mathbf{imp}}\, s_2 \Leftrightarrow \exists p \bullet p\, \mathbf{imp}\, s_1 \wedge p\, \mathbf{imp}\, s_2$.

3. $\forall s \bullet$ **stop** $\leq_{tr} s$, hence **stop** is the required bottom element.
4. Define a process Run, that can perform all possible traces and never refuses any action, as follows[2]:

   $\mathsf{Run} := \Sigma\{\mathsf{a};\, \mathsf{Run} \mid \mathsf{a} \in \mathbf{L}\}$

   Observe that, $\forall \sigma \in \mathbf{L}^* \bullet Ref(\mathsf{Run}, \sigma) = \{\emptyset\}$. Therefore, $\forall s \bullet \mathsf{Run}\, \mathbf{conf}\, s$.
5. The process Run, defined above, also has more traces than any other process, i.e. $\forall s \bullet Tr(\mathsf{Run}) = \mathbf{L}^* \supseteq Tr(s)$. Therefore, $\forall s \bullet \mathsf{Run}\, \mathbf{ext}\, s$. □

The following theorem gives a sufficient condition for two specifications (say s_1 and s_2) to be consistent with respect to reduction. The condition requires that s_1 and s_2 can at least refuse all the actions they may not both do after a certain trace.

Theorem 8. *Let $s_1, s_2 \in \mathcal{P}$ be two specifications, then $s_1\, C^2_{\mathbf{red}}\, s_2$ if:*

$$\forall \sigma \in Tr(s_1) \cap Tr(s_2) \bullet \mathbf{L} \backslash (Out(s_1, \sigma) \cap Out(s_2, \sigma)) \in Ref(s_1, \sigma) \cap Ref(s_2, \sigma)$$

Proof. See [18].

3.2 Unbalanced Consistency

Unbalanced consistency is more complicated than the balanced case. First of all, there are many more cases of unbalanced consistency. Moreover, unlike balanced consistency relations, unbalanced ones are not symmetric. However, there is a close relationship between $C_{\mathbf{imp}_1,\mathbf{imp}_2}$ and $C_{\mathbf{imp}_2,\mathbf{imp}_1}$.

Proposition 9. *For any two implementation relations* $\mathbf{imp}_1, \mathbf{imp}_2$,

$$C_{\mathbf{imp}_2,\mathbf{imp}_1} = C^{-1}_{\mathbf{imp}_1,\mathbf{imp}_2}.$$

Since it is easy to derive the inverse of a relation (just swap the arguments), this proposition gives an easy recipe for deriving $C_{\mathbf{imp}_2,\mathbf{imp}_1}$ from the relation with the implementation relations reversed $C_{\mathbf{imp}_1,\mathbf{imp}_2}$. It halves our problem of finding 30 consistency conditions.

For the remaining 15 cases, observe that all implementation relations are *reflexive*. The following proposition therefore allows us to derive at least a sufficient condition for consistency to hold in each of these cases.

[2] The operator Σ generalises the choice operator ($\_ \,[\,]\, \_$).

Proposition 10. *Given a consistency relation* $\mathcal{C}_{\mathbf{imp}_1,\mathbf{imp}_2}$*, such that* $\mathbf{imp}_1$ *is reflexive,*

$$\mathbf{imp}_2 \subseteq \mathcal{C}_{\mathbf{imp}_1,\mathbf{imp}_2} .$$

Proof. From reflexivity of $\mathbf{imp}_1$, it follows that $\mathbf{Id} \subseteq \mathbf{imp}_1^{-1}$. And, by monotonicity of $\circ$, $\mathbf{Id} \subseteq \mathbf{imp}_1^{-1} \Rightarrow \mathbf{imp}_2 \subseteq \mathbf{imp}_1^{-1} \circ \mathbf{imp}_2 = \mathcal{C}_{\mathbf{imp}_1,\mathbf{imp}_2}$. □

Under the condition that the inverse of $\mathbf{imp}_1$ is stronger than $\mathbf{imp}_2$ and $\mathbf{imp}_2$ is a transitive relation, $\mathbf{imp}_2$ is both a necessary and sufficient condition. This result applies to six of the remaining cases.

Theorem 11. *Given a consistency relation* $\mathcal{C}_{\mathbf{imp}_1,\mathbf{imp}_2}$*, such that*

- $\mathbf{imp}_1$ *is reflexive,*
- $\mathbf{imp}_2$ *is transitive, and*
- $\mathbf{imp}_1^{-1} \subseteq \mathbf{imp}_2$,

then $\mathcal{C}_{\mathbf{imp}_1,\mathbf{imp}_2} = \mathbf{imp}_2$.

Proof. By Prop. 10, we have $\mathbf{imp}_2 \subseteq \mathcal{C}_{\mathbf{imp}_1,\mathbf{imp}_2}$. In the other direction, we derive by monotonicity of $\circ$ and transitivity of $\mathbf{imp}_2$, that $\mathbf{imp}_1^{-1} \subseteq \mathbf{imp}_2 \Rightarrow \mathbf{imp}_1^{-1} \circ \mathbf{imp}_2 \subseteq \mathbf{imp}_2 \circ \mathbf{imp}_2 \subseteq \mathbf{imp}_2$. □

Corollary 12.

1. $\mathcal{C}_{\approx_{\mathrm{te}},\leq_{\mathrm{tr}}} = \leq_{\mathrm{tr}}$
2. $\mathcal{C}_{\approx_{\mathrm{te}},\approx_{\mathrm{tr}}} = \approx_{\mathrm{tr}}$
3. $\mathcal{C}_{\approx_{\mathrm{te}},\mathbf{red}} = \mathbf{red}$
4. $\mathcal{C}_{\approx_{\mathrm{te}},\mathbf{ext}} = \mathbf{ext}$
5. $\mathcal{C}_{\approx_{\mathrm{tr}},\leq_{\mathrm{tr}}} = \leq_{\mathrm{tr}}$
6. $\mathcal{C}_{\mathbf{ext},\leq_{\mathrm{tr}}} = \leq_{\mathrm{tr}}$

Since testing equivalence is stronger than all other implementation relations, and because it is an equivalence, we almost always have $\mathcal{C}_{\approx_{\mathrm{te}},\mathbf{imp}_2} = \mathbf{imp}_2$. The only case that is missing, is when $\mathbf{imp}_2 = \mathbf{conf}$. Even though $\mathbf{conf}$ is not transitive, we still have the same result.

Theorem 13. $\mathcal{C}_{\approx_{\mathrm{te}},\mathbf{conf}} = \mathbf{conf}$

Proof. By Prop. 10 we have $\mathbf{conf} \subseteq \mathcal{C}_{\approx_{\mathrm{te}},\mathbf{conf}}$. For inclusion in the other direction, observe that, by Prop. 5 and symmetry of $\approx_{\mathrm{te}}$, $\mathcal{C}_{\approx_{\mathrm{te}},\mathbf{conf}} = \approx_{\mathrm{te}} \circ \mathbf{conf}$. We now prove $\approx_{\mathrm{te}} \circ \mathbf{conf} \subseteq \mathbf{conf}$ by extensionality: $\forall s_1, s_2 \in \mathcal{P}$,

$$\begin{aligned}
& s_1 \approx_{\mathrm{te}} \circ \,\mathbf{conf}\, s_2 \\
\Leftrightarrow\ & \exists p \bullet s_1 \approx_{\mathrm{te}} p \wedge p\, \mathbf{conf}\, s_2 \\
\Leftrightarrow\ & \exists p \bullet (\forall \sigma \in \mathbf{L}^* \bullet \mathit{Ref}(s_1,\sigma) = \mathit{Ref}(p,\sigma)) \\
& \quad \wedge (\forall \sigma \in \mathit{Tr}(s_2) \bullet \mathit{Ref}(p,\sigma) \subseteq \mathit{Ref}(s_2,\sigma)) \\
\Rightarrow\ & \forall \sigma \in \mathit{Tr}(s_2) \bullet \mathit{Ref}(s_1,\sigma) \subseteq \mathit{Ref}(s_2,\sigma) \\
\Leftrightarrow\ & s_1\, \mathbf{conf}\, s_2
\end{aligned}$$

□

Of the remaining consistency relations, one holds for any two specifications.

Theorem 14. $\mathcal{C}_{\mathbf{ext},\mathbf{conf}} = \mathcal{P} \times \mathcal{P}$

Proof. Use the same witness as in the proofs of $\mathcal{C}^2_{\mathbf{conf}} = \mathcal{C}^2_{\mathbf{ext}} = \mathcal{P} \times \mathcal{P}$. □

The remaining two consistency relations with **ext** coincide with trace refinement.

Theorem 15.

1. $\mathcal{C}_{\approx_{\mathrm{tr}},\mathbf{ext}} = \geq_{\mathrm{tr}}$
2. $\mathcal{C}_{\mathbf{red},\mathbf{ext}} = \geq_{\mathrm{tr}}$

Proof. In one direction, inclusion follows by a simple monotonicity argument:

1. Since $\mathbf{ext} \subseteq \geq_{\mathrm{tr}}$, it follows that $\mathcal{C}_{\approx_{\mathrm{tr}},\mathbf{ext}} = \approx_{\mathrm{tr}} \circ\, \mathbf{ext} \subseteq \approx_{\mathrm{tr}} \circ \geq_{\mathrm{tr}} = \geq_{\mathrm{tr}}$.
2. Since $\mathbf{red}^{-1} \subseteq \geq_{\mathrm{tr}}$ and $\mathbf{ext} \subseteq \geq_{\mathrm{tr}}$, it follows that $\mathcal{C}_{\mathbf{red},\mathbf{ext}} = \mathbf{red}^{-1} \circ \mathbf{ext} \subseteq \geq_{\mathrm{tr}} \circ \geq_{\mathrm{tr}} = \geq_{\mathrm{tr}}$.

In the other direction, we need to exhibit a common implementation for any two specifications s_1, s_2 such that $s_1 \geq_{\mathrm{tr}} s_2$. In both cases, such a common implementation is given by the deterministic process with the same traces as s_1. □

In an earlier version of this paper, we defined a relation **cons** $\subseteq \mathcal{P} \times \mathcal{P}$ at this point (see definition 18) and proposed that being in this relation provided a sufficient and necessary condition for four of the remaining consistency relations, viz. $\mathcal{C}_{\leq_{\mathrm{tr}},\mathbf{conf}}$, $\mathcal{C}_{\leq_{\mathrm{tr}},\mathbf{red}}$, $\mathcal{C}_{\mathbf{red},\mathbf{conf}}$, and $\mathcal{C}_{\approx_{\mathrm{tr}},\mathbf{conf}}$. However, we now know this not to be the case. Although **cons** is indeed a precise characterisation of $\mathcal{C}_{\approx_{\mathrm{tr}},\mathbf{conf}}$ (see theorem 19) and it plays a role in the characterisation of $\mathcal{C}_{\approx_{\mathrm{tr}},\mathbf{red}}$ (see theorem 20), $\mathcal{C}_{\approx_{\mathrm{tr}},\mathbf{conf}}$ does not coincide with the other three aforementioned consistency relations. We can, however, establish a relative ordering between the four relations.

Proposition 16.

1. $\mathcal{C}_{\leq_{\mathrm{tr}},\mathbf{red}} = \mathcal{C}_{\leq_{\mathrm{tr}},\mathbf{conf}}$
2. $\mathcal{C}_{\approx_{\mathrm{tr}},\mathbf{conf}} \subset \mathcal{C}_{\leq_{\mathrm{tr}},\mathbf{conf}}$
3. $\mathcal{C}_{\mathbf{red},\mathbf{conf}} \subset \mathcal{C}_{\leq_{\mathrm{tr}},\mathbf{conf}}$

Proof.

1. Firstly, since $\mathbf{red} \subseteq \mathbf{conf}$, it follows that $\mathcal{C}_{\leq_{\mathrm{tr}},\mathbf{red}} = \geq_{\mathrm{tr}} \circ\, \mathbf{red} \subseteq \geq_{\mathrm{tr}} \circ\, \mathbf{conf} = \mathcal{C}_{\leq_{\mathrm{tr}},\mathbf{conf}}$. Secondly, suppose $\exists p \bullet p \leq_{\mathrm{tr}} s_1 \wedge p\, \mathbf{conf}\, s_2$, but $p \not\leq_{\mathrm{tr}} s_2$. There must then be a $\sigma \in \mathit{Tr}(p) \cap \mathit{Tr}(s_2)$ such that $a \in \mathit{Out}(p,\sigma) \setminus \mathit{Out}(s_2,\sigma)$ for some $a \in \mathbf{L}$. However, then $\{a\} \in \mathit{Ref}(s_2,\sigma)$ so we can remove the a-transition from p without invalidating that $p \leq_{\mathrm{tr}} s_1$ and $p\, \mathbf{conf}\, s_2$. Now, let p' be the process constructed from p by removing all these violating transition and we clearly have $p' \leq_{\mathrm{tr}} s_1$ and $p'\, \mathbf{red}\, s_2$.
2. Since $\approx_{\mathrm{tr}} \subseteq \geq_{\mathrm{tr}}$, it follows that $\mathcal{C}_{\approx_{\mathrm{tr}},\mathbf{conf}} = \approx_{\mathrm{tr}} \circ\, \mathbf{conf} \subseteq \geq_{\mathrm{tr}} \circ\, \mathbf{conf} = \mathcal{C}_{\leq_{\mathrm{tr}},\mathbf{conf}}$. Moreover, there exist specifications s_1, s_2 such that $s_1\, \mathcal{C}_{\leq_{\mathrm{tr}},\mathbf{conf}}\, s_2$, but $\neg(s_1\, \mathcal{C}_{\approx_{\mathrm{tr}},\mathbf{conf}}\, s_2)$ (see example 17).

3. Since $\mathbf{red}^{-1} \subseteq \geq_{\mathbf{tr}}$, it follows that $\mathcal{C}_{\mathbf{red},\mathbf{conf}} = \mathbf{red}^{-1} \circ \mathbf{conf} \subseteq \geq_{\mathbf{tr}} \circ \mathbf{conf} = \mathcal{C}_{\leq_{\mathbf{tr}},\mathbf{conf}}$. Moreover, there exist specifications s_1, s_2 such that $s_1 \, \mathcal{C}_{\leq_{\mathbf{tr}},\mathbf{conf}} \, s_2$, but $\neg(s_1 \, \mathcal{C}_{\mathbf{red},\mathbf{conf}} \, s_2)$ (see example 17). □

Example 17. Consider the following specifications:

$$s_1 := \mathsf{a};\ \mathbf{stop} \ [] \ \mathsf{b};\ \mathbf{stop}$$
$$s_2 := \tau;\ \mathsf{a};\ \mathbf{stop} \ [] \ \mathsf{b};\ \mathsf{c};\ \mathbf{stop}$$

then we have $s_1 \, \mathcal{C}_{\leq_{\mathbf{tr}},\mathbf{conf}} \, s_2$, because a; **stop** is a common implementation, but not $s_1 \, \mathcal{C}_{\approx_{\mathbf{tr}},\mathbf{conf}} \, s_2$ and not $s_1 \, \mathcal{C}_{\mathbf{red},\mathbf{conf}} \, s_2$. In the latter two cases, any common implementation would have to perform b initially and then refuse c to be an implementation of s_1, but such a process can never be conformant to s_2, which requires c after b. □

Definition 18. *Define a relation* **cons** $\subseteq \mathcal{P} \times \mathcal{P}$ *as follows:*

$$p \, \mathbf{cons} \, q \overset{def}{\iff} \forall \sigma \in Tr(p) \cap Tr(q) \bullet (\mathbf{L} \setminus Out(p, \sigma)) \in Ref(q, \sigma).$$

The relation **cons** characterises $\mathcal{C}_{\approx_{\mathbf{tr}},\mathbf{conf}}$, as is shown in the following theorem. In order for a process p to be '*trace-conf consistent*' with a process q, q must be able to refuse everything that p cannot do after a certain trace σ common to both p and q.

Theorem 19. $\mathcal{C}_{\approx_{\mathbf{tr}},\mathbf{conf}} = \mathbf{cons}$

Proof. Firstly from left to right. Assuming that $\exists p \bullet p \approx_{\mathbf{tr}} s_1 \ \wedge \ p \, \mathbf{conf} \, s_2$ we need to show that $s_1 \, \mathbf{cons} \, s_2$. Suppose not. By definition of **cons** this means that $\mathbf{L} \setminus Out(s_1, \sigma) \notin Ref(s_2, \sigma)$ for some trace $\sigma \in Tr(s_1) \cap Tr(s_2)$. From the assumption that $p \approx_{\mathbf{tr}} s_1$ it follows that $Out(p, \sigma) = Out(s_1, \sigma)$ and therefore that $\mathbf{L} \setminus Out(p, \sigma) \notin Ref(s_2, \sigma)$. However, for p to be a valid process (e.g., see [12, p. 62]), we must have $\mathbf{L} \setminus Out(p, \sigma) \in Ref(p, \sigma)$, which contradicts that $p \, \mathbf{conf} \, s_2$.

Secondly, from right to left. Assume $s_1 \, \mathbf{cons} \, s_2$. Next, construct a process p with the following traces and refusals:

$$Tr(p) = Tr(s_1)$$
$$Ref(p, \sigma) = Ref(s_2, \sigma), \text{ if } \sigma \in Tr(s_2)$$
$$Ref(p, \sigma) = \wp(\mathbf{L} \setminus Out(p, \sigma)), \text{ if } \sigma \in Tr(p) \setminus Tr(s_2)$$

It immediately follows that $p \approx_{\mathbf{tr}} s_1$ and $p \, \mathbf{conf} \, s_2$. However, it still needs to be verified that the combination of traces and refusals satisfy certain properties in order for them to define a valid process (e.g., see [12, p. 62]). Most of these properties follow trivially from the given definitions, but the following may require some formal justification:

$$X \in Ref(p, \sigma) \Rightarrow X \cup (L \setminus Out(p, \sigma)) \in Ref(p, \sigma)$$

By contradiction: suppose $X \cup (L \setminus Out(p,\sigma)) \notin Ref(p,\sigma)$ for some $\sigma \in Tr(p)$ such that $X \in Ref(p,\sigma)$. If $\sigma \notin Tr(s_2)$, then we have a straightforward contradiction, because then $\mathbf{L} \setminus Out(p,\sigma) \in Ref(p,\sigma)$ by definition. Otherwise, there must be some $a \in \mathbf{L} \setminus Out(p,\sigma)$ such that $\{a\} \notin Ref(p,\sigma)$, since $X \in Ref(p,\sigma)$. From the fact that $Tr(p) = Tr(s_1)$, we also know that $a \notin Out(s_1,\sigma)$. However, then it follows by $s_1 \, \mathbf{cons} \, s_2$, that $\{a\} \in Ref(s_2,\sigma)$, which contradicts that $\{a\} \notin Ref(p,\sigma)$, because $Ref(p,\sigma) = Ref(s_2,\sigma)$ by definition. □

Theorem 20. $\mathcal{C}_{\approx_{tr},\mathbf{red}} = \leq_{tr} \cap \, \mathbf{cons}$

Proof. In one direction, a simple calculation suffices:

$$\begin{aligned} \mathcal{C}_{\approx_{tr},\mathbf{red}} &= \approx_{tr} \circ \, \mathbf{red} = \approx_{tr} \circ (\leq_{tr} \cap \, \mathbf{conf}) \\ &\subseteq (\approx_{tr} \circ \leq_{tr}) \cap (\approx_{tr} \circ \, \mathbf{conf}) = \leq_{tr} \cap \, \mathbf{cons} \end{aligned}$$

In the other direction, assume $s_1 \leq_{tr} s_2$ and $s_1 \, \mathbf{cons} \, s_2$ for some s_1, s_2. By $s_1 \, \mathbf{cons} \, s_2$, we have $p \approx_{tr} s_1$ and $p \, \mathbf{conf} \, s_2$ for some p. By $s_1 \leq_{tr} s_2$, it then follows that $p \leq_{tr} s_2$ and therefore that $p \, \mathbf{red} \, s_2$. □

3.3 Summary of Consistency Results

By instantiating the general definition of binary consistency with the implementation relations defined in section 2.3, 36 different notions of consistency were obtained. For most of these notion of consistency a necessary and sufficient condition has been derived, in the form of a characterising relation, under which two specifications can be considered consistent. Eventhough we did not yet find such characterising relations for $\mathcal{C}_{\leq_{tr},\mathbf{conf}} = \mathcal{C}_{\leq_{tr},\mathbf{red}}$ and $\mathcal{C}_{\mathbf{red},\mathbf{conf}}$, we conjecture that they exist nevertheless. In the following we denote these two unknown relations $\mathbf{cs}_1$ and $\mathbf{cs}_2$, resp. The obtained results are summarised in Table 3.

In order to verify the consistency of two specifications s_1, s_2 interpreted via implementation relations $\mathbf{imp}_1, \mathbf{imp}_2$, respectively, look up the relation in the row labelled by $\mathbf{imp}_1$ and the column labelled by $\mathbf{imp}_2$. Say this is a relation C. Now, if $s_1 \, C \, s_2$, then $s_1 \, \mathcal{C}_{\mathbf{imp}_1,\mathbf{imp}_2} \, s_2$ holds.

Table 3. Consistency conditions

	$\leq_{tr}$	$\approx_{tr}$	**conf**	**red**	**ext**	$\approx_{te}$
$\leq_{tr}$	$\mathcal{P} \times \mathcal{P}$	$\geq_{tr}$	$\mathbf{cs}_1$	$\mathbf{cs}_1$	$\geq_{tr}$	$\geq_{tr}$
$\approx_{tr}$	$\leq_{tr}$	$\approx_{tr}$	**cons**	$\leq_{tr} \cap \, \mathbf{cons}$	$\geq_{tr}$	$\approx_{tr}$
conf	$\mathbf{cs}_1^{-1}$	$\mathbf{cons}^{-1}$	$\mathcal{P} \times \mathcal{P}$	$\mathbf{cs}_2^{-1}$	$\mathcal{P} \times \mathcal{P}$	$\mathbf{conf}^{-1}$
red	$\mathbf{cs}_1^{-1}$	$\geq_{tr} \cap \, \mathbf{cons}^{-1}$	$\mathbf{cs}_2$	$\mathcal{C}^2_{\mathbf{red}}$	$\geq_{tr}$	$\mathbf{red}^{-1}$
ext	$\leq_{tr}$	$\leq_{tr}$	$\mathcal{P} \times \mathcal{P}$	$\leq_{tr}$	$\mathcal{P} \times \mathcal{P}$	$\mathbf{ext}^{-1}$
$\approx_{te}$	$\leq_{tr}$	$\approx_{tr}$	**conf**	**red**	**ext**	$\approx_{te}$

Fig. 1 relates the consistency relations in terms of their relative strength. The strongest consistency relation ($\approx_{te}$) can be found at the bottom of the spectrum;

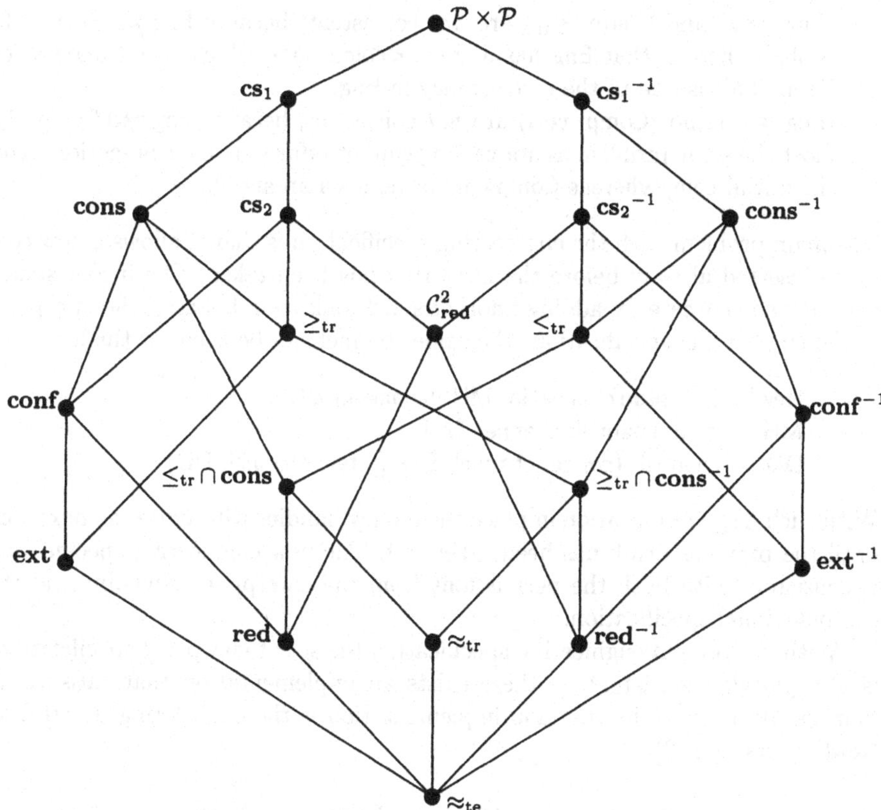

Fig. 1. The spectrum of consistency relations

the weakest relation ($\mathcal{P} \times \mathcal{P}$) at the top. A line between two relations indicates that the lower one is included in the higher one. It is always sufficient to verify a strictly stronger relation rather than the required notion of consistency.

The relationships depicted in the bottom half of Fig. 1 are mostly well-known results from the literature [13,12,3]. The other relationships between consistency relations usually follow from a straightforward monotonicity argument as in Prop. 16 or directly from the definitions.

3.4 Consistency Checking Example

Using the results obtained above, we can now verify the pair-wise consistency of the specifications in Sect. 2.4:

- (Obl, **ext**) and (Perm, $\leq_{tr}$) are consistent, because Obl $\leq_{tr}$ Perm.
- (Obl, **ext**) and (Comp, **red**) are consistent, because Obl $\leq_{tr}$ Comp.
- (Eng, $\approx_{te}$) and (Obl, **ext**) are consistent, because Eng **ext** Obl.
- (Perm, $\leq_{tr}$) and (Comp, **red**) are consistent, because Perm **red** Comp, which is a sufficient condition for consistency by Prop. 10.

- (Eng, $\approx_{te}$) and (Perm, $\leq_{tr}$) are *not* consistent, because Eng $\not\leq_{tr}$ Perm. The problem here is that Eng has a trace <coin.coin>, which is not allowed by Perm. This is due to the concurrency in Eng.
- (Eng, $\approx_{te}$) and (Comp, **red**) are *not* consistent, because Eng r̸e̸d Comp. Almost the same problem as above. Eng cannot refuse to do a coin-action, after the initial coin, whereas Comp cannot do such an action.

The main problem with the engineering specification is that it allows a new coin to be inserted already before the last drink has been taken. The inconsistency can be resolved here by adding another synchronisation between the two parts of the engineering specification (the same channel can be used for this):

```
NewEng := hide channel in  MH |[channel]| DD
MH := coin; channel; channel; MH
DD := channel; (coffee; channel; DD [] tea; channel; DD)
```

With such a synchronisation in place the money handler will refuse the next coin until the previous drink has been taken out. The new engineering specification is consistent with both the permissions from the enterprise viewpoint and the computational specification.

With the revised engineering specification the set of viewpoint specifications is also *globally* consistent — there exists an implementation that satisfies all four specifications. The common implementation is the engineering description NewEng (see Fig. 2).

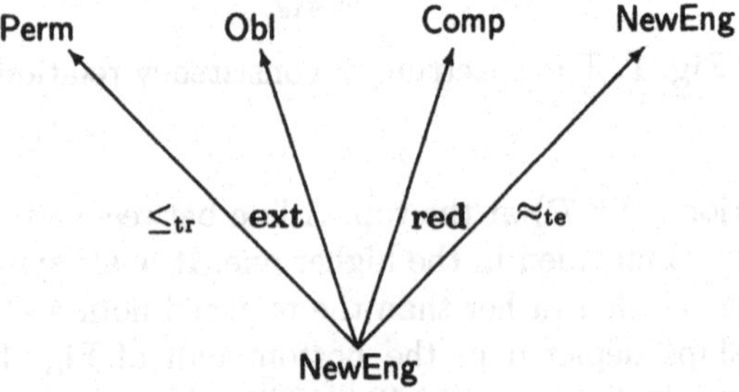

Fig. 2. Global consistency

4 Conclusion

We have presented characterisations of all possible, i.e., balanced and unbalanced, binary consistency relations between six different trace and/or refusal based specification formalisms for process behaviour. These consistency relations are vital if formal specifications are to be used in a multiple viewpoint approach to specification, as is advocated, e.g., by the RM-ODP [9].

Various other approaches to partial process specification have been suggested in the literature [4,11,12], some with associated consistency conditions. However, those authors do not consider, what we have called, unbalanced consistency relations.

Ongoing research at the University of Kent focuses on the 'translation' of the consistency relations to consistency checking techniques and tools for more 'user-friendly', graphical specification notations. The main question here is "what implementation relations are (implicitly) assumed by specifiers of State Charts, Sequence Diagrams, etc?"

Another topic for further study is how to deal with specifications at different levels of abstraction. A single action in an enterprise specification may correspond to a more complicated behaviour in the computational specification. In order to support consistency checking between such specifications, we need to consider also implementation relations that incorporate some form of action refinement.

Acknowledgements

We would like to thank Guy Leduc for his comments on an earlier version of this paper.

References

1. T. Bolognesi and E. Brinksma. Introduction to the ISO specification language LOTOS. *Computer Networks and ISDN Systems*, 14:25–59, 1987.
2. G. Booch. *Object oriented design with applications*. Benjamin/Cummings, 1991.
3. E. Brinksma, G. Scollo, and C. Steenbergen. LOTOS specifications, their implementations and their tests. In *Protocol Specification, Testing and Verification VI*, pages 349–360. IFIP, 1987.
4. R. Cleaveland and B. Steffen. A preorder for partial process specifications. In J. C. M. Baeten and J. W. Klop, editors, *CONCUR '90: Theories of Concurrency: Unification and Extension*, LNCS 458, pages 141–151. Springer-Verlag, 1990.
5. A. Finkelstein, J. Kramer, B. Nuseibeh, L. Finkelstein, and M. Goedicke. Viewpoints: a framework for integrating multiple perspectives in system development. *International Journal on Software Engineering and Knowledge Engineering, Special issue on Trends and Research Directions in Software Engineering Environments*, 2(1):31–58, March 1992.
6. C. Fischer and G. Smith. Combining CSP and Object-Z: Finite or infinite trace semantics. In T. Mizuno, N. Shiratori, T. Higashino, and A. Togashi, editors, *FORTE/PSTV'97*, pages 503–518, Osaka, Japan, November 1997. Chapman & Hall.
7. R. J. van Glabbeek. The linear time – branching time spectrum II; the semantics of sequential systems with silent moves (extended abstract). In E. Best, editor, *CONCUR'93*, LNCS 715, pages 66–81. Springer-Verlag, 1993.
8. C. A. R. Hoare. *Communicating Sequential Processes*. Prentice Hall, 1985.
9. ISO/IEC JTC1/SC21/WG7. Basic Reference Model of Open Distributed Processing. ISO 10746, 1993. Parts 1–4.

10. K. G. Larsen. Ideal specification formalism = expressivity + compositionality + decidability + testability + $\cdots$. In *CONCUR'90. Theories of Concurrency: Unification and Extensions*, LNCS 458, pages 33–56. Springer-Verlag, 1990.
11. K. G. Larsen. Modal specifications. In J. Sifakis, editor, *Automatic Verification Methods for Finite State Systems: Proceedings*, LNCS 407, pages 232–246. Springer-Verlag, 1990.
12. G. Leduc. *On the Role of Implementation Relations in the Design of Distributed Systems using LOTOS.* PhD thesis, University of Liège, Belgium, June 1991.
13. G. Leduc. A framework based on implementation relations for implementing LOTOS specifications. *Computer Networks and ISDN Systems*, 25:23–41, 1992.
14. P. F. Linington. RM-ODP: The Architecture. In K. Raymond and L. Armstrong, editors, *Open Distributed Processing II*, pages 15–33. Chapman & Hall, February 1995.
15. R. Milner. *Communication and Concurrency.* Prentice Hall, 1989.
16. R. Sinnott and K. J. Turner. Modeling ODP viewpoints. In H. Kilov, W. Harvey, and H. Mili, editors, *Workshop on Precise Behavioral Specifications in Object-Oriented Information Modeling, OOPSLA 1994*, pages 121–128. OOPSLA, October 1994.
17. M. W. A. Steen. *Consistency and Composition of Process Specifications.* PhD thesis, University of Kent at Canterbury, May 1998. Submitted for examination.
18. M. W. A. Steen, H. Bowman, and J. Derrick. Composition of LOTOS specifications. In P. Dembiński and M. Średniawa, editors, *Protocol Specification, Testing and Verification XV*, pages 87–102. Chapman & Hall, 1995.

Observational Logic

Rolf Hennicker*, Michel Bidoit**

*Institut für Informatik, Ludwig-Maximilians-Universität München
Oettingenstr. 67, D-80538 München, GERMANY
**Laboratoire Spécification et Vérification, CNRS & Ecole Normale Supérieure de Cachan
61, Avenue du Président Wilson, 94235 Cachan Cedex, FRANCE

Abstract. We present an institution of observational logic suited for state-based systems specifications. The institution is based on the notion of an observational signature (which incorporates the declaration of a distinguished set of observers) and on observational algebras whose operations are required to be compatible with the indistinguishability relation determined by the given observers. In particular, we introduce a homomorphism concept for observational algebras which adequately expresses observational relationships between algebras. Then we consider a flexible notion of observational signature morphism which guarantees the satisfaction condition of institutions w.r.t. observational satisfaction of arbitrary first-order sentences. From the proof theoretical point of view we construct a sound and complete proof system for the observational consequence relation. Then we consider structured observational specifications and we provide a sound and complete proof system for such specifications by using a general, institution-independent result of [6].

1 Introduction

In this paper we study a logical framework for the specification of the observable behaviour of software systems which is particularly suited for state-based systems but may also be used for specifying infinite data and behavioural properties of abstract data types. Formally, we introduce an institution of observational logic and we study proof methods for first-order observational properties of structured specifications built over this institution.

Although our approach is novel, it is influenced by previous behavioural approaches, in particular of [4, 11, 16, 21, 23]. The important difference to [4] is that in the present approach we use a built-in observational semantics which previously led to problems w.r.t. the encapsulation of observational properties of parts of a system specification (cf. [1, 15]). In the hidden sorted algebra approach (cf. e.g. [11]) encapsulation is achieved at the cost of a rather restrictive notion of signature morphism which recently was generalized in [8] (see the discussion in Section 5). Another popular formalism which deals with a state-based view of systems is provided in the framework of coalgebras (cf. e.g. [23, 16]). These approaches, however, have problems to deal with n-ary operations working on several non-observable (hidden) argument sorts which frequently occur in practice (see also Section 5). Moreover, coalgebraic approaches are based on terminal semantics while we are interested in a loose semantics in order to obtain sufficient freedom for the choice of implementations.

A.M. Haeberer (Ed.): AMAST'98, LNCS 1548, pp. 263-277, 1998.

The starting point of our approach is a methodological consideration: A well-known method for specifying abstract data types (or, more concretely, for writing functional programs) is to determine first a set of constructor symbols which describe how the elements of the data type are constructed and then to define functions which can be applied to the data (usually by case distinction over the given constructors). An analogous method can be used for specifying state-based systems. First, a set of observer symbols is declared which determines an indistinguishability relation (also called observational equality) for the non-observable elements (i.e. for the states). Then the operations are specified, usually by describing their effects w.r.t. the given observers.

Formally, these considerations lead to our notion of an observational signature which contains a distinguished set of observer symbols. A similar idea was presented in [21] (and recently in [9] and [8]).[1] Based on the notion of an observational signature we define observational algebras as those structures whose operations are compatible with the observational equality (determined by the observers of the signature). In this way we obtain, in Section 3, a category of observational algebras with a notion of observational homomorphism that is suited to express observational relationships between algebras. Moreover, we establish a full and faithful functor from the category of observational algebras to the category of standard algebras which is compatible with the observational satisfaction relation defined in Section 4.

In Section 5, we introduce the institution of observational logic. It turns out that our general notion of an observer (used in observational signatures) allows us to define a powerful notion of observational signature morphism which guarantees, nevertheless, that the (observational) satisfaction condition of the institution is valid. Then, in Section 6, we define a sound and complete proof system for observational logic.

The results obtained so far allow us, first, to apply a generic construction of structured specifications over an arbitrary institution (cf. [24]), thus obtaining a basic language of structured observational specifications. Secondly, we can also apply a generic construction of a sound and complete proof system for structured specifications (cf. [6]) which leads to a corresponding proof system for structured observational specifications.

2 Algebraic Preliminaries

We assume the reader to be familiar with the basic notions of algebraic specifications (cf. e.g. [18]), like the notions of (many-sorted) *signature* $\Sigma = (S, OP)$ (with a set S of *sorts* and a set OP of *operation symbols* op: $s_1,\ldots,s_n \to s$), *signature morphism* $\sigma\colon \Sigma \to \Sigma'$, *total Σ-algebra* $A = ((A_s)_{s\in S}, (f^A)_{f\in F})$, *$\Sigma$-congruence*, *$\Sigma$-term algebra* $T(\Sigma, X)$, *valuation* $\alpha : X \to A$ and *interpretation* $I_\alpha\colon T(\Sigma, X) \to A$. Throughout this paper we assume that the carrier sets A_s of a Σ-algebra are not empty and that $X = (X_s)_{s\in S}$ is a family of countably infinite sets X_s of variables of sort $s \in S$. The class of all Σ-algebras is denoted by $Alg(\Sigma)$. Together with Σ-homomorphisms this class forms a category, for simplicity also denoted by $Alg(\Sigma)$.

---

[1] Indeed our notion of an "observer" is a generalization of an "action" in the sense of [21] and of a "behavioural operation" in the sense of [8].

For any signature morphism $\sigma: \Sigma \rightarrow \Sigma'$ the *reduct functor* $\_|_\sigma : Alg(\Sigma') \rightarrow Alg(\Sigma)$ is defined as usual. The *reduct* of a relation $\varphi' \subseteq A' \times B'$ w.r.t. $\sigma: \Sigma \rightarrow \Sigma'$ is denoted by $\varphi'|_\sigma$ where $\varphi'|_\sigma \subseteq A'|_\sigma \times B'|_\sigma$ is defined by $(\varphi'|_\sigma)_s =_{def} \varphi'_{\sigma(s)}$ for all $s \in S$.
The set of (many-sorted) *first-order* Σ*-formulas* is defined as usual whereby we will also admit infinitary Σ-formulas built by countably infinite conjunctions (or disjunctions). A *finitary* Σ-formula is a Σ-formula which contains no infinitary conjunction (disjunction resp.) and a Σ-*sentence* is a Σ*-formula* which contains no free variable. The *(standard) satisfaction relation*, denoted by $A \models \phi$, is defined as usual in the first-order predicate calculus (with a straightforward extension to infinitary formulas, cf. e.g. [17]). The notation $A \models \phi$ is extended in the usual way to classes of algebras and sets of formulas. A Σ-sentence ϕ is a *semantic consequence* of a set Φ of Σ-sentences, also denoted by $\Phi \models \phi$, if for any Σ-algebra A with $A \models \Phi$ we have $A \models \phi$.

3 The Category of Observational Algebras

An observational signature is a generalization of a standard algebraic signature with a distinguished set of observable sorts (determining the carrier sets of the observable values) and with a distinguished set of observer operations (determining the experiments that can be used to distinguish non-observable elements, often called "states"). An n-ary operation op: $s_1, \ldots, s_n \rightarrow s$ with several non-observable argument sorts may also be used as an observer (which is not the case in [21] and in [8]). In this case op is equipped with a "position number" $1 \leq i \leq n$ which indicates the argument sort of the states to be observed by op. For instance, if op: $s_1, s_2 \rightarrow s$ is a binary operation then we can declare either (op, 1) or (op, 2) or both, (op, 1) and (op, 2), as observer thus obtaining as much flexibility as needed in practical examples.

Definition 3.1 *(Observational signature)* Let $\Sigma = (S, OP)$ be a signature and $S_{Obs} \subseteq S$ be a set of *observable sorts*. An *observer* is a pair (op, i) where (op: $s_1, \ldots, s_n \rightarrow s$) $\in$ OP is an operation symbol such that $1 \leq i \leq n$ and $s_i \notin S_{Obs}$. (op, i) is a *direct observer* of s_i if $s \in S_{Obs}$; otherwise it is an *indirect observer*. If op: $s_1 \rightarrow s$ is a unary observer we will simply write op instead of (op, 1). An *observational signature* $\Sigma_{Obs} = (\Sigma, S_{Obs}, OP_{Obs})$ consists of a signature $\Sigma = (S, OP)$, a set $S_{Obs} \subseteq S$ of observable sorts and a set OP_{Obs} of observers (op, i) with op $\in$ OP. ♦

Convention We implicitly assume in the following (if not stated otherwise) that whenever we consider an observational signature Σ_{Obs}, then $\Sigma_{Obs} = (\Sigma, S_{Obs}, OP_{Obs})$ with $\Sigma = (S, OP)$ and similarly for Σ'_{Obs} etc.

Example 3.2 The following is a simple observational signature for bank accounts with observer "bal" determining the balance of an account and an operation "update" subsuming the usual credit and debit operations. Here and in the following examples we use postfix notation for unary operations and infix notation for binary operations.

sorts {account, int}
observable sorts {int}
observers {_.bal: account → int}
operations {new: → account, _.update_ : account, int → account}

A more advanced signature for bank accounts may be obtained by introducing also an indirect observer "_.undo: account → account" intended to reconstruct the previous state of an account after having performed an action. ♦

Any observational signature determines a set of observable contexts which represent those experiments which allow us to distinguish elements by the given observers.

Definition 3.3 *(Σ_{Obs}-context)* Let Σ_{Obs} be an observational signature, let $X = (X_s)_{s \in S}$ be the generally assumed family of variable sets and let $Z = (\{z_s\})_{s \in S}$ be a disjoint S-sorted family of singleton sets. For all s, s' ∈ S the set $C(\Sigma_{Obs})_{s \to s'}$ of *Σ_{Obs}-contexts* with "application sort" s and "result sort" s' is inductively defined as follows:

(1) For each $s \in S$, $z_s \in C(\Sigma_{Obs})_{s \to s}$.

(2) For each $(op, i) \in OP_{Obs}$ with $op: s_1, \ldots, s_n \to s'$, for each $c \in C(\Sigma_{Obs})_{s \to s_i}$ and pairwise disjoint variables $x_1, \ldots, x_n$ (not occurring in c) of sort $s_1, \ldots, s_n$,

$$op(x_1, \ldots, x_{i-1}, c, x_{i+1}, \ldots, x_n) \in C(\Sigma_{Obs})_{s \to s'}.$$

Each context $c \in C(\Sigma_{Obs})_{s \to s'}$ contains, besides variables in X, exactly one occurrence of the "context variable" z_s. The application of a context $c \in C(\Sigma_{Obs})_{s \to s'}$ to a term t of sort s, denoted by c[t], is the term obtained by substituting the term t for z_s.
An *observable Σ_{Obs}-context* is a Σ_{Obs}-context with observable result sort $s' \in S_{Obs}$.
We denote by $C(\Sigma_{Obs})_{s \to S_{Obs}}$ the set of observable contexts with application sort s. ♦

In Example 3.2 the only observable context is "$z_{account}$.bal". If we additionally use the indirect observer "undo" then there are infinitely many observable contexts of the form "$z_{account}$.undo.undo … .bal".
Elements which cannot be distinguished by the experiments of an observational signature are considered to be observationally equal, formally defined as follows.

Definition 3.4 *(Σ_{Obs}-equality)* Let Σ_{Obs} be an observational signature. For any Σ-algebra $A \in Alg(\Sigma)$ the *observational Σ_{Obs}-equality* on A is denoted by $\approx_{\Sigma_{Obs},A}$ and defined by:
For all $s \in S$, two elements $a, b \in A_s$ are observationally equal w.r.t. Σ_{Obs} i.e. $a \approx_{\Sigma_{Obs},A} b$, if and only if for all observable contexts $c \in C(\Sigma_{Obs})_{s \to S_{Obs}}$ and for all valuations $\alpha, \beta: X \cup \{z_s\} \to A$ with $\alpha(x) = \beta(x)$ if $x \in X$, $\alpha(z_s) = a$, $\beta(z_s) = b$, we have $I_\alpha(c) = I_\beta(c)$. Obviously, if s is an observable sort, then for all $a, b \in A$, $a \approx_{\Sigma_{Obs},A} b$ is equivalent to $a = b$. ♦

For any Σ-algebra A, $\approx_{\Sigma_{Obs},A}$ is an equivalence relation on A. But it is important to note that for an arbitrary Σ-algebra A there may exist (non-observer) operations which are not compatible with the observational equality $\approx_{\Sigma_{Obs},A}$, i.e. $\approx_{\Sigma_{Obs},A}$ is in general not a Σ-congruence on A.
In this paper we follow the loose semantics approach to algebraic specifications where a specification can be considered as a description of all admissible implementations (represented by the models of the specification). The basic assumption of the present approach is that, having declared a set of observers, an implementation can only be admissible if all its operations respect the observational equality determined by the

given observers. Formally, this is expressed by the following notion of an observational algebra.

Definition 3.5 *(Observational algebra)* Let Σ_{Obs} be an observational signature. An *observational Σ_{Obs}-algebra* is a Σ-algebra A such that $\approx_{\Sigma Obs,A}$ is a Σ-congruence on A.[2] The class of all observational Σ_{Obs}-algebras is denoted by $Alg_{Obs}(\Sigma_{Obs})$. ♦

Note that in the special case where all operations op ∈ OP are declared as observers (for each non-observable argument sort) any Σ-algebra is an observational Σ_{Obs}-algebra and $\approx_{\Sigma Obs,A}$ is just the (total) observational equality of elements defined in [4][3] and similarly in other approaches in the literature like, for instance, in [22, 11]. Let us now point out the relationship to the coalgebraic framework (cf. e.g. [16, 23]). For this purpose assume that $\Sigma_{Obs} = (\Sigma, S_{Obs}, OP_{Obs})$ is an observational signature such that for any observer (op, i) ∈ OP_{Obs}, s_i is the only non-observable argument sort of op. Moreover, assume that any observable sort s ∈ S_{Obs} is interpreted in any observational Σ_{Obs}-algebra by the same fixed set of observable values (e.g. integers, booleans etc.). Then a (polynomial) functor T: Set → Set can be associated to OP_{Obs} which captures the functionality of the observer symbols.[4] Any observational Σ_{Obs}-algebra A is an extension of a T-coalgebra C which has the same carrier sets as C and defines the non-observer operations op ∈ $OP \backslash OP_{Obs}$ on top of C. The fact that the extension A is an observational algebra is equivalent to the fact that each operation of OP preserves bisimilarity of elements (cf. e.g. [16]).

In order to obtain a category of observational algebras we still need an appropriate morphism notion. Of course, since any observational algebra is a Σ-algebra, one could simply use standard homomorphisms between Σ-algebras. But this does not reflect the relationships between the *observable behaviour* of algebras. Therefore, we have chosen another definition where an observational homomorphism is defined as an appropriate relation which is compatible with observational equalities.

Definition 3.6 *(Observational homomorphism)* Let A, B ∈ $Alg_{Obs}(\Sigma_{Obs})$. An *observational Σ_{Obs}-homomorphism* φ: A → B is an S-sorted family $(\varphi_s)_{s\in S}$ of relations $\varphi_s \subseteq A_s \times B_s$ with the following properties for all s ∈ S:

(1) For all $a \in A_s$ there exists $b \in B_s$ such that $a \ \varphi_s \ b$.
(2) For all $a \in A_s$, $b, b' \in B_s$, if $a \ \varphi_s \ b$ then ($a \ \varphi_s \ b'$ if and only if $b \approx_{\Sigma Obs,B} b'$).
(3) For all $a, a' \in A_s$, $b \in B_s$, if $a \ \varphi_s \ b$ and $a \approx_{\Sigma Obs,A} a'$ then $a' \ \varphi_s \ b$.
(4) For all (op: $s_1,\ldots, s_n \to s$) ∈ OP and $a_i \in A_{s_i}$, $b_i \in B_{s_i}$,
if $a_i \ \varphi_{s_i} \ b_i$ for $i = 1,\ldots, n$ then $op^A(a_1,\ldots, a_n) \ \varphi_s \ op^B(b_1,\ldots, b_n)$. ♦

2 Obviously, for this it is sufficient that all non-observer operations are compatible with $\approx_{\Sigma Obs,A}$.

3 In [4] also partial observational equalities are considered. It should be straightforward to extend our approach to this case.

4 If there is more than one non-observable sort in $S \backslash S_{Obs}$ then a set of functors has to be associated to Σ_{Obs}.

Theorem 3.7 *(The category of observational algebras)*
For each observational signature Σ_{Obs}, the class $Alg_{Obs}(\Sigma_{Obs})$ together with observational Σ_{Obs}-homomorphisms is a category which, by abuse of notation, will also be denoted by $Alg_{Obs}(\Sigma_{Obs})$. Thereby the composition of observational homomorphisms is the usual composition of relations and, for each observational Σ_{Obs}-algebra A, the identity id_A: A → A in the category $Alg_{Obs}(\Sigma_{Obs})$ is the observational equality $\approx_{\Sigma_{Obs},A}$. (The proof is straightforward; see [14].) ♦

Since for any observational Σ_{Obs}-algebra A the observational equality $\approx_{\Sigma_{Obs},A}$ is a Σ-congruence, we can construct the quotient algebra $A/\approx_{\Sigma_{Obs},A}$ which identifies all elements of A which are indistinguishable "from the outside". $A/\approx_{\Sigma_{Obs},A}$ can be considered as the "black box view" of A thus representing the "observable behaviour" of A w.r.t. Σ_{Obs}. Using this behaviour construction we obtain (for any observational signature Σ_{Obs}) a functor from the category $Alg_{Obs}(\Sigma_{Obs})$ of observational algebras into the category Alg(Σ) of (standard) Σ-algebras which establishes a one to one correspondence between observational homomorphisms φ: A → B and standard homomorphisms h: $A/\approx_{\Sigma_{Obs},A} \to B/\approx_{\Sigma_{Obs},B}$, i.e. the functor is full and faithful.

Theorem 3.8 *(Behaviour functor)* For any observational signature Σ_{Obs}, $\mathcal{F}_{\Sigma_{Obs}}$: $Alg_{Obs}(\Sigma_{Obs}) \to Alg(\Sigma)$ is a full and faithful functor where $\mathcal{F}_{\Sigma_{Obs}}$ is defined by:
For each $A \in Alg_{Obs}(\Sigma_{Obs})$, $\mathcal{F}_{\Sigma_{Obs}}(A) =_{def} A/\approx_{\Sigma_{Obs},A}$ and
for each observational Σ_{Obs}-homomorphism φ: A → B,

$$\mathcal{F}_{\Sigma_{Obs}}(\varphi): A/\approx_{\Sigma_{Obs},A} \to B/\approx_{\Sigma_{Obs},B} \text{ is defined by } \mathcal{F}_{\Sigma_{Obs}}(\varphi)([a]) =_{def} [b] \text{ if } a\ \varphi\ b.$$

(For $A \in Alg_{Obs}(\Sigma_{Obs})$, $\mathcal{F}_{\Sigma_{Obs}}(A)$ is called the *observational behaviour* of A. The proof of the theorem is straightforward; see [14].) ♦

Remark 3.9 Since $\mathcal{F}_{\Sigma_{Obs}}$ is full and faithful, it is obvious that two observational algebras are observationally isomorphic if and only if they have isomorphic behaviours. Hence, as a consequence of a result in [4], observational isomorphism coincides with usual notions of observational equivalence between algebras (cf. e.g. [22]).[5] This also points out the adequacy of our morphism notion. ♦

4 Observational Satisfaction

The underlying idea of the observational satisfaction relation is to interpret the equality symbol "=" occurring in a first-order formula ϕ not by the set-theoretic equality but by the observational equality of elements.

Definition 4.1 The *observational satisfaction relation* between observational Σ_{Obs}-algebras and Σ-formulas is denoted by $\models_{\Sigma_{Obs}}$ and defined as follows:

(1) For any two terms t, r ∈ $T(\Sigma, X)_s$ of the same sort s and for any valuation α: X → A, A, α $\models_{\Sigma_{Obs}}$ t = r holds if $I_\alpha(t) \approx_{\Sigma_{Obs},A} I_\alpha(r)$.

[5] To our knowledge [20] is the first paper where observational equivalence of algebras is characterized by isomorphism of some category.

(2) For any arbitrary Σ-formula ϕ and for any valuation $\alpha: X \to A$, $A, \alpha \models_{\Sigma Obs} \phi$ is defined by induction over the structure of the formula ϕ in the usual way.
(3) For any arbitrary Σ-formula ϕ, $A \models_{\Sigma Obs} \phi$ holds if for all valuations $\alpha: X \to A$, $A, \alpha \models_{\Sigma Obs} \phi$ holds.

The notation $A \models_{\Sigma Obs} \phi$ is extended in the usual way to classes of observational algebras and sets of formulas. ♦

Technically the observational satisfaction relation could be defined in the same way for arbitrary Σ-algebras which do not necessarily belong to $Alg_{Obs}(\Sigma_{Obs})$. This is the approach of extended hidden algebra (cf. [8]) where a special predicate symbol "~" is introduced for representing the observational equality of (non-observable) elements. But then the congruence rule of the equational calculus is only sound w.r.t. "~" if one can prove that all operations of a specification are "behaviourally coherent", i.e. are compatible with the given observational equality.[6]

Definition 4.2 *(Observational consequence)* A Σ-sentence ϕ is an *observational consequence* of a set Φ of Σ-sentences, also denoted by $\Phi \models_{\Sigma Obs} \phi$, if for any observational Σ_{Obs}-algebra A, $A \models_{\Sigma Obs} \Phi$ implies $A \models_{\Sigma Obs} \phi$. ♦

The next proposition shows that the behaviour functor defined in Theorem 3.8 is compatible with the observational and the standard satisfaction relations. (For the proof see [14].)

Proposition 4.3 For any $A \in Alg_{Obs}(\Sigma_{Obs})$ and any Σ-formula ϕ,
$A \models_{\Sigma Obs} \phi$ if and only if $\mathcal{F}_{\Sigma Obs}(A) \models \phi$.[7] ♦

As a consequence of Remark 3.9 and Proposition 4.3 we can generalize Scott's theorem (cf. e.g. [17]) to observational algebras and observational satisfaction (taking into account that Σ-formulas may be infinitary; cf. Section 2).

Corollary 4.4 *(Observational version of Scott's theorem)*[8]
Let $A, B \in Alg_{Obs}(\Sigma_{Obs})$ be two observational Σ_{Obs}-algebras such that $\mathcal{F}_{\Sigma Obs}(A)$ and $\mathcal{F}_{\Sigma Obs}(B)$ are countable. The following conditions are equivalent:
(1) A and B are observationally isomorphic.
(2) For all (possibly infinitary) Σ-formulas ϕ, $A \models_{\Sigma Obs} \phi$ if and only if $B \models_{\Sigma Obs} \phi$. ♦

We are now able to define the syntax and semantics of flat observational specifications. Structured specifications will be considered in Section 7.

---

6 The idea of introducing a denotation "~" for observational equalities is suggested in [2, 3] as a proof-theoretic means for proving behavioural theorems and implementation correctness.

7 In an abstract category-theoretic setting this fact can be used as a definition of behavioural satisfaction as in [7] and [5].

8 A related result, but formulated in terms of observational equivalence of algebras instead of observational isomorphism, is given in [4].

Definition 4.5 A *flat observational specification* SP = $\langle\Sigma_{Obs}, Ax\rangle$ consists of an observational signature $\Sigma_{Obs} = (\Sigma, S_{Obs}, OP_{Obs})$ and a set Ax of Σ-sentences, called the *axioms* of SP. The semantics of SP is given by its signature $Sig_{Obs}(SP)$ and by its class of models $Mod_{Obs}(SP)$ which are defined by

$$Sig_{Obs}(SP) =_{def} \Sigma_{Obs},\ Mod_{Obs}(SP) =_{def} \{A \in Alg_{Obs}(\Sigma_{Obs}) \mid A \models_{\Sigma_{Obs}} Ax\}.$$ ♦

For any observational specification SP, the class $Mod_{Obs}(SP)$ is closed under observational isomorphisms.

Example 4.6 The following specification of bank accounts has additionally to the account operations of Example 3.2 an operation "paycharge" which reduces the balance of an account by a constant monthly fee.

```
spec ACCOUNT =
      sorts {account, int}
      observablesorts {int}
      observers {_.bal: account → int, _.undo: account → account}
      operations "operations for the integers" ∪
                 {new: → account, _.update_ : account, int → account,
                 _.paycharge: account → account}
      axioms     "axioms for the integers" ∪
                 {∀x: int, s: account.
                 new.bal = 0, new.undo = new,
                 s.update(x).bal = s.bal+x, s.update(x).undo = s,
                 s.paycharge.bal = s.bal-10, s.paycharge.undo = s}
```

A possible model of the specification ACCOUNT which satisfies the axioms even literally can be defined in terms of lists of integers. Another model which satisfies the axioms observationally (but not literally) can be constructed by using the well-known array with pointer realization of lists.
In the above specification the behaviour of the operations is uniquely specified w.r.t. the given observers. A proper loose specification can be obtained, for instance, by removing the equations for the "paycharge" operation. Then the semantics of the specification is still restricted to those models where the interpretation of "paycharge" is compatible with the given observational equality (since only observational algebras are admissible models). ♦

5 The Institution of Observational Logic

The category of observational algebras is the basis for defining an institution (cf. [10]) of observational logic which captures the model-theoretic view of the observable behaviour of systems. An essential ingredient to build an institution is an appropriate morphism notion for observational signatures which is defined as follows.

Definition 5.1 *(Observational signature morphism)* Let $\Sigma_{Obs} = (\Sigma, S_{Obs}, OP_{Obs})$ and $\Sigma'_{Obs} = (\Sigma', S'_{Obs}, OP'_{Obs})$ be two observational signatures with $\Sigma = (S, OP)$ and $\Sigma' = (S', OP')$. An *observational signature morphism* $\sigma: \Sigma_{Obs} \to \Sigma'_{Obs}$ is a signature morphism $\sigma: \Sigma \to \Sigma'$ such that the following conditions are satisfied:

(1) For all $s \in S$, $s \in S_{Obs}$ if and only if $\sigma(s) \in \sigma(S_{Obs})$.
(2) If $(op, i) \in OP_{Obs}$ then $(\sigma(op), i) \in OP'_{Obs}$.
(3) If $(op', i) \in OP'_{Obs}$ such that $op': s_1',\ldots,s_n' \to s'$ and $s_i' = \sigma(s_i)$ for some $s \in S$ then there exists $(op, i) \in OP_{Obs}$, $op: s_1,\ldots,s_n \to s$ such that $op' = \sigma(op)$. ♦

Condition (1) is standard. It requires that observable and non-observable sorts are preserved by σ. Condition (2) requires that also observers are preserved by σ. Condition (3) is essential for the satisfaction condition presented below. It says that whenever the image $\sigma(s)$ of some "old" sort s of Σ_{Obs} is observed by an observer op' of Σ'_{Obs} then there must be a corresponding observer op of Σ_{Obs} which observes s and which is mapped to op'. Thus no "new" observations can be introduced for "old" sorts. However, it is important to note that there is still sufficient flexibility for the following:

1. We can introduce new operation symbols in $OP' \backslash OP'_{Obs}$ which are *not* in the image of σ but nevertheless may have argument sorts which *are in the image* of σ. A standard example may be given by the signature of a specification of a bank which is based on (i.e. imports) a specification of accounts. Then the bank specification may introduce a new operation "_.add_: bank, account → bank" for adding an account to a bank. This is not a problem as long as "add" is not used as an observer for accounts (i.e. (add, 2) is not an observer which indeed would be strange). Examples like this, where some argument sorts of an operation are imported from another signature frequently occur in practice, in particular, in object-oriented programming. This situation cannot be dealt with by hidden signature morphisms; cf. e.g. [19]. Extended hidden algebra, however, solves this problem; cf. [8].
2. We can introduce new observers (op', i) in OP'_{Obs} as long as the observed sort s_i' is not in the image of σ. For instance, in the bank example one has definitely to introduce some observer(s) for the new sort "bank". This can neither be done in hidden algebra nor in extended hidden algebra.

Definition 5.2 *(Observational reduct functor)* For any observational signature morphism $\sigma: \Sigma_{Obs} \to \Sigma'_{Obs}$, $Alg_{Obs}(\sigma): Alg_{Obs}(\Sigma'_{Obs}) \to Alg_{Obs}(\Sigma_{Obs})$ is defined by:
For each $A' \in Alg_{Obs}(\Sigma'_{Obs})$, $Alg_{Obs}(\sigma)(A') =_{def} A'|_\sigma$
For each observational Σ'_{Obs}-homomorphism $\varphi': A' \to B'$,

$$Alg_{Obs}(\sigma)(\varphi'): A'|_\sigma \to B'|_\sigma \text{ is defined by } Alg_{Obs}(\sigma)(\varphi') =_{def} \varphi'|_\sigma$$

(See Section 2 for the definition of the reducts $A'|_\sigma$ and $\varphi'|_\sigma$.) ♦

The following lemma is essential for proving that $Alg_{Obs}(\sigma)$ is indeed a well-defined functor (cf. Theorem 5.4) and also for checking the (observational) satisfaction condition (cf. Theorem 5.5). It says that the observational equality is preserved by observational reducts. (The proof of the lemma and of the subsequent theorems is given in [14].)

Lemma 5.3 For any observational signature morphism $\sigma: \Sigma_{Obs} \to \Sigma'_{Obs}$ and observational Σ'_{Obs}-algebra $A' \in Alg_{Obs}(\Sigma'_{Obs})$, $(\approx_{\Sigma_{Obs},A'})|_\sigma = \approx_{\Sigma_{Obs},(A'|_\sigma)}$. ♦

Theorem 5.4 For any observational signature morphism $\sigma: \Sigma_{Obs} \to \Sigma'_{Obs}$, $Alg_{Obs}(\sigma): Alg_{Obs}(\Sigma'_{Obs}) \to Alg_{Obs}(\Sigma_{Obs})$ is a well-defined functor. ♦

We are now able to state the satisfaction condition for observational logic which generalizes the satisfaction condition for hidden algebra. It guarantees encapsulation in the sense that observational properties are respected when composing specifications. In Section 7 we will consider structured specifications and we will see how a straightforward sound and complete proof system for structured specifications can be constructed which needs the validity of the satisfaction condition.

Theorem 5.5 *(Observational satisfaction condition)* Let $\sigma: \Sigma_{Obs} \to \Sigma'_{Obs}$ be an observational signature morphism. For any $A' \in Alg_{Obs}(\Sigma'_{Obs})$ and Σ-sentence ϕ,

$$A' \models_{\Sigma'_{Obs}} \sigma(\phi) \text{ if and only if } Alg_{Obs}(\sigma)(A') \models_{\Sigma_{Obs}} \phi$$

where $\sigma(\phi)$ is the usual extension of a signature morphism to Σ-sentences. ♦

Corollary 5.6 *(The institution of observational logic)*
The quadruple $INS_{Obs} = (Sig_{Obs}, Sen_{IFOLEQ}, Alg_{Obs}, \models_{Obs})$ is an institution whereby:

- Sig_{Obs} is the category of observational signatures and observational signature morphisms.
- The functor $Sen_{IFOLEQ}: Sig_{Obs} \to Set$ maps
  - each observational signature $\Sigma_{Obs} = (\Sigma, S_{Obs}, OP_{Obs})$ to the set of (possibly infinitary) many-sorted first-order Σ-sentences (cf. Section 2) and
  - each observational signature morphism $\sigma: \Sigma_{Obs} \to \Sigma'_{Obs}$ to the obvious translation function which transforms Σ-sentences into Σ'-sentences.
- The functor $Alg_{Obs}: (Sig_{Obs})^{op} \to Cat$ maps
  - each observational signature Σ_{Obs} to the category $Alg_{Obs}(\Sigma_{Obs})$ of observational Σ_{Obs}-algebras and observational Σ_{Obs}-homomorphisms and
  - each observational signature morphism $\sigma: \Sigma_{Obs} \to \Sigma'_{Obs}$ to the observational reduct functor $Alg_{Obs}(\sigma): Alg_{Obs}(\Sigma'_{Obs}) \to Alg_{Obs}(\Sigma_{Obs})$.
- $\models_{Obs} = (\models_{\Sigma_{Obs}})_{\Sigma_{Obs} \in |Sig_{Obs}|}$ where, for each observational signature Σ_{Obs}, $\models_{\Sigma_{Obs}}$ is the observational satisfaction relation of a Σ-sentence by an observational Σ_{Obs}-algebra. ♦

According to Proposition 4.3 the family $(\mathcal{F}_{\Sigma_{Obs}})_{\Sigma_{Obs} \in |Sig_{Obs}|}$ of (full and faithful) functors $\mathcal{F}_{\Sigma_{Obs}}: Alg_{Obs}(\Sigma_{Obs}) \to Alg(\Sigma)$ can be extended to an institution morphism (cf. [10]) which maps the institution of observational logic to the institution of (standard) infinitary first order logic.

6 A Proof System for Observational Logic

In this section we study the proof theory for observational logic. For defining an appropriate proof system we first associate to any observational signature Σ_{Obs} the following set $FA_{\Sigma_{Obs}}$ of Σ-sentences.

Definition 6.1 Let $\Sigma_{Obs} = (\Sigma, S_{Obs}, OP_{Obs})$ be an observational signature with $\Sigma = (S, F)$. $FA_{\Sigma Obs} =_{def} \{FA_{\Sigma Obs}(s) \mid s \in S \backslash S_{Obs}\}$ where for any $s \in S \backslash S_{Obs}$,

$$FA_{\Sigma Obs}(s) =_{def} \forall x_L, x_R{:}s. \; (\bigwedge_{c \in C(\Sigma_{Obs})_{s \to S_{Obs}}} \forall Var(c).\; c[x_L] = c[x_R]) \Rightarrow x_L = x_R.$$

Thereby $Var(c)$ denotes the set of all variables occurring in c besides the context variable z_s. ♦

The underying idea for considering $FA_{\Sigma Obs}$ stems from a result in [4] where it is shown that the behavioural theory of a class C of Σ-algebras coincides with the standard theory of the fully abstract algebras of C. The following theorem shows that indeed the sentences $FA_{\Sigma Obs}$ allow us to characterize the observational consequence relation in terms of the standard consequence relation. (For the proof see [14].)

Theorem 6.2 Let Σ_{Obs} be an observational signature, Φ be a set of Σ-sentences and φ be a Σ-sentence. $\Phi \models_{\Sigma Obs} \phi$ if and only if $\Phi \cup FA_{\Sigma Obs} \models \phi$. ♦

In the sequel we assume given, for each signature Σ, a sound and complete proof system $\Pi(\Sigma)$ for (many-sorted) infinitary first-order logic (see the discussion below). The proof system $\Pi(\Sigma_{Obs})$ for observational logic is then constructed by adding to the axioms and rules of $\Pi(\Sigma)$ the sentences $FA_{\Sigma Obs}$ as further axioms.

Definition 6.3 *(Proof system for observational logic)*
For any observational signature Σ_{Obs}, $\Pi(\Sigma_{Obs}) =_{def} \Pi(\Sigma) \cup FA_{\Sigma Obs}$.
We write $\Phi \vdash_{\Sigma Obs} \phi$ ($\Phi \vdash_{\Sigma} \phi$ resp.) if φ is a Σ-sentence that can be deduced from a set Φ of Σ-sentences by the axioms and rules of $\Pi(\Sigma_{Obs})$ ($\Pi(\Sigma)$ resp.). ♦

Corollary 6.4 *(Soundness and completeness)* For any observational signature Σ_{Obs}, set Φ of Σ-sentences and Σ-sentence φ, $\Phi \vdash_{\Sigma Obs} \phi$ if and only if $\Phi \models_{\Sigma Obs} \phi$.

Proof: $\Phi \vdash_{\Sigma Obs} \phi$ iff, by definition of $\Pi(\Sigma_{Obs})$, $\Phi \cup FA_{\Sigma Obs} \vdash_{\Sigma} \phi$ iff, by soundness and completeness of $\Pi(\Sigma)$, $\Phi \cup FA_{\Sigma Obs} \models \phi$ iff, by Theorem 6.2, $\Phi \models_{\Sigma Obs} \phi$. ♦

The axioms $FA_{\Sigma Obs}$ can be considered as a coinductive proof principle (cf. e.g. [16]) which, together with $\Pi(\Sigma)$, allows us to prove the observational validity not only of equations but of arbitrary first-order formulas. If Σ_{Obs} contains only direct observers there exist (up to α-conversion) only finitely many observable contexts and hence $FA_{\Sigma Obs}$ is finitary. In this case $\Pi(\Sigma)$ can be chosen as a formal (i.e. finitary) proof system and any available theorem prover for first-order logic can be used to prove that φ is an observational consequence of Φ.[9]
If Σ_{Obs} contains indirect observers there may be infinitely many observable contexts and then $FA_{\Sigma Obs}$ contains infinitary conjunctions. In this case we can choose for $\Pi(\Sigma)$ a proof system for infinitary first-order logic (for instance, the many-sorted variant of

[9] For instance, using the Larch Prover one can directly implement the axioms $FA_{\Sigma Obs}$ by the "partitioned by" construct of LP; cf. [12].

the proof system in [17]). Then the above completeness result is mainly of theoretical interest. However, it is important to note that the infinitary formulas $FA_{\Sigma Obs}$ can still be very useful because in practical examples the validity of (an instantiation of) the infinitary conjunction of $FA_{\Sigma Obs}$ can often be verified by an induction proof (cf. Example 6.5 below). Using a result of [3] it is even possible to encode the infinitary formulas $FA_{\Sigma Obs}$ by finitary ones if one introduces auxiliary symbols and reachability constraints. Hence the problem of the non-completeness of finitary proof systems for observational logic corresponds exactly to the non-completeness of finitary proof systems for inductively defined data types (in particular of arithmetic).

Example 6.5 Consider the signature of the ACCOUNT specification of Example 4.6. It induces the infinitary sentence FA(account) $=_{def}$

$$\forall s_L, s_R\text{: account. } (\bigwedge_{i \in N} s_L.\text{undo}^i.\text{bal} = s_R.\text{undo}^i.\text{bal}) \Rightarrow s_L = s_R.$$

Now consider the implicitly universally quantified equation

$$s.\text{paycharge} = s.\text{update}(-10).$$

It is easy to prove by induction that for all $i \in N$,

$$s.\text{paycharge}.\text{undo}^i.\text{bal} = s.\text{update}(-10).\text{undo}^i.\text{bal}$$

can be derived from the axioms of ACCOUNT. Then, using FA(account), we deduce

$$s.\text{paycharge} = s.\text{update}(-10)$$

and therefore, by Corollary 6.4, this equation is an observational consequence of the ACCOUNT specification. ♦

7 Structured Observational Specifications

In [6] (and similarly in [24]) a basic set of specification-building operations is defined which allows one to build structured specifications over an arbitrary institution. We will now apply these operators to the particular institution of observational logic thus obtaining the following set of operations for constructing structured observational specifications. The semantics of such a specification SP is determined by its (observational) signature, denoted by $Sig_{Obs}(SP)$, and by its class of models, denoted by $Mod_{Obs}(SP)$. In the following definition we assume that $\sigma: \Sigma_{Obs} \rightarrow \Sigma'_{Obs}$ is an injective observational signature morphism.[10]

basic: Any presentation $\langle \Sigma_{Obs}, Ax \rangle$ is an observational specification. Its semantics is defined in Definition 4.5.

union: For any two observational specifications SP1 and SP2 with $Sig_{Obs}(SP1) = Sig_{Obs}(SP2)$, the expression $SP1 \cup SP2$ is an observational specification with semantics

$$Sig_{Obs}(SP1 \cup SP2) =_{def} Sig_{Obs}(SP1),$$
$$Mod_{Obs}(SP1 \cup SP2) =_{def} Mod_{Obs}(SP1) \cap Mod_{Obs}(SP2).$$

---

10 The injectivity requirement ensures that the interpolation property for institutions (cf. [6]) needed for the completeness proof holds. Whether the interpolation property holds without this assumption seems to be an open question.

translate: For any observational specification SP with $Sig_{Obs}(SP) = \Sigma_{Obs}$, the expression **translate** SP **by** σ is an observational specification with semantics

$$Sig_{Obs}(\textbf{translate } SP \textbf{ by } \sigma) =_{def} \Sigma'_{Obs},$$
$$Mod_{Obs}(\textbf{translate } SP \textbf{ by } \sigma) =_{def} \{A' \in Alg_{Obs}(\Sigma'_{Obs}) \mid Alg_{Obs}(\sigma)(A') \in Mod_{Obs}(SP)\}.$$

derive: For any observational specification SP' with $Sig_{Obs}(SP') = \Sigma'_{Obs}$, the expression **derive from** SP' **by** σ is an observational specification with semantics

$$Sig_{Obs}(\textbf{derive from } SP' \textbf{ by } \sigma) =_{def} \Sigma_{Obs},$$
$$Mod_{Obs}(\textbf{derive from } SP' \textbf{ by } \sigma) =_{def} \{Alg_{Obs}(\sigma)(A') \mid A' \in Mod_{Obs}(SP')\}.$$

Definition 7.1 Let SP be an observational specification with signature Σ_{Obs}. A Σ-sentence ϕ is called an *observational theorem* of SP, written $SP \models_{\Sigma_{Obs}} \phi$, if $Mod_{Obs}(SP) \models_{\Sigma_{Obs}} \phi$. ♦

In the following we are interested in a proof system which allows us to prove observational theorems of structured (observational) specifications. For this purpose we instantiate the institution-independent proof system of [6] and obtain the following rules which generate, for each observational signature Σ_{Obs}, a relation $\vdash_{\Sigma_{Obs}}$ between observational specifications SP and Σ-sentences ϕ.

(pi-obs)
$$\frac{SP \vdash_{\Sigma_{Obs}} \phi_i \text{ for } i \in I,\ \{\phi_i \mid i \in I\} \vdash_{\Sigma_{Obs}} \phi}{SP \vdash_{\Sigma_{Obs}} \phi}$$

(basic)
$$\frac{\phi \in Ax}{\langle \Sigma_{Obs}, Ax\rangle \vdash_{\Sigma_{Obs}} \phi}$$

(union-1)
$$\frac{SP1 \vdash_{\Sigma_{Obs}} \phi}{SP1 \cup SP2 \vdash_{\Sigma_{Obs}} \phi}$$

(union-2)
$$\frac{SP2 \vdash_{\Sigma_{Obs}} \phi}{SP1 \cup SP2 \vdash_{\Sigma_{Obs}} \phi}$$

(translate)
$$\frac{SP \vdash_{\Sigma_{Obs}} \phi}{\textbf{translate } SP \textbf{ by } \sigma \vdash_{\Sigma'_{Obs}} \sigma(\phi)}$$

(derive)
$$\frac{SP' \vdash_{\Sigma'_{Obs}} \sigma(\phi)}{\textbf{derive from } SP' \textbf{ by } \sigma \vdash_{\Sigma_{Obs}} \phi}$$

According to the rule (pi-obs) the proof system for structured specifications is based on the proof system $\Pi(\Sigma_{Obs})$ for observational logic (cf. Section 6). The other rules correspond to the specification-building operations and hence proofs of observational theorems can be performed according to the structure of a given specification. An institution-independent proof of the soundness of the above rules is presented in [24]. The completeness can be checked by applying the results of [6] to the insti-tution INS_{Obs} of observational logic. For this purpose one has to show that INS_{Obs} satisfies the amalgamation and interpolation properties which is detailed in [14].

Theorem 7.2 *(Soundness and completeness)* Let SP be an observational specification with signature Σ_{Obs} and let ϕ be a Σ-sentence.

$$SP \models_{\Sigma_{Obs}} \phi \text{ if and only if } SP \vdash_{\Sigma_{Obs}} \phi. \ ♦$$

8 Conclusion

Observational logic provides a formal foundation for an observational specification methodology for state-based systems which works quite analogously to functional specifications of reachable data structures. In the latter case one usually starts by declaring a set of data type constructors. Similarly, in the observational case one starts by declaring a set of observers which do not tell us how elements are constructed but how elements can be observed. While data type constructors induce a generation principle which restricts the admissible models of a specification to reachable algebras, observer operations induce an observational equality which restricts the admissible models to observational algebras. Moreover, the operations on reachable data structures can be specified by inductive definitions while the operations on non-observable elements (i.e. states) can be defined (coinductively) by describing their effect w.r.t. the given observers. Analogously to abstract data type specifications, a loose observational specification describes a class of observational algebras which is closed under observational isomorphisms. Such a class can be considered as an "abstract behaviour type". If it contains only one observational isomorphism class its specification can be regarded as an "observationally monomorphic" specification of an object-oriented program.
The main topic of our next research steps is the consideration of refinement relations between (structured) observational specifications with an emphasis on refinement proofs. We hope that we can reuse several results of [2] and [13] but we are aware that these approaches do not deal with a built-in (internalised) observational semantics of structured specifications as considered in this paper. Another important direction of future research is concerned with an extension of observational logic to take into account concurrent systems specifications.

Acknowledgement We would like to thank Andrzej Tarlecki and the referees of this paper for several valuable remarks.

References

1. G. Bernot, M. Bidoit: Proving the correctness of algebraically specified software: modularity and observability issues. Proc. AMAST '91, 216-242, Springer-Verlag Works. in Comp. Series, 1992.
2. M. Bidoit, R. Hennicker: Proving the correctness of behavioural implementations. In Proc. AMAST '95, LNCS 936, 152-168, 1995. An extended version entitled "Modular correctness proofs of behavioural implementations" will appear in Acta Informatica.
3. M. Bidoit, R. Hennicker: Behavioural theories and the proof of behavioural properties. Theoretical Computer Science 175, 3-55, 1996.
4. M. Bidoit, R. Hennicker, M. Wirsing: Behavioural and abstractor specifications. Science of Computer Programming 25, 149-186, 1995.
5. M. Bidoit, A. Tarlecki: Behavioural satisfaction and equivalence in concrete model categories. Proc. CAAP '96, Trees in Algebra and Progr., LNCS 1059, 241-256, 1996.

6. T. Borzyszkowski: Completeness of a logical system for structured specifications. In: F. Parisi Presicce (ed.): Recent Trends in Algebraic Development Techniques, LNCS 1376, 107-121, 1998.
7. R. Burstall, R. Diaconescu: Hiding and behaviour: an institutional approach. In: A. W. Roscoe (ed.): A Classical Mind: Essays in Honour of C.A.R. Hoare, 75-92, Prentice-Hall, 1994.
8. R. Diaconescu: Behavioural coherence in object-oriented algebraic specification. Japan Advanced Institute for Science and Technology, IS-RR-98-0017F, 1998.
9. R. Diaconescu, K. Futatsugi: CafeOBJ Report: The Language, Proof Techniques, and Methodologies for Object-Oriented Algebraic Specification, AMAST Series in Computing, Vol. 6, World Scientific, 1998.
10. J. Goguen, R. Burstall: Institutions: abstract model theory for specification and programming. Journal of the Association for Computing Machinery 39 (1), 95-146, 1992.
11. J. Goguen, G. Malcolm: A hidden agenda. Report CS97-538, Univ. of Calif. at San Diego, 1997.
12. J. Guttag, J. Horning: Larch: Languages and Tools for Formal Specification. Texts and Monographs in Computer Science, Springer, 1993.
13. R. Hennicker: Structured specifications with behavioural operators: semantics, proof methods and applications. Habilitation thesis, Institut für Informatik, Ludwig-Maximilians-Universität München, 1997.
14. R. Hennicker, M. Bidoit: Observational logic (long version). www.lsv.ens-cachan.fr/Publis/RAPPORTS_LSV/, 1998.
15. R. Hennicker, F. Nickl: A behavioural algebraic framework for modular system design with reuse. In: H. Ehrig, F. Orejas (eds.): Recent Trends in Data Type Specification, LNCS 785, 220-234, 1994.
16. B. Jacobs, J. Rutten: A Tutorial on (Co)Algebras and (Co)Induction. EATCS Bulletin 62, 222-259, 1997.
17. H. J. Keisler: Model theory for infinitary logic. North-Holland, 1971.
18. J. Loeckx, H.-D. Ehrich, M. Wolf: Specification of Abstract Data Types. Wiley and Teubner, 1996.
19. G. Malcolm, J. A. Goguen: Proving correctness of refinement and implementation. Technical Monograph PRG-114, Oxford University Computing Laboratory, 1994.
20. P. Nivela, F. Orejas: Initial behaviour semantics for algebraic specifications. In: D. T. Sannella, A. Tarlecki (eds.): Recent Trends in Data Type Specification, Springer Lecture Notes in Computer Science 332, 184-207, 1988.
21. P. Padawitz: Swinging data types: syntax, semantics, and theory. In: M. Haveraaen, O. Owe, O.-J. Dahl (eds.): Recent Trends in Data Type Specification, LNCS 1130, 409-435, 1996.
22. H. Reichel: Initial computability, algebraic specifications, and partial algebras. International Series of Monographs in Computer Science No. 2, Oxford: Clarendon Press, 1987.
23. H. Reichel: An approach to object semantics based on terminal co-algebras. Math. Struct. Comp. Sci., 5, 129-152, 1995.
24. D. T. Sannella, A. Tarlecki: Specifications in an arbitrary institution. Information and Computation 76, 165-210, 1988.

Scheduling Algebra

Rob van Glabbeek[1], Peter Rittgen[2]

[1]Computer Science Department, Stanford University, CA 94305-9045, USA
rvg@cs.stanford.edu
[2]Institut für Wirtschaftsinformatik, Universität Koblenz-Landau, Rheinau 1, 56075 Koblenz, Germany
rittgen@uni-koblenz.de

Abstract. Our goal is to develop an algebraic theory of process scheduling. We specify a syntax for denoting processes composed of actions with given durations. Subsequently, we propose axioms for transforming any specification term of a scheduling problem into a term of all valid schedules. Here a *schedule* is a process in which all (implementational) choices (e.g. precise timing) are resolved. In particular, we axiomatize an operator restricting attention to the *efficient* schedules. These schedules are representable as trees, because in an efficient schedule actions start only at time zero or when a resource is released, i.e. upon termination of the action binding a required resource. All further delay is useless. Nevertheless, we do not consider resource constraints explicitly here. We show that a normal form exists for every term of the algebra and establish both soundness of our axioms with respect to a schedule semantics and completeness for efficient processes.

Introduction

The problem of scheduling entails assigning an execution time (and sometimes a processor) to each of a set of actions with given durations, depending on a certain goal function (e.g. shortest schedule) and certain causal and resource constraints. The theory of scheduling has been investigated since the early 50s. The research so far can be divided into two categories: partitioning the set of all scheduling problems into (complexity) classes, and finding efficient algorithms for each class. Most of the problems being NP-complete, substantial effort has been spent on problem relaxations and heuristics to obtain algorithms that compute near-optimal schedules in polynomial time. But an axiomatization of the theory of scheduling still remains to be given. Here we outline such a calculus abstracting from both resource constraints and goal function and limiting the structure of the causal order. Our aim is to provide an axiom system that turns a process specification into a set of *efficient* (or *semi-active*) schedules [7]. These are the schedules that are potentially optimal for certain constraints and goal functions, provided the latter do not favor spending idle time. For simplicity, we consider only specifications in which no action has more than one immediate causal predecessor, i.e. the precedence graph has a multi-tree order.

Viewed from a different angle, our approach can be described as applying concurrency theory to scheduling. A lot of algebras of concurrent processes have been given, e.g. ACP [3], CCS [12] or CSP [11]. Hence, one might ask why we develop a

A.M. Haeberer (Ed.): AMAST'98, LNCS 1548, pp. 278-292, 1998.

new calculus instead of using an existing one. The answer lies in the specific requirements of processes needed for scheduling: A scheduling calculus must include the notion of time, the concept of durational actions and the possibility to delay an action arbitrarily.

Time has been incorporated into the calculi mentioned above (ACP [2], TCCS [13], TCSP [14]), but generally employing instantaneous actions. Durational actions are treated in [1, 10], for example, but these approaches do not allow actions to be delayed (or in the case of [10] only to wait for a synchronization partner). This means that in "$a \parallel b$" actions a and b are always executed simultaneously whereas in scheduling they might as well be run sequentially in either order. In e.g. [4, 6], and in various references quoted in [6], semantics are given that allow for arbitrary delays, but they are not accompanied by an axiomatization. Moreover, nowhere in the process algebra literature a concept of efficiency is formalized resembling the one of [7]. Hence, a new algebra is necessary. We call it *scheduling algebra* and develop its theoretical framework in the following sections.

An example process is given in fig. 1. Its algebraic term is: $a; (b \parallel c; d) \parallel e$.

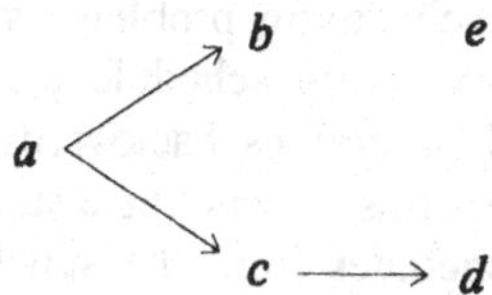

Fig. 1. Example process

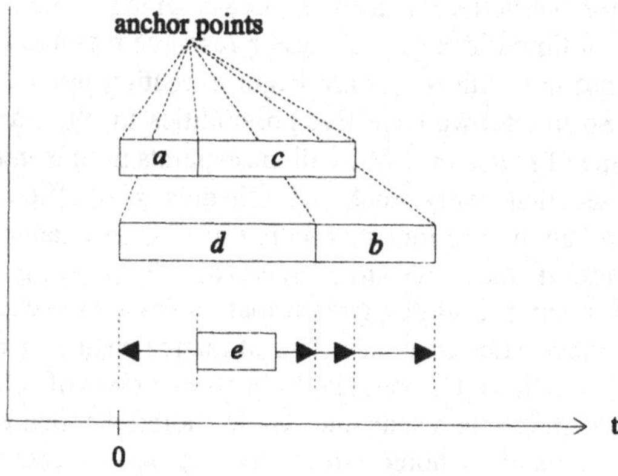

Fig. 2. GANTT diagram

To visualize a schedule, the so-called GANTT chart [5] is used (see fig. 2). It depicts every action as a box labeled with its identifier. The length of the box indicates the duration of the action, and the placement of the box shows exactly when it takes place. In the scenario of fig. 2, actions a and d start simultaneously (on different processors), whereas c is executed immediately after a (and b after d). For the additional action e, the arrows indicate possible placements.

Anchoring and Efficiency

A full-fledged scheduling problem consists of a *specification*, determining a set of valid schedules, some *resource constraints*, saying that certain actions cannot overlap in time, and a *goal function*, specifying how satisfactory certain schedules are. The objective is then to find an optimal schedule among the valid schedules that satisfy the resource constraints. Here we deal with scheduling problems in which the goal function and resource constraints are unspecified. We are interested in describing the valid schedules, and in finding, among the possibly infinite assortment of valid schedules, a finite subset, as small as possible, that for any choice of resource constraints and goal function contains an optimal schedule. In this quest, as is usual in scheduling theory, we limit attention to goal functions that do not favor spending idle time: a valid schedule, meeting the resource constraints, cannot improve by moving an action forwards in time.

A schedule is called *efficient* (or *semi-active*) if it is valid and no actions can be moved backwards in time without violating validity or some hypothetical resource constraints. For the types of scheduling problems studied in this paper, as well as elsewhere in scheduling theory, every schedule can be "improved" to an efficient schedule by moving some of its actions backwards in time (cf. [7, theorem 2.1]). Depending on the goal function, this may not be a strict improvement, but it is never a regress. Thus the efficient schedules form the subclass of schedules sought above. However, in order to be of practical use, a more explicit characterization of the efficient schedules is needed.

We call a schedule *anchored* if each action starts either at time 0 or at the termination time of some other process. Fig. 2 depicts an anchored schedule. If the execution times for the actions *a* to *d* is already given as indicated, an additional action *e* may start at time 0 if it does not use a resource required by either *a* or *d*. If *e* conflicts with *a* but not with *d*, it may begin execution upon termination of *a* (as drawn in fig. 2). So in total we have five possibilities for the starting time of *e*: 0 or the termination time of *a*, *b*, *c* or *d*. We call these points in time *anchor points*.

It is easy to see that every anchored schedule *s* is efficient. If the resource constraints say that any two actions that do not overlap in *s* cannot overlap at all, it is impossible to preserve these resource constraints by moving actions backwards. Moreover, a goal function could be chosen that makes every schedule in which even one action takes place later than in *s* less attractive than *s* itself, no matter what happens to the other actions. Conversely, for a special class of scheduling problems it is shown in [7] that all efficient schedules are anchored, i.e. that every valid schedule can be turned into a valid anchored one by moving actions backwards in time. Here we are interested in scheduling problems for which the same holds. For these the efficient schedules are exactly the anchored ones, and our efficiency operator can be implemented as an anchoring operator.

Syntax

Let *A* be a set of *atomic actions* (external actions), and let *T* be the set of positive reals including zero. A pair (*a*, *t*) of $A \times T$ written as *a*(*t*) is called a time-stamped action or

activity. It indicates that action a takes t units of time (the duration of a). The elements t of T also denote internal time actions which can be seen as silent steps (τ, t).
Process terms can be constructed using the constants

ε	denoting the empty process,
$a(t)^{+} \in A\times T$	denoting the starting of action a with duration t,

the unary operators

$a(t)\cdot$	*Sequence*, $a(t)\cdot\ x$ meaning that x starts immediately after the execution of a, i.e. at time t,
$t\cdot$	*Time offset* for $t\in T$; in $t\cdot\ x$ the starting time of the actions in x is delayed by t time units and an anchor point at time t is added,
Δ	Arbitrary *delay operator*, meaning that the prefixed process can be postponed indefinitely,
[.]	*Delay elimination*, deleting every occurrence of the delay operator from the enclosed term by attaching the processes prefixed by Δ to every possible anchor point,

and the binary operators

$\cup$	*Choice*; $x \cup y$ denotes the execution of either x or y exclusively,
$\curlywedge$	*forking of time* ('Hühnerfuß' or 'chicken claw'), an urgent parallel composition; $x \curlywedge y$ means that the initial actions of both x and y start at the same time.

In addition, we use the following abbreviations:

$a(t);$	(*causal*) *precedence*, a unary operator expressing linear order, so $a(t); x$ says that x has to be executed some time (but not necessarily immediately) after completion of a. We have: $a(t); x = a(t)\cdot \Delta x$.
$\|\|$	*concurrence*, a binary operator expressing independence of processes. We have: $x \| y = \Delta x \curlywedge \Delta y$.

The binding precedence of the binary operators (from loose to tight) is: $\cup$, $\|$ and $\curlywedge$. The unary operators bind tightest.

Terms built from ε, $a(t)^{+}$, $a(t)\cdot$, $t\cdot$, $\curlywedge$ and [.] denote unique schedules. Choice between various schedules is introduced by the operators $\cup$ and Δ, and hence also by $a(t)$; and $\|$. The delay elimination simply prunes some choices introduced by Δ.

We take a, b to range over $A\times T\cup A$, t, u, v to range over T and x, y, z to range over all process expressions. For convenience, we sometimes abbreviate the expression $a(t)$ by a. Moreover, we leave out trailing ε's in terms $a(t)\cdot\ \varepsilon$ and $t\cdot\ \varepsilon$.

Example: The expression $a; b \| c; d$ denotes two concurrent tasks, one executing first a and then b, and the other running c followed by d. Example schedules for the efficient process $[a; b \| c; d]$ are shown in fig. 3.

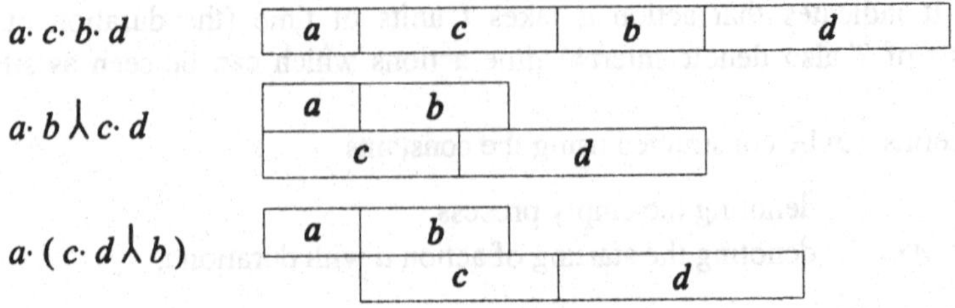

Fig. 3. Example schedules

As *specification terms* we consider processes P or $[P]$ that contain no further occurrences of [.] and no direct occurrences of the sequencing operator $a(t)\cdot$, although occurrences of $a(t);$ are permitted. Let us call an occurrence of $a(t)\cdot$ which can not be regarded as an occurrence of $a(t);$ *tight sequencing*. Tight sequencing can be encoded with the delay elimination operator, namely $a(t)\cdot b(t') = [a(t)\cdot \Delta b(t')] = [a(t);\ b(t')]$. For these specification terms the argument of [7] showing that all efficient schedules are anchored applies, so that the delay elimination operator indeed describes efficiency. This need not be true for general processes. Consider for example the process $\Delta\, a(1)\cdot b(1) \parallel c(2)$. A valid schedule for this process specification is shown in fig. 4. This schedule is efficient because it is the shortest schedule for the given specification if we assume that b and c may not overlap due to a resource conflict. But at the same time it is not anchored because the starting time of action a does not coincide with the finishing time of any other action.

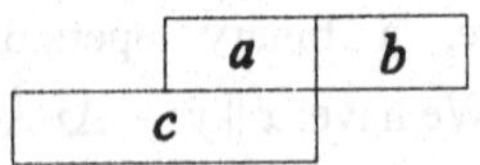

Fig. 4. An efficient schedule that is not anchored

Axiomatic System

The axioms of scheduling algebra are listed below. Please note that "+" refers to the real number addition.

Basic Axioms

$$x \cup (y \cup z) = (x \cup y) \cup z \tag{A1}$$

$$x \cup y = y \cup x \tag{A2}$$

$$x \cup x = x \tag{A3}$$

$$(x \curlywedge y) \curlywedge z = x \curlywedge (y \curlywedge z) \tag{A4}$$

$$x \curlywedge y = y \curlywedge x \tag{A5}$$

Zero Axioms

$$0 \cdot x = x \tag{Z1}$$

$$x \curlywedge \varepsilon = x \tag{Z2}$$

Distributivity of choice

$$t \cdot (x \cup y) = t \cdot x \cup t \cdot y \tag{D1}$$

$$\Delta (x \cup y) = \Delta x \cup \Delta y \tag{D2}$$

$$x \curlywedge (y \cup z) = x \curlywedge y \cup x \curlywedge z \tag{D3}$$

Normal Form Axioms

$$a(t) \cdot x = a(t)^{+} \curlywedge t \cdot x \tag{N1}$$

$$t \cdot x \curlywedge (t+u) \cdot y = t \cdot (x \curlywedge u \cdot y) \tag{N2}$$

$$\Delta x \curlywedge \Delta y = \Delta(x \curlywedge \Delta y \cup y \curlywedge \Delta x) \tag{N3}$$

Elimination of delay (anchor axioms)

$$[x \cup y] = [x] \cup [y] \tag{AA1}$$

$$[a(t)^{+} \curlywedge x] = a(t)^{+} \curlywedge [x] \tag{AA2}$$

$$[t \cdot x] = t \cdot [x] \tag{AA3}$$

$$[\Delta x] = [x] \tag{AA4}$$

$$[t \cdot x \curlywedge \Delta y] = [t \cdot x \curlywedge y] \cup t \cdot [x \curlywedge \Delta y] \tag{AA5}$$

$$[\varepsilon] = \varepsilon \tag{AA6}$$

Proposition
The following equation follows from applying N1, D1, D3 and N1 in the given order:

$$a\cdot (x \cup y) \;=\; a\cdot x \cup a\cdot y \qquad \text{P1}$$

The distributivity axioms and P1 indicate that we work in a linear-time setting. This is because we treat scheduling as a planning problem where all decisions (choices) are made prior to the execution of the process.

Note that we do not have the axioms $x \cup \varepsilon = x$ and $t\cdot u\cdot x = (t+u)\cdot x$. They are not sound because the terms on the left side introduce an additional empty schedule and an anchor point at t, respectively.

Example
Applying the axioms, the definitions and P1 in a suitable order, we can compute the schedules for an example process specification. The penultimate equality shows the term in *normal form* (see below); the last equality gives the term in *anchor form* (explained later). Recall that we abbreviate $a(t)$ by a, and that we leave out trailing ε's.

$$[\, a(1); b(3) \,\|\, c(2); d(1) \,] =$$

$$[\, \Delta(\, a(1)\cdot \Delta b(3)\,) \curlywedge \Delta(\, c(2)\cdot \Delta d(1)\,)\,] =$$

$$\begin{array}{lll}
a^+ \curlywedge 1\,(b^+ \curlywedge 3\,(c^+ \curlywedge 2\,(d^+ \curlywedge 1))) & \cup\; a^+ \curlywedge 1\,(c^+ \curlywedge b^+ \curlywedge 2\,(d^+ \curlywedge 1)) & \cup \\
a^+ \curlywedge 1\,(c^+ \curlywedge b^+ \curlywedge 2\cdot 1\,(d^+ \curlywedge 1)) & \cup\; a^+ \curlywedge 1\,(c^+ \curlywedge 2\,(b^+ \curlywedge 3\,(d^+ \curlywedge 1))) & \cup \\
a^+ \curlywedge 1\,(c^+ \curlywedge 2\,(b^+ \curlywedge (d^+ \curlywedge 1\cdot 2))) & \cup\; a^+ \curlywedge 1\,(c^+ \curlywedge 2\,(d^+ \curlywedge 1\,(b^+ \curlywedge 3))) & \cup \\
c^+ \curlywedge 2\,(a^+ \curlywedge 1\,(b^+ \curlywedge 3\,(d^+ \curlywedge 1))) & \cup\; c^+ \curlywedge 2\,(a^+ \curlywedge 1\,(d^+ \curlywedge (b^+ \curlywedge 1\cdot 2))) & \cup \\
c^+ \curlywedge 2\,(a^+ \curlywedge 1\,(d^+ \curlywedge 1\,(b^+ \curlywedge 3))) & \cup\; c^+ \curlywedge 2\,(d^+ \curlywedge a^+ \curlywedge (1\,(b^+ \curlywedge 3))) & \cup \\
c^+ \curlywedge 2\,(d^+ \curlywedge 1\,(a^+ \curlywedge (1\,(b^+ \curlywedge 3)))) & \cup\; c^+ \curlywedge a^+ \curlywedge 1\,(b^+ \curlywedge 1\cdot 2\,(d^+ \curlywedge 1)) & \cup \\
c^+ \curlywedge a^+ \curlywedge 1\,(b^+ \curlywedge 1\,(d^+ \curlywedge 1\cdot 1)) & \cup\; c^+ \curlywedge a^+ \curlywedge 1\cdot 1\,(b^+ \curlywedge 3\,(d^+ \curlywedge 1)) & \cup \\
c^+ \curlywedge a^+ \curlywedge 1\cdot 1\,(d^+ \curlywedge 1\,(b^+ \curlywedge 3)) & \cup\; c^+ \curlywedge a^+ \curlywedge 1\cdot 1\,(d^+ \curlywedge (b^+ \curlywedge 1\cdot 2)) & =
\end{array}$$

$$\begin{array}{l}
a\cdot b\cdot c\cdot d \cup a\cdot c\cdot b\cdot d \cup a\cdot c\cdot d\cdot b \cup c\cdot a\cdot b\cdot d \cup c\cdot a\cdot d\cdot b \cup c\cdot d\cdot a\cdot b \cup \\
a\cdot (c\cdot d \curlywedge b) \cup a\cdot (b\cdot d \curlywedge c) \cup c\cdot (a \curlywedge d\cdot b) \cup a\cdot c\cdot (b \curlywedge d) \cup a\cdot b \curlywedge c\cdot d \cup \\
c\cdot a\cdot (b \curlywedge d) \cup a \curlywedge c\cdot (b \curlywedge d) \cup a \curlywedge c\cdot b\cdot d \cup a \curlywedge c\cdot d\cdot b \cup c \curlywedge a\cdot b\cdot d
\end{array}$$

Normal Form Theorem

All terms can be written in the form $\bigcup N$ where

$$\begin{array}{lcll}
N & ::= & a(t)^+ \curlywedge N & \\
 & \mid & t\cdot N & \text{for } t \neq 0 \\
 & \mid & \Delta N & \\
 & \mid & t\cdot N \curlywedge \Delta N & \text{for } t \neq 0 \\
 & \mid & \varepsilon &
\end{array}$$

This normal form theorem is proved by structural induction on terms. We show for every operator that the composition of terms in normal form again yields a term that can be expressed in normal form.

The choice operator distributes over every other operator (D1 - D3, P1, AA1) so that it can always be moved to the top level. This and associativity (A1) and commutativity (A2) allow us to denote every term as a union of choice-free subterms. Prefixing a normal term with Δ or $t\cdot$ (for $t\neq 0$) is by definition normal, as is ε. Z1 takes care of terms $0\cdot N$. Using Z2, the constant $a(t)^+$ can be written as the normal form $a(t)^+ \curlywedge \varepsilon$. For $a(t)\cdot N$, axiom N1 leads to the normal form

$$a(t)\cdot N = a(t)^+ \curlywedge t\cdot N.$$

Delay elimination is straightforward with AA1 - AA6. It remains to be shown that *forking* of normal terms can be normalized. Let K, L, M and N be normal terms. Then

$$(a(t)^+ \curlywedge M) \curlywedge N \xrightarrow{\text{A4}} a(t)^+ \curlywedge (M \curlywedge N) \xrightarrow{\text{IH}} a(t)^+ \curlywedge N',$$
$$\varepsilon \curlywedge N \xrightarrow{\text{A5,Z2}} N,$$

where IH abbreviates induction hypothesis, and N' denotes the normalization of $M \curlywedge N$. For the remaining 3 normal forms we have to prove normalization only for the upper triangle of the following matrix (due to commutativity of forking), in which $t, u>0$:

$\curlywedge$	$u\cdot K$	ΔK	$uK \curlywedge \Delta L$
$t\cdot M$	N2 ($t\leq u$), IH	√	A4, N2 ($t\leq u$), IH
ΔM		N3, IH, IH	A5, A4, N3, IH, IH
$tM \curlywedge \Delta N$			A4, A4, A5, A4, A4, N2 ($t\leq u$), N3, IH, IH, IH

Table 1. Normal form proofs

The proof proceeds applying the laws stated modulo commutativity (√ means that the term is already in normal form).

Semantics

As a semantic model, we define schedules similar to the real-time execution sequences in [14]. Each expression maps to a set of possible schedules:

$$[\![.]\!]: Expr \rightarrow 2^{\mathbb{N}^{A\times T\times T} \times 2^{T} \times 2^{T-\{0\}}}$$

A schedule is a triple $\sigma = \langle\sigma_0; \sigma_1; \sigma_2\rangle$ where σ_0 is a multi-set over triples (action, duration, starting time), σ_1 is a set of starting and finishing times and σ_2 a set of anchor points (or finishing times), excluding 0. Because 0 is always an anchor point, there is no need to record it explicitly; suppressing it turns out to have technical advantages. Typically, we leave out the brackets delimiting σ_0, σ_1 and σ_2, i.e.

$$[\![a(1) \cdot b(2) \curlywedge a(1)]\!] = \{ \langle (a, 1, 0), (a, 1, 0), (b, 2, 1) ; 0, 1, 3 ; 1, 3 \rangle \}.$$

This schedule is interpreted as follows:

- $a(1)$ is started twice at time 0,
- $b(2)$ at time 1 (immediately after a),
- the starting times are 0 and 1 (for a and b respectively)
- and the anchor points are 1 and 3 (the finishing times of a and b respectively).

Because Z1 yields $a(1) \cdot b(2) = a(1) \cdot b(2) \cdot 0$ and the latter term has a starting time 3 (for the internal action of duration 0), 3 can be regarded as a potential starting time of the process $a(1) \cdot b(2)$ as well. For this reason we include in σ_1 not only the starting times, but also the finishing times of actions.

For two schedules σ and ρ, we write $\sigma \cup \rho$ for $\langle \sigma_0 \cup \rho_0 ; \sigma_1 \cup \rho_1 ; \sigma_2 \cup \rho_2 \rangle$. The offset of a schedule by a fixed amount of time is defined by:

$$\sigma + u = \langle \{\!\{ (a, t, p+u) \mid (a, t, p) \in \sigma_0 \}\!\} ; \{ p+u \mid p \in \sigma_1 \} ; \{ p+u \mid p \in \sigma_2 \} \rangle.$$

The general definition of the semantics is as follows:

$$[\![\varepsilon]\!] = \{ \langle \varnothing ; 0 ; \varnothing \rangle \} \tag{1}$$

$$[\![a(t)^{+}]\!] = \{ \langle (a, t, 0) ; 0 ; \varnothing \rangle \} \tag{2}$$

$$[\![a(t) \cdot P]\!] = \{ \langle (a, t, 0) ; 0 ; \{t\} - \{0\} \rangle \cup (\sigma + t) \mid \sigma \in [\![P]\!] \} \tag{3}$$

$$[\![t \cdot P]\!] = \{ \langle \varnothing ; 0 ; \{t\} - \{0\} \rangle \cup (\sigma + t) \mid \sigma \in [\![P]\!] \} \tag{4}$$

$$[\![\Delta P]\!] = \{ \langle \varnothing ; 0 ; \varnothing \rangle \cup (\sigma + u) \mid \sigma \in [\![P]\!], u \in T \} \tag{5}$$

$$[\![[P]]\!] = \{ \sigma \in [\![P]\!] \mid \sigma_1 \subset \sigma_2 \cup \{0\} \} \tag{6}$$

$$[\![P \cup Q]\!] = [\![P]\!] \cup [\![Q]\!] \tag{7}$$

$$[\![P \curlywedge Q]\!] = \{ \sigma \cup \rho \mid \sigma \in [\![P]\!], \rho \in [\![Q]\!] \} \tag{8}$$

Note that this definition ensures that $0 \in \sigma_1$, $0 \notin \sigma_2$ and $\sigma_2 \subset \sigma_1$, for all schedules σ.

The Role of the Elimination Operator [.]

A term not containing [.] represents a set of schedules, both efficient and inefficient ones. Delay elimination restricts attention to the efficient schedules, thereby eliminating all schedules known not to be optimal under any goal function or constraints. On the syntactical level, this corresponds to the elimination of Δ's, which is achieved by the operator [.]. One might ask why this is not done implicitly, i.e. by giving every term p the semantics of $[p]$. The answer is that the following implication does not hold:

$$[\![[x]]\!] = [\![[y]]\!] \Rightarrow [\![[x \| z]]\!] = [\![[y \| z]]\!]. \quad ↯$$

A counterexample to the implication can be constructed easily by setting:

$$x = a(1) \cdot b(1),$$
$$y = a(1) \cdot \Delta\, b(1),$$
$$z = c(1).$$

We then get:

$$[\![[a(1) \cdot b(1)]]\!] = \{ \langle (a, 1, 0), (b, 1, 1) \,;\, 0, 1, 2 \,;\, 1, 2 \rangle \} = [\![[a(1) \cdot \Delta\, b(1)]]\!],$$

but:

$$\langle (a, 1, 0), (b, 1, 2), (c, 1, 1) \,;\, 0, 1, 2, 3 \,;\; 1, 2, 3 \rangle \in [\![[a(1) \cdot \Delta\, b(1) \| c(1)]]\!],$$
$$\langle (a, 1, 0), (b, 1, 2), (c, 1, 1) \,;\, 0, 1, 2, 3 \,;\; 1, 2, 3 \rangle \notin [\![[a(1) \cdot b(1) \| c(1)]]\!].$$

Correctness

Axioms A1 to A3 are correct w.r.t. our schedule semantics because set union is associative, commutative and idempotent. Associativity and commutativity of multi-set union yield A4 and A5. Z1 and Z2 are explained by 0 and the empty set being the neutral elements of addition and (multi-)set union respectively, taking into account that 0 is already in σ_1 for every schedule σ. The correctness proofs for the remaining axioms can be found in [9].

Completeness w.r.t. Δ-free Terms

In this section, we will prove completeness of the axiomatic system for Δ-free terms. The normal form theorem allows restriction to Δ-free normal forms, which, for example, arise from normalizing arbitrary terms of the form $[P]$. Each choice-free subterm of such a term corresponds uniquely to a set of only one schedule.

Lemma 1. Let N be a term of the form

$$N \;::=\; a(t)^{+} \curlywedge N \;\mid\; t \cdot N \;\mid\; \varepsilon, \text{ where } t \neq 0.$$

Then $| [\![N]\!] | = 1$.

Proof by structural induction:

$$| [\![\varepsilon]\!] | = | \{ \langle \varnothing \,;\, 0 \,;\, \varnothing \rangle \} | = 1.$$
$$[\![t \cdot N]\!] = \{ \langle \varnothing \,;\, 0 \,;\, \{t\} - \{0\} \rangle \cup (\sigma + t) \mid \sigma \in [\![N]\!] \}.$$

So if $| [\![N]\!] | = 1$, then $| [\![t \cdot N]\!] | = 1$.

The case $a(t)^{+} \curlywedge N$ proceeds likewise.

Lemma 2. Let N, M be of the form stipulated above, then

$$[\![N]\!] = [\![M]\!] \;\Rightarrow\; \text{A4, A5} \vdash N = M.$$

Proof by structural induction:

$$N = a_1(t_1)^+ \curlywedge a_2(t_2)^+ \curlywedge \ldots \curlywedge a_n(t_n)^+ \curlywedge N' \text{ with } N' = \varepsilon \text{ or } N' = t\cdot N'' \text{ with } t{>}0, \text{ and}$$
$$M = b_1(u_1)^+ \curlywedge b_2(u_2)^+ \curlywedge \ldots \curlywedge b_m(u_m)^+ \curlywedge M' \text{ with } M' = \varepsilon \text{ or } M' = u\cdot M'' \text{ with } u{>}0.$$

Let σ be the unique schedule in $[\![N]\!] = [\![M]\!]$. We have $N' = \varepsilon \Leftrightarrow \sigma_2 = \varnothing \Leftrightarrow M' = \varepsilon$.

In case $N' = M' = \varepsilon$ we have

$$\sigma_0 = \{\!|(a_1, t_1, 0), (a_2, t_2, 0), \ldots, (a_n, t_n, 0)|\!\} = \{\!|(b_1, u_1, 0), (b_2, u_2, 0), \ldots, (b_m, u_m, 0)|\!\}.$$

Hence $\{\!|\, a_1(t_1)^+, a_2(t_2)^+, \ldots, a_n(t_n)^+ \,|\!\} = \{\!|\, b_1(u_1)^+, b_2(u_2)^+, \ldots, b_m(u_m)^+ \,|\!\}$, so in particular $n = m$. It follows that A4, A5 $\vdash N = M$.

In case $N' = t\cdot N''$ and $M' = u\cdot M''$ we have

$$[\![N]\!] = \{\, \langle (a_1, t_1, 0), (a_2, t_2, 0), \ldots, (a_n, t_n, 0) \,;\, 0 \,;\, t \rangle \cup (\sigma{+}t) \mid \sigma \in [\![N'']\!] \,\} \text{ and}$$
$$[\![M]\!] = \{\, \langle (b_1, u_1, 0), (b_2, u_2, 0), \ldots, (b_m, u_m, 0) \,;\, 0 \,;\, u \rangle \cup (\rho{+}u) \mid \rho \in [\![M'']\!] \,\}.$$

Again it follows that $\{\!|\, a_1(t_1)^+, a_2(t_2)^+, \ldots, a_n(t_n)^+ \,|\!\} = \{\!|\, b_1(u_1)^+, b_2(u_2)^+, \ldots, b_m(u_m)^+ \,|\!\}$, so in particular $n = m$. Furthermore, we have $u = t$ because t is the least anchor point in the unique schedule of $[\![N]\!]$, and so is u for $[\![M]\!]$. Finally, $[\![N'']\!] = [\![M'']\!]$, and therefore by induction A4, A5 $\vdash N'' = M''$. Hence A4, A5 $\vdash N = M$.

Theorem. Let K, L be Δ-free normal forms with $[\![K]\!] = [\![L]\!]$ then

$$\text{A1}, \ldots, \text{A5} \vdash K = L.$$

Proof: Using A1, A2, A3, it suffices to show that for every choice-free subterm N of K there is a choice-free subterm M of L such that A4, A5 $\vdash N = M$ (and vice versa). By lemma 1, $[\![N]\!] = \{\sigma\}$, so $\sigma \in [\![K]\!] = [\![L]\!]$. Again by lemma 1, there is a choice-free subterm M of L such that $[\![M]\!] = \{\sigma\} = [\![N]\!]$. Now, by lemma 2, A4, A5 $\vdash N = M$.

Anchor Form

For every t-free term $[P]$ we can eliminate $[.]$ without introducing time offsets.

$$P \text{ is } t\text{-free} \;\Rightarrow\; \exists Q\colon Q = [P] \text{ where } Q \text{ contains neither } [.] \text{ nor } t\cdot$$

We call this the *anchor form* of a term because all schedules are given by just anchoring actions to the endpoints of others. The GANTT chart can be drawn easily from this form.

Using a straightforward structural induction, one can establish that for every schedule σ of a t-free term we have

$$\forall t \in \sigma_2\colon \exists (a, u, v) \in \sigma_0\colon t = u + v \text{ with } u > 0. \tag{*}$$

Furthermore, employing N2, Z2 and again N2, we obtain

$$t\cdot (u\cdot x \curlywedge v\cdot y) \;=\; t\cdot \varepsilon \curlywedge (t{+}u)\cdot x \curlywedge (t{+}v)\cdot y.$$

Every term $[P]$ can be written in a Δ-free normal form, which, using structural induction, Z1 and the law above, can be converted to a sum of terms of the form

$$u_1\cdot a_1(t_1)^+ \curlywedge u_2\cdot a_2(t_2)^+ \curlywedge \ldots \curlywedge u_k\cdot a_k(t_k)^+ \curlywedge u_{k+1}\cdot \varepsilon \curlywedge u_{k+2}\cdot \varepsilon \curlywedge \ldots \curlywedge u_n\cdot \varepsilon.$$

Assuming P is t-free, (*) implies that for every $u_i > 0$ there exists a $j \in \{1, \ldots k\}$ such that $u_i = u_j + t_j$ and $t_j > 0$. We will remove step by step the time offsets $u_i\cdot$ in the term above, starting with the largest, until only time offsets $0\cdot$ remain. The intermediate terms encountered during this process will always remain in the form

$$u_1\cdot a_1(t_1)^+ \curlywedge u_2\cdot a_2(t_2)^+ \curlywedge \ldots \curlywedge u_k\cdot a_k(t_k)^+ \curlywedge u_{k+1}\cdot N_{k+1} \curlywedge u_{k+2}\cdot N_{k+2} \curlywedge \ldots \curlywedge u_n\cdot N_n \qquad (**)$$

where for every $u_i > 0$ there exists a $j \in \{1, \ldots k\}$ such that $u_i = u_j + t_j$, $t_j > 0$ and the N_i contain neither time offsets nor occurrences of [.] or Δ.

Let $v > 0$ be the maximum of $\{ u_1, u_2, \ldots u_n \}$, and let $j \in \{1, \ldots k\}$ be such that $v = u_j + t_j$ and $t_j > 0$. Using A4-5, we write all subterms $u_i\cdot a_i(t_i)$ and $u_i\cdot N_i$ with $u_i = v$ as well as $u_j\cdot a_j(t_j)$ at the right of the formula, thus obtaining an expression of the form

$$u'_1\cdot a'_1(t'_1)^+ \curlywedge u'_2\cdot a'_2(t'_2)^+ \curlywedge \ldots \curlywedge u'_l\cdot a'_l(t'_l)^+ \curlywedge u'_{l+1}\cdot N'_{l+1} \curlywedge u'_{l+2}\cdot N'_{l+2} \curlywedge \ldots \curlywedge u'_m\cdot N'_m \curlywedge K$$

with $K = u_j\cdot a_j(t_j)^+ \curlywedge v\cdot M_1 \curlywedge v\cdot M_2 \curlywedge \ldots \curlywedge v\cdot M_h$. Using axioms N2 and Z1, K can be rewritten as $u_j\cdot a_j(t_j)^+ \curlywedge v\cdot (M_1 \curlywedge M_2 \curlywedge \ldots \curlywedge M_h)$ and, with N1, as $u_j\cdot a_j(t_j)\cdot (M_1 \curlywedge M_2 \curlywedge \ldots \curlywedge M_h)$. This brings the term again in the form (**) and we can continue with the next-largest member of $\{ u_1, u_2, \ldots u_n \}$. In the end, all remaining u_i's will be 0 and Z1 turns the term into the required form.

Example

Consider the term [a; (b || c) || d] where all actions consume one unit of time. Transformation of this term into its anchor form yields 19 possible schedules. They are shown in fig. 5 grouped into the three basic orders "a; (b || c)", "a; b; c" and "a; c; b" with the independent action d attached to an anchor point or inserted in front of or in between the other actions.

This example might suggest that it is possible to convert a t-free term $[P]$ into its anchor form without considering the duration of its actions. However, this is not possible in general: the anchor form of the term [a(1); b(3) || c(2); d(1)] that has been calculated in a previous example contains the schedule $a(1) \curlywedge c(2)\cdot (b(3) \curlywedge d(1))$, but not $a(1)\cdot (b(3) \curlywedge d(1)) \curlywedge c(2)$.

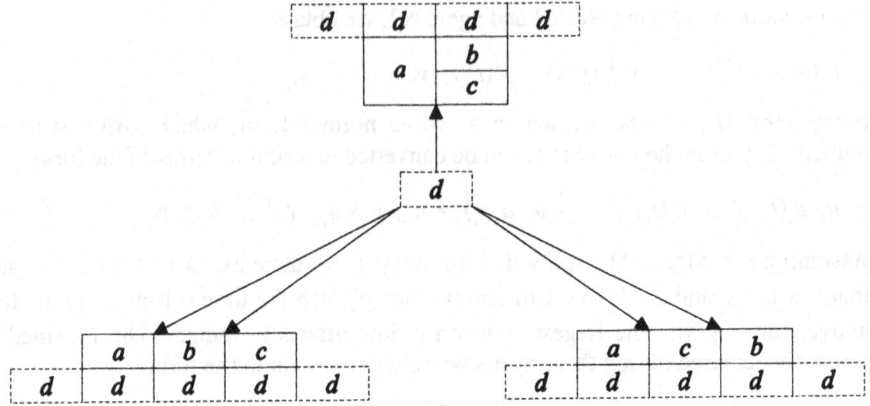

Fig. 5. All schedules

Term Rewriting

We show that modulo associativity and commutativity of choice and Hühnerfuß our axioms, when read from left to right, form a confluent and (strongly) normalizing rewrite system, provided that we add the rewrite rule $[a(t)^{+}] \rightarrow a(t)^{+}$ (which is derivable from the axiom system). If we drop the rewrite rule A3 (the idempotence of choice), the normal forms are exactly the normal forms in the sense of our normal form theorem, except that all subterms $a(t)^{+} \curlywedge \varepsilon$ are replaced by $a(t)^{+}$ (so $a(t)^{+}$ is a normal form for the rewrite system). In the presence of A3 the same normal forms apply, but without pairs of summands that are equal modulo A4 and A5.

It is easy to check that such normal forms allow no further rewriting; the proof of the normal form theorem (but omitting the reverse application of Z2) shows essentially that all other terms can be rewritten into a normal form. Thus the rewrite system is normalizing. Here we omit the proof that it is even strongly normalizing; however, it involves only standard term rewriting techniques.

Confluence for terms of the form $[P]$ follows immediately from the proof of our completeness theorem for such terms (because only the axioms A1-5 are involved in the completeness proof for Δ-free normal forms). Confluence for the entire rewrite system can be established by the Knuth-Bendix method, skipping pairs of overlapping redexes that lay within a term of the form $[P]$. In the table 2, we review the relevant pairs of overlapping redexes.

An implementation of the system without A3 can be found at the URL

`http://www.uni-koblenz.de/~rittgen/SA.html.`

This applet generates the normal form for the axiom system given any term of your chosing.

A3 with D1, D2 and D3	trivial
Z1 with D1 and N2 (in two ways)	trivial
Z2 with D3 and Z2	trivial
D1 with N2 (in two ways)	trivial with D3
D2 with N3 (in two ways)	trivial with D3
D3 with D3	trivial
N2 with itself	OK
N3 with itself	OK

Table 2. Overlapping redexes

Conclusion

In this paper, we have proposed a calculus for durational actions, namely scheduling algebra. The constants and operators of this algebra allow to specify process terms as a multi-tree order over actions with a certain duration. A normal form exists for each term which for Δ-free terms is unique modulo A1-5. Using the axioms of this calculus, the set of all efficient schedules $[P]$ can be "computed" for any process P without tight sequencing. The operator [.] does so by generating a subterm (schedule) for any possible attachment of actions to anchor points, thus eliminating all time delays Δ.

We established soundness of the axiomatic system w.r.t. a schedule semantics and showed that the algebra is complete for Δ-free terms (sets of schedules), such as arise from terms of the form $[P]$.

Currently, our calculus is designed for multi-tree precedence. To extend it to arbitrary orders (with more than one predecessor to an action), synchronization must be present in the algebra (as follows from the existence of partial orders which are not "series-parallel" [8]). This is left for future research.

References

1. Aceto, L.; Murphy, D.: *Timing and Causality in Process Algebra*, Acta Informatica 33, 1996, pp. 317-350
2. Baeten, J.C.M.; Bergstra, J.A.: *Real Time Process Algebra*, Formal Aspects of Computing 3(2), 1991, pp. 142-188
3. Bergstra, J.A.; Klop, J.W.: *Process Algebra for Synchronous Communication*, Information and Control 60, 1984, pp. 109-137
4. Chen, X.J.; Corradini, F.; Gorrieri, R.: *A Study on the Specification and Verification of Performance Properties*, Proceedings of AMAST '96, Lecture Notes in Computer Science 1101, Springer, Berlin, 1996, pp. 306-320

5. Clark, W.: *The Gantt Chart: A Working Tool of Management*, Pitman, London, 1935
6. Corradini, F.: *On Performance Congruences for Process Algebras*, Information and Computation 145(2), 1998, pp. 191-230
7. French, S.: *Sequencing and Scheduling: An Introduction to the Mathematics of the Job-Shop*, Ellis Horwood, Chichester, 1982
8. Gischer, J.L.: *The Equational Theory of Pomsets*, Theoretical Computer Science 61, 1988, pp. 199-224
9. Glabbeek, R.J. van; Rittgen, P.: *Scheduling Algebra*, Arbeitsberichte des Instituts für Wirtschaftsinformatik Nr. 12, Universität Koblenz-Landau, Germany, 1998 http://www.uni-koblenz.de/~iwi/publicfiles/Arbeitsberichte/Nr12.ps.zip
10. Gorrieri, R.; Roccetti, M.; Stancampiano, E.: *A Theory of Processes with Durational Actions*, Theoretical Computer Science 140(1), 1995, pp. 73-94
11. Hoare, C.A.R.: *Communicating Sequential Processes*, Prentice Hall, Englewood Cliffs, 1985
12. Milner, R.: *A Calculus of Communicating Systems*, Lecture Notes in Computer Science 92, Springer, Berlin, 1980
13. Moller, F.; Tofts, C.: *A Temporal Calculus of Communicating Systems*, in: Baeten, J.C.M.; Klop, J.W.: Proceedings of CONCUR '90, Lecture Notes in Computer Science 458, Springer, Berlin, 1990, pp. 401-415
14. Reed, G.M.; Roscoe, A.W.: *A Timed Model for Communicating Sequential Processes*, Theoretical Computer Science 58, 1988, pp. 249-261

Algebraic Semantics of Coordination or What Is in a Signature[(*)]

José Luiz Fiadeiro[‡] and Antónia Lopes

Department of Informatics
Faculty of Sciences, University of Lisbon,
Campo Grande, 1700 Lisboa, PORTUGAL
{llf,mal}@di.fc.ul.pt

Abstract. We propose an algebraic characterisation of the notion of coordination in the sense of recently proposed languages and computational models that provide a clear separation between the modelling of individual software components and their interaction in the overall software organisation. We show how this separation can be captured in Goguen's categorical approach to General Systems Theory and borrow examples from specification logics, program design languages, mathematical models of behaviour, and coordination languages to illustrate the applicability of our algebraic characterisation.

1 Introduction

Several recently proposed languages and computational models, e.g. those discussed in [4], support the separation between what, in the definition of a system, is responsible for its computational aspects and what is concerned with *coordinating* the interaction between its different components. As explained in [12]: "(A) computation model allows programmers to build a single computational activity: a single-threaded, step-at-a-time computation; (a) coordination model is the glue that binds separate activities into an ensemble".

The clean separation that is achieved between individual software components and their interaction in the overall software organisation makes large applications more tractable, supports global analysis, and enhances reuse of software. Hence, it is not surprising that the significance of this original work on "coordination languages" has now been recognised in areas of Software Engineering concerned with system configuration and architectural description languages [11].

(*) This work was partially supported by LMF-DI (PUC-Rio, Brazil) through the ARTS project, and through PRAXIS XXI contracts PCSH/OGE/1038/95 (MAGO), 2/2.1/TIT/1662/95 (SARA) and PCEX/P/MAT/46/96 (ACL).

‡ On leave at Laboratório de Métodos Formais, Departamento de Informática, Pontifícia Universidade Católica do Rio de Janeiro, Brazil, as a grantee of CNPq (Brazil) and ICCI (Portugal).

A.M. Haeberer (Ed.): AMAST'98, LNCS 1548, pp. 293-307, 1998.

In this paper, we show how the separation between computation and coordination can be captured in the framework of Goguen's categorical approach to General Systems Theory [13]. We capitalise on our previous work on the formalisation of architectural principles in software design [7], which was based on a formal notion of "coordination" that we wish to revise, motivate, discuss and put forward in a more comprehensive way. Examples are drawn from specification logics, concurrency models, parallel program design languages and, of course, coordination languages.

2 Coordination in the Context of General Systems

We start by illustrating the categorical framework that we have been adopting for modelling the development of complex systems, and then motivate the formalisation of "coordination".

2.1 The Categorical Approach to Systems Modelling – An Example

The basic motto of the categorical approach to systems, as explained in [13], is that morphisms can be used to express interaction between components, so that "given a category of widgets, the operation of putting a system of widgets together to form some super-widget corresponds to taking the colimit of the diagram of widgets that shows how to interconnect them" [15]. As shown in [5,16], these categorical principles can be used to formalise process models for concurrent systems such as transition systems, synchronisation trees, event structures, etc. We shall illustrate the approach using a trace-based model.

A process alphabet is a finite set, and a process is a pair $<A,\Lambda>$ *where A is an alphabet and* $\Lambda \subseteq 2^{A^\omega}$ *is the language of the process, where* 2^{A^ω} *denotes the set of infinite sequences over* 2^A.

The alphabet models the set of actions in which the process may involve itself. Each sequence of events in the language of the process captures a possible behaviour of the process, where each event consists of a set of actions that occur concurrently during that event. The empty set of actions models an event of the environment.

We take a morphism of process alphabets to be a total function, and a process morphism $f{:}<A_1,\Lambda_1>\rightarrow<A_2,\Lambda_2>$ *as an alphabet morphism* $f{:}A_1\rightarrow A_2$ *such that, for every* $\lambda\in\Lambda_2$, $f^{-1}(\lambda)\in\Lambda_1$, *where* $f^{-1}(\lambda)(i)=f^{-1}(\lambda(i))$.

The idea is that a morphism of processes captures the relationship that exists between a system (target of the morphism) and any of its components (source). That is, every morphism identifies a component within a system. Hence, an alphabet morphism identifies each action of a component with an action of the system. Each such morphism f defines a contravariant mapping between the sets of events associated with each process $f^{-1}: 2^{A_2}\rightarrow 2^{A_1}$. That is, each event in the life of the system is mapped to an event in the life of the component. The empty event arises when the component is not involved in that specific event of the system, which then acts as the environment for the component. Finally, each behaviour of the system is mapped to one of the possible behaviours of the component.

Diagrams express how a complex system is put together through the interconnection of simpler components. The colimit of such a configuration diagram returns the process that results from the interconnection. The simplest configuration diagram expresses the interconnection between two components via a third one:

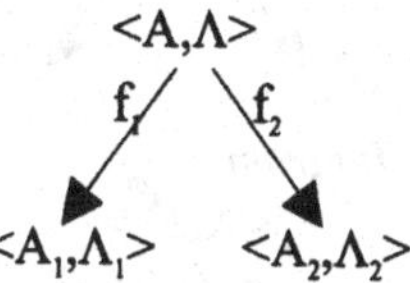

The colimit (pushout) of this diagram is calculated as follows: the pushout (amalgamated sum) of the underlying diagram of process alphabets is calculated, returning

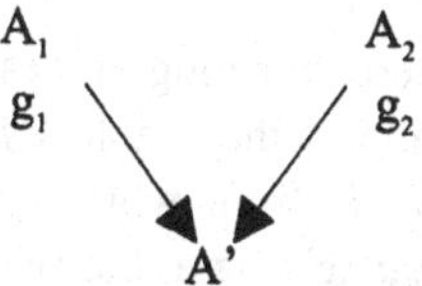

The alphabet A' is obtained from the disjoint union of A_1 and A_2 through the quotient that results from the equivalence relation generated by the set of all pairs $<f_1(a),f_2(a)>$ where $a\in A$. That is to say, each action of A establishes a synchronisation point for the component processes $<A_1,\Lambda_1>$ and $<A_2,\Lambda_2>$.

The resulting process is then calculated over the alphabet thus computed by taking the intersection of the inverse images of the component behaviours:

$$\Lambda'=\{\lambda\in 2^{A'} : g_1^{-1}(\lambda)\in\Lambda_1 \text{ and } g_2^{-1}(\lambda)\in\Lambda_2\}$$

That is to say, the system thus configured can execute all the actions that its components can, subject to the synchronisations specified by the interconnection, and exhibits the behaviours that are allowed by both its components.

2.2 Separating Coordination

How can we talk about computation and coordination in the example above, and in what sense can they be separated?

It seems intuitive to associate the "computational" part of the model to the set Λ of traces in the sense that this set is what captures the local behaviour of the process.

It seems also clear that interconnection between processes, the "coordination" part of the model, is achieved through the alphabets. Indeed, in the proposed model, processes interact by synchronising at designated actions identified via morphisms from what we could call *channels* (or points of *rendez-vous*). These channels correspond to the middle process that we used in the previous example.

It is not difficult to see that only alphabets are involved in interconnections. On the one hand, as we have seen, the alphabet of the process resulting from an interconnection is obtained from the pushout of the underlying diagram of alphabets. The behaviours of the processes involved do not interfere in this calculation. On the other hand, the behaviour of the middle process is not relevant for determining the behaviour of the resulting process: the interconnection is expressed, completely, in the alphabet morphisms that result from the pushout. This property is illustrated by the fact that *every diagram of the form*

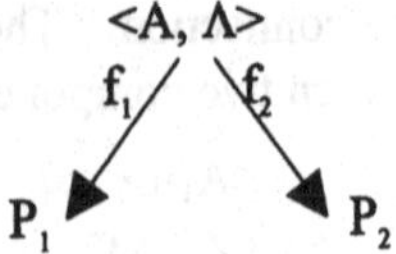

admits the same pushouts as the diagram

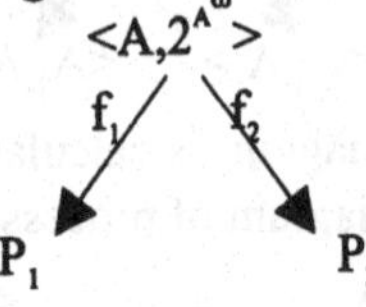

Hence, for the purposes of interconnecting processes, it is sufficient to use middle processes whose set of behaviours is the whole language. That is to say, we can identify a precise class of *channels* from which any interconnection can be built. Notice that the fact that such channels have the whole language for their set of behaviours means that they have no behaviour of their own, i.e. are "passive" and just transmit signals between components. Hence, they can be identified with alphabets.

In this sense, alphabets represent the coordination part of the model. They provide the *interfaces* over which interconnections between components are established. We can then say that coordination is separated from computation, and that the trace model we described is "coordinated over alphabets".

2.3 Coordinated Formalisms

Let us see how, from the example above, we can generalise a set of requirements for considering a category of systems, or abstractions of systems, to be "coordinated" over a given notion of *interface*.

We shall take the separation between coordination and computation to be materialised through a functor mapping systems to interfaces. We require this functor to be faithful (injective over each hom-set), meaning that morphisms of systems cannot induce more relationships between systems than between their underlying interfaces.

Consider given a category ***SYS*** *(of systems) and a category* ***INT*** *(of interfaces) together with a faithful functor* ***int:SYS→INT***.

Which properties should we require of ***int*** that make ***SYS*** coordinated over ***INT***? Basically, we have to capture the fact that any interconnection of systems is established via their interfaces. Part of this intuition can be expressed by the property that *the coordination functor* ***int*** *lifts colimits*.

That is to say, given any diagram ***dia:I→SYS*** and colimit $(int(S_i)\rightarrow C)_{i:I}$ of ***(dia;int)*** there exists a colimit $(S_i\rightarrow S)_{i:I}$ of ***dia*** such that $int(S_i\rightarrow S)=(int(S_i)\rightarrow C)$. In other words, if we interconnect system components through a diagram, then any colimit of the underlying diagram of interfaces can be lifted to a colimit of the original diagram of system components.

There are two aspects in this requirement that should be noted.

On the one hand, lifting means that if the configuration of interfaces is "viable", in the sense that it has a colimit (i.e. gives rise to a system), then so is the corresponding configuration of components. Indeed, although the category of processes defined in 2.1 is cocomplete, meaning that every diagram admits a colimit and, hence, represents the configuration of a "real system", not every category of systems needs to satisfy this property. The requirement that ***int*** lifts colimits means that the computations assigned to the components cannot interfere with the viability of the underlying configuration of interfaces.

On the other hand, lifting of colimits also requires that the system that results from the colimit in ***SYS*** be mapped to the underlying colimit of interfaces. That is to say, the computations assigned to the components cannot restrict the interface of the resulting system, which is calculated through the colimit in ***INT***.

The inverse property requires that every interconnection of system components is an interconnection of the underlying interfaces. In particular, it requires that the computations do not make viable a configuration of system components whose underlying configuration of interfaces is not viable. This property is verified when *the coordination functor **int** preserves colimits*.

That is to say, given any diagram ***dia**:**I**→**SYS*** and colimit $(S_i \to S)_{i:I}$ of ***dia***, $(\boldsymbol{int}(S_i) \to \boldsymbol{int}(S))_{i:I}$ is a colimit of ***(dia;int)***. This property means, using the terminology of [1], that all colimits in ***SYS*** are *concrete*.

These two properties together imply that any colimit in ***SYS*** can be computed by first translating the diagram to ***INT***, then computing the colimit in ***INT***, and finally lifting the result back to ***SYS***.

Preservation and lifting of colimits are two properties that relate diagrams in ***INT*** and diagrams in ***SYS***. We would now like that, similarly to what we showed for processes, interconnections of components can be achieved by using interfaces, or system components that "correspond" to interfaces (channels), as "middle objects". The property that we have in mind is the *existence of discrete structures* for ***int*** as a concrete category in the sense of [1]:

*For every interface C:**INT** there exists s(C):**SYS** such that, for every morphism f:C→**int**(S), there is a morphism g:s(C)→S such that **int**(g)=f.*

That is to say, every interface C has a "realisation" (a discrete lift) as a system component *s(C)* in the sense that, using *C* to interconnect a component *S*, which is achieved through a morphism *f:C→**int**(S)*, is tantamount to using *s(C)* through any *g:s(C)→S such that **int**(g)=f*. Notice that, because ***int*** is faithful, there is only one such g, which means that *f* and g are, essentially, the same. That is, sources of morphisms in diagrams in ***SYS*** are, essentially, interfaces.

This property allows us to use graphical representations in which interfaces are used as connectors between components, a kind of "hybrid diagrams" that are more economical. For instance, in the previous section, discrete lifts are given by the pairs $\langle A,2^{A^\omega}\rangle$. Indeed, morphisms between alphabets A and B are exactly the same as morphisms between $\langle A,2^{A^\omega}\rangle$ and any process $\langle B,\Lambda\rangle$. Hence, we could have used

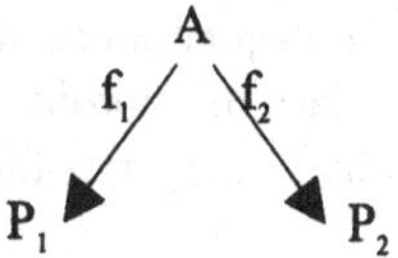

to express the interconnection between P_1 and P_2.

Because ***int*** is faithful, the existence of discrete structures implies that ***int*** admits a left adjoint ***sys:INT→SYS*** such that (1) ***sys(C)=s(C)*** for every ***C:INT***, and (2) ***sys;int=id***$_{INT}$. Hence, ***sys*** is a full embedding which means that, as illustrated in the previous section, interfaces can be identified with a particular subclass of system components: the subcategory of ***channels***.

In [7,8], we characterised coordinated formalisms precisely in terms of the existence of a full embedding that is a left adjoint for the forgetful functor ***int***. We feel, however, that the additional properties of preservation and lifting of colimits are equally important. They are the ones that establish that colimits in ***INT*** and in ***SYS*** have the same expressive power as far as interconnections of components are concerned.

The existence of discrete lifts allows us to simplify the way in which we interconnect components by limiting "middle objects" to channels. Another consequence of this fact is that, being faithful, ***int*** preserves colimits. Therefore, the characterisation of coordination can be reduced to lifting of colimits and existence of discrete structures.

There is an observation that sheds additional light on the nature of the formalisms that we have been characterising. The fact that ***int*** lifts colimits and has discrete structures implies that ***SYS*** is "almost" topological over ***INT***. To be topological [1], ***int*** would have to lift colimits uniquely, which would make the concrete category amnestic, i.e. the fibres of interfaces would have to be partial orders. As far as the algebraic properties of the underlying formalism are concerned, this is not a problem because every concrete category can be modified to produce an amnestic, concretely equivalent version. However, and although the process category discussed in 2.1 is amnestic, we shall see two examples of concrete categories that are not topological but which we would still like to consider to be coordinated.

2.4 Summary

Definition: *A functor **int:SYS→INT** is said to be coordinated, and **SYS** is said to be coordinated over **INT**, iff*

- ***int*** *is faithful;*
- ***int*** *lifts colimits;*
- ***int*** *has discrete structures.*

Proposition: *Let **int:SYS→INT** be coordinated. The following properties hold:*

- ***int*** *admits a left adjoint **sys:INT→SYS** which is a full embedding and satisfies **sys;int=id***$_{INT}$*;*
- ***int*** *preserves colimits;*
- *if **INT** is (finitely) cocomplete then so is **SYS**.*

Proposition: *Every topological category is coordinated.*

This property tells us that the class of coordinated categories has many "interesting" categories. It also includes many categories that are "interesting" in Computing:

***Proposition:** Let $\mathbf{THE}_I$ and $\mathbf{PRE}_I$ be the categories of theories and theory presentations of an institution I [14]. Both $\mathbf{THE}_I$ and $\mathbf{PRE}_I$ are coordinated over the underlying category of signatures.*

The same result holds if we work with •-institutions [6]. Recall that, in both cases of institutions and •-institutions, the objects of $\mathbf{THE}_I$ consist of pairs $<\Sigma,\Phi>$ where Σ is a signature and Φ is a set of sentences over the language of Σ that is closed under consequence. Theory morphisms are signature morphisms that induce inclusions between the sets of theorems. Such categories are topological.

However, if we take the usual definition of $\mathbf{PRE}_I$ as having for objects pairs $<\Sigma,\Phi>$ where Σ is a signature and Φ is a set of sentences over the language of Σ, not necessarily closed under consequence, and for morphisms all signature morphisms that induce theory morphisms between the presented theories (i.e. preserve theorems), then we obtain a coordinated category that is not topological. Indeed, the class of presentations over a given signature is not a partial order because any two presentations of the same theory are isomorphic but not necessarily identical. In practical terms, this means that colimits are not lifted uniquely from signatures to presentations.

We can also find plenty of examples of coordinated categories among models of concurrency, of which the model presented in section 2.1 is a particularly simple case.

2.5 An Example from Coordination Languages

In this section, we briefly discuss an example borrowed from coordination formalisms: the language Gamma [2] based on the chemical reaction paradigm.

A Gamma program P consists of

- a signature $\Sigma=<S,\Omega,\Pi>$, where S is a set of sorts, Ω is a set of operation symbols and Π is a set of relation symbols, representing the data types that the program uses;
- a set of reactions, where a reaction R has the following structure:

$$R \equiv X, t_1, ..., t_n \rightarrow t'_1, ..., t'_m \Leftarrow c$$

where

- X is a set (of variables); each variable is typed by a data sort in S;
- $t_1, ..., t_n \rightarrow t'_1, ..., t'_m$ is the action of the reaction – a pair of sets of terms over X;
- c is the reaction condition – a proposition over X.

An example of a Gamma program is the following producer of burgers and salads from, respectively, meat and vegetables:

PROD ≡	*sorts*	meat, veg, burger, salad
	ops	vprod: veg→salad, mprod: meat→burger
	reactions	m:meat, m → mprod(m)
		v:veg, v → vprod(v)

Parallel composition of Gamma programs, defined in [2], is a program consisting of all the reactions of the component programs. Its behaviour is obtained by executing the reactions of the component programs in any order, possibly in parallel. This leads us to the following notion of morphism. A morphism σ between Gamma programs P_1 and P_2 is a morphism between the underlying data signatures s.t. $\sigma(P_1)\subseteq P_2$, i.e., P_2 has more reactions than P_1.

Concerning system configuration in Gamma, let us consider that we want to interconnect the producer with the following consumer:

CONS ≡ *sorts* food, waste
ops cons: food →waste
reactions f:food, f → cons(f)

The interconnection of the two programs is based on the identification of the food the consumer consumes, that is, the interconnection is established between their data types. For instance, the coordination of the producer and the consumer based on meat is given by the following interconnection:

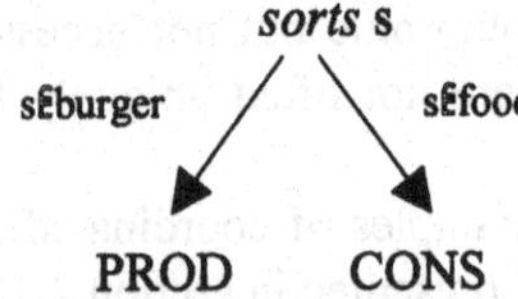

Gamma is, indeed, coordinated over the category of data types:

- the forgetful functor *dt* from Gamma programs to data types is faithful;
- given any diagram in the category Gamma, a colimit $\sigma_i{:}(dt(P_i)\rightarrow\Sigma)_{i\in I}$ of the corresponding diagram in the category of data types is lifted to the following colimit of programs $\sigma_i{:}(P_i\rightarrow<\Sigma, \cup\sigma_j(R_j)>)_{i\in I}$;
- the discrete lift of a data type is the program with the empty set of reactions.

Notice, however, that we have extended the way in which Gamma programs are traditionally put together. Gamma assumes a global data space whereas we have made it possible for Gamma programs to be developed separately and put together by matching the features that they are required to have in common. This localisation further enhances the reusability of coordinated programs.

3 An Example from Parallel Program Design

In order to consolidate the definitions put forward in the previous section we shall now discuss an example drawn from parallel program design.

The language COMMUNITY [9] is similar to IP [10] and UNITY [3]. Its definition has evolved over the years through experience gained in using it in different contexts. It is precisely the changes that were required to make it coordinated that we shall illustrate in this section. On the one hand, we feel that these changes reveal more of our intuition of what it means to be "coordinated". On the other hand, they reflect some of the typical hesitations that one faces when designing formalisms, and for which the need to establish coordinated frameworks helps to make a decision.

3.1 COMMUNITY

We assume a fixed algebraic specification $<\Sigma,\Phi>$ representing the data supported by the language. That is to say, $\Sigma=<S,\Omega>$ is a signature in the usual algebraic sense and Φ is a set of (first-order) axioms over Σ defining the properties of the operations. Data types can be made local to each program but assuming them to be fixed simplifies considerably the presentation.

A COMMUNITY program P has the following structure:

$$\begin{array}{lll} P \equiv & var & V \\ & read & R \\ & init & I \\ & do & \mathop{[]}\limits_{g \in \Gamma} g: [B(g) \rightarrow \mathop{\|}\limits_{a \in D(g)} a{:=}F(g,a)] \end{array}$$

where

- V is the set of local attributes (i.e. the program "variables"); each attribute is typed by a data sort in S;
- R is the set of read-only attributes used by the program (i.e. attributes that are to be instantiated with local attributes of other components in the environment); each attribute is typed by a data sort in S;
- Γ is the set of *action names*; each action name has an associated statement (see below) and can act as a *rendez-vous* point for program synchronisation;
- I is a condition on the attributes – the initialisation condition;
- for every action $g\in\Gamma$, D(g) is the set of attributes that g can change (its domain or write frame); we also denote by D(a) the set of actions in Γ that have the attribute *a* in their write frame;
- for every action $g\in\Gamma$, B(g) is a condition on the attributes – its *guard*;
- for every action $g\in\Gamma$ and attribute $a\in D(g)$, F(g,a) is a term denoting the value that *g* assigns to *a*.

An example of a COMMUNITY program is the following vending machine:

$$\begin{array}{lll} VM \equiv & var & \text{ready, eat, drink: bool} \\ & do & \text{coin}: [\neg ready \wedge (eat \vee drink) \rightarrow ready{:=}tt \,\|\, eat{:=}ff \,\|\, drink{:=}ff] \\ & [] & \text{cake}: [ready \wedge \neg(eat \vee drink) \rightarrow eat{:=}tt \,\|\, drink{:=}ff] \\ & [] & \text{coke}: [ready \wedge \neg(eat \vee drink) \rightarrow drink{:=}tt \,\|\, eat{:=}ff] \\ & [] & \text{reset}: [ready \wedge (eat \vee drink) \rightarrow ready{:=}ff] \end{array}$$

The machine is initialised so as to accept only coins. Once it accepts a coin it can deliver either a cake or a coke (but not both). After delivering a cake or a coke it can only be reset, after which it is ready to accept more coins.

A morphism σ between COMMUNITY programs P_1 and P_2 consists of:

- a map $\sigma_\alpha: V_1 \cup R_1 \rightarrow V_2 \cup R_2$;
- a map $\sigma_\gamma: \Gamma_1 \rightarrow \Gamma_2$

such that,

1. For every $a\in V_1\cup R_1$, $sort(a)=sort(\sigma_\alpha(a))$;
2. For every $a\in V_1$, $\sigma_\alpha(a)\in V_2$;
3. For every $a\in V_1$, $\sigma_\gamma(D_1(a))=D_2(\sigma_\alpha(a))$;
4. $\Phi \vdash (I_2 \supset \sigma(I_1))$;

5. Forall $g_1 \in \Gamma_1, a_1 \in D_1(g_1)$,
$\Phi\,\hat{\imath}\,B_2(\sigma(g_1)) \supset (F_2(\sigma(g_1),\sigma(a_1))=\sigma(F_1(g_1,a_1)))$;
6. For every $g_1 \in \Gamma_1$, $\Phi\,\hat{\imath}\,(B_2(\sigma(g_1)) \supset \sigma(B_1(g_1)))$

where $\hat{\imath}$ means consequence in the first-order sense, and σ is also used to denote the translation induced by the morphism over the language of the source signature.

Condition 1 indicates that morphisms have to respect the sorts of the attributes. Condition 2 means that local attributes of a component must also be local within the system. It also allows read attributes of a component to become local in the system: this is the typical situation when the attribute being read by the component is local to some other component within the same system. Condition 3 does not allow actions of the system that do not belong to the component to change the local attributes of the component. Condition 4 means that the initialisation condition of the component must be respected by the system. Condition 5 means that assignments made by the component are preserved. Condition 6 means that guards cannot be weakened. These conditions capture what in the literature is known as superposition [3].

An example of a morphism is the identity mapping the program below to the vending-machine defined above:

SW ≡	*var*	ready: bool
	init	¬ready
	do	coin : [¬ready → ready:=tt]
	[]	reset : [ready → ready:=ff]

The morphism identifies a component within the vending machine, namely the mechanism that sets and resets it. Notice how new actions can be introduced which use the old attributes in the guards but cannot update them. The guards of the old actions can be strengthened and so can the initialisation condition.

Is COMMUNITY coordinated? Over what notion of interface?

3.2 Lifting Colimits

When one is defining a logic, or a model for concurrency, the nature of interfaces seems pretty obvious because there is a clear separation between "syntax", i.e. the identification of the symbols over which language is generated, and "semantics" in the sense of what is defining the "contents" of the individual components.

In COMMUNITY, the choice is perhaps less clear. It seems obvious that a program signature will have to include the set of attributes (read and local) and the set of actions. But what about the other features?

One criterion for deciding what to place in a signature is the need to be able to lift colimits. Naturally, the more we put in signatures the easier it is for colimits to be lifted. However, we want to put in signatures as little as possible so that we end up with interfaces that are as simple as possible. As shown, for instance, in [8], this is important for facilitating the establishment of relationships like adjunctions between the interface categories of different coordinated formalisms.

It is not difficult to see that, if we consider that program signatures are triples <V,R,Γ>, we are not able to lift colimits. For instance, we cannot interconnect the following programs P_1 and P_2 by synchronising actions g_1 and g_2 in P_1.

P ≡ *do* g_1: [tt → *skip*]
[] g_2: [tt → *skip*]

id ↙ ↘ g_1,g_2£g

P_1 ≡ *var* a: bool
do g_1: [tt → a:= tt]
[] g_2: [tt → *skip*]

P_2 ≡ *do* g: [tt → *skip*]

Such an interconnection of programs does not admit a colimit although the corresponding diagram of signatures clearly admits a pushout consisting of a local attribute *a* and an action *g*. This happens because the restriction on domains (3) applied to g_1 requires that the resulting synchronised action belongs to the write frame of *a* whereas, when applied to g_2, it requires that it does not belong to the write frame of *a*.

This example shows that, without including domains in signatures, it is not possible to lift colimits. More concretely, it shows that what is being left in the computation side of programs interferes with the interconnections. Indeed, action domains enforce locality of attributes and, therefore, constrain the interference that can exist between programs. That is to say, action domains are part of what in COMMUNITY is responsible for coordination and, therefore, must be part of interfaces.

The suggestion, then, is that program signatures are triples <V,R,Γ> where Γ, rather than a set, is a 2^V indexed family of sets (the index of a set is the domain of the actions in that set) and signature morphisms satisfy the equality of domains expressed in condition 3 of program morphisms.

Notice that, in this case, the diagram of signatures obtained from the diagram above does not admit a colimit, meaning that the configuration is not viable. Indeed, we are attempting to synchronise two actions within a program, which may not be feasible due to conflicting types.

3.3 Existence of Discrete Structures

Consider now the need to define, for every program signature, its discrete lift, i.e. the program over that signature that can replace it when establishing interconnections with other programs. The condition that we discussed in section 2.3 basically means that such discrete lifts need to be "neutral" with respect to the computational aspects so as not to compromise the establishment of relationships (morphisms). A neutral initialisation condition is any tautology. The same holds for action guards.

Assignments raise a more interesting case. Let σ:<V,R,Γ>→<V',R',Γ'> be a signature morphism, where <V',R',Γ'> is the signature of some program P'. For σ to be a morphism from the discrete lift of <V,R,Γ> to P' it is necessary that Φ î B'(σ(g)) ⊃ (F'(σ(g),σ(a))=σ(F(g,a)), for every a∈D(g).

Clearly, given $g \in \Gamma$ and $a \in D(g)$, we cannot find a value F(g,a) that satisfies that property for any possible F'. That is to say, given an action and an attribute in its domain, it is not always possible to find a value for the assignment such that we can match any other assignment.

Does this mean that we should shift assignments into signatures?

Shifting assignments into signatures would mean that they are one of the factors that restrict the kind of interconnections allowed in COMMUNITY. This is indeed the case. For instance, we cannot interconnect the following two programs

$P_1 \equiv$ *var* a: {1,-1}
do g: [$b_1 \rightarrow$ a:=1]

$P_2 \equiv$ *var* a: {1,-1}
do g: [$b_2 \rightarrow$ a:=-1]

by a middle object of the form

P ≡ *var* a: {1,-1}
do g: [tt → a:=e]

in order to make them share attribute *a* and synchronise at action *g*. Indeed, the local assignments on *a* are conflicting, i.e. that there is no term *e* that can be mapped to both *1* and *-1*.

Hence, it is only natural that we recognise that assignments are one of the instruments of coordination. Notice that we can, however, interconnect the signatures of the two programs through ***int(P)*** so as to produce the desired synchronisation. The problem is that the middle signature, consisting of local attribute *a* and action *g* with domain *{a}* cannot be lifted to a middle program.

On the other hand, recognising this fact may make us feel uncomfortable about the model of coordination that we have defined. For instance, we might feel that the interference between the assignments is only a problem if the synchronised action occurs. The practical effect of guarding the equality between assigned values in condition 5 of morphisms should be to forbid the execution of the system action whenever it is required to perform conflicting assignments. Hence, we might be interested in a model of coordination that would postpone the resolution of interfering assignments to execution time, allowing the configuration to be established.

This means that we need to change our notion of program to allow for discrete lifts! This is exactly what happened between [9] and [7]. The solution we found was to introduce non-deterministic assignments. The assignment lifted from a signature is the universal one, i.e. it assigns the whole range of possible values, thus ensuring "neutrality". In the case of the example above, we would use the program (channel)

P ≡ *var* a: {1,-1}
do g: [tt → a:∈{1,-1}]

together with identity morphisms for the interconnections. The program resulting from the interconnection is (see the summary section for details on the construction)

P' ≡ *var* a: {1,-1}
do g: [$b_1 \wedge b_2 \rightarrow$ a:∈Ø]

which is idle for as long as b_1 or b_2 are false, and deadlocks when they both are true.

3.4 Summary

The resulting coordinated category can be defined as follows:

Definition: *A* program signature *is a triple* $<V,R,\Gamma>$ *where*

- *V and R are S-indexed families of sets where S is the set of sorts.*
- Γ *is a* 2^V*-indexed family of sets. We denote by D(g) the type of each g in* Γ.

All these sets of symbols are assumed to be finite and mutually disjoint.

Definition/Proposition: *Given signatures* $\theta_1=<V_1,R_1,\Gamma_1>$ *and* $\theta_2=<V_2,R_2,\Gamma_2>$, *a* signature morphism σ *from* θ_1 *to* θ_2 *is a pair* $<\sigma_\alpha: V_1\cup R_1 \rightarrow V_2\cup R_2, \sigma_\gamma: \Gamma_1 \rightarrow \Gamma>$ *of functions such that,*

1. For every $a\in V_1\cup R_1$, $sort(a)=sort(\sigma_\alpha(a))$.
2. For every $a\in V_1$, $\sigma_\alpha(a)\in V_2$.
3. For every $a\in V_1$, $\sigma_\gamma(D_1(a))=D_2(\sigma_\alpha(a))$.

Program signatures and morphisms constitute a category ***SIG****.*

Definition: *A* program *is a pair* $<\theta,\Delta>$ *where* θ *is a signature* $<V,R,\Gamma>$ *and* Δ, *the* body *of the program, is a triple* $<I,F,B>$ *where*

- *I is a proposition over the local attributes (V);*
- *F assigns to every action* $g\in\Gamma$ *a non-deterministic command, i.e. F maps every attribute a in D(g) to a set expression F(a);*
- *B assigns to every action* $g\in\Gamma$ *a proposition over the attributes (V and R).*

Definition/Proposition: *A* program morphism $\sigma: <\theta_1,\Delta_1> \rightarrow <\theta_2,\Delta_2>$ *is a signature morphism* $\sigma:\theta_1\rightarrow\theta_2$ *such that*

1. $\Phi \models (I_2 \supset \sigma(I_1))$.
2. For all $g_1\in\Gamma_1, a_1\in D_1(g_1), \Phi \models (F_2(\sigma(g_1),\sigma(a_1))\subseteq\sigma(F_1(g_1,a_1)))$.
3. For every $g_1\in\Gamma_1$, $\Phi \models (B_2(\sigma(g_1)) \supset \sigma(B_1(g_1)))$.

where $\models$ *means validity in the first-order sense. Programs and superposition morphisms constitute a category* ***PRO****.*

Proposition: *The forgetful functor* ***sig*** *mapping programs to the underlying signatures lifts colimits as follows: let* ***dia:X→PRO*** *be a diagram and* $(\sigma_i: sig(S_i)\rightarrow\theta)_{i:X}$ *a colimit of* ***(dia;sig)****; the colimit* $(\sigma_i: S_i\rightarrow S)_{i:X}$ *of* ***dia*** *lifted by* ***sig*** *is characterised by:*

- *the initialisation condition I is* $\wedge \{\sigma_i(I_i) \mid i:X\}$;
- *given any action* $g\in\Gamma$, *B(g) is* $\wedge \{\sigma_i(B_i(g')) \mid \sigma_i(g')=g, g'\in\Gamma_i, i:X\}$;
- *given any action* $g\in\Gamma$ *and* $a\in D(g)$,
  $F(g,a)= \cap \{\sigma_i(F_i(g',a')) \mid \sigma_i(g')=g, \sigma_i(a')= a, i:X\}$.

Proposition: *The functor* ***sig*** *has discrete structures. The discrete lift for a signature* $<V,R,\Gamma>$ *is the program defined by:*

var V
read R
init tt
do $\underset{g\ \Gamma}{[]}$ g: [tt $\rightarrow$ $\underset{a\ D(g)}{\|}$ a:$\in s_a$]

where s_a *is a term expression denoting the whole set of elements of sort a.*

4 Concluding Remarks

In this paper, we proposed a formalisation for the property according to which a framework for system design supports the separation between computation and coordination. We used Goguen's categorical approach to systems design [13,15] as a platform for the formalisation. The perceived advantages of the proposed notion of coordination are the following.

On the one hand, it provides us with a way of checking whether a given formalism supports the separation between computation and coordination, which we take as being a good measure of the ability of the formalism to cope with the complexity of systems. In the paper, we borrowed examples from specification logics, mathematical models of behaviour, parallel program design languages and coordination languages to illustrate these points, which shows that "coordination" is more that a property of "programming languages", i.e. it applies to other levels of specification and design.

On the other hand, because such an algebraic characterisation of coordination is independent of specific languages and models, it provides us with a framework for the integration of different formalisms for software specification and design that is based on relationships between their interaction models rather than their computational paradigms (the latter being recognisably much harder to integrate). For instance, the earlier work reported in [7] provides a formal account of some of the contributions of "coordination" to the architectural approach to software design [17]. It also suggests ways of extending the expressive power of current architectural description languages by supporting heterogeneous connectors, i.e. connectors in which the roles and the glue are not necessarily described in the same formalism. In a related context, the work reported in [8] shows that interconnections between programs can be synthesised from interconnections between their specifications in "coordinated" frameworks. It also characterises a stronger notion of compositionality in which implementation of computations is decoupled from coordination aspects.

Work is now in progress towards studying the impact that coordination may have on the analysis of behavioural properties of systems, as well as on the characterisation and analysis of the behavioural properties of configurations of systems.

Acknowledgements

Thanks are due to Tom Maibaum and Michel Wermelinger for many discussions that have helped to focus the proposed characterisation of coordination.

References

1. J.Adámek, H.Herrlich and G.Strecker, *Abstract and Concrete Categories*, John Wiley & Sons 1990.

2. J.P.Banâtre and D.Le Métayer, "Programming by Multiset Transformation", *Communications ACM* 16(1), 1993, 55-77.
3. K.Chandy and J.Misra, *Parallel Program Design - A Foundation*, Addison-Wesley 1988.
4. P.Ciancarini and C.Hankin, *Coordination Languages and Models*, LNCS 1061, Springer-Verlag 1996.
5. J.F.Costa, A.Sernadas, C.Sernadas and H.-D.Ehrich, "Object Interaction", *Mathematical Foundations of Computer Science*, LNCS 629, Springer-Verlag 1992, 200-208.
6. J.L.Fiadeiro and A.Sernadas, "Structuring Theories on Consequence", in D.Sannella and A.Tarlecki (eds), *Recent Trends in Data Type Specification*, LNCS 332, Springer-Verlag 1988, 44-72.
7. J.L.Fiadeiro and A.Lopes, "Semantics of Architectural Connectors", *Theory and Practice of Software Development*, M.Bidoit and M.Dauchet (eds), LNCS 1214, Springer-Verlag 1997, 505-519.
8. J.L.Fiadeiro, A.Lopes and T.Maibaum, "Synthesising Interconnections", in R.Bird and L.Meertens (eds), *AlgorithmicLanguagesandCalculi*,ChapmanHall1997,240-264.
9. J.L.Fiadeiro and T.Maibaum, "Categorical Semantics of Parallel Program Design", *Science of Computer Programming* 28(2-3), 1997, 111-138.
10. N.Francez and I.Forman, *Interacting Processes*, Addison-Wesley 1996.
11. D.Garlan and D.Le Metayer, *Coordination Languages and Models*, LNCS 1282, Springer-Verlag 1997.
12. D.Gelernter and N.Carriero, "Coordination Languages and their Significance", *Communications ACM* 35(2), 1992, 97-107.
13. J.Goguen, "Categorical Foundations for General Systems Theory", in F.Pichler and R.Trappl (eds) *Advances in Cybernetics and Systems Research*, Transcripta Books 1973, 121-130.
14. J.Goguen and R.Burstall, "Institutions: Abstract Model Theory for Specification and Programming", *Journal of the ACM* 39(1), 1992, 95-146.
15. J.Goguen, "A Categorical Manifesto", *Mathematical Structures in Computer Science* 1(1), 1991, 49-67.
16. V.Sassone, M.Nielsen and G.Winskel, "A Classification of Models for Concurrency", in E.Best (ed) *CONCUR'93*, LNCS 715, Springer-Verlag, 82-96.
17. M.Shaw and D.Garlan, Software Architecture: Perspectives on an Emerging Discipline, Prentice Hall, 1996.

An Algebraic Approach to Combining Processes in a Hardware/Software Partitioning Environment

Leila Silva, Augusto Sampaio, Edna Barros, and Juliano Iyoda

Depto. de Informática - UFPE - Caixa Postal 7851
Cidade Universitária - CEP 50740-540 Recife - PE - Brazil
{lmas,acas,ensb,jmi}@di.ufpe.br

Abstract. In this paper we present a strategy for combining processes belonging to the same hardware or software component (cluster), in the context of hardware/software partitioning of a system. The strategy takes as input an occam description of a system. This description is the parallel composition of the system components in a predefined form, together with annotations that indicate how the processes, in each cluster, must be combined: by serialisation or by parallelisation. The description given as input can be seen as a binary tree. The strategy to combine processes is based on the reduction of possible configurations in that tree, by using transformation rules which are provable from an algebraic semantics of occam.

1 Introduction

Hardware/Software Co-design is the design of systems comprising two kinds of components: specific application (hardware) components and general programmable ones (software components). The decision about which parts of the system will be implemented in hardware or in software is characterised as the partitioning problem. Partitioning is a well-known NP-complete problem, and thus, some heuristic algorithms to perform the hardware/software partitioning have been developed [8,6,2,11,15]. Recently, some works [5,9,1] have suggested the use of formal methods to validate the partitioning process. However, none of them includes a formal verification that the partitioning preserves the semantics of the original description.

In [3] Barros and Sampaio present some initial ideas towards a partitioning approach whose emphasis is correctness. The proposed approach uses occam[13] as the description and reasoning language, and suggests that partitioning can be characterised as a program transformation task. This work was the seed of the PISH project [4], whose goal was to develop an environment for hardware/software co-design that comprises all the steps from the partitioning of the system into hardware and software components to the layout generation of the hardware.

A.M. Haeberer (Ed.): AMAST'98, LNCS 1548, pp. 308–324, 1998.

The ideas suggested in [3] were illustrated through a case study but no formal strategy to perform the partitioning was presented. In [17] Silva *et al.* give a more precise characterisation of the partitioning process, which clearly separates correctness from efficiency issues. The proposed approach comprises four phases: *splitting*, *classification*, *clustering* and *joining*. The major contribution of the work reported in [17] is the complete formalisation of the splitting phase, which transforms the input description into a set of parallel processes in a normal form, suitable for classification and clustering analysis. The formalism employed is occam and the algebraic laws that define its semantics [14].

This work further develops the ideas described in [4,17] and presents a strategy for the joining phase, in which the processes belonging to the same cluster are effectively combined. This strategy is based on algebraic transformations, and deals with a subset of the occam language which does not include iteration.

This paper is organised as follows: after presenting the relevant subset of occam (Section 2), the partitioning process is described (Section 3). Then we present the strategy for the joining phase (Section 4). Finally, Section 5 summarises the contribution of this paper, briefly describes the environment that implements our partitioning strategy, and discusses topics of further research.

2 A Language of Communicating Processes

The goal of this section is to present the language which is used both to describe the applications and to reason about the partitioning process itself. This language is a subset of occam, defined by the BNF-style syntax given below. For convenience, we

sometimes linearise occam syntax in this paper. For example,

we may write SEQ(P_1, P_2,..., P_n) instead of the standard vertical style.

```
P ::= SKIP | STOP | x := e | ch ? x | ch ! e
    | IF (c1 P1, c2 P2, ..., cn Pn) | ALT (c1&g1 P1, c2&g2 P2,..., cn&gn Pn)
    | SEQ (P1, P2, ..., Pn) | PAR (P1, P2, ..., Pn)
    | VAR x: P | CHAN ch: P
```

In what follows we give a short description of these commands (for more details see, for example, [13]). The SKIP construct has no effect and always terminates successfully. STOP is the canonical deadlock process which can make no further progress. The commands x := e, ch ? x and ch ! e are assignment, input and output commands, respectively; the communication in occam is synchronous. The commands IF and ALT select a process to execute, based on a condition (IF) or on a guard (ALT). The commands SEQ and PAR denote the sequential and parallel composition of processes, respectively. Processes within a PAR constructor run concurrently, with the possibility of communication between them, and cannot share variables. The constructs VAR and CHAN declare local variables and channels, respectively. Here we avoid mentioning a particular type for the declared variables or channels.

The main reason for choosing occam as the description language is that occam obeys a set of algebraic laws [14] which can be used to carry out program transformation with the preservation of semantics. For example, in what follows, Law 1 and Law 2 define the symmetry and the associativity of the PAR constructor, respectively; Law 3 expresses the associativity of the SEQ constructor.

Law 1: PAR(P_1, P_2) = PAR(P_2, P_1)
Law 2: PAR(P_1, P_2, ..., P_n) = PAR(P_1, PAR(P_2, ..., P_n))
Law 3: SEQ(P_1, P_2, ..., P_n) = SEQ(P_1, SEQ(P_2, ..., P_n))

3 The Hardware/Software Partitioning Approach

The general structure of the partitioning approach adopted in this work is depicted in Figure 1. The target architecture underlying this approach includes a single software component and an arbitrary number of hardware components, which can exhibit distinct degrees of parallelism.

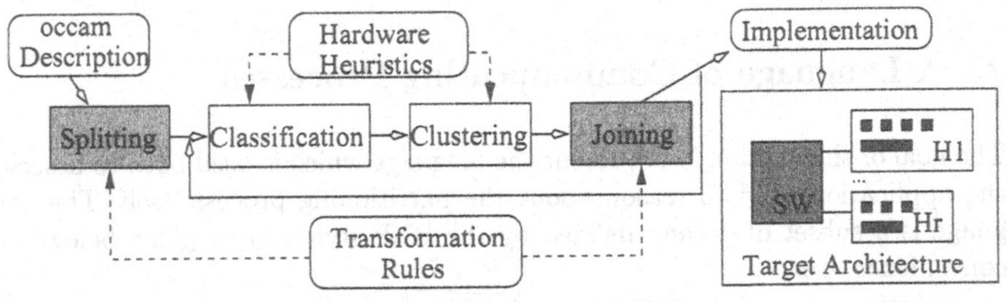

Fig. 1. The partitioning phases.

As mentioned in Section 1, our partitioning approach accepts as input an occam description of a system and carries out the partitioning in four phases: *splitting*, *classification*, *clustering*, and *joining*. There is a clear orthogonality between efficiency and correctness issues in our approach. The phases represented by white boxes (classification and clustering) are concerned with the efficiency of the partitioning process and are based on the work proposed by Barros [2]. The phases represented by gray boxes (splitting and joining) are related to the correctness of the partitioning process (in the sense of preserving the semantics of the original description) and have been originally suggested in [3]. The splitting phase has been completely formalised in [17] and the aim of this work is to present a strategy for the joining phase (Section 4).

To carry out partitioning, we extend the subset of occam given in Section 2 to include six new constructors: BOX, CON, PARsw, PARhw, PARpar, and PARser. These constructors can be regarded as annotations useful for the partitioning process and have no semantic effect. For example, BOX P = P and PARsw P= PAR P. The BOX constructor allows user interaction and is used to indicate that part of description must be considered as an *atomic* process, into which the splitting rules should not be applied. The cost of all processes included into a BOX constructor is analised as a whole by the clustering phase. This constructor is very useful, for example, when the designer knows beforehand that part of the description will necessarily be implemented in hardware, say to make use of existing hardware components. The CON constructor is an annotation for *controlling processes*, whose usefulness will became clear in the next section. The constructors PARsw and PARhw are used to indicate the software and hardware clusters, respectively. The constructors PARpar and PARser specify which processes should be parallelised or serialised, respectively. In what follows we give an overview of each partitioning phase.

3.1 The Splitting Phase

The goal of the splitting phase is to transform the original description into a set of *simple* parallel processes, with the *normal form* below:

$$\texttt{CHAN } ch_1, ch_2, \ldots, ch_n\texttt{: PAR}(P_1, P_2, \ldots, P_k) \qquad (1)$$

The definition of simple process is given in [17]. Here it is enough to mention that each simple process has at most one *atomic* process to execute, which can be either an assignment, a communication command or a BOX constructor. This level of granularity allows the analysis of every possibility of combining the commands of the original program, exploring the different ways of sharing resources. Furthermore, since PAR is a commutative operator in occam, the normal form above allows the classification and clustering phases to analise all permutations for combining processes in hardware or in software components.

To transform the original program into the above normal form a reduction strategy, given in [17], is performed. This strategy applies, to the original description, algebraic rules which are derived from the basic laws of occam.

The description generated by the splitting phase has two major characteristics. First, each process is *closed* in the sense that all free[1] variables used and assigned in this process are declared locally. Moreover, for all pair of processes, a controlling process is introduced. The controlling process acts as an interface between each process under its control and the environment.

To give an idea of some steps of the splitting strategy, consider the example of Figure 2(a). For this simple example, the splitting strategy begins by applying Law 3 to transform this description into one in binary form, as shown in Figure 2(b). After that, each named process is turned into one which is simple and closed, controlling processes are added, and the description is transformed into one in the form of Equation (1) (see Figure 2(e)). Observe that the process CS7 is the controlling process of P8 and P9. Thus, although P8 and P9 are in parallel, in fact their execution is sequential: first CS7 synchronises with P8 through channels ch13 and ch14, and only after that CS7 synchronises with P9 through channel ch15. CP2 is the controlling process of P3 and P4, CS5 is the controlling process

[1] If P is some occam term and x is a variable, we say that an occurrence of x in P is *free* if it is not in the scope of any declaration of x in P, and *bound* otherwise.

of P6 and CS7, and so on. It might seem strange that processes originally in parallel (like P3' and P4') need a controlling process, but the introduction of this kind of controlling process is essential to maintain a desirable uniformity during the splitting strategy.

For didactic purposes, we can represent the description generated by the splitting phase by using binary trees. The nodes of this tree are parallel processes and the edges represent the introduced communication between the controlling process and each process under its control. To distinguish processes of the original description from controlling processes, we represent the former ones by circles and the latter ones by boxes if they are sequential controllers, and by lozenges if they are parallel controllers. The leaves of this tree are always processes of the original description. This kind of representation will be used in the reminder of this paper. Considering Figure 2, the diagrammatic representation of the description shown in (e) is depicted in (c).

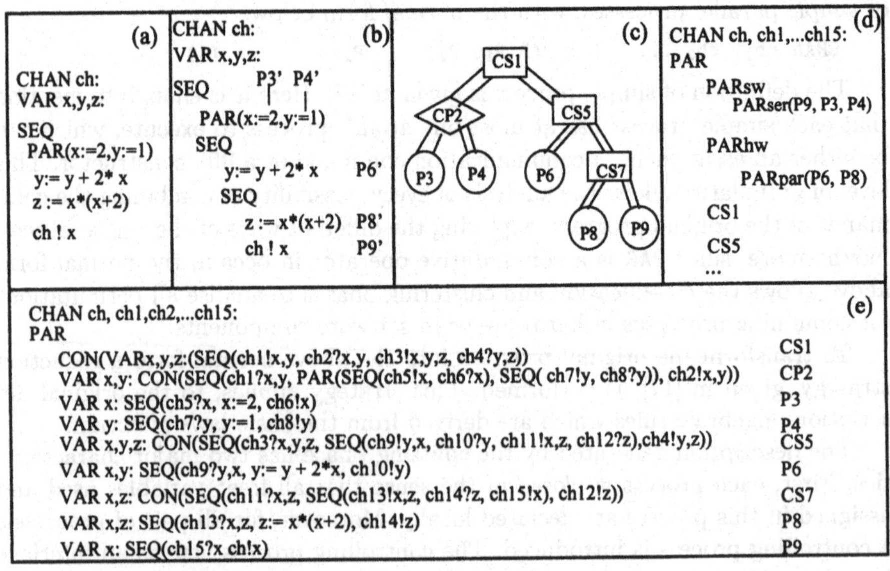

Fig. 2. An example to illustrate the partitioning strategy.

We use some auxiliary notations, useful for the next partitioning phases. USED and ASS stand for the lists of used and assigned free variables of the considered processes, respectively. INPUT and OUTPUT stand for the lists of input and output channels of the considered process, respectively. In the case of a controlling process, these lists are the union of the lists of the processes under its control. In addition, we give a label and a name for each process Q. The name is the concatenation of the identification and the depth of Q. The identifications are CS, CP and P, standing for sequential controller, parallel controller and process of the original description, respectively. The depth is an integer number that expresses the order in which Q is visited, when a depth-first search is applied on the splitting tree. The label expresses the *history* of Q, by the concatenation of the names of all its ancestors in the splitting tree, beginning with the name of its father. For example, the complete representation of process P9 is given below.

```
USED(x):ASS():INPUT():OUTPUT(ch):CS7CS5CS1:P9:VAR x:SEQ(ch15 ?x, ch!x)
```

Observe that the names of the processes in the Figure 2(e) attend the mentioned notation. To refer to an arbitrary process, we use P (possibly with subscripts) to range over processes.

3.2 The Classification and Clustering Phases

After the splitting phase, the classification and clustering take place. The classification phase establishes a set of implementation alternatives, for each simple process, whereas the clustering phase maps the simple processes to a hardware or software cluster, based on hardware heuristics which include the consideration of communication costs and the area/delay tradeoff.

Figure 2(d) illustrates a possible clustering result for the previous example. It indicates that the original description will now be implemented partially in software and partially in hardware. Moreover, it is required that processes P6 and P8 be implemented in parallel, instead of their original sequential order. On the other hand, P3 and P4 should be implemented in sequence.

3.3 The Joining Phase

The clustering phase establishes the clusters composition, but does not effectively combine the processes into the same cluster. This is carried out at the joining phase, by applying algebraic rules to transform the description generated by the clustering phase into one in the form:

$$\mathtt{CHAN}\ ch_1, ch_2, \ldots, ch_m\mathtt{:}\ \mathtt{PAR}(SW, H_1, H_2, \ldots, H_r) \qquad (2)$$

where SW and each H_s, $1 \leq s \leq r$, are the generated clusters. Each P_i generated by the splitting phase (see Equation (1)) is in exactly one of these clusters. Note that in this way we capture the precise mathematical notion of partition. The SW, by convention, stands for the software process and each H_i for one hardware process.

4 The Joining Strategy

There are two major tasks to be performed during the joining phase: combining processes belonging to the same cluster in sequence or in parallel. The strategy to perform the combination should eliminate the local communication among processes, as well as auxiliary variables introduced during the splitting phase.

Related to these tasks, there are two major problems to consider. Firstly, the serialisation of processes originally in parallel can introduce deadlock. On the other hand, processes originally in sequence can have data-dependency and the parallelisation of these processes is not allowed in occam. Also, the parallelisation of processes can eliminate deadlock, therefore changing the semantics of the input description. So, the strategy must check some conditions to avoid these two problems. We assume that the input description is deadlock free; we consider that deadlock absence of the input description is a separate concern which should be checked at an earlier stage - as part of the validation of the specification. Thus, we should be concerned only with avoiding introducing deadlock.

Although the splitting strategy deals with iteration, the joining strategy described in this paper consider the subset of the occam language presented in Section 2.

The joining strategy is based on the transformation and reduction of configurations in a binary tree which represents the current description, by applying algebraic rules. There are two kinds of configurations to consider: basic and intermediary. Sections 4.1 and 4.2 present these configurations and some of the rules necessary to transform them, and Section 4.3 describes the algorithm which guides the application of the rules. For didactic purposes, the joining rules are expressed in a diagrammatic version. We select one of these rules and show how it can be expressed as an equation relating occam terms. The occam description of the complete set of rules is given in [18]. In addition to circles, boxes and lozenges, a triangle is used to represent an arbitrary subtree. Moreover, if processes P_x and P_y are combined, we use the notation $P_{x.y}$ as the symbolic name of the new combined process.

4.1 Basic Configurations

Basic configurations are the ones that, after being transformed, reduce the size (number of nodes) of the splitting tree. A reduction is performed through the elimination of one controlling process and the combination of two brother processes, say P_1 and P_2. Figure 3(a) shows the two possible basic configurations.

Processes structured as in a basic configuration can be combined in sequence or in parallel. If, for example, these processes were originally in sequence and must be combined in sequence, the inverse procedure of the splitting phase can be directly applied. Otherwise, if they must be combined in parallel, it is necessary to guarantee that they have no data-dependency, as parallel processes in occam cannot share variables. Rule 1 in Figure 3(b) is used to parallelise two processes originally in sequence. The boolean function `dependency`, when applied to two arbitrary processes, say Q_1 and Q_2, verifies the condition

$$(\mathtt{ASS}(Q_1) \cup \mathtt{USED}(Q_1)) \cap \mathtt{ASS}(Q_2) \neq \emptyset \text{ or } (\mathtt{ASS}(Q_2) \cup \mathtt{USED}(Q_2)) \cap \mathtt{ASS}(Q_1) \neq \emptyset$$

and returns `TRUE` if these conditions are satisfied (Q_1 and Q_2 have data-dependency), and `FALSE` otherwise. The textual version of this rule is showed in (c) of the same figure, where the lists, the label and the name associated with each process are intentionally omitted. Notice that a function `F` is used to represent an arbitrary context which includes P_1 and P_2 as subcomponents, as they can belong to a `PARpar` constructor in any level of the cluster hierarchy. The matching of the controlling processes is immediate; all controlling processes are in the

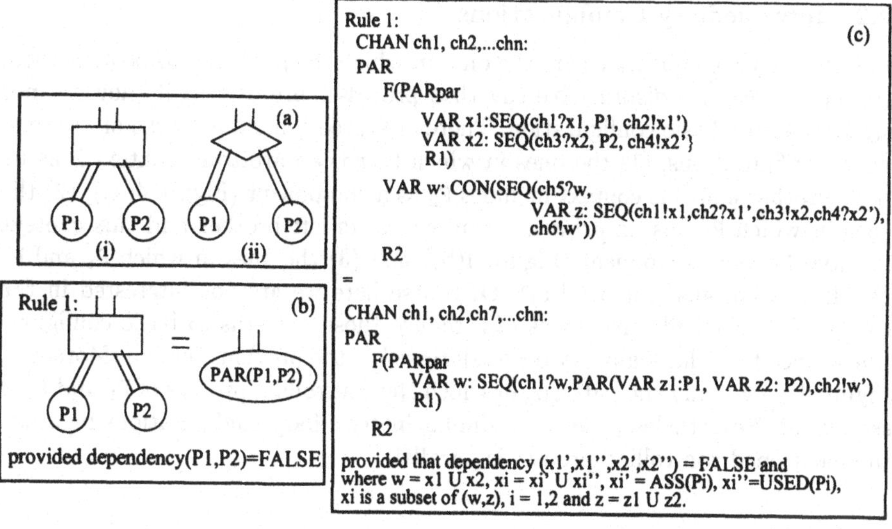

Fig. 3. Basic configurations.

same level, the most external one (see Figure 2(d)). The processes P_1 and P_2 are combined if the conditions are satisfied, and during this combination the controlling process of P_1 and P_2 disappears. On the left-hand side of the equation of Figure 3(c), R_1 stands for the remaining processes into the **PARpar** constructor under consideration and R_2 stands for the remaining clusters and controlling processes of the whole description. It is important to notice that every time the current tree is being transformed, it is necessary to update the labels and the names of the processes affected by this transformation.

In the case of serialising two processes originally in parallel, the introduction of deadlock must be avoided. If P_1 or P_2 does not communicate directly with the environment through user declared channels (that is, one of them communicates with the environment only through their controller), there is no problem to serialise P_1 and P_2, in any order, since they operate on disjoint data spaces. This is syntactically ensured because occam does not allow parallel processes to share variables. Nevertheless, if P_1 and P_2 have communication commands introduced by the user in the original description, it is necessary to check if there is any possible order in which these processes can communicate with the environment without introducing deadlock. This is performed by the function **sequence** [18], which uses the information about the order of occurrence of all events in the system. Also, it may be the case in which P_1 and P_2 synchronise. In this case, the serialisation of P_1 and P_2 involves the elimination of internal channels, and can be achieved following the strategy suggested in [14].

4.2 Intermediary Configurations

Intermediary configurations are the ones in which the processes to be combined, say P_1 and P_2, are *distant*. We say that processes are distant if they are not structured in a basic configuration. There are three general patterns of intermediary configurations: (1) the ones in which P_1 has an arbitrary context F as its right brother and this context includes P_2 as a component (Figure 4(a)); (2) the ones in which P_1 has an arbitrary context as its left brother and this context includes P_2 as a component (Figure 4(b)) and (3) the ones in which P_1 and P_2 are distant cousins (Figure 4(c)). Of course here we are not interested in the cases where particular instances of F reduce those patterns to basic configurations, since these have already been addressed in the previous section. Moreover, Figure 4 shows only the patterns in which the controlling process of P_1 and P_2 is sequential. Nevertheless, there are similar intermediary configurations in which the mentioned controlling process is parallel.

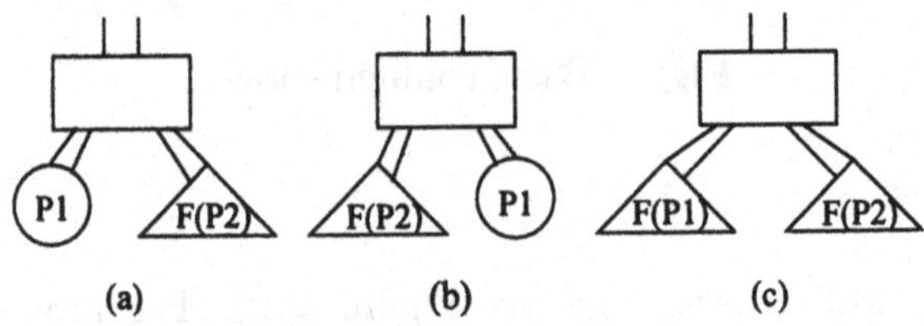

Fig. 4. Intermediary configurations.

The aim of the transformations applied to these configurations is to approximate P_1 and P_2. However, unlike in the case of basic configurations, these transformations do not reduce the tree. The rules applied in Cases (1) and (2) try to approximate P_1 and P_2 by moving P_2 upwards and/or P_1 downwards, until they can be structured as in a basic configuration. In Case (3), rules are applied to move P_1 and/or P_2 upwards, until one of them becames the grandson of the common ancestor of P_1 and P_2. (The sons of the common ancestor must be controlling processes, otherwise we are considering Case (1) or (2). These promotion rules are very similar to the ones applied in the moving upwards procedure of Cases (1) and (2).) Then, a transformation rule is applied to structure P_1 and P_2 as in Cases (1) or (2). For conciseness reasons, in what follows we will focus on the approximation procedure of Case (1). The other cases are extensively described in [18].

4.2.1 The Approximation Procedure of Case (1). The approximation of processes P_1 and P_2 when they obey Case (1) may involve three procedures: the transformation procedure of *stop configurations*, the *moving downwards* procedure of P_1 and the *moving upwards* procedure of P_2. Notice that at this point it does not matter whether P_1 and P_2 will be combined in sequence or in parallel; only the reduction of basic configurations deals with this issue.

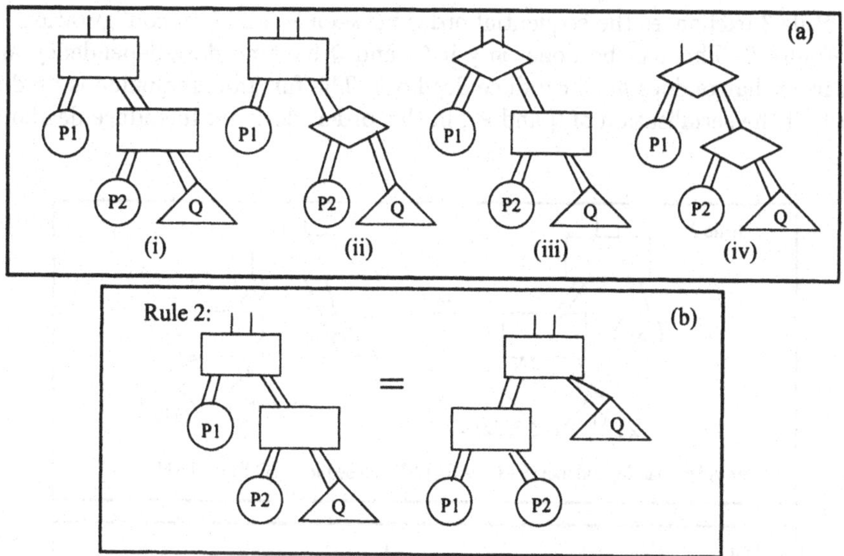

Fig. 5. Stop configurations of Case (1).

The Transformation Procedure of Stop Configurations for Case (1). Stop configurations are configurations which can be directly transformed in a basic configuration. They can be considered as target configurations for the moving upwards/downwards procedure. The stop configurations of Case (1) are depicted in Figure 5(a). Rule 2 in Figure 5(b) is an example of the rules applied to transform stop configurations (in this case it transforms the stop configuration (i) of the same figure). This rule is justified by the associativity of the `SEQ` operator (see Law 3 of Section 2).

The Moving Downwards Procedure for Case (1). The configurations showed in Figure 6 can occur during the moving downwards procedure applied to $\mathtt{P}_1$. In addition to these configurations, it must be considered the ones in which $(\mathtt{F}(\mathtt{P}_2))$ exchanges role with `Q`. Two of the rules applied to transform some of these configurations is given in Figure 7.

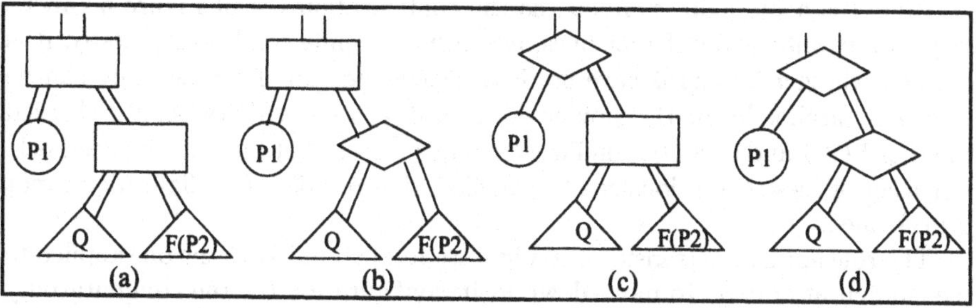

Fig. 6. Configurations for the moving downwards procedure in Case (1).

Rule 3 exchanges the sequential order between P_1 and Q in configuration (a) of Figure 6. This can be done only if P_1 and Q have no data-dependency and if this exchange does not introduce deadlock. The function **sequence** is used to verify if the serialisation of Q and P_1, in this order, does not introduce deadlock.

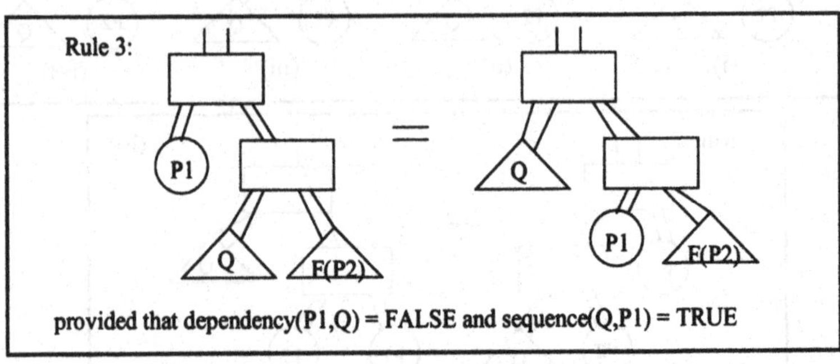

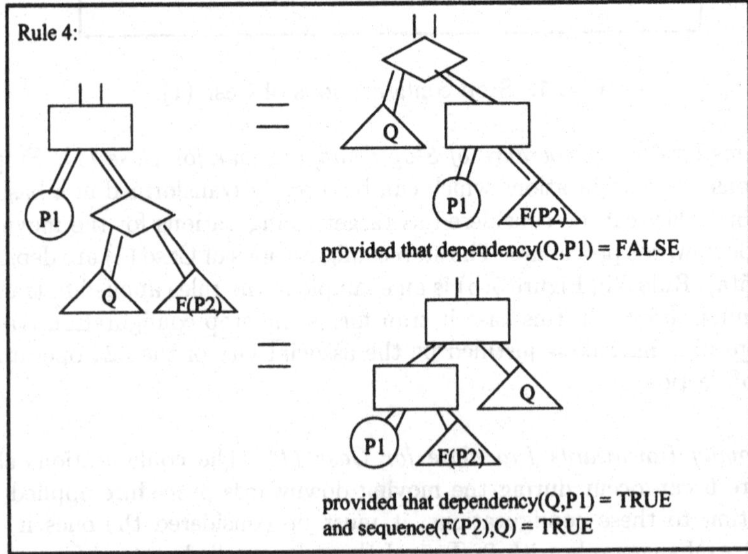

Fig. 7. Four Rules for the moving downwards of process P_1.

Rule 4 is a bit more complex, as the configuration (b) of Figure 6 can be transformed into two different ones, according to some conditions. Firstly, it is checked whether P_1 and Q have no data-dependency. In this case, they can be placed in parallel during the moving downwards process. Otherwise, it is checked whether F(P_2) and Q can be combined in sequence without introducing deadlock. If neither of these conditions can be satisfied, the considered configuration is not transformed.

The rule applied to transform configuration (c) is similar to the one applied to the configuration (b). To move down P_1 in configuration (d) the commutativity of the PAR operator is used.

The Moving Upwards Procedure for Case (1). The configurations for the moving upwards procedure are very similar to the ones for the moving downwards procedure. The only difference between the rules for this case and the ones for the moving upwards procedure is that the controlling processes transformed in the moving upwards procedure are the father and the grandfather of P_2, whereas in the moving downwards procedure are the father and the brother of P_1.

4.3 The Algorithm

The algorithm to combine processes at the joining phase applies the rules presented in Sections 4.1 and 4.2 and is composed of three main steps: resolution of the basic configurations of each cluster, resolution of the intermediary configurations of each cluster and some final transformations. The first two steps (which are the major ones) of the algorithm are described in what follows, while a more detailed version is given in [18]. Before presenting the algorithm of each step, we will describe the data structures and the procedures used by the algorithm.

Data structures: Cl_i – cluster variable; $S_i = (S_{i,1}, S_{i,2}, ..., S_{i,m})$ – sequence where each $S_{i,j}$ is a a **PARpar** or **PARser** process of Cl_i. The order of the elements in this sequence follows the order of the nested levels of the cluster hierarchy, where the first element is the most internal **PARpar** or **PARser** process in the cluster hierarchy, and the last element is the most external **PARpar** or **PARser** process. The order of the elements in the same level of the hierarchy is irrelevant. If a cluster has no **PARpar** or **PARser** constructor, it has only one process P_i and so, S_i is empty; $S_{i,j} = (P_1, P_2, ..., P_k)$ – is a sequence of processes belonging to a **PARser** or **PARpar** process; L_1 – sequence of pair of processes (P_x, P_y) from $S_{i,j}$, where P_x is the brother of P_y; L_2, L_3 and L_4– sequences are all composed of pair of processes (P_x, P_y) such as the depth of any process in the current $S_{i,j}$ is not in between the depth of P_x and that of P_y. (This guarantees that closest processes are first considered during the combination procedure.) Furthermore, in L_2 the processes P_x and P_y must be structured as in Case (1) of intermediary configuration, in L_3 the processes must be structured as in Case (2) and in L_4 as in Case (3); **tryAgain** and **moveUp** – auxiliary boolean variables.

Global Procedures

```
function Reduce_Basic_Configuration(Px,Py): boolean
 begin
   apply the appropriate rule for combining Px and Py
   if the combination is successful then
     delete Px and Py from Si,j and add Px.y to Si,j
     return TRUE
   else return FALSE
 end

procedure Unitary_Constructor(Si,j)
 begin
   if Si,j becames unitary then
   apply either the rule PARpar P=P or the rule PARser P=P
 end
```

Step 1: Resolution of Basic Configurations. The purpose of this step is, for each cluster, to combine all processes which are structured as in a basic configuration. Observe that we consider any pair of processes only once during the algorithm. If the combination of these processes is possible, it is performed. Otherwise, the processes remain without any change.

Algorithm of Step 1

```
for each cluster Cli, i = 1,2,...,n do
  if Si is not empty then
    for each Si,j, j = 1,2,...,m do
      generate L1 by finding the pair of brothers of Si,j
      while L1 is not empty do
        let (Px, Py) be the first element of L1
        delete (Px, Py) from L1
        if Reduce_Basic_Configuration(Px,Py) then
          add to L1 each new pair of brothers (Px.y, Pz)
      Unitary_Constructor(Si,j)
```

Step 2: Resolution of Intermediary Configurations. The purpose of this step is, for each cluster, to combine distant processes. These processes are structured as in cases (1), (2) or (3) of Section 4.2. As mentioned before, we will concentrate the description in the resolution of Case (1). The other cases are fully described in [18]. To combine processes structured as in Case (1), first it is checked if the processes are structured as in a stop configuration, in which case a transformation rule, and after, a reduction rule are applied. Otherwise, the moving downwards and/or the moving upwards procedures are applied to approximate the processes.

Algorithm of Step 2

```
for each cluster Cli, i = 1,2,...,n do
  if Si is not empty then
    for each Si,j, j = 1,2,...,m' do
      generate L2, L3 and L4
      -- resolution of case 1
      while L2 is not empty do
        let (Px, Py) be the first element of L2
        delete (Px, Py) from L2
        set the variables tryAgain to TRUE and moveUp to FALSE
        --  moving downwards
        while tryAgain and not moveUp do
          if Px and Py are structured as in a stop configuration then
            apply the appropriate stop configuration rule to
             structure Px and Py in a basic configuration
            if the transformation is successful then
              if Reduce_Basic_Configuration(Px,Py) then
                replace each occurrence of Px and Py in L2,L3 and L4 by Px.y
            set the variable tryAgain to FALSE
          else
            apply the appropriate rule for moving downwards, say Px
            if the moving downwards is not successful then
```

```
        set moveUp to TRUE
    -- moving upwards
    while moveUp do
      apply the appropriate rule for moving upwards, say Py
      if the moving upwards is not successful then
        set moveUp to FALSE
      else
        if Px and Py are structured as in a stop configuration then
          apply the appropriate stop configuration rule to
           structure Px and Py in a basic configuration
          if the transformation is successful then
            if Reduce_Basic_Configuration(Px,Py) then
              replace each occurrence of Px and Py in L2, L3
               and L4 by Px.y
          set the variable moveUp to FALSE
  -- resolution of case (2)
  -- resolution of case (3)
  Unitary_Constructor(Si,j)
```

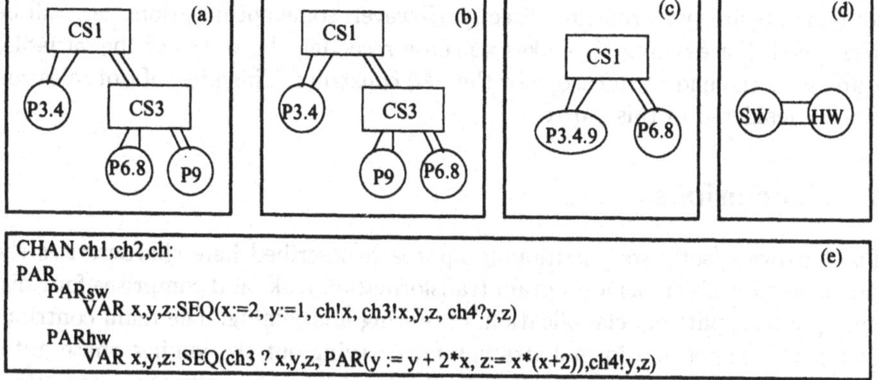

Fig. 8. The application of the joining strategy to the example of Figure 2.

Considering the clustering result of Figure 2(d), during the first step of the algorithm processes P3 and P4 are combined, as they are structured as in a basic configuration. At the second step, Rule 2 is applied to structure P6 and P8 in a basic configuration and after that Rule 1 is applied to combine these processes. The reduced tree is shown in Figure 8(a), where, for didactic purposes, we use a symbolic name for the combined processes. Observe that it is not possible to put $P_{3.4}$ and P9 closer by moving downwards $P_{3.4}$, as this process has data-dependency with process $P_{6.8}$. However, it is possible to exchange P9 with $P_{6.8}$, because they do not have data-dependency (see Figure 8(b)). After that $P_{3.4}$ and P9 are structured in the form of the stop configuration (i) of Figure 5(a) and Rule 2 is again applied. Finally, the processes are combined by applying a serialisation rule for the basic configuration (i) of Figure 3(a), resulting in the tree showed in Figure 8(c).

Final Considerations. These two steps of the algorithm are responsible for combine (when possible) all processes of a given cluster. However, to guarantee that the description generated after this step follows the clustering result, it is necessary some additional work, detailed in [18]. For example, if two processes into a PARser construct have not been combined in the previous steps (because they did not satisfy the conditions associated to the rules attempted), it is necessary to check if these processes really execute in sequence, as required. The labels of these processes are used for this purpose. Moreover, there might still be opportunity for further transformations: inside each process, the inverse rules of the splitting phase are applied to re-compose the IF's and ALT's commands, which might have been broken at the splitting phase, and to linearise the SEQ and PAR constructors, which are in binary form. After that, the final description is generated, by adopting as a design decision the condensation of the remaining controlling processes with the software cluster.

Considering the example of Figure 8, the partitioned system (Figure 8(d,e)) is generated through the condensation of the remaining controlling process CS1 with the software process ($P_{3.4.9}$). Observe that the hardware and the software components are synchronising directly. However, some optimisations can still be performed. For example, it makes no sense receiving the values of the variables y and z at the end of the scope of the SEQ construct. This kind of optimisation is not considered in this paper.

5 Conclusions

The hardware/software partitioning approach described here characterises the partitioning problem as a program transformation task, and comprises four distinct phases: splitting, classification, clustering and joining. The main contribution of this paper is a formal strategy for carrying out the joining phase automatically. The processes belonging to the hardware and software components are effectively combined in this phase, by serialisation or by parallelisation. By checking some conditions, the strategy presented here avoids the introduction of deadlock.

Because the output of the partitioning process in our approach is still an occam program, it is possible to carry out simulations at a very early stage of the design. In other approaches to co-design [6,8], the output of the partitioning is only an indication of what should be done, but the final program is not automatically generated.

One important point of our approach is the orthogonality between the efficiency and the correctness issues of the partitioning, as mentioned in Section 3. An immediate consequence of this orthogonality is that we can reuse the same splitting and joining phases as part of several different strategies for hardware/software partitioning, provided of course the splitting granularity is adequate for the heuristics employed. Nevertheless, even for partitioning heuristics that consider coarser granularity, the proposed BOX constructor may be used to encapsulate pieces of code for hardware/software partitioning analysis.

We have briefly justified a few rules (like Rule 2 in Figure 5) based on some of the basic laws of occam, such as the associativity of sequential composition (Law 3 of Section 2). One immediate topic for further work is to formally derive each rule of the joining phase from the basic laws of occam, like has been done for the splitting phase [17]. We also need to extend the joining strategy to consider replicated constructs and to deal with the optimisations mentioned at the very end of Section 4.3. The ultimate goal is to guarantee, by construction, that the splitting and joining strategies preserve the semantics of the original description.

We are not aware of any other work which presents a formal characterisation of the partitioning problem as done here and in [17]. Nevertheless, it is worth mentioning that the kind of algebraic framework used here has been used previously to characterise and reason about a number of other applications [14,16,10]. All these works can be regarded as applications of refinement algebra.

We have developed an environment, the Partitioning Transformation System (ParTS), to carry out the partitioning automatically. It has been developed as an extension of OTS (the *Occam Transformation System*) [7], which performs general transformations of occam programs. While the basic laws of occam [14] implemented in OTS are useful for program transformation in general, they express only simple transformations, and are not suitable to capture the partitioning problem. The aim of ParTS is to extend OTS with transformation rules for the splitting and joining phases, as well as the splitting and joining strategies. The current version of ParTS includes the implementation of all splitting rules, of the splitting strategy and of some of the joining rules. Each transformation rule is captured in ParTS by a function in the SML [12] language. The splitting and joining strategies are also coded as functions, taking advantage of the pattern matching facilities of SML.

Channels and local variables introduced during the splitting phase can interfere on the efficiency of the partitioned system. We are developing some optimisations to guarantee that all local communication and local variables will be eliminated from the final description of the system. Moreover, the classification and clustering phases, during the cost analysis, deal with introduced variables and channels in a different way from the original variables and channels. An accurate analysis of the efficiency of the final system will be possible only after developing some large case studies. In this paper we are emphasising the correctness issue of the partitioning process.

Acknowledgements: The authors acknowledge Prof. He Jifeng, Dr. Geraint Jones and Dr. George Justo for discussing this work and making some useful suggestions. The authors also acknowledge the financial support given by the Brazilian research agencies, CAPES and CNPq. Leila Silva acknowledges the Federal University of Sergipe, for financial support.

References

1. A. Balboni,W. Fornaccari, D. Sciuto. TOSCA: A Pragmatic Approach to Co-Design Automation of Control-Dominated Sustems. In *Hardware/Software Codesign*, (1996) 265–294, Kluwer Academic Publishers.
2. E. Barros. *Hardware/Software Partitioning using UNITY*. PhD thesis, Universität Tübingen, 1993.
3. E. Barros and A. Sampaio. Towards Probably Correct Hardware/Software Partitioning Using Occam. In *Proceedings of the Third International Workshop on Hardware/Software Codesign, Codes/CASHE94*, (1994) 210-217, IEEE Press.
4. E. Barros *et al.* The PISH Methodology for Hardware/Software Codesign. In *Proceedings of the Workshop of ProTem-CC, CNPq*, (1998) 65–98.
5. M. Chiodo, P. Giusto, H. Hsieh, A. Jurecska, L. Lavagno and A. Sangiovanni-Vicentelli. A Formal Methodology for Hardware/Software Co-design of Embedded Systems. In *IEEE Micro*, August 1994.
6. R. Ernst and J. Henkel. Hardware-Software Codesign of Embedded Controllers Based on Hardware Extraction. In *Handouts of the International Workshop on Hardware-Software Co-Design*, October 1992.
7. M. Goldsmith. *The Oxford* **occam** *Transformation System*. Technical report, Oxford University Computing Laboratory, January 1988.
8. R. Gupta and G. De Micheli. System-level Synthesis Using Re-programmable Components. In *Proceedings of EDAC*, (1992) 2–7, IEEE Press.
9. R. B. Highes and G. Musgrave. The Lambda Approach to System Verification. In Hardware/Software Co-design, G. De Micheli and M. Sami Editors, Kluwer Academeic Publishers, 1996.
10. Jifeng He, I. Page, and J. Bowen. A Provable Hardware Implementation of occam. In *Correct Hardware Design and Verification Methods (Advanced Research Working Conference, CHARME' 93), Lecture Notes in Computer Science*, Springer Verlag, **683**, (1993) 214–225.
11. A. Kalavade and E. Lee. The Extended Partitioning Problem: Hardware/Software Mapping, Scheduling and Implementation-bin Selection. In *Design Automation for Embedded Systems*, 1997, 2(2): 125–163.
12. L. Paulson. *ML for the working programmer*. Cambridge University Press, 1991.
13. D. Pountain and D. May. *A Tutorial Introduction to OCCAM Programming*. Inmos BSP Professional Books, (1987).
14. A. Roscoe and C. A. R. Hoare. The laws of **occam** programming. In *Theoretical Computer Science*, **60**, (1988) 177–229.
15. D. Saha, R. S. Mitra and A. Basu. Hardware/Software Partitioning Using Genetic Algorithm. In *Proc. of 10th International Conference on VLSI Design*, India, 1997, 155–160.
16. A. Sampaio. *An Algebraic Approach to Compiler Design*. Volume 4 of Algebraic Methodology and Software Technology (AMAST) Series in Computing, World Scientific, 1997.
17. L. Silva, A. Sampaio and E. Barros. A Normal Form Reduction Strategy for Hardware/Software Partitioning. In *Formal Methods Europe (FME) 97. Lecture Notes in Computer Science 1313*, (1997) 624–643.
18. L. Silva, A. Sampaio and E. Barros. *A Reduction Strategy for Combining Processes in a Hardware/Software Partitioning Environment*. Technical Report, RT-DI/UFPE, Federal University of Pernambuco, Recife, Brazil. (draft)

An Algebraic View of Program Composition

Pietro Cenciarelli

Ludwig–Maximilians–Universität München
cenciare@informatik.uni-muenchen.de

Abstract We propose a general categorical setting for modeling program composition in which the call-by-value and call-by-name disciplines fit as special cases. Other notions of composition arising in denotational semantics are captured in the same framework: our leading examples are nondeterministic call-by-need programs and nonstrict functions with side effects. Composition of such functions is treated in our framework with the same degree of abstraction that Moggi's categorical approach based on monads allows in the treatment of call-by-value programs. By virtue of such abstraction, interesting program equivalences can be validated axiomatically in mathematical models obtained by means of modular constructions.

1 Introduction

In denotational semantics programs are interpreted in domains with suitable computational structure. For example, a domain for interpreting integer programs with exceptions must include (besides integers) denotations for exceptions and allow case analysis. In the categorical semantics proposed in [Mog91], the concrete structure of such domains is hidden behind the structure of a strong monad T, where TX is the domain of programs of type X. The advantage of describing program denotations in terms of the abstract structure of a monad is that a language can be extended with new computational features (e.g. a mechanism for exceptions or side-effects) and reinterpreted by just adopting a "more powerful" monad, without rewriting the old semantic equations. The *computational lambda calculus* (or computational *metalanguage*), the formal system associated in [Mog91] with this semantics, features a type constructor T and an operator let_T to compose programs of the form $A \to TB$, parametric in A, with programs of type A, which live in the domain TA.

The notion of composition implemented by let_T corresponds to a *call-by-value* parameter evaluation in that programs are modelled by morphisms of the form $A \to TB$, indexed by values in A, and they satisfy only a restricted form of substitution. On the other hand, *call-by-name* programs, which accept unevaluated expressions as inputs, are modelled by morphisms $TA \to TB$, indexed by "computations" in TA. In the metalanguage composition of such programs works according to β-reduction. Categorically, call-by-value programs compose in the Kleisli category of a monad T, while call-by-name programs compose in the base category.

A.M. Haeberer (Ed.): AMAST'98, LNCS 1548, pp. 325–340, 1998.

Other notions of composition arise in computer science. For nondeterministic partial functions, for example, the *call-by-need* discipline differs from call-by-value in that it is nonstrict, and from call-by-name in that different occurrences of a parameter are always assigned the same value. How do call-by-need programs compose? In the computational metalanguage one has no choice but treating nondeterministic call-by-need programs as a special kind of call-by-name programs (the "additive" ones) and interpret them as morphism $\mathcal{P}A \to \mathcal{P}B$, for some power construction $\mathcal{P}$. However, a more finely grained semantics can be obtained by interpreting programs as morphisms of the form $A_\perp \to \mathcal{P}B$ and exploiting the relation between $\mathcal{P}$ and the lifting construction $(\_)_\perp$ to get an operation $let^{\perp}_{\mathcal{P}}$ for composing such morphisms just like $let_{\mathcal{P}}$ composes strict programs $A \to \mathcal{P}B$. Similar operations let^{R}_{T} are available for monads R and T when the structure of T extends, in a suitable sense to be explained below, the structure of R. Such operations and the categorical setting in which they arise are studied in this paper.

We propose a general categorical framework for modeling program composition in which the call-by-value and call-by-name disciplines fit as special cases. In view of the relation between monads and algebraic theories, different notions of composition are obtained by distinguishing the algebraic structure with respect to which programs behave as homomorphisms. This approach gives a uniform account of different strategies of parameter evaluation capturing notions of composition which do not accommodate naturally in the monadic setting of [Mog91]. Common programming constructs such as exception handlers, pipes etc. can be interpreted in the proposed framework without exposing the concrete structure of the semantic domains. The benefits are twofold: On the one hand our framework allows an axiomatic approach to validation of program equivalences in large classes of models. On the other hand it allows property-preserving reinterpretation of program constructs under model extensions, thus supporting a modular approach to denotational semantics in the spirit of [Mog90a, Cen95].

Synopsis. Section 2 discusses a motivating example. Section 3 gives a general categorical explanation of the constructions of Section 2 and presents a semantic framework which gives a uniform account of different disciplines of program composition in terms of the algebraic notion of homomorphism. The setting of Section 3 is further generalised in Section 4, where a weak theory of program composition is proposed; the theory features two operations, similar to the unit and lifting of Kleisli triples, of which simple equational properties are proven. Applications are described in Section 5 where these operations are used to define the semantics of common program constructs. Then properties of such constracts are derived axiomatically and shown to be preserved when models are suitably extended with new computational features.

2 A Motivating Example

The viewpoint proposed in this paper is that different strategies of parameter evaluation can be described in terms of how programs preserve computational structure. In this section we discuss an example where semantic domains are provided by a composite monad $T = Q \circ R$. In such cases, an operation of program composition let^R_T is available, where only the structure of Q is preserved. Using such an operation for defining the denotational semantics of programs with side-effects (modeled by Q) and failure (modeled by R) we are able to validate program equivalences axiomatically. Observing that similar operations are available, with the same benefits, for monads which are not of the form $Q \circ R$, we look for the general semantic setting, subsuming monad composition, where such operations arise. This is done in the next section.

In a language with side effects and a mechanism for aborting computation, for example, a construct *handle* (M, N) runs the program M and, if a failure occurs, makes a second attempt to produce a value by running N. Such a language may be interpreted in a cartesian closed category by mapping terms of type τ to elements of $((\llbracket\tau\rrbracket + 1) \times S)^S$, where S is some object of states. In particular:

$$\llbracket handle(M,N)\rrbracket = \lambda s : S.\ case\ \pi_0(\llbracket M\rrbracket s)\ of\ inl(v).\ \langle inl(v), \pi_1(\llbracket M\rrbracket s)\rangle \\ inr(u).\ \llbracket N\rrbracket(\pi_1(\llbracket M\rrbracket s)).$$

By using lambda abstraction and projections, this equation exposes the concrete structure of the domains of interpretation. Hence, it works fine for a toy language but not for more realistic ones where domains of the form $((X + 1) \times S)^S$ may be inadequate to host programs. A more general presentation of the semantics of failure handling can be given by using the computational lambda calculus as metalanguage. In any model where programs are interpreted in domains of the form QRX, where Q is an arbitrary monad and $RX = X + 1$ we define:

$$\llbracket handle(M,N)\rrbracket \;=\; let_Q\ z \Leftarrow \llbracket M\rrbracket\ in\ case\ z\ of\ inl(v).\ val_Q(inl(v)) \\ inr(u).\ \llbracket N\rrbracket. \tag{1}$$

The case above is obtained when $QX = (X \times S)^S$. Adopting interpretation (1), one can work formally in a suitable theory of the computational lambda calculus and validate program equivalences for any model of the above class. The following equation, for example, can be easily derived from the axioms of the calculus.

$$handle\,(handle\,(L, M), N) = handle\,(L, handle\,(M, N)). \tag{2}$$

Unfortunately, there are perfecly reasonable models for exceptions where no val_Q and let_Q operations are available to implement the handling of an exception. The monad $TX = ((X \times S) + 1)^S$, for example, models a "dramatic" form of failure, in which the state is lost upon occurrence of an exception. Equation (2) should also hold for programs of this form, but we have no formal (i.e. axiomatic) means of proving this equivalence without exposing again the concrete structure of T.

However, the monads R and T are related by two operations $val^R_T : R \to T$ and let^R_T, the latter feeding programs of the form $RA \to TB$ with arguments of type TA, which do for T what val_Q and let_Q do for the monad QR. Given $L : RA$, $M : TA$ and $N : RA \to TB$, define:

$$val^R_T(L) = \lambda s : S.\, case\ L\ of\ inl(v).\ inl\langle v, s\rangle \\ \qquad\qquad\qquad\qquad\qquad\qquad inr(u).\, inr(u)$$

$$let^R_T\ x \Leftarrow M\ in\ N(x) = \lambda s : S.\, case\ M(s)\ of\ inl\langle v, s'\rangle.\, N(inl(v))s' \\ \qquad\qquad\qquad\qquad\qquad\qquad\qquad inr(u).\quad N(inr(u))s_0,$$

where s_0 is some recovery state from which computation is resumed if a dramatic failure occurs. We can now define the semantics of dramatic exception handlers by just replacing val_Q with val^R_T and let_Q with let^R_T in (1), and the given proof of (2) goes through unchanged (see application 51). This approach is shown in Section 5 to yield a uniform interpretation of *handle* in a large class of models obtained by modular constructions: let H be an arbitrary monad and let $\mathcal{F}H$ be the monad $(\mathcal{F}H)X = (H(X \times S))^S$; assuming that suitable operations val^R_H and let^R_H are given for interpreting failure in the computational setting of H, one obtains operations $val^R_{\mathcal{F}H}$ and $let^R_{\mathcal{F}H}$ for reinterpreting failure in the more elaborate setting of $\mathcal{F}H$.

For which monads R and T can we find suitable operations val^R_T and let^R_T lifting R-computation to T-computation? What equations should one expect such operations to satisfy? Associativity seems a reasonable assumption. Moreover, in the above example, val^R_T is a *left* unit for let^R_T, that is: $let^R_T(val^R_T) = id$. On the other hand, it is not a *right* unit, that is, $let^R_T(f) \circ val^R_T = f$ does not hold. If let^R_T is to model a nonstrict form of program composition and the view is adopted that programs should form a category, this is a rather odd state of affairs. In the next section we look for a categorical picture to give us a convincing set of axioms for a general theory of program composition.

3 An Algebraic View of Program Composition

In this section we propose an abstract categorical setting, called *extension setting*, for interpreting program composition. The underlying algebraic intuition is explained by discussing the example of non-deterministic call-by-need programs.

In the functional programming language Haskell programs are said to evaluate their parameters "by-need." Call-by-need differs from call-by-value in that application is nonstrict: A typical Haskell implementation of the Ackermann function, for example, would include a clause `ack 0 n =1`. Then, for a nonterminating program `loop`, the term `ack(0,loop)` evaluates to `1`, while it would fail to produce a result in Standard ML, where parameters are called by-value.

Call-by-need also differs from call-by-name in the presence of nondeterminism. A sequential program may exhibit nondeterministic behaviour when interacting with the operating system. For example, many programming languages, including Haskell, feature a library function `GetTime` which returns nondeterministically the current value of the system clock. Let the call `ack(2,GetTime)` match the clause `ack n m = ack (ack (n-1) m) (m-1)`. With a call-by-name discipline, as in the Algol-like language of [Ten91], this call would result in evaluating the second argument at different times, thus producing nonsensical results. Conversely, arguments that are called by-need are evaluated only once, if ever. The discriminating notion here is *additivity*. Let `p` and `q` be programs and let `p or q` be the program which runs either `p` or `q`, nondeterministically. A program `f` is called additive when `f(p or q)=f(p) or f(q)`. Then, call-by-need and call-by-value programs are additive while call-by-name are not.

The above discussion suggests an "algebraic" explanation of these three calling mechanisms. Consider an interpretation in the category *Set* of small sets of a simple nondeterministic language, where programs producing values in X are interpreted as elements of the finite powerset $\mathcal{P}X$ of X. Two operations are fundamental in the finite powerset construction: binary union, which we can use to interpret `or`, and emptyset, which we can use to interpret `loop`. In this setting, one can view call-by-value programs as homomorphisms with respect to both operations, call-by-need with respect to union only, and call-by-name with respect to neither. The following interpretation of the three calling mechanisms is based upon this observation.

A nondeterministic program `p(x)`, with a call-by-name parameter `x`, expects an unevaluated expression as input. Therefore such programs correspond (roughly) to functions of the form $\mathcal{P}A \to \mathcal{P}B$, and `p(q)` is obtained by straight composition. On the other hand, if `p` is call-by-value, it must run on the *results* of its argument's evaluation and produce nothing if `q` produces none. Therefore, such programs correspond to morphisms of the form $A \to \mathcal{P}B$. Composition of such programs is obtained by exploiting the operation of Kleisli lifting $(\_)^{*\mathcal{P}}$ of the monad $\mathcal{P}$, which maps morphisms $A \to \mathcal{P}B$ to morphisms $\mathcal{P}A \to \mathcal{P}B$. In particular, $[\![\texttt{p(q)}]\!] = [\![\texttt{p}]\!]^{*\mathcal{P}}[\![\texttt{q}]\!]$. In the computational metalanguage this is written:

$$[\![\texttt{p(q)}]\!] = let_{\mathcal{P}}\, x \Leftarrow [\![\texttt{q}]\!] \textit{ in } [\![\texttt{p}]\!].$$

Morphisms of the form $f^{*\mathcal{P}}$ are strict and additive precisely because finite powersets are the free construction associated with the theory of *semilattices*. Semilattices are algebraic structures with a nullary operation 0 and a binary operation $\vee$ satisfying the following axioms:

$$\begin{aligned} x \vee x &= x \\ x \vee y &= y \vee x \\ x \vee (y \vee z) &= (x \vee y) \vee z \\ x \vee 0 &= x. \end{aligned}$$

A monad providing the free construction associated with an algebraic theory is said to *classify* the theory. The correspondence between monads and algebraic theories in enriched categories is studied in [KP93, Rob95].

To model a call-by-need p, not only must we say how it behaves on values, but also what it can do when no value is produced in input. This can be done by interpreting p as a morphism $A_\perp \to \mathcal{P}B$, where $A_\perp = \{X \in \mathcal{P}A \mid card(X) \leq 1\}$ and $card(X)$ is the cardinality of X. Then, to interpret p(q), we look for an operation $(\_)^*$ to return an additive, possibly nonstrict extension of $[\![p]\!]$ to $\mathcal{P}A$.

The idea is to split the finite powerset monad into two constructions, one for each operation of the theory of semilattices. First we consider the theory of 0, with no axioms. The free models of this theory in *Set* are given by the lifting monad $(\_)_\perp$. To "finish" the construction, we cannot use the finite *nonempty* powerset monad $\mathcal{P}^\vee$ which classifies the theory of $\vee$, as $\mathcal{P}A \neq \mathcal{P}^\vee(A_\perp)$. In fact, we must consider this theory not in *Set* but in $Set^\perp$, the category of algebras of the monad $(\_)_\perp$. An algebra (A, a) for this monad consists of a set A and a distingiushed element $a \in A$. Homomorphisms from (A, a) to (B, b) are functions $f : A \to B$ such that $f(a) = b$. The free models of the theory of $\vee$ in $Set^\perp$ are given by the monad $\langle \mathcal{P}^+, \eta, \mu \rangle$, where $\mathcal{P}^+$ maps (A, a) to $(\{X \in \mathcal{P}A \mid a \in X\}, \{a\})$, $\eta_{(A,a)}(x) = \{x, a\}$ and $\mu_{(A,a)}(X) = \bigcup_{W \in X} W$. Clearly, writing $A_\perp$ for the free algebra $(A_\perp, \emptyset)$, the underlying set of $\mathcal{P}^+(A_\perp)$ is (isomorphic to) $\mathcal{P}A$.

In fact, $\mathcal{P}$ is the extension of the monad $\mathcal{P}^+$ along the forgetful functor $Set^\perp \to Set$ in the sense explained below. Similarly, the operation $(\_)^{*_{\mathcal{P}^+}}$, which lives in $Set^\perp$, extends to an operation $(\_)^*$ in *Set*. In particular, $(\_)^*$ maps functions $A_\perp \to \mathcal{P}B$ to functions $\mathcal{P}A \to \mathcal{P}B$ where $f^*X = \bigcup\{f(x) \mid x \in X_\perp\}$. Then, call-by-need is modelled as: $[\![p(q)]\!] = [\![p]\!]^* [\![q]\!]$. Pretty-printing:

$$[\![p(q)]\!] = let^{\perp}_{\mathcal{P}}\ x \Leftarrow [\![q]\!]\ in\ [\![p]\!].$$

Functions of the form f^* are strict only when f is strict. Moreover, they are additive because $(\_)^*$ extends the Kleisli composition of the monad $\mathcal{P}^+$ which classifies the theory of $\vee$.

The situation just described generalises as follows. We call *extension setting* a categorical picture

$$\mathcal{C} \underset{G}{\overset{F}{\rightleftarrows}} \mathcal{X} \xrightarrow{M} \mathcal{X}$$

where $F \dashv G$ are adjoint functors and $M = \langle M, \eta^M, \mu^M \rangle$ is a monad on $\mathcal{X}$. Let $R = \langle GF, \eta^R, \mu^R \rangle$ be the monad induced by the adjunction $F \dashv G$ on $\mathcal{C}$, let ϵ be the counit of this adjunction and let T be the functor $GMF : \mathcal{C} \to \mathcal{C}$. The latter is the right Kan extension of GM along G. The natural transformations $\eta^T = G\eta^M F \circ \eta^R$ and $\mu^T = G\mu^M F \circ GM\epsilon MF = G(\epsilon MF)^{*_M}$ endow T with the structure of a monad. Following [Str72], we call such a monad the extension of M along G. This extension is unique in the sense of [Str72, §2].

We write settings like the one above as triples $\langle F, G, M \rangle$ and indicate with R, T and ϵ respectively the monad GF, the monad GMF and the counit of the adjunction $F \dashv G$. We call $\mathcal{C}$ the *base category* of the setting.

Given an extension setting $\langle F, G, M\rangle$ on a base category $\mathcal{C}$, we intend to interpret program composition by means of a family $(\_)^*$ of associative operations of the form:

$$(\_)^*_{A,B} : \mathcal{C}(RA, TB) \to \mathcal{C}(TA, TB),$$

extending M-lifting along G. More formally: we require that, for all morphisms $h : FA \to MFB$, $f : RB \to TC$ and $g : RA \to TB$, the following holds:

$$(Gh)^* = G(h^{*_M}) \tag{3}$$
$$f^* \circ g^* = (f^* \circ g)^* \tag{4}$$

When the functor G is monadic, (3) requires $(\_)^*$ to behave like $(\_)^{*_M}$ on R-homomorphisms. Since T extends M along G, this is to say that f^* should preserve T structure whenever f preserves R structure. Note that the natural transformation $\iota = G\eta^M F : GF \to GMF$ in a setting $\langle F, G, M\rangle$ is a monad morphism from R to T, and that the equation $\iota^* = id$ follows immediately from (3).

Example 31 *Call-by-value and call-by-name.*

Given a monad T, an interpretation $[\![\mathtt{p(q)}]\!] = [\![p]\!]^*[\![q]\!]$ of call-by-value program composition is obtained in the setting $\langle Id, Id, T\rangle$, where R is the identity and $(\_)^* = (\_)^{*_T}$. On the other hand, a call-by-name interpretation is obtained in any setting $\langle F, G, Id\rangle$, where $R = GF = T$ and $(\_)^*$ is the identity. These examples are the "extreme" cases where R possesses all or nothing of the structure of T. The following examples show that intermediate cases are also interesting.

Example 32 *Call-by-need.*

Here we describe a setting, analogous to the finite powerset example developed earlier, relating the lifting and Hoare powerdomain monads in the category of cpos. A similar picture can be drawn for algebraic cpos.

Let *Cpo* be the category of possibly bottomless cpos. The Hoare powerdomain $\mathcal{P}(A)$ of such a cpo A is the set of downward closed subsets of A ordered by inclusion. Empty set and union are the universal operations on $\mathcal{P}(A)$ satisfying the theory of semilattices, together with the axiom: $x \vee y \geq x$.

One can split this construction in two steps as done for powersets. In particular, let $Cpo^\perp$ be the category of cpos with bottom element and strict continuous functions. This is the category of algebras of the lifting monad. If X is an object of $Cpo^\perp$, let $\mathcal{P}^+X$ be the cpo of nonempty downwards closed subsets of X ordered by inclusion and let $\eta_X : X \to \mathcal{P}^+X$ map x to $\{y \mid y \leq x\}$. The union operation makes of $\mathcal{P}^+X$ the free $\{\vee\}$-algebra generated by X in $Cpo^\perp$. That is: for any map $f : A \to B$ in $Cpo^\perp$, where B is endowed with an operation $\vee$ satisfying the given equations, there is a unique $\vee$-homomorphism $f^\dagger : \mathcal{P}^+A \to B$ such that $f^\dagger \circ \eta_A = f$. This gives to $\mathcal{P}^+$ the structure of a monad which extends to $\mathcal{P}$ along the forgetful functor $G^\perp = Cpo^\perp \to Cpo$. The operation $(\_)^*$ such that $f^*X = \bigcup\{f(x) \mid x \in X_\perp\}$ extends $(\_)^{*_{\mathcal{P}^+}}$ along $G^\perp$.

Example 33 *Composition of monads.*

Any two monads which compose give rise to an operation $(\_)^\star$. Let R and Q be monads on a category $\mathcal{C}$ and let the natural transformation $\lambda : RQ \dot{\rightarrow} QR$ be a distributive law of Q over R. The functor $\hat{Q} : \mathcal{C}^R \rightarrow \mathcal{C}^R$ mapping R-algebras α to $Q\alpha \circ \lambda$ has the structure of a monad which forms an extension setting $\langle F^R, G^R, \hat{Q}\rangle$. The monad $\hat{Q}$ is the *lifting* of Q to $\mathcal{C}^R$ associated with λ (see [Bec69]). In particular, we have $G^R\hat{Q} = QG^R$. The extension of $\hat{Q}$ along the forgetful functor G^R is the composite monad QR. In fact, we have $G^R\hat{Q}F^R = QG^RF^R = QR$. Note that the pair (F^R, λ) is a monad morphism $Q \rightarrow \hat{Q}$.

Writing $h^{*_{\hat{Q}}} = \mu^{\hat{Q}} \circ \hat{Q}h$ and noticing that μ^Q is the underlying natural transformations of $\mu^{\hat{Q}}$, we see that the operation $(\_)^{*_Q}_{RA,RB}$, which is obviously associative, extends $(\_)^{*_{\hat{Q}}}$ along the forgetful functor G^R.

4 Notions of Composition

In the previous section we developed some intuition on how a general operation $(\_)^\star$ to interpret program composition should look like and we wrote axioms to support our intuition formally. We assumed that such an operation, similar to composition in the Kleisli category of a monad, lives in an extension setting. Here we develop the theory of more general notions of composition, which need not belong to an extension setting. When they do, we prove that the equations of Section 3 are satisfied. However, it is in the more general theory that we derive the properties that we expected to hold from our earlier discussion.

Let $R : \mathcal{C} \rightarrow \mathcal{C}$ be a functor, let T be a monad on $\mathcal{C}$ and let $\sigma : R \rightarrow T$ be a natural transformation. Given $h : A \rightarrow TB$, we write $h_\sigma : RA \rightarrow TB$ the morphism

$$h_\sigma = h^{*_T} \circ \sigma.$$

Note that, when R has the structure of a monad and σ is a monad morphism, there is a forgetful functor $G_\sigma : \mathcal{C}^T \rightarrow \mathcal{C}^R$ mapping T-algebras (A, α) to R-algebras $(A, \alpha \circ \sigma)$. In this case we have $h_\sigma = G^R h^\dagger$, where $h^\dagger : F^R A \rightarrow G_\sigma F^T B$ corresponds bijectively to h by the adjunction $F^R \dashv G^R$.

Definition 41 *Let $R : \mathcal{C} \rightarrow \mathcal{C}$ be a functor and let T be a monad on a category $\mathcal{C}$; a* weak notion of composition *is a pair $(\iota, (\_)^\star)$, where $\iota : R \rightarrow T$ is a natural transformation and $(\_)^\star$ is a family of operations:*

$$(\_)^\star_{A,B} : \mathcal{C}(RA, TB) \rightarrow \mathcal{C}(TA, TB)$$

satisfying the following equations: for all $f : RB \rightarrow TC$, $g : RA \rightarrow TB$ and $h : A \rightarrow TB$,

$$f^\star \circ g^\star = (f^\star \circ g)^\star \tag{4}$$

$$h^\star_\iota = h^{*_T}. \tag{5}$$

We write $(\iota, (\_)^\star) : R \rightarrow T$ for a weak notion of composition as above to make R and T understood.

Proposition 42 *The operation $(\_)^\star_{A,B}$ of a weak notion of composition is natural in B.*

Proof. Let $(\iota, (\_)^\star) : R \to T$ be a weak notion of composition; naturality of $(\_)^\star_{A,B}$ in B is expressed by the equation $Tf \circ g^\star = (Tf \circ g)^\star$, where $f : B \to C$ and $g : RA \to TB$. Then,

$$Tf \circ g^\star = (\eta \circ f)^* \circ g^\star = (\eta \circ f)^\star_\iota \circ g^\star = ((\eta \circ f)^\star_\iota \circ g)^\star = (Tf \circ g)^\star.$$

Proposition 43 *Weak notions of composition $(\iota, (\_)^\star) : R \to T$ satisfy the following equations:*

$$\iota^\star = id_T \tag{6}$$
$$f^* \circ g^\star = (f^* \circ g)^\star \tag{7}$$
$$Th = (\iota \circ Rh)^\star. \tag{8}$$

Proof. Note that $\iota = \eta_\iota$. Let $f : B \to TC$, $g : RA \to TB$ and $h : A \to B$,

$$\begin{aligned} \iota^\star &= \eta_\iota^\star = \eta^* = id_T; \\ f^* \circ g^\star &= f_\iota^\star \circ g^\star = (f_\iota^\star \circ g)^\star = (f^* \circ g)^\star; \\ Th &= (\eta \circ h)^* \circ \iota^\star = ((\eta \circ h)^* \circ \iota)^\star = (Th \circ \iota)^\star = (\iota \circ Rh)^\star. \end{aligned}$$

Weak notions of composition $(\iota, (\_)^\star) : R \to T$ in which R is a monad on $\mathcal{C}$ and ι is a monad morphism, often live in an extension setting. In fact, the forgetful functor $G_\iota : \mathcal{C}^T \to \mathcal{C}^R$ induced by ι often has left adjoint. This is always the case when $\mathcal{C}$ is *Set* [BW85, 9.3]. In general, it is well known that G_ι has left adjoint when $\mathcal{C}^T$ has all coequalisers of reflexive pairs [Lin69, coroll. 1]. A sufficient condition for that to happen is that $\mathcal{C}$ has such coequalisers and T preserves them [Lin69, coroll. 3].

If G_ι has left adjoint F_ι, we obtain an extension setting $\langle F^R, G^R, M\rangle$ where M is the monad induced on $\mathcal{C}^R$ by the adjunction $F_\iota \dashv G_\iota$. In this setting, $G^R M F^R = G^R G_\iota F_\iota F^R = G^T F^T = T$.

The following theorem shows the correspondence between weak notions of composition and operations satisfying (3) and (4) as in Section 3.

Theorem 44 *Let $\langle F, G, M\rangle$ be an extension setting; a family of operations $(\_)^\star_{A,B} : \mathcal{C}(RA, TB) \to \mathcal{C}(TA, TB)$ satisfies (3) and (4) if and only if $(\iota, (\_)^\star)$ is a weak notion of composition, for some ι such that $\iota \circ \eta^R = \eta^T$.*

Proof. [(3)⇒(5)] Let $\iota = G\eta^M F$ and let $h : A \to TB$. Using the naturality of $(\_)^{*_M}$, we have:

$$\begin{aligned} h_\iota^\star &= (\mu_B^T \circ \iota_{TB} \circ Rh)^\star = (G\epsilon^{*_M}_{MFB} \circ G\eta^M_{FTB} \circ GFh)^\star = (G(\epsilon^{*_M}_{MFB} \circ \eta^M_{FTB} \circ Fh))^\star \\ &= G(\epsilon^{*_M}_{MFB} \circ \eta^M_{FTB} \circ Fh)^{*_M} = G\epsilon^{*_M}_{MFB} \circ G(\eta^M_{FTB} \circ Fh)^{*_M} \\ &= \mu_B^T \circ G((\eta^M_{FTB})^{*_M} \circ MFh) = \mu_B^T \circ Th = h^{*_T}. \end{aligned}$$

[(5)⇒(3)] Let $(\iota, (\_)^*) : R \to T$, with $\iota \circ \eta^R = \eta^T$, be a weak notion of composition. For any $f : RA \to TB$ we have $(f \circ \eta^R)_\iota = f$ by easy calculations. Let $h : FA \to MFB$, we have:

$$\begin{aligned}(Gh)^* &= (Gh \circ \eta^R_A)^*_\iota = (Gh \circ \eta^R_A)^{*_T} = \mu^T \circ T(Gh \circ \eta^R_A) \\ &= G\mu^M_{FB} \circ GM\epsilon_{MFB} \circ GMFGh \circ T\eta^R_A \\ &= G\mu^M_{FB} \circ GMh \circ GM\epsilon_{FA} \circ GMF\eta^R_A \\ &= G(\mu^M_{FB} \circ Mh) \circ GM(\epsilon_{FA} \circ F\eta^R_A) = Gh^{*_M}.\end{aligned}$$

Definition 45 *A* notion of composition *is a* weak notion of composition *$(\iota, (\_)^*)$ such that, for all $f : RA \to TB$ the following equation holds:*

$$f^* \circ \iota = f. \tag{9}$$

When working with sets, the inclusions $\iota_A : A_\perp \to \mathcal{P}A$ of Section 3 do not satisfy (9) while they do in the case of cpos (example 32). Another *strictly* weak notion of composition is the pair of operation val^R_T and let^R_T defined in the introduction to model dramatic failure. On the other hand, (9) is satisfied in the models of interleaving of Application 54.

The following results are used in the next section:

Proposition 46 *The operation $(\_)^*_{A,B}$ of a notion of composition is natural in A.*

Proof. Let $(\iota, (\_)^*) : R \to T$ be a notion of composition; naturality of $(\_)^*_{A,B}$ in A is expressed by the equation $f^* \circ Tg = (f \circ Rg)^*$, where $f : RA \to TB$ and $g : C \to A$. Then, from (8) and (9) we have:

$$f^* \circ Tg = f^* \circ (\iota_A \circ Rg)^* = (f^* \circ \iota_A \circ Rg)^* = (f \circ Rg)^*.$$

Proposition 47 *Let $(\iota, (\_)^*) : R \to T$ be a weak notion of composition and let $\iota' : S \to T$ be a natural transformation such that $\iota = \iota' \circ \nu$ for some natural transformation ν. The pair $(\iota', (\_)^{*\nu})$, where $f^{*\nu} = (f \circ \nu)^*$, is a weak notion of composition $S \to T$.*

Proof. The associativity of $(\_)^{*\nu}$ is an immediate consequence of the associativity of $(\_)^*$. Moreover, let $h : A \to TB$,

$$h^{*\nu}_{\iota'} = (h^{*_T} \circ \iota')^{*\nu} = (h^{*_T} \circ \iota' \circ \nu)^* = (h^{*_T} \circ \iota)^* = h^{*_T} \circ \iota^* = h^{*_T}.$$

Proposition 48 *Let $(\iota, (\_)^*) : R \to T$ be a notion of composition and let $\nu : R \to S$ be a natural transformation with a right inverse, that is a natural transformation ν' such that $\nu \circ \nu' = id$. The pair $(\iota', (\_)^{*\nu})$, where $\iota' = \iota \circ \nu'$ and $f^{*\nu}$ is as above, is a notion of composition $S \to T$.*

Proof. It is a weak notion of composition by the previous proposition. Moreover:

$$f^{*\nu} \circ \iota' = (f \circ \nu)^* \circ \iota \circ \nu' = f \circ \nu \circ \nu' = f.$$

5 Applications to Modular Semantics

In [Mog90a], a modular approach to denotational semantics is proposed, where mathematical models of computation are obtained by stepwise application of *monad constructors*. These are functions $\mathcal{F}$ mapping monads to monads and satisfying certain naturality conditions. Intuitively, the monad $\mathcal{F}T$ augments the structure of T with the machinery to interpret a new computational feature. For example, the constructor $\mathcal{F}$ such that $(\mathcal{F}T)A = (T(A \times S))^S$ adds to T the capability of modelling side-effects. Monad constructors are studied in [CM93, Cen95].

In [Mog90b] the notion of *uniform redefinition* is introduced to lift operations defined in a computational setting M to a new setting $\mathcal{F}M$. Let $\zeta(\_)$ be some type scheme, let $op_A : \zeta(MA)$ be an operation defined for a monad M in a category $\mathcal{C}$, let $H : \mathcal{C} \to \mathcal{C}$ be an endofunctor and let $\mathcal{F}$ be a monad constructor of the form $(\mathcal{F}T)A = THA$; op can be uniformly redefined for the monad $\mathcal{F}M$ as follows:

$$(\mathcal{F}op)_A = op_{HA}.$$

This technique is not always applicable: when either $\mathcal{F}$ or op are not of the appropriate form, ad-hoc redefinitions must be sought. The above constructor for side-effects, for example, does not fulfill the requirements. Neither does the operation $C_{A,B} : (A \to MB) \times (MA \to MB) \times MA \to MB$ used in [CM93, example 2.10] to perform case analysis on interleaving programs. In this section we propose a technique based on notions of composition which yields well behaved redefinitions of operations in both cases.

We show that two benefits derive from using notions of composition to define operations in a computational setting M: on the one hand it allows properties of the operations to be formally derived without exposing (all of) the structure of M (thus for a large class of models); on the other hand it allows the operations to be redefined in cases where uniform redefinitions are not available, and their properties automatically preserved.

Application 51 *Reinterpreting failure in state models.*

Let R be the monad $RA = A + 1$. We say that a monad $\langle H, \eta, (\_)^*\rangle$ has a *structure for failure* when it is equipped with a weak notion of composition $(\iota, (\_)^\star) : R \to H$ and with a natural transformation $\rho : H \to H$ such that:

$$\iota \circ inl = \eta \quad (10)$$

$$f^\star \circ \eta = f \circ inl \quad (11)$$

$$f^\star \circ fail = \rho \circ f \circ inr \quad (12)$$

$$f^* \circ fail = fail \quad (13)$$

where $fail_A : HA$ is the natural transformation $\iota \circ inr$. Intuitively, $\rho(N)$ is the program running N after some recovery action. For example, ρ would be the identity for $HA = ((A+1) \times S)^S$ while it would feed its argument with some recovery state for $HA = ((A \times S) + 1)^S$.

An operation $handle_A : HA \times HA \to HA$ running its first argument and handling a possible failure with its second can be defined as follows:

$$handle\,(M, N) = [\eta, N]^\star \circ M.$$

The following equations are satisfied by *fail* and *handle* in any structure for failure:

$$\begin{aligned} handle\,(\eta, N) &= \eta \\ handle\,(fail, N) &= \rho(N) \\ handle\,(L, handle\,(M, N)) &= handle\,(handle\,(L, M), N). \end{aligned}$$

In fact, $handle\,(\eta, N) = [\eta, N]^\star \circ \eta = [\eta, N] \circ inl = \eta$ and similarly for the other equations.

The signature $HA \times HA \to HA$ is indeed of the form $\zeta(HA)$, which makes *handle* qualify for uniform redefinition. Not so however for the monad constructor $\mathcal{F}$ mapping H to $(\mathcal{F}H)A = (H(A \times S))^S$. This constructor, however, extends to structures for failure as follows:

$$\begin{aligned} &(\mathcal{F}\rho)w = \lambda\, s.\, \rho(w(s_0)) \\ &(\mathcal{F}\iota)z = \lambda\, s.\, let_H\ a \Leftarrow \iota(z)\ in\ val_H\langle a, s\rangle \\ &f^{\mathcal{F}\star}w = \lambda\, s.\, let^R_H\ z \Leftarrow w(s)\ in\ case\ z\ of\ inl\langle a, s'\rangle.\, f(inl(a))s' \\ &\qquad\qquad\qquad\qquad\qquad\qquad\qquad\qquad\quad inr(u).\quad f(inr(u))s_0. \end{aligned}$$

These operations are easily shown to satisfy the axioms (10-13). Thus, by suitably extending the action of the constructor $\mathcal{F}$ to weak notions of composition $R \to H$, operations such as *fail* and *handle* are automatically redefined in models of computation with side-effects, in such a way that the relevant properties are also preserved.

Remark. The operations val^R_T and let^R_T of Section 1 are obtained by applying the constructuion just described to the identity notion of composition $R \to R$.

Application 52 Inwards *monad constructors.*

Here we describe a class of monad constructors $\mathcal{F}$ which have a canonical lifting of (weak) notions of composition $R \to T$ to $\mathcal{F}R \to \mathcal{F}T$. In the next application we use this construction to obtain a reinterpretation interleaving in models of exceptions.

We call *inwards* a monad constructor $\mathcal{F}$ such that:

$$\begin{aligned} &(\mathcal{F}M)A = M(HA) \ \text{ for some functor } H, \text{ and} \\ &\quad \mu^{\mathcal{F}M} = \mu^M H \circ Mp \ \text{ for some natural transformation } p : HMH \to MH. \end{aligned}$$

Remark. The above condition on $\mu^{\mathcal{F}M}$ arises when composing monads with functors. Let H be a functor, let $\eta^H : Id \to H$ be a natural transformation and let η^{MH} be $\eta^M H \circ \eta^H$. There is a one to one correspondence between natural transformations $p : HMH \to MH$ satisfying

$$\begin{aligned} p \circ \eta^H MH &= id \\ p \circ H\eta^{MH} &= \eta^M H \\ p \circ H\mu^{MH} &= \mu^{MH} \circ pMH, \end{aligned}$$

where μ^{MH} is the natural transformation $\mu^M H \circ Mp$, and monads $\langle MH, \eta^{MH}, \mu\rangle$ such that $\mu \circ \mu^M HMH = \mu^M H \circ M\mu$ (see [JD93]).

Proposition 53 *Let the monad constructor $(\mathcal{F}M)A = M(HA)$ be inwards, and let $(\iota, (\_)^\star) : R \to T$ be a (weak) notion of composition. The pair of operations $(\iota H, (\_)^{\star H})$, where $(\_)^{\star H}_{A,B} = (\_)^\star_{HA,HB}$, is a (weak) notion of composition $\mathcal{F}R \to \mathcal{F}T$.*

Proof Associativity (and right unit) are inherited immediately form $(\iota, (\_)^\star)$. Let $f : A \to THB$, noticing that $f^{*_{TH}} = (p \circ Hf)^{*_T}$, we have:

$$\begin{aligned} f^{\star H}_{\iota H} &= (f^{*_{TH}} \circ \iota)^\star = ((p \circ Hf)^{*_T} \circ \iota)^\star = (p \circ Hf)^{*_T} \circ \iota^\star \\ &= \mu^T \circ Tp \circ THf = \mu^{TH} \circ THf = f^{*_{TH}}. \end{aligned}$$

Application 54 *Reinterpreting interleaving in models of exceptions.*

In [CM93], the semantics of computation with interleaving is described in terms of the "resumptions monad" $TA = \mu X.Q(A + X)$ and two families of operations

$$\begin{aligned} \tau_A &: TA \to TA \\ C_{A,B} &: (A \to TB) \times (TA \to TB) \times TA \to TB. \end{aligned}$$

from which interesting programming constructs can be defined, such as the operator *pand* of parallel composition described in [Cen95, 7.3]. The operations τ and C arise from a notion of composition $(\iota, (\_)^\star) : Id + T \to T$. Let α_A be the isomorphism $Q(A + TA) \to TA$ and let γ_A be its inverse. We define:

$$\begin{aligned} \iota(z) &= \mathit{case}\ z\ \mathit{of}\ \mathit{inl}(a).\ \alpha(\mathit{val}_Q(\mathit{inl}(a))) \\ &\qquad\qquad\qquad \mathit{inr}(u).\alpha(\mathit{val}_Q(\mathit{inr}(u))) \\ f^\star(w) &= \alpha(\mathit{let}_Q\ z \Leftarrow \gamma(w)\ \mathit{in}\ \gamma(f(z))) \end{aligned}$$

The associativity of $(\_)^\star$ follows easily from the associativity of $(\_)^{*_Q}$. Similarly, $f^\star \circ \iota = f$ follows from $f^{*_Q} \circ \eta^Q = f$. As for (5), note that, for $h : A \to TB$, we have:

$$\begin{aligned} h_\iota(z) \;&=\; \mathit{case}\ z\ \mathit{of}\ \mathit{inl}(a).\ h(a) \\ &\qquad\qquad \mathit{inr}(u).\alpha(\mathit{val}_Q(\mathit{inr}(h^{*_T}u))) \end{aligned}$$

and hence

$$h_\iota^\star(w) = \alpha(let_Q\ z \Leftarrow \gamma(w)\ in\ case\ z\ of\ inl(a).\ \gamma(h(a)) \\ inr(u).\ val_Q(inr(h^{*_T}u))) \\ = h^{*_T}w.$$

Now, the operations τ and C of [CM93] can be defined as follows:

$$\tau(w) = \iota(inr(w)) \\ C(f,g,w) = (\lambda z.\ case\ z\ of\ inl(a).\ f(a) \\ inr(u).\ g(u))^\star w.$$

Noticing that $\eta^T = \iota \circ inl$, from (6) and (9) one can easily derive the equations

$$C(f,g,\eta) = f \\ C(f,g,\tau) = g \\ C(\eta,\tau,w) = w$$

showing that τ and C behave respectively as right injection and case analysis. The commutativity of the operator *pand* of [Cen95, 7.3] can be easily derived from these equation and from the commutativity of an operation of nondeterministic choice.

Let H be the functor $HA = A + E$, where E is some object "of exceptions." Given a monad M there is a *unique* monad $\langle MH, \eta, \mu\rangle$ such that $\eta = \eta^M H \circ inl$ and $\mu \circ \mu^M HMH = \mu^M H \circ M\mu$. This follows from the remark in 52. We write $\mathcal{F}$ the monad constructor for exceptions, mapping monads M to MH.

The constructor $\mathcal{F}$ is inwards. Hence, applying Proposition 53 to the notion of composition $Id + T \to T$ defined above, we get a notion of composition $H + TH \to TH$ satifying (9). Then, noticing that the natural transformation $[id + \eta^T H \circ inr, inr] : H + TH \to Id + TH$ has a right inverse $inl + id$, we obtain, by Proposition 48, a notion of composition $Id + TH \to TH$ to interpret interleaving in models constructed by $\mathcal{F}$. Again, C and τ are automatically reinterpreted in such models and the relevant properties are preserved.

6 Conclusions

We proposed a general categorical setting for modeling program composition in which the call-by-value and call-by-name disciplines fit as special cases. Call-by-need is also captured in this framework for nondeterministic programs; it is an interesting question whether call-by-need programs with side effects can be captured similarly. The proposed theory of program composition features two operations ι and $(\_)^\star$, reminiscent of the unit and lifting of Kleisli triples, of which only weak properties are assumed. These are however enough to derive simple equational properties of common program constructs such exception handling and parallel composition. The paper argues that, by defining program constructs in terms of ι and $(\_)^{\star}$, not only can one validate program equivalences axiomatically for large classes of models, but also reinterpret the constructs when models are extended, preserving the truth of the relevant axioms. Since we are able to do this in cases where the uniform redefinition proposed in [Mog90b] are not available, our technique makes one step forward towards a modular approach to denotational semantics. The proposed technique is applied in [Cen98] to the semantics of Java, where we seek a modular proof of computational adequacy with respect to the operational semantics of [CKRW98].

References

[Bec69] Jon Beck. Distributive laws. In B. Eckman, editor, *Seminar on triples and categorical homology theory*, pages 119–140, Berlin, 1969. Springer LNM 80.

[BW85] M. Barr and C. Wells. *Toposes, triples and theories*. Springer-Verlag, New York, 1985.

[Cen95] P. Cenciarelli. *Computational applications of calculi based on monads*. PhD thesis, Department of Computer Science, University of Edinburgh, 1995. CST-127-96. Also available as ECS-LFCS-96-346.

[Cen98] P. Cenciarelli. Objects and computation. Presented at the Dagstuhl seminar on "The Semantic Challenge of Object-Oriented Programming", Schloss Dagstuhl, Wadern, Germany, 28/6 - 3/7 1998.

[CKRW98] P. Cenciarelli, A. Knapp, B. Reus, and M. Wirsing. An Event-Based Structural Operational Semantics of Multi-Threaded Java. In J. Alves-Foss, editor, *Formal Syntax and Semantics of Java*, 1523 LNCS. Springer, 1998.

[CM93] P. Cenciarelli and E. Moggi. A syntactic approach to modularity in denotational semantics. In *Proceedings of 5th Biennial Meeting on Category Theory and Computer Science*. CTCS-5, 1993. CWI Tech. Report.

[JD93] M.P. Jones and L. Duponcheel. Composing monads. Technical Report YALEU/DCS/RR-1004, Department of Computer Science, Jale University, 1993.

[KP93] G.M. Kelly and A.J. Power. Adjunctions whose counits are coequalizers, and presentations of finitary monads. *Journal of Pure and Applied Algebra*, 89:163–179, 1993.

[Lin69] F.E.J. Linton. Coequalizers in Categories of Algebras. In *Seminar on Triples and Categorical Homology Theory*, pages 75–90. Springer LNM 80, 1969.

[Mog90a] E. Moggi. An abstract view of programming languages. Technical Report ECS-LFCS-90-113, University of Edinburgh, Comp. Sci. Dept., 1990.

[Mog90b] E. Moggi. Modular approach to denotational semantics. Unpublished manuscript, November 1990.

[Mog91] E. Moggi. Notions of computation and monads. *Information and Computation*, 93(1), 1991.

[Rob95] E.P. Robinson. Note on the presentation of enriched monads. Unpublished manuscript, available by ftp from `theory.doc.ic.ac.uk`, 1995.

[Str72] R. Street. The formal theory of monads. *Journal of Pure and Applied Algebra*, 2:149–168, 1972.

[Ten91] R.D. Tennent. *Semantics of programming languages.* Prentice Hall, 1991.

Architectural Specifications in CASL

Michel Bidoit[1], Donald Sannella[2], and Andrzej Tarlecki[3]

[1] Laboratoire Spécification et Vérification, CNRS & ENS de Cachan, France
[2] Laboratory for Foundations of Computer Science, University of Edinburgh, UK
[3] Institute of Informatics, Warsaw University and Institute of Computer Science, Polish Academy of Sciences, Warsaw, Poland.

Abstract. One of the novel features of CASL, the Common Algebraic Specification Language, is the provision of so-called *architectural specifications* for describing the modular structure of software systems. A discussion of refinement of CASL specifications provides the setting for a presentation of the rationale behind architectural specifications. This is followed by details of the features provided in CASL for architectural specifications, hints concerning their semantics, and simple results justifying their usefulness in the development process.

1 Introduction

A common feature of present-day algebraic specification languages (see e.g. [SW83], [EM85], [GH93], [CoFI96], [SW98]) is the provision of *specification-building operations* [BG77] for building large specifications in a structured fashion from smaller and simpler ones. Less usual are features for describing the modular structure of software systems under development. This paper is about the facilities for this that are provided in CASL, the new *Common Algebraic Specification Language* [CoFI98b] that has been developed under the auspices of the Common Framework Initiative [Mos97, CoFI98a] in an attempt to create a focal point for future joint work on algebraic specifications and a platform for exploitation of past and present work on methodology, support tools, etc.

Following practical experiences [FJ90] and foundational work [Bid88], [ST89], [SST92], [BH93], we argue that mechanisms to structure specifications cannot suffice for describing the modular structure of software under development. CASL therefore provides a separate kind of specifications, so-called *architectural specifications*, for this purpose. An architectural specification consists of a list of *unit declarations*, indicating the component modules required with specifications for each of them, together with a *unit term* that describes the way in which these modules are to be combined. Such architectural specifications are aimed at the "implementation" modular structure of the system rather than at the "interaction" relationships between modules in the sense of [AG97] (the latter to be considered when specifications of "reactive" modules are introduced in a CASL extension).

The aim of this paper is to present motivation, intuition and technicalities related to this concept. We provide some information about CASL in Sect. 2, discuss the development of programs from specifications by stepwise refinement in

A.M. Haeberer (Ed.): AMAST'98, LNCS 1548, pp. 341–357, 1998.

Sect. 3 and then introduce architectural specifications in Sect. 4. The semantics and correctness issues of architectural specifications are discussed in Sects. 5, 6 and 7. The development process in the presence of architectural specifications is briefly discussed in Sect. 8.

Even though we present architectural specifications in the context of CASL, the ideas apply in any specification and development framework, as we mention in Sect. 9. We also briefly mention there the issue of *behavioural refinement.*

2 CASL Preliminaries

CASL is a formalism to describe CASL *structures*: many-sorted algebras with subsorts, partial operations and predicates. Structures are classified by *signatures*, which give *sort* names (with their subsorting relation), partial/total *operation* names, and *predicate* names, together with *profiles* of operations and predicates. For each signature Σ, the class of all Σ-structures is denoted $\mathbf{Mod}[\Sigma]$.

The basic level of CASL includes *declarations* to introduce components of signatures and *axioms* to give properties of structures that are to be considered as *models* of a specification. The logic used to write the axioms is essentially first-order logic built over *atomic formulae* which include strong and existential equalities, definedness formulae and predicate applications. A basic CASL specification SP amounts to a definition of a signature Σ and a set of axioms Φ. It denotes the class $[\![SP]\!] \subseteq \mathbf{Mod}[\Sigma]$ of its *models*, which are those Σ-structures that *satisfy* all the axioms in Φ: $[\![SP]\!] = \{A \in \mathbf{Mod}[\Sigma] \mid A \models \Phi\}$.

CASL provides ways of building complex specifications out of simpler ones by means of various *structuring constructs.* These include translation, hiding, union, and both free and loose forms of extension. *Generic specifications* and their *instantiations* with pushout-style semantics [EM85] are also provided. Structured specifications built using these constructs can be given a compositional semantics where each specification SP determines a signature $Sig[SP]$ and a class $[\![SP]\!] \subseteq \mathbf{Mod}[Sig[SP]]$ of models.

2.1 Example

Here is a sequence of definitions of CASL specifications.

spec NUM =
  sort *Num*
  ops *0* : *Num*;
    succ : *Num* $\rightarrow$ *Num*
end

spec ADDNUM = NUM **then op** $plus : Num \times Num \rightarrow Num$
  vars $x, y : Num$
  axiom $plus(x, succ(y)) = succ(plus(x, y))$

spec ORDNUM = NUM **then pred** $\_ < \_ : Num \times Num$
  axiom $\forall x : Num \bullet 0 < succ(x)$

spec CODENUM =
  ADDNUM **and** ORDNUM
  then **op** $code : Num \to Num$
    axiom $\forall x : Num \bullet 0 < code(x)$

We start with a signature for natural numbers, and then extend it in two ways: by a binary operation with a simple axiom and by a loosely specified binary predicate. In CODENUM we put both extensions together and then add a unary operation on *Num* with another simple axiom.

spec ELEM = **sort** *Elem* **end**

spec PARTCONTAINER [ELEM] =
  generated type $Cont ::= empty \mid add(Elem; Cont)?$
  pred $addable : Elem \times Cont$
  vars $x, y : Elem;\ C : Cont$
  axiom $def\ add(x, C) \Leftrightarrow addable(x, C)$
  pred $\_\_ \in \_\_ : Elem \times Cont$
  axioms $\neg\ (x \in empty)$;
    $(x \in add(y, C) \Leftrightarrow x = y \lor x \in C)$ *if* $addable(y, C)$
end

This is a generic (in ELEM) specification of "partial containers", which introduces a datatype *Cont* generated by a constant *empty* and a partial constructor *add* that adds an element to a container. An element x may be added to a container C if and only if $addable(x, C)$ is satisfied. But *addable* is left unspecified at this stage. The usual membership predicate is provided as well.

spec PARTNUMCONT =
  PARTCONTAINER[CODENUM **fit** $Elem \mapsto Num$]

We instantiate PARTCONTAINER to CODENUM, with an appropriate fitting of the parameter. The result contains all the components of CODENUM together with those added by PARTCONTAINER with their profiles adjusted accordingly.

spec UNIQUENUMCONT =
  PARTNUMCONT
  then **vars** $x : Num;\ C : Cont$
    axiom $addable(x, C) \Leftrightarrow \neg\ (x \in C) \land \neg\ (code(x) \in C)$

Finally, we constrain the addability condition, requiring that a number is addable to a container if and only if neither it nor its code are already included there.

3 Program Development and Refinement

The intended use of CASL is to specify programs. Each CASL specification should determine a class of programs that realize the specified requirements. It follows that programs must be written in a language having a semantics which assigns[1] to each program its *denotation* as a CASL structure. Then each program P determines a signature $Sig[P]$ and a structure $[\![P]\!] \in \mathbf{Mod}[Sig[P]]$. The denotation $[\![SP]\!]$ of a specification SP is a description of its admissible realizations: a program P is a *(correct) realization* of SP if $Sig[P] = Sig[SP]$ and $[\![P]\!] \in [\![SP]\!]$.

In an idealized view of program development, we start with an initial loose requirements specification SP_0 and refine it step by step until some easily-realizable specification SP_{last} is obtained:

$$SP_0 \rightsquigarrow SP_1 \rightsquigarrow \cdots \rightsquigarrow SP_{last}$$

Stepwise refinement only makes sense if the above chain of refinements guarantees that any correct realization of SP_{last} is also a correct realization of SP_0: for any P, if $[\![P]\!] \in [\![SP_{last}]\!]$ then $[\![P]\!] \in [\![SP_0]\!]$. This is ensured by the definition of refinement: for any SP and SP' with the same signature, we define

$$SP \rightsquigarrow SP' \iff [\![SP']\!] \subseteq [\![SP]\!].$$

The construction of a program to realize SP_{last} is outside the scope of CASL. Furthermore, there is no construct in CASL to explicitly express refinement between specifications. All this is a part of the meta-level, though firmly based on the formal semantics of CASL specifications.

A more satisfactory model of refinement allows for modular decomposition of a given development task into several tasks by refining a specification to a sequence of specifications, each to be further refined independently. (Of course, a development may branch more than once, giving a tree structure.)

$$SP \rightsquigarrow BR \begin{cases} SP_1 \rightsquigarrow \cdots \rightsquigarrow SP_{1,last} \\ \vdots \\ SP_n \rightsquigarrow \cdots \rightsquigarrow SP_{n,last} \end{cases}$$

Once we have realizations $P_1, \ldots, P_n$ of the specifications $SP_{1,last}, \ldots, SP_{n,last}$, we should be able to put them together with no extra effort to obtain a realization of SP. So for each such branching point we need an operation to combine arbitrary realizations of $SP_1, \ldots, SP_n$ into a realization of SP. This may be thought of as a linking procedure $LINK_{BR}$ attached to the branching point BR, where for any $P_1, \ldots, P_n$ realizing $SP_1, \ldots, SP_n$, $LINK_{BR}(P_1, \ldots, P_n)$ realizes SP: if $[\![P_1]\!] \in [\![SP_1]\!], \ldots, [\![P_n]\!] \in [\![SP_n]\!]$ then $[\![LINK_{BR}(P_1, \ldots, P_n)]\!] \in [\![SP]\!]$.

The nature of $LINK_{BR}$ depends on the nature of the programs considered. Our preferred view is that the programming language in use has reasonably powerful and flexible modularization facilities, such as those in Standard ML or Ada. Then $P_1, \ldots, P_n$ are program modules (structures in Standard ML, packages in Ada) and $LINK_{BR}$ is a module expression (or a *generic module* on its own) with formal parameters for which the actual modules $P_1, \ldots, P_n$ may be

[1] This may be rather indirect, and in general involves a non-trivial abstraction step. It has not yet been attempted for any real programming language.

substituted. Note that if we later replace a module P_i by another realization P_i' of SP_i, "recompilation" of $LINK_{BR}(P_1,\ldots,P_i',\ldots,P_n)$ might be required but in no case will it be necessary to modify the other modules.

One might expect that BR above is just a *specification-building operation* OP (or a specification construct expressible in CASL), and branching could be viewed as "ordinary" refinement $SP \rightsquigarrow OP(SP_1,\ldots,SP_n)$. Further refinement of $OP(SP_1,\ldots,SP_n)$ might then consist of separate refinements for $SP_1,\ldots,SP_n$ as above. Then we need at least that OP is "monotonic" w.r.t. inclusion of model classes.[2] This view is indeed possible provided that the specification-building operation OP is *constructive*: for any realizations $P_1,\ldots,P_n$ of $SP_1,\ldots,SP_n$, we must be able to construct a realization $LINK_{OP}(P_1,\ldots,P_n)$ of $OP(SP_1,\ldots,SP_n)$. However, simple examples show that some standard specification-building operations (like the union of specifications) do not have this property. (See [HN92] for a different approach to this problem.)

Another problem with the refinement step $SP \rightsquigarrow OP(SP_1,\ldots,SP_n)$ is that it does not explicitly indicate that subsequent refinement is to proceed by independently refining each of $SP_1,\ldots,SP_n$, so preserving the structure imposed by the operation OP. The structure of the specification $OP(SP_1,\ldots,SP_n)$ in no way prescribes the structure of the final program. And this is necessarily so: while preserving this structure in the subsequent development is convenient when it is natural to do so, refinements that break this structure must also be allowed. Otherwise, at very early stages of the development process we would have to fix the final structure of the resulting program: any decision about structuring a specification would amount to a decision about the structure of the final program. This is hardly practical, as the aims of structuring specifications in the early development phases (and at the requirements engineering phase) are quite distinct from those of structuring final programs. Simple examples are mentioned below, cf. [FJ90].

On the other hand, at certain stages of program development we need to fix the structure of the system under development: the design of the architecture of the system is often among the most important design decisions in the development process. In CASL, this is the role of *architectural specifications*, see Sect. 4.

3.1 Example

Consider the task of realizing UNIQUENUMCONT from Sect. 2.1. Its structure does not provide useful guidance to the structure of its realization. For instance, the last extension of PARTNUMCONT by an axiom for *addable* cannot be a directive to first realize PARTNUMCONT and then somehow miraculously ensure

[2] The specification-building operations we use here, hence all derived specification constructs, are monotonic, as are most of the constructs of CASL and other specification languages. The few exceptions — like imposing the requirement of freeness — can be viewed as operations which add "constraints" to specifications rather than as fully-fledged specification-building operations.

that the predicate *addable* does indeed satisfy the axiom. One might change this specification, so that a realization of PARTNUMCONT would be required for any choice of *addable* — but this would be quite a different specification with quite a different structure. Moreover, it would not enable the implementor to take advantage of the fact that the axiom for *addable* ensures that an element need never be added to a container more than once.

We might re-structure the above specification instead by introducing some new "constructive" compositions or exposing some existing ones. For instance:

spec UNIQUECONTAINER [CODENUM] =
  PARTCONTAINER[CODENUM **fit** $Elem \mapsto Num$]
  then vars $x : Num$; $C : Cont$
    axiom $addable(x, C) \Leftrightarrow \neg (x \in C) \wedge \neg (code(x) \in C)$

spec UNIQUENUMCONT' = UNIQUECONTAINER[CODENUM]

Then we have that UNIQUENUMCONT $\rightsquigarrow$ UNIQUENUMCONT' (in fact, the two specifications are equivalent) and the instantiation in the latter specification is "constructive", which indicates a possible split of further development to a part where a realization of CODENUM is developed and another part where UNIQUECONTAINER is implemented. See Sect. 4.1 below for details.

4 Architectural Specifications

The conclusion from Sect. 3 is that there are two different kinds of structuring mechanisms needed in the specification and development process.

On one hand we need the standard mechanisms to structure specifications to facilitate their construction, reading, understanding and re-use. These are provided by the specification-building operations of CASL, disregarding whether these operations are "constructive" or not. On the other hand, at a certain stage of program development we need to design the structure of the final program, and consider these decisions binding in the subsequent development process. Such a design is given by refining a specification to a "constructive" combination of specified components. The essence here is not so much the use of a constructive specification-building operation, as rather some specific construction (linking procedure) that builds a realization of the original specification once given realizations of the component specifications.

The latter structuring facility, although quite standard in modular programming languages, is rarely explicitly provided in specification formalisms. In many approaches, the structure of the specification is regarded as determining the structure of the final program, examples like those in Sect. 3.1 notwithstanding, see e.g. [GB80, MA91]. Or else *ad hoc* informal mechanisms are used to indicate that a certain part of the structure of a specification (given by a constructive specification-building operation) is to remain fixed throughout the rest of the development. We consider this unsatisfactory and likely to be confusing.

Therefore CASL provides an explicit notation whereby one specifies the components required together with a way to combine them to build the resulting program. Such *architectural specifications* can be used to refine ordinary specifications, whether structured or not, explicitly introducing branching into the development process and structure into the final program:

$$SP \rightsquigarrow \boxed{BR \begin{cases} SP_1 \\ \vdots \\ SP_n \end{cases}}$$

The corresponding architectural specification is written as follows:

units $U_1 : SP_1$;
...
$U_n : SP_n$
result $LINK_{BR}(U_1, \ldots, U_n)$

Notice that we provide names for program *units* to be implemented according to the component specifications given, and we give a "linking procedure" $LINK_{BR}$ to combine these units rather than an operation to combine their specifications. The component specifications $SP_1, \ldots, SP_n$ are ordinary CASL specifications. The "linking procedure" $LINK_{BR}(U_1, ..., U_n)$ is just a *unit term* that might involve the units named $U_1, \ldots, U_n$. It builds a new unit when given actual units $U_1, \ldots, U_n$ correctly realizing the specifications $SP_1, \ldots, SP_n$. Typically $SP_1, \ldots, SP_n$ (and so, units that realize them) will contain shared parts, or some of them will rely on others. For instance, we might start by implementing some simple specification SP_1. Then, given an implementation U_1 of SP_1, build an implementation U_2 of some "larger" specification SP_2 using U_1, etc. The last stage is to build an implementation U_n of SP_n using U_{n-1}, and the final result is U_n. The corresponding architectural specification is:

units $U_1 : SP_1$;
$U_2 : SP_2$ **given** U_1;
...
$U_n : SP_n$ **given** U_{n-1}
result U_n

Of course, this is just the simplest case. In particular, it does not cover multiple dependencies (where a unit might use several other units), sharing between various units in a more flexible way than just having each unit use the previous one, or reusability (whereby a unit may be used more than once). Still, it illustrates the idea of splitting a development task into subtasks, clearly indicating their interfaces and the flow of information between them. In the extreme, such a split may be done step by step, each time splitting the work into just two parts:

$SP \rightsquigarrow$

units $U_1 : SP_1$;
$U_2 : SP$ **given** U_1
result U_2

The task of providing a realization U_1 for SP_1 is *independent* from the task of providing a realization U_2 for SP using U_1. It follows that no properties of U_1 may be exploited in the development of U_2 other than those explicitly ensured by the specification SP_1. This requires a realization of SP for *any* realization of SP_1, which is tantamount to requiring a *generic* realization F of SP which takes the particular realization of SP_1 as parameter. Then we obtain U_2 by simply feeding U_1 to F.

Genericity here arises from the independence of the developments of U_1 and U_2, rather than from the desire to build multiple realizations of SP using different realizations of SP_1. This is reflected in the fact that F is not named in the architectural specification above. If it is desired to indicate the potential for re-use explicitly, we may give F "first-class" status as a so-called *generic unit* with a specification $SP_1 \rightarrow SP$ which indicates that it will realize SP when given a realization of SP_1:

units $U_1 : SP_1$;
$F : SP_1 \rightarrow SP$;
$U_2 = F[U_1]$
result U_2

Here, $U_2 = F[U_1]$ is a so-called *unit definition.*

The earlier specification is equivalent to this version except that F is anonymous there. This shows how to explain architectural specifications involving "**given**" by translation to architectural specifications involving generic units. A key insight is the use of genericity to control the flow of information between developments of independent units, as well as for multiple instantiation. Despite this, it seems useful to retain both notations as they convey different pragmatic intuitions.

Generic unit specifications correspond to functor headings in Extended ML [ST89] and to a restricted form of Π-specifications in [SST92], cf. Spectral [KS91]. Generic unit specifications and generic specifications coincide in ACT ONE [EM85], which the above discussion argues is inappropriate.

4.1 Example

Recall the specifications built in Sect. 2.1 and the further comments on them in Sect. 3.1. We ended up there with a specification

spec UNIQUENUMCONT' = UNIQUECONTAINER[CODENUM]

which indicates a way of decomposing the task of implementing UNIQUENUMCONT. This may be turned into a design decision by refining this specification to an architectural specification that captures the decomposition meant here:

arch spec UCNUM =
units N : CODENUM;
UCN : UNIQUENUMCONT' **given** N
result UCN

Then UNIQUENUMCONT $\rightsquigarrow$ UCNUM.

We might, however, be a bit more clever in our design and require a realization of containers with the specified "uniqueness" property for arbitrary elements equipped with the operations that allow one to express this property. For instance:[3]

spec TRANSELEM =
  sort *Elem*
  op *transform* : *Elem* $\rightarrow$ *Elem*
end

spec ABSTRACTUNIQUECONT =
  PARTCONTAINER[TRANSELEM]
  then **vars** $x : Elem;\ C : Cont$
    axiom $addable(x, C) \Leftrightarrow \neg (x \in C) \wedge \neg (transform(x) \in C)$

arch spec ABSTRACTUCNUM =
  units N : CODENUM;
    AUC : TRANSELEM $\rightarrow$ ABSTRACTUNIQUECONT
  result $AUC[N$ **fit** $Elem \mapsto Num,\ transform \mapsto code]$

We still have UNIQUENUMCONT $\rightsquigarrow$ ABSTRACTUCNUM.

The required generic unit AUC here is more abstract and more general than the "anonymous" unit to build UCN as required in UCNUM. AUC has to work for arbitrary structures fitting the abstract TRANSELEM specification; it could be re-used in the future for arguments other than N.

5 Semantics of Unit Specifications

Consider a unit specification of the form $SP' \rightarrow SP$. In CASL, SP is implicitly viewed as an extension of SP'. We therefore assume that in each specification of the form $SP' \rightarrow SP$, SP extends SP', that is: $Sig[SP'] \subseteq Sig[SP]$ and $[\![SP]\!]|_{Sig[SP']} \subseteq [\![SP']\!]$.

To realize the specification $SP' \rightarrow SP$, we should provide a "program fragment" ΔP for $SP \setminus SP'$ that extends any realization P' of SP' to a realization $\Delta P(P')$ of SP. For all programs P' such that $[\![P']\!] \in [\![SP']\!]$, $\Delta P(P')$ must be a program that extends P' and realizes SP. Hence, semantically ΔP determines a function $[\![\Delta P]\!] \colon [\![SP']\!] \rightarrow [\![SP]\!]$ that "preserves" its argument. Consequently:

$$[\![SP' \rightarrow SP]\!] = \{F \colon [\![SP']\!] \rightarrow [\![SP]\!] \mid \text{for all } A' \in [\![SP']\!], F(A')|_{Sig[SP']} = A'\}$$

This view of program fragments as functions naturally leads to further generalisations. The most obvious one is to admit multi-argument functions, providing for the possibility that the realization of some specification might depend

[3] The reader is kindly asked to rely on her/his intuition and the obvious analogy with the instantiation of generic specifications to grasp the meaning of instantiation of generic units with non-trivial fitting of arguments.

on realizations of more than one (sub-)specification. Specifications of multiply-dependent units will have the form $SP_1 \times \ldots \times SP_n \rightarrow SP$. As with singly-dependent units, we assume that SP extends each of $SP_1, \ldots, SP_n$ (or equivalently, their union). We then have:

$$[\![SP_1 \times \ldots \times SP_n \rightarrow SP]\!] = \{F : [\![SP_1 \times \ldots \times SP_n]\!] \rightarrow [\![SP']\!] \mid \text{for all } \langle A_1, \ldots, A_n \rangle \in [\![SP_1 \times \ldots \times SP_n]\!], F(A_1, \ldots, A_n)|_{Sig[SP_i]} = A_i, \text{ for } i = 1, \ldots, n\}$$

We have not yet defined $[\![SP_1 \times \ldots \times SP_n]\!]$. In general, not all tuples $\langle A_1, \ldots, A_n \rangle$ of structures $A_1 \in [\![SP_1]\!], \ldots, A_n \in [\![SP_n]\!]$ can be extended to structures in $[\![SP]\!]$: if a symbol in SP is inherited from one or more of $SP_1, \ldots, SP_n$, then its interpretation in the resulting structure must be the same as in each corresponding argument structure. So, if such a symbol occurs in several arguments then it is impossible to expand a tuple of arguments to a result unless all of the relevant arguments interpret this symbol in the same way.

A tuple $\langle A_1, \ldots, A_n \rangle$ of structures $A_1 \in \mathbf{Mod}[\Sigma_1], \ldots, A_n \in \mathbf{Mod}[\Sigma_n]$ is *compatible* if any symbol that occurs in both Σ_i and Σ_j is interpreted in the same way in A_i and A_j, for $1 \leq i, j \leq n$. Then we take $\mathbf{Mod}[\Sigma_1 \times \ldots \times \Sigma_n]$ to be the class of all compatible tuples of structures from $\mathbf{Mod}[\Sigma_1], \ldots, \mathbf{Mod}[\Sigma_n]$, respectively, and define:

$$[\![SP_1 \times \ldots \times SP_n]\!] = \{\langle A_1, \ldots, A_n \rangle \in \mathbf{Mod}[\Sigma_1 \times \ldots \times \Sigma_n] \mid A_1 \in [\![SP_1]\!], \ldots, A_n \in [\![SP_n]\!]\}$$

6 Sharing and Well-Formedness

The definitions at the end of the previous section convey important methodological concepts. Namely, we now have a way to require that a number of units (fed to a unit dependent on them) *share* some of their parts. Even though they might be developed independently, certain parts of the argument units must be identical. In CASL, this requirement is imposed by the use of the same names in argument signatures for symbols which are to be shared between the argument units. An application of a generic unit to a tuple of arguments is *well-formed* only if the arguments share their commonly-named parts. In a programming language like Standard ML, this is a part of the "type discipline" and the required sharing is (type-)checked statically.

Consider the following simple example:

spec SP_0 = **sort** s **end**
spec SP_a = **sort** s **op** $a : s$ **end**
spec SP_b = **sort** s **op** $a, b : s$ **end**
spec SP_c = **sort** s **op** $a, c : s$ **end**
spec SP_d = **sort** s **op** $a, b, c, d : s$ **axiom** $d = b \vee d = c$ **end**

Then the generic unit specification $SP_b \times SP_c \to SP_d$ imposes a constraint on the arguments for the generic unit: they are required to share a common realization of the sort s and constant a. Consequently, given the following unit declarations:

$$\begin{aligned} \textbf{units}\ & U_b : SP_b; \\ & U_c : SP_c; \\ & F_d : SP_b \times SP_c \to SP_d \end{aligned}$$

the instantiation $F_d[U_b, U_c]$ cannot be allowed, since we have no way to ensure that the units U_b and U_c do indeed share s and a. On the other hand, consider the following unit declarations:

$$\begin{aligned} \textbf{units}\ & U_a : SP_a; \\ & F_b : SP_a \to SP_b; \\ & F_c : SP_a \to SP_c; \\ & F_d : SP_b \times SP_c \to SP_d \end{aligned}$$

The unit term $F_d[F_b[U_a], F_c[U_a]]$ is well-formed in the context of these declarations. The required sharing between the two arguments for F_d, namely between $F_b[U_a]$ and $F_c[U_a]$, is ensured. In both $F_b[U_a]$ and $F_c[U_a]$ the sort s and constant a come from U_a, and so must be the same.

The situation becomes a bit less clear if components of instantiations of generic units are involved. For instance, consider:

$$\begin{aligned} \textbf{units}\ & U_0 : SP_0; \\ & F_a : SP_0 \to SP_a \end{aligned}$$

and declarations of F_b, F_c, F_d as above. Is $F_d[F_b[F_a[U_0]], F_c[F_a[U_0]]]$ well-formed? One might expect so: the sort s in the two arguments for F_d can be traced to U_0, and the constant a to the two occurrences of $F_a[U_0]$. But the argument that the two occurrences of $F_a[U_0]$ share the constant a cannot be carried too far. In general, to decide if two instantiations of F_a, say $F_a[U_0]$ and $F_a[U_0']$, share the constant a, we would have to check if the two argument units U_0 and U_0' are identical. Clearly, this is too complicated for static analysis, even if in trivial cases it can be seen to hold immediately, as above. Moreover, in some programming languages (Standard ML, Ada) the new items introduced by instantiation of generic modules are distinct for each such instantiation.

Therefore, for safety, we assume that new symbols introduced by a generic unit are not shared between its instantiations, even when its arguments are the same in each case. (For programming languages with "applicative" rather than "generative" modules, this treatment is sound albeit marginally more awkward than necessary.) Auxiliary unit definitions may be used in CASL to avoid repetition of unit instantiation. For instance, we can rewrite the previous example:

$$\begin{aligned} \textbf{units}\ & U_0 : SP_0; \\ & F_a : SP_0 \to SP_a; \\ & U_a' = F_a[U_0]; \\ & F_b : SP_a \to SP_b; \end{aligned}$$

$$F_c : SP_a \rightarrow SP_c;$$
$$F_d : SP_b \times SP_c \rightarrow SP_d$$

In this context, $F_d[F_b[U'_a], F_c[U'_a]]$ is well-formed and captures the intention behind $F_d[\,F_b[F_a[U_0]],\ F_c[F_a[U_0]]\,]$.

To sum up: in the context of a sequence of unit declarations and definitions, symbols in two units *share* if they can be traced to a common symbol in a non-generic unit. The "tracing procedure" can be broken down according to the constructs available for forming unit terms. For applications of generic units to arguments, symbols in the result are new if they do not occur in the argument signatures. Otherwise they can be traced to the same symbols in the arguments (and, transitively, to the symbols those can be traced to). The symbols of a declared unit can be traced only to themselves. The symbols of a defined unit may be traced according to the definitional term for the unit.

7 Semantics of Unit Terms

An architectural specification comprises a sequence of unit declarations and definitions followed by a unit term which shows how the named units can be put together to build the result. Obviously, it is not possible to put together units in completely arbitrary ways; they must fit together properly, as in modular programming languages. Then given an *environment* which maps the declared unit names to particular (possibly generic) structures, the result term denotes a structure.

The static analysis of unit terms, with sharing analysis etc., is just the beginning of checking their correctness. The most crucial step is to check that when a unit (or tuple of units) is fed to a generic unit then the interfaces match, making sure that the requirements imposed on the parameter(s) of the generic unit by its specification are fulfilled by the argument (tuple). To take a simple example:

units $U : SP;$
$F : SP' \rightarrow SP''$

Can we now feed the unit U to the generic unit F? Or in other words: is the unit term $F[U]$ *correct*? In order for it to be well-formed, the signatures of U and of the argument of F must coincide: $Sig[SP] = Sig[SP']$. And if F were multiply-dependent with symbols in common between different arguments, then sharing would also have to be checked. But also, F is required to work only for arguments that realize SP', including the requirements imposed by any axioms SP' may contain. So, for $F[U]$ to be correct, we must make sure that what we know about U is sufficient to establish what is required of the argument for F. Clearly, everything we know about U is recorded in SP — no other information is available. Even later on, when the unit U has been developed, the whole point of its declaration here — which decomposes the development task into developing U and F separately — is to limit the knowledge about U at this level to what is provided by SP. So, what we know about the unit U is that it denotes a structure in $[\![SP]\!]$. The argument of F is required to denote a structure in $[\![SP']\!]$. Consequently, the term $F[U]$ is correct provided $[\![SP]\!] \subseteq [\![SP']\!]$.

We have used different words to describe different aspects of "good" unit terms. *Well-formedness* is a static property, expected to be decidable so that it can be checked automatically. To check whether a unit term is well-formed we need information about the signatures of the units available as well as sharing information about them. In such a context, well-formedness of a term is determined as sketched in Sect. 6. *Correctness* requires verification: it is not decidable in general. To check whether a unit term is correct we need full semantic information about the available units, as explained below.

The last example was perhaps misleadingly simple: the argument U of F came equipped with an explicit specification that provided all the information that was available about U. In general, the argument may be more complex than this, and still we have to be able to gather all the information about it that is available. So, for instance, what do we know about $F[U]$, assuming that $Sig[SP] = Sig[SP']$ and $[\![SP]\!] \subseteq [\![SP']\!]$? Clearly, we know that the result realizes SP''. Is this all? Not quite: we also know that U, and hence the reduct of $F[U]$ to $Sig[SP]$, realizes SP, which may carry more information than SP' does.

Given an environment ρ which maps unit names to particular (possibly generic) structures, a unit term T denotes a structure $[\![T]\!]_\rho$, defined inductively as follows:

- If T is a unit name U then $[\![T]\!]_\rho = \rho(U)$.
- If T is an instantiation $F[T_1, \ldots, T_n]$ where F is an n-ary generic unit and $T_1, \ldots, T_n$ are unit terms, then $[\![F[T_1, \ldots, T_n]]\!]_\rho = \rho(F)([\![T_1]\!]_\rho, \ldots, [\![T_n]\!]_\rho)$.

Some unit terms will not denote. A trivial reason for this might be the application of a generic unit to the wrong number of arguments, or to arguments with wrong signatures, or the use of an unbound unit name. Less trivially, there might be an attempt to apply a generic unit to a non-compatible tuple of structures. These cannot happen if the term is well-formed in the sense discussed above. Finally, a term will not denote if it involves application of a generic unit to a structure outside its domain; this cannot happen if the term is correct.

Correctness is defined in a *context* γ where unit names are associated with specifications. We say that an environment ρ *matches* a context γ if they bind the same unit names and for each unit name U in their domain, the structure $\rho(U)$ realizes the specification $\gamma(U)$: $\rho(U) \in [\![\gamma(U)]\!]$.[4] For any unit term T that is well-formed in the context γ, we write $[T]_\gamma$ for the class of all structures $[\![T]\!]_\rho$ that T denotes in environments ρ that match γ. Intuitively, $[T]_\gamma$ captures the properties of the unit built by T using unit declarations and definitions that determine γ.

Correctness of a well-formed unit term is defined inductively as follows:

- A unit name U is correct. (By well-formedness, U is declared in γ.) It follows that $[U]_\gamma = [\![\gamma(U)]\!]$.

---

[4] Moreover, the units in ρ share the components indicated by the sharing information in γ.

- An instantiation $F[T_1,\ldots,T_n]$ is correct, where $\gamma(F)$ is $SP_1\times\ldots\times SP_n\rightarrow SP$, if $T_1,\ldots,T_n$ are so and $[T_1]_\gamma \subseteq [\![SP_1]\!],\ldots,[T_n]_\gamma \subseteq [\![SP_n]\!]$. It follows that $[F[T_1,\ldots,T_n]]_\gamma = \{A \in [\![SP]\!] \mid A|_{Sig[SP_1]} \in [T_1]_\gamma,\ldots,A|_{Sig[SP_n]} \in [T_n]_\gamma\}$.

This omits the use of defined units in unit terms, treated in the obvious way with information about these units extracted from their definitional terms and stored in the context as well. Some further constructs for unit terms (amalgamation, reduct/renaming, pushout-style instantiation using a fitting morphism, local unit definitions, λ-notation for generic units) are available in CASL, but these are not discussed here for lack of space.

The above statements defining the correctness of unit terms also provide a more direct way to compute $[T]_\gamma$, without referring to the class of all environments that match γ. This can be proved by induction on the structure of unit terms, and can be used to directly calculate the ensured properties of T, and to validate its correctness.

Theorem 1. *Let γ be a context and let T be a unit term that is well-formed and correct in γ. Then for any environment ρ that matches γ, $[\![T]\!]_\rho$ is defined (and $[\![T]\!]_\rho \in [T]_\gamma$).*

This means that once we have finished the development process and so have provided realizations of each of the units declared, a correct result term will successfully combine these realizations to give a structure which satisfies the properties we can calculate directly from the architectural specification. Correctness of the result term of an architectural specification can be checked before realizations of its component units are provided. No *a posteriori* checking is necessary!

8 Refinements of Architectural Specifications

Section 4 indicated how a specification may be refined to an architectural specification. Architectural specifications themselves can in turn be refined by refining each of the specifications for its declared units separately. One remaining issue is to define refinements between specifications of generic units:

$$SP_1\rightarrow SP_2 \quad \rightsquigarrow \quad SP'_1\rightarrow SP'_2$$

To begin with, we need the signatures to agree, that is: $Sig[SP_1] = Sig[SP'_1]$ and $Sig[SP_2] = Sig[SP'_2]$. Furthermore, we need that every generic unit that realizes $SP'_1\rightarrow SP'_2$ must correctly realize $SP_1\rightarrow SP_2$, but allowing for restrictions of mappings between structures to smaller domains. This amounts to requiring $[\![SP_1]\!] \subseteq [\![SP'_1]\!]$ and $[\![SP'_2$ **and** $SP_1]\!] \subseteq [\![SP_2]\!]$. Notice that the latter condition is slightly weaker than the most obvious $[\![SP'_2]\!] \subseteq [\![SP_2]\!]$ — we can take advantage of the fact that we are expected to apply the unit to arguments that realize SP_1.

This allows for linear development of individual units declared in an architectural specification. To allow further decomposition here, we can refine unit specifications to architectural specifications. For closed units this is covered above.

Specifications of generic units may be refined to architectural specifications with generic result units.

The overall effect is that we have a development tree, rather than just a sequence of refinement steps. This was indeed the target from the very beginning. Each leaf of such a tree may be developed independently from the others, using the full machinery of further decomposition via architectural design etc. The development subtree beginning at any given node may be replaced by another development tree without affecting the other parts as long as the new development subtree is correct with respect to the specification at its root.

9 Further Comments

We have discussed the issue of designing the structure of a system to be developed from a specification. Our conclusion has been that apart from the usual mechanisms for structuring requirements specifications, we need a separate mechanism to describe the modular structure of the system to be developed. CASL provides this in the form of *architectural specifications.* We presented the basic ideas behind this concept. The semantics of architectural specifications has been sketched as well, but see [CoFI98c] for all the details. This was sufficient to state a few basic facts about the semantics, as well as to argue that properties of architectural specifications ensure that the basic goals of their design have been achieved. Namely, architectural specifications make it possible to describe the structure of the system to be developed by listing the units to be built, providing their specifications and indicating the way they are to be combined. Once such an architectural specification is given then its internal correctness can be checked and the ensured properties of the resulting module can be calculated (to check if the original requirements specification has been fulfilled by this design). Moreover, further developments of the units required may proceed independently from each other, which brings in all the benefits of modular development.

The above ideas have been presented in the specific context of CASL. However, both the overall idea and the constructs for architectural specifications are largely independent from the details of the underlying CASL logical system. In fact, everything here can be presented in the context of an arbitrary *institution* [GB92] equipped with some extra structure — see [Mos98] for details.

One issue which we have omitted above is that of *behavioural implementation* [Sch87, ST89, NOS95, ST97, BH9?]. The idea is that when realizing a specification it is sufficient to provide a structure that is *behaviourally equivalent* to a model. Intuitively, two structures are *behaviourally equivalent* if they cannot be distinguished by computations involving only the predicates and operations they provide.

When using a structure that was built to realize a specification up to behavioural equivalence, it is very convenient to pretend that it actually is a true model of the specification. This is sound provided all the available constructions on structures (hence all the generic units that can be developed) map behaviourally equivalent arguments to behaviourally equivalent results. More precisely: a generic unit is *stable* if for any behaviourally equivalent arguments provided for it via a fitting morphism, the overall results of instantiations of this unit on them are behaviourally equivalent as well. If all units are stable, it is sufficient to check *local behavioural correctness* of unit terms only: this is defined like correctness in Sect. 7, but allows the arguments for generic units to fit their formal requirement specifications only up to behavioural equivalence. Then the ensured properties $[T]_\gamma$ of any well-formed and locally behaviourally correct unit term T in a context γ can still be calculated exactly as in Sect. 7, as justified by the following theorem:

Theorem 2. *Let γ be a context and let T be a unit term that is well-formed and locally behaviourally correct in γ. Then for any environment ρ that matches γ up to behavioural equivalence, $[\![T]\!]_\rho$ is in $[T]_\gamma$ up to behavioural equivalence.*

Acknowledgements Our thanks to the whole of CoFI, and in particular to the Language Design Task Group, for many discussions and opportunities to present and improve our ideas on architectural specifications. Thanks to Till Mossakowski for comments on a draft. This work has been partially supported by KBN grant 8 T11C 018 11, the LoSSeD workpackage of CRIT-2 funded by ESPRIT and INCO (AT), a French-Polish project within the CNRS-PAS cooperation programme (MB, AT), and by EPSRC grant GR/K63795, an SOEID/RSE Support Research Fellowship and the FIREworks working group (DS).

References

[AG97] R. Allen and D. Garlan. A Formal Basis for Architectural Connection. *ACM Transactions on Software Engineering and Methodology*, July 1997.

[Bid88] M. Bidoit. The stratified loose approach: a generalization of initial and loose semantics. *Selected Papers from the 5th Workshop on Specification of Abstract Data Types*, Gullane. Springer LNCS 332, 1–22 (1988).

[BH93] M. Bidoit and R. Hennicker. A general framework for modular implementations of modular systems. *Proc. 5th Joint Conf. on Theory and Practice of Software Development*, Orsay. Springer LNCS 668, 199–214 (1993).

[BH9?] M. Bidoit and R. Hennicker. Modular correctness proofs of behavioural implementations. *Acta Informatica*, to appear (199?).

[BG77] R. Burstall and J. Goguen. Putting theories together to make specifications. *Proc. 5th Intl. Joint Conf. on Artificial Intelligence*, Cambridge, 1045–1058 (1977).

[CoFI96] The Common Framework Initiative. Catalogue of existing frameworks. `http://www.brics.dk/Projects/CoFI/Catalogue.html` (1996).

[CoFI98a] The Common Framework Initiative. CoFI: The Common Framework Initiative for algebraic specification and development (WWW pages). `http://www.brics.dk/Projects/CoFI/` (1998).

[CoFI98b] CoFI Task Group on Language Design. CASL – The CoFI algebraic specification language – Summary (version 1.0). `http://www.brics.dk/Projects/CoFI/Documents/CASL/Summary/` (1998).

[CoFI98c] CoFI Task Group on Semantics. CASL – The CoFI algebraic specification language – Semantics (version 1.0). To appear (1998).

[EM85] H. Ehrig and B. Mahr. *Fundamentals of Algebraic Specification I: Equations and Initial Semantics.* Springer (1985).

[FJ90] J. Fitzgerald and C. Jones. Modularizing the formal description of a database system. *Proc. VDM'90 Conference*, Kiel. Springer LNCS 428, 198–210 (1990).

[GB80] J. Goguen and R. Burstall. CAT, a system for the structured elaboration of correct programs from structured specifications. Technical report CSL-118, SRI International (1980).

[GB92] J. Goguen and R. Burstall. Institutions: abstract model theory for specification and programming. *Journal of the Assoc. for Computing Machinery* 39:95–146 (1992).

[GH93] J. Guttag and J. Horning. *Larch: Languages and Tools for Formal Specification.* Springer (1993).

[HN92] R. Hennicker and F. Nickl. A behavioural algebraic framework for modular system design and reuse. *Recent Trends in Data Type Specifications. Proc. 9th Workshop on Specification of Abstract Data Types* ADT'92, Caldes de Mavella, Springer LNCS 785, 220–234.

[KS91] B. Krieg-Brückner and D. Sannella. Structuring specifications in-the-large and in-the-small: higher-order functions, dependent types and inheritance in SPECTRAL. *Proc. Colloq. on Combining Paradigms for Software Development*, Joint Conf. on Theory and Practice of Software Development (TAPSOFT), Brighton. Springer LNCS 494, 313–336 (1991).

[MA91] J. Morris and S. Ahmed. Designing and refining specifications with modules. *Proc. 3rd Refinement Workshop*, Hursley Park, 1990. Springer Workshops in Computing, 73–95 (1991).

[Mos98] T. Mossakowski. Institution-independent semantics for CASL-in-the-large. CoFI note S-8 (1998).

[Mos97] P. Mosses. CoFI: The Common Framework Initiative for algebraic specification and development. *Proc. 7th Intl. Joint Conf. on Theory and Practice of Software Development*, Lille. Springer LNCS 1214, 115–137 (1997).

[NOS95] M. Navarro, F. Orejas and A. Sanchez. On the correctness of modular systems. *Theoretical Computer Science* 140:139–177 (1995).

[SST92] D. Sannella, S. Sokołowski and A. Tarlecki. Toward formal development of programs from algebraic specifications: parameterisation revisited. *Acta Informatica* 29:689–736 (1992).

[ST89] D. Sannella and A. Tarlecki. Toward formal development of ML programs: foundations and methodology. *Proc. 3rd Joint Conf. on Theory and Practice of Software Development*, Barcelona. Springer LNCS 352, 375–389 (1989).

[ST97] D. Sannella and A. Tarlecki. Essential concepts of algebraic specification and program development. *Formal Aspects of Computing* 9:229–269 (1997).

[SW83] D. Sannella and M. Wirsing. A kernel language for algebraic specification and implementation. *Proc. 1983 Intl. Conf. on Foundations of Computation Theory*, Borgholm. Springer LNCS 158, 413–427 (1983).

[SW98] D. Sannella and M. Wirsing. Specification languages. Chapter 8 of *Algebraic Foundations of Systems Specification* (eds. E. Astesiano, H.-J. Kreowski and B. Krieg-Brückner). Springer, to appear (1998).

[Sch87] O. Schoett. Data Abstraction and the Correctness of Modular Programming. Ph.D. thesis, report CST-42-87, Dept. of Computer Science, Univ. of Edinburgh (1987).

Pi-Congruences as CCS Equivalences

Paola Quaglia*

BRICS**, Aarhus University

Abstract. We show the coincidence of non-ground π-calculus and $\pi\xi$-calculus, a CCS-like calculus that allows processes to be explicitly represented as temporary functions of input parameters, and as permanent functions of their free names.
As intermediate results, strong and weak π-calculus full congruences are characterized as finitary closures of the corresponding π-bisimilarities.
In this paper we consider only late π-calculus, but all of the characterizations can be easily adapted to deal with non-ground semantics of the early family.

0 Introduction

The π-calculus is a well-known process algebra that allows the description of mobile systems [MPW92]. The calculus, that has been shown to have great flexibility and expressive power, exploits a name-passing interaction paradigm. Names, synonyms of channels, can be sent around and received, possibly changing the local/global acquaintances of the inputting process.

By this, name substitution has a fundamental role in the operational semantics of the calculus. Substitutions are most important also at the bisimulation level. This is essentially due to the fact that they can increase process move capabilities. For instance, unless the notion of name instantiation is moved inside the bisimulation definition itself (like, *e.g.*, in the open semantics [San96]), bisimulation fails to be a congruence w.r.t. input prefix. The canonical example to see this is in terms of the interleaving interpretation of parallel composition. Taking $\overline{x}z$ and $x(u)$ to denote, respectively, an output and an input action at channel x, such interpretation equates the processes $(\overline{x} \mid y)$ and $(\overline{x}.y + y.\overline{x})$ (unnecessary syntactic details are omitted). Each of the – either strong or τ-forgetting – *late* or *early* bisimulation semantics, here generically denoted by $\dot{\asymp}$, is such that

$$\overline{x} \mid y \;\dot{\asymp}\; \overline{x}.y + y.\overline{x}$$

$$x(y).(\overline{x} \mid y) \;\dot{\not\asymp}\; x(y).(\overline{x}.y + y.\overline{x})$$

The reason for the above inequivalence is that, whenever the leading input action $x(y)$ causes y to be instantiated by x, the left-hand process $x(y).(\overline{x} \mid y)$ can perform a τ-move that the right-hand process is not able to match. The usual

* Funded by the EU, under the Marie Curie TMR Programme.

** Basic Research in Computer Science, Centre of the Danish National Research Foundation.

A.M. Haeberer (Ed.): AMAST'98, LNCS 1548, pp. 358–372, 1998.

way to extract the full congruence $\asymp$ from the *ground* relation $\dot{\asymp}$ passes through a closure over name substitutions, so that classical definitions read as follows:

$$P \asymp Q \text{ iff } P\sigma \dot{\asymp} Q\sigma \text{ for all substitutions } \sigma$$

Full congruences, also called *non-ground* relations, are pragmatically much more appealing than their ground duals, although the universal quantification over substitutions makes prohibitive to verify them.

The main concern of this paper is to present finitary characterizations of late, either strong or weak, π-calculus full congruences. The investigation is carried on in both the π-calculus itself and the $\pi\xi$-calculus, a generalization of the process algebras presented in [FMQ96, FMQ95].

As for the π-calculus we show that, by adequately quotienting name substitutions, $P \asymp Q$ can be expressed in terms of a finite number of bisimilarity checks $P\sigma \dot{\asymp} Q\sigma$.

Turning to the $\pi\xi$-calculus, we characterize late π-congruences as corresponding CCS-like equivalences of one suitable single pair of $\pi\xi$-processes.

The coincidence between π-calculus and $\pi\xi$-calculus makes it possible to easily export to the π-calculus some well-know results about the automated verification and the mathematical theory (logical, axiomatic, and denotational) of CCS equivalences (see, *e.g.*, [CPS93, HM85, ABV94, Abr91]).

The π-calculus meta-syntactic operation of substitution is explicitly handled in the $\pi\xi$-calculus. The main operational idea underpinning the calculus is to avoid to directly apply name substitutions to π-processes, and let a distinct explicit component ξ to serve the same purpose. Such a component acts as some kind of environment and represents associations among names. The generic $\pi\xi$-process looks like $\xi :: P$, where P is an agent obtained by the usual π-calculus syntax added with abstraction prefixes '$\lambda y.$'. These prefixes are used to associate a concrete operational counterpart to the actualization of formal parameters. The idea is to interpret each input action $x(y)$ as a non atomic event. First a commitment at x takes place, coercing the inputting process to temporarily become a function of y. The subsequent move actualizes the name y, and the functional dependency on it is definitely dropped away.

An analogous approach was used in [FMQ96, FMQ95] to characterize ground π-calculus semantics in terms of CCS-like bisimilarities. However, the proposed calculi were not general enough to capture the flavour of non-groundness, and the challenging issue of characterizing full congruences was left open.

Here we fill that gap by stressing the correspondence between π-calculus name substitutions and $\pi\xi$-calculus environments, and by emphasizing the significance of abstraction prefixes. We interpret π-calculus processes as actual, permanent, functions of their free names, and show that suitable λ-closures of P and Q induce, in an effective way, the generation of those environments that correspond to the finite set of substitutions sufficient to infer $P \asymp Q$ from the ground bisimilarity of $P\sigma$ and $Q\sigma$. After this we prove the coincidence of non-ground π-calculus and $\pi\xi$-calculus. Axiomatic characterizations of finite processes are provided, too.

In this paper we only deal with late π-calculus, however, by minor changes, all of the characterizations stated for late congruences do hold of non-ground semantics of the early family [Qua96]. We only present the finite fragment of the $\pi\xi$-calculus. Nevertheless the coincidence with π-calculus non-ground semantics does extend to the full language.

A final remark is about the fact that the peculiar interpretation of input actions and abstraction prefixes ensures that, but for the unguarded behaviour of the replication operator, $\pi\xi$-processes can be always modelled by finitely branching structures, which is generally not the case for data-dependent agents. So, our results could be particularly useful in the perspective of verifying mobile systems.

The rest of the paper is organized as follows. In Section 1 we prove that each π-calculus non-ground late semantics can be expressed as the conjunction of a finite number of corresponding ground bisimilarities. In Section 2 we present the $\pi\xi$-calculus, a simplified and generalized version of the calculi in [FMQ96, FMQ95]. Then we characterize late π-congruences as standard strong and weak bisimilarities of one single suitable pair of $\pi\xi$-processes. An equational characterization of late π-calculus non-groundness is also provided. Section 3 contains some concluding remarks.

Due to space constraints, proofs are omitted or briefly sketched. For a full account on them the reader is referred to [Qua97].

1 Finitary Reformulation of π-Congruences

This section is devoted to show that any late π-calculus congruence $\asymp$ can be expressed by closing its corresponding ground relation $\dot{\asymp}$ over a finite – vs. infinite – set of carefully chosen substitutions σ. Although our investigation is referred to semantics of the late family, exactly the same results do hold for early relations [Qua96].

Familiarity with the π-calculus is assumed. We only recall a few main issues about its semantics. Letting $\mathcal{N}$ be a denumerably infinite set of names (ranged over by x, y, z, ...), the syntax of π-calculus processes (ranged over by P, Q, ...) is defined by the following grammar:

$$P ::= \text{nil} \;\big|\; x(y).P \;\big|\; \overline{x}y.P \;\big|\; \tau.P \;\big|\; P+P \;\big|\; P \mid P \;\big|\; [x=y]P \;\big|\; (y)P \;\big|\; !P$$

In $x(y).P$ and in $(y)P$ the displayed occurrences of y are bindings with scope P. Free and bound names of a process P (of an action α) are denoted by $\text{fn}(P)$ ($\text{fn}(\alpha)$), and by $\text{bn}(P)$ ($\text{bn}(\alpha)$) respectively. The symbol '$\equiv$' denotes α-conversion.

Name substitutions (ranged over by σ, σ', ...) are functions from $\mathcal{N}$ to $\mathcal{N}$ defined almost everywhere as the identity. Sometimes, when the substitution σ differs from the identity for the names in $\{x_1, \ldots, x_n\}$, σ is simply written $\{x_1\sigma/x_1, \ldots, x_n\sigma/x_n\}$. The set $\{x_1, \ldots, x_n\}$ is then referred to as domain of σ.

Definition 1 (late π-calculus semantics)
Assuming $\xrightarrow{\alpha}$ to be the π-calculus operational transition relation, let $\Longrightarrow$ be the reflexive and transitive closure of $\xrightarrow{\tau}$, and $\overset{\alpha}{\Longrightarrow}$ be $\Longrightarrow\xrightarrow{\alpha}\Longrightarrow$, and $\overset{\hat{\alpha}}{\Longrightarrow}$ be $\Longrightarrow$

if $\alpha = \tau$, $\stackrel{\alpha}{\Longrightarrow}$ otherwise. A binary symmetric relation $\mathcal{S}$ over π-calculus processes is a

- *late strong ground bisimulation* if $P \mathcal{S} Q$ implies that: *(i)* if $P \stackrel{\alpha}{\longrightarrow} P'$ with $\alpha \neq x(y)$ and $\mathrm{bn}(\alpha) \notin \mathrm{fn}(P, Q)$, then for some Q', $Q \stackrel{\alpha}{\longrightarrow} Q'$ and $P' \mathcal{S} Q'$; *(ii)* if $P \stackrel{x(y)}{\longrightarrow} P'$ with $y \notin \mathrm{fn}(P, Q)$, then for some Q', $Q \stackrel{x(y)}{\longrightarrow} Q'$ and, for all w, $P'\{w/y\} \mathcal{S} Q'\{w/y\}$;
- *late weak ground bisimulation* if $P \mathcal{S} Q$ implies that: *(i)* if $P \stackrel{\alpha}{\longrightarrow} P'$ with $\alpha \neq x(y)$ and $\mathrm{bn}(\alpha) \notin \mathrm{fn}(P, Q)$, then for some Q', $Q \stackrel{\widehat{\alpha}}{\Longrightarrow} Q'$ and $P' \mathcal{S} Q'$; *(ii)* if $P \stackrel{x(y)}{\longrightarrow} P'$ with $y \notin \mathrm{fn}(P, Q)$, then for some Q', $Q \Longrightarrow\stackrel{x(y)}{\longrightarrow} Q'$ and, for all w, $P'\{w/y\} \mathcal{S} Q'\{w/y\}$.

P and Q are: *(i) late strong ground bisimilar*, written $P \dot{\sim}_L Q$, if $P \mathcal{S} Q$ for some strong late ground bisimulation $\mathcal{S}$; *(ii) late strong bisimilar*, written $P \sim_L Q$, if $P\sigma \dot{\sim}_L Q\sigma$ for all substitutions σ; *(iii) late weak ground bisimilar*, written $P \dot{\approx}_L Q$, if $P \mathcal{S} Q$ for some late weak ground bisimulation $\mathcal{S}$; *(iv) late weak bisimilar*, written $P \approx_L Q$, if $P\sigma \dot{\approx}_L Q\sigma$ for all substitutions σ; *(v) late weak ground equal*, written $P \dot{\simeq}_L Q$, iff $P \dot{\approx}_L Q$ and whenever $P \stackrel{\tau}{\longrightarrow} P'$ then for some Q', $Q \stackrel{\tau}{\Longrightarrow} Q'$ with $P' \dot{\approx}_L Q'$, and symmetrically; *(vi) late weak equal*, written $P \simeq_L Q$, if $P\sigma \dot{\simeq}_L Q\sigma$ for all substitutions σ. □

The universal quantification over substitutions in the definition of non-ground bisimilarities gives raise to a heavy requirement to check in practice. Nevertheless, not all of the infinite name substitutions are either always or equally relevant to infer process equivalence. For instance, given an arbitrary process P and any name $x \notin \mathrm{fn}(P)$, for all w it holds that $P\{w/x\} = P$. Hence, a first simple improvement in deciding whether or not P is congruent to Q is to consider only those substitutions whose domains are given by $\mathrm{fn}(P, Q)$. To set further improvements it is crucial to examine how the application of name substitutions can change process move potentials. At least in the absence of a mismatching operator, substitutions do not decrease those potentials. However they might increase performance capabilities.

For instance, assuming $x \neq y$, let $R_1 = (\overline{x} \mid y)$ and $\sigma_x = \{x/x, x/y\}$. The parallel components of $R_1\sigma_x$ may be involved in a communication which is forbidden to the subprocesses of R_1. A τ-move is equally possible for $R_1\sigma_y$ when $\sigma_y = \{y/x, y/y\}$, and, more generally, a communication occurs between the subprocesses $\overline{x}\sigma_z$ and $y\sigma_z$ of $R_1\sigma_z$ for any $\sigma_z = \{z/x, z/y\}$. Now suppose to want to check the congruence of R_1 and R_2, with R_2 such that $\mathrm{fn}(R_2) = \mathrm{fn}(R_1) = \{x, y\}$. Comparable action capabilities are expected for $R_1\sigma_x$ and $R_1\sigma_y$. This suggests that the bisimilarity of $R_1\sigma_x$ and $R_2\sigma_x$ might be related to that of $R_1\sigma_y$ and $R_2\sigma_y$ in a very precise sense. Our investigation leads to the proof of the following claim.

Claim 2 Whenever $\mathrm{fn}(R_1, R_2) = \{x, y\}$ and $u \neq w$, the late congruence $R_1 \asymp R_2$ can be simply characterized as conjunction of the two following ground relations:

$$R_1\{u/x, u/y\} \dot{\asymp} R_2\{u/x, u/y\} \qquad R_1\{u/x, w/y\} \dot{\asymp} R_2\{u/x, w/y\}$$ □

The comparability of the move potentials of the above processes $R_1\sigma_x$ and $R_1\sigma_y$ depends on the fact that either σ_x or σ_y map both x and y into the same target name. From this we gain the intuition that substitutions can be quotiented according to the way their domains are partitioned into subsets of names sharing the same image, no matter what such an image is.

Definition 3 The name substitution $\sigma : \mathcal{N} \to \mathcal{N}$ is said to *represent the partition of* $N \subseteq \mathcal{N}$ into the k disjoint and non-empty sets $N_1, \ldots, N_k$ iff $\forall j, h \in \{1, \ldots, k\} : j \neq h, \forall x, y \in N_j, \forall z \in N_h$ it holds that $x\sigma = y\sigma$ and $x\sigma \neq z\sigma$. □

Given two substitutions σ, σ' representing the same partition of $\text{fn}(P, Q)$, our next goal is to relate the ground (in)equivalence of $P\sigma$ and $Q\sigma$ to the (in)equivalence of $P\sigma'$ and $Q\sigma'$. As auxiliary intermediate results, we prove statements on the relationship between the strong and the weak behaviour of $P\sigma$ and $P\sigma'$ with σ and σ' representing the same partition of $\text{fn}(P)$.

Lemma 4 *Let* $\sigma, \sigma' : \mathcal{N} \to \mathcal{N}$ *represent the same partition of* $\text{fn}(P)$. *If* $P\sigma \xrightarrow{\alpha} P_1$ *with* $\text{bn}(\alpha) \notin \text{fn}(P\sigma, P\sigma')$ *and such that* $(\text{bn}(\alpha))\sigma = (\text{bn}(\alpha))\sigma' = \text{bn}(\alpha)$, *then by an inference of equal depth* $P\sigma' \xrightarrow{\beta} P_2$ *where, for some action* γ *and some process* P' *with* $\text{fn}(P') \subseteq \text{fn}(P) \cup \text{bn}(\alpha)$, *it holds that* $\alpha = \gamma\sigma$, $\beta = \gamma\sigma'$ *and* $P_1 \equiv P'\sigma$, $P_2 \equiv P'\sigma'$. □

Lemma 5 *Let* $\sigma, \sigma' : \mathcal{N} \to \mathcal{N}$ *represent the same partition of* $\text{fn}(P)$. *If* $P\sigma \Longrightarrow P_1$ *then by a derivation of equal length* $P\sigma' \Longrightarrow P_2$ *where, for some process* P' *with* $\text{fn}(P') \subseteq \text{fn}(P)$, *it holds that* $P_1 \equiv P'\sigma$ *and* $P_2 \equiv P'\sigma'$. □

Theorem 6 *Let* $\sigma, \sigma' : \mathcal{N} \to \mathcal{N}$ *represent the same partition of* $\text{fn}(P, Q)$ *and let* $\dot{\asymp} \in \{\dot{\sim}_L, \dot{\approx}_L, \dot{\simeq}_L\}$. *Then* $P\sigma \dot{\asymp} Q\sigma$ *iff* $P\sigma' \dot{\asymp} Q\sigma'$.

PROOF (SKETCH): In the case of strong semantics, we set $\mathcal{S} = \bigcup_n \mathcal{S}_n$ where

$$\begin{aligned} \mathcal{S}_0 &= \dot{\sim}_L \\ \mathcal{S}_{n+1} &= \{ (P\sigma, Q\sigma) \mid P\sigma' \, \mathcal{S}_n \, Q\sigma' \text{ and} \\ &\qquad \sigma, \sigma' \text{ represent the same partition of } \text{fn}(P, Q) \} \end{aligned}$$

Then $\mathcal{S}$ is shown to be a strong late ground bisimulation by proving, by induction on n, that $P \, \mathcal{S}_n \, Q$ implies that: *(i)* if $P \xrightarrow{\alpha} P'$ with $\alpha \neq x(y)$ and $\text{bn}(\alpha) \notin \text{fn}(P, Q)$, then for some Q', $Q \xrightarrow{\alpha} Q'$ and $P' \, \mathcal{S} \, Q'$; *(ii)* if $P \xrightarrow{x(y)} P'$ with $y \notin \text{fn}(P, Q)$, then for some Q', $Q \xrightarrow{x(y)} Q'$ and, for all w, $P'\{w/y\} \, \mathcal{S} \, Q'\{w/y\}$. An analogous proof schema is used for τ-forgetting semantics. In those cases, by hypothesis, $(P\sigma', Q\sigma') \in \mathcal{U}$ for some weak late ground bisimulation $\mathcal{U}$ that possibly ensures $P\sigma' \dot{\simeq}_L Q\sigma'$. We set $\mathcal{T} = \bigcup_n \mathcal{T}_n$ where

$$\mathcal{T}_0 = \mathcal{U}$$

$$\mathcal{T}_{n+1} = \{ (P\sigma, Q\sigma) \mid P\sigma' \, \mathcal{T}_n \, Q\sigma' \text{ and } \sigma, \sigma' \text{ represent the same partition of } \mathrm{fn}(P,Q) \}$$

and show that $\mathcal{T}$ is a weak late ground bisimulation that possibly guarantees $P\sigma \stackrel{.}{\simeq}_L Q\sigma$. □

The above theorem ensures that lots of checks may be saved when deciding the non-ground bisimilarity of two processes P and Q. In fact, once $P\sigma \stackrel{.}{\asymp} Q\sigma$ has been proved, any other test on $P\sigma' \stackrel{.}{\asymp} Q\sigma'$ is superfluous if σ and σ' represent the same partition of $\mathrm{fn}(P,Q)$. It is then a matter of isolating a kernel of relevant substitutions. To this end, substitutions are grouped into families representing a given set of names.

Definition 7 Let $N \subseteq \mathcal{N}$ be a set of names and $\{\sigma_i\}_{i \in I}$ be a family of name substitutions $\sigma_i : \mathcal{N} \to \mathcal{N}$. Then $\{\sigma_i\}_{i \in I}$ is a *partition family of* N iff the following holds:

- if $N = \emptyset$ then $\{\sigma_i\}_{i \in I}$ contains only the identity substitution;
- for each partition of $N \neq \emptyset$ into k disjoint and non-empty sets $N_1, \ldots, N_k$, there is exactly one substitution in $\{\sigma_i\}_{i \in I}$ that represents $N_1, \ldots, N_k$. □

Notice that an infinite number of distinct partition families of $N \neq \emptyset$ there exists. However, any partition family of N is unredundant in a very precise sense: it contains one and only one representative of each of the possible partitions of N. Relying on the notion of partition family, we can eventually justify Claim 2 by proving the main result of this section.

Theorem 8 *Let P, Q be π-calculus processes and let $\stackrel{.}{\asymp} \in \{\stackrel{.}{\sim}_L, \stackrel{.}{\approx}_L, \stackrel{.}{\simeq}_L\}$. Also, assume $\asymp$ to be the non-ground relation corresponding to the chosen $\stackrel{.}{\asymp}$. Then $P \asymp Q$ iff $P\sigma \stackrel{.}{\asymp} Q\sigma$ for all $\sigma \in \{\sigma_i\}_{i \in I}$ with $\{\sigma_i\}_{i \in I}$ partition family of* $\mathrm{fn}(P,Q)$. □

As $\mathrm{fn}(P,Q)$ is a finite set, any partition family of $\mathrm{fn}(P,Q)$ is finite. Hence Theorem 8 provides a finite upper bound to the number of ground bisimilarity checks which are necessary to infer full congruence.

2 Characterizing π-Congruences in $\pi\xi$-Calculus

In the following we present the $\pi\xi$-calculus, a CCS-like process algebra that simplifies and generalizes the calculi proposed in [FMQ96, FMQ95]. For $\asymp \in \{\sim_L, \approx_L, \simeq_L\}$, we show that $P \asymp Q$ coincides with the corresponding CCS-bisimilarity of two single $\pi\xi$-processes. Equational characterizations of late non-ground π-calculus semantics are also provided. Although only the finite fragment of the $\pi\xi$-calculus is presented here, its semantic coincidence with non-ground π-calculus does hold of the full language.

$$\tau.P \xrightarrow{\langle\tau,\mathtt{true}\rangle} P \qquad \overline{x}y.P \xrightarrow{\langle\overline{x}y,x\downarrow\rangle} P \qquad x(y).P \xrightarrow{\langle x(y),x\downarrow\rangle} P \qquad \lambda y.P \xrightarrow{\langle[y],\mathtt{true}\rangle} P$$

$$\frac{P \xrightarrow{\omega} P'}{P+Q \xrightarrow{\omega} P'} \qquad \frac{P \xrightarrow{\omega} P'}{P\,|\,Q \xrightarrow{\omega} P'\,|\,Q} \qquad \frac{P \xrightarrow{\omega} P' \quad Q \xrightarrow{\omega'} Q'}{P\,|\,Q \xrightarrow{\omega\|\omega'} P'\,|\,Q'}$$

$$\frac{P \xrightarrow{\omega} P'}{(y)P \xrightarrow{o_y\omega} P'} \qquad \frac{P \xrightarrow{\omega} P'}{(y)P \xrightarrow{\nu_y\omega} (y)P'} \qquad \frac{P \xrightarrow{\omega} P'}{[x=y]P \xrightarrow{\mu_y^x(\omega)} P'}$$

where

$$\mu_y^x\langle\alpha, C\rangle = \langle\alpha, C \wedge x = y\rangle$$

$$\langle\alpha_1, C_1'\rangle \parallel \langle\alpha_2, C_2'\rangle = \begin{cases} \langle\tau[y/w], C_1 \wedge C_2 \wedge x = z\rangle & \text{if } \alpha_1 \in \{\overline{x}y, \overline{x}(y)\},\ \ C_1' = C_1 \wedge x\downarrow \\ & \text{and } \alpha_2 = z(w),\ \ C_2' = C_2 \wedge z\downarrow \\ & \text{or symmetrically} \\ \langle\tau, \mathtt{false}\rangle & \text{otherwise} \end{cases}$$

$$\nu_y\langle\alpha, C\rangle = \begin{cases} \langle\alpha, C \wedge y \neq z\rangle & \text{if } \alpha = \overline{x}z \\ \langle\alpha, C\rangle & \text{otherwise} \end{cases} \qquad o_y\langle\alpha, C\rangle = \begin{cases} \langle\overline{x}(z), C \wedge y = z\rangle & \text{if } \alpha = \overline{x}z \\ \langle\alpha, \mathtt{false}\rangle & \text{otherwise} \end{cases}$$

Table 1. symbolic operational semantics

Processes of the $\pi\xi$-calculus are written $\xi :: P$ and ranged over by $S, S_1, \ldots$. The right component of the state operator '_ :: _' is (essentially) a π-calculus process, while ξ keeps track of the associations among names carried out in the past of the ongoing computation. Name substitutions are never applied to the right component of $\xi :: P$, hence ξ can be viewed as an environment giving the actual associations of names. As for the right component P, reasoning up to α-conversion, we assume that there is no homonymy either among bound names or among free and bound names. But for this requirement, P is defined by the usual π-calculus syntax added with the new family of prefixes '$\lambda y.$'. For brevity, in spite of these new prefixes, we often refer to the right component of any $\pi\xi$-calculus process as to a π-calculus process.

The operational semantics of the $\pi\xi$-calculus follows the SOS style [Plo81] and is based on a two-stage approach. The first stage consists of the definition of a symbolic semantics where transition labels record requirements on names. The evaluation of those requirements is one of the main concerns of the top level transition system.

The symbolic operational semantics is given by the axioms and rules reported in Tab. 1 together with symmetric rules for choice and asynchronous parallel composition. Labels, ranged over by $\omega, \omega', \ldots$, are pairs of the form $\langle\alpha, C\rangle$. The first component is an action; the second component, called *obligation*, is a logical formula that codes requirements on names.

The execution of the prefix λy results in the action $[y]$ that calls for an actualization of the formal parameter y. Differently from the π-calculus, communication is characterized by one single inference rule. Complete information about privacy of names is captured by environments, hence, at this first level, communicating private names is considered to be the same as sending public channels. Before plunging processes into environments, we only impose a consistency requirement: no process must be allowed to commit on a link that is not known outside. To this end, input and output transition labels include the obligation $x\downarrow$. It demands for a delayed check against the privacy of channel x.

Left components of $\pi\xi$-processes are sets of equations on two distinct entities: names and constants. Constants are taken from a denumerably infinite set $\mathcal{D}$ that is ranged over by $c, c_1, c_2, \ldots$, and disjoint from the set $\mathcal{N}$ of names.

Definition 9 An *environment* ξ is an equivalence relation over $\mathcal{N} \cup \mathcal{D}$ which is:

- *consistent*: $c_i \; \xi \; c_j$ implies $c_i = c_j$
- *finitely active*: the set $\{(a,b) \mid a \; \xi \; b \text{ and } a \neq b\}$ is finite

We denote by $[a]_\xi$ the equivalence class of ξ containing a. A constant c is *active* in ξ iff there exists y such that $y \in [c]_\xi$, it is *inactive* otherwise. The identity environment is denoted by $\mathrm{Id}_\mathcal{E}$. Also, letting R_1, R_2 be relations over $\mathcal{N} \cup \mathcal{D}$, the smallest equivalence relation including $(R_1 \cup R_2)$ is denoted by $R_1 + R_2$. □

We let ξ sometimes assume the reading of a partial function. In particular, whenever $(y \; \xi \; c)$, we write $\xi(y)$ for c, and say that $\xi(\_\_)$ is defined on y, denoted by $\xi(y)\downarrow$. Also, in order to guarantee the run-time generation of fresh constants, we assume the existence of two specialized functions: all$\mathcal{D}$ and new$\mathcal{D}$. The application all$\mathcal{D}(\xi)$ returns the finite set of all the constants that are active in the argument, while new$\mathcal{D}(\xi)$ yields a constant which is inactive in ξ. We suppose that new$\mathcal{D}$ only depends on the active constants in the argument, so that all$\mathcal{D}(\xi_1)$ = all$\mathcal{D}(\xi_2)$ implies new$\mathcal{D}(\xi_1)$ = new$\mathcal{D}(\xi_2)$.

The late $\pi\xi$-calculus operational semantics is described by the transition relation '$\longrightarrow\!\!\triangleright$' defined in Tab. 2. The behaviour of $\xi :: P$ is inferred from a symbolic transition of P by invoking suitable functions on environments. The possibly many-valued update function η takes care of extending the environment ξ with the name associations activated by the transition. The first step in computing η consists in checking the satisfiability of the obligation. Whenever the requirements expressed by C are not met in the environment ξ, the application η returns the empty set, so that the $\pi\xi$-process at hand is unable to move. Otherwise, depending on the structure of the action α, the update function yields a set of environments obtained by possibly adding a pair to ξ.

The case when α is a placeholder instantiation is particularly interesting: the function η yields as many environments as the possible choices of c in all$\mathcal{D}(\xi)$, plus a new constant. This corresponds to instantiate y with (possibly a superset of) all the free names of the process at hand, plus a new fresh one. The intuitive reason for this relies on the following. Given P', the processes $P'\{z/y\}$ and $P'\{u/y\}$ have analogous move potentials whenever $z, u \notin \mathrm{fn}(P')$. More precisely

$$\frac{P \xrightarrow{\omega} P' \quad \omega \neq \langle x(y), C\rangle \quad \xi' \in \eta(\xi,\omega)}{\xi :: P \overset{\delta(\xi',\omega)}{\longrightarrow\!\!\!\!\triangleright} \xi' :: P'} \qquad \frac{P \xrightarrow{\langle x(y),C\rangle} P' \quad \xi' \in \eta(\xi, \langle x(y), C\rangle)}{\xi :: P \overset{\delta(\xi',\langle x(y),C\rangle)}{\longrightarrow\!\!\!\!\triangleright} \xi' :: \lambda y.P'}$$

where

$$\begin{array}{l} \delta\xi\alpha C = \mathbf{case}\ \alpha\ \mathbf{in} \\ \qquad \tau, \tau[x/y] : \tau \\ \qquad \overline{x}(y), \overline{x}y : \overline{\xi(x)}\xi(y) \\ \qquad x(y) : \xi(x) \\ \qquad [y] : [\xi(y)] \\ \qquad \mathbf{end\_case} \end{array} \qquad \begin{array}{l} \eta\xi\alpha C = \neg[\![C]\!]\xi \longrightarrow \emptyset, \\ \quad \mathbf{case}\ \alpha\ \mathbf{in} \\ \qquad \tau : \{\xi\} \\ \qquad \tau[x/y] : \{\xi + (y,x)\} \\ \qquad \overline{x}(y), \overline{x}y : \xi(y)\downarrow \longrightarrow \{\xi\},\ \{\xi + (y,\ \mathrm{new}\mathcal{D}(\xi))\} \\ \qquad x(y) : \{\xi\} \\ \qquad [y] : \bigcup_{c \in (\mathrm{all}\mathcal{D}(\xi)\ \cup\ \mathrm{new}\mathcal{D}(\xi))} \{\xi + (y,c)\} \\ \quad \mathbf{end\_case} \end{array}$$

with

$$\begin{array}{l} [\![C]\!]\xi = \mathbf{case}\ C\ \mathbf{in} \\ \qquad \mathbf{true} : \mathtt{tt} \\ \qquad \mathbf{false} : \mathtt{ff} \\ \qquad x\downarrow : \xi(x)\downarrow \longrightarrow \mathtt{tt},\ \ \mathtt{ff} \\ \qquad x = y : x\ \xi\ y \longrightarrow \mathtt{tt},\ \ \mathtt{ff} \\ \qquad x \neq y : x\ \xi\ y \longrightarrow \mathtt{ff},\ \ \mathtt{tt} \\ \qquad C_1 \wedge C_2 : [\![C_1]\!]\xi\ \mathrm{and}\ [\![C_2]\!]\xi \\ \qquad \mathbf{end\_case} \end{array}$$

Table 2. definition of $\longrightarrow\!\!\!\!\triangleright$

either $P'\{z/y\}$ or $P'\{u/y\}$ have the same action capabilities as P' has. Then, whenever P performs the input $x(y)$ transforming into P', the relevant instantiations of y in $P'\{w/y\}$ are given by $w \in \mathrm{fn}(P') \subseteq \mathrm{fn}(P) \cup \{y\}$. The definition of $\eta(\xi, [y], C)$ is meant to mimic those instantiations. The function all$\mathcal{D}(\xi)$ plays the role of fn(P) while new$\mathcal{D}(\xi)$ stays for the set $\{y\}$. At any time during execution only finitely many constants are active. Hence finite $\pi\xi$-processes (or, in the full calculus, processes with at most guarded occurrences of the replication operator) can always be represented by finitely branching transition systems.

The result function δ yields either τ or the constant(s) associated with the relevant name(s). Observation labels are given by the grammar $\rho ::= \tau \mid c \mid [c] \mid cc' \mid \overline{c}c'$.

A last remark on the operational semantics of the calculus is about the first-class role of name instantiation. Correspondingly to the execution of the symbolic action $x(y)$, process $\xi :: P$ evolves to $\xi' :: \lambda y.P'$ whose right component is an explicit temporary function of the parameter y. The next – compulsory – move of $\xi' :: \lambda y.P'$ is the instantiation of y.

Notice that, although the first level operational semantics is symbolic, the $\pi\xi$-calculus transition system is labelled by concrete events. This allows us to set, as extensional semantics of the calculus, the usual *strong*, *weak*, and *observational* CCS equivalences [Par81, Mil83, HM85], denoted by $\sim$, $\approx$, and $\approx^c$, respectively.

2.1 Main Results

In [FMQ96, FMQ95] calculi analogous to the one introduced above were used to characterize ground π-calculus semantics. Those calculi were not general enough to encompass coincidence with π-congruences. Here we fill this gap, and characterize π-calculus non-ground semantics, either strong or τ-forgetting.

Relying on the results of Section 1, we show that λ-abstractions can be exploited to capture the intrinsic and permanent functional dependency of π-processes on their free names. This is mainly due to the fact that environments encode name substitutions in a genuine way. In particular they naturally represent, via equivalence classes, partitions of sets of names.

Definition 10 The environment ξ *represents the partition* of $N \subseteq \mathcal{N}$ into the disjoint and non-empty sets $N_1, \ldots, N_k$ iff the following holds:

- $N = \bigcup_{c \in \text{all}\mathcal{D}(\xi)} ([c]_\xi \cap \mathcal{N})$;
- for all $j \in \{1, \ldots, k\}$ there exists $c \in \text{all}\mathcal{D}(\xi)$ such that $N_j = ([c]_\xi \cap \mathcal{N})$. $\square$

A more effective feature can be modelled in $\pi\xi$-calculus. Consider the process $S_0 = (\text{Id}_\mathcal{E} :: \lambda x.\lambda y.\lambda z.\,\text{nil})$. No constant is active in $\text{Id}_\mathcal{E}$. Then, when the leading prefix λx fires, the name x is deterministically associated with $c_1 = \text{new}\mathcal{D}(\text{Id}_\mathcal{E})$. Correspondingly, S_0 is transformed into $S_1 = (\text{Id}_\mathcal{E} + (x, c_1) :: \lambda y.\lambda z.\,\text{nil})$ with $(\text{Id}_\mathcal{E} + (x, c_1))$ representing the only possible partition of $\{x\}$. Since the actual environment contains one active constant, the symbolic step labelled by $\langle [y], \mathtt{true} \rangle$ induces now two distinct transitions from S_1. The one is labelled by $[c_1]$, the other is labelled by the concretion of a new constant c_2. The two derivatives of S_1 have a common right component $(\lambda z.\,\text{nil})$, but their environments are distinct because of the association of y with different constants. Such environments represent the partitioning of the set $\{x, y\}$ into $\{x, y\}$ and into $\{x\} \cup \{y\}$, respectively. Each computation goes on by firing the λz prefix, and the environments of the derivative processes represent all the possible partitions of $\{x, y, z\}$. This sample argument is made precise by the following statement.

Proposition 11 *Suppose that* $\xi :: \lambda y.P \xrightarrow{\rho} \xi' :: P$, *then the following holds.*

1. *If* $\xi = \text{Id}_\mathcal{E}$ *then* $\rho = [\text{new}\mathcal{D}(\text{Id}_\mathcal{E})]$ *and* ξ' *represents the only possible partition of* $\{y\}$.
2. *If* ξ *represents the partition of* N *into* $N_1, \ldots, N_k$, *with* $y \notin N$, *then*
   - *if* $\rho = [c]$ *with* $([c]_\xi \cap \mathcal{N}) = N_j$, *then* ξ' *represents the partition of* $N \cup \{y\}$ *into the* k *sets* $N_1, \ldots, N_j \cup \{y\}, \ldots, N_k$;
   - *if* $\rho = [\text{new}\mathcal{D}(\xi)]$ *then* ξ' *represents the partition of* $N \cup \{y\}$ *into the* $k+1$ *sets* $N_1, \ldots, N_k, \{y\}$. $\square$

Proposition 11 highlights that environments representing partitions of sets can be effectively generated by sequential compositions of λ-abstractions. Coincidence between π-calculus and $\pi\xi$-calculus is obtained by suitably combining this feature with the fact that the congruence $P \asymp Q$ is expressible as conjunction of $P\sigma \dot{\asymp} Q\sigma$ for all σ in a partition family of $\mathrm{fn}(P, Q)$ (Theorem 8).

Definition 12 Let P be a π-calculus process, and $\mathcal{L}$ be a list of names. The set of names that occur in $\mathcal{L}$ is denoted by $\mathrm{n}(\mathcal{L})$. The *λ-closure of P w.r.t. $\mathcal{L}$* is defined as $\lambda\_\mathrm{clos}(P, \mathcal{L})$ where

$$\lambda\_\mathrm{clos}(P, [\,]) = P$$
$$\lambda\_\mathrm{clos}(P, [y, \mathcal{L}]) = \lambda\_\mathrm{clos}(\lambda y.P, \mathcal{L})$$

with $[\,]$ denoting the empty list, and $[\_, [\_]]$ denoting the 'cons' list operator. □

Theorem 13 *(coincidence with late non-ground semantics)*
Let P, Q be π-calculus processes and let $\mathcal{L}$ be a list s.t. $\mathrm{n}(\mathcal{L}) = \mathrm{fn}(P, Q)$. Then

1. $P \sim_L Q$ *iff* $\mathrm{Id}_\mathcal{E} :: \lambda\_\mathrm{clos}(P, \mathcal{L}) \sim \mathrm{Id}_\mathcal{E} :: \lambda\_\mathrm{clos}(Q, \mathcal{L})$
2. $P \approx_L Q$ *iff* $\mathrm{Id}_\mathcal{E} :: \lambda\_\mathrm{clos}(P, \mathcal{L}) \approx \mathrm{Id}_\mathcal{E} :: \lambda\_\mathrm{clos}(Q, \mathcal{L})$
3. $P \simeq_L Q$ *iff* $\mathrm{Id}_\mathcal{E} :: \lambda\_\mathrm{clos}(P, \mathcal{L}) \approx^c \mathrm{Id}_\mathcal{E} :: \lambda\_\mathrm{clos}(Q, \mathcal{L})$ □

Reasoning about congruences compels to take into account that putting a process into an arbitrary context might cause arbitrary instantiations of its free names. Taking λ-closed terms corresponds to this intuition: processes are interpreted as permanent functions of their free names.

Assuming that $\mathrm{card}(\mathrm{fn}(P, Q)) = n$, there are $n!$ distinct permutations of the elements in $\mathrm{fn}(P, Q)$. Hence there exist $n!$ distinct lists built out of $\mathrm{fn}(P, Q)$. Theorem 13 asserts that the congruence of P and Q can be checked relying on one of those lists, chosen at random. The role played by $\mathcal{L}$ in the definition of $\lambda\_\mathrm{clos}(P, \mathcal{L})$ is to generate the environments that represent a partition family of $\mathrm{fn}(P, Q)$. The wanted effect is independent on the actual choice of the list. Whenever the same $\mathcal{L}$ is used in the definition of either the λ-closure of P or the λ-closure of Q, we are guaranteed that the bisimilarity of the pair of processes $(\mathrm{Id}_\mathcal{E} :: \lambda\_\mathrm{clos}(P, \mathcal{L}), \mathrm{Id}_\mathcal{E} :: \lambda\_\mathrm{clos}(Q, \mathcal{L}))$ is factorized into the bisimilarity of all the pairs in $\bigcup_p (\xi_p :: P, \xi_p :: Q)$ where p ranges over the possible partitions of $\mathrm{fn}(P, Q)$ and ξ_p represents the partition p.

The proof of Theorem 13 relies on the definition of encodings of each calculus in the other, so that, for instance, environments are translated into proper substitutions and consistent information about the privacy of names is retrieved. A key point of the proof is the characterization of operational correspondences between related terms. The observability of the actualization of input parameters plays a central role, too. On the π-calculus side, late bisimulation input clauses use a sequence of quantifiers of the shape '$\forall\exists\forall$', with the last quantification relating to the placeholder instantiation w. On their side, CCS-like bisimulations can express universal quantifications only on the set of the possible next moves. Hence forcing the parameter actualization to become a distinguished operational step is the only way to mimic the above '$\forall w$' in a CCS-like setting.

$\mathcal{A}_s : (U)\ \xi :: \omega.P = \bigoplus_{\xi' \in \eta_L(\xi,\omega)} \delta(\xi',\omega); (\xi' :: P) \quad \text{if } \omega \neq \langle x(y), C\rangle$

$(V)\ \xi :: \langle x(y), C\rangle.P = \bigoplus_{\xi' \in \eta_L(\xi,\langle x(y),C\rangle)} \delta(\xi', \langle x(y), C\rangle); (\xi' :: \langle [y], \mathbf{true}\rangle.P)$

$(D)\ \xi :: (P_1 + P_2) = \xi :: P_1 \oplus \xi :: P_2$

$(C1)\ P_1 + P_2 = P_2 + P_1$	$(C1')\ S_1 \oplus S_2 = S_2 \oplus S_1$
$(C2)\ (P_1 + P_2) + P_3 = P_1 + (P_2 + P_3)$	$(C2')\ (S_1 \oplus S_2) \oplus S_3 = S_1 \oplus (S_2 \oplus S_3)$
$(C3)\ P + P = P$	$(C3')\ S \oplus S = S$
$(C4)\ P + \text{nil} = P$	$(C4')\ S \oplus \xi :: \text{nil} = S$

$(R)\quad (x)P = (\nu_x)P + (o_x)P$

$(EX)\ P_1 \mid P_2 = P_1 \rfloor\!\rfloor P_2 + P_2 \rfloor\!\rfloor P_1 + P_1 \parallel P_2$

$(A1)\ [x = y]\ \omega.P = \mu^x_y(\omega).P$	$(D1)\ [x = y](P_1 + P_2) = [x = y]P_1 + [x = y]P_2$
$(A2)\ \omega.P_1 \rfloor\!\rfloor P_2 = \omega.(P_1 \mid P_2)$	$(D2)\ (P_1 + P_2) \rfloor\!\rfloor P = P_1 \rfloor\!\rfloor P + P_2 \rfloor\!\rfloor P$
$(A3)\ \omega_1.P_1 \parallel \omega_2.P_2 =$	$(D3)\ (P_1 + P_2) \parallel P = P_1 \parallel P + P_2 \parallel P$
$\quad (\omega_1 \parallel \omega_2).(P_1 \mid P_2)$	$(D4)\ P \parallel (P_1 + P_2) = P \parallel P_1 + P \parallel P_2$
$(A4)\ (\nu_x)(\omega.P) = \nu_x(\omega).(x)P$	$(D5)\ (\nu_x)(P_1 + P_2) = (\nu_x)P_1 + (\nu_x)P_2$
$(A5)\ (o_x)(\omega.P) = o_x(\omega).P$	$(D6)\ (o_x)(P_1 + P_2) = (o_x)P_1 + (o_x)P_2$

$(IN)\ [x = y]\,\text{nil} = (\nu_x)\,\text{nil} = (o_x)\,\text{nil} = \text{nil} \rfloor\!\rfloor P = P \parallel \text{nil} = \text{nil} \parallel P = \text{nil}$

$\mathcal{A}_w : (T1)\ \rho; \tau; S = \rho; S$

$(T2)\ S \oplus \tau; S = \tau; S$

$(T3)\ \rho; (S_1 \oplus \tau; S_2) \oplus \rho; S_2 = \rho; (S_1 \oplus \tau; S_2)$

Table 3. axiom systems $\mathcal{A}_s$, $\mathcal{A}_w$

2.2 Equational Characterizations

In [Lin95], π-calculus congruences have been characterized in terms of symbolic bisimulation [HL95]. The perspective of our encoding is orthogonal to Lin's symbolic view. Reasoning in terms of $\pi\xi$-calculus semantics moves to the π-calculus a number of verification algorithms and general results that have been developed, over the years, for CCS and its theory.

A significant example of the relevance of inheriting the general CCS metatheory is the following equational characterization, based on the axiom system drawn in Tab. 3.

Theorem 14 *(equational characterization of late non-ground semantics)* *Let P, Q be finite π-calculus processes and let $\mathcal{L}$ be a list s.t.* $\text{n}(\mathcal{L}) = \text{fn}(P, Q)$. *Then*

1. $P \sim_L Q$ *iff* $\mathcal{A}_s \vdash \text{Id}_\mathcal{E} :: \lambda\_\text{clos}(P, \mathcal{L}) = \text{Id}_\mathcal{E} :: \lambda\_\text{clos}(Q, \mathcal{L})$
2. $P \simeq_L Q$ *iff* $\mathcal{A}_s, \mathcal{A}_w \vdash \text{Id}_\mathcal{E} :: \lambda\_\text{clos}(P, \mathcal{L}) = \text{Id}_\mathcal{E} :: \lambda\_\text{clos}(Q, \mathcal{L})$ □

The axiom system $\mathcal{A}_s$ in Tab. 3 is defined after the following observations. The symbolic operational semantics of Tab. 1 fits in a simple generalization of the

De Simone format [DS85]. By this, a head-normalizing axiom system for the π-component of any $\pi\xi$-process can be stated by imposing the monoidal laws for summation $(C1 - C4)$ and exploiting the procedure presented in [ABV94]. The idea is to introduce auxiliary operators in order to break down those process constructor whose operational behaviour is described by more than one inference rule (R, EX). Auxiliary operators are then pushed as deep as possible inside the structure of the term by means of action axioms $(A1 - A5)$ and distributive laws $(D1 - D6)$, till they actually disappear, due to inaction equations (IN).

Once its right component has been reduced into head normal form, any finite $\pi\xi$-process can be transformed into a finite labelled tree given by the grammar $S ::= \xi :: P \mid \rho; S \mid S \oplus S$, where ';' is a prefixing operator, and '$\oplus$' is a non-deterministic choice constructor, and $\bigoplus_\emptyset$ is assumed to be the neutral element of $\oplus$, *i.e.* the $\pi\xi$-process $\xi ::$ nil for an arbitrary instantiation of ξ. Now, reducing a $\pi\xi$-calculus process into a sum of prefixes only requires: two axioms corresponding of the operational rules that define the top level transition relation (U, V); one equation describing the distributivity of the low level choice operator over the state constructor (D); and the monoidal laws for the top level summation $(C1' - C4')$.

Eventually, $\mathcal{A}_w$ simply consists of the Milner's τ-laws for observational congruence.

3 Concluding Remarks

The main contribution of the paper is the finitary nature of the characterizations of late, either strong or weak, π-calculus congruences. Target languages of those characterizations are either the π-calculus itself or the $\pi\xi$-calculus, a generalization of the calculi proposed in [FMQ96, FMQ95].

A characterization of weak π-calculus non-groundness already appeared in [Lin95]. Lin's formulation is in terms of symbolic bisimulation [HL95]. Our perspective is orthogonal to that symbolic view. Non-ground π-calculus semantics are rephrased as CCS equivalences that come equipped with well-established mathematical properties and verification tools. The axiomatizations we presented are relevant examples of the significance of re-using the general CCS meta-theory.

Relating to the full π-calculus, it was shown that each late non-ground equivalence can be expressed by closing the corresponding ground bisimilarity under the substitutions of a suitable finite family $\{\sigma_i\}_i$. This result can be directly instantiated to early semantics, either strong or τ-forgetting. So it gives insights on the complexity of checking a large spectrum of π-calculus congruences.

The $\pi\xi$-calculus view allowed processes to be explicitly interpreted as temporary functions of input parameters and as permanent functions of their free names. One of the advantages of adopting this perspective is that the mere evolution of a λ-closed term induces the generation of (the environments corresponding to) the substitutions $\{\sigma_i\}_i$ needed to check non-groundness. This adds effectiveness to the first alternative characterization we provided.

Here we dealt only with the finite fragment of the $\pi\xi$-calculus, and tackled only its late semantics. However, the coincidence of non-ground π-calculus and $\pi\xi$-calculus does hold of the full language [Qua97]. Further, π-calculus congruences of the early family can be characterized using an approach analogous to that we adopted for the late case. The early view can be retrieved as a special case of late $\pi\xi$-semantics by minor changes to the top level transition system [Qua96]. In the same spirit as the free input actions of [MPW93], those changes essentially amount to make atomic any input step and the subsequent instantiation move.

Acknowledgments We would like to thank Gian-Luigi Ferrari and Ugo Montanari for the joint work that provided the basis of the $\pi\xi$-calculus, and for helpful comments relating to this.

References

[Abr91] S. Abramsky. A Domain Equation for Bisimulation. *Information and Computation*, 92(2):161–218, 1991.

[ABV94] L. Aceto, B. Bloom, and F. Vaandrager. Turning SOS Rules into Equations. *Information and Computation*, 111(1):1–52, 1994.

[CPS93] R. Cleaveland, J. Parrow, and B. Steffen. The Concurrency Workbench: A Semantics-Based Tool for the Verification of Concurrent Systems. *ACM Trans. on Programming Languages and Systems*, 15(1):36–72, 1993.

[DS85] R. De Simone. Higher level synchronizing devices in MEIJE-SCCS. *Theoretical Computer Science*, 37(3):245–267, 1985.

[FMQ95] G.-L. Ferrari, U. Montanari, and P. Quaglia. The Weak Late π-calculus Semantics as Observation Equivalence. In I. Lee and S.A. Smolka, editors, *Proc. 6th International Conference on Concurrency Theory, CONCUR '95*, volume 962 of *LNCS*, pages 57–71. Springer-Verlag, 1995.

[FMQ96] G.-L. Ferrari, U. Montanari, and P. Quaglia. A π-calculus with Explicit Substitutions. *Theoretical Computer Science*, 168(1):53–103, 1996.

[HL95] M. Hennessy and H. Lin. Symbolic Bisimulations. *Theoretical Computer Science*, 138:353–389, 1995.

[HM85] M. Hennessy and R. Milner. Algebraic Laws for Nondeterminism and Concurrency. *Journal of the ACM*, 32(1):137–161, 1985.

[Lin95] H. Lin. Complete Inference Systems for Weak Bisimulation Equivalences in the π-Calculus. In P.D. Mosses, M. Nielsen, and M.I. Schwartzbach, editors, *Proc. 6th International Joint Conference CAAP/FASE, TAPSOFT '95*, volume 915 of *LNCS*. Springer-Verlag, 1995.

[Mil83] R. Milner. Calculi for synchrony and asynchrony. *Theoretical Computer Science*, 25:267–310, 1983.

[MPW92] R. Milner, J. Parrow, and D. Walker. A Calculus of Mobile Processes, Part I and II. *Information and Computation*, 100(1):1–77, 1992.

[MPW93] R. Milner, J. Parrow, and D. Walker. Modal logics for mobile processes. *Theoretical Computer Science*, 114(1):149–171, 1993.

[Par81] D. Park. Concurrency and automata on infinite sequences. In *Proc. 5th GI-Conference*, volume 104 of *LNCS*. Springer-Verlag, 1981.

[Plo81] G. D. Plotkin. A Structural Approach to Operational Semantics. Technical Report DAIMI-FN-19, Computer Science Department, Aarhus University, 1981.

[Qua96] P. Quaglia. *The π-calculus with explicit substitutions*. PhD thesis, Università degli Studi di Pisa, Dipartimento di Informatica, 1996. Report TD-09/96.

[Qua97] P. Quaglia. On the finitary characterization of π-congruences. BRICS Report RS-97-52, Computer Science Department, Aarhus University, 1997.

[San96] D. Sangiorgi. A Theory of Bisimulation for the π-calculus. *Acta Informatica*, 33(1):69–97, 1996.

Algebraic Specifications, Higher-Order Types, and Set-Theoretic Models

Hélène Kirchner[1] and Peter D. Mosses[23]

[1] LORIA-CNRS
B.P. 239
F-54506 Vandœuvre-lès-Nancy Cedex, France
e-mail: Helene.Kirchner@loria.fr

[2] Computer Science Laboratory, SRI International
333 Ravenswood Avenue, Menlo Park, CA 94025, USA
e-mail: mosses@csl.sri.com.

[3] On leave from BRICS, Department of Computer Science
University of Aarhus, Ny Munkegade bldg. 540
DK-8000 Aarhus C, Denmark
http://www.brics.dk/~pdm/

Abstract. In most algebraic specification frameworks, the type system is restricted to sorts, subsorts, and first-order function types. This is in marked contrast to the so-called model-oriented frameworks, which provide higher-order types, interpreted set-theoretically as Cartesian products, function spaces, and power-sets. This paper presents a simple framework for algebraic specifications with higher-order types and set-theoretic models. It may be regarded as the basis for a Horn-clause approximation to the Z framework, and has the advantage of being amenable to prototyping and automated reasoning. Standard set-theoretic models are considered, and conditions are given for the existence of initial reducts of such models. Algebraic specifications for various set-theoretic concepts are considered.

1 Introduction

There are two main schools of thought regarding the formal specification of abstract data types: the model-oriented and the property-oriented. In a model-oriented specification, the emphasis is on specifying data types as set-theoretic structures (products, power sets, etc.), the operations of the data types then being defined as particular functions on these structures. The underlying logic for reasoning about such a specification is a powerful higher-order logic, e.g., based on ZF set theory. In a property-oriented specification, one generally tries to avoid choosing an explicit representation: types are left abstract as so-called sorts—sometimes equipped with a subsort inclusion relation, but otherwise unstructured. The operations are specified by axioms that relate them to each other, usually including the main intended algebraic properties. The underlying logic is often a modest Horn-clause fragment of equational first-order logic—supplemented by an induction rule when dealing with initial algebra semantics

A.M. Haeberer (Ed.): AMAST'98, LNCS 1548, pp. 373–388, 1998.

rather than loose semantics. In practice, some model-oriented specification languages (such as Z) do allow types to be left abstract (or 'given'), with the operations on them specified by axioms. Moreover, the use of auxiliary ('hidden') sorts and operations in property-oriented specifications can give these a model-oriented flavour. There are also some wide-spectrum languages (e.g., RSL, Spectrum) which encompass both approaches, allowing model- and property-oriented specifications to be mixed together.

It seems to us that both the model- and property-oriented approaches have their advantages *and* disadvantages. In particular, we regard the restriction to Horn-clause logic in the latter as beneficial, since not only are the consequences of a specification much more obvious than in full higher-order logic, but also automated reasoning and prototyping are feasible. The resulting existence of initial models is useful (but does not preclude considering loose semantics of specifications). The usual restriction to first-order functions and unstructured sorts in property-oriented specifications, however, we regard as a definite disadvantage.

This has led us to investigate an intermediate or hybrid approach:

- Types may be polymorphic, and include abstract types as well as the concrete set-theoretical product, power-set, and function types.
- Operations may be higher-order and partial.
- The only built-in relations are equality, set membership and definedness (the last merely abbreviates an equality).
- Formulae are restricted to Horn clauses (no disjunction, variables are universally quantified).
- Models have set-theoretic foundations.
- Specifications have initial models (when consistent).
- Specifications are amenable to prototyping and reasoning using rewriting and saturation techniques.

The main novelty of our approach is its treatment of set-theoretic concepts using the rather weak logical framework of (equational) Horn clauses. It will be helpful to consider altogether three kinds of models of specifications: arbitrary algebraic models, where values of set-types do not necessarily have any concrete set-theoretic structure at all; labelled-set models, where they are pairs of arbitrary labels and ordinary sets; and standard set-theoretic models, where sets have neither labels nor extra elements. For the algebraic and labelled-set models we obtain initial models, and we give conditions such that the standard set-theoretic models have initial reducts. The conclusion is that when sets are used essentially as types, and their equality is of no concern, one need only consider the standard set-theoretic models of our specifications.

Thus a framework based on our approach could be attractive for those who prefer the concrete, higher-order, set-theoretic nature of B or Z, but who also like the possibility of automated reasoning and prototyping for exploring the consequences of (requirements or design) specifications. The price to be paid for the latter is the restriction to Horn clauses, and the avoidance of conditions involving equality between sets that do not necessarily have the same members in all models. As a compensation, the models of our specifications are of a much simpler nature than those of Z specifications.

Plan of the Paper. Section 2 explains the syntax of our proposal for algebraic specifications with (higher-order, polymorphic) set-theoretic types, giving some simple examples. Section 3 defines the notion of a presentation. Section 4 gives a deductive system for type judgements. Section 5 presents our main proof system for reasoning about the consequences of specifications. Section 6 investigates the three kinds of models mentioned above. Section 7 outlines how the language presented here can be extended to cater for explicit subset inclusions, and indicates how accurately other familiar set-theoretic constructs can be specified in this framework. Finally, Section 8 discusses the relationship of our framework to Membership Equational Logic [5] and other work, and considers possible directions for future developments. Due to space limitations all proofs of results are either sketched or deferred to the extended version of this paper.

2 Specifications

Some simple examples of specifications are given in Tables 1–6. They are intended mainly to illustrate the basic form of our specifications; this is reminiscent of (non-imperative) B and Z, although there are also some significant differences, such as our treatment of partiality, polymorphism, and overloading. The rest of this section gives an informal explanation of the syntax and semantics of the specification language.

Table 1. Natural Numbers

Naturals

$type\ \mathbf{N}$
$0 \in \mathbf{N}$
$succ \in \mathbf{N} \nrightarrow \mathbf{N}$
$pred \in \mathbf{N} \nrightarrow \mathbf{N}$

$\forall n \in \mathbf{N} \bullet$
$pred(succ(n)) = n$
$pred(n)\downarrow \Rightarrow succ(pred(n)) = n$
$pred(0)\downarrow \Rightarrow \bot$

As illustrated in Table 1, a specification here is mainly just a set of Horn clauses where the atomic formulae are equations $t_1 = t_2$, set memberships $t \in s$, or definedness assertions $t\downarrow$. A Horn clause with conditions $A_1, \ldots, A_n$ and conclusion A is written $A_1 \wedge \ldots \wedge A_n \Rightarrow A$; when $n = 0$ it is written $\top \Rightarrow A$, or simply A. Negative clauses, asserting that the conditions cannot all hold together, are written $A_1 \wedge \ldots \wedge A_n \Rightarrow \bot$.

Terms t are formed from constants and variables by the two binary operations of function application $t_1(t_2)$ and pairing (t_1, t_2). Only one pair of parentheses is needed in $f((x, y))$, and $(x, (y, z))$ may be written (x, y, z). Mixfix notation is allowed for application: when a function constant f has n place-holders (written '_'), the application $f(t_1, \ldots, t_n)$ (or $f(t_1) \ldots (t_n)$) may be written as f with each place-holder replaced by the corresponding argument term t_i. E.g., $\_+\_(m, succ(n))$ may be written $m + succ\ n$, as illustrated in Table 2.

Table 2. Natural Numbers with addition

NaturalsPlus

Naturals

$\_+\_ \in \mathbb{N} \times \mathbb{N} \nrightarrow \mathbb{N}$

$\forall m, n \in \mathbb{N} \bullet$

$m + 0 = m$

$m + succ(n) = succ(m + n)$

$pred(n) \downarrow \Rightarrow m + pred(n) = pred(m + n)$

The function constants $\_\times\_$, $\_\nrightarrow\_$, $\mathbb{P}\_$, used for expressing sets, have interpretations such that for any sets S, T:

- the members of $S \times T$ are all pairs (s, t) with $s \in S$, $t \in T$;
- all members of $S \nrightarrow T$ represent partial functions which, when defined, map values in S to values in T;
- all members of $\mathbb{P}\, S$ are subsets of S.

By letting $S \nrightarrow T$ and $\mathbb{P}\, S$ return subsets of the usual results, we admit models with countable universes. Meinke [3, 4] uses a similar technique for the type constructors in his approach to higher-order algebra.

We distinguish some terms denoting sets as *types*, using them in checking the consistency of use of symbols in specifications. An abstract 'given' type S is simply a set constant whose members remain to be specified; it is introduced by writing *type* S, as with $\mathbb{N}$ in Table 1. Applications of the function constants $\_\times\_$, $\_\nrightarrow\_$, $\mathbb{P}\_$ then generate concrete set-theoretic types based on the abstract types. E.g., $\mathbb{N} \nrightarrow \mathbb{N}$ and $\mathbb{N} \times \mathbb{N} \nrightarrow \mathbb{N}$ are types in Tables 1 and 2. (We let $\_\times\_$ have higher precedence than $\_\nrightarrow\_$, and group both of them to the right; $\mathbb{P}\_$ has higher precedence than $\_\times\_$.) The type of the constant $\mathbb{N}$ itself is implicitly $\mathbb{P}\,\mathbb{N}$; the membership $\mathbb{N} \in \mathbb{P}\,\mathbb{N}$, used later, might look a bit strange, but it merely corresponds to the inclusion $\mathbb{N} \subseteq \mathbb{N}$, implying that $\mathbb{N}$ must denote a set.

All the constants c used in a specification are required to have declared types written as (unconditional) memberships of the form $c \in T$, where T is a type term. The assertion $c \in T$ implies the definedness of c (and of T, but in fact type terms T always have defined values).

We follow the Z style of collecting all the declarations concerning types at the head of the specification, separating them from the body by a short line. This style has the benefit of clearly exhibiting all the used constants, together with their types. Furthermore, since declarations are generally just membership assertions, they may be considered as ordinary clauses, which allows one to ignore the separation between the head and the body of the specification when considering its logical consequences.

We also follow Z in the way that a specification may be named. A reference to its name in the head of another specification is equivalent to inserting a copy of the named specification (putting the head and body in the right places, and renaming any clashing variables). Thus the specification in Table 2 extends that in Table 1.

Although all types are sets, not all sets need be types. Table 3 illustrates the specification of the sets *Even* and *Odd* as *subsets* of $\mathbb{N}$. Whereas checking whether a term has a particular type is decidable (as we shall see in Section 4), it is in general undecidable to check membership of a set, since a membership assertion may have equations as conditions, as well as other memberships.

Table 3. Even and Odd Numbers

```
┌─ EvenOdd ─────────────────────────────
│ Naturals
│     Even ∈ ℙℕ
│     Odd ∈ ℙℕ
├──────────
│ ∀n ∈ ℕ •
│     0 ∈ Even
│     n ∈ Even ⇒ succ(n) ∈ Odd
│     n ∈ Odd ⇒ succ(n) ∈ Even
└───────────────────────────────────────
```

The clauses in the body of a specification generally involve variables, and we require the types of the variables to be declared at the top of the body, again writing them as memberships, e.g., $n \in \mathbb{N}$. For conciseness, several variables of the same type may be declared together, e.g., $m, n \in \mathbb{N}$. We follow the Z style by enclosing the list of type memberships in $\forall \ldots \bullet$, and by using semicolons as separators. Apart from being used in type-checking, each variable membership is regarded as an implicit condition of all the clauses in the body of the specification (or of just those in which the variable occurs, if one prefers, as the choice here doesn't affect the consequences or satisfaction of the extended clauses). Note that the value of a variable can never be undefined.

Type variables are allowed too, and used as arguments of the type constructors $\_\times\_$, $\_\leftrightarrow\_$ (partial function arrow), $\mathbb{P}\_$ to provide concrete *polymorphic* types. E.g., the first projection function *first* for pairs has type $X \times Y \nrightarrow X$, where X and Y are both type variables. Since type variables are needed for use in declarations, they are listed before them, at the top of the head of the specification.

Table 4 gives a specification of the predefined constructors for polymorphic set-theoretic types: $V \times W$, $X \nrightarrow Y$, $\mathbb{P}\,Z$, and of the projection functions for pairs: *first*, *second*. This anonymous specification is assumed to be implicitly referenced by every other specification, so that its declarations and clauses are always available.

Table 4. Predefined constants

$$
\begin{array}{l}
\forall\, \mathit{type}\ V, W, X, Y, Z \bullet \\
\quad \mathit{type}\ V \times W \\
\quad \mathit{type}\ X \nrightarrow Y \\
\quad \mathit{type}\ \mathbb{P}\,Z \\
\quad \mathit{first} \in V \times W \nrightarrow V \\
\quad \mathit{second} \in V \times W \nrightarrow W
\end{array}
$$

$$
\begin{array}{l}
\forall p \in V \times W;\ f \in X \nrightarrow Y; \\
\quad v \in V;\ w \in W;\ x \in X;\ z \in Z; \\
\quad Q \in \mathbb{P}\,V;\ R \in \mathbb{P}\,W;\ S \in \mathbb{P}X;\ T \in \mathbb{P}\,Y;\ U, U' \in \mathbb{P}Z; \bullet \\
\qquad v \in Q \wedge w \in R \Rightarrow (v, w) \in Q \times R \\
\qquad p \in Q \times R \Rightarrow \mathit{first}(p) \in Q \\
\qquad p \in Q \times R \Rightarrow \mathit{second}(p) \in R \\
\qquad (\mathit{first}(p), \mathit{second}(p)) = p \\
\qquad \mathit{first}(v, w) = v \\
\qquad \mathit{second}(v, w) = w \\
\qquad f \in S \nrightarrow T \wedge x \in S \wedge f(x) \downarrow \Rightarrow f(x) \in T \\
\qquad z \in U \wedge U \in \mathbb{P}\,U' \Rightarrow z \in U'
\end{array}
$$

Type variables are also used to declare abstract polymorphic types, such as $Seq(X)$ in Table 5. In fact *Seq* is interpreted as a (total) function from sets to sets, but in general we only need to consider its application to particular argument type terms, e.g., $Seq(\mathbb{N})$, $Seq(Seq(X))$. Notice that we only need to reference *GenericSequences*, after which we may use $Seq(X)$ with arbitrary arguments X; the treatment is different in Z, where a specification using *GenericSequences* would explicitly instantiate X to a particular type.

Table 5 illustrates also the possibility of overloading, or *ad hoc* polymorphism, by declaring a second type for the function (constant) $\_ + \_$. In models of this

Table 5. Generic Sequences

GenericSequences

NaturalsPlus
$\forall$ *type* $X \bullet$
$\quad$ *type* $Seq(X)$
$\quad \langle\rangle \in Seq(X)$
$\quad \langle\_\rangle \in X \twoheadrightarrow Seq(X)$
$\quad \_+\_ \in Seq(X) \times Seq(X) \twoheadrightarrow Seq(X)$
$\quad |\_| \in Seq(X) \twoheadrightarrow \mathbb{N}$

$\forall s, t, u \in Seq(X);\ x \in X \bullet$
$\quad \langle\rangle + s = s$
$\quad s + \langle\rangle = s$
$\quad s + (t + u) = (s + t) + u$
$\quad |\langle\rangle| = 0$
$\quad |\langle x\rangle| = succ(0)$
$\quad |s + t| = |s| + |t|$

specification, $\_+\_$ is interpreted as a single partial function on the entire universe, returning only results in $\mathbb{N}$ when applied to arguments in $\mathbb{N}$, and only results in $Seq(X)$ when applied to arguments in $Seq(X)$ (for any set X). Had we followed Z in interpreting $S \twoheadrightarrow T$ as a subset of $\mathbb{P}(S \times T)$, we would have needed a more complicated interpretation of overloaded functions (as functions from types to the actual functions of interest).

The concrete types $V \times W$, $X \twoheadrightarrow Y$, and $\mathbb{P}\,Z$ never have any values in common. Thus a specification where an overloaded constant is declared, say, with both a pair type and a set type, is simply inconsistent, with no models at all. Similarly, abstract types never have any values in common with concrete types (although they may have values in common with other abstract types).

The use of higher-order functions in our framework is illustrated in Table 6. The somewhat tedious definedness conditions are needed because equations are interpreted existentially, implying that their terms have defined values. They would not be needed if f were restricted to total functions (types of total functions are considered briefly in Section 7). One could also eliminate the need for definedness conditions by introducing so-called strong equations, which hold also when the values of both terms are undefined.

This completes the informal explanation of our syntax and its intended interpretation. We believe that many existing (non-imperative) Z specifications could be reformulated in our language. The following sections focus on the for-

Table 6. Mapping Generic Sequences

MappingGenericSequences
GenericSequences
$\forall\, type\ X, Y \bullet$
$\quad mapseq\ \_\ \_ \in (X \rightarrow\!\!\!\!\rightarrow Y) \rightarrow\!\!\!\!\rightarrow Seq(X) \rightarrow\!\!\!\!\rightarrow Seq(Y)$

$\forall f \in X \rightarrow\!\!\!\!\rightarrow Y;\ s, t \in Seq(X);\ x \in X \bullet$
$\quad mapseq\ f\ \langle\rangle = \langle\rangle$
$\quad f(x) \downarrow \Rightarrow mapseq\ f\ \langle x \rangle = \langle f(x) \rangle$
$\quad mapseq\ f\ s \downarrow \wedge\ mapseq\ f\ t \downarrow \Rightarrow$
$\qquad mapseq\ f\ (s + t) = mapseq\ f\ s + mapseq\ f\ t$

mal foundations: type-checking, proof system, and models. But first, let us strip away the concrete syntax of specifications to obtain abstract *presentations*.[3]

3 Presentations

As explained in the preceding section, a specification *SP* (always extending Table 4) is mainly a set of Horn clauses, separated into two parts. It determines an abstract presentation $(\mathcal{F}, \mathcal{X}, \mathcal{M}, \mathcal{H})$ as follows:

- $\mathcal{F}$ is the set of all the (untyped) constant symbols declared by memberships $c \in T$ or by type declarations *type* c in the head of *SP*. The type symbols form a distinguished subset $\mathcal{F}_t$ of $\mathcal{F}$. For a polymorphic type declaration *type* $F(X_1, \ldots, X_n)$ we let $\mathcal{F}_t$ include the function symbol F.
- $\mathcal{X}$ is the set of all the (untyped) variables declared by variable memberships $x \in T$ in the body of *SP* or by type variable declarations *type* X in the head of *SP*. The type variables form a distinguished subset $\mathcal{X}_t$ of $\mathcal{X}$.
- $\mathcal{M}$ is the set of all the membership atoms specified in the head of the specification, together with implicit memberships determined by type declarations: a declaration of an abstract type constant c gives rise to $c \in \mathbb{P}\, c$; and a polymorphic type $F(X_1, \ldots, X_n)$ provides $F \in \mathbb{P}\, X_1 \times \ldots \times \mathbb{P}\, X_n \rightarrow\!\!\!\!\rightarrow \mathbb{P}(F(X_1, \ldots, X_n))$, as well as $F(X_1, \ldots, X_n) \downarrow$, the latter ensuring the totality of F. Moreover, for each type variable X in $\mathcal{X}_t$, $\mathcal{M}$ includes the membership atom $X \in \mathbb{P}\, X$.
- $\mathcal{H}$ is the set of Horn clauses in the body of *SP* with the declarations of variables in the body added as extra conditions.

[3] Presentations could be divided into the signatures and sentences of an institution, but the details are not relevant here.

The following negative clauses are added to $\mathcal{H}$:

$$p \in Q \times R \wedge p \in S \nrightarrow T \Rightarrow \bot$$
$$f \in S \nrightarrow T \wedge f \in \mathbb{P}\, U \Rightarrow \bot$$
$$z \in \mathbb{P}\, U \wedge z \in Q \times R \Rightarrow \bot$$

together with similar clauses expressing the impossibility of values common to both concrete and abstract types, such as $z \in \mathbb{N} \wedge z \in \mathbb{P}\, U \Rightarrow \bot$. This reflects the intention of keeping pairs, functions, and sets distinct from other values in our models.

Let $\overline{\mathcal{F}}$ (resp. $\overline{\mathcal{F}_t}$) be the set composed of all user-defined and predefined constants $\mathcal{F}$ (resp. $\mathcal{F}_t$) and of two binary operators, namely application $\cdot(\cdot)$ and pairing $(\cdot,\cdot)$. $\mathcal{T}(\overline{\mathcal{F}}, \mathcal{X})$ denote the set of terms built from the set of variables $\mathcal{X}$ and $\overline{\mathcal{F}}$. The set of *type terms* $\mathcal{T}(\overline{\mathcal{F}_t}, \mathcal{X}_t)$ is similarly built from $\overline{\mathcal{F}_t}$.

Notice that whereas a specification *SP* involves declarations of *typed* symbols, the declarations $(\mathcal{F}, \mathcal{X})$ of the corresponding abstract presentation are *untyped*, and the original type information has been replaced by membership axioms. Thus standard results and tools pertaining to *unsorted* Horn clause logic may be applied to our presentations. Nevertheless, as we shall see in the next section, we can still check that our presentations are well-typed.

Substitutions are mappings from $\mathcal{X}$ to $\mathcal{T}(\overline{\mathcal{F}}, \mathcal{X})$, mapping variables to terms, such that type variables are mapped only to type terms. They are denoted by Greek letters. $SUBST(\mathcal{F}, \mathcal{X})$ denotes the set of such substitutions defined on $\mathcal{T}(\overline{\mathcal{F}}, \mathcal{X})$. The domain of σ, written $\mathcal{D}om(\sigma)$, is defined as the set $\{x \in \mathcal{X} \mid \sigma(x) \neq x\}$, and is assumed finite.

4 Type Checking

In this and the following sections we assume given a fixed presentation $\mathcal{P} = (\mathcal{F}, \mathcal{X}, \mathcal{M}, \mathcal{H})$, determined by some specification as explained in Section 3.

Let us introduce type judgements of the form $t : T$, where t is any term but T is restricted to type terms. Our deduction rules for such judgements are given in Table 7. For each membership axiom of the form $t \in T$ in $\mathcal{M}$, we have $t : T$ as an axiom too. However, note that $t \in T$ is only a consequence of $t : T$ when the value of t is defined. The axioms **Pair**, **Appl** in Table 7 reflect that the judgement $t : T$ is merely a decidable approximation to $t \in T$, disregarding whether the values of terms are defined, in contrast to the corresponding clauses given in Table 4 above. Note that we deliberately leave out the potential typing axiom: $x : T \wedge T : \mathbb{P}\, T' \Rightarrow x : T'$. Thinking of $T : \mathbb{P}\, T'$ as a type inclusion $T \subseteq T'$, such an axiom would merely let us deduce larger types from smaller ones.

Definition 1. *A term t is well-typed if one can deduce with the typing deduction rules in Table 7 that $t : T$. Then T is called a type of t.*

Table 7. Deduction rules for type-checking

Axioms:	$t : T$ if $(t \in T)$ is in $\mathcal{M}$
Pair:	$x : X \wedge y : Y \Rightarrow (x, y) : (X \times Y)$
Appl:	$f : X \rightarrow\!\!\!\!+ Y \wedge x : X \Rightarrow f(x) : Y$
Subst:	$\dfrac{G \Rightarrow L}{\sigma(G) \Rightarrow \sigma(L)}$ if σ is in $SUBST(\mathcal{F}, \mathcal{X})$
Cut:	$\dfrac{G \wedge L' \Rightarrow L, \quad G' \Rightarrow L'}{G \wedge G' \Rightarrow L}$

Due to overloading of constants, and to substitutions for type variables in polymorphic types, a term may have many types. Checking whether a term t is well-typed amounts to find solutions of the goal $t : X$ where X is a variable, in the Horn theory given by the clauses in Table 7, which can be done by resolution.

Proposition 1. *Assuming that the declarations in a presentation provide a finite number of membership axioms for each constant and variable used, there exists a procedure to enumerate the set of types of any term. This procedure terminates, so it is decidable to check whether a term is well-typed.*

Proof: By resolution in the Horn clause theory defined by the axioms of Table 7, we can enumerate the solutions of a goal $t : X$ where X is a variable. In this process, variables in t are considered as constants and cannot be instantiated. The procedure terminates since there is a finite number of typing axioms for constants, and, in each resolution step using a non-atomic clause, the structural complexity of terms to be typed strictly decreases. □

5 Proof System

The proof rules shown in Table 8 are for a deduction relation $\mathcal{P} \vdash \Phi$ taking two arguments: a presentation $\mathcal{P}$ and a formula Φ. $\mathcal{P} \vdash \Phi$ means that one can derive Φ by applying rules in Table 8.

The same set of formulas could be deduced with a slightly different set of deduction rules where **Subst**, **Cut** and **Paramod** are replaced by a resolution rule and a paramodulation rule involving unification of atomic formulas and terms.

The meta-variables G, G' range over possibly-empty conjunctions of atomic formulae (the empty conjunction is written $\top$) and L, L' may be a single atomic formula or $\bot$. Recall that an atom L is identified with the Horn clause $\top \Rightarrow L$.

Thanks to the paramodulation rule **Paramod**, it is possible to deduce in this proof system formulas with apparently non well-typed terms. Let us consider types A, B, C, membership declarations $a \in A$, $a \in B$, $b \in B$, $b \in C$, $f \in A \rightarrow\!\!\!\!+ A$

Table 8. Deduction rules for presentations

WellDef:	$\frac{L[t]}{t\downarrow}$ if $L[t]$ is an atom containing t as a sub-term
PartialRef:	$\frac{t\downarrow}{t=t}$
Axioms:	$G \Rightarrow L$ if $(G \Rightarrow L)$ is in $\mathcal{H} \cup \mathcal{M}$
Subst:	$\frac{G \Rightarrow L,\quad \sigma(x)\downarrow \textit{ for all } x \text{ in } \mathcal{D}om(\sigma)}{\sigma(G) \Rightarrow \sigma(L)}$ if σ is in $SUBST(\mathcal{F},\mathcal{X})$
Cut:	$\frac{G \wedge L' \Rightarrow L,\quad G' \Rightarrow L'}{G \wedge G' \Rightarrow L}$
Paramod:	$\frac{G \Rightarrow L[s],\quad G' \Rightarrow (s=t)}{G \wedge G' \Rightarrow L[t]}$

and an equality $a = b$. We get $f(a) : A$ and $f(a) \in A$, then $f(b) \in A$ using **Paramod**, although $f(b)$ is not well-typed.

The next propositions state relations between type judgements and formulas deduced in the proof system.

Proposition 2. *Let $\mathcal{P}$ be a presentation, and T any* type *term in $\mathcal{T}(\overline{\mathcal{F}},\mathcal{X})$ other than a variable. Then $\mathcal{P} \vdash T\downarrow$.*

Proof: (Sketch) Existence follows from the memberships in $\mathcal{M}$ determined by type declarations, using the rules **WellDef** and **Subst.** □

Proposition 3. *Let $\mathcal{P}$ be a presentation, and t, T any terms in $\mathcal{T}(\overline{\mathcal{F}},\mathcal{X})$. Suppose $\mathcal{P} \vdash t : T$ (from Table 7), as well as $\mathcal{P} \vdash t\downarrow$ (from Table 8). Then $\mathcal{P} \vdash t \in T$.*

Proof: (Sketch) Any proof of $t : T$ can be converted into a proof of $t \in T$, using **WellDef** to provide the required existence of subterms. □

6 Models

We are interested primarily in standard set-theoretic models, where values of set-types are actually ordinary sets. But in order to obtain our initiality results, we shall consider other classes of models as well.

6.1 Algebraic Models

The first class of models we consider is a class of algebraic models $\mathcal{I}$, namely unsorted partial first-order structures. In these models, $=$ is interpreted as identity,

$\downarrow$ as existence of a value, $\in$ by a binary relation $\in^{\mathcal{I}}$, and constants c by values $c^{\mathcal{I}}$. Homomorphisms are classical ones that, by construction, preserve type values and definedness.

Formula satisfaction is written $\mathcal{I} \models \Phi$, or $\mathcal{P} \models \Phi$ in case all models of $\mathcal{P}$ satisfy the formula Φ. Let $\mathbf{Alg}(\mathcal{P})$ denote the class of all partial $\overline{\mathcal{F}}$-structures that satisfy all formulae in $\mathcal{P}$.

Since axioms in $\mathcal{M} \cup \mathcal{H}$ of a presentation $\mathcal{P}$ are Horn clauses, the class of unsorted partial $\overline{\mathcal{F}}$-structures satisfying $\mathcal{M} \cup \mathcal{H}$ has initial models (provided that the set of axioms is consistent).

Let $\mathcal{T}(\mathcal{P})$ denote the initial $\overline{\mathcal{F}}$-structure of $\mathbf{Alg}(\mathcal{P})$. It may be constructed from equivalence classes $[t]$ of those terms whose existence follows from $\mathcal{P}$, where two terms t_1, t_2 are deemed equivalent when their equality follows from $\mathcal{P}$.

Theorem 1. *Let $\mathcal{P}$ be a presentation. The deduction rules in Table 8 are sound and complete, i.e. for any atomic formula ψ, $\mathcal{P} \models \psi$ iff $\mathcal{P} \vdash \psi$.*

Proof: Since $\mathcal{T}(\mathcal{P})$ is an algebraic model, if $\mathcal{P} \models \psi$, then $\mathcal{T}(\mathcal{P}) \models \psi$. By definition of $\mathcal{T}(\mathcal{P})$, for any atomic formula ψ, $\mathcal{T}(\mathcal{P}) \models \psi$ iff $\mathcal{P} \vdash \psi$. □

The following definition plays an important rôle in relating algebraic models to models with set-theoretic structure:

Definition 2. *Let $\mathcal{I}$ be an $\overline{\mathcal{F}}$-structure in* $\mathbf{Alg}(\mathcal{P})$. *The* set-like *values of $\mathcal{I}$ are those values s in $\mathcal{I}$ such that $s \in^{\mathcal{I}} \mathbb{P}^{\mathcal{I}}(v)$ for some value v in $\mathcal{I}$.*

It follows from Proposition 3 that in a model of a presentation, the value of any ground term of type $\mathbb{P}\ T$ for some T is set-like, when defined.

6.2 Labelled-Sets Models

Now we restrict the class of algebraic models by imposing that set-like values (cf. Definition 2) are interpreted as labelled sets of values. A labelled set is denoted by S^l where ^{l} is the label.

Two labelled sets are equal if their underlying sets are equal and their labels are identical. The membership predicate $\in$ is interpreted by ordinary membership between values and the underlying sets, thus ignoring the label. Let $\mathbf{LSAlg}(\mathcal{P})$ be the class of all labelled-set models that satisfy $\mathcal{P}$. From $\mathcal{T}(\mathcal{P})$, let us now build a labelled-set model $\mathcal{S}(\mathcal{P})$ isomorphic to $\mathcal{T}(\mathcal{P})$. First let us choose the set of labels as the set of equivalence classes of $\mathcal{T}(\mathcal{P})$. Then the idea is to associate to any term $t : \mathbb{P}(T)$ a labelled set with label $[t]$, and whose elements are values that are provably members of t.

Let us now formally define $\mathcal{S}(\mathcal{P})$. The carrier of $\mathcal{S}(\mathcal{P})$ contains equivalence classes of $\mathcal{T}(\mathcal{P})$ as atomic objects and labelled sets of values. Let us consider the mapping $h : \mathcal{T}(\mathcal{P}) \to \mathcal{S}(\mathcal{P})$ defined as follows: if $t : \mathbb{P}(T)$, then $h([t]) = \{h([u]) \mid \mathcal{P} \vdash u \in t\}^{[t]}$, else $h([t]) = [t]$.

Proposition 4. *In $\mathcal{S}(\mathcal{P})$, two labelled sets are equal iff their labels are equal.*

Proof: If two labelled sets are equal, their labels are equal by definition. Conversely assume that two labels are equal: $[t_1] = [t_2]$, i.e. $\mathcal{P} \vdash t_1 = t_2$. Let us define $s_1 = \{h([u]) \mid \mathcal{P} \vdash u \in t_1\}$ and $s_2 = \{h([v]) \mid \mathcal{P} \vdash v \in t_2\}$. For all $h([u])$ in s_1, from $u \in t_1$ and $t_1 = t_2$, $u \in t_2$ is deducible by **Paramod**. So $h([u])$ is in s_2 for all u, hence s_1 is included in s_2. Similarly, s_2 is included in s_1. □

Thus h is an isomorphism and we get:

Proposition 5. *$\mathcal{S}(\mathcal{P})$ is initial in the class of labelled set models* **LSAlg**$(\mathcal{P})$.

Proof: (Sketch) $\mathcal{S}(\mathcal{P})$ is isomorphic to $\mathcal{T}(\mathcal{P})$ which is initial in the class of algebraic models of $\mathcal{P}$. $\mathcal{S}(\mathcal{P})$ is also a labelled set model and is thus initial in the subclass of labelled set models. □

6.3 Set-Theoretic Models

Let us finally consider the main class of models of interest here, namely set-theoretic models.

A set-theoretic model is one where all set-like values (cf. Definition 2) are ordinary, unlabelled sets. The interpretation of $=$ and $\in$ is now exactly as in standard set theory. Let **SAlg**$(\mathcal{P})$ be the class of all standard set-theoretic models that satisfy $\mathcal{P}$.

Set-theoretic models can be obtained from labelled-set models by forgetting the labels on the sets. This may however map two different labelled sets to the same unlabelled set. In particular, the set-theoretic model obtained by forgetting labels from $\mathcal{S}(\mathcal{P})$ need not be initial in **SAlg**$(\mathcal{P})$. E.g., suppose that $\mathcal{P}$ declares two constants of set type, but does not require them to have any members at all, so they are interpreted as distinctly-labelled empty sets in $\mathcal{S}(\mathcal{P})$; forgetting the labels identifies the interpretations of the two constants, preventing homomorphisms to those set-theoretic models of $\mathcal{P}$ where they have distinct members.

Suppose however that one forgets not only the labels, but also the sets themselves! To do this, let us first define the following subset of type terms:

Definition 3. *The* $\mathbb{P}$-less types *are those type terms that do not contain any occurrence of the concrete type-constructor* $\mathbb{P}$ *at all.*

Let us consider then the restriction of $\mathcal{F}$ by keeping only those constants that have a declared $\mathbb{P}$-less type; call the result $\mathcal{G}$. Notice that we hereby eliminate also our predefined type constructors from $\mathcal{F}$ in $\mathcal{G}$. Now taking any model of $\overline{\mathcal{F}}$ and retaining only those values that are members of $\mathbb{P}$-less types gives us an ordinary algebraic model for $\overline{\mathcal{G}}$. This removes not only all the set-like values, but also pairs of set-like values, and functions that return set-like values.

Even though this removal of all the values involving sets may seem rather drastic, the important point is that we retain all values of all types such as $\mathbb{N}$, $\mathbb{N} \times \mathbb{N}$, $Seq(\mathbb{N})$, etc. We even retain the values of function types, e.g., $succ \in \mathbb{N} \times \mathbb{N} \nrightarrow \mathbb{N}$ (since we have avoided identifying functions with their graphs, and since the existence of a value does not require the existence of a set that contains it—in contrast to many-sorted frameworks, where forgetting a sort requires forgetting all operations whose profiles include it).

Now let $\mathcal{R}(\mathcal{P})$ be the algebraic $\mathcal{G}$-model obtained as a reduct of $\mathcal{S}(\mathcal{P})$ by keeping only values of $\mathbb{P}$-less types.

Proposition 6. *$\mathcal{R}(\mathcal{P})$ is initial in the class of all $\mathcal{G}$-models arising as reducts of models in* $\mathbf{SAlg}(\mathcal{P})$.

Proof: (Sketch) We show that the unique homomorphism h from $\mathcal{S}(\mathcal{P})$ to an arbitrary $\mathcal{L}$ in $\mathbf{LSAlg}(\mathcal{P})$ cuts down to a unique homomorphism from $\mathcal{R}(\mathcal{P})$ to the reduct of $\mathcal{L}$. Let x be any value of $\mathcal{R}(\mathcal{P})$; then there exists a $\mathbb{P}$-less type term T such that $x \in^{\mathcal{S}(\mathcal{P})} T^{\mathcal{S}(\mathcal{P})}$ holds. Therefore $h(x) \in^{\mathcal{L}} h(T^{\mathcal{S}(\mathcal{P})}) = T^{\mathcal{L}}$, so $h(x)$ is in the reduct of $\mathcal{L}$. Thus h cuts down to a homomorphism between the reducts; a simple induction proves uniqueness. □

We leave it to future work to find sufficient syntactic restrictions on specifications such that $\mathcal{R}(\mathcal{P})$ is unaffected by the labels on sets, i.e., adding equations between sets to the clauses leaves $\mathcal{R}(\mathcal{P})$ unchanged.

7 Algebraic Specification of Set Theory

Our predefined notation for sets consists merely of the function constants $\_\times\_$, $\_\nrightarrow\_$, $\mathbb{P}\_$, *first*, *second*, together with the atomic formulae for membership and equality. But we may go much further. We have investigated an extension of our framework to allow subset inclusions as atomic formulae. When used as a declaration in the head of a specification, an inclusion may indicate a subtype relationship, much as in order-sorted algebra [1]; when used as a clause, it corresponds to a membership implication—but the inclusion may also be used as a condition in a clause, in contrast to the implication. The essential properties of inclusion are that it is a partial order, preserved by most operations. Polymorphic type declarations are generalized to allow both monotonic and anti-monotonic dependency on the type variables, giving rise to corresponding axioms for set arguments. The definition of a labelled set has also to be generalized to take account of an inclusion order on the labels themselves, but in the set-theoretic models the interpretation of inclusion is completely standard.

Further familiar set-theoretic notation can be specified using our framework: sets of relations and total functions, set union and intersection, singletons and (finite) set comprehensions. Relations are represented simply as partial functions to a singleton set; since homomorphisms preserve definedness, they also preserve the holding of relations. A set of total functions between two sets is a subset of the partial functions between the same sets, but not necessarily containing all such functions that happen to be total; a total-function type constructor is provided so that each function required to be total can simply be declared as such. Union and intersection are specified to have the properties of a distributive lattice, as in unified algebras [6]. A finite set comprehension merely lists the elements that are its members, leaving it to initiality to ensure that there are no further members.

Our intention, however, is not to try to provide a full "Mathematical Toolkit" like the ones available for Z and B, but rather to test the limits of our modest Horn-clause specification framework. It is admittedly harder to work out the various Horn-clause properties of an operation like union, instead of defining it extensionally using a disjunctive formula. But such properties are perhaps often needed in proofs about sets in any case, so work in this direction should be of significant practical relevance.

8 Conclusion

We have presented a framework for algebraic specifications with higher-order types and set-theoretic models. It embodies significant simplifications, compared to our original proposal for such a framework [2]. We have studied three classes of models (algebraic, labelled-set, and standard set-theoretic) and obtained initiality results. We refer to[2] for a discussion of the connections between our original proposal and such frameworks as R^n/G^n logics, ETL, and unified algebras. Let us here consider the relationship between our work and two other frameworks that have been developed in recent years: Meinke's higher-order initial algebra specifications, and Meseguer's first-order Membership Equational Logic.

Meinke [3, 4] has studied the theoretical properties and practical applications of a different framework with higher-order (non-polymorphic) types and set-theoretic models. Although product and (total) function types are allowed, power-set types are not, and it appears that it would not be easy to incorporate them [3, page 388]. The lack of power-set types precludes considering set membership in formulae, and the use of types as values. Higher-order algebras with pairing and application operations are considered. The values of the product and function types are the standard set-theoretic objects, but, as in our own framework, the types themselves are allowed to be subsets of the usual types. Higher-order algebras are shown to be isomorphic to first-order algebras that satisfy the usual extensionality axiom for functions. Specified axioms are restricted to conditional equations involving terms of base type. Meinke provides a complete finitary (conditional) equational logic for the class of all extensional models, but needs an infinitary logic for the construction of initial models as quotients of term models. In contrast to Meinke, we have focussed on the power-set type, and kept to a finitary Horn-clause logic, getting a useful expressiveness but keeping tractability by not requiring extensional equality of sets.

Meseguer [5] has developed Membership Equational Logic (MEL), a first-order algebraic specification framework with set-theoretic models. The types are abstract types (called kinds), first-order n-ary total function types (for declaring operations), and first-order power-set types (called sorts); positive conditional clauses are allowed as axioms. Subsort inclusions can be declared. Atomic formulae are equations between values of the same kind, and memberships of such values in sorts. Equality between sets (or functions) is not directly expressible, so extensionality is not an issue for obtaining a complete proof system and initial models.

In an extended version of the present paper we intend to explain our treatment of subtype declarations, and to give specifications of further familiar set-theoretic concepts and notations. We will also investigate sufficient conditions for the independence of reducts and labels on sets. We intend also to clarify the relationship between our framework and MEL. We believe that any MEL specification can be straightforwardly translated into a presentation $\mathcal{P}$ in our framework, such that the initial model in MEL corresponds exactly to our initial labelled-set model $\mathcal{S}(\mathcal{P})$. The image of this translation would identify a sub-framework where we could exploit the term-rewriting techniques that have already been developed for MEL. It would be interesting then to see how far these techniques could be extended to allow more general (e.g., partial, higher-order) specifications. Finally, along the same lines as in our previous paper [2], we will further consider the use of saturation techniques for obtaining a refutationally-complete automatic theorem-prover for consequences of our specifications.

Acknowledgements. The authors are grateful to the anonymous referees for helpful comments. Hélène Kirchner has been partially supported by the Esprit Basic Research Working Group 22457 - Construction of Computational Logics II. Peter D. Mosses has been supported by BRICS (Centre for Basic Research in Computer Science), established by the Danish National Research Foundation in collaboration with the Universities of Aarhus and Aalborg, Denmark; by an International Fellowship from SRI International; and by DARPA-ITO through NASA-Ames contract NAS2-98073.

References

[1] J. A. Goguen and J. Meseguer. Order-sorted algebra I: Equational deduction for multiple inheritance, overloading, exceptions and partial operations. *Theoretical Computer Science*, 105:217–273, 1992.

[2] C. Hintermeier, H. Kirchner, and P. D. Mosses. Combining algebraic and set theoretic specifications. In M. Haveraaen, O. Owe, and O.-J. Dahl, editors, *Recent Trends in Data Type Specification, Proc. 11th Workshop on Specification of Abstract Data Types joint with the 9th general COMPASS workshop. Oslo, Norway, September 1995. Selected papers*, volume 1130 of *Lecture Notes in Computer Science*, pages 255–273. Springer-Verlag, 1996.

[3] K. Meinke. Universal algebra in higher types. *Theoretical Computer Science*, 100:385–417, 1992.

[4] K. Meinke. Higher-order equational logic for specification, simulation and testing. In *HOA'95, Proc. Second Int. Workshop on Higher-Order Algebra, Logic and Term Rewriting*, volume 1074 of *Lecture Notes in Computer Science*, pages 124–143. Springer-Verlag, 1996.

[5] J. Meseguer. Membership algebra as a semantic framework for equational specification. In *Recent Trends in Algebraic Development Techniques*, volume 1376 of *Lecture Notes in Computer Science*, pages 18–61. Springer-Verlag, 1998.

[6] P. D. Mosses. Unified algebras and institutions. In *Proceedings 4th IEEE Symposium on Logic in Computer Science, Pacific Grove*, pages 304–312, 1989.

Type Analysis for CHIP*

Włodzimierz Drabent[1] and Paweł Pietrzak[2]

[1] IPI PAN, Polish Academy of Sciences, Ordona 21, Pl - 01-237 Warszawa and IDA, Linköpings universitet
[2] IDA, Linköpings universitet, S - 581 83 Linköping, Sweden.
{wlodr,pawpi}@ida.liu.se.

Abstract. This paper proposes a tool to support reasoning about (partial) correctness of constraint logic programs. The tool infers a specification that approximates the semantics of a given program. The semantics of interest is an operational "call-success" semantics. The main intended application is program debugging. We consider a restricted class of specifications, which are regular types of constrained atoms.
Our type inference approach is based on bottom-up abstract interpretation, which is used to approximate the declarative semantics (c-semantics). By using "magic transformations" we can describe the call-success semantics of a program by the declarative semantics of another program. We are focused on CLP over finite domains. Our prototype program analyzer works for the programming language CHIP.

1 Introduction and Motivation

In this paper we are interested in supporting reasoning about program correctness in the context of CLP (constraint logic programming). Speaking informally, a program is correct if it behaves as expected by the user. But user expectations are seldom well documented. This paper describes an analyzer that for a given CLP program produces a characterization of the form of calls and successes in any execution of the program starting from a given class of goals. The user may inspect the description produced to see whether it conforms to her expectations. We deal with partial correctness, the given program is partially correct w.r.t. the obtained description.

The starting point are well-known verification conditions for partial correctness of logic programs wrt to a specification, which gives a set of procedure calls and a set of procedure successes. (Such verification conditions were proposed in [DM88, Dra88]; a useful special case was given in [BC89, AM94]). We generalize the verification conditions for the case of CLP.

Generally the conditions are undecidable. But they become decidable for a restricted class of specifications. For the case of LP (logic programming) it was shown [Boy96] that it is sufficient to consider specifications describing regular tree sets. In the literature this kind of specifications is often called regular types [YS91, DZ92]. While successes and calls in LP are atoms, their counterpart in

* This work has been supported by the ESPRIT 4 Project 22532 DiSCiPl.

A.M. Haeberer (Ed.): AMAST'98, LNCS 1548, pp. 389–405, 1998.

CLP are constrained atoms. Therefore this paper adapts regular types for CLP so that one can describe sets of constrained terms and atoms. This includes adaptation of certain operations on regular types.

To compute semantic approximations of programs, we need static analysis techniques. We show that the verification conditions for a CLP program P constitute another CLP program Q whose declarative semantics describes the calls and successes. (Such approach is often called "magic transformation"). For this purpose we introduce a generalization for CLP of c-semantics [Cla79, FLMP89]; this results in more precise descriptions than using the standard $\mathcal{D}$-model semantics. We adapt then the technique of Gallagher and de Waal [GdW92, GdW94] of bottom-up abstract interpretation to synthesize an approximation of the c-semantics of Q; it also is an approximation of the call-success semantics of P. As a side effect we obtain a tool to approximate the declarative semantics of CLP programs.

We are particularly interested in CLP over finite domains (CLP(FD)) [Hen89], especially the language CHIP [Cos96]. We have implemented a prototype type analysis system for CHIP. It is a major modification of the system described in [GdW92, GdW94]. A preliminary version of our work was presented in [DP98b].

The use of types, as in our work, to approximate the semantics of programs in an untyped language is usually called *descriptive* typing. Another approach is *prescriptive* typing. In that approach the type information, provided by the programmer, influences the semantics of a program. In particular, variables are typed and may only be bound to the values from the respective types. Usually the programmer is required to provide types for function symbols and/or for predicates. Prescriptive typing is a basis of a few programming languages (e.g. TypedProlog [LR91], Gödel [HL94], Mercury [SHC96]).

Experience with languages like Gödel shows that their mechanism of types is able to find numerous errors at compile time. This is an immense advantage in comparison to finding them during testing and run-time debugging. Our work adds a similar potential of static checking to any typeless CLP language (by comparing the types obtained from the analysis with the intended ones).

The paper is organized as follows. The next section summarizes basic concepts of CLP and presents the declarative and the operational semantics. Then we propose a system of regular types for CLP. Section 4 describes the type inference method used in this work. Then we present an example of type analysis for CHIP.

2 Semantics of CLP

In this work we employ two semantics of CLP. We need a semantics providing information about the form of procedure calls and successes during the execution of CLP programs; this is the role of a call-success semantics. The analysis method employs magic transformation, so we also need a declarative semantics. Both semantics are introduced below in this section.

Most of implementations of CLP use syntactic unification[1]. In this paper we are interested in CLP with syntactic unification, we believe however that our work can be adapted to the "standard" CLP.

2.1 Basic Concepts

We consider a fixed constraint domain. It is given by fixing a signature and a structure $\mathcal{D}$ over this signature. Predicate symbols of the signature are divided into *constraint predicates* and (non-constraint) *predicates*. The former have a fixed interpretation in $\mathcal{D}$, the interpretation of the latter is defined by programs. Similarly, the function symbols are divided into *interpreted function symbols* and *constructors*. All the function symbols have a fixed interpretation. It is assumed that the interpretations of constructors are bijections with disjoint co-domains. So the elements of structure $\mathcal{D}$ can be seen as terms built from some elementary values by means of constructors[2]. That is why we will often call them $\mathcal{D}$-terms. An *atomic constraint* is an atomic formula with a constraint predicate symbol. Throughout this paper by a *constraint* we will mean an atomic constraint or $c_1 \wedge c_2$ or $c_1 \vee c_2$ or $\exists x c_1$, where c_1 and c_2 are constraints and x is a variable. A CLP clause is of the form: $h \leftarrow c, b_1, \ldots, b_n$ where $h, b_1, \ldots, b_n$ are atoms (i.e. atomic formulae built up from non-constraint predicate symbols) and c is a conjunction of atomic constraints. A CLP program is a finite set of CLP clauses.

2.2 Declarative Semantics

The standard least $\mathcal{D}$-model semantics is insufficient for our purposes. We are interested in the actual form of computed answers[3]. Two programs with the same least $\mathcal{D}$-model semantics may have different sets of computed answers. For instance take the following two CLP(FD) programs

$$P_1 = \{p(1).;\ p(2).\} \qquad P_2 = \{p(x) \leftarrow x \in \{1,2\}.\}$$

and a goal $p(x)$. Constraint $x \in \{1,2\}$ is an answer for P_2 but not for P_1. In order to describe such differences, we generalize the c-semantics [Cla79, FLMP89]. For logic programs, this semantics is given by the set of (possibly non ground) atomic logical consequences of a program. The c-semantics for CLP will be expressed by means of constrained atoms.

---

[1] In CLP with syntactic unification, function symbols occurring outside of constraints are treated as constructors. So, for instance in CLP over integers, the goal $p(4)$ fails with the program $\{p(2{+}2) \leftarrow\}$, but the goal $X \#{=} 4, p(X)$ succeeds (where $\#{=}$ is the constraint of arithmetical equality).

[2] Notice that in many CLP languages function symbols play also the role of constructors. For instance, the interpretation of $2+3$ may be a number, while that of $a+3$ (where a is a 0-ary constructor) is a $\mathcal{D}$-term with the main symbol $+$.

[3] $\mathcal{D}$-model semantics can be used to describe CLP with syntactic unification, one has to made $\mathcal{D}$ to be a a Herbrand domain. (No element of the carrier of such a domain is a value of two distinct ground terms).

Definition 1. A *constrained expression* (atom, term, ...) is a pair $c[]E$ of a constraint c and an expression E such that each free variable of c occurs (freely) in E.

If ν is a valuation such that $\mathcal{D} \models \nu(c)$ then $\nu(E)$ is called an $\mathcal{D}$-*instance* of $c[]E$.

A constrained expression $c'[]E'$ is an *instance* of a constrained expression $c[]E$ if c' is satisfiable in $\mathcal{D}$ and there exists a substitution θ such that $E' = E\theta$ and $\mathcal{D} \models c' \rightarrow c\theta$ ($c\theta$ means here applying θ to the free variables of c, with a standard renaming of the non-free variables of c if a conflict arises).

If $c[]E$ is an instance of $c'[]E'$ and vice versa then $c[]E$ is a *variant* of $c'[]E'$

By the *instance-closure* $cl(E)$ of a constrained expression E we mean the set of all instances of E. For a set S of constrained expressions, its instance-closure $cl(S)$ is defined as $\bigcup_{E \in S} cl(E)$.

Note that, in particular, $c\theta[]E\theta$ is an instance of $c[]E$ and that $c'[]E$ is an instance of $c[]E$ whenever $\mathcal{D} \models c' \rightarrow c$. The relation of being an instance is transitive. (Take an instance $c'[]E\theta$ of $c[]E$ and an instance $c''[]E\theta\sigma$ of $c'[]E\theta$. As $\mathcal{D} \models c'' \rightarrow c'\sigma$ and $\mathcal{D} \models c' \rightarrow c\theta$, we have $\mathcal{D} \models c'' \rightarrow c\theta\sigma$).

Notice also that if c is not satisfiable then $c[]E$ does not have any instance (it is not an instance of itself).

We will often not distinguish E from $true[]E$ and from $c[]E$ where $\mathcal{D} \models \forall c$. Similarly, we will also not distinguish $c[]E$ from $c'[]E$ when c and c' are equivalent constraints ($\mathcal{D} \models c \leftrightarrow c'$).

Example 2. $a + 7$, $Z + 7$, $1{+}7$ are instances of $X + Y$, but 8 is not.

$f(X){>}3[]f(X){+}7$ is an instance of $Z{>}3[]Z{+}7$, which is an instance of $Z + 7$, provided that constraints $f(X){>}3$ and $Z{>}3$, respectively, are satisfiable.

Assume a numerical domain with the standard interpretation of symbols. Then $4 + 7$ is an instance of $X{=}2{+}2[]X{+}7$ (but not vice versa), the latter is an instance of $Z{>}3[]Z{+}7$.

Consider CLP(FD) [Hen89]. A domain variable with the domain S, where S is a finite set of natural numbers, can be represented by a constrained variable $x{\in}S\,[]\,x$ (with the expected meaning of the constraint $x{\in}S$).

If $Vars(c) \not\subseteq Vars(E)$ then $c[]E$ will denote $(\exists_{-Vars(E)}c)[]E$ (where $\exists_{-V}$ stands for quantification over the variables not in V).

Two notions of groundness arise naturally for constrained expressions. $c[]E$ is *syntactically ground* when E contains no variables. $c[]E$ is *semantically ground* if it has exactly one $\mathcal{D}$-instance.

Now we define the c-semantics for CLP with syntactic unification. In the next definition we apply substitutions to program clauses. So let us define $\downarrow P$ as $\{\, C\theta \mid C \in P,\ \theta \text{ is a substitution} \,\}$.

Definition 3 (Immediate consequence operator for c-semantics). Let P be a CLP program. $T_P^{\mathcal{C}}$ is a mapping over sets of constrained atoms, defined by

$$
\begin{aligned}
T_P^{\mathcal{C}}(I) = \{\ c[]h \mid\ & (h \leftarrow c', b_1, \ldots, b_n) \in {\downarrow}P,\ n \geq 0, \\
& c_i[]b_i \in I, \ \text{ for } i = 1, \ldots, n, \\
& c = \exists_{-\,Vars(h)}(c', c_1, \ldots, c_n), \\
& \mathcal{D} \models \exists c \ \ \}
\end{aligned}
$$

(where $Vars(E)$ is the set of free variables occurring in E).

Notice that in the definition syntactic unification is used for parameter passing, but terms occurring in constraints are interpreted w.r.t. $\mathcal{D}$.

$T_P^{\mathcal{C}}$ is continuous w.r.t. $\subseteq$. So it has the least fixpoint $T_P^{\mathcal{C}} \uparrow \omega = \bigcup_{i=0}^{\infty} (T_P^{\mathcal{C}})^i(\emptyset)$. By the *declarative semantics* (or *c-semantics*) $M(P)$ of P we mean the instance-closure of the least fixpoint of $T_P^{\mathcal{C}}$:

$$M(P) = cl(T_P^{\mathcal{C}}{\uparrow}\omega).$$

Speaking informally, cl is used here only to add new constraints but not new (non-constraint) atoms: As $T_P^{\mathcal{C}}{\uparrow}\omega$ is closed under substitution, for every $c[]u \in M(P)$ there exists a $c'[]u \in T_P^{\mathcal{C}}{\uparrow}\omega$ such that $\mathcal{D} \models c \rightarrow c'$.

Example 4. Consider programs P_1 and P_2 from the beginning of this section. $M(P_1) = \{p(1), p(2)\}$. $T_{P_2}^{\mathcal{C}}{\uparrow}\omega$ contains $p(1)$, $p(2)$ and $x{\in}\{1,2\}[]p(x)$. (It also contains variants of the latter constrained atom, obtained by renaming variable x). $M(P_2)$ contains additionally all the instances of $x{\in}\{1,2\}[]p(x)$, like $y{=}1[]p(y)$.

The traditional least $\mathcal{D}$-model semantics and the c-semantics are related by the fact that the set of $\mathcal{D}$-instances of the elements of $M(P)$ is a subset of the least $\mathcal{D}$-model of P. If we take a least $\mathcal{D}$-model semantics for CLP with syntactic unification (where $\mathcal{D}$ is a Herbrand domain) then the set of $\mathcal{D}$-instances of the elements of $M(P)$ and the least $\mathcal{D}$-model of P coincide.

2.3 Call-Success Semantics

We are interested in the actual form of procedure calls and successes that occur during the execution of a program. We assume the Prolog selection rule. Such semantics will be called the *call-success semantics.*

Without loss of generality we can restrict ourselves to atomic initial goals. Given a program and a class of initial goals, we want to provide two sets of constrained atoms corresponding to the calls and to the successes. For technical reasons it is convenient to have just one set. So for each predicate symbol p we introduce two new symbols ${}^\bullet p$ and $p^\bullet$; we will call them *annotated predicate symbols.* They will be used to represent, respectively, call and success instances of atoms whose predicate symbol is p. For an atom $A = p(\tilde{t})$, we will denote ${}^\bullet p(\tilde{t})$ and $p^\bullet(\tilde{t})$ by ${}^\bullet A$ and $A^\bullet$ respectively. We will use analogous notation for constrained atoms. (If $A = c[]p(\tilde{t})$ then ${}^\bullet A = c[]{}^\bullet p(\tilde{t})$, etc).

The call-success semantics is defined in terms of the computations of the program. For a given operational semantics, which specifies what the computations of a program are, one defines what are the *procedure calls* and the *procedure successes* of these computations. For logic programs and LD-resolution this is done for instance in [DM88]. It is rather obvious how to generalize it to CLP, we omit the details.

Definition 5. Let P be a CLP program and $\mathcal{G}$ a set of constrained atoms. Their *call-success semantics* $CS(P, \mathcal{G})$ is a set of constrained atoms (with annotated predicate symbols) such that

1. $c[]^{\bullet}p(\tilde{t}) \in CS(P, \mathcal{G})$ iff there exists an LD-derivation for P with the initial goal in $\mathcal{G}$ and in which $c[]p(\tilde{t})$ is a procedure call;
2. $c[]p^{\bullet}(\tilde{t}) \in CS(P, \mathcal{G})$ iff there exists an LD-derivation for P with the initial goal in $\mathcal{G}$ and in which $c[]p(\tilde{t})$ is a procedure success.

We will characterize the call-success semantics of a program P as the declarative semantics of some other program P^{CS}. In logic programming this approach is often called "magic transformation". Program P^{CS} can also be viewed as the verification conditions of the proof method of [BC89] or an instance of the verification conditions of the proof method of [DM88].

Proposition 6. Let P be a CLP program and $\mathcal{G}$ a set of constrained atoms. Then

$$cl(CS(P, \mathcal{G})) = cl\left((T^{\mathcal{C}}_{P^{CS}})^{\omega}(\mathcal{G})\right)$$

where P^{CS} is a program that for each clause $H \leftarrow c, B_1, \ldots, B_n$ from P contains clauses:

$$
\begin{array}{l}
c, {}^{\bullet}H \rightarrow {}^{\bullet}B_1 \\
\ldots \\
c, {}^{\bullet}H, B_1^{\bullet}, \ldots, B_{i-1}^{\bullet} \rightarrow {}^{\bullet}B_i \\
\ldots \\
c, {}^{\bullet}H, B_1^{\bullet}, \ldots, B_{n-1}^{\bullet} \rightarrow {}^{\bullet}B_n \\
c, {}^{\bullet}H, B_1^{\bullet}, \ldots, B_n^{\bullet} \rightarrow H^{\bullet}
\end{array}
$$

PROOF (outline) One shows that all the procedure calls and successes occurring in (a prefix of) an SLD-derivation of length j are in $(T^{\mathcal{C}}_{P^{CS}})^j(\mathcal{G})$. Conversely, for any member of $(T^{\mathcal{C}}_{P^{CS}})^j(\mathcal{G})$ the corresponding call/success occurs in a derivation. Both proofs are by induction on j. □

Assume that the set of initial constrained goals is characterized by a CLP program P': $\mathcal{G} = \{\, A \mid {}^{\bullet}A \in M(P') \,\}$. Assume that no predicate $p^{\bullet}$ occurs in P'. From the last proposition it follows that the declarative semantics of $P^{CS} \cup P'$ describes the call-success semantics of P:

$$cl(CS(P, \mathcal{G})) = M(P^{CS} \cup P') \cap \mathcal{A}$$

where $\mathcal{A}$ is the set of all constrained atoms with annotated predicate symbols. (The role of the intersection with $\mathcal{A}$ is to remove auxiliary predicates that may originate from P').

3 Types

We are interested in computing approximations of the call-success semantics of programs. A program's semantics is an instance closed set of constrained atoms, an approximation is its superset. The approximations are to be manipulated by an analysis algorithm and communicated to the user.

We need a suitable class of approximations and a language to specify them. We extend for that purpose the formalism of regular unary logic programs [YS91] used in LP to describe regular sets of terms/atoms.[4] We call such sets regular (constraint) types. So we use (a restricted class of) CLP programs and their declarative c-semantics to describe approximations of the call-success semantics of CLP programs.

3.1 Regular Unary Programs

Our approach to defining types is a generalization of canonical regular unary logic (RUL) programs [YS91]. We begin this section with presenting RUL programs. Then we introduce our generalization, called RULC programs. We conclude with several examples.

To define types we will use a restricted kind of programs, with unary predicates only. In such a program R a predicate symbol t is considered to be a *name* of a type and $[\![t]\!]_R := \{\, c[]u \mid c[]t(u) \in M(R) \,\}$ is the corresponding type.

Definition 7. A (canonical) regular unary logic program (**RUL program**) is a finite set of clauses of the form:

$$t_0(f(x_1, \ldots, x_n)) \leftarrow t_1(x_1), \ldots, t_n(x_n). \tag{1}$$

(where $n \geq 0$ and $x_1, \ldots, x_n$ are distinct variables) such that no two clause heads have a common instance.

Notice that the types defined by a RUL program are sets of ground terms. (For such programs there is no difference between the c-semantics and the least Herbrand model semantics).

RUL programs were introduced in [YS91]. In [FSVY91] they are called reduced regular unary-predicate programs. The formalism defines tuple distributive [Mis84, YS91] sets of terms. So if $f(u_1, u_2)$ and $f(u_1', u_2')$ are members of such a set then also $f(u_1, u_2')$ and $f(u_1', u_2)$ are. (For exact definitions the reader is referred to [Mis84, YS91]).

We will write $F[x_1, \ldots, x_n]$ to stress that F is a formula such that $Vars(F) \subseteq \{x_1, \ldots, x_n\}$. $F[u_1, \ldots, u_n]$ will denote F with each x_i replaced by the term u_i.

---

[4] The formalism is equivalent to deterministic root-to-frontier tree automata [GS97], to (deterministic) regular term grammars (see e.g. [DZ92] and references therein) and to type graphs of [JB92, HCC95].

Definition 8. A constraint $c[x]$ in a constraint domain $\mathcal{D}'$ will be called a **regular constraint** if there exists a RUL program R and a predicate symbol t such that for any ground term u, $\mathcal{D}' \models c[u]$ iff $u \in [\![t]\!]_R$. Constraint c will be called the **corresponding constraint** for t and R.

Notice that a constraint corresponding to a RUL program may be not regular (if $\mathcal{D}'$ is a non Herbrand domain). For instance consider domain $\mathcal{D}'$ of integers, where $+$ is an interpreted function symbol. Take a program $R = \{t(4).\}$. The set of terms satisfying the corresponding constraint contains for instance $1+3$ and $3+1$ but not $3+3$. So it cannot be described by a RUL program.

The next definition provides a CLP generalization of RUL programs. From now on we assume that the constraint domain $\mathcal{D}$ contains the regular constraints.

Definition 9. By an *instance of the head* of a clause $h \leftarrow c, b_1, \ldots, b_n$ (where c is a constraint and $b_1, \ldots, b_n$ are non constraint atoms) we mean an instance of $c[]h$. A regular unary constraint logic program (**RULC program**) is a finite set of clauses of the form (1) or of the form

$$t_0(x) \leftarrow c[x]. \tag{2}$$

(where $c[x]$ is a regular constraint) such that no two clause heads have a common instance.

Example 10. The type t described by the RUL program $\{\, t(2).,\ t(3).,\ t(4).\,\}$ is the set $\{2,3,4\}$ of ground terms.

Consider CLP(FD) [Hen89]. To describe type t extended by a domain variable, with $\{2,3,4\}$ as its domain, we use a regular constraint $x \in \{2,3,4\}$ in a RULC program $R' = \{\, t'(x) \leftarrow x \in \{2,3,4\}\,\}$. Indeed, $[\![t']\!]_{R'} = cl(x \in \{2,3,4\} \,[]\, x)$.

Example 11. A type of lists with (possibly nonground) elements satisfying a constraint c can be expressed by the following RULC program R:

$$\begin{array}{l} list([\,]) \leftarrow . \\ list([x|xs]) \leftarrow elem(x), list(xs). \\ elem(x) \leftarrow c[x] \end{array}$$

The c-semantics of this program is
$M(R) = cl\left(\{\, c[x_1], \ldots, c[x_n] \,[]\, list([x_1, \ldots, x_n]) \mid n \geq 0 \,\} \cup \{\, c[x][]elem(x) \,\}\right)$.

Let Q be a RUL program such that $c[x]$ is the corresponding constraint for *elem* and Q. Replacing in R the last clause by (the clauses of) Q results in a RUL program R' describing the set of ground lists from the previous type.

Let $c_{list}[x]$ be the corresponding constraint for *list* and R'. A type of possibly non-ground lists with elements of the type *elem* can be defined by a one clause RULC program R''

$$list(x) \leftarrow c_{list}[x].$$

The type contains unbound variables whose further bindings are restricted to be lists (i.e. constrained variables of the form $c_{list}[y][]y$). It also contains all their instances. Thus our approach makes it possible to express prescriptive types like those of programming language Gödel [HL94].

Comparing the three list types presented here, we obtain $[\![list]\!]_{R'} \subseteq [\![list]\!]_{R} \subseteq [\![list]\!]_{R''}$.

Example 12. The type of all ground terms (over the given signature) is defined by predicate *ground* and a (RUL) program containing the clause $ground(f(x_1,\ldots,x_n)) \leftarrow ground(x_1),\ldots,ground(x_n)$ for each function symbol f of arity $n \geq 0$.

The type of all constrained terms is defined by predicate *any* and program $\{\,any(x) \leftarrow true\,\}$.

3.2 Operations on Types

In type analysis some basic operations on types are employed. One has to perform a check for type emptiness and inclusion. One has to compute the intersection and (an approximation of) the union of two types[5]. One has to find type $\{\,c_1,\ldots,c_n[]f(u_1,\ldots,u_n) \mid c_i[]u_i \in [\![t_i]\!],\ i=1,\ldots,n\,\}$ for given types $t_1,\ldots,t_n$, and for a given type t and an i find type $\{\,(\exists_{-Vars(u_i)}c)[]u_i \mid c[]f(u_1,\ldots,u_n) \in [\![t]\!]\,\}$. These operations for RULC are generalization of those for RUL [GdW94], and are described in [DP98a]. Here we present only an example. To find the intersection of the types t_1, t_2 defined by

$$\begin{array}{l} t_1(f(x_1,\ldots,x_n)) \leftarrow r_1(x_1),\ldots,r_n(x_n) \\ t_2(x) \leftarrow c[x] \end{array}$$

we construct clauses

$$\begin{array}{l} (t_1 \sqcap t_2)(f(x_1,\ldots,x_n)) \leftarrow (r_1 \sqcap s_1)(x_1),\ldots,(r_n \sqcap s_n)(x_n). \\ s_1(x_1) \leftarrow \exists_{-\{x_1\}}\, c[f(x_1,\ldots,x_n)]. \\ \ldots \\ s_n(x_n) \leftarrow \exists_{-\{x_n\}}\, c[f(x_1,\ldots,x_n)]. \end{array}$$

Here $r \sqcap s$ is a new type, it is the intersection of types r, s. $s_1,\ldots,s_n$ are new types. Notice that $\exists_{-\{x_i\}}\, c[f(x_1,\ldots,x_n)]$ is a regular constraint.

3.3 Regular Programs as an Abstract Domain

In this section we present how RULC programs are used to approximate the semantics of CLP programs. We also show that it is a rather unusual case of abstract interpretation, as most of the commonly required conditions [CC92] are not satisfied.

In our approach, the concrete domain **C** is that of the semantics of programs. So **C** is the set of sets of constrained atoms over the given language. (We do not

[5] The union of two types defined by RULC programs may be not definable by RULC programs.

need to make the domain more sophisticated by removing from **C** those elements that are not the meaning of any program). $(\mathbf{C}, \subseteq)$ is a complete lattice.

We want to approximate sets of constrained atoms by RULC programs. Following [GdW92, GdW94] we introduce a distinguished (unary) predicate symbol *approx*. The type corresponding to *approx* in a RULC program R is understood as the set of constrained atoms specified by R. Notice that the arguments of *approx* are treated both as atoms and as terms, we use here the ambivalent syntax [AB96]. So R *approximates* a set I of constrained atoms iff $I \subseteq [\![approx]\!]_R$. We will call such a program R a *regular approximation* of I.

Example 13. Let P be the following CLP(R) program

$$rev([\,], Y, Y).$$
$$rev([f(V,X)|T], Y, Z) \leftarrow V{*}V + X{*}X < 9,\ rev(T, Y, [f(V,X)|Z]).$$

Then the following program is a regular approximation of $M(P)$.

$$approx(rev(X,Y,Z)) \leftarrow t1(X), any(Y), any(Z).$$
$$t1([\,]).$$
$$t1([X|Xs]) \leftarrow t2(X),\ t1(Xs).$$
$$t2(f(X,Y)) \leftarrow t3(X),\ t3(Y).$$
$$t3(X) \leftarrow -3 < X,\ X < 3.$$

So the abstract domain **A** is the set of RULC programs (over the given language). The concretization function $\gamma : \mathbf{A} \to \mathbf{C}$ is defined as the meaning of *approx*:

$$\gamma(R) := [\![approx]\!]_R.$$

The ordering of the concrete domain induces the relation $\preceq$ on **A**:

$$R \preceq R' \text{ iff } \gamma(R) \subseteq \gamma(R').$$

$\preceq$ is a pre-order but not a partial order.

This is a case of abstract interpretation, in which an abstraction function *does not exist.* The reason is, roughly speaking, that there may exist an infinite decreasing sequence of regular approximations (of some $I \in \mathbf{C}$) which does not have a g.l.b. in **A**[6] [DP98a].

We want also to mention that the abstract immediate consequence function T_P^A, defined later on and used in type inference, may be not monotonic. So its least fixpoint may not exist. The properties outlined above hold already for the approach of [GdW92, GdW94]; this contradicts some claims of [GdW92, GdW94].

[6] This property also holds when the pre-order $(\mathbf{A}, \preceq)$ is replaced by the induced partial order on the set $\mathbf{A}/_{\preceq}$. Also, using another natural pre-order on **C** ($R \sqsubseteq R'$ iff $M(R) \subseteq M(R')$) does not improve the properties discussed in this section.

3.4 Types for CLP(FD)

The concept of *finite domains* was introduced to logic programming by [Hen89]. We will basically follow this framework, including the terminology. So within this section "domain" stands for a finite domain in the sense of [Hen89]. We assume that a domain is a finite set of natural numbers (including 0). This is the case in most of CLP(FD) languages. To any domain S there corresponds a *domain constraint* $x \in S$, with the expected meaning. Usually a variable involved in such a constraint is called a domain variable.

In our type analysis for CHIP we use some types that correspond to restrictions on the form of arguments of finite domain constraint predicates. We need the type of natural numbers, the type of integers, the type of finite domains (the l.u.b. of the types of the form $cl(x{\in}S[]x)$), the type of arithmetical expressions and its subset of so called linear terms.

Defining the first three of them by a RULC program would require an infinite set of clauses. So we extend RULC programs by three "built-in" types[7]. We introduce unary predicate symbols *nat*, *neg* and *anyfd*, which cannot occur in the left hand side of a RULC clause. We assume that (independently from a RULC program) $[\![nat]\!]$ is the set of all non-negative integer constants, $[\![neg]\!]$ is the set of all negative integer constants and $[\![anyfd]\!]$ is $cl(\{\, x{\in}S[]x \mid S \subseteq \mathbb{N},\ S \text{ is finite}\,\})$.[8] We allow clauses of the form $t(x) \leftarrow builtin(x)$ to occur in RULC programs (where *builtin* is one of the three symbols). By an instance of the head of such clause we mean any element of $[\![builtin]\!]$.

The type *int* of integers and the type of arithmetical expressions are defined by means of these special types by a RULC program. The type of linear terms cannot be defined by a RULC program. (For instance, for domain variables x, y and a natural number n, it contains $x * n$ and $n * y$ but not $x * y$). So we use a RULC description of a superset of it.

4 Type Inference

The core of our method is computing a regular approximation of the c-semantics of a program. It is described in [DP98a], here we present an outline. Our approach is based on [GdW92, GdW94], it can be seen as a bottom-up abstract interpretation. We use a function $T_P^{\mathcal{A}} : \mathbf{A} \to \mathbf{A}$, which approximates the immediate consequence operator $T_P^{\mathcal{C}}$. The program semantics $M(P)$ is approximated by a fixpoint of $T_P^{\mathcal{A}}$. A technique of widening, similar to that of [CC92], is applied to assure that a fixpoint is reached in a finite number of steps.

For a CLP program P and an RULC program R, $T_P^{\mathcal{A}}(R)$ is defined as

$$T_P^{\mathcal{A}}(R) = norm\left(R \amalg \coprod_{C \in P} solve(C, R) \right).$$

[7] Alternatively we can assume that the type of integers is finite. A similar solution is taken in constructing a semantics for CLP with interval constraints [BO97].

[8] If all the finite domains are the subset of some maximal domain $0..max$, then this type may be defined by a RULC clause $anyfd(x) \leftarrow x \in 0..max$.

Here *norm* [GdW94, DP98a] is a widening function; $R \preceq norm(R)$ for any R. For RULC programs Q and Q', $Q \amalg Q'$ is a RULC program such that $Q \preceq Q \amalg Q'$ and $Q' \preceq Q \amalg Q'$. It is computed using the type union operation of Sect. 3.2.

The main function is *solve*, which gives a regular approximation of $T^{\mathcal{C}}_{\{C\}}(\gamma(R))$: $T^{\mathcal{C}}_{\{C\}}(\gamma(R)) \subseteq \gamma(solve(C, R))$. Due to lack of space we only briefly outline its definition. It is based on that of [GdW92, GdW94]. The main difference is that we take into account the constraints occurring in clause C. Let $C = h \leftarrow c, b_1, \ldots, b_m$, where $c[x_1, \ldots, x_n]$ is a conjunction of elementary constraints. We approximate c by computing a "projection" of c. The projection consists of one argument constraints $c_1[x_1], \ldots, c_n[x_n]$ such that

$$\mathcal{D} \models c[x_1, \ldots, x_n] \rightarrow c_1[x_1], \ldots, c_n[x_n].$$

It is computed using the constraint solver of the underlying CLP implementation (or possibly some more powerful solver). So the types defined in the RULC program $R' = \{\, t_i(x_i) \leftarrow c_i[x_i] \mid i = 1, \ldots, n \,\}$ approximate the sets of possible values of the variables in c. Now clause $C' = h \leftarrow t_1(x_1), \ldots, t_n(x_n), b_1, \ldots, b_m$ is submitted as an argument to the function **solve** of [GdW92, GdW94], together with $R \cup R'$ as the second argument. It computes an approximation of $T^{\mathcal{C}}_{\{C'\}}(\gamma(R \cup R'))$, thus of $T^{\mathcal{C}}_{\{C\}}(\gamma(R))$.

As $T^{\mathcal{C}}_{\{C\}}(\gamma(R)) \subseteq \gamma(solve(C, R))$ and $R \preceq norm(R)$, we have that $T^{\mathcal{A}}_{P}$ approximates the concrete semantic function $T^{\mathcal{C}}_{P}$:

$$T^{\mathcal{C}}_{P}(\gamma(R)) \subseteq \gamma(T^{\mathcal{A}}_{P}(R))$$

and thus $\forall n \; T^{\mathcal{C}}_{P} \uparrow n \subseteq \gamma(T^{\mathcal{A}}_{P} \uparrow n)$.

Due to widening, a fixed point of $T^{\mathcal{A}}_{P}$ is found in a finite number of iterations (conf. [GdW94]); $T^{\mathcal{A}}_{P} \uparrow n = T^{\mathcal{A}}_{P} \uparrow \omega$, for some n. We call it the *computed fixpoint*. Function $T^{\mathcal{A}}_{P}$ is in general not monotonic w.r.t. $\preceq$ (as *norm* is not monotonic [DP98a] and $\amalg$ is not required to be). Thus we cannot claim that the computed fixpoint is the least fixpoint.

The result $T^{\mathcal{A}}_{P} \uparrow \omega$ of the computation approximates $M(P)$ as $M(P) = lfp(T^{\mathcal{C}}_{P}) \subseteq \gamma(T^{\mathcal{A}}_{P} \uparrow \omega) = [\![approx]\!]_{T^{\mathcal{A}}_{P} \uparrow \omega}$

5 Examples

This section presents a type analysis of two example programs. The user interface of our prototype analyser employs, instead of RULC programs, a more convenient formalism. So we explain it before coming to the examples.

To provide a more compact and more readable notation, we use *regular term grammars with constraints*. They can be seen as an abbreviation for RULC programs. A clause $t_0(f(x_1, \ldots, x_n)) \leftarrow t_1(x_1), \ldots, t_n(x_n)$ is represented by the *grammar rule* $t_0 \rightarrow f(t_1, \ldots, t_n)$, a clause $t(x) \leftarrow c[x]$ by the rule $t \rightarrow c[x]$.

The formalism includes parametric types. It uses type symbols of arity ≥ 0 and type variables; terms built out of them are called *type terms*. A *parametric grammar rule* is of the form $t(\alpha_1, \ldots, \alpha_k) \rightarrow f(t_1, \ldots, t_n)$ where t is a k-ary type symbol, t_j are type terms and α_i are type variables. (One requires that $Vars(t_1, \ldots, t_n) \subseteq \{\alpha_1, \ldots, \alpha_k\}$). Such a rule stands for a family of RULC clauses represented by the (non parametric) rules $t(s_1, \ldots, s_k) \to f(t_1, \ldots, t_n)\theta$, where s_i are arbitrary types and θ is the substitution $\{\alpha_i/s_i \mid i = 1, \ldots, k\}$. [9]

For example, rules

$$list(\alpha) \to [\,] \qquad\qquad list(\alpha) \to [\alpha|list(\alpha)]$$

correspond to a family of RULC programs

$$list(t)([\,]). \qquad\qquad list(t)([x_1|x_2]) \leftarrow t(x_1), list(t)(x_2).$$

which for any type term t define the type $list(t)$ of lists of elements of type t.

The user may declare some types by providing (possibly parametric) grammar rules.[10] Whenever possible, the system uses the declared types in its output. Thus the output may be expressed (partially) in terms of types familiar to the user; this can substantially improve the readability of the results of the analysis. For instance, assume that the system derives a type t with the corresponding fragment of a RULC program:

$$t([\,]). \qquad\qquad t([x|y]) \leftarrow nat(x), t(y).$$

Then, instead of displaying the RULC clauses (or actually the corresponding grammar) the system informs that the type is *list*(*nat*). Notice that the system does not infer parametric polymorphic types, the polymorphism comes only from user declarations.

As the first example we use the following program, which solves the well-known N-queens problem. The current version of our analyzer treats all the finite domains in a uniform way, namely as *anyfd* (the types of the form $cl(x \in S[]x)$ are not yet implemented).

```
:- entry nqueens(nat,any).
nqueens(N,List) :- length(List,N), List::1..N,
                   constraint_queens(List), labeling(List).
labeling([]).
labeling([X|Y]) :- indomain(X), labeling(Y).
constraint_queens([]).
constraint_queens([X|Y]) :- safe(X,Y,1), constraint_queens(Y).
safe(_,[],_).
safe(X,[Y|T],K) :- noattack(X,Y,K), K1 is K+1, safe(X,T,K1).
```

[9] So now the predicate symbols of RULC are type terms. We allow only such grammars for which no two corresponding clauses have a common head instance (conf. Def. 9). We should deal with finite RULC programs. But the program corresponding to a set of parametric rules may be infinite. So a condition on grammars is imposed: in the obtained RULC program any type should depend on a finite set of types. For details see [DP98a, DZ92].

[10] The widely used type $list(\alpha)$, declared as above, is predefined in the system.

The **entry** declaration indicates the top goal and its call patterns for the call-success analysis. Types inferred by the system are presented below.

```
call    : nqueens(nat,any)
success : nqueens(nat,list(nat))
-------------------------
call    : labeling(list(anyfd))
success : labeling(list(nat))
-------------------------
call    : constraint_queens(list(anyfd))
success : constraint_queens(list(anyfd))
-------------------------
call    : safe(anyfd,list(anyfd),int)
success : safe(anyfd,list(anyfd),int)
-------------------------
call    : noattack(anyfd,anyfd,int)
success : noattack(anyfd,anyfd,int)
```

Assume now that the second clause defining **safe/3** contains a bug:

```
safe(X,[Y|T],K):-noattack(X,Y,K),K1 is K+1,safe(X,t,K1).% bug here
```

Types inferred by the analyzer look like follows (we show only those which differ from ones generated previously):

```
success : nqueens(nat,t102)
t102 --> [nat|t78]
t102 --> []
t78 --> []
-------------------------
call    : labeling(t90)
t90 --> []
t90 --> [anyfd|t78]
success : labeling(t102)
-------------------------
success : constraint_queens(t90)
-------------------------
call    : safe(anyfd,t71,int)
t71 --> []
t71 --> [anyfd|list(anyfd)]
t71 --> t
success : safe(anyfd,t78,int).
```

The types inferred are obviously suspicious and should be helpful in localizing the bug in the program. For instance, the second argument of success of **nqueens/2** (type **t102**) is an empty list or a one-element list of naturals. A similar problem is with **constraint_queens**. The problem may be traced down to **safe/3** which succeeds with the empty list as the second argument.

The next example illustrates inferring non-trivial constraints in the approximation of a program. The predicate `split5(Xs,Ls,Gs)` splits an input list `Xs` of finite domain variables (or natural numbers) into lists of elements less and greater or equal to 5 (`Ls` and `Gs` respectively).

```
:-entry split5(list(anyfd),any,any).
split5([],[],[]).
split5([X|Xs],[X|Ls],Gs) :- X #< 5, split5(Xs,Ls,Gs).
split5([X|Xs],Ls,[X|Gs]) :- X #>= 5, split5(Xs,Ls,Gs).
```

The inferred types are presented below.

```
call    : split5(list(anyfd),any,any)
success : split5(list(anyfd),list(t1),list(t2))
t1 --> X #< 5
t2 --> X #>= 5
```

6 Conclusions and Future Work

In this paper we propose a method of computing semantic approximations for CLP programs. Our aim is a practical tool that would be helpful in debugging. We are mainly interested in CLP(FD), particularly in the language CHIP. Our approach considers the (operational) call-success semantics and the (declarative) c-semantics.

As a specification language to express the semantic approximations we propose a system of regular types for CLP, which is an extension of an approach used for logic programs. The types are defined by (a restricted class of) CLP programs, called RULC programs. We present an algorithm for computing regular approximations of the declarative semantics. This algorithm can also be used for approximating the call-success semantics, due to a characterization of this semantics by the c-semantics of a transformed program.

We have adopted a regular approximation system (described in [GdW92, GdW94]) to constraint logic programming over finite domains. The current version analyzes programs in the language CHIP. We expect it to be easily portable to work with other CLP languages, as we have isolated its parts responsible for the built-ins of CHIP. The prototype has been implemented in CHIP and has been ported to SICStus Prolog and CIAO [CLI97]. The latter implementation is a part of an assertion-based framework for debugging in CLP [PBM98].

The system presents types to the user as regular term grammars, which are more easily comprehensible than RULC programs. This provides a restricted but useful kind of polymorphism (conf. Section 5)

A subject for future work is obtaining more precise analysis by using a more sophisticated treatment of constraints. We also plan to evaluate the method experimentally by applying it to non-toy programs.

Another direction of further work is relating our technique to abstract debugging [CLMV98]. A clear relationship between these two techniques should be established. The first step is a diagnosis method [CDP98, CDMP98] which finds the clauses responsible for a program being incorrect w.r.t. a type specification. That work uses the type system presented here as the class of specifications. Computing an approximation of T_C^C, as discussed in Sect. 4, is at the core of the diagnosis algorithm.

ACKNOWLEDGMENT

The authors want to thank Jan Małuszyński for discussions and suggestions.

References

[AB96] K.R. Apt and R. Ben-Eliyahu. Meta-variables in Logic Programming, or in Praise of Ambivalent Syntax. *Fundamenta Informaticae*, 28(1-2):22–36, 1996.

[AM94] K.R. Apt and E. Marchiori. Reasoning about Prolog programs: from modes through types to assertions. *Formal Aspects of Computing*, 6(6A):743–764, 1994.

[BC89] A. Bossi and N. Cocco. Verifying correctness of logic programs. In *Proceedings of the International Joint Conference on Theory and Practice of Software Development TAPSOFT '89, vol. 2*, pages 96–110. Springer-Verlag, 1989. Lecture Notes in Computer Science.

[BO97] F. Benhamou and W. Older. Applying Interval Arithmetic to Real, Integer and Boolean Constraints. *Journal of Logic Programming*, 32(1):1–24, July 1997.

[Boy96] J. Boye. *Directional Types in Logic Programming.* Linköping studies in science and technology, dissertation no. 437, Linköping University, 1996.

[CC92] P. Cousot and R. Cousot. Abstract Interpretation and Application to Logic Programming. *Journal of Logic Programming*, 13(2–3):103–179, 1992.

[CDMP98] M. Comini, W. Drabent, J. Małuszyński, and P. Pietrzak. A type-based diagnoser for CHIP. ESPRIT DiSCiPl deliverable, September 1998.

[CDP98] M. Comini, W. Drabent, and P. Pietrzak. Diagnosis of CHIP programs using type information. In *proceedings of Types for Constraint Logic Programming, post-conference workshop of JICSLP'98*, 1998.

[Cla79] K. L. Clark. Predicate logic as computational formalism. Technical Report 79/59, Imperial College, London, December 1979.

[CLI97] The CLIP Group. *CIAO System Reference Manual.* Facultad de Informática, UPM, Madrid, August 1997. CLIP3/97.1.

[CLMV98] M. Comini, G. Levi, M. C. Meo, and G. Vitiello. Abstract diagnosis. *Journal of Logic Programming*, 1998. To appear.

[Cos96] Cosytec SA. *CHIP System Documentation*, 1996.

[DM88] W. Drabent and J. Małuszyński. Inductive Assertion Method for Logic Programs. *Theoretical Computer Science*, 59:133–155, 1988.

[DP98a] W. Drabent and P. Pietrzak. Inferring call and success types for CLP programs. ESPRIT DiSCiPl deliverable, September 1998.

[DP98b] W. Drabent and P. Pietrzak. Type analysis for CHIP. In *proceedings of Types for Constraint Logic Programming, post-conference workshop of JICSLP'98*, 1998.

[Dra88] W. Drabent. On completeness of the inductive assertion method for logic programs. Unpublished note (available from `www.ipipan.waw.pl/~drabent`), Institute of Computer Science, Polish Academy of Sciences, May 1988.

[DZ92] P. Dart and J. Zobel. A regular type language for logic programs. In F. Pfenning, editor, *Types in Logic Programming*, pages 157–187. MIT Press, 1992.

[FLMP89] M. Falaschi, G. Levi, M. Martelli, and C. Palamidessi. Declarative modelling of the operational behaviour of logic languages. *Theoretical Computer Science*, 69(3):289–318, 1989.

[FSVY91] T. Fruewirth, E. Shapiro, M. Vardi, and E. Yardeni. Logic programs as types for logic programs. In G. Kahn, editor, *Annual IEEE Symposium on Logic in Computer Science (LICS)*, pages 300–309, Amsterdam, July 1991. IEEE Computer Society Press. Corrected version available from `http://WWW.pst.informatik.uni-muenchen.de/~fruehwir`.

[GdW92] J. Gallagher and D. A. de Waal. Regular Approximations of Logic Programs and Their Uses. Technical Report CSTR-92-06, Department of Computer Science, University of Bristol, 1992.

[GdW94] J. Gallagher and D. A. de Waal. Fast and Precise Regular Approximations of Logic Programs. In P. Van Hentenryck, editor, *Proc. of the Eleventh International Conference on Logic Programming*, pages 599–613. MIT Press, 1994.

[GS97] F. Gécseg and M. Steinby. Tree languages. In G. Rozenberg and A. Salomaa, editors, *Handbook of Formal Languages*, volume 3, Beyond Words. Springer-Verlag, 1997.

[HCC95] P. Van Hentenryck, A. Cortesi, and B. Le Charlier. Type analysis of Prolog using type graphs. *Journal of Logic Programming*, 22(3):179–209, March 1995.

[Hen89] P. Van Hentenryck. *Constraint Satisfaction in Logic Programming*. MIT Press, 1989.

[HL94] P.M. Hill and J.W. Lloyd. *The Gödel Programming Language*. MIT Press, 1994.

[JB92] G. Janssens and M. Bruynooghe. Deriving descriptions of possible values of program variables by means of abstract interpretation. *Journal of Logic Programming*, 13(2 & 3):205–258, 1992.

[LR91] T.K. Lakshman and U.S. Reddy. Typed Prolog: A semantic reconstruction of the Mycroft-O'Keefe type system. In V. Saraswat and K. Ueda, editors, *Proc. of the 8th International Logic Programming Symposium*, pages 202–217. MIT Press, 1991.

[Mis84] P. Mishra. Towards a theory of types in Prolog. In *Proceedings of the IEEE International Symposium on Logic Programming*, pages 289–298, 1984.

[PBM98] G. Puebla, F. Bueno, and Hermenegildo M. A framework for assertion-based debugging in constraint logic programming. In *proceedings of Types for Constraint Logic Programming, post-conference workshop of JICSLP'98*, 1998.

[SHC96] Z. Somogyi, F. Hederson, and T. Conway. The execution algorithm of Mercury: an efficient purely declarative logic programming language. *Journal of Logic Programming*, 29(1–3):14–64, 1996.

[YS91] E. Yardeni and E. Y. Shapiro. A type system for logic programs. *Journal of Logic Programming*, 10(2):125–153, 1991.

Categorical Programming with Abstract Data Types

Martin Erwig

FernUniversität Hagen, Praktische Informatik IV
58084 Hagen, Germany
erwig@fernuni-hagen.de

Abstract. We show how to define fold operators for abstract data types. The main idea is to represent an ADT by a bialgebra, that is, an algebra/coalgebra pair with a common carrier. A program operating on an ADT is given by a mapping to another ADT. Such a mapping, called *metamorphism*, is basically a composition of the algebra of the second with the coalgebra of the first ADT. We investigate some properties of metamorphisms, and we show that the metamorphic programming style offers far-reaching opportunities for program optimization that cover and even extend those known for algebraic data types.

1 Introduction

Expressing recursion over data types in terms of catamorphisms, or fold operations, has been successfully employed by Bird and Meertens to calculate programs from specifications [3, 21]. They formulated laws expressing algebraic identities of programs and used them to derive algorithms in a sequence of simple transformation steps. Their work was primarily focused on lists, but it has been extended to regular algebraic data types [20, 22, 25]: a data type is given by a morphism which is a fixed point of a functor defining the signature of the data type. Since fixed points are initial objects, homomorphisms to other data types are uniquely defined, and this makes it possible to specify a program on a data type by simply selecting an appropriate target data type. Along with these generalizations a lot of work on program optimization has emerged that essentially relies on programs being expressed as catamorphisms, for example, [13, 25, 26, 16, 17]. The strong interest in program fusion is certainly due to the fact that catamorphisms encapsulate a class of recursion over data types that enjoys some nice mathematical properties.

Besides the original idea of having a framework for calculating programs from their specifications, avoiding general recursion is also important from a programming methodology point of view. For instance, Meijer et al. [22] stress in their often cited paper the aspect that using folds in functional languages is truly in the spirit of the structured programming methodology. Similar beliefs, that is, avoiding general recursion and using a fixed set of higher order functions, had been already emphasized before by Backus [1]. The programming languages CPL

A.M. Haeberer (Ed.): AMAST'98, LNCS 1548, pp. 406–421, 1998.

[14] and Charity [11] are designed thoroughly on the basis of this programming methodology.

It is striking that the categorical framework has been applied only sporadically to data types that are not just given by free term structures, such as abstract data types. Although the catamorphism approach principally works also for sub-algebras that satisfy certain laws, one cannot map into algebras with less structure [9, 10]. This innocent looking restriction means a rather severe limitation of expressiveness: for instance, such a simple task as counting the elements of a set *cannot* be expressed by a catamorphism.

Therefore we propose to base mappings between ADTs not just on constructors, but on explicitly defined destructors. Formally, this means to represent an ADT by a bialgebra, that is, a pair (algebra, coalgebra) with a common carrier, and to define a mapping between two ADTs D and D' by composing the algebra of D' with the coalgebra of D. This offers much freedom in specifying ADTs and mappings between them. It also provides a new programming style encouraging the compositional use of ADTs. The proposed approach essentially uses existing concepts, such as algebra and coalgebra, on a higher level of abstraction, and this is the reason that all the optimization rules developed for algebraic data types are still valid in this extended framework. But in addition to this, the "programming by ADT composition" style offers some new optimization opportunities: for example, since intermediate ADTs are intrinsically used in a single-threaded way, a compiler can automatically insert efficient imperative update-in-place implementations for them.

The rest of the paper is structured as follows: after reviewing the categorical definition of algebraic data types in Sect. 2, we show how to represent abstract data types as bialgebras in Sect. 3. In Sect. 4 we introduce metamorphisms as mappings between ADTs, we provide various example programs, and we show some basic properties of ADTs and metamorphisms. In Sect. 5 we investigate several aspects of program transformation. Related work is described in Sect. 6, and conclusions follow in Sect. 7.

2 Data Types and Homomorphisms

In this section we briefly review the categorical framework for modeling data types. More detailed introductions are given, for example, in [4, 9]. Specific information about coalgebras can be found in [15, 24], and hylomorphisms are explained and used in [22, 26, 16]. We assume some basic knowledge about category theory (an understanding of *category*, *functor*, and *natural transformation* should be sufficient), for an introduction see, for example, [2] or [4].

In Sect. 2.1 we briefly recall the notion of algebraic data types of functional languages. In Sect. 2.2 we show how to express signatures by functors. This is the basis for the definition of algebras and coalgebras in Sect. 2.3. We demonstrate how algebra homomorphisms can express programs on data types in Sect. 2.4.

2.1 Algebraic Data Types in Functional Languages

In modern functional languages, like ML or Haskell, data structures are represented by terms that are built by typed constructors. All constructors of one type are introduced in one definition, and the defined type is called an *algebraic data type*. For example, a term representation for natural numbers and a polymorphic list data type are given by:

$$\begin{aligned} nat &= Zero \mid Succ\ nat \\ list\ A &= Nil \mid Cons(A, list\ A) \end{aligned}$$

This introduces the four constructors $Zero : \mathbf{1} \to nat$, $Succ : nat \to nat$, $Nil : \mathbf{1} \to list\ A$, and $Cons : A \times list\ A \to list\ A$, where $\mathbf{1}$ denotes the *unit type* that contains just the one element (), that is, constants of type A are identified with functions of type $\mathbf{1} \to A$. Thus, a data type can be viewed as an algebra whose operations are given by the data type constructors. Note that we consider non-strict constructors.

2.2 Polynomial Functors

In this paper the default category $\mathcal{C}$ is **CPO**, whose objects are complete partially ordered sets with a least element $\perp$ and whose morphisms are continuous functions. Working in **CPO** guarantees the existence of least fixed points for certain recursive equations needed in Sect. 2.4 and in Sect. 4. The terminal object in $\mathcal{C}$ is the one-element type $\mathbf{1}$. In the sequel we consider only endofunctors on $\mathcal{C}$ (that is, functors from $\mathcal{C}$ to $\mathcal{C}$) which are built by the four basic functors I (*identity*), $\underline{A}$ (*constants*), $\times$ (*product*), and $+$ (*separated sum*).

The effect of the identity functor I on types and functions is $I\ A = A$ and $I\ f = f$, and the constant functor for type A, denoted by $\underline{A}$, operates on types and functions by $\underline{A}\ B = A$ and $\underline{A}\ f = \mathrm{id}_A$ where id_A denotes the identity morphism on A. For an object x we denote its constant function by $\underline{x}$, that is, $\underline{x}\ y = x$. The *product* of two types A and B and its operation on functions is defined as:

$$\begin{aligned} A \times B &= \{(x,y) \mid x \in A, y \in B\} \\ (f \times g)\ (x,y) &= (f\ x, g\ y) \end{aligned}$$

Related operations are left and right projection and tupling (also called split):

$$\begin{aligned} \pi_1\ (x,y) &= x \\ \pi_2\ (x,y) &= y \\ \langle f,g \rangle\ x &= (f\ x, g\ x) \end{aligned}$$

Finally, the *separated sum* of two types A and B and its operation on functions is defined as:

$$
\begin{aligned}
A + B &= \{1\} \times A \cup \{2\} \times B \cup \{\bot\} \\
(f + g)\,(1, x) &= (1, f\ x) \\
(f + g)\,(2, y) &= (2, g\ y) \\
(f + g)\ \bot &= \bot
\end{aligned}
$$

Related operations are left and right injection and case analysis (also called junc):

$$
\begin{aligned}
\iota_1\ x &= (1, x) \\
\iota_2\ y &= (2, y) \\
[f, g]\,(1, x) &= f\ x \\
[f, g]\,(2, y) &= g\ y \\
[f, g]\ \bot &= \bot
\end{aligned}
$$

The use of seperated sum is convenient for the treatment of, for example, infinite lists. However, algebras for, say, natural numbers, are usually better modeled with coalesced sum (which identifies bottom elements). These issues are discussed in some detail in [9, 19].

Separated sum and product are bifunctors that map from the product category $\mathcal{C} \times \mathcal{C}$ to $\mathcal{C}$. Fixing one parameter of a bifunctor yields a monofunctor: the (left) *section* of a bifunctor F and an object A is defined as $F_A(B) = F(A, B)$. Thus, for example, $\times_A$ is a monofunctor which takes an object B and maps it to the product $A \times B$.

Now *polynomial functors* are inductively defined as follows: (i) I and $\underline{A}$ are polynomial functors, and (ii) if F and G are polynomial functors, then so are their composition FG, their sum $F+G$, and their product $F \times G$ where: $(F+G)(X) = F(X) + G(X)$ and $(F \times G)(X) = F(X) \times G(X)$ (for both types and functions X). Two examples of polynomial functors are:

$$
\begin{aligned}
N &= \underline{1} + I \\
L_A &= \underline{1} + \underline{A} \times I
\end{aligned}
$$

Here L_A is actually a left section of a bifunctor.

(A note on operator precedence: function application binds strongest, and $\times$ binds stronger than $+$, which in turn binds stronger than composition "$\circ$".)

2.3 Algebras and Coalgebras

Let endofunctor $F : \mathcal{C} \to \mathcal{C}$ represent a signature. Then an *F-algebra* is a morphism $\alpha : F(A) \to A$. Object A is called the *carrier* of the algebra. We can extract the carrier of an algebra with the forgetful functor U, that is, $U(\alpha) = A$. Dually, an *F-coalgebra* is a morphism $\overline{\alpha} : A \to F(A)$. An *F-homomorphism* from algebra $\alpha : F(A) \to A$ to algebra $\beta : F(B) \to B$ is a morphism $h : A \to B$ in $\mathcal{C}$ that satisfies $h \circ \alpha = \beta \circ F(h)$. As a shorthand for this condition we write:

$h : \alpha \to_F \beta$. The category of F-algebras $\mathbf{Alg}(F)$ has as objects F-algebras and as arrows F-homomorphisms. (Composition and identities in $\mathbf{Alg}(F)$ are taken from $\mathcal{C}$.) Dually, $\mathbf{CoAlg}(F)$ is the category of F-coalgebras with F-cohomomorphisms (where a morphism $h : A \to B$ from coalgebra $\overline{\alpha} : A \to F(A)$ to coalgebra $\overline{\beta} : B \to F(B)$ is a *cohomomorphism* if it satisfies $F(h) \circ \overline{\alpha} = \overline{\beta} \circ h$).

If F is a polynomial functor on **CPO**, $\mathbf{Alg}(F)$ has an initial object, which is denoted by in_F. This means that $in_F : F(T) \to T$ is an F-algebra with carrier $T = U(in_F)$ (the "T" reminds of "term algebra"). Dually, $\mathbf{CoAlg}(F)$ has a terminal object, denoted by out_F, and $out_F : T \to F(T)$ is an F-coalgebra with the same carrier T as in_F. Moreover, in_F and out_F are each other's inverses, and they thus define an isomorphism $T \cong F(T)$ in **CPO**.

Now the definitions of Sect. 2.1, written categorically as:

$$\begin{aligned} [\mathit{Zero}, \mathit{Succ}] &: N(\mathit{nat}) \to \mathit{nat} \\ [\mathit{Nil}, \mathit{Cons}] &: L_A(\mathit{list}\ A) \to \mathit{list}\ A \end{aligned}$$

with $N(\mathit{nat}) = \mathbf{1} + \mathit{nat}$ and $L_A(\mathit{list}\ A) = \mathbf{1} + A \times \mathit{list}\ A$, define the data types as initial objects in the category of N-algebras, respectively, L_A-algebras, that is, $\mathit{nat} = [\mathit{Zero}, \mathit{Succ}] := in_N$ and $\mathit{list}\ A = [\mathit{Nil}, \mathit{Cons}] := in_{L_A}$.

With sums we can also define conditionals. First, we define a type for booleans by:

$$\mathit{bool} = \mathit{True} \mid \mathit{False}$$

(which is just syntax for $\mathit{bool} = [\mathit{True}, \mathit{False}] := in_B$ with $B = \underline{1} + \underline{1}$.) Now for each predicate $p : A \to \mathit{bool}$ a morphism $p? : A \to A + A$ is defined by [22]:

$$p?(a) = \begin{cases} \bot & \text{if } p(a) = \bot \\ \iota_1\ a & \text{if } p(a) = \mathit{True} \\ \iota_2\ a & \text{if } p(a) = \mathit{False} \end{cases}$$

The conditional is then simply defined by **if** p **then** f **else** $g = [f, g] \circ p?$. (A more detailed categorical exposition of this can be found in [4].)

Those coalgebras that are the inverses of initial algebras defined by algebraic data type definitions just undo the term construction. So we can think, for example, of out_N and out_{L_A} as being defined by:

$$\begin{aligned} out_N = \lambda n.\mathbf{case}\ n\ \mathbf{of}\ & \mathit{Zero} \to (1, ()) \\ \mid\ & \mathit{Succ}(m) \to (2, m) \\ out_{L_A} = \lambda l.\mathbf{case}\ l\ \mathbf{of}\ & \mathit{Nil} \to (1, ()) \\ \mid\ & \mathit{Cons}(x, l') \to (2, (x, l')) \end{aligned}$$

2.4 Catamorphisms, Anamorphisms, and Hylomorphisms

Initial and terminal objects are unique up to isomorphism, and they are characterized by having exactly one morphism to, respectively, from, all other objects.

This means that for each F-algebra α in the category $\mathbf{Alg}(F)$ there is exactly one F-homomorphism $h : in_F \rightarrow_F \alpha$. Since h is uniquely determined by α, it is conveniently denoted by $(\!|\alpha|\!)_F$, or just $(\!|\alpha|\!)$ when F is clear from the context, and h is called a *catamorphism* [22]. Accordingly, for each F-coalgebra $\overline{\alpha}$ in the category $\mathbf{CoAlg}(F)$ there is exactly one F-cohomomorphism $h : \overline{\alpha} \rightarrow_F out_F$, which is denoted by $[\!(\overline{\alpha})\!]_F$ (or just $[\!(\overline{\alpha})\!]$) and which is called an *anamorphism*.

Programs mapping from an initial algebra to another data type can be succinctly expressed as catamorphisms. An F-catamorphism $(\!|\alpha|\!)$ can be thought of as a function replacing the constructors of in_F by the functions/constructors of α; catamorphisms offer a canonical way of consuming a data structure. Similarly, mappings to terminal algebras can be expressed by anamorphisms, which provide a canonical way of constructing data structures. It is clear that the identity is the unique morphism from the initial F-algebra to itself (respectively, to the terminal F-coalgebra from itself):

$$(\!|in_F|\!)_F \;=\; [\!(out_F)\!]_F \;=\; \mathrm{id} \qquad \text{(CataId, AnaId)}$$

Finally, a *hylomorphism* is essentially the composition of a catamorphism with an anamorphism. Formally, a hylomorphism $[\![\alpha, \overline{\alpha}]\!]_F$ is defined as the least morphism h satisfying:

$$h \;=\; \alpha \circ F(h) \circ \overline{\alpha} \qquad \text{(HyloDef)}$$

Hylomorphisms are related to cata- and anamorphisms in an obvious way:

$$\begin{aligned} [\![\alpha, \overline{\alpha}]\!]_F &= (\!|\alpha|\!)_F \circ [\!(\overline{\alpha})\!]_F && \text{(HyloSplit)} \\ [\![\alpha, out_F]\!]_F &= (\!|\alpha|\!)_F && \text{(HyloCata)} \\ [\![in_F, \overline{\alpha}]\!]_F &= [\!(\overline{\alpha})\!]_F && \text{(HyloAna)} \end{aligned}$$

Hylomorphisms enjoy a number of useful laws [22, 26], for example,

$$\begin{aligned} [\![\alpha \circ \eta, \overline{\alpha}]\!]_F &= [\![\alpha, \eta \circ \overline{\alpha}]\!]_G \;\Leftarrow\; \eta : F \dot{\rightarrow} G && \text{(HyloShift)} \\ [\![\alpha, \overline{\alpha}]\!]_F \circ [\![\beta, \overline{\beta}]\!]_F &= [\![\alpha, \overline{\beta}]\!]_F \;\Leftarrow\; \overline{\alpha} \circ \beta = \mathrm{id} && \text{(HyloFusion)} \end{aligned}$$

A hylomorphism $[\![\alpha, \overline{\alpha}]\!]_F$ defines a recursive function whose recursion follows that of the functor F.

3 Abstract Data Types as Bialgebras

The main hindrance for expressing certain catamorphisms on ADTs is that homomorphisms are not able to map to less constrained structures.

One solution is to decouple the decomposition of ADT values from their construction to gain more flexibility. This can be achieved by modeling an ADT by a pair $(\alpha, \overline{\alpha})$ where α is an F-algebra, $\overline{\alpha}$ is a G-coalgebra, and $U(\alpha) = U(\overline{\alpha})$. Such an algebra/coalgebra-pair with a common carrier is called an F, G-*bialgebra* [10]. An F, G-*bialgebra homomorphism* from $(\alpha, \overline{\alpha})$ to $(\beta, \overline{\beta})$ is a morphism satisfying

$h \circ \alpha = \beta \circ F(h)$ and $G(h) \circ \overline{\alpha} = \overline{\beta} \circ h$. The F,G-algebras and homomorphisms form a category $\mathbf{BiAlg}(F,G)$, which is built upon $\mathcal{C}$.[1] In the sequel we use the terms "ADT" and "bialgebra" as synonyms. Given an ADT $D = (\alpha, \overline{\alpha})$, we call α the *constructor* of D and $\overline{\alpha}$ the *destructor* of D.

Note that, in general, we have to provide the carrier of an ADT explicitly, since it is not determined by a universal property (like for initial algebras or terminal coalgebras). If not stated otherwise, we will always implicitly take the carrier of the initial algebra (or terminal coalgebra) whenever it is used as a constructor (respectively, as a destructor). For example, the carrier of the bialgebras *Nat*, *Range*, and *Prod* is always $U(in_N)$, and the carrier of the bialgebras *List*, *Queue*, and *Set* is $U(in_{L_A})$.

Let us consider some examples. First of all, algebraic data types can be regarded as ADTs by taking the initial algebra as constructor and its inverse as destructor. For example, ADT $List = (in_{L_A}, out_{L_A})$ is an L_A, L_A-bialgebra; *List* can also be used as a stack ADT. Similarly, we can define: $Nat = (in_N, out_N)$. But we can define many more different ADTs for natural numbers. We can consider, for instance, binary destructors, that is, $L_{U(in_N)}$-coalgebras. One example (that will be used later) is the ADT *Range* which decomposes a number by returning the number itself in addition to the predecessor:

$$Range \;=\; (in_N, [I, \langle succ, I\rangle] \circ out_N)$$

Note that using *succ* is indeed correct here, since out_N gives the predecessor (which is preserved by I) and *succ* re-builds the original number value. This also shows that the constructor and the destructor of an ADT need not have the same signature. Another example for this is the $L_{U(in_N)}, N$-bialgebra *Prod* that constructs numbers by multiplication:

$$Prod \;=\; ([1, *], out_N)$$

Our next example is an ADT for queues. The constructors of a queue are the same as for *List*. The destructor is also an L_A-coalgebra, but it is different from out_{L_A}, since elements are taken from the end. There are different ways to define the queue destructor. First of all, we can give a recursive function definition (which is possible, since we are working in **CPO**).

$$\begin{array}{rll} dequeue \;=\; \mu[\, f = \lambda l.\textbf{case}\ l\ \textbf{of} & & \\ & Nil & \rightarrow (1, ()) \\ & \mid\ Cons(x, Nil) & \rightarrow (2, (x, Nil)) \\ & \mid\ Cons(x, l') & \rightarrow (I + I \times (Cons \circ \langle \underline{x}, I\rangle))\ (f\ l')\ ] \end{array}$$

(This definition could be written more conveniently if **let**-expressions were available.) A more categorical style is to use only combinators and catamorphisms.

[1] An F,G-bialgebra is just a special case of an F,G-*dialgebra*, that is, $\mathbf{BiAlg}(F,G) = \mathbf{DiAlg}([F,I],[I,G])$ [10, 9, 7]. Working with bialgebras is sufficient for our purposes and makes the separation of constructors and destructors more explicit.

With the aid of a function *snoc* for appending a single element at the end of a list and a function *rev* for reversing a list – both defined by L_A-catamorphisms:[2]

$$\begin{aligned} snoc(x, l) &= (\![Cons(x, Nil), Cons]\!)_{L_A}\ l \\ rev &= (\![Nil, snoc]\!)_{L_A} \end{aligned}$$

we can define the queue destructor as follows.

$$dequeue = I + I \times rev \circ out_{L_A} \circ rev$$

This means, a queue is represented by a list where elements are enqueued at the front and dequeued from the rear. In particular, dequeueing from a list l works as follows: reverse l, take first element x and tail l' from $rev\ l$, and finally reverse l' to get the standard queue representation. Now we can define:

$$Queue = (in_{L_A}, dequeue).$$

As our final example we define a set ADT, again based on the "cons"-view given by L_A. This can be done in two principally different ways. One possibility that quickly comes to mind is to define equations E expressing idempotence and commutativity and work with the quotient algebra "in_{L_A}/E".[3] The problem with this approach is that homomorphisms are forced to stay within L_A/E-algebras, and it is not obvious how to define destructors into different algebras. Thus, it is not clear how to define a function for counting the elements of a set.

To define sets as bialgebras, we can use a list carrier and normalize lists in constructors or destructors. The second option means to take in_{L_A} as constructor. The destructor must then be defined so that a value is retrieved from a set at most once. This can be realized by splitting an arbitrary element off (for example, the one that was inserted last) and removing all occurrences of this element in the returned set. We need the following functions:

$$\begin{aligned} append(l, l') &= (\![l', Cons]\!)_{L_A}\ l \\ flatten &= (\![Nil, append]\!)_{L_{list}\ A} \\ map\ f &= (\![Nil, Cons \circ f \times I]\!)_{L_A} \\ filter(p, l) &= (flatten \circ map\ (\lambda y.\mathbf{if}\ p(y)\ \mathbf{then}\ Cons(y, Nil)\ \mathbf{else}\ Nil))\ l \\ remove(x, l) &= filter\ ((\neq x), l) \end{aligned}$$

Now we can define the set destructor and the set ADT by:

$$\begin{aligned} deset &= I + \langle \pi_1, remove \rangle \circ out_{L_A} \\ Set &= (in_{L_A}, deset) \end{aligned}$$

Note that the definition works only for types A on which equality is defined.

[2] For readability we omit the junc-brackets inside catamorphisms.

[3] Equations can be expressed categorically by *transformers* [10], which are mappings between algebras. Examples can be found in [10, 9, 18].

We have actually given one concrete implementation of sets based on lists, and strictly we have still to prove that this implementation is correct. So the presented bialgebra approach to programming with ADTs is definitely not as high-level as equational specifications. However, we believe it is more flexible. In particular, the bialgebra approach encourages to combine different F-algebras and G-coalgebras, which makes it easy to adapt ADTs to changing requirements. For example, instead of splitting off single elements with *deset* we can also use a P-coalgebra *split* (where $P = \underline{1} + I \times I$) to partition a set into two equally sized sets. This can be useful, for example, for divide-and-conquer algorithms.

4 Programming by Metamorphisms

¿From now on let $D = (\alpha, \overline{\alpha})$ be an F, G-bialgebra, let $D' = (\beta, \overline{\beta})$ be an H, J-bialgebra, and let $C = (\varphi, \overline{\varphi})$ be a K, M-bialgebra.

Metamorphisms and Data Type Filters. If $f : G \dot{\rightarrow} H$ is a natural transformation, the *f-metamorphism* from D to D' is defined as the least solution of the equation

$$h = \beta \circ f \circ G(h) \circ \overline{\alpha} \qquad \text{(MetaDef)}$$

and is denoted by $D \overset{f}{\rightsquigarrow} D'$ (we write $D \rightsquigarrow D'$ if $f = \text{id}$). We call D/D' the *source/target* and f the *map* of the metamorphism. This definition says that a metamorphism from D to D' is essentially a hylomorphism:

$$D \overset{f}{\rightsquigarrow} D' = [\![\beta \circ f, \overline{\alpha}]\!]_G \qquad \text{(MetaHylo)}$$

As an important special case, metamorphisms from algebraic data types reduce to catamorphisms, that is,

$$D \rightsquigarrow D' = (\![\beta]\!)_G \quad \Leftarrow \quad D = (in_G, out_G) \qquad \text{(MetaAlg)}$$

This can be seen as follows. First we know $H = G$, since $f = \text{id}$. Then

$$\begin{aligned} D \rightsquigarrow D' &= [\![\beta, out_G]\!]_G && \{ \text{ MetaHylo } \} \\ &= (\![\beta]\!)_G \circ [\!(out_G)\!]_G && \{ \text{ HyloSplit } \} \\ &= (\![\beta]\!)_G \circ \text{id} && \{ \text{ AnaId } \} \\ &= (\![\beta]\!)_G \end{aligned}$$

As an abbreviation for the composition of two metamorphisms we introduce the notion of an *ADT-filter*. The *C-filter* from D to D' is defined as:

$$D \overset{f}{\rightsquigarrow} C \overset{g}{\rightsquigarrow} D' = C \overset{g}{\rightsquigarrow} D' \circ D \overset{f}{\rightsquigarrow} C \qquad \text{(FilterDef)}$$

Here D and D' are called the *source* and *target* of the filter, and C is called the *filter data type*. Again, we omit f and g if they are just identities.

A New Programming Style for ADTs. Let us now consider some metamorphic programs. First of all, examples for algebraic data types translate directly from the corresponding catamorphisms. For instance, the length of a list can be computed by the metamorphism

$$\mathit{length} \;=\; \mathit{List} \overset{I+\pi_2}{\rightsquigarrow} \mathit{Nat}$$

We can always save the metamorphism map, here $I + \pi_2$, by selecting a target ADT whose constructor functor agrees with the functor of the source ADT destructor. We can actually calculate the desired ADT as follows.

$$\begin{aligned} \mathit{List} \overset{I+\pi_2}{\rightsquigarrow} \mathit{Nat} &= [\![[\mathit{Zero}, \mathit{Succ}] \circ (I + \pi_2), \mathit{out}_{L_A}]\!]_{L_A} && \{\ \text{MetaHylo}\ \} \\ &= [\![[\mathit{Zero} \circ I, \mathit{Succ} \circ \pi_2], \mathit{out}_{L_A}]\!]_{L_A} && \{\ \text{sum}\ \} \end{aligned}$$

Thus, we can define the L_A, N-bialgebra $\mathit{Count} = ([\mathit{Zero}, \mathit{Succ} \circ \pi_2], \mathit{out}_N)$, and we obtain for *length* the modified form:

$$\mathit{length} \;=\; \mathit{List} \rightsquigarrow \mathit{Count}$$

In an actual programming environment there will be lots of different ADTs representing many algebra/coalgebra-combinations. We envision a system that supports the writing of metamorphisms by automatically offering sets of functor-matching target ADTs and/or sets of natural transformations that can be used as maps in metamorphisms.

Now let us consider the more interesting case for non-algebraic ADTs. We are eventually able to count the number of elements in a set by:

$$\mathit{count} = \mathit{Set} \rightsquigarrow \mathit{Count}$$

Mapping a function f to all elements of a set can be expressed by:

$$\mathit{mapset} \;=\; \mathit{Set} \overset{I+f\times I}{\rightsquigarrow} \mathit{Set} \qquad (= \mathit{Set} \overset{L_A(f)}{\rightsquigarrow} \mathit{Set})$$

And we must not forget the factorial function, which can be computed by:

$$\mathit{fac} \;=\; \mathit{Range} \rightsquigarrow \mathit{Prod}$$

Filters are very handy in expressing certain algorithms, for example,

$\mathit{List} \rightsquigarrow \mathit{Set} \rightsquigarrow \mathit{List}$	Remove duplicates
$\mathit{List} \rightsquigarrow \mathit{Set} \overset{I+\pi_2}{\rightsquigarrow} \mathit{Nat}$	Number of different list elements
$\mathit{List} \rightsquigarrow \mathit{Queue} \rightsquigarrow \mathit{List}$	List reverse
$\mathit{List} \rightsquigarrow \mathit{PQueue} \rightsquigarrow \mathit{List}$	Heapsort

We have not defined the ADT *PQueue* for priority queues yet. This can be done similar to *Queue*, except that the destructor selects the smallest instead of the last list element. The reader might wonder whether *Stack* (= *List*) instead of *Queue* should be used in the list reverse example. *Queue* is indeed the proper filter here, since metamorphisms proceed in a "bottom-up" manner (which means for L_A "right-to-left"), that is, the last element of the first list will be inserted first into the queue and will thus be consed last to the target list. One might also object that the given program for reverse is unacceptably inefficient because during the decomposition phase each intermediate queue value is reversed twice thus resulting in a quadratic running time. We will address two ways for optimization in Sect. 5.

Until now we have only worked with types of linear functors. It is clear that all presented concepts also apply to, say, tree-like structures. For these the ADT approach gives a very nice view on divide-and-conquer algorithms.

With a final example we give an impression of the power the metamorphic programming style can offer when the right abstractions are chosen. Suppose we have defined a representation for graphs (based on a suitable functor *Gr*) and a function *roots* for computing and removing the list of a graph's roots.[4] We can then define an ADT $RootGraph = (in_{Gr}, roots)$, which can be used for realizing topological sorting as follows:[5]

$$topsort \;=\; flatten \circ RootGraph \rightsquigarrow List$$

5 Program Transformation

One interesting property of ADTs is *invertability*:

$$D \text{ is } invertible \iff \overline{\alpha} \circ \alpha = \mathrm{id}$$

Invertible data types are important, since they can be fused away (see Theorem 1 below). In particular, all algebraic data types are invertible. It is clear that a data type can be invertible only if destructor and constructor have compatible signatures:

Lemma 1. *An F, G-algebra D can be invertible only if $F = G$.* □

Next we consider how program transformation and optimization present themselves in the framework of bialgebras and metamorphisms. First of all, we stress that we can use all of the existing results developed for algebraic data types: we show that fusion of algebraic data types is still possible, and we demonstrate the use of well-known laws in the optimization of metamorphic programs. In addition, we show that the fixed recursion pattern enables specific optimizations for ADTs, and finally we show that the filter programming style offers optimization opportunities that go beyond fusion and even promise asymptotic speed-ups.

[4] We will show a simple relational representation for graphs in Sect. 5. Based on that a function for computing roots is given by $- \circ \langle dom, rng \rangle$.

[5] Note that this does *not* work with the representation of Sect. 5, since it is not possible to represent isolated nodes, which might occur during the graph decomposition. The extension is not difficult, but it is not needed here to understand the point.

A Fusion Law for ADTs. An important property of invertible data types is that they do not have an effect as filter data types, that is, they can be safely omitted from filters.

Theorem 1 (Filter Fusion). *C is invertible* $\Longrightarrow\ D{\leadsto}C{\leadsto}D' = D{\leadsto}D'$.

Proof.

$$
\begin{aligned}
D{\leadsto}C{\leadsto}D' &= C{\leadsto}D' \circ D{\leadsto}C && \{\text{ FilterDef }\}\\
&= [\![\beta,\overline{\varphi}]\!]_H \circ [\![\varphi,\overline{\alpha}]\!]_G && \{\text{ MetaHylo }\}\\
&= [\![\beta,\overline{\varphi}]\!]_G \circ [\![\varphi,\overline{\alpha}]\!]_G && \{\text{ Lemma 1 }\}\\
&= [\![\beta,\overline{\alpha}]\!]_G && \{\text{ Assumption, HyloFusion }\}\\
&= D{\leadsto}D' && \{\text{ MetaHylo }\}
\end{aligned}
$$

□

This is a reformulation of the well-known fusion law for algebraic data types. Its importance lies in the fact that the extension to ADTs and metamorphisms is conservative in the sense that the fusion optimization for algebraic data types is not affected and can still be applied in the extended framework.

Applying Classical Transformations. Assume we represent a graph by a binary relation on integers. We can use the already defined ADT *Set* for this; we call it *Rel* here just for clarity. A simple method for computing the set of all nodes in a graph is then to take the union of the domain and the codomain of the relation, which are defined by two simple metamorphisms:

$$
\begin{aligned}
\mathit{dom} &= \mathit{Rel} \overset{I+\pi_1\times I}{\leadsto} \mathit{Set}\\
\mathit{rng} &= \mathit{Rel} \overset{I+\pi_2\times I}{\leadsto} \mathit{Set}\\
\mathit{nodes} &= \cup \circ \langle \mathit{dom}, \mathit{rng}\rangle
\end{aligned}
$$

With this implementation the relation must be traversed twice, once for computing the left components of all pairs and once for computing the right components. With the aid of the so-called *banana-split* law [4]

$$\langle (\!|\alpha|\!)_F, (\!|\beta|\!)_F\rangle \quad = \quad (\!|\langle \alpha \circ F(\pi_1), \beta \circ F(\pi_2)\rangle|\!)_F$$

we can obtain an improved version of *nodes* in which *dom* and *rng* are computed by a single scan over the relation. First, we expand *dom* and *rng* by (MetaHylo) and (HyloSplit) to:

$$
\begin{aligned}
\mathit{dom} &= (\!|\mathit{in}_{L_A} \circ I + \pi_1 \times I|\!) \circ [\!(\mathit{deset})\!]\\
\mathit{rng} &= (\!|\mathit{in}_{L_A} \circ I + \pi_2 \times I|\!) \circ [\!(\mathit{deset})\!]
\end{aligned}
$$

By factorizing the anamorphism from the split we get:

$$\mathit{nodes} = \cup \circ \langle (\!|\mathit{in}_{L_A} \circ I + \pi_1 \times I|\!), (\!|\mathit{in}_{L_A} \circ I + \pi_2 \times I|\!)\rangle \circ [\!(\mathit{deset})\!]$$

Now we can apply the banana-split law (with $F = L_A$) and obtain the following optimized version for *nodes*:

$$\begin{aligned} nodes &= \cup \circ (\!|\langle (in_{L_A} \circ I + \pi_1 \times I) \circ L_A(\pi_1), \\ &\qquad (in_{L_A} \circ I + \pi_2 \times I) \circ L_A(\pi_2)\rangle|\!) \circ [\![deset]\!] \end{aligned}$$

This can be simplified by evaluating L_A and applying laws for product and sum yielding:

$$nodes = \cup \circ (\!|\langle in_{L_A} \circ I + \pi_1 \times \pi_1, in_{L_A} \circ I + \pi_2 \times \pi_2\rangle|\!) \circ [\![deset]\!]$$

which can be finally written as a metamorphism:

$$nodes = \cup \circ Rel \overset{\langle I+\pi_1\times\pi_1, I+\pi_2\times\pi_2\rangle}{\rightsquigarrow} (Set \times Set)$$

Depending on the definition of the function $\cup$, we can possibly optimize further. For example, if $\cup$ is itself defined by a catamorphism, we can fuse that definition with the metamorphism just obtained. We do not elaborate on this here, the goal of this part was just to show that optimizations and transformations can be well performed using already existing laws.

Exploiting Fixed Recursion Scheme. We have already noted that the filter for implementing list reverse is unacceptably inefficient, since actually each tail of the list is reversed twice. This gives a quadratic running time, and, no doubt, a direct use of the function *rev* would be much better.[6] But if we look at how the queue is used in a metamorphism, we observe that in each step one element is taken from the queue and the (intermediate) queue values themselves are never needed, except for decomposing/dequeueing. In order to exploit this knowledge we formulate equations for different versions of the queue (q_i) and the dequeued elements (x_i). We abbreviate $\pi_1 \circ out_{L_A}$ by *hd* and $\pi_2 \circ out_{L_A}$ by *tl*. Recall the definition of $dequeue = I + I \times rev \circ out_{L_A} \circ rev$. Now given an (non-empty) intermediate queue q_{i-1}, we have:

$$\begin{aligned} q_i &= (rev \circ \pi_2 \circ out_{L_A} \circ rev)\ q_{i-1} &= (rev \circ tl \circ rev)\ q_{i-1} \\ x_i &= (\pi_1 \circ out_{L_A} \circ rev)\ q_{i-1} &= (hd \circ rev)\ q_{i-1} \end{aligned}$$

Since $q_{i-1} = (rev \circ tl \circ rev)\ q_{i-2}$ we have $q_i = (rev \circ tl \circ rev \circ rev \circ tl \circ rev)\ q_{i-2} = (rev \circ tl^2 \circ rev)\ q_{i-2}$. By induction it follows (given an initial queue q_0) that

$$q_i = (rev \circ tl^i \circ rev)\ q_0$$

[6] In general, however, we do not know about the implementation of an ADT, and thus we might not have access to a function like *rev*.

Now we observe that the last queue value is *Nil* and that all other queue values are only used for dequeueing. This means that we can work inside the decomposition with reversed queues, that is, using $r_i = rev\ q_i$ we get:

$$\begin{aligned} r_1 &= rev\ q_0 \\ r_i &= rev\ ((rev \circ tl \circ rev)\ q_{i-1}) = (tl \circ rev)\ q_{i-1} = tl\ r_{i-1} \\ x_i &= hd\ r_{i-1} \end{aligned}$$

This gives a much more efficient implementation for the *Queue* ADT. In particular, the representing list has to be reversed only once.

Single-Threaded Analysis for Free! Consider the filters *List*⇝*Queue*⇝*List* and *List*⇝*PQueue*⇝*List*. First, from the definition of filter it is clear that: (i) the filter ADTs *Queue* and *PQueue* are completely built up before they are decomposed. Second, from the definition of metamorphism it can be seen that (ii) an ADT is constructed from one generator (the source ADT) where only one version exists at any time, and (iii) an ADT is destructed just from one consumer (the target ADT) thus also maintaining only one version at any time.

Hence at any time only one version of the filter ADT is referenced, and this means that the update operations to be performed on the filter can be safely implemented in an imperative way. This can increase the efficiency of programs much more than fusion is ever able to achieve. We are faced with a twisted situation here: it is not the elimination of data structures that improves the running time of programs, but rather the introduction of filter structures.

The nice thing is that a compiler does not need a sophisticated analysis technique to determine single-threadedness. Selecting update-in-place implementations is particularly important for data types like arrays or graphs, since persistent (= functional) implementations for these can become quite complex [23, 6], and as demonstrated in [8], predefined imperative implementations of fold operations can speed up computations considerably.

6 Related Work

Much of the work concerning catamorphisms on algebraic data types has already been mentioned in the introduction. There is surprisingly little work addressing structured recursion on non-algebraic data types, that is, data types satisfying equational laws. In particular, most approaches deal with specific data types, and there is almost no general framework available that could be used for a large class of abstract data types.

Chuang presents in [5] essentially three different views of arrays and defines for each view corresponding fold operations. Gibbons [12] defines a data type for directed acyclic multi-graphs. With a careful choice of operations, which obey certain algebraic laws, the definition of graph catamorphisms becomes feasible, and some functions on graphs, such as reversing the edges of a graph (graph reversal) or determining shortest paths (measured in number of edges), can be expressed as graph catamorphisms. However, the whole approach is very limited, since it applies only to acyclic graphs having no edge labels.

We have presented a more general view of graphs in [8]. In that paper an important aspect was the definition of a couple of fold operations that can be used to express operations, such as graph reversal, depth first search, evaluation of expression DAGs, or computing all simple paths in a graphs. Two theorems for program fusion were presented that allow the removal of intermediate search trees as well as intermediate graph structures.

The only general approach for expressing catamorphisms over non-free data types we know of is the work of Fokkinga [10, 9]. The idea is to represent terms by combinators called *transformers* and to represent an equation by a pair of transformers. Several properties of transformers are investigated, and it is shown how transformers can be combined to yield new transformers thus resulting in a variable-free language for expressing equations. The use of transformers is demonstrated in showing the equivalence of two different stack implementations. However, the whole approach suffers from the already mentioned restrictions caused by the constraints that homomorphisms must map to quotients.

7 Conclusions

We have demonstrated how the structured recursion programming discipline can be applied to abstract data types. The main idea was to represent ADTs by bialgebras and to express mappings between ADTs by metamorphisms.

Our approach demands the explicit definition of destructors. However, this additional effort pays off, since it offers much freedom in the design of ADTs, in particular, the separation into algebra and coalgebra provides a high degree of modularity. Moreover, it also provides with metamorphisms a much more general computing device than homomorphisms, since we can map into types with less structure.

Nevertheless, metamorphisms on bialgebras are a conservative extension of homomorphisms: the fusion law for algebraic data types is still valid and can be applied for invertible ADTs. Moreover, a very promising property of filter ADTs is that they can be safely implemented in a destructive way without loosing referential transparency, since metamorphisms (and filters) use them in a single-threaded way.

References

[1] J. Backus. Can Programming be Liberated from the von Neumann Style? A Functional Style and its Algebra of Programs. *Communications of the ACM*, 21:613–641, 1978.

[2] M. Barr and C. Wells. *Category Theory for Computing Science*. Prentice-Hall International, 1996.

[3] R. S. Bird. Lectures on Constructive Functional Programming. In M. Broy, editor, *Constructive Methods in Computer Science*, NATO ASI Series, Vol. 55, pages 151–216, 1989.

[4] R. S. Bird and O. de Moor. *The Algebra of Programming*. Prentice-Hall International, 1997.

[5] T.-R. Chuang. A Functional Perspective of Array Primitives. In *2nd Fuji Int. Workshop on Functional and Logic Programming*, pages 71–90, 1996.

[6] P. F. Dietz. Fully Persistent Arrays. In *Workshop on Algorithms and Data Structures*, LNCS 382, pages 67–74, 1989.

[7] H. Dybkjær. *Category Theory, Types, and Programming Laguages*. PhD thesis, University of Copenhagen, 1991.

[8] M. Erwig. Functional Programming with Graphs. In *2nd ACM Int. Conf. on Functional Programming*, pages 52–65, 1997.

[9] M. M. Fokkinga. *Law and Order in Algorithmics*. PhD thesis, University of Twente, 1992.

[10] M. M. Fokkinga. Datatype Laws without Signatures. *Mathematical Structures in Computer Science*, 6:1–32, 1996.

[11] T. Fukushima and C. Tuckey. *Charity User Manual*, January 1996.

[12] J. Gibbons. An Initial Algebra Approach to Directed Acyclic Graphs. In *Mathematics of Program Construction*, LNCS 947, pages 282–303, 1995.

[13] A. Gill, J. Launchbury, and S. L. Peyton Jones. A Short Cut to Deforestation. In *Conf. on Functional Programming and Computer Architecture*, pages 223–232, 1993.

[14] T. Hagino. *A Categorical Programming Language*. PhD thesis, University of Edinburgh, 1987.

[15] T. Hagino. Codatatypes in ML. *Journal of Symbolic Computation*, 8:629–650, 1993.

[16] Z. Hu, H. Iwasaki, and M. Takeichi. Deriving Structural Hylomorphisms from Recursive Definitions. In *1st ACM Int. Conf. on Functional Programming*, pages 73–82, 1996.

[17] Z. Hu, H. Iwasaki, M. Takeichi, and A. Takano. Tupling Calculation Eliminates Multiple Data Traversals. In *2nd ACM Int. Conf. on Functional Programming*, pages 164–175, 1997.

[18] J. T. Jeuring. *Theories for Algorithm Calculation*. PhD thesis, University of Utrecht, 1993.

[19] D. J. Lehmann and M. B. Smyth. Algebraic Specification of Data Types: A Synthetic Approach. *Mathematical Systems Theory*, 14:97–139, 1991.

[20] G. Malcolm. Homomorphisms and Promotability. In *Mathematics of Program Construction*, LNCS 375, pages 335–347, 1989.

[21] L. Meertens. Algorithmics – Towards Programming as a Mathematical Activity. In *CWI Symp. on Mathematics and Computer Science*, pages 289–334, 1986.

[22] E. Meijer, M. Fokkinga, and R. Paterson. Functional Programming with Bananas, Lenses, Envelopes and Barbed Wire. In *Conf. on Functional Programming and Computer Architecture*, pages 124–144, 1991.

[23] M. E. O'Neill and F. W. Burton. A New Method for Functional Arrays. *Journal of Functional Programming*, 7(5):487–513, 1997.

[24] J.J.M.M. Rutten. Universal Coalgebra: a Theory of Systems. Report CS-R9652, CWI, Department of Software Technology, 1996.

[25] T. Sheard and L. Fegaras. A Fold for all Seasons. In *Conf. on Functional Programming and Computer Architecture*, pages 233–242, 1993.

[26] A. Takano and E. Meijer. Shortcut Deforestation in Calculational Form. In *Conf. on Functional Programming and Computer Architecture*, pages 306–313, 1995.

Condensing Lemmas for Pure Type Systems with Universes

Blas C. Ruiz Jiménez

Dpto. de Lenguajes y Ciencias de la Computación. E.T.S.I. Informática.
Campus Teatinos, 29071–Málaga, Spain
e-mail: blas@lcc.uma.es, fax: +34–5–2131397

Abstract. Pure Type Systems with universes (γPTS) provide a right frame to model programming languages features and are the core of Logical Frameworks, widely used in theorem proof systems. For these systems some authors propose a single rule, which includes both typing under β–conversion and the relation between universes. Our proposal adds an independent rule parameterized over a relation γ between sorts. Non trivial properties of the PTS like the *weak strengthening lemma* can be obtained in γPTS by extending a method proposed by van Benthem Jutting and using weak-closure for γ–reduction. This lemma is important due to two main reasons: (1) it provides a condensing lemma that determines in the underlying logic system a *cut* rule that simplifies the task in proof assistant systems; (2) the proof of type checking decidability can be eased in some normalizing systems.

1 Introduction

As the literature reflects, the Pure Type Systems (PTS) [2] introduced by Terlouw and Berardi are a right frame to model programming languages features and logic systems using Howard-Curry isomorphism (propositions-as-types).

The inclusion of an explicit hierarchy between universes without needing a type for every type can be used to work in a similar way without sacrificing the consistency. For these systems, some authors [9,11] propose a single rule (the *abstract reduction* rule), which includes both typing under β–conversion and the relation between generalized universes (dependent products). Our proposal captures the typing under universes with an independent additional rule. This rule is parameterized over a relation γ between sorts (basic universes) that determines another hierarchical level (apart from the hierarchy determined by the usual axioms and rules in PTS) and allows the definition of a relation $\twoheadrightarrow_\gamma$ between generalized universes. These systems, named γPTS in [12], are a natural generalization of PTS and other theories (as CC^ω [6,8]) and are a special case of γPTS for a proper initial relation γ.

Elementary properties of the PTS (subject β–reduction, substitution lemma, etc.) can be expressed in γPTS without imposing restrictions neither to γ or to the axioms and rules. Other properties, as type checking decidability or the strengthening and condensing properties, need the addition of restrictions to

A.M. Haeberer (Ed.): AMAST'98, LNCS 1548, pp. 422–437, 1998.

the system. These restrictions have been captured in an uniform way using the concept of weak–closure for γ–reduction:

$$\Gamma \vdash c : \Box \qquad c \twoheadrightarrow_\gamma c' \;\Rightarrow\; \Gamma \vdash c' : \Box' \qquad [S_s^\partial \gamma]$$

The technique shown in this paper is an extension to the method proposed by L. van Benthem Jutting in [3]. Thus, the set Λ of terms will be divided into two disjoint classes: $\Lambda = T_v \cup T_s$. In class T_v we have uniqueness of types for a relation $\simeq$ (defined in terms of relations $\twoheadrightarrow_\gamma$ and $\twoheadrightarrow_\beta$), and we also get another interesting property, the *strengthening lemma:*

$$\Gamma_1 \vdash a : \_ \qquad \Gamma_2 \vdash a : A \qquad \Gamma_1 \subseteq \Gamma_2 \;\Rightarrow\; \Gamma_1 \vdash a : A' \;{}_\beta\!\twoheadleftarrow A$$

Class T_s verifies another essential property, the *weak strengthening lemma:*

$$\Gamma_1 \vdash a : \_ \qquad \Gamma_1 \subseteq \Gamma_2 \;\Rightarrow\; \Gamma_2 \underline{\partial} a \subseteq \Gamma_1 \overline{\partial} a$$

where $\overline{\partial}$ and $\underline{\partial}$ denote the operators

$$\Gamma \overline{\partial} a = \{\, s \mid \exists d, \Delta[a \twoheadrightarrow_\beta \lambda\Delta.d \wedge \Gamma, \Delta \vdash d : s]\}$$
$$\Gamma \underline{\partial} a = \{\, s \mid \exists \Delta[\Gamma \vdash a : \Pi\Delta.s]\}$$

which allow us to generalize some results already obtained in [3]. This property directly provides the standard condensation lemma, i.e.

$$\Gamma_1, y : D, \Gamma_2 \vdash a : A \qquad y \notin \mathrm{FV}(\Gamma_2) \cup \mathrm{FV}(a : A) \;\Rightarrow\; \Gamma_1, \Gamma_2 \vdash a : A$$

The remainder of the paper is organized as follows. In section 2 we formalize the γPTS, state the elementary properties and characterize the possible types. Section 3 studies the concept of weak–closure for γ–reduction and some of its consequences, like the *Church-Rosser property* for class T_v. Section 4 is devoted to the *strengthening lemmas* and to one of its corollaries: the standard condensation lemma. The essential properties of $\overline{\partial}$ and $\underline{\partial}$ operators are stated, and will be used in the following section when proving the type checking decidability for a class of normalizing systems.

We have also included in this paper some relevant examples to show that some of the conditions in the results cannot be eliminated or even relaxed. On the other hand, most of the proofs have been omitted due to the lack of space. A final appendix summarizes the notation used throughout the paper to improve its readability.

2 Pure Type Systems with Universes (γPTS)

Considering a set of variables $\mathcal{V}$ ($x, y, \ldots \in \mathcal{V}$) and a set of constants or *sorts* $\mathcal{S}$ ($p, s, \Box, \triangle, \ldots \in \mathcal{S}$), the set Λ of terms for a γPTS is inductively defined as:

$$a \in \mathcal{V} \cup \mathcal{S} \qquad \Rightarrow\; a \in \Lambda$$
$$A, B, a, b \in \Lambda \;\Rightarrow\; a\ b,\ \lambda x : A.b,\ \Pi x : A.B \in \Lambda$$

We will denote by $\rightarrow_\beta$ a one step β–reduction, $\twoheadrightarrow_\beta$ its reflexive and transitive closure, and by $=_\beta$ the equality generated by $\twoheadrightarrow_\beta$ [1] . The set of β–normal forms will be denoted by β_n, $\mathrm{FV}(a)$ the set of free variables and $\mathrm{FV}(a : A) = \mathrm{FV}(a) \cup \mathrm{FV}(A)$. A *context* Γ is a sorted sequence with the following form $x_1 : A_1, \ldots, x_n : A_n$. We will denote $x : A \in \Gamma$ iff $\exists i[x \equiv x_i \wedge A \equiv A_i]$, $\Gamma \subseteq \Gamma'$ iff $\forall x[x : A \in \Gamma \;\Rightarrow\; x : A \in \Gamma']$, $\mathrm{Var}(\Gamma) = \{x_1, \ldots, x_n\}$; $x \in \Gamma$ means $x \in \mathrm{Var}(\Gamma)$; $\mathrm{FV}(\Gamma) \stackrel{def}{=} \bigcup_{1 \le i \le n} \mathrm{FV}(A_i)$, $(\Gamma_1, x : A, \Gamma_2) \downarrow_x \equiv \Gamma_1$. For the context $\Delta \equiv x : A, y : B, z : C$ we denote by $\Pi\Delta.P$ the term $\Pi x : A.\Pi y : B.\Pi z : C.P$.

This notation will also be used for $\lambda\Delta.b$. For a relation $\rightarrow_\eta$ from terms, we extend the concept of reduction context $\Gamma \rightarrow_\eta$ as usual.

A γPTS is defined by a tuple $\lambda S = (\mathcal{S}, \gamma, \mathcal{A}, \mathcal{R})$, where γ is a relation over $\mathcal{S}$ (its objects are called classes or basic universes), $\mathcal{A}(\subseteq \mathcal{S} \times \mathcal{S})$ is the set of *axioms* and $\mathcal{R}(\subseteq \mathcal{S} \times \mathcal{S} \times \mathcal{S})$ is the set of *rules*. We will abbreviate $\Box : \Delta \in \mathcal{A}, s_1 : s_2 : s_3 \in \mathcal{R}$. $(s, s') \in \gamma$ means that s *is a subclass or subuniverse of* s'; $\rightarrow_\gamma$ denotes right Π–compatible closure:

$$\frac{(Q,R)\in\gamma}{Q\rightarrow_\gamma R} \qquad \frac{Q\rightarrow_\gamma R}{\Pi x{:}M.Q\rightarrow_\gamma \Pi x{:}M.R}$$

We will denote by $\twoheadrightarrow_\gamma$ the reflexive and transitive closure of $\rightarrow_\gamma$ and by $=_\gamma$ the equivalence (not necessarily compatible) generated by $\twoheadrightarrow_\gamma$; we will make similar assumptions over the $\gamma\beta$ relation ($\rightarrow_{\gamma\beta}\equiv\rightarrow_\gamma \cup \rightarrow_\beta$). The symbol $\cdot$ denotes the composition of two relations: $x\ R \cdot R'\ y \iff \exists z[xRz \wedge zR'y]$. We will omit the symbol $\cdot$ if there is not possible confusion: $\twoheadrightarrow_\beta \cdot {}_\gamma\twoheadleftarrow \equiv \twoheadrightarrow_\beta\ {}_\gamma\twoheadleftarrow$. We will write $\rightarrow_\gamma \models \Diamond$ as $\gamma \models \Diamond, \beta \Diamond \gamma$ for the commutativity of β and γ relations, and CR for the Church-Rosser's property:

$$\beta \in CR \equiv\ {}_\beta\twoheadleftarrow\twoheadrightarrow_\beta \subseteq \twoheadrightarrow_\beta\ {}_\beta\twoheadleftarrow \qquad \beta \Diamond \gamma \equiv\ {}_\beta\twoheadleftarrow\twoheadrightarrow_\gamma \subseteq \twoheadrightarrow_\gamma\ {}_\beta\twoheadleftarrow$$

The notion of derivation $\Gamma \vdash_{\lambda S} a : A$, that we will also write as $\Gamma \vdash a : A$, is defined by the inductive system in the attached figure. The first seven rules are the usual ones in the PTS. The last one captures typing under the presence of classes or universes. When γ is empty the system is an standard PTS. In the

γPTS

(ax)	$\dfrac{}{\vdash p{:}s}$	$p : s \in \mathcal{A}$
(var)	$\dfrac{\Gamma\vdash A{:}s}{\Gamma,x{:}A\vdash x{:}A}$	$x \notin \Gamma$
$(weak)$	$\dfrac{\Gamma\vdash b{:}B \qquad \Gamma\vdash A{:}s}{\Gamma,x{:}A\vdash b{:}B}$	$x \notin \Gamma$
(apl)	$\dfrac{\Gamma\vdash f{:}(\Pi x{:}A.F) \qquad \Gamma\vdash a{:}A}{\Gamma\vdash fa{:}F[x{:=}a]}$	
(Π)	$\dfrac{\Gamma\vdash A{:}s_1 \qquad \Gamma,x{:}A\vdash B{:}s_2}{\Gamma\vdash \Pi x{:}A.B\ :\ s_3}$	$\overline{s} \equiv s_1 : s_2 : s_3 \in \mathcal{R}$
(λ)	$\dfrac{\Gamma\vdash A{:}s_1 \qquad \Gamma,x{:}A\vdash B{:}s_2 \qquad \Gamma,x{:}A\vdash b{:}B}{\Gamma\vdash \lambda x{:}A.b\ :\ \Pi x{:}A.B}$	$\overline{s} \in \mathcal{R}$
(β)	$\dfrac{\Gamma\vdash a{:}A \qquad \Gamma\vdash A'{:}s}{\Gamma\vdash a{:}A'}$	$A =_\beta A'$
(γ)	$\dfrac{\Gamma\vdash a{:}A \qquad \Gamma\vdash A'{:}s}{\Gamma\vdash a{:}A'}$	$A \twoheadrightarrow_\gamma A'$

derivation $\Gamma \vdash a : A$, term a is the *subject* and A the *predicate*. A context Γ is *legal* (we will use $\Gamma \vdash$) when there are terms a and A such that $\Gamma \vdash a : A$. We will write $\Gamma \vdash a : A : s$ for the pair of derivations $\Gamma \vdash a : A \wedge \Gamma \vdash A : s$, and $\Gamma \vdash a : A, A'$ for $\Gamma \vdash a : A \wedge \Gamma \vdash a : A'$. A free variable to the right of an implication is implicitly quantified in the correct way. For example, the consequent of the implication $\Gamma \vdash \Box : R \Rightarrow \Box : s \in \mathcal{A}$ is read as $\exists s \in \mathcal{S}[\ \Box : s \in \mathcal{A}\ ]$, and will also be written as $\Box : \_ \in \mathcal{A}$.

The $\vdash_w$ system obtained when substituting the constraint in the (*weak*) rule for $b \in \mathcal{S} \cup \mathcal{V}, x \notin \Gamma$ is equivalent to the former: $\Gamma \vdash b : B \iff \Gamma \vdash_w b : B$. This rule is the one exposed in [4] and it is interesting on it own because the generation lemma does not depend on the (*weak*) rule.

Lemma 1.
(*i*) $\beta \diamond \gamma$
(*i'*) $A \twoheadrightarrow_\gamma B \in \beta_n \;\Rightarrow\; A \in \beta_n$
(*ii*) $\twoheadrightarrow_\gamma \cdot \twoheadrightarrow_\beta \subseteq \twoheadrightarrow_\beta \cdot \twoheadrightarrow_\gamma$
and for terms following the pattern $\Pi\Delta.s : \twoheadrightarrow_\beta \cdot \twoheadrightarrow_\gamma \subseteq \twoheadrightarrow_\gamma \cdot \twoheadrightarrow_\beta$
(*iii*) $A \twoheadrightarrow_{\gamma\beta} B \;\Rightarrow\; A[x := N] \twoheadrightarrow_{\gamma\beta} B[x := N]$
(*iv*) $A \twoheadrightarrow_\gamma B \;\Rightarrow\; \mathrm{FV}(A) = \mathrm{FV}(B)$
(*v*) $\gamma \models \diamond \quad\Rightarrow\; \gamma\beta \in CR$

The next lemma resumes the basic properties of the relation that will appear in the generation lemma.

Lemma 2 (Order generated by γ). (*a*) *Let $\rhd_1, \rhd_2, \rhd_3$ be three reflexive and transitive relations in Λ verifying $\forall i, j[i > j][\rhd_i \cdot \rhd_j \subseteq \rhd_j \cdot \rhd_i]$. Then*

$$(\rhd_1 \cup \rhd_2 \cup \rhd_3)^* = \rhd_1 \cdot \rhd_2 \cdot \rhd_3$$

In particular, when $\succeq\!\!> \;\equiv\; \twoheadrightarrow_\beta \cdot \twoheadrightarrow_\gamma \cdot {}_\beta\!\twoheadleftarrow$,

(*b*) $\succeq\!\!> \;=\; (\twoheadrightarrow_\beta \cup \twoheadrightarrow_\gamma \cup {}_\beta\!\twoheadleftarrow)^* \;=\; (=_\beta \cup \twoheadrightarrow_\gamma)^* \;=\; (=_\beta \cdot \twoheadrightarrow_\gamma \cdot =_\beta)$

(*c*) *The $\succeq\!\!>$ relation is the least reflexive, transitive and right Π-compatible relation that includes the relation* $(=_\beta \cup \twoheadrightarrow_\gamma)$

(*d*) *If $\twoheadrightarrow_\gamma$ is an (partial) order relation in $\mathcal{S} \times \mathcal{S}$, then $\succeq\!\!>$ is an order relation with respect to $=_\beta$; i.e.: $A \succeq\!\!> A' \succeq\!\!> A \;\Rightarrow\; A =_\beta A'$*

Proof. (*a*) is trivial; (*b*) follows from (*a*) taking $\rhd_1 \equiv \twoheadrightarrow_\beta, \rhd_2 \equiv \twoheadrightarrow_\gamma, \rhd_3 \equiv {}_\beta\!\twoheadleftarrow$; by $\beta \diamond \gamma$, $\rhd_3 \cdot \rhd_2 \subseteq \rhd_2 \cdot \rhd_3$. By $\beta \in CR$, $\rhd_3 \cdot \rhd_1 \subseteq \rhd_1 \cdot \rhd_3$, and by Lemma 1(*ii*), $\rhd_2 \cdot \rhd_1 \subseteq \rhd_1 \cdot \rhd_2$. Similarly, by $\beta \in CR, \beta \diamond \gamma$ and by Lemma 1(*ii*): $=_\beta \cdot \twoheadrightarrow_\gamma \cdot =_\beta \subseteq \twoheadrightarrow_\beta \cdot \twoheadrightarrow_\gamma \cdot {}_\beta\!\twoheadleftarrow$, and the reciprocal inclusion $\supseteq$ is trivial. The rest of the equalities follow from $(=_\beta \cup \twoheadrightarrow_\gamma)^* = (\twoheadrightarrow_\beta \cup \twoheadrightarrow_\gamma \cup {}_\beta\!\twoheadleftarrow)^*$ that is obvious because of $\beta \in CR$. In order to prove (*d*) it is enough to show that

$$B^* \twoheadrightarrow_\gamma Q^* \;{}_\beta\!\twoheadleftarrow A' \wedge A' \twoheadrightarrow_\beta Z \twoheadrightarrow_\gamma B^* \qquad \Rightarrow \qquad B^* =_\beta A' \qquad (1)$$

and observe the next diagram on the left (the non annotated arrows correspond to the $\twoheadrightarrow_\beta$ relation). The part of the diagram [1] is due to $\beta \in CR$, and the part [2] to $\beta \diamond \gamma$.

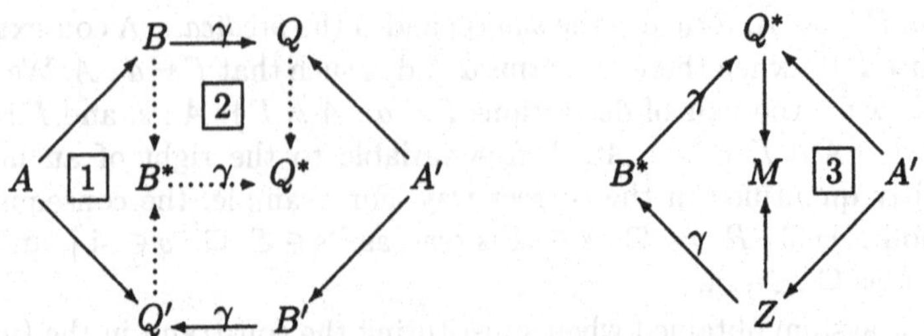

In order to prove (1), using Lemma 1(i, ii), it is enough to prove

$$B^* \twoheadrightarrow_\gamma Q^* \twoheadrightarrow_\beta M \;{}_\beta\!\twoheadleftarrow Z \twoheadrightarrow_\gamma B^* \quad \Rightarrow \quad B^* =_\beta Z \qquad (2)$$

and look at the former diagram on the right (the part $\boxed{3}$ of the diagram follow from $\beta \in CR$); (2) is proved by induction on the derivation of $Z \twoheadrightarrow_\gamma B^*$. □

If we include Σ–types, these results also hold, and when the $\twoheadrightarrow_\gamma$ relation is an order in $\mathcal{S} \times \mathcal{S}$ the $\succeq\!\!>$ relation is the $\preceq$ relation of [9].

Example 3 (The Generalized Calculus of Constructions as a γPTS). Let us consider the numerable set of constants $\mathcal{S} \equiv \{*, \Box_0, \ldots, \Box_i, \ldots\}_{i \in \omega}$, and the initial relation over constants $\gamma = \{* \rightarrow_\gamma \Box_0, \Box_i \rightarrow_\gamma \Box_{i+1}\}_{i\in\omega}$. Then, $\twoheadrightarrow_\gamma$ is an order relation and, by Lemma 2, $\succeq\!\!>$ is an order with respect to $=_\beta$. Let γCC^ω be the γPTS generated by $(\mathcal{S}, \gamma, \mathcal{A}, \mathcal{R})$, γ the former relation and let $\mathcal{A}$ and $\mathcal{R}$ be the following set of axioms and rules

$$\mathcal{A} = \{* : \Box_0,\ \Box_i : \Box_{i+1}\}_{i\in\omega}$$
$$\mathcal{R} = \{* : * : *,\ \Box_i : * : *,\ \Box_i : \Box_i : \Box_i,\ * : \Box_i : \Box_i\}_{i\in\omega}$$

Then, this system is equivalent to Thierry Coquand's CC^ω system [8].

Lemma 4. *Let $\Gamma \equiv x_1 : A_1, \ldots, x_n : A_n \vdash c : C$. Then*

(i) $\mathrm{FV}(c : C) \subseteq \mathrm{Var}(\Gamma)$

(ii) $\forall i, j[1 \le i, j \le n][x_i \equiv x_j \Rightarrow i = j]$

(iii) $p : q \in \mathcal{A} \Rightarrow \Gamma \vdash p : q$

(iv) $y : D \in \Gamma \Rightarrow \Gamma \vdash y : D$

(a) *Substitution lemma: If $\Gamma_1 \vdash d : D$, then*

$$\Gamma_1, y : D, \Gamma_2 \vdash c : C \;\Rightarrow\; \Gamma_1, \Gamma_2[y := d] \vdash c[y := d] : C[y := d]$$

(b) *Thinning lemma:*

$$\Gamma_2 \vdash \qquad \Gamma_1 \vdash b : B \qquad \Gamma_1 \subseteq \Gamma_2 \;\Rightarrow\; \Gamma_2 \vdash b : B$$

(c) $\Delta \vdash \lambda x : A.b : R \vee \Delta \vdash x : R \vee \Delta \vdash ab : R \;\Rightarrow\; \Delta \vdash R : \Box$

Lemma 5 (Generation lemma).

(i) $\Gamma \vdash p : R \;\Rightarrow\; p : q \in \mathcal{A} \wedge q \succeq\!\!> R$

(ii) $\Gamma \vdash x : R \;\Rightarrow\; x : R' \in \Gamma \wedge \Gamma \downarrow_x\!\vdash R' : s' \wedge R' \succeq\!\!> R$

(iii) $\Gamma \vdash \Pi x : A.B : R \;\Rightarrow\; \Gamma, x : A \vdash B : s_2 \wedge \Gamma \vdash A : s_1 \wedge s_3 \succeq\!\!> R$

(iv) $\Gamma \vdash \lambda x : A.b : R \;\Rightarrow\; \Gamma \vdash A : s_1 \wedge \Gamma, x : A \vdash b : B : s_2 \wedge \Pi x : A.B \succeq\!\!> R$

(v) $\Gamma \vdash bc : R \;\Rightarrow\; \Gamma \vdash b : \Pi x : C.F, c : C, F[x := c] : s \wedge F[x := c] \succeq\!\!> R$

Corollary 6.

(a) $\Gamma \vdash s : \Box \wedge \Delta \vdash \;\Rightarrow\; \Delta \vdash s : \Box$

(b) *Correctness of types:* $\Gamma \vdash c : C \;\Rightarrow\; s \equiv C \;\vee\; \Gamma \vdash C : s$

(c) *If* $\Gamma \vdash \lambda x : A.b : R$, *there is a sequence of terms* $\{X_i\}_{0 \leq i \leq n}$ *verifying*
$X_0 \equiv \Pi x : A.B \qquad X_n \equiv R \qquad \Gamma \vdash A : s_1 \qquad \Gamma, x : A \vdash b : B : s_2$
$\forall i [\, 0 \leq i < n \,][\, \Gamma \vdash \lambda x : A.b : X_i \wedge (X_i =_\beta X_{i+1} \vee X_i \twoheadrightarrow_\gamma X_{i+1})\,]$

(d) *If* $\Gamma \vdash \Pi x : A.B : Q$, *for a certain* $\overline{s}$ *rule*
$\Gamma \vdash A : s_1 \qquad \Gamma, x : A \vdash B : s_2 \qquad (s_3 \equiv Q \vee \Gamma \vdash Q : s, s_3 \rightsquigarrow Q)$

The reciprocal to (b) does not hold: $\vdash C : \Box$ could hold, but $\forall c[\not\vdash c : C]$; i.e., there are correct types *without inhabitants* (Corollary 16). In γCC^ω and in other systems a stronger result can be obtained

$$\Gamma \vdash c : C \quad \Rightarrow \quad \Gamma \vdash C : s \qquad [SCT]$$

and that property eliminates many technical problems. This result does not hold for every PTS, and hence, for every γPTS. However, this property can be obtained by adding some additional properties to γ (some cumulative properties).

Corollary 7. (a) *If* Γ *legal, then* $\Box : \mathcal{A}_\gamma = \{\, s \mid \Gamma \vdash \Box : s \,\}$

(b) *If* $\Gamma \vdash \Pi x : A.B : \_$, *then*
$$(\Gamma \vdash A) : (\Gamma, x : A \vdash B) : \mathcal{R}_\gamma = \{\, s \mid \Gamma \vdash \Pi x : A.B : s \,\}$$

where $\Gamma \vdash A$ *denotes the set of* Γ*-types of* A, *and*

$$\begin{aligned} \Box : \mathcal{A}_\gamma &= \{\, \Delta \mid \Box : \Delta' \in \mathcal{A}, \Delta' \equiv \Delta \vee \Delta' \twoheadrightarrow_\gamma \Delta : \_ \in \mathcal{A} \,\} \\ \mathcal{M} : \mathcal{N} : \mathcal{R}_\gamma &= \{\, \Box \mid \overline{s} \in \mathcal{R}, s_1 \in \mathcal{M}, s_2 \in \mathcal{N}, s_3 \equiv \Box \vee s_3 \twoheadrightarrow_\gamma \Box : \_ \in \mathcal{A} \,\} \end{aligned}$$

In order to achieve a decidable type checking the sets $\Box : \mathcal{A}_\gamma$ and $\mathcal{M} : \mathcal{N} : \mathcal{R}_\gamma$ must be (effectively) decidable if $\mathcal{M}$ and $\mathcal{N}$ sets are decidable.

In functional PTS [2] the uniqueness of types is obtained for the $=_\beta$ equality. Because of the (γ) rule, the corresponding result is weaker in γPTS. We now introduce a natural generalization of *functional* PTS. So, a γPTS is γ*-functional* when:

$$\begin{aligned} &p : q,\; p' : q' \in \mathcal{A} \wedge p =_\gamma p' &&\Rightarrow\; q =_\gamma q' \\ &s_1 : s_2 : s,\; s_1' : s_2' : s' \in \mathcal{R} \wedge s_1 =_\gamma s_1' \wedge s_2 =_\gamma s_2' &&\Rightarrow\; s =_\gamma s' \end{aligned}$$

Lemma 8 (Uniqueness of types for γ–functional and confluent systems). *If* λS *is* γ*-functional and* $\gamma \models \diamond$, *then* $\Gamma \vdash c : C, C' \Rightarrow C =_{\gamma\beta} C'$

This lemma can be used to prove the condensation lemma for γ–functional systems. If the system is not γ–functional the uniqueness of types is not assured, not even for the ordinary PTS. For PTS, van Benthem Jutting [3] introduces a class of terms (T_v) where the uniqueness of types under $=_\beta$ is assured. Also, if $\gamma \not\models \diamond$, the following lemma may be used to study the least class of terms with uniqueness of types for the following relation:

$$A \simeq A' \iff \exists X, Z, Z'[A \twoheadrightarrow_\beta Z \;{}_\gamma\!\twoheadleftarrow X \twoheadrightarrow_\gamma Z' \;{}_\beta\!\twoheadleftarrow A']$$

We want to assure that when $\Gamma \vdash a : A, A'$, the types A and A' must be *connected* through the relations $\twoheadrightarrow_\beta$ y $\twoheadrightarrow_\gamma$.

Lemma 9 (Properties for the $\simeq$ relation).
(i) $\simeq$ *is substitutive, reflexive and simetric*
(i′) If $\gamma^{-1} \models \diamond$*, then* $\simeq$ *is an equivalence*
(ii) $B \simeq B' \,\underline{\simeq\!\!>}\, A' \;\Rightarrow\; B \simeq A'$
(ii′) $A \,\underline{<\!\!\simeq}\, Z \,\underline{\simeq\!\!>}\, A' \;\Rightarrow\; A \simeq A'$
(iii) $\Pi x : C.F \simeq \Pi x : C'.F' \iff C =_\beta C' \wedge F \simeq F'$
(iv) Let $\mathcal{U} = \{\, a \mid \forall \Gamma, A, A' [\, \Gamma \vdash a : A, A' \;\Rightarrow\; A \simeq A']\}$*, then*
(a) $\mathcal{V} \subseteq \mathcal{U}$
(b) $b \in \mathcal{U} \;\Rightarrow\; b\,c,\; \lambda x : A.b \in \mathcal{U}$

Definition 10. *Let* T_v *be the least class of terms verifying the properties (a) and (b) in the last lemma, and* T_s *the complementary class* $(\Lambda - T_v)$*. Using Lemma 9,* T_v *and* T_s *are the inductively defined classes by the axioms*
(i) $\mathcal{V} \subseteq T_v$ *(ii)* $b \in T_v \;\Rightarrow\; b\,c,\; \lambda x : A.b \in T_v$
(i′) $\Box,\; \Pi x : A.B \in T_s$ *(ii′)* $b \in T_s \;\Rightarrow\; b\,c,\; \lambda x : A.b \in T_s$

The T_s class is closed under β-reduction, while T_v verifies the dual property:

$$a \twoheadrightarrow_\beta a' \in T_v \qquad \Rightarrow \qquad a \in T_v$$

Lemma 11 (Uniqueness of types in T_v class). *If* $a \in T_v$*,*
(a) $\Gamma \vdash a : A, A' \;\Rightarrow\; \Gamma \vdash A : \Box \qquad A \simeq A'$
(b) In any PTS, $\Gamma \vdash a : A, A' \;\Rightarrow\; A =_\beta A'$

Proof. (*a*) follows from Lemma 4 and Lemma 9; for (*b*) we take $\gamma = \emptyset$. □

The idea of dividing the set of terms into two classes ($\Lambda = T_s \cup T_v$) is due to L.S. van Benthem Jutting, and that partition is used to prove two important results: the condensation lemma and the decidability of type checking. We will also use that division for the same purpose. The result (*b*) above is the one obtained in [3](page 33). There is a better result that generalizes the Lemma 4.4 in [3](page 34):

Lemma 12 (Characterization of types in T_s class). *The relation*

$$B \overset{s}{\simeq} B' \iff \exists \Delta, \Box, \Box' [B =_\beta \Pi\Delta.\Box \wedge B' =_\beta \Pi\Delta.\Box']$$

verifies the following properties
(i) $\overset{s}{\simeq}$ *is a substitutive equivalence*
(ii) $B \overset{s}{\simeq} B' \wedge B' \,\underline{\simeq\!\!>}\, C \Rightarrow B \overset{s}{\simeq} C$
(iii) $B \overset{s}{\simeq} B' \iff \Pi x : A.B \overset{s}{\simeq} \Pi x : A.B'$
(iv) $c \in T_s \wedge \Gamma \vdash c : C, C' \Rightarrow C \overset{s}{\simeq} C'$
(v) Uniqueness of domains:

$$\Gamma \vdash a : \Pi x : A_1.B_1,\; \Pi x : A_2.B_2 \qquad \Rightarrow \qquad A_1 =_\beta A_2$$

We see that β–normal types in a term $a \in T_s$ follow the pattern $\Pi\Delta.s$. If $\Gamma \vdash a : \Pi\Delta.s, \Pi\Delta.s'$ the constants s and s' will not always be related. This result also happens in *PTS*.

3 Contexts Reductions, Subjects, and Predicates

Definition 13. *A relation* $\triangleright$ *on* Λ *verifies the following properties when*

subject reduction:	$\Gamma \vdash c : C \wedge c \triangleright c' \Rightarrow \Gamma \vdash c' : C$	$[S\triangleright]$
context reduction:	$\Gamma \vdash c : C \wedge \Gamma \triangleright \Gamma' \Rightarrow \Gamma' \vdash c : C$	$[C\triangleright]$
predicate reduction:	$\Gamma \vdash c : C \wedge C \triangleright C' \Rightarrow \Gamma \vdash c : C'$	$[P\triangleright]$

We will simplify the notation by writing $S\beta$ instead of $S \twoheadrightarrow_\beta$, and similarly for $P\gamma$. It may be interesting the not conservation of the type when the subject is reduced in γ. For this purpose, we introduce the concept of *weak subject* γ*-reduction*

$$\Gamma \vdash c : \Box \qquad c \twoheadrightarrow_\gamma c' \qquad \Rightarrow \qquad \Gamma \vdash c' : \Box' \qquad [S^{\partial}_{s}\gamma]$$

Lemma 14. *If* $\Gamma \vdash c : C$*, then*

$(i)\ c \twoheadrightarrow_\beta c' \Rightarrow \Gamma \vdash c' : C$	$[S\beta]$
$(ii)\ \Gamma \twoheadrightarrow_\beta \Gamma' \Rightarrow \Gamma' \vdash c : C$	$[C\beta]$
$(iii)\ C \twoheadrightarrow_\beta C' \Rightarrow \Gamma \vdash c : C'$	$[P\beta]$

Proof. (i) and (ii) can be proved simultaneously by induction on the derivation $\psi \equiv \Gamma \vdash c : C$ and for a single step of reduction ($c \to_\beta c'$ and $\Gamma \to_\beta \Gamma'$). The proof is identical to the one exposed in [2,7], except when ψ is inferred from the (apl)–rule. For this case, the Corollary 6 can be applied; (iii) follows from the correctness of types, $S\beta$ and from the (β)–rule. $\Box$

Corollary 15. *Let* $\vdash_{\mathcal{L}}$ *be the system obtained by replacing the* (β) *and* (γ) *rules with* [1]

$$(\beta\gamma) \qquad \frac{\Gamma \vdash_{\mathcal{L}} a : A \qquad \Gamma \vdash_{\mathcal{L}} A' : s}{\Gamma \vdash_{\mathcal{L}} a : A'}\ A \mathrel{\Rightarrow\!\!\!>} A'$$

Then, the $\vdash$ *and* $\vdash_{\mathcal{L}}$ *systems are equivalent:* $\Gamma \vdash_{\mathcal{L}} c : C \iff \Gamma \vdash c : C$

Proof. $(\Leftarrow)$: by induction on $\Gamma \vdash c : C$ using i.h. and $(=_\beta \cup \twoheadrightarrow_\gamma) \subseteq \mathrel{\Rightarrow\!\!\!>}$. $(\Rightarrow)$: by induction on $\Gamma \vdash_{\mathcal{L}} c : C$ using i.h. and $SP\beta$. $\Box$

Corollary 16.

(a) *Context substitution: If* $\Gamma \vdash Y_0 : s_0$*, then*
$\Gamma, y : Y, \Gamma' \vdash c : C \qquad Y_0 \mathrel{\Rightarrow\!\!\!>} Y \Rightarrow \Gamma, y : Y_0, \Gamma' \vdash c : C$

(b) $\Delta \vdash \qquad \Gamma \vdash s : R \Rightarrow \Delta \vdash s : R'\ {}_\beta\!\twoheadleftarrow R$

(c) *If* $S^{\partial}_{s}\gamma \vee P\gamma$*, then* $\Gamma \vdash y : R \Rightarrow \Gamma \downarrow_y \vdash R' : s \wedge R'\ {}_\beta\!\twoheadleftarrow R$

(d) $\Gamma \vdash \lambda x : A.b : R \Rightarrow \Gamma \vdash P : s_3, Q : s', \Pi x : A.B : s_3 \qquad \Gamma, x : A \vdash b : B$
and $\Pi x : A.B \twoheadrightarrow_\beta P \twoheadrightarrow_\gamma Q\ {}_\beta\!\twoheadleftarrow R$

(e) $\Gamma \vdash \lambda x : A.b : \Pi x : A'.B' \Rightarrow \Gamma, x : A \vdash b : B'$

(f) $\lambda\mathbf{S} \models N \Rightarrow \nexists a, s[\vdash a : \Pi x : s.x]$

---

[1] the $(\beta\gamma)$ rule is named elsewhere as the *abstract reduction* rule [11]

If we consider $Y_0 \twoheadrightarrow_{\gamma\beta} Y$, the property in the last corollary is a property of *$\beta\gamma$–weak context expansion* in the sense that $\Gamma \vdash Y_0 : s$ is necessary. The proof for (f) is similar to the one presented by [2](5.2.31) for *PTS*. If $\Box : \triangle \in \mathcal{A}$ and $\triangle : \Box : \Box \in \mathcal{R}$, then $\vdash (\Pi x : \Box.x) : \Box$, so, if the system is normalizing, there are correct types without inhabitants. If $\vdash a : \Pi x : s.x$ hold, the system would not be normalizing.

Lemma 17 (Predicate γ–reduction in strong cumulative systems). *If $\lambda\mathbf{S} \models S_s^{\partial}\gamma$, $[AC]$ and $[RC]$, then $\lambda\mathbf{S} \models P\gamma$, where $[AC]$ and $[RC]$ are the following strong cumulative conditions:*

$$\_ : q \in \mathcal{A} \wedge q \rightarrow_\gamma q' \;\Rightarrow\; q' : \_ \in \mathcal{A} \qquad [AC]$$
$$\_ : \_ : \Box \in \mathcal{R} \wedge \Box \rightarrow_\gamma \Box' \;\Rightarrow\; \Box' : \_ \in \mathcal{A} \qquad [RC]$$

In γCC^ω the strong correctness of types $[SCT]$ holds, hence $P\gamma$ also holds. Let us study the $S_s\gamma^{-1}$ property in γCC^ω . This property holds for constants,

$$\Gamma \vdash s' : \Box \wedge s \twoheadrightarrow_\gamma s' \;\Rightarrow\; \Gamma \vdash s : \Box$$

It is enough prove it for a single reduction step. So, if $* \rightarrow_\gamma \Box_0$ and $\Gamma \vdash \Box_0 : p$, because of generation lemma $\Box_1 \twoheadrightarrow_\gamma p$, hence $\Gamma \vdash * : \Box_0$ and we can use $P\gamma$. The other case is treated in a similar way. Now, we can apply the following result:

Lemma 18. *In order to verify $S_s\gamma^{-1}$ it is enough, for every constants, to prove:*

$$\Gamma \vdash s' : \Box \qquad s \twoheadrightarrow_\gamma s' \;\Rightarrow\; \Gamma \vdash s : \Box$$

This result does not hold if we replace $S_s\gamma^{-1}$ by $S_s^{\partial}\gamma^{-1}$, i.e, $S_s^{\partial}\gamma^{-1}$ may hold for constants but not for every term (see Example 20).

We saw that, if $\Gamma \vdash a : A, A'$, then:

$$a \in T_v \;\Rightarrow\; A \twoheadrightarrow_\beta \twoheadrightarrow_\gamma X \;_\gamma\!\twoheadleftarrow\; _\beta\!\twoheadleftarrow A'$$

but $\Gamma \vdash a : X$ does not always holds. Let us study now these two problems

$$\Gamma \vdash a : A, A' \;\Rightarrow\; \Gamma \vdash a : X \qquad A \Leftarrow X \Rightarrow A'$$
$$\Gamma \vdash a : A \qquad \Rightarrow\; \Gamma \vdash a : A^* \qquad \forall A'[\Gamma \vdash a : A' \;\Rightarrow\; A^* \Rightarrow A'\,]$$

Such A^*, if it exists, is named a *principal type* (*PT*) for a. The first property is named *CR property* by [9]. We obtain a better result with the following

Lemma 19 (*CR* property for the T_v class). *Let us consider a γPTS where the property $S_s^{\partial}\gamma^{-1}$ holds over the terms in the class Π; then, for each $a \in T_v$,*

$$\Gamma \vdash a : A, A' \;\Rightarrow\; \Gamma \vdash a : X \qquad A \twoheadrightarrow_\beta \twoheadrightarrow_\gamma X \;_\gamma\!\twoheadleftarrow\; _\beta\!\twoheadleftarrow A'$$

The condition $S_s^{\partial}\gamma^{-1}$ can not be eliminated as the next example shows

Example 20. Let us consider the set of axioms, rules and relations γ

$$\mathcal{A} = \{\triangle : \Box, \qquad \triangle_1 : \Box_1, \qquad \triangle_2 : \Box_2\}$$
$$\mathcal{R} = \{\Box : \Box_1 : \Box_1, \qquad \Box : \Box_2 : \Box_2\}$$
$$\gamma = \{\triangle \rightarrow_\gamma \triangle_1, \qquad \triangle \rightarrow_\gamma \triangle_2\}$$

Then, $\vdash \Pi x : \triangle.\triangle_1 : \Box_1$ and $\vdash \Pi x.\triangle.\triangle_2 : \Box_2$, so $\vdash \lambda x : \triangle.x : \Pi x : \triangle.\triangle_1$ and $\vdash \lambda x : \triangle.x : \Pi x.\triangle.\triangle_2$. But, $\not\vdash \lambda x : \triangle.x : \Pi x : \triangle.\triangle$ (if $\vdash \Pi x : \triangle.\triangle : s$, by generation $\Box : \Box : \_ \in \mathcal{R}$ – absurd). Let us note that $\lambda\mathbf{S} \not\models S_s^{\partial}\gamma^{-1} \wedge S_s\gamma^{-1}$, but $\lambda\mathbf{S} \models S_s^{\partial}\gamma$, because it holds for the constants, and it is a trivial task to prove that

$$\Gamma \vdash \Pi x : A.B : s \wedge B \twoheadrightarrow_\gamma B' \quad \Rightarrow \quad B \equiv B'$$

So, $\lambda x : \Delta.x$ does not have a principal type. If we now consider

$$\mathcal{A} = \{\Delta : \Box, \quad \Delta_1 : \Box_1, \quad \Delta_2 : \Box_2, \quad \Delta : \Box_1, \quad \Delta : \Box_2\}$$

$S_s^\partial\gamma^{-1}$ will hold over T_v but will not over T_s. In fact: $\vdash \Pi x : \Delta.\Delta : \Box_1$ and $\vdash \Pi x : \Delta.\Delta : \Box_2$ but $\Box_1 \not\simeq \Box_2$, so there are terms in class Π without PT. The same may happen for class λ:

$$\vdash \lambda y : \Delta.(\Pi x : \Delta.\Delta) : \Pi y : \Delta.\Box_1$$

$$\vdash \lambda y : \Delta.(\Pi x : \Delta.\Delta) : \Pi y : \Delta.\Box_2 \qquad \Pi y : \Delta.\Box_1 \not\simeq \Pi y : \Delta.\Box_2$$

4 Strengthening and Condensing Lemmas

Theorem 21 (Strengthening lemma for T_v class). *Let us consider a γPTS with ${}_{T_v}P\gamma$ and $S_s\gamma^{-1}$ (or $S_s^\partial\gamma\gamma^{-1}$) [2], then for each $a \in T_v$,*

$$\Gamma_1 \vdash a : \_ \qquad \Gamma_2 \vdash a : A \qquad \Gamma_1 \subseteq \Gamma_2 \Rightarrow \Gamma_1 \vdash a : A' \;{}_\beta\!\twoheadleftarrow A$$

Proof. induction on a, using Lemma 5 and Lemma 19. □

Let us note that A could contain free variables appearing in Γ_2 but not in Γ_1, so $\Gamma_1 \vdash a : A$ would not hold. The corollary above is trivial if β–uniqueness of types holds in T_v.

¿From the substitution lemma, if $\Gamma_1 \vdash d : D$, then:

$$\Gamma_1, y : D, \Gamma_2 \vdash a : A \Rightarrow \Gamma_1, \Gamma_2[y := d] \vdash a[y := d] : A[y := d]$$

hence, if $y \notin \mathrm{FV}(\Gamma_2) \cup \mathrm{FV}(a : A)$, then

$$\Gamma_1, y : D, \Gamma_2 \vdash a : A \Rightarrow \Gamma_1, \Gamma_2 \vdash a : A$$

Under certain conditions, the condition $\Gamma_1 \vdash d : D$ may be eliminated. Even more, it may happen that $\nexists d[\Gamma_1 \vdash d : D]$ (Lemma 16). This property will be named the *condensation property* as in [2] (we will use *strengthening* to denote the property of Theorem 21). We will prove the condensation lemma using special techniques that we analyze in the next subsections.

We know (Lemma 12) that types for terms in class T_s are β–reducible to terms in the form $\Pi\Delta.s$. It is natural thinking that a term of the form $\Pi\Delta.s$ is β–equivalent to an abstraction such as $\lambda\Delta.d$. This property is related to the strengthening property. Extending the property of Theorem 21 to class T_s is difficult. We will obtain for the T_s class a weaker property that we will name *weak strengthening*

$$\begin{array}{l} \Gamma_1 \vdash a : \_ \qquad \Gamma_2 \vdash a : \Pi\Delta.s \qquad \Gamma_1 \subseteq \Gamma_2 \\ \Rightarrow \\ \Gamma_1, \Delta' \vdash d : s \wedge a \twoheadrightarrow_\beta \lambda\Delta'.d \end{array} \qquad [ST^\partial]$$

Besides, Theorem 26 shows that this property also holds for class T_v if we add an additional condition ($\Gamma_2 \vdash a : s_1$). This property was observed by [3] and its proof is based in the use of a computable operator that we will study later on. Now, let us study two operators that capture the final constants that may appear in the $[ST^\partial]$ property.

[2] ${}_{T_v}P\gamma$ denotes $P\gamma$ over the T_v class, and $S_s^\partial\gamma\gamma^{-1}$ means $S_s^\partial\gamma \wedge S_s^\partial\gamma^{-1}$

Definition 22. *Let $\overline{\partial}$ and $\underline{\partial}$ be the following operators*

$$\Gamma\overline{\partial}a = \{\ s \mid \exists d, \Delta[a \twoheadrightarrow_\beta \lambda\Delta.d \wedge \Gamma, \Delta \vdash d : s]\}$$
$$\Gamma\underline{\partial}a = \{\ s \mid \exists\Delta[\Gamma \vdash a : \Pi\Delta.s]\}$$

From this operators we can express the $[ST^\partial]$ property as

$$\Gamma_1 \vdash a : \_ \qquad \Gamma_1 \subseteq \Gamma_2 \qquad \Rightarrow \qquad \Gamma_2\underline{\partial}a \subseteq \Gamma_1\overline{\partial}a \qquad [ST^\partial]$$

In order to study this property, we will show the operators $\underline{\partial}$ and $\overline{\partial}$ have similar properties and some dual ones.

Lemma 23 (Properties for the $\underline{\partial}$ operator).

$$\begin{array}{llrcl}
(ii)\ \Gamma \vdash a : s_1 & \Rightarrow & \{s|\Gamma \vdash a : s\} & = & \Gamma\underline{\partial}a \\
(iii)\ \Gamma \textit{ legal} & \Rightarrow & \Gamma\underline{\partial}\Box & = & \Box : \mathcal{A}_\gamma \\
(iv)\ \Gamma \vdash \Pi x : A.B : _ & \Rightarrow & \Gamma\underline{\partial}(\Pi x : A.B) & = & \Gamma\underline{\partial}A : (\Gamma, x : A)\underline{\partial}B : \mathcal{R}_\gamma \\
(v) & & \Gamma\underline{\partial}(\lambda x : A.b) & \subseteq & (\Gamma, x : A)\underline{\partial}b \\
(vi)\ b \in T_s, (S^\partial_s\gamma \vee P\gamma) & \Rightarrow & \Gamma\underline{\partial}(bc) & \subseteq & \Gamma\underline{\partial}b
\end{array}$$

The condition $b \in T_s$ in the property (vi) is essential; by example, if we consider a PTS where $\vdash \Pi x : \triangle.x : \Box$ and $\vdash \Box : \triangle$, then, taking $\Gamma \equiv y : \Pi x : \triangle.x$, and using (apl)–rule and Lemma 23(ii): $\Box \in \Gamma\underline{\partial}(y\,\Box)$ but $\Gamma\underline{\partial}y = \emptyset$ (if $\Gamma \vdash y : \Pi\Psi.s$, then, by uniqueness of types in T_v, $\Pi x : \triangle.x \simeq \Pi\Psi.s$, and this is impossible because $x \not\simeq s$).

Example 24. Neither the condition $P\gamma$ nor $S^\partial_s\gamma$ can be eliminated for the property (vi). For example, let us consider $\lambda\mathbf{S}$ *as*

$$\begin{array}{l}
\mathcal{A} = \{\triangle : \Box, \quad \Box : \triangle, \quad \Box : \triangle', \quad \triangle' : \nabla\} \\
\mathcal{R} = \{\Box : \Box : \Box\} \\
\gamma\ = \{\triangle \rightarrow_\gamma \triangle'\}
\end{array}$$

then,

$$\begin{array}{lll}
\textit{for } b \equiv \lambda x : \triangle.x (\in T_v) : & \triangle' \in \underline{\partial}(b\,\Box) & \triangle' \notin \underline{\partial}(b) \\
\textit{for } b' \equiv \lambda x : \triangle.\Box (\in T_s) : & \triangle' \in \underline{\partial}(b'\,\Box) & \triangle' \notin \underline{\partial}(b') \\
\lambda\mathbf{S} \not\models S^\partial_s\gamma; \textit{i.e.}: & \vdash \Pi x : \triangle.\triangle : \Box & \not\vdash \Pi x : \triangle.\triangle' : _ \\
\lambda\mathbf{S} \not\models P\gamma; \textit{i.e.}: & \vdash b : \Pi x : \triangle.\triangle & \not\vdash b : \Pi x : \triangle.\triangle'
\end{array}$$

Lemma 25 (Properties for the $\overline{\partial}$ operator).

$$\begin{array}{llrcl}
(i)\ a \twoheadrightarrow_\beta a' & \Rightarrow & \Gamma\overline{\partial}a' & = & \Gamma\overline{\partial}a \\
(ii)\ \Gamma \vdash a : s_1 & \Rightarrow & \Gamma\overline{\partial}a & = & \{s|\Gamma \vdash a' : s, a \twoheadrightarrow_\beta a'\} \\
(iii)\ \Gamma \textit{ legal} & \Rightarrow & \Gamma\overline{\partial}\Box & = & \Box : \mathcal{A}_\gamma \\
(iv)\ \Gamma \vdash \Pi x : A.B : _ & \Rightarrow & \Gamma\overline{\partial}(\Pi x : A.B) & = & \Gamma\overline{\partial}A : (\Gamma, x : A)\overline{\partial}B : \mathcal{R}_\gamma \\
(v) & & (\Gamma, x : A)\overline{\partial}b & \subseteq & \Gamma\overline{\partial}(\lambda x : A.b) \\
(vi)\ \Gamma \vdash bc : _ & \Rightarrow & \Gamma\overline{\partial}b & \subseteq & \Gamma\overline{\partial}(bc)
\end{array}$$

The condition "bc is Γ-typable" in (vi) is essential. For the system in Exampl 24: $\overline{\partial}((\lambda x : \triangle.x)\underline{y}) = \emptyset$ and $\overline{\partial}(\lambda x : A.x) \neq \emptyset$. The reciprocal inclusion in (vi does not hold: $\Gamma\overline{\partial}(x\,\Box) \neq \emptyset$ and $\Gamma\overline{\partial}x = \emptyset$

Let us observe that $a \twoheadrightarrow_\beta a' \Rightarrow \Gamma\underline{\partial}a \subseteq \Gamma\underline{\partial}a' \subseteq \Gamma\overline{\partial}a' = \Gamma\overline{\partial}a$: "with th β-reduction $\underline{\partial}$ increases but its bound depends on a", but not on any redex a *We conjecture* that for β–normal forms $\Gamma\underline{\partial}a = \Gamma\overline{\partial}a$.

Theorem 26 (Weak strengthening property). *Let $\Gamma_1 \subseteq \Gamma_2$ be two contexts in a γPTS verifying $S_s\gamma^{-1}P\gamma$* [3] *(or $S_s^{\partial}\gamma\gamma^{-1}$). Then, for every term a Γ_1-typable*

$$a \in T_v \wedge \Gamma_2 \vdash a : s_1 \vee a \in T_s \Rightarrow \Gamma_2\underline{\partial}a \subseteq \Gamma_1\overline{\partial}a \qquad [ST^{\partial}]$$

Proof. Induction on a using Lemmas 23–25, Theorem 21 and Corollary 6. □

Example 27. In $[ST^{\partial}]$ the inclusion $\subseteq$ can be proper. Take ($\gamma = \emptyset$)

$$\mathcal{A} = \{\, \triangle : \square, \quad \square : \triangle, \quad \square : \nabla \,\}$$
$$\mathcal{R} = \{\, \square : \square : \square \,\}$$

Then, if $a \equiv (\lambda x : \triangle.x)\,\square$, we will have that $\vdash a : \square$ and $\not\vdash a : \nabla$, so

$$\overline{\partial}a = \{\triangle, \nabla\} \qquad \nabla \in \overline{\partial}a \qquad \nabla \notin \underline{\partial}a$$
$$\vdash a : \triangle \qquad a \twoheadrightarrow_\beta \square \qquad \vdash \square : \nabla \qquad \not\vdash a : \nabla$$

This example also shows that subject expansion *property does not hold.*

Corollary 28. *If $\lambda\mathbf{S} \models S_s\gamma^{-1}P\gamma \vee S_s^{\partial}\gamma\gamma^{-1}$, then*

$$\Gamma_1 \vdash a : \triangle \qquad \Gamma_2 \vdash a : \square \qquad \Gamma_1 \subseteq \Gamma_2 \;\Rightarrow\; \Gamma_1 \vdash a \twoheadrightarrow_\beta a' : \square$$

Proof. Follows from thinning, Lemmas 23(ii)–25(ii') and Theorem 26. □

Theorem 29. *If $\lambda\mathbf{S} \models S_s\gamma^{-1}P\gamma \vee S_s^{\partial}\gamma\gamma^{-1}$,*

(*a*) *Weak condensation lemma:*

$$\Gamma_1, y : D, \Gamma_2 \vdash a : A \qquad y \notin \mathrm{FV}(\Gamma_2) \cup \mathrm{FV}(a) \;\Rightarrow\; \Gamma_1, \Gamma_2 \vdash a : A' \;{}_{\beta}\!\twoheadleftarrow A$$

(*b*) *Condensation lemma:*

$$\Gamma_1, y : D, \Gamma_2 \vdash a : A \qquad y \notin \mathrm{FV}(\Gamma_2) \cup \mathrm{FV}(a : A) \;\Rightarrow\; \Gamma_1, \Gamma_2 \vdash a : A$$

Proof. We denote $\Gamma_y \equiv \Gamma_1, y : D, \Gamma_2$ and $\Gamma \equiv \Gamma_1, \Gamma_2$. ($b$): follows from ($a$), type correctness and the (β)–rule. (a): by induction on the derivation $\psi \equiv \Gamma_y \vdash a : A$. Let us proof some cases:

$\psi : -(\gamma)$ with $\Gamma_y \vdash a : A_1, A : s \qquad A_1 \twoheadrightarrow_\gamma A$; then, by i.h., $\Gamma \vdash a : A_1' \;{}_{\beta}\!\twoheadleftarrow A_1$, and by $\beta \diamond \gamma$, for a A', we have that $A \twoheadrightarrow_\beta A'$ and $A_1' \twoheadrightarrow_\gamma A'$.

- If $A_1' \in \mathcal{S}$, then $A' \equiv \square \in \mathcal{S}$, so, by $S\beta$ we obtain $\Gamma_y \vdash \square : s$, because it is Γ-legal, and by Lemma 6: $\Gamma \vdash \square : s$, and we can use the ($\gamma$)–rule to obtain $\Gamma \vdash a : A'$.
- If $A_1' \notin \mathcal{S}$, by type correctness, $\Gamma \vdash A_1' : s$, and by $S_s^{\partial}\gamma$, we will have $\Gamma \vdash A' : s'$, and using (γ)–rule again.
- If $P\gamma$ the last two steps are direct.

$\psi : -(\lambda)$, $a \equiv \lambda x : C.b$, with

$$\Gamma_y \vdash C : s_1 \qquad (0)$$
$$\Gamma_y, x : C \vdash B : s_2 \qquad (1)$$
$$\Gamma_y, x : C \vdash b : B \qquad (2)$$

We can apply i.h. to (0) and (2) to obtain

$$\Gamma \vdash C : s_1 \qquad \Gamma, x : C \vdash b : B' \;{}_{\beta}\!\twoheadleftarrow B \qquad (4)$$

and besides (1) and $S\beta$, we will have

$$\Gamma_y, x : C \vdash B' : s_2 \qquad (5)$$

[3] is an simplification of $S_s\gamma^{-1} \wedge P\gamma$

But, because $y \in \mathrm{FV}(B)$ is possible, we can not apply i.h. to (1). This is the real problem, because if we had

$$\Gamma, x : C \vdash B' : s_2 \tag{6}$$

we could apply the (λ) rule, in addition to (4) to obtain

$$\Gamma \vdash (\lambda x : C.b) : \Pi x : C.B' \ {}_{\beta}\!\twoheadleftarrow \Pi x : C.B$$

In order to obtain something similar to (6) we reason as follows: by the type correctness property applied to (4), we can distinguish two cases:
- If $B' \equiv \Box \in \mathcal{S}$, then, by (5) and Lemma 6: $\Gamma, x : C \vdash \Box : s_2$.
- If $\Gamma, x : C \vdash B' : s_2'$, then we apply Corollary 28 to this derivation, and (5) to obtain $\Gamma, x : C \vdash B'' : s_2$, where $B'' \ {}_{\beta}\!\twoheadleftarrow B' \ {}_{\beta}\!\twoheadleftarrow B$, then apply $P\beta$ to (4), and finally apply the (λ)–rule to obtain

$$\Gamma \vdash \lambda x : C.b : \Pi x : C.B'' \ {}_{\beta}\!\twoheadleftarrow \Pi x : C.B$$ □

The first formulation of the condensation lemma is due to van Daalen (for AUTOMAT); [9] proves an analogous result for the system **ECC** but he obtains $A' \precsim A$ instead of $A' \ {}_{\beta}\!\twoheadleftarrow A$. But, because in **ECC** we have $\forall i, j, [i < j][\ \vdash \Box_i : \Box_j\ ]$, then the conditions of the weak condensation theorem hold. Thus, our result is stronger. [3] proves the same lemma for PTS, but it is a particular instance of our lemma when $\gamma = \emptyset$.

5 Decidability

If we substitute in Lemma 23(v, vi) the symbol $\subseteq$ by an equality, a new operator ∂ is obtained, and we can extend it to any term:

Definition 30. *Let ∂ be the following operator*

$$\Gamma\partial\Box = \Box : \mathcal{A}_\gamma$$
$$\Gamma\partial(\Pi x : A.B) = \Gamma\partial A : (\Gamma, x : A)\partial B : \mathcal{R}_\gamma$$
$$\textit{If } b \in T_v, \qquad \Gamma\partial b = \{s | \Gamma \vdash b : s\}$$
$$\textit{If } b \in T_s, \quad \Gamma\partial(\lambda x : A.b) = (\Gamma, x : A)\partial b$$
$$\Gamma\partial(b\,c) = \Gamma\partial b$$

It is easy to realize that ∂ extends *in a natural way* the $\Sigma(\Gamma, a)$ operator from [3]. In addition, we will admit the following conditions hold in our system (they are trivial when $\mathcal{S}$ is finite)

$$\Box : \mathcal{A}_\gamma \text{ is decidable} \qquad [\mathcal{A} - dec]$$
$$\mathcal{M}, \mathcal{N} \text{ decidable} \quad \Rightarrow \quad \mathcal{M} : \mathcal{N} : \mathcal{R}_\gamma \text{ decidable} \qquad [\mathcal{R} - dec]$$

Then, ∂ is computable if it is computable for the terms in T_v. So, the computability of the set $\{s|\Gamma \vdash a : s\}$ is assured if we prove that

$$\{s|\Gamma \vdash a : s\} \subseteq \Gamma\underline{\partial}a \subseteq \Gamma\partial a \subseteq \Gamma\overline{\partial}a \subseteq \{s|\Gamma \vdash a : s\}$$

The following results generalize the results 5.3 – 5.4 in [3](page 34).

Lemma 31. (*a*) *If a is Γ-typable, $\Gamma\partial a \subseteq \Gamma\overline{\partial}a$*
(*b*) *If $\lambda\mathbf{S} \models S_s^{\partial}\gamma \vee P\gamma$,*

$$a \in T_v \wedge \Gamma \vdash a : s_1 \vee a \in T_s \Rightarrow \Gamma\underline{\partial}a \subseteq \Gamma\partial a$$
$$a \in \beta_n \wedge \Gamma \vdash a : s_1 \qquad \Rightarrow \Gamma\partial a = \{s|\Gamma \vdash a : s\}$$

Proof. Induction on a using Lemma 23 and Lemma 25. □

Lemma 32 (Generation lemma for β–normal types). *If $\lambda\mathbf{S} \models P\gamma \vee S_s^{\partial}\gamma$, the Φ operator, $\Gamma\Phi a = \{A | A \in \beta_n \wedge \Gamma \vdash a : A\}$, verifies the following properties:*

(i) $\Gamma\Phi(\Pi x : A.B) = \Gamma\Phi A : (\Gamma, x : A)\Phi B : \mathcal{R}_\gamma$

(ii) $\Gamma\Phi(\lambda x : A.b) = \Pi x : A_\beta.\{ B \in (\Gamma, x : A)\Phi b \,|\, \Gamma\Phi A : (\Gamma, x : A)\Phi B : \mathcal{R} \neq \emptyset\}$

(iii) $\Gamma\Phi(b\,c) = \{A' | \Pi x : C.A \in \Gamma\Phi b, C \in \Gamma\Phi c, A[x := c]_\beta \twoheadrightarrow_\gamma A'\}$

(iv) $$(\Gamma, y : A)\Phi e = \begin{cases} \Gamma\Phi e & if\ (e \equiv s \vee e \equiv x \not\equiv y) \wedge \Gamma\Phi A \cap \mathcal{S} \neq \emptyset \\ \{B | A_\beta \twoheadrightarrow_\gamma B\} & if\ e \equiv y \wedge \Gamma\Phi A \cap \mathcal{S} \neq \emptyset \\ \emptyset & if\ \Gamma\Phi A \cap \mathcal{S} = \emptyset \end{cases}$$

(v) $\Phi\Box = \Box : \mathcal{A}_\gamma$

(vi) $\Phi x = \emptyset$

Proof. (i) follows from Corollary 6 and the rules (Π) and (γ); (ii)– $\subseteq$ follows from Lemma 16 and type correctness; (ii)– $\supseteq$ follows from (λ)–rule; (iii)– $\subseteq$ follows from *apl*–generation; in order to prove (iii)– $\supseteq$ we use the (*apl*)–rule, $SP\beta$, $S_s^{\partial}\gamma$ and the (γ)–rule, or immediately if we have $P\gamma$; (iv) follows from Corollary 16, *var*–generation, thinning, (*weak*)–rule and contexts substitution; (v) follows from Lemma 4(i) and Corollary 7. □

Theorem 33. *If the system is normalizing, $\mathcal{S}$ is finite and $S_s\gamma^{-1}P\gamma \vee S_s^{\partial}\gamma\gamma^{-1}$,*
(i) The operator Φ is computable
(ii) $\Gamma \vdash a : A$ is decidable

Proof. (ii): by type correctness and (β)–rule:

$$\Gamma \vdash a : A \Longleftrightarrow A \in \mathcal{S} \cap \Gamma\Phi a \vee (\mathcal{S} \cap \Gamma\Phi A \neq \emptyset \wedge A_\beta \in \Gamma\Phi a)$$

and by (i) the predicate on the right is computable. In order to prove (i) we consider the lexicographical relation $\prec$ defined for the pairs $\Gamma \dashv a$ (i.e.: $\Gamma' \dashv c \prec \Gamma \dashv b\,c \quad \Gamma \dashv b \prec \Gamma, x : A \dashv b \quad \Gamma \dashv a \not\prec \,<>\dashv x$, etc.). We will prove (i) by induction on $\prec$ in two phases:

• *Phase 1: For T_v class.-* We use induction on $\prec$ and inductive definition of T_v. From generation, uniqueness of domains, CR and Lemma 32, we have for typable terms in T_v class that

$$\Gamma\Phi(b\,c) = \{A' | \Pi x : C.A \in \Gamma\Phi b, A[x := c]_\beta \twoheadrightarrow_\gamma A'\}$$
$$\Gamma\Phi(\lambda x : A.b) = \Pi x : A_\beta.(\Gamma, x : A)\Phi b$$

Using induction on $\prec$ and Lemma 32(iv) we have the computability of Φ for (typable) terms in T_v class. ¿From Corollary 31 and definition of ∂ we obtain the computability of $\{ s \,|\, \Gamma \vdash a : s\}$ for $a \in \beta_n$.

• *Phase 2: For T_s class.-* We use induction on $\prec$ and inductive definition of T_s. For base case $(\Phi\Box)$ we apply Lemma 32(v) (and $[\mathcal{A} - dec]$); the inductive steps follow from Lemma 32$(i - iv)$. The only complication is the implication:

Both $(\Gamma, x : A)\Phi b$ and $\Gamma\Phi A$ computable

$\Rightarrow$

$\{B \in (\Gamma, x : A)\Phi b \,|\, \Gamma\Phi A : (\Gamma, x : A)\Phi B : \mathcal{R} \neq \emptyset\}$ computable

Let $\Delta \equiv \Gamma, x : A$. By phase 1, if $\Delta \vdash B : s_2$ and $B \in \beta_n$, the set $\Delta\partial B = \{s | \Delta \vdash B : s\}$ is computable. But if $\Delta \vdash B : M$ then $M \simeq s_2$, or $M \overset{s}{\simeq} s_2$. Hence, if $M \in \beta_n$ then $M \in \mathcal{S}$. Summarizing: $\Delta\Phi B = \{s | \Delta \vdash B : s\}$ is computable. □

6 Related Work, Future Work, and Conclusions

Some authors study the problem of decidability using syntax directed type systems [4,11], while other authors use the *typed abstract reduction* concept [5,10]. Our study is closer to the original PTS. Obviously, our results are theoretical, since it is quite hard to obtain a practical algorithm for type checking using normal forms. While the general properties of γPTS can be generalized when introducing Σ-types (dependent sums), it is difficult to obtain a condensing lemma keeping the partition proposed by [3]. We are currently working on generalizing these results, and studying the impact of the properties of weak–closure for γ–reduction in the normalization method of Girard–Tait.

Acknowledgments We would like to thank the referees for their valuable comments and corrections. And to my colleagues José Gallardo and Antonio Vallecillo for their help in the preparation of the final version.

References

1. Barendregt, H.: The Lambda Calculus (Its syntax and semantics). North-Holland (1985) (sec. ed.)
2. Barendregt, H.: Lambda Calculi with Types. HandBook of Logic in Computer Science (Abramsky, S. et al., eds.). Oxford University Press (1992) 117–309
3. van Benthem Jutting, L.S.: Typing in Pure Type Systems. Information and Computation **105 (3)** (1993) 30–41
4. van Benthem Jutting, L.S., McKinna, J., Pollack, R.: Checking Algorithms for Pure Type Systems. Types for Proofs and Programs: International Workshop TYPES'93, Nijmegen (Barendregt, H. and Nipkow, T., eds.). LNCS **806** (1993) 19–61
5. Castagna, G., Chen, G.: Dependent type with subtyping and lately–binded overloading. DMI-LIENS. Ecole Normale Supérieure, Paris (1997) (submitted to TLCA'97)
6. Coquand, T., Huet, G.: The Calculus of Constructions. Information and Computation **76(2/3)** (1988) 95–120
7. Geuvers, H., Nederhof, M.: Modular proof of strong normalization for the calculus of constructions. Journal of Functional Programming **1 (2)** (1991) 155–189
8. Harper, R., Pollack, R.: Type Checking with Universes. Theor. Computer Science **89** (1991) 107–136
9. Luo, Z.: An Extended Calculus of Constructions. Ph. D. University of Edinburgh (1990)
10. Luo, Z.: Coercive subtyping. Journal of Logic and Computation (1997) (to appear)
11. McKinna, J., Pollack, R.: Some Lambda Calculus and Type Theory Formalized. Laboratory for Foundations of Computer Science, University of Edinburgh (1997)
12. Ruiz, B.C., Santos, A.: Decidibilidad en sistemas γSTP. III Jornadas de Informática, Cádiz (Spain) (Torres, J.C., ed.) (1997) 217–226

Appendix (Special Notation Used)

Sets:

β_n	β–normal forms
$\mathrm{FV}(a)$	free variables of a
$\mathrm{FV}(a:A)$	$\mathrm{FV}(a) \cup \mathrm{FV}(A)$
$\mathrm{Var}(\Delta)$	$\{x_1, \ldots, x_n\}$, for context $\Delta \equiv x_1 : A_1, \ldots, x_n : A_n$
$x \in \Delta$	$x \in \mathrm{Var}(\Delta)$, for $\Delta \equiv x_1 : A_1, \ldots, x_n : A_n$
$\mathrm{FV}(\Delta)$	$\mathrm{FV}(A_1) \cup \ldots \cup \mathrm{FV}(A_n)$, for $\Delta \equiv x_1 : A_1, \ldots, x_n : A_n$
$\Box : \mathcal{A}$	$\{\, \triangle \mid \Box : \triangle \in \mathcal{A} \,\}$
$\mathcal{M} : \mathcal{N} : \mathcal{R}$	$\{\, s_3 \mid \overline{s} \in \mathcal{R}, s_1 \in \mathcal{M}, s_2 \in \mathcal{N} \,\}$
$\Box : \mathcal{A}_\gamma$	$\{\, \triangle \mid \Box : \triangle' \in \mathcal{A}, \triangle' \equiv \triangle \vee \triangle' \twoheadrightarrow_\gamma \triangle : \_ \in \mathcal{A} \,\}$
$\mathcal{M} : \mathcal{N} : \mathcal{R}_\gamma$	$\{\, \Box \mid \overline{s} \in \mathcal{R}, s_1 \in \mathcal{M}, s_2 \in \mathcal{N}, s_3 \equiv \Box \vee s_3 \twoheadrightarrow_\gamma \Box : \_ \in \mathcal{A} \,\}$

Classes:

T_v	$\mathcal{V} \subseteq T_v$; $\quad b \in T_v \Rightarrow bc, \lambda x : A.b \in T_v$
T_s	$\Box$, $\Pi x : A.B \in T_s$; $\quad b \in T_s \Rightarrow bc,\ \lambda x : A.b \in T_s$

Operators:

$\Delta \downarrow_{x_{i+1}}$	$x_1 : A_1, \ldots, x_i : A_i$
$\Pi \Delta.B$	$\Pi x_1 : A_1.\Pi x_2 : A_2. \cdots .\Pi x_n : A_n.B$
$\lambda \Delta.b$	$\lambda x_1 : A_1.\lambda x_2 : A_2. \cdots .\lambda x_n : A_n.b$
$\Pi \Delta.\mathcal{N}$	$\{\, \Pi \Delta.N \mid N \in \mathcal{N} \,\}$
$\cdot$	composition: $x\ R \cdot R'\ y \iff \exists z[xRz \wedge zR'y]$
$\overline{\partial}$	$\Gamma \overline{\partial} a = \{\ s \mid \exists d, \Delta[a \twoheadrightarrow_\beta \lambda \Delta.d \wedge \Gamma, \Delta \vdash d : s]\}$
$\underline{\partial}$	$\Gamma \underline{\partial} a = \{\ s \mid \exists \Delta[\Gamma \vdash a : \Pi \Delta.s]\}$

Relations:

$\rightarrow_\gamma$	right Π–compatible closure of γ: $\quad \frac{(Q,R) \in \gamma}{Q \rightarrow_\gamma R} \qquad \frac{Q \rightarrow_\gamma R}{\Pi x:M.Q \rightarrow_\gamma \Pi x:M.R}$
$\twoheadrightarrow_\gamma$	reflexive and transitive closure of $\rightarrow_\gamma$
$\stackrel{\sim}{\Rightarrow}$	$\twoheadrightarrow_\beta \cdot \twoheadrightarrow_\gamma \cdot {}_\beta\!\twoheadleftarrow \ \equiv\ (=_\beta \cdot \twoheadrightarrow_\gamma \cdot =_\beta)$
$\simeq$	$\twoheadrightarrow_\beta \cdot {}_\gamma\!\twoheadleftarrow \cdot \twoheadrightarrow_\gamma \cdot {}_\beta\!\twoheadleftarrow$
$\stackrel{s}{\simeq}$	$B \stackrel{s}{\simeq} B' \ \equiv\ \exists \Delta, \Box, \Box'[B =_\beta \Pi \Delta.\Box \wedge B' =_\beta \Pi \Delta.\Box']$

Properties:

CR	Church-Rosser: ${}_\beta\!\twoheadleftarrow \cdot \twoheadrightarrow_\beta \subseteq \twoheadrightarrow_\beta \cdot {}_\beta\!\twoheadleftarrow$
$\beta \diamond \gamma$	commutativity of β and γ: ${}_\beta\!\twoheadleftarrow \cdot \twoheadrightarrow_\gamma \subseteq \twoheadrightarrow_\gamma \cdot {}_\beta\!\twoheadleftarrow$

– predicate γ–reduction in class T_v:

${}_{T_v}P\gamma$	$c \in T_v \wedge \Gamma \vdash c : C \wedge C \twoheadrightarrow_\gamma C' \Rightarrow \Gamma \vdash c : C'$

– subject and weak subject γ–reduction:

$S_s\gamma^{-1}$	$\Gamma \vdash c : \Box \wedge c' \twoheadrightarrow_\gamma c \Rightarrow \Gamma \vdash c' : \Box$
$S_s^{\partial}\gamma$	$\Gamma \vdash c : \Box \wedge c \twoheadrightarrow_\gamma c' \Rightarrow \Gamma \vdash c' : \Box'$

Improving Computations in a Typed Functional Logic Language*

Jesús Manuel Almendros-Jiménez

Dpto. Sistemas Informáticos y Programación
Escuela Superior de Informática, Univ. Complutense, Madrid, Spain
jesusmal@eucmos.sim.ucm.es

Abstract. We propose an implementation of a functional logic language with parametric order sorted types by translating typed lazy narrowing into Prolog, and establishing a reasonable strategy of combination of data and type constraints solving. By means of this translation, we describe how types can be used at run-time to prune the search space of solutions of a goal. Moreover, we present type inference and checking compile-time tools for the language showing its usefulness for improving computations.

1 Introduction

The combination of different declarative (specially functional and logic) paradigms has been widely investigated during the last decade (see [13] for a survey) and as a consequence, several well-known functional logic languages arose, using *lazy narrowing* as goal solving mechanism. More recently, the language CURRY [11,14] integrates the most important features from functional, logic and concurrent programming, providing a common platform for the research and the application of this kind of languages.

One of the traditional topics of research in declarative paradigms has been the incorporation of *type systems* as an attempt to provide more structured readable compile-time type-safe programs and run-time optimizations. In the functional [8,18] and the logic paradigm [20,12,17,15,4,5] the combination of *parametric types* and *order sorted hierarchies*, so-called *subtyping*, has been adopted as a simple type system allowing to subdivide and parametrize the universe of discourse. In particular, the logic paradigm has benefited from the type system by replacing syntactic unification for *typed unification*, which reduces the search space of solutions.

In this paper we describe the implementation of a functional logic language with subtyping whose operational semantics is based on typed lazy narrowing (other implementations of typed logic languages are [19], based on Prolog, and [5], based on a WAM extension with typed unification [6]). Typed lazy narrowing was presented in [1] and defines a rewriting calculus for solving data and type

* This research has been partially supported by the Spanish National Projects TIC95-0433-C03-01 "CPD"and TIC98-0445-C03-02 "TREND".

A.M. Haeberer (Ed.): AMAST'98, LNCS 1548, pp. 438–454, 1998.

constraints, but it does not state any concrete strategy for combining these solvings. The interest of this combination is the use of type constraints for pruning the search space of data constraints. It implies to establish how to keep type information and how to use it at run-time for solving lazy unification and looking for values for logic variables.

As a part of the implementation of the language, the type analysis at compile-time improves computations in several aspects (avoiding useless computations, reducing the search space of logic variables). In the literature, the problem of type analysis in logic programming with subtyping has been studied for type inference in [4,15,20]. The approaches based on type inference are focused in the inference of an environment defining one type for every data variable in every clause of a logic program. Following this line, [4] presents an algorithm solving the incompleteness of the previous algorithms [20,15]; however, it can not be adapted to our language because we use type conditions over expressions instead of type conditions over terms.

The organization of the paper is as follows. Section 2 presents the syntax of our language, and the semantics by means of a calculus. The operational semantics is defined in Section 3 and specified in Prolog in Section 4; Section 5 describes the details of the representation. Finally Section 6 studies type analysis.

2 The Programming Language

A *polymorphic signature* Σ consists of (1) *a type specification* containing a ranked set of polymorphic type constructors, partially ordered by $\leq$. For instance, the type specification $\mathsf{opnat}\#\mathsf{nat} \leq \mathsf{int}$, $\mathsf{posint}\#\mathsf{zero} \leq \mathsf{nat}$, $\mathsf{negint}\#\mathsf{zero} \leq \mathsf{opnat}$, $\mathsf{elist}(\alpha)\#\mathsf{nelist}(\alpha) \leq \mathsf{list}(\alpha)$ defines subsort relations for integer numbers and parametric lists. Following [15], $\leq$ is a lower quasi-lattice (and therefore upper quasi-lattice) defined for type constructors of the same arity. *Types* $\tau, \tau', \ldots$ are built up from type constructors $(K, L, \ldots)$ and type variables $(\alpha, \beta, \ldots)$, and they can be compared by extending $\leq$ to types. For instance, $\mathsf{nelist}(\mathsf{elist}(\alpha)) \leq \mathsf{list}(\mathsf{list}(\alpha))$ and $\mathsf{elist}(\mathsf{nat}) \leq \mathsf{list}(\mathsf{int})$. And (2) *a set of type declarations* for data constructors (CON) and functions (FUN). For instance, $0 : \mathsf{zero}$, $\mathsf{suc} : \mathsf{nat} \rightarrow \mathsf{posint}$, $\mathsf{pred} : \mathsf{opnat} \rightarrow \mathsf{negint}$, $[] : \mathsf{elist}(\alpha)$, $[\_|\_] : \alpha \rightarrow \mathsf{list}(\alpha) \rightarrow \mathsf{nelist}(\alpha)$, $\mathsf{head} : \mathsf{nelist}(\alpha) \rightarrow \alpha$, $\mathsf{tail} : \mathsf{nelist}(\alpha) \rightarrow \mathsf{list}(\alpha)$. Data constructor declarations $c : \tau_1 \ldots \tau_n \rightarrow \tau_0$ with extra type variables in τ_0 are completed by adding variables new arguments in the type declaration, needed to ensure the existence of the *least type* for terms [12]. *Expressions* $e, e', \ldots$ are built up from data constructor and function symbols and data variables $(X, Y, \ldots)$ and *Terms* $t, t', \ldots$ only uses data constructors and data variables.

Programs consist of a polymorphic signature Σ and a set of non-overlapping conditional constructor based rewriting rules defining the behaviour of function symbols. For instance, $\mathsf{filter} : (\alpha \rightarrow \mathsf{bool}) \rightarrow \mathsf{list}(\alpha) \rightarrow \mathsf{list}(\alpha)$ and $\leq : \mathsf{int} \rightarrow \mathsf{int} \rightarrow \mathsf{bool}$ are defined by the rules:

filter P [ ] := [ ] ⇐ P : $\alpha \rightarrow$ bool
filter P [X|L] := [X|filter P L] ⇐ (P X) == true□P : $\alpha \rightarrow$ bool, X : α, L : list(α)
filter P [X|L] := filter P L ⇐ (P X) == false□P : $\alpha \rightarrow$ bool, X : α, L : list(α)

≤ X Y := true ⇐ X : opnat, Y : nat
≤ X Y := false ⇐ X : posint, Y : opnat
≤ X Y := false ⇐ X : zero, Y : negint
≤ suc(X) suc(Y) :=≤ X Y ⇐ X : nat, Y : nat
≤ pred(X) pred(Y) :=≤ X Y ⇐ X : opnat, Y : opnat

Formally, a *program rule* has the form $f\ t_1 \ldots t_n := r \Leftarrow C_D \Box C_T$, where the body r of the rule is an expression, and the rule condition, which establishes the applicability conditions, is composed of a set of data (C_D) and type (C_T) conditions. The elements of C_D are strict equalities $l == r$ between expressions whereas the elements of C_T are type conditions $e : \tau$[1]. Furthermore, (i) the tuple $(t_1, \ldots, t_n)$ of terms is linear (i.e., no variable occurs more than once), and (ii) $dvar(r) \subseteq dvar(t_1 \ldots t_n)$. Remark that (ii) does not exclude the existence of extra variables in the rule condition. (i) and (ii) together with non-overlapping natural conditions ensure the functionality of definitions. A program rule is said *static well-typed* if $(\Sigma, C_T) \vdash_{SW} t_i : \tau_i, r : \tau_0$ and there are σ_j such that $(\Sigma, C_T) \vdash_{SW} l_j : \sigma_j, r_j : \sigma_j$, for every $l_j == r_j \in C_D$, where $f : \tau_1 \ldots \tau_n \rightarrow \tau_0 \in FUN$, where the relation $(\Sigma, C_T) \vdash_{SW} e : \tau$ holds if $e : \tau' \in C_T$ and $\tau' \leq \tau$ or $e \equiv h\ e_1 \ldots e_n$, $\tau_0 \leq \tau$ and $(\Sigma, C_T) \vdash_{SW} e_i : \tau_i$, $1 \leq i \leq n$, for an instance $\tau_1 \ldots \tau_n \rightarrow \tau_0$ of the type declaration of $h \in CON \cup FUN$.

Goals are as conditions of a well-typed rule. *Answers* consist of triples (V, θ, ρ), where V is an environment, θ (resp. ρ) is a data (resp. type) substitution satisfying the well-typedness property $(\Sigma, V) \vdash_{SW} X\theta : \tau$, for every $X : \tau \in V$. For instance:

X==filter (≤ Y)[0, pred(0), suc(suc(0)), suc(0), pred(pred(0))]□X: nelist(int), Y: nat

has the following answers:

==> X : nelist(nat), Y : cero, X = [0, suc(suc(0)), suc(0)], Y = 0
==> X : nelist(posint), Y : posint, X = [suc(suc(0)), suc(0)], Y = suc(0)
==> X : nelist(posint), Y : posint, X = [suc(suc(0))], Y = suc(suc(0))

The *most general answers* of a goal consist of the most general data and type substitutions and the greatest (resp. smallest) environments V w.r.t. the free (resp. bound) variables (see [1] for more details).

The *semantics* of the language is presented by means of a Typed Rewriting Calculus (in short *TRC*) in which formulas φ are derived from a program P and an environment V (denoted by $(P, V) \vdash_{TRC} \varphi$). In [2,1] the soundness and

[1] Environments (exactly one type condition for every data variable) are not strong enough to guarantee the static well-typedness of rules in our language. For instance, the function second : nelist(α) $\rightarrow \alpha$ defined by second X := head(tail X) ⇐ X : nelist(α), tail X : nelist(α) needs the condition tail X : nelist(α) to ensure the existence of a second element in the list.

completeness of *TRC* w.r.t. the model semantics and typed lazy narrowing was proved. In the sequel we restrict the presentation to a first order fragment of the language, but it can be extended to higher order as in [10]. *TRC* provides meaning to the following formulas φ: *non-strict equalities* $e \rhd t$ –representing that the *partial* (i.e. possibly containing $\perp$[2]) term t approximates the value of e (i.e. e is lazily unified to the partial term t)–, *strict equalities* $e == e'$ –which will be derived by reducing e and e' to the same *total* (without $\perp$) term– and as a novelty w.r.t. untyped calculi (e.g. [9]) *type conditions* $e : \sigma$ –which semantically define the type of an expression. *TRC* is composed of the rules below[3].

Bottom $\dfrac{e:\tau, t:\tau}{e \rhd \perp(t)}$ **Reflexivity** $\dfrac{}{t \rhd t}$ if $t \equiv X:\tau \in V$ or $t \equiv c:\tau \in CON$

Decomposition $\dfrac{e_1 \rhd t_1, \ldots, e_n \rhd t_n, t_1:\tau_1, \ldots, t_n:\tau_n}{c(e_1, \ldots, e_n) \rhd c(t_1, \ldots, t_n)}$ if $\tau_1 \ldots \tau_n \rightarrow \tau_0$ is an instance of the type declaration of $c \in CON$

Reduction I $\dfrac{e_1 \rhd t_1, \ldots, e_n \rhd t_n, C_D, C_T, r \rhd t}{f(e_1, \ldots, e_n) \rhd t}$ if $f(t_1, \ldots, t_n) := r \Leftarrow C_D \Box C_T \in [P]$ and $t \not\equiv \perp(t')$

Strict Equality $\dfrac{e \rhd t, e' \rhd t}{e == e'}$ if t is total **Type Assumption** $\dfrac{\sigma \leq \tau}{X:\tau}$ if $X:\sigma \in V$

Type Declaration $\dfrac{e_1:\tau_1, \ldots, e_n:\tau_n, \tau_0 \leq \tau}{h(e_1, \ldots, e_n):\tau}$ if $\tau_1..\tau_n \rightarrow \tau_0$ is an instance of the declaration of $h \in CON \cup FUN$

Reduction II $\dfrac{e_1 \rhd t_1, \ldots, e_n \rhd t_n, C_D, C_T, r:\tau}{f(e_1, \ldots, e_n):\tau}$ if $f(t_1, \ldots, t_n) := r \Leftarrow C_D \Box C_T \in [P]$

The novelty of the typed lazy calculus we present is the proof of type conditions about functional expressions. In order to derive $f(\bar{e}) : \tau$ there are two possibilities: to consult the type declaration of f or to use a program rule of f. For instance, from(0) : nelist(nat) can be proved directly from the type declaration of from : nat → nelist(nat). However, second(from(0)) : nat needs the reduction of second(from(0)) to be proved. The lazy combination of both mechanisms is the so-called *lazy type checking*. Notice that lazy type checking is not necessarily a head normal form reduction; for instance, we can prove f(0) : nat, where f : nat → int is defined by f(X) := g(X) ⇐ X : nat, from g(0) : nat, where g : nat → nat.

TRC mixes in a coherent way equalities and type conditions, in the sense that it can be proved that *the TRC-derivable (strict and non-strict) equalities are TRC-well-typed.* An equality is *TRC-well-typed* if there exists a *TRC*-provable common type for both sides of the equality (see [2] for more details).

3 A Typed Lazy Narrowing Calculus

Operationally, typed lazy narrowing is expressed through a Typed Lazy Narrowing Calculus (TLNC) which consists of a set of transformation rules Tr transforming goals $C \bullet\!\!\rightarrow^{Tr} C'$ or producing failure. A TLNC-successful derivation $C \bullet\!\!\rightarrow^{*} C'$ is a sequence of applications of transformation rules such that

[2] To model the behaviour of non-strict or partial functions, we introduce in the signature the data constructor $\perp : \alpha \rightarrow \alpha$ denoting the undefined value.

[3] $[P]$ is the set of program rule instances of the form $f(\bar{t})\theta := r\theta \Leftarrow C_D\theta \Box C_T\theta\rho$.

$\mathcal{C}'$ is in solved form (i.e. TLNC-irreducible) representing a TLNC-answer for $\mathcal{C}$. A TLNC-answer (solution) (V, θ, ρ) for a given goal $\mathcal{C}$, denoted by $Sol(\mathcal{C})$, corresponds (via soundness and completeness results, cfr. [1]) to a *TRC*-proof of $(\mathcal{P}, V) \vdash_{TRC} \mathcal{C}\theta\rho$. During the transformation process, $\mathcal{C}$ can also contain statements $e \rhd t$ and $\tau \leq \tau'$, whose meaning was (implicitly) given in *TRC*. Beside this, new statements are added to represent the answer (V, θ, ρ); so $X = t$ and $\alpha = \tau$ are used in the construction of θ and ρ respectively, and $Y : \alpha_Y$ is used for V[4]. Below, we summarize the transformation rules[5].

Rules for $\approx$

$\approx_1$) $f(\bar{e}) \approx e, \mathcal{C} \bullet\!\!\to e_1 \rhd t_1, \ldots, e_n \rhd t_n, \mathcal{C}_D, \mathcal{C}_T, r \approx e, \bar{X} : \alpha_{\bar{X}}, \mathcal{C}$

$\approx_2$) $X \approx t, X : \alpha_X, \mathcal{C} \bullet\!\!\to X = t, X : \alpha_X, t : \alpha_X, \mathcal{C}[X/t]$

$\approx_3$) $X \approx c(\bar{e}), X : \alpha_X, \mathcal{C} \bullet\!\!\to X = c(\bar{X}), X : \alpha_X, c(\bar{X}) : \alpha_X, X_1 \approx e_1, \ldots, X_n \approx e_n,$
$\bar{X} : \alpha_{\bar{X}}, \mathcal{C}[X/c(\bar{X})]$, if X is not safe in $c(\bar{e})$, and failure, otherwise

$\approx_4$) $e \rhd X, X : \alpha_X, \mathcal{C} \bullet\!\!\to e : \alpha_X, \mathcal{C}$

$\approx_5$) $X == X, \mathcal{C} \bullet\!\!\to \mathcal{C}$

$\approx_6$) $c(\bar{e}) \approx d(\bar{e'}), \mathcal{C} \bullet\!\!\to e_1 \approx e'_1 \ldots e_n \approx e'_n, \mathcal{C}$, if $c \equiv d$, and failure, otherwise

When a program rule is applied using *Narrowing* ($\approx_1$ rule), data $\mathcal{C}_D$ and type $\mathcal{C}_T$ conditions of the rule are incorporated to the constraint system. *Binding* ($\approx_2$ rule), *Imitation* ($\approx_3$ rule[6]) and *Eager variable elimination* ($\approx_4$ rule) differ from untyped versions in that lazy unification is substituted by typed lazy unification. When $X \approx e$ is solved, $e : \alpha_X$ is introduced in the constraint system in order to check that e and X have the same type (forcing the well-typedness of the equality). *Identity* ($\approx_5$ rule) and *Decomposition* ($\approx_6$ rule) remain as in untyped versions. Furthermore, in order to achieve laziness of the unification, it is supposed that a demand driven strategy [16] is followed, suspending the lazy unification of $e \rhd X$ meanwhile X does not demand the evaluation of e in the rule conditions.

Any given implementation of these transformation rules must determine the order for solving data and type constraints. When a program rule is applied ($\approx_1$ rule), lazy unification of $e_i \rhd t_i$ is solved before rule conditions $\mathcal{C}_D, \mathcal{C}_T$ are solved. With respect to the demand driven strategy, instead of suspending $e \rhd X$, X is bound to e and the effect of this substitution is propagated over the goal,achieving the sharing of e by considering a suspended form of e [7]. In $\approx_2, \approx_3, \approx_4$, if $X \approx e$ is solved, the pruning of the search space is achieved by checking then $e : \alpha_X$.

[4] We add a type variable α_Y to compute the type of every data variable Y in $\mathcal{C}$.

[5] In order to simplify the notation, we use $\approx$ for representing both $\rhd$ and $==$. In *TLNC* we suppose that type declarations and program rules are used with fresh variables. In the case of program rules $X_i : \alpha_{X_i}$ is introduced for every new data variable X_i (shortly $\bar{X} : \alpha_{\bar{X}}$).

[6] $\approx_3$ needs occur check of X in $c(\bar{e})$. Occur check in lazy unification differs from syntactic occur-check. Occur check in $\approx_3$ checks that X is not safe in $c(\bar{e})$ (i.e. X is in $c(\bar{e})$ in the scope of a function symbol).

Rules for :

$:_1)\ X : \tau, C \bullet\!\!\to \alpha_X \leq \tau, C$

$:_2)\ h(\bar{e}) : \tau, C \bullet\!\!\to e_1 : \tau_1, \ldots, e_n : \tau_n, \tau_0 \leq \tau, C$, where $h \in CON \cup FUN$

$:_3)\ f(\bar{e}) : \tau, C \bullet\!\!\to e_1 \rhd t_1, \ldots, e_n \rhd t_n, C_D, C_T, r : \tau, \bar{X} : \alpha_{\bar{X}}, C$

According to the semantics of the language, type constraint solving may involve the application of a program rule ($:_3$ rule) when $f(\bar{e}) : \tau$ can not be checked by using the type declaration of f ($:_2$ rule). Since syntactical conditions of applicability can not be established, *don't know nondeterminism* arises between $:_2$ and $:_3$ rules.

Any given implementation of these transformation rules should solve $\tau_0 \leq \tau$ in $:_2$ rule as soon as it is incorporated to the goal because its failure will avoid the solving of $e_i : \tau_i$. Moreover, $\alpha_X \leq \tau$ should be checked in $:_1$ rule to be compatible with other type constraints about α_X. On other hand, to get laziness in type conditions solving, a *lazy type checking strategy* should be followed, by applying $:_3$ whenever $:_2$ is not successful.

Rules for $\leq$

$\leq_1)\ K(\bar{\tau}) \leq L(\bar{\sigma}), C \bullet\!\!\to \tau_1 \leq \sigma_1, \ldots, \tau_n \leq \sigma_n, C$, if $K \leq L$, and failure, otherwise

$\leq_2)\ \alpha \leq \alpha, C \bullet\!\!\to C$

$\leq_3)\ \alpha \leq K(\bar{\tau}), \alpha \leq L(\bar{\sigma}), C \bullet\!\!\to \alpha \leq I(\bar{\alpha}), \alpha_1 \leq \tau_1, \ldots, \alpha_n \leq \tau_n, \alpha_1 \leq \sigma_1, \ldots, \alpha_n \leq \sigma_n, C$
if there exists the infimum I for K and L, and failure, otherwise

$\leq_4)\ \alpha \leq L(\bar{\tau}), C \bullet\!\!\to \alpha = K(\bar{\alpha}), \alpha_1 \leq \tau_1, \ldots, \alpha_n \leq \tau_n, C[\alpha/K(\bar{\alpha})]$, where $K \leq L$

$\leq_5)\ \alpha \leq \beta, C \bullet\!\!\to \alpha = \beta, C[\alpha/\beta]$

$\leq_6)\ \alpha \leq \beta, C \bullet\!\!\to \alpha = K(\bar{\alpha}), \beta = L(\bar{\beta}), \alpha_1 \leq \beta_1, \ldots, \alpha_n \leq \beta_n, C[\alpha/K(\bar{\alpha}), \beta/L(\bar{\beta})]$, where $K \leq L$

These rules for solving subtype conditions are similar to those presented in [4,18,8]. Rules obtaining a supremum $\leq_3'$ and binding lower bound type variables $\leq_4'$, are defined in a similar way to their corresponding $\leq_3$ and $\leq_4$. There is *don't know nondeterminism* in the choice of $K \leq L$ in $\leq_4$ rule and also between $\leq_5$ and $\leq_6$ rules, when $\alpha \leq \beta$ is solved; in the second case, to get completeness, there are two possibilities, either to bind α and β or to enumerate solutions for α and β.

Any given implementation of these rules should collect lower and upper bounds of every type variable ($\leq_3$, $\leq_3'$ rules) in order to reduce the nondeterminism of $\leq_4$, $\leq_4'$ rules. The search space is also reduced if we suppose that C is closed under transitivity.

4 Prolog Specification

In this section we propose the implementation of the operational semantics of the language. To this end we translate every program P (resp. goal G) into a set of Prolog clauses $PC(P)$ (resp. a Prolog goal $PG(G)$). The specification of the operational semantics $PC(TLNC)$ together with $PC(P)$, $PG(G)$ and the control regime of Prolog yields a search tree which defines a set of solutions of the goal G w.r.t. the program P.

Data variables X are represented by Prolog terms including a Prolog variable NX, used as internal name, and the representation of the type variable α_X. Type variables α are represented by Prolog terms containing the internal name of the type variable and their upper and lower bounds. We can suppose that every type variable α is constrained as $\beta_1, \ldots, \beta_l, K(\alpha_1, \ldots, \alpha_n) \leq \alpha \leq \gamma_1, \ldots \gamma_u, L(\delta_1, \ldots, \delta_n)$, where $\beta_i, \alpha_j, \gamma_k, \delta_l$ are type variables, and $\beta_1, \ldots, \beta_l, K(\alpha_1, \ldots, \alpha_n)$ (resp. $\gamma_1, \ldots \gamma_u$, $L(\delta_1, \ldots, \delta_n)$) are the lower (resp. upper) bounds of α. K and L are the so-called *current* lower and upper bounds of α. They are obtained by computing infimum and supremum ($\leq_3$, $\leq_3'$ rules) of type constructors. We call *quasi-solved form* to this representation and it will be guaranteed is *consistent*, that is, it defines *at least a solution* for α.

When a type condition is solved, and new subtype constraints arise, new upper and lower bounds in the representations of the involved type variables are added. If the checking of the consistence of the new representation of some type variable fails, this produces backtracking. For instance, the new constraint $nat \leq \beta$ becomes inconsistent the representation $\alpha \leq \beta, bool$.

Matching of type variables ($\leq_4, \leq_4', \leq_5, \leq_6$ rules) must be delayed until it is sound. A matching will be sound when no new type constraints involving its type variables will appear. Note that unsound matchings lead to backtracking in later steps; for instance, binding α to nat when $\alpha \leq nat$, is unsound if $\alpha \leq opnat$ is required afterwards. When a program rule is applied, the fresh type variables of C_T can be soundly matched, if the incorporated data and type constraints have been solved. However, type variables α_X of the fresh data variables of the rule can not be solved until X is solved.

4.1 Solving a Goal

A goal contains $C_D \square C_T$, however in practice, data and type constraints can be written in any order (we write C instead $C_D \square C_T$). In fact, it would be better to write them following an order allowing to prune the search space. There is a significantly improvement in the typed lazy unification if type conditions $X : \tau$ in C are collected in advance in the representation of type variables α_X; so, when $e : \alpha_X$ is solved the constraint $\alpha_X \leq \tau$ is checked in the representation of α_X. In our specification we will suppose that goals are nested pairs $(Constr, Rest)$[7].

$solve\_goal(C) :- solve\_conditions(C), postprocessed(C, AX), matching\_free(AX).$
$solve\_conditions(C)\text{:-}preprocessed(C, TVar), solve\_constr(C), matching\_free(TVar)$
$solve\_constr((Constr, Rest)) :- solve\_constr(Constr), solve\_constr(Rest).$
$solve\_constr(L == R) :- equals(L, R).$
$solve\_constr(E : T) :- type\_cond(E, T).$

[7] Furthermore, in the goal solver we suppose a preprocess which collects in $TVar$, type variables occurring in the goal, and a postprocess which collects in AX, type variables α_X of data variables X remaining free in the answer (remark that α_X has not been bound because X is free). All these variables are matched via *matching_free* predicate.

4.2 Solving Data Constraints

The predicate *equals* solves strict equalities $e{=}{=}e'$. To this end, computes head normal forms of e and e' via *hnf*. This strategy corresponds with the application of $\approx_1$ rule until someone of $\approx_2, \approx_3, \approx_5, \approx_6$ rules can be applied. *equals_hnf* and *equal_vars* correspond with $\approx_5, \approx_2, \approx_3, \approx_6$ rule applications. The failure cases ($\approx_6$ rule for clashing of constructors, and $\approx_3$ rule if occur check in *equal-vars* succeeds) are implemented as Prolog failure branches.

$equals(L, R) : -hnf(L, HL), hnf(R, HR), equals\_hnf(HL, HR).$
$equals\_hnf(X, R) : -is\_data\_var(X), !, equal\_vars(X, R).$
$equals\_hnf(L, X) : -is\_data\_var(X), !, equal\_vars(X, L).$
$equals\_hnf(L, R) : -is\_c\_apply(L, c(L_1, \ldots, L_n), T_c), is\_c\_apply(R, c(R_1, \ldots, R_n), T_c),$
$\quad equals(L_1, R_1), \ldots, equals(L_n, R_n).$
% $is\_data\_var(E)$ succeeds if E is a data variable.
% $is\_c\_apply(E, c(E_1, \ldots, E_n), T)$ (resp. $is\_c\_apply(E, c(E_1, \ldots, E_n), T)$ succeeds
% if E represents an expression $c(e_1, \ldots, e_n)$ (resp. $f(e_1, \ldots, e_n)$),
% where E_i represents e_i and T represents the declared type of c (resp. f).
$equal\_vars(X, Y) : -is\_data\_var(Y), is\_same\_data\_var(X, Y), !.$
$equal\_vars(X, Y) : -is\_data\_var(Y), !, bind\_data(X, Y).$
$equal\_vars(X, E) : -is\_c\_apply(E, c(E_1, \ldots, E_n), T), not\_occur(X, c(E_1, \ldots, E_n)),$
$\quad imitation\_data(X, c(X_1, \ldots, X_n)), equals(X_1, E_1), \ldots, equals(X_n, E_n).$
% $is\_same\_data\_var(X, Y)$ determines if X and Y represent the same data variable.
$hnf(E, H) : -is\_data\_var(E), !, H = E.$
$hnf(E, H) : -is\_c\_apply(E, c(E_1, \ldots, E_n), T_c), !, H = c(E_1, \ldots, E_n).$
$hnf(E, H) : -is\_f\_apply(E, f(E_1, \ldots, E_n), T_f), \#f(E_1, \ldots, E_n, R), hnf(R, H).$
$not\_occur(X, Y) : -is\_data\_var(Y), !, name\_data\_var(X, NX),$
$\quad name\_data\_var(Y, NY), NX = \backslash = NY.$
$not\_occur(X, E) : -is\_c\_apply(E, c(E_1, \ldots, E_n), T_c), !,$
$\quad not\_occur(X, E_1), \ldots, not\_occur(X, E_n).$
$not\_occur(X, E) : -is\_f\_apply(E, f(E_1, \ldots, E_n), T_f).$
% $name\_data\_var(X, NX)$ returns the Prolog variable NX associated to X.

$\#f$ represents the Prolog predicate associated to f. Every program rule $f(t_1, \ldots, t_n) := r \Leftarrow C$ is translated to $\#f(E_1, \ldots, E_n, r) : -unify(E_1, t_1), \ldots, unify(E_n, t_n), solve\_conditions(C)$. This translation of a program rule differs from other translations in the literature [3]. In effect, instead of the computation of the head normal form of $f(\bar{e})$, the $\#f$ predicate computes a first approximation of $f(\bar{e})$(the body r of a rule) via the application of the rule. This is so because type constraints solving does not need head normal forms.

In general, when a program rule is applied, type variables in C_T are collected and solved as if they were part of a goal. However the matching of the remaining type variables α_X corresponding to fresh free variables X of the program rule is delayed until the binding of X. The lazy unification is specified as follows:

$unify(E, T) : -is\_data\_var(T), !, bind\_data(T, E).$
$unify(E, T) : -hnf(E, H), unify\_hnf(H, T).$
$unify\_hnf(E, T) : -is\_data\_var(E), !, bind\_data(E, T).$

$unify\_hnf(E,T) : -is\_c\_apply(E, c(E_1,\ldots,E_n), T_c), is\_c\_apply(T, c(T_1,\ldots,T_n), T_c),$
$\quad unify(E_1,T_1),\ldots, unify(E_n,T_n).$

When e is lazily unified to t then we distinguish two cases. If t is a data variable, t is bound to e ($\approx_2$, $\approx_4$ rules), otherwise t is a data constructor term and it demands its head normal form ($\approx_1$, $\approx_6$ rules). $\approx_3$ rule has not been considered in the case of non strict equalities, once the demand driven strategy has been substituted by sharing. For the sake of presentation we will consider suspended forms and sharing below; in this preliminary version, multiple copies will be eventually evaluated. Finally, binding and imitation rules are specified as follows:

$bind\_data(X,T)\text{:-}type\_var(X,AX), infer(T,AX), name\_data\_var(X,NX), NX{=}T.$
$imitation\_data(X, c(X_1,\ldots,X_n))\text{:-}type\_var(X,AX), infer(c(X_1,\ldots,X_n), AX),$
$\quad name\_data\_var(X,NX), NX = c(X_1,\ldots,X_n).$

% $type\_var(X,AX)$ returns AX representing α_X.

4.3 Solving Type Constraints

The predicate *type_cond* solves type conditions. The don't know nondeterminism of $:_2$ and $:_3$ is achieved through the backtracking of Prolog.

$type\_cond(X,T) : -is\_data\_var(X), !, type\_var(X,AX), add\_constraints(AX \leq T).$
$type\_cond(E,T) : -is\_c\_apply(E, c(E_1,\ldots,E_n), T_1 \ldots T_n \to T_0), !,$
$\quad lower\_matching(T_0,T), solve\_constr(E_1 : T_1,\ldots,E_n : T_n).$
$type\_cond(E,T) : -is\_f\_apply(E, f(E_1,\ldots,E_n), T_1 \ldots T_n \to T_0),$
$\quad lower\_matching(T_0,T), solve\_constr(E_1 : T_1,\ldots,E_n : T_n).$
$type\_cond(E,T)\text{:-}is\_f\_apply(E, f(E_1,\ldots,E_n), T_f), \#f(E_1,\ldots,E_n,R), type\_cond(R,T).$

If $h(\bar{e}) : \tau$ is solved by consulting the type declaration ($:_2$ rule) then *lower_matching* (τ_0,τ) matches type variables of τ_0 to the infimum of their upper bounds in τ. These matchings are sound because type variables of τ_0 are bound to the greatest types, which guarantees soundness by monotonicity of the type declarations. Type variables of τ (which possibly are elsewhere in the goal) are matched in a later step, when the solving of C_T finishes. Type variables of $\cup_{i=1,\ldots,n} tvar(\tau_i) \backslash tvar(\tau_0)$ will not be matched, since their matching is not relevant to the computation; only the consistence of their constraints will be checked.

Type conditions $e : \alpha_X$ obtained from solving $X \approx e$ are solved by binding α_X to the least type of e through $infer$ predicate. $Infer$ is analogous to the *type_cond* predicate, but using *matching* predicate instead of *lower_matching* predicate. To get the more general solution for X in V, $matching(\tau,\tau')$ matches type variables of τ to the greatest types in τ', as in *lower_matching*, and type variables of τ' to the least types in τ, that is, the supremum of their lower bounds in τ'.

Due to the lack of space we omit the specification of subtype constraints solving, including *lower_matching*, *matching* and *matching_free* predicates. Supposed given a quasi-solved form $\beta_1,\ldots,\beta_l$, $K(\alpha_1,\ldots,\alpha_n) \leq \alpha \leq \gamma_1, \ldots \gamma_u,$

$L(\delta_1, \ldots, \delta_n)$ for a type variable α, (1) *lower_matching*(τ_0, τ) matches every type variable α of τ_0 to his corresponding $L(\delta_1, \ldots, \delta_n)$ ($\leq_4$ rule), (2) *matching*(τ, τ') matches every type variable α of τ to his corresponding $L(\delta_1, \ldots, \delta_n)$ ($\leq_4$ rule) and every type variable α of τ' to his corresponding $K(\alpha_1, \ldots, \alpha_n)$ ($\leq_4'$ rule), and (3) *matching_free* gets more general solutions, matching every type variable α to its greatest type, either $L(\beta_1, \ldots, \beta_n)$ ($\leq_4$ rule) or γ_j ($\leq_5$ rule) or any maximal type w.r.t. $\leq$ ($\leq_6$ rule).

Whenever a type variable is matched (e.g. $\alpha = L(\delta_1, \ldots, \delta_n)$)the binding is propagated to the upper and lower bounds, adding new constraints ($\beta_i \leq L(\delta_1, \ldots, \delta_n)$, $\alpha_j \leq \delta_j$ -as new upper bounds of β_i, α_j- and $L(\delta_1, \ldots, \delta_n) \leq \gamma_k$ -as new lower bounds of γ_k), checking the consistence of every involved type variable.

5 Representation

The representation of data variables X is defined as $R[X] = dvar(NX, R[\alpha_X])$ where NX is a Prolog variable or eventually the representation of any data variable to which X has been bound. The representation of type variables α (supposed is in quasi-solved form) is defined as $R[\alpha] = tvar(R[\{\beta_1, \ldots, \beta_m\}], R[K(\alpha_1, \ldots, \alpha_n)], NA, R[\{\gamma_1, \ldots, \gamma_k\}], R[L(\delta_1, \ldots, \delta_n)])$, where NA is a Prolog variable or eventually the representation of any data variable to which α has been bound.

Upper and lower bound type variables are represented by partial lists of lists allowing to add new upper and lower bounds along the computation. More precisely, $R[\{\beta_1, \ldots, \beta_m\}] = [V_1, \ldots, V_n, [R[\beta_1], \ldots, R[\beta_m]]|Xs]$, where Xs is a Prolog variable and $V_1, \ldots, V_n$ are lists of type variables. The current upper and lower bounds are represented by partial lists of non-variable types, so $R[K(\alpha_1, \ldots, \alpha_n)] = [I_1, \ldots, I_n, K(R[\alpha_1], \ldots, R[\alpha_n])|Xs]$, where Xs is a Prolog variable and $I_1, \ldots, I_n$ are non-variable types.

The most important predicate which depends on the representation is *add_constraints*. This predicate supports the representation of every type variable. Each time a new constraint involving α is added, the representation of α is modified; for instance, if the new constraint is of the form $\epsilon \leq \alpha$ and ϵ is a type variable, ϵ is added to $\{\beta_1, \ldots, \beta_m\}$, and if is of the form $M(\tau_1, \ldots, \tau_n) \leq \alpha$, then the supremum S of K and M is computed and $S(\lambda_1, \ldots, \lambda_n)$, where λ_i are new type variables, is added as new current lower bound for α, generating the new constraints $\tau_1 \leq \lambda_1, \ldots, \tau_n \leq \lambda_n$, $\alpha_1 \leq \lambda_1, \ldots, \alpha_n \leq \lambda_n$ ($\leq_3'$ rule) and checking that the involved type variables are consistent. Notice that, in the case of introducing $\epsilon \leq \alpha$, simultaneously α is added as upper bound of ϵ. This kind of circular references could produce problems when type variables α and ϵ are mutually bound. We exploit the lack of occur check in Prolog, accepting cyclic terms in our representation.

By simplicity we have substituted demand driven strategy for propagating substitutions over the goal. This produces as a side effect that multiple copies of expressions are evaluated in data and/or type conditions. As some kind of sharing would optimize this non desirable situation, we adopt (following ideas in [7]) a better representation of functional expressions and consider $f(e_1,\ldots,e_n,S)$, where S is a Prolog variable if $f(e_1,\ldots,e_n)$ has not been evaluated, and otherwise it represents the evaluation state of $f(e_1,\ldots,e_n)$, that is, if $f(e_1,\ldots,e_n)$ has been evaluated to e' (not necessarily in head normal form) then S represents e'. This differs from previous works [3], which focus on head normal forms and not on intermediate steps.

6 Type Analysis

Essentially, the idea of type analysis is to require a stronger static well-typedness condition to program rules and queries. The static well-typedness condition we have required so far allows to type program rules and goals straightforwardly; in effect, given $f(\bar{t}) := r \Leftarrow C_D \square \emptyset$, where $f : \tau_1 \ldots \tau_n \to \tau_0 \in FUN$, to get static well-typedness, it is enough to include in C_T the type conditions $t_1 : \tau_1, \ldots, t_n : \tau_n, r : \tau_0$ and, for every $l == r \in C_D$, $l : \alpha, r : \alpha$, where α is a new type variable (this is so even if we have $C_T \neq \emptyset$).

A stronger static well-typedness condition will allow to improve the execution.In order to get a better search space reduction, C_T should only contain *the essential* type information to be checked and this information should be *as precise as possible.* For this reason, we are interested in simplifying C_T and obtaining the closest type information for each program rule. Even more, we are interested in detecting goals or program rules leading to useless branches, either due to illtyped C_D's or unsatisfiable C_T's.

6.1 Strong Static Well-Typedness

In order to define a notion of strong static well-typedness, we need to require an additional condition to program rules. In the following, we suppose that for all $f(\bar{e})$, program rule $f(\bar{t}) := r \Leftarrow C_D \square C_T \in [P]$ and environment V, $\{\sigma/(\mathcal{P},V) \vdash_{TRC} f(\bar{e}) : \sigma\} = \{\sigma/(\mathcal{P},V) \vdash_{TRC} r : \sigma\}$, whenever $(\mathcal{P},V) \vdash_{TRC} e_i \rhd t_i$, $1 \leq i \leq n$, and $(\mathcal{P},V) \vdash_{TRC} C_D, C_T$. We call this condition *non-ambiguity w.r.t. types*, likewise the condition of *non-ambiguity w.r.t. data* defined in [2]. Notice that for this set equality, the inclusion $\supseteq$ derives from $\vdash_{TRC}$, while $\subseteq$ is ensured whenever program rules are not referred to *a particular case of the type declaration for* f, and assuming that non-ambiguity w.r.t. data holds. Non-ambiguity w.r.t. types yields a natural property over expressions, namely that every expression TRC-well-typed w.r.t. V has a *TRC-least type* w.r.t. V. The TRC-least type σ of e w.r.t. V satisfies $(\mathcal{P},V) \vdash_{TRC} e : \sigma$ and $\sigma \leq \tau$, for every τ such that $(\mathcal{P},V) \vdash_{TRC} e : \tau$. From now on we associate a type variable α_e to represent the TRC-least type of e.

C_T leads to unsuccessful branches when C_T is unsatisfiable. At compile-time it is possible to detect the unsatisfiability by checking if C_T contains type conditions which are not compatible w.r.t. the type declarations or if there is a subset of type conditions of an expression without infimum.

We say C_T is *compatible* if there exists ρ such that C_T is compatible via ρ and C_T is *compatible via* ρ if the following conditions holds:

- every *type condition* $h(\bar{e}) : \tau \in C_T$ *is compatible w.r.t. the type declarations* via ρ, which means that if $h : \tau_1 \ldots \tau_n \to \tau_0$ then $\{e_1 : \tau_1, \ldots, e_n : \tau_n\}$ is compatible via ρ and, if $h \equiv c$ then $\tau_0\rho \leq \tau\rho$, and if $h \equiv f$ then $\alpha_{f(\bar{e})}\rho$ is a lower bound of $\tau_0\rho$ and $\tau\rho$
- every *subset* $\{e : \tau_1, \ldots, e : \tau_n\} \subseteq C_T$ *is lower bound* via ρ, that is, $\alpha_e\rho \leq \tau_1\rho, \ldots, \tau_n\rho$.

On other hand, C_T should only contain the minimal type information for assuring the well-typedness. Therefore we require that C_T can not be simplified to a simpler or smaller set of type conditions.

We will say that C_T is *non-simplifiable* if there not exist ρ and C'_T such that C_T is simplifiable to C'_T via ρ such that $Sol(C_D, C_T) = Sol(C_D, C'_T)$, and C_T is *simplifiable to* C'_T *via* ρ if some of the following conditions holds:

- there exists a *type condition* $h(\bar{e}) : \tau \in C_T$ *simplifiable via* ρ to $\{e_1 : \tau_1, \ldots, e_n : \tau_n\}$, that is, $\tau_1 \ldots \tau_n \to \tau_0 \in [h]$, $\tau_0 \leq \tau\rho$ and $C'_T = (C_T \backslash \{h(\bar{e}) : \tau\}) \cup \{e_1 : \tau_1, \ldots, e_n : \tau_n\}$
- there exists a *subset* $\{e : \tau_1, \ldots, e : \tau_n\} \subseteq C_T$ *simplifiable via* ρ to $e : \tau$, that is, $\tau \leq \tau_1\rho, \ldots, \tau_n\rho$ and $C'_T = (C_T \backslash \{e : \tau_1, \ldots, e : \tau_n\}) \cup \{e : \tau\}$.

C_T is simplifiable if either (1) it contains a type condition simplifiable to their arguments, for some instance of the type declaration, or (2) it contains a set of type conditions for some expression and this set can be simplified to a unique type condition, for some lower bound. However, in order to preserve soundness, the simplification of C_T have to consider that the applicability conditions of a program rule and the type constraints of a goal depends on C_T, therefore C_T can be simplified only if the simplification guarantees the semantic equivalence of C_D, C_T and C_D, C'_T.

With respect to C_D, we can also consider a stronger well-typedness condition for equalities $l == r$. The static well-typedness condition required to find a common type for both sides of every equality $l == r$ of C_D. Now, we strength this condition by requiring the existence of a common *minimal type* for l and r deduced from C_T. Notice that at compile-time we only can refer to *the set of minimal types* of l and r w.r.t. C_T (every C_T defines a finite set of more general minimal types for every well-typed expression in C_T) and not to a unique least type. These minimal types can be seen as approximations at compile-time to the TRC-least type. Given $(V, \theta, \rho) \in Sol(C_D, C_T)$ and an expression e, one of the minimal types will be an upper bound of the TRC-least type of $e\theta$ w.r.t. V.

We will say that σ is a *minimal type* of e in C_T if $(\Sigma, C_T) \vdash_{SW} e : \sigma$ and there not exist τ such that $\tau \leq \sigma$, $\tau \neq \sigma$ and $(\Sigma, C_T) \vdash_{SW} e : \tau$.

We will say that C_D is *strongly well-typed* w.r.t. C_T if for every $l{==}r \in C_D$ there is at least a common minimal type σ of l and r w.r.t. C_T, and for every σ, σ' minimal types of l and r w.r.t. C_T respectively, either $\sigma = \sigma'$ or there is no infimum of σ and σ'.

This notion of strong well-typedness of C_D is based on the existence of TRC-least type for expressions and terms. Indeed if $l\theta \rhd t$ and $r\theta \rhd t$, for a given $(V, \theta, \rho) \in Sol(C_D, C_T)$, then some common minimal type of l and r w.r.t. C_T is an upper bound of the TRC-least type of t w.r.t. V.

Combining all the required conditions, we will say that $f(\bar{t}) := r \Leftarrow C_D \Box C_T$ is *strongly static well-typed* if it is static well-typed w.r.t. C_T, C_T is compatible and non-simplifiable and C_D is strongly well-typed w.r.t. C_T.

Examples

(1) $C_D = \{X == Y, Y == head(tail([0, Z]))\}$ is strongly well-typed w.r.t. $C_T = \{X : int, Y : int, tail([0, Z]) : nelist(int)\}$. In effect, $tail([0, Z]) : nelist(int)$ is compatible with $tail : list(\alpha) \rightarrow list(\alpha)$, but can not be simplified to $Z : int$.

(2) $C_D = \{X == head([0, suc(Y)]), Z == second([X, suc(X)])\}$ is strongly well-typed w.r.t. $C_T = \{X : nat, Y : nat, Z : nat\}$, but not w.r.t. $C_T = \{X : int, Y : nat, Z : int\}$, because the unique minimal type of $head([0, suc(Y)])$ w.r.t. C_T is nat, the unique minimal type of X w.r.t. C_T is int, $nat \neq int$ and they have infimum.

Two aspects must be considered yet. The first one is how to know if a given C_T is simplifiable or not to a C'_T in such a way that $Sol(C_D, C_T) = Sol(C_D, C'_T)$. For instance, if $C_D = \{Z == head([X, Y])\}$ then $C_T = \{Z : nat, head([X, Y]) : nat\}$ can not be simplified to $C'_T = \{Z : nat, X : nat, Y : nat\}$ because $Sol(C_D, C_T) \neq Sol(C_D, C'_T)$. As we will see, to guarantee this semantic equivalence it will be needed to consider simplifications w.r.t. an *upper more general matching* ρ.

The second aspect to be specified is how to compute minimal types w.r.t. a given C_T. The set of minimal types of an expression w.r.t. C_T corresponds with the smallest types which are compatible w.r.t. the type declarations, therefore they can be computed using *lower more general matchings* ρ. For instance, nat is the smallest type such that $head([0, suc(Y)])$ is compatible w.r.t. type declarations and $C_T = \{Y : nat\}$.

6.2 Matching Algorithms

Type Analysis is based on the subtype constraint matching that can be deduced from C_T. Given C_T, type analysis transforms C_T into a set of type constraints R_{C_T} as follows. Starting with $R_{C_T} := C_T$, repeat the following process until no rule can be applied:

- for every $X : \sigma \in R_{C_T}$ and $\sigma \not\equiv \alpha_X$, $R_{C_T} := (R_{C_T} \backslash \{X : \sigma\}) \cup \{X : \alpha_X, \alpha_X \leq \sigma\}$
- for every $c(\bar{e}) : \sigma \in R_{C_T}$, $R_{C_T} := (R_{C_T} \backslash \{c(\bar{e}) : \sigma\}) \cup \{e_1 : \tau_1, \ldots, e_n : \tau_n, \tau_0 \leq \sigma\}$, where $\tau_1 \ldots \tau_n \rightarrow \tau_0$ is a fresh variant of the type declaration of c
- for every $f(\bar{e}) : \sigma \in R_{C_T}$ and $\sigma \not\equiv \alpha_{f(\bar{e})}$, $R_{C_T} := (R_{C_T} \backslash \{f(\bar{e}) : \tau\}) \cup \{e_1 : \tau_1, \ldots, e_n : \tau_n, \alpha_{f(\bar{e})} \leq \tau_0, \alpha_{f(\bar{e})} \leq \sigma, f(\bar{e}) : \alpha_{f(\bar{e})}\}$, where $\tau_1 \ldots \tau_n \rightarrow \tau_0$ is a fresh variant of the type declaration of f.

C_T is transformed in this way into a set R_{C_T} which includes: (1) a set C_{C_T} of type conditions of the form $e : \alpha_e$, for data variables and functional expressions , and (2) subtype constraints S_{C_T} of the form $\sigma \leq \sigma'$. Notice that $Sol(R_{C_T}) = Sol(C_T)$ can be proved from the semantics of type conditions.

Once $R_{C_T}(= S_{C_T} \cup C_{C_T})$ is obtained, the type analysis studies the upper and lower more general matchings of S_{C_T}. Given a set of subtype constraints $C = \{\tau_1 \leq \sigma_1, \ldots, \tau_n \leq \sigma_n\}$, we say that ρ is a *type matching* of C (written $\rho \in TMatch(C)$) if $\tau_1\rho \leq \sigma_1\rho, \ldots, \tau_n\rho \leq \sigma_n\rho$. We say that $\{\rho_i, i \in I\}$

is a *complete set of upper (resp. lower) more general matchings* of C if $\rho_i \in TMatch(C)$, for all $i \in I$, and for each $\rho \in TMatch(C)$ there exists $i \in I$ and a type substitution λ such that $\rho \leq \rho_i\lambda$ (resp. $\rho_i\lambda \leq \rho$). It can be proved that if $TMatch(C) \neq \emptyset$ then there exists a complete and finite set of upper (resp. lower) more general matchings for C. For instance, a complete set of upper more general matchings of $C = \{\alpha \leq list(\beta), nelist(\alpha) \leq \gamma, \beta \leq nat\}$ only contains $\rho = \{\beta/nat, \alpha/list(nat), \gamma/list(list(int))\}$ and a complete set of lower more general matchings is $\{\rho_1, \rho_2, \rho_3, \rho_4\}$ where $\rho_1 = \{\beta/zero, \alpha/elist(zero), \gamma/nelist(elist(zero))\}$, $\rho_2 = \{\beta/posint, \alpha/elist(posint), \gamma/nelist\ (elist\ (posint))\}$, $\rho_3 = \{\beta/zero, \alpha/\ nelist(zero), \gamma/nelist(nelist(zero))\}$ and $\rho_4 = \{\beta/posint, \alpha\ /nelist\ (posint), \gamma/nelist(nelist(posint))\}$.

The following algorithm, written as transformation rules of the form $(C, \mu) \rightarrow (C', \mu')$, computes the matchings ρ defining a complete set of upper more general matchings for a set of subtype constraints C. To this end, the transformation process starts with $(C, \emptyset)$ and ends with $(\emptyset, \rho)$, if no failure arises, by applying the following rules:

- (DEC): $(\{K(\bar{\tau}) \leq L(\bar{\sigma})\} \cup C, \rho) \rightarrow (\{\tau_1 \leq \sigma_1, \ldots, \tau_n \leq \sigma_n\} \cup C, \rho)$
  % $K \leq L$, and failure, otherwise
- (IMI I): $(\{\alpha \leq \tau, L(\bar{\sigma}) \leq \beta\} \cup C, \rho) \rightarrow (\{\alpha \leq \tau, L(\bar{\beta}) \leq \beta, \sigma_1 \leq \beta_1, \ldots, \sigma_n \leq \beta_n\} \cup C, \rho)$
  % α occurs in $L(\bar{\sigma})$ and $\bar{\beta}$ are new type variables
- (IMI II): $(\{\alpha \leq K(\bar{\tau}), \tau \leq \beta\} \cup C, \rho) \rightarrow (\{\alpha \leq K(\bar{\alpha}), \tau \leq \beta, \alpha_1 \leq \tau_1, \ldots, \alpha_n \leq \tau_n\} \cup C, \rho)$
  % β occurs in $K(\bar{\tau})$ and $\bar{\alpha}$ are new type variables
- (INF): $(\{\alpha \leq K_i(\bar{\tau}_i), 1 \leq i \leq m\} \cup C, \rho) \rightarrow (\{\alpha_j \leq \tau_{ij}/1 \leq i \leq m, 1 \leq j \leq n\} \cup C, \rho)[\alpha/I(\bar{\alpha})]$
  % $\bar{\alpha}$ are new type variables, α only occurs in C in constraints $\alpha \leq \beta$ and there exists the infimum I of $K_1, \ldots, K_m$, and failure, otherwise
- (MAX): $(\{K_i(\bar{\tau}_i) \leq \alpha, 1 \leq i \leq m\} \cup C, \rho) \rightarrow (\{\tau_{ij} \leq \alpha_i/1 \leq i \leq m, 1 \leq j \leq n\} \cup C, \rho)[\alpha/M(\bar{\alpha})]$
  % $\bar{\alpha}$ are new type variables, α only occurs in C in constraints $\beta \leq \alpha$ and there exists a maximal M such that $M \geq K_1, \ldots, K_m$, and failure, otherwise
- (VAR I): $(\{\alpha \leq \beta\} \cup C, \rho) \rightarrow (C, \rho)[\beta/\alpha]$
- (VAR II): $(\{\alpha \leq \beta_1, \ldots, \alpha \leq \beta_n\} \cup C, \rho) \rightarrow (C, \rho)[\alpha/I(\bar{\alpha}), \beta_1/M_1(\bar{\alpha}), \ldots, \beta_n/M_n(\bar{\alpha})]$
  % $\bar{\alpha}$ are new type variables, M_j/n are maximals, I is the infimum ($I \neq M_j$, for some j) of $M_1, \ldots, M_n$ and α does not occur in C

(DEC), (IMI I) and (IMI II) collect lower and upper bounds of every type variable. Then (INF) and (MAX) are applied to match type variables to as great as possible types. (MAX) provides different upper more general matchings by binding to maximal elements. When no lower and upper non-variable bounds can be found, (VAR I) matches to type variables and (VAR II) matches to maximal types. We exploit the non determinism of these transformation rules ((MAX), (VAR I), (VAR II)), computing a set of upper more general matchings. We write $UGM(C) = \{\rho/(C, \emptyset) \rightarrow (\emptyset, \rho)\}$.

We have proved that $UGM(C)$ is a complete set of upper more general matchings of C. There is a similar algorithm to compute a complete set $LGM(C)$ of lower more general matchings of C, taking infima instead of suprema and minimal types instead of maximal types. Our algorithms are based in the algorithm of Hill and Topor [15] to compute the more general infimum of a set of types.

6.3 Inferring a C_T

The type analysis we present is expressed as transformation rules $C_T \rightarrow C'_T$ of sets of type conditions. Given a program rule $f(\bar{t}) := r \Leftarrow C_D \Box C_T$ (C_T establishes the applicability conditions of the rule) we start with C_T for which the rule is statically well-typed and we finish either with a C'_T w.r.t. which the rule is strongly well-typed, or with a failure, meaning that the program rule is not strongly typable w.r.t. C_T. The transformation rules are the following:

- **(SWE) Strong Well-Typedness of Equalities**
  $C_T \rightarrow_{SWE} \{l : \tau, r : \tau\} \cup C_T$
  % $l == r \in C_D$, α, β and γ are new type variables, $CS = LGM(S_{\{l:\beta, r:\gamma, l:\alpha, r:\alpha\} \cup C_T} |_{\{\beta,\gamma,\alpha\}})$, there exists $\tau = \mathit{infimum}(\beta\rho, \gamma\rho)$, for all $\rho \in CS$, and failure when CS is empty or there not exist infimum for $\beta\rho, \gamma\rho$, for all $\rho \in CS$.

- **(SCTC) Simplification and Compatibility of Type Conditions**
  $\{e : \tau\} \cup C_T \rightarrow_{SCTC_1} C'_T \cup C_T$
  % $CS = UGM(S_{\{e:\tau\}} \cup S_{C_T}|_{tvar(R_{\{e:\tau\}})})$, there is $\rho = \{\alpha/\alpha\rho_i, \rho_i \in CS, \alpha \in tvar(C_{\{e:\tau\}})\}$, and failure, when CS is empty; moreover if $e \equiv f(\bar{e})$, $\alpha_{f(\bar{e})}\rho$ is an instance of the range of f, $C'_T = C_{\{e:\tau\}}\rho$

  $\{e : \tau_1, \ldots, e : \tau_n\} \cup C_T \rightarrow_{SCTC_2} \{e : \tau\} \cup C_T$
  % $CS = UGM(\{\alpha_e \leq \tau_1, \ldots, \alpha_e \leq \tau_n\} \cup S_{\{e:\tau_1,\ldots,e:\tau_n\} \cup C_T}|_{\cup_{i=1,\ldots,n} tvar(\tau_i)})$, $\alpha_e\rho = \tau$, for all $\rho \in CS$, and failure, when CS is empty.

Since type variables can occur in different type conditions, the soundness in the matching of type variables is preserved if we collect in $S_{C_T}(\subseteq R_{C_T})$, all the subtype constraints for them; for instance, in (SWE), the subtypes constraints for α, β, γ.

(SWE) rule includes in C_T type conditions $l : \tau$ and $r : \tau$, for every $l == r \in C_D$. The type conditions $l : \alpha, r : \alpha$ are used to force the well-typedness of $l == r$, while the type conditions $l : \beta, r : \gamma$ are used to compute in β (resp. γ) minimal types of l (resp. r). There are two possible cases of failure: either l and r are ill-typed (CS is empty) or l and r have no compatible types (no infimum for minimal types). In both cases $l == r$ is not strongly typable. When there exist two or more infima, the type of $l == r$ can not be precise and C_T does not change (see Example (3) below).

($\mathsf{SCTC_1}$) rule first computes a complete set CS of upper more general matchings of $S_{\{e:\tau\}} \cup S_{C_T}|_{tvar(R_{\{e:\tau\}})}$. Then $e : \tau$ is simplifiable to $C_{\{e:\tau\}}\rho$ when there exists an extension ρ of CS over the type variables of $C_{\{e:\tau\}}$. If e is a functional expression, then the type condition can be simplified whenever ρ is matched to the range of the type declaration of f. ($\mathsf{SCTC_2}$) rule simplifies a set of type conditions of an expression by computing the more general infimum. To this end, it computes a complete set CS of upper more general matchings of $\{\alpha_e \leq \tau_1, \ldots, \alpha_e \leq \tau_n\} \cup S_{\{e:\tau_1,\ldots,e:\tau_n\} \cup C_T} |_{\cup_{i=1,\ldots,n} tvar(\tau_i)}$. The set of type conditions is simplifiable if CS defines a more general infimum for α_e. If CS either in ($\mathsf{SCTC_1}$) or in ($\mathsf{SCTC_2}$) results the empty set, we can conclude that C_T is not compatible.

Examples

(1) Let $C_D = \{X == Y, Y == head(tail([0, Z]))\}$, (SWE) rule obtains $C_T = \{X : int, Y : int, head(tail([0, Z])) : int\}$, then (SCTC$_1$) simplifies $head(tail([0, Z])) : int$ to $tail([0, Z]) : nelist(int)$, and therefore $C_T = \{X : int, Y : int, tail([0, Z]) : nelist(int)\}$ is obtained.

(2) Let $C_D = \{X == head([0, suc(Y)]), Z == second([X, suc(X)])\}$, (SWE) obtains $C_T = \{X : nat, head([0, suc(Y)]) : nat, Z : nat, second([X, suc(X)]) : nat\}$. (SCTC$_1$) simplifies to $C_T = \{X : nat, Y : nat, Z : nat\}$.

(3) Let $C_D = \{X == head([A, Y])\}$ and $C_T = \{X : \alpha, head([A, Y]) : \alpha\}$, where $A : a, a \leq b, a \leq c$ (without supremum for b and c). In this case, it is not possible to type $X == head([A, Y])$ because we deduce that $Y : b$ or $Y : c$ and therefore $X : b$ or $X : c$. Notice that, the strong well-typedness condition is ensured by considering $C_T = \{X : \alpha, head([A, Y]) : \alpha\}$.

Theorem (Strong Well-Typedness)
Given a program rule $f(\bar{t}) := r \Leftarrow C_D \Box C_T$:
(1) If $C_T \rightarrow^* C'_T$ and C'_T is irreducible by the rules, then $f(\bar{t}) := r \Leftarrow C_D \Box C'_T$ is strongly well-typed and $Sol(C_D, C_T) = Sol(C_D, C'_T)$
(2) If $C_T \rightarrow^* failure$ then $f(\bar{t}) := r \Leftarrow C_D \Box C_T$ is not strongly typable w.r.t. C_T.

7 Conclusions and Future Work

In this paper we have presented an implementation of a functional logic language with subtyping. We have shown how types can be used in typed lazy narrowing to prune the search space of solutions, avoiding otherwise necessary functional reductions. We have also studied how type analysis at compile-time can be used for improving computations. Most of the optimizations and ideas we have presented have been tested in a *prototype* realized in SICSTUS Prolog. As future work we plan to improve the implementation, by considering more refined translations to Prolog, following ideas from [16,3].

Acknowledgments. I would like to thank to Antonio Gavilanes and Ana Gil for their comments to this work.

References

1. J.M.Almendros-Jiménez, A. Gil-Luezas. *Lazy Narrowing with Parametric Order-Sorted Types*, Procs. ALP-HOA'97, LNCS 1298, pp. 159-173, 1997.
2. J.M. Almendros-Jiménez, A. Gavilanes-Franco, A. Gil-Luezas. *Algebraic Semantics for Functional Logic Programs with Polymorphic Order-Sorted Types*, Procs. ALP'96, LNCS 1139, pp. 299-313, 1996.
3. P.Arenas-Sánchez, A.Gil-Luezas, F.J.López-Fraguas. *Combining Lazy Narrowing with Disequality Constraints*, Procs. PLILP'94, LNCS 844, pp. 385-399, 1994.
4. C. Beierle. *Type Inferencing for Polymorphic Order-Sorted Logic Programs*, Procs. ICLP'95, The MIT Press, pp. 765-779, 1995.

5. C. Beierle. *Concepts, Implementation and Applications of a Typed Logic Programming Language,* Logic Programming: Formal Methods and Practical Applications. C. Beierle and L. Plümer Eds. Elsevier Science, pp. 139-167, 1995.
6. C. Beierle, G. Meyer. *Run-Time Type Computations in the Warren Abstract Machine,* JLP, 18, pp. 123-148, 1994.
7. P.H. Cheong. L. Fribourg. *Implementation of Narrowing: The Prolog-Based Approach,* In K.R. Apt, J.W. Bakker, J.J.M.M. Rutten, Eds. Logic Programming Languages: Constraints, Functions and Objects. The MIT Press, pp. 1-20. 1993.
8. Y. Fuh, P. Mishra. *Type Inference with Subtypes,* TCS 73, pp. 155-175, 1990.
9. J.C. González-Moreno, T. Hortalá-González, F. J. López-Fraguas, M. Rodríguez-Artalejo. *A Rewriting Logic for Declarative Programming,* Procs. ESOP'96, LNCS 1058, pp. 156-172, 1996.
10. J.C.González-Moreno, T.Hortalá-González, M.Rodríguez-Artalejo. *A Higher Order Rewriting Logic for Functional Logic Programming,* Procs.ICLP'97,The MIT Press, pp.153-167, 1997.
11. M. Hanus, S. Antoy, H. Kuchen, F.J. López-Fraguas, F. Steiner *Curry: An Integrated Functional Logic Language,* Draft June 1998.
12. M. Hanus. *Parametric Order-Sorted Types in Logic Programming,* Procs. TAPSOFT'91, LNCS 494, pp. 181-200, 1991.
13. M. Hanus. *The Integration of Functions into Logic Programming: A Survey,* JLP (19,20), Special issue "Ten Years of Logic Programming", pp. 583-628, 1994.
14. M. Hanus. *A Unified Computation Model for Functional and Logic Programming,* Procs. POPL'97, pp. 80-93, 1997.
15. P.M. Hill, R.W. Topor. *A Semantics for Typed Logic Programming,* in [17], Chapter 1, pp. 1-58, 1992.
16. R.Loogen, F.J.López-Fraguas, M.Rodríguez-Artalejo. *A Demand Driven Computation Strategy for Lazy Narrowing,*Procs.PLILP'93,LNCS 714,pp.184-200,1993.
17. F.Pfenning. *Types in Logic Programming,* The MIT Press, 1992.
18. G.S. Smith. *Principal Type Schemes for FunctionalPrograms with Overloading and Subtyping,* Science of Computer Programming 23, pp. 197-226, 1994.
19. G. Smolka. *TEL (version 0.9), Report and User Manual,* SEKI-Report SR, 87-17, FB Informatik, Universität Kaiserslautern, Germany, 1988.
20. G. Smolka. *Logic Programming over Polymorphically Order-Sorted Types,* PhD thesis, FB Universität Kaiserslautern, Germany, 1989.

Abstract Interpretation of Prolog Programs

Fausto Spoto and Giorgio Levi

Dipartimento di Informatica - Università di Pisa
Corso Italia, 40, 56100 Pisa, Italy
{spoto,levi}@di.unipi.it
Ph.: +39-50-887246 Fax: +39-50-887226

Abstract. In this paper we propose an abstract version of the denotational semantics defined in [8]. This leads to a precise goal–independent abstract interpretation of Prolog programs. We deal with the control rules of Prolog and the cut operator. Moreover, we get a simple denotation for negation as finite failure. The abstract analysis is proposed both for computed answers analysis and for call patterns analysis. In both cases the abstract semantics is finitely computable.

Keywords: *Abstract interpretation, static analysis, logic programming.*

1 Introduction

In [8] we proposed a denotational semantics which models the set of computed answers of a Prolog interpreter in a goal–independent fashion. In this paper, we show how that semantics can be used as an effective base for program analysis. Handling Prolog control features allows us to get a more precise analysis. In particular our approach deals with the leftmost selection rule, the depth first search rule and the cut operator. Moreover, it is a simple task to "implement" the Prolog `not/1` operator once we have the cut operator. Hence even negation as finite failure is modeled denotationally. Similarly we can implement some other built-in's like `if_then/2` and `if_then_else/3`.

A previous approach to the abstract interpretation of logic programs with control rules of Prolog and the cut operator was presented in [6]. The main difference between our approach and theirs is that we delay the decision divergence/no divergence or cut/no cut to the actual evaluation of a denotation in a given constraint store. They take instead this decision just when they compose denotations. It is not clear how their approach could lead to a goal-independent semantics. In any case their approach makes impossible to compute the denotations of the most general goals only, like we do in this paper. This is important because it leads to a simpler and more efficient computation of the semantics.

The problems arising in the abstract interpretation framework, when we deal with control, were already tackled in [7, 8]. We will give a simple and effective solution to the problem of "downward approximation" of constraints. Another problem with [7, 8] was that an abstract analysis built on that semantics was not

A.M. Haeberer (Ed.): AMAST'98, LNCS 1548, pp. 455–470, 1998.

necessarily finite even when defined on a finite abstract domain. This important problem is solved in the present work (see section 4.2).

Another approach strongly related to ours is the one described in [1], where a semantics for logic programs with Prolog control was obtained by "compiling" a Prolog program into an ask/tell language, so that the semantics of the Prolog program can be viewed as the semantics of a constraint logic program. The main problem with [1] is that the transformation uses constraints like "the execution of this goal terminates", whose abstraction is not trivial. The problem related to the abstraction of these goals, in a finite abstract analysis framework, is the same as the problem of finding an (upward or downward) approximation of SLD trees with control faced in [2], or even as the problem of the approximation of the consistency of constraints faced in [7, 8]. Up to now there was no sensible proposal for control approximation. Hence all these approaches are still theoretical rather than practical. In this paper we follow the approach of [1]. However we use a new kind of constraints, in particular we use "observability" constraints which check whether they are consistent with the constraint store. Moreover, we suggest how to handle these constraints in the abstract case, and we propose a general approach for approximating them, by keeping only the information which is relevant from the abstract point of view, rather than too concrete information like termination.

In order to better understand the ideas underlying the following sections, consider this Prolog program:

```
select_vars_in_term(X, [X]) :- var(X), !.
select_vars_in_term(A, []) :- atom(A), !.
select_vars_in_term(F, L) :- F = ..[_Name|Args],
        select_vars_in_list(Args, L).

select_vars_in_list([], []).
select_vars_in_list([H|T], [H1|T1]) :- select_vars_in_term(H, H1),
        select_vars_in_list(T, T1).
```

Assume we know from some global analysis that the procedure `select_vars_in_term` is always called with its second argument free. If we take into account the control information, it is easy to guess that in the third clause F can never be a free variable nor an atom. This would allow us to optimize the compilation of the procedure.

Roughly speaking, dealing with control allows us to collect not only the information related to the successful execution of a branch of computation, but even that related to the simple observability of it, which should not be hidden by divergence or cut; we will not try to guess when a branch of a computation is observable or not (this would lead to control approximation) but merely what abstract information can we derive from the observability of a branch of computation. We will try to solve the problem: "if I arrived here, what do I know?" and not the problem "when do I arrive here?". Hence we will add to any constraint, in the following called "kernel" constraint, an observability condition in the form of an "observability" constraint, which is to be satisfied in order to make the kernel constraint observable.

2 Preliminaries

We assume the reader familiar with basic algebraic structures [5]. A sequence is an ordered collection of elements with repetitions. We will write $\mathcal{SEQ}(\mathcal{E})$ for the set of sequences of elements of $\mathcal{E}$. :: denotes sequence concatenation.

Abstract interpretation [3, 4] is a technique which allows us to statically ("at compile time") determine some dynamic ("at run time") properties of a program. The idea is that of executing the program on an "abstract" domain, where every element represents a set of elements of the "concrete" domain. This technique is widely used in computer science both for reasoning about the relationships between different semantics and for program analysis.

Note that in the program analysis case the abstract domain should be chosen with the aim of making the computation effective. The abstraction (and related approximation) is needed because we know from Rice's theorem that most "interesting" properties of programs are not effectively computable.

One way of formalizing abstract interpretation is by means of Galois connections

Definition 1. *A Galois connection between the posets $\langle P, \sqsubseteq\rangle$ and $\langle P^a, \sqsubseteq^a\rangle$ is a pair $\langle \alpha, \gamma\rangle$ of total maps such that the following condition holds:*

$$\forall p \in P, p^a \in P^a : \alpha(p) \sqsubseteq^a p^a \quad \text{if and only if} \quad p \sqsubseteq \gamma(p^a) .$$

α and γ are the abstraction and the concretization maps of the connection.

We know from a theorem in [4] that, given two complete lattices $\langle P, \sqsubseteq\rangle$ and $\langle P^a, \sqsubseteq^a\rangle$, if $\langle \alpha, \gamma\rangle$ is a Galois connection between them, $\phi : P \mapsto P$ and $\phi^a : P^a \mapsto P^a$ are two monotonic operators and $\alpha(\bot) = \bot^a$, then the local correctness condition implies the global one, i.e.:

- $\alpha \circ \phi \sqsubseteq^a \phi^a \circ \alpha$ implies $\alpha(lfp(\phi)) \sqsubseteq^a lfp(\phi^a)$,
- $\alpha \circ \phi = \phi^a \circ \alpha$ implies $\alpha(lfp(\phi)) = lfp(\phi^a)$.

We recall that two posets can be extended to complete lattices in such a way that any monotonic map α between them can be extended to a continuous map $\tilde{\alpha}$ between their extensions. The same extension leads to the following result which will allow us to get simpler proofs.

Proposition 1. *Let $\langle F, \sqsubseteq\rangle$ and $\langle F^a, \sqsubseteq^a\rangle$ be two posets and $T : F \mapsto F$, $T^a : F^a \mapsto F^a$ and $\alpha : F \mapsto F^a$ be three monotonic maps such that $\alpha(\bot) = \bot^a$ and $\alpha \circ T \sqsubseteq^a T^a \circ \alpha$. Then we can extend F and F^a to two complete lattices $\widetilde{F}$ and $\widetilde{F^a}$ respectively and α, T and T^a to continuous maps $\widetilde{\alpha} : \widetilde{F} \mapsto \widetilde{F^a}$, $\widetilde{T} : \widetilde{F} \mapsto \widetilde{F}$ and $\widetilde{T^a} : \widetilde{F^a} \mapsto \widetilde{F^a}$ such that $\langle \widetilde{\alpha}, \alpha^{-1}\rangle$ is a Galois connection between $\widetilde{F}$ and $\widetilde{F^a}$, $\widetilde{\alpha}(\bot) = \bot^a$ and the correctness condition $\widetilde{\alpha} \circ \widetilde{T} \sqsubseteq^a \widetilde{T^a} \circ \widetilde{\alpha}$ holds.*

The relevance of the above proposition is that we do not need to be concerned with the infinite elements of the two lattices, neither we have to define the semantic operators and the abstraction map on them. Roughly speaking, F consists of the finite elements of $\widetilde{F}$ and F^a consists of the finite elements of $\widetilde{F^a}$.

In the following we will use an "abstract" syntax for Prolog programs, which simplifies the semantic operators. The translation from Prolog into our syntax is straightforward and can be understood by noting that the Prolog clause: `q(X):-p(X),!,s(X).` is translated into $\mathtt{q}(x) : -\mathtt{cut}(\mathtt{p}(x))$ **and** $\mathtt{s}(x)$.. Finally, our abstract syntax assumes all predicates to be unary. This constraint simplifies the definition of the semantics without loss of generality. The extension of that definition to the general case is anyway straightforward.

A clause will be of the form $\mathtt{p}(x) : -G_1$ **or** $\cdots$ **or** G_n., where $G_1, \ldots, G_n$ are goals, defined by the grammar: $G ::= c|G$ **and** $G|\mathtt{exists}\ x.G|\mathtt{cut}(G)$, where c is a constraint and x is a variable symbol.

3 The Denotational Semantics

We recall here the basic definitions of the denotational semantics defined in [8], where further details and explanations can be found.

Definition 2. *Given a program P, its immediate consequence operator is defined as* $T_P(I)(\mathtt{p}) = \exists_x([\![\delta_{\alpha,x}]\!] \otimes \mathsf{i}(\mathcal{E}[\![B]\!]I))$*, where* $\mathtt{p}(x) : -B.$ *is the definition of* p *in P, I is an interpretation, i.e., a map that tells us what we already know about the semantics of the more general goals,* α *is a distinguished variable (not present in programs) and*

$$\mathcal{E}[\![c]\!]I = [\![c]\!] \qquad \mathcal{E}[\![G_1 \text{ and } G_2]\!]I = \mathcal{E}[\![G_1]\!]I \otimes \mathcal{E}[\![G_2]\!]I$$

$$\mathcal{E}[\![G_1 \text{ or } G_2]\!]I = \mathcal{E}[\![G_1]\!]I \oplus \mathcal{E}[\![G_2]\!]I \qquad \mathcal{E}[\![\mathtt{p}(\mathtt{x})]\!]I = \exists_\alpha([\![\delta_{\mathtt{x},\alpha}]\!] \otimes I(\mathtt{p}))$$

$$\mathcal{E}[\![\mathtt{cut}(G)]\!]I = !(\mathcal{E}[\![G]\!]I) \qquad \mathcal{E}[\![\mathtt{exists}\ \mathtt{x}.G]\!]I = \exists_{\mathtt{x}}(\mathcal{E}[\![G]\!]I)\ .$$

The semantics of a program P is defined as: $\mathcal{S}_P = \bigsqcup_{i \geq 0}(T_P^i(I^0))$.

We now define the semantic domain. The lattice of basic constraints could be thought of as the domain of analysis, for instance equations over rational trees, though more abstract "domains" could be used (and they actually will in section 4). The diagonal elements $\delta_{x,y}$ represent unification of x and y. For instance, in the case of rational trees, they represent the equation $x = y$. They are needed to perform parameter passing. The cylindrification operators $\exists_x$ are used to remove from a constraint all the information related to x, and are a simple way for avoiding all renaming problems.

Observability constraints are a new concept; roughly speaking, an observability constraint o is satisfied in a given constraint store S if and only if it is consistent with it, i.e., if and only if $S \wedge o$ is satisfiable. Note that we do not require o to be entailed by S. From an alternative point of view, we can look at observability constraints as constraints which give us some information on the constraint store. If o is satisfied in S then S is consistent with o. This remark will be useful when we will consider abstract observability constraints.

In the following, we will consider pairs consisting of a "kernel" constraints and its "observability" part. A kernel constraint k is observable in a given constraint store if and only if its observability part o is satisfied. The kernel part represents the contribution of the constraint to the constraint store in the case the constraint is observable.

Definition 3. *A basic constraint is an element of a lattice* $\langle \mathcal{B}, \leq, \vee, \wedge, true, false\rangle$, *where* $\wedge$ *is the greatest lower bound operator,* $\vee$ *is the least upper bound operator,* $true$ *is the top of the lattice and* $false$ *is the bottom of the lattice. We assume there exist elements* $\delta_{x,y} \in \mathcal{B}$: *for instance* $\delta_{x,y}$ *represents the constraint identifying the variables* x *and* y. *Moreover, we assume there is a family of monotonic operators* $\exists_x$ *on the set of constraints, representing the restriction of a constraint obtained by hiding all the information related to the variable* x.

The set of kernel constraints is defined as $\mathcal{K} = \mathcal{B}$.

The set $\mathcal{O}$ *of observability constraints is defined as the least set containing basic constraints and closed w.r.t. the binary operations* $\sqcap$ *ans* $\sqcup$ *and the unary operation* $-$. *We assume an injection map from kernel constraints into observability constraints defined as* $k \propto obs = k$. *This map will have maximum precedence. Hence,* $o \sqcap k \propto obs$ *will mean* $o \sqcap (k \propto obs)$. *Cylindrification is defined on observability constraints as the elementwise extension of cylindrification on kernel constraints: if* $o = k \propto obs$ *then* $\exists_x o = \exists_x k$. *Otherwise:*

$$\exists_x(o_1 \sqcap o_2) = \exists_x o_1 \sqcap \exists_x o_2 \qquad \exists_x(o_1 \sqcup o_2) = \exists_x o_1 \sqcup \exists_x o_2$$
$$\exists_x(-o) = -\exists_x o \ .$$

A conditional convergent constraint (convergent constraint for short) is an element of $C = \mathcal{O} \times \mathcal{K}$. $o + k$ *will denote* $\langle o, k\rangle \in C$, *where* o *is the observability part and* k *is the kernel part of the constraint. We will use the usual notation for field selection:* $(o+k)_1 = o$ *and* $(o+k)_2 = k$.

We define the following sets of constraints:

- $\widetilde{C} = \{\widetilde{o+k} \mid o+k \in C\}$ *(divergent constraints);*
- $\underline{C} = \{\underline{o+k} \mid o+k \in C\}$ *(open convergent constraints);*
- $\underline{\underline{C}} = \{\underline{\underline{o+k}} \mid o+k \in C\}$ *(open internal constraints).*

A constraint is an element of the set $\mathfrak{C} = C \cup \widetilde{C} \cup \underline{C} \cup \underline{\underline{C}}$. *The module operator* $|\cdot| : \mathfrak{C} \mapsto C$ *is defined as:* $|o+k| = o+k$, $|\widetilde{o+k}| = o+k$, $|\underline{o+k}| = o+k$ *and* $|\underline{\underline{o+k}}| = o+k$.

Given a goal G, $\mathrm{con}(G)$ *is the constraint part of* G, *i.e., the kernel constraint already computed by* G *(up to the first procedure call in* G*) in a left to right scanning of the goal. Formally (we assume* $c \wedge \widetilde{c'} = \widetilde{c \wedge c'}$*)*

$$\mathrm{con}(c) = c \qquad \mathrm{con}(\mathbf{p}(\mathbf{x})) = \widetilde{true}$$
$$\mathrm{con}(G_1 \ \mathbf{and}\ G_2) = \mathrm{con}(G_1) \wedge \mathrm{con}(G_2) \qquad \mathrm{con}(\mathtt{exists}\ x.G) = \exists_x \mathrm{con}(G)$$
$$\mathrm{con}(\mathtt{cut}(G)) = \mathrm{con}(G) \ .$$

Note that $\mathrm{con}(G) \in \widetilde{C}$ *if and only if* G *contains a procedure call, and* $\mathrm{con}(G) \in C$ *if and only if it does not.*

Definition 4. *The usual notion of satisfiability is defined on kernel constraints. A kernel constraint k is satisfiable in a constraint store S and in a structure interpretation $\mathfrak{I}$, if and only if there exists an environment ρ such that $\models^{\rho}_{\mathfrak{I}} (S \wedge k)$.*

Satisfiability for observability constraints is different in that we allow different environments to be used in different proofs.

- *if $o \in \mathcal{B}$, then o is satisfiable in S and $\mathfrak{I}$, if and only if o is satisfiable in S and $\mathfrak{I}$ as a kernel constraint;*
- *$o_1 \sqcap o_2$ is satisfiable in S and $\mathfrak{I}$, if and only if both o_1 and o_2 are satisfiable in S and $\mathfrak{I}$;*
- *$o_1 \sqcup o_2$ is satisfiable in S and $\mathfrak{I}$, if and only if o_1 or o_2 is satisfiable in S and $\mathfrak{I}$;*
- *$-o$ is satisfiable in S and $\mathfrak{I}$, if and only if o is not satisfiable in S and $\mathfrak{I}$.*

Definition 5. *The lifting of an observability constraint with respect to a kernel constraint is defined as follows:*

$$k \bullet k' = k \wedge k' \qquad k \bullet (o_1 \sqcap o_2) = (k \bullet o_1) \sqcap (k \bullet o_2)$$

$$k \bullet (o_1 \sqcup o_2) = (k \bullet o_1) \sqcup (k \bullet o_2) \qquad k \bullet -o = -(k \bullet o)\ .$$

A preorder on the set of observability constraints is defined as $o_1 \leq o_2$ if and only if, for all constraint stores S, if $S \bullet o_1$ is satisfiable in $\emptyset$ and $\mathfrak{I}$, then $S \bullet o_2$ is satisfiable in $\emptyset$ and $\mathfrak{I}$. Obviously $\leq$ is reflexive and transitive. However, it is not antisymmetric. Hence we define an equivalence relation $\sim$, such that $o_1 \sim o_2$, if and only if $o_1 \leq o_2$ and $o_2 \leq o_1$. In such a way, the extension of $\sim$ on $\sim$-equivalence classes is an ordering relation. In the following, we will always consider an observability constraint as its equivalence class, and therefore $=$ will denote $\sim$.

Given a sequence of constraints s, we define its divergence condition as $\delta(s) = \bigsqcup_{\widetilde{o+k} \in s} o \sqcap k \propto obs$, its cut condition as $\kappa(s) = \bigsqcup_{\underline{o+k} \in s} (o \sqcap k \propto obs) \sqcup \bigsqcup_{\overline{o+k} \in s} (o \sqcap k \propto obs)$, its block condition as $\beta(s) = \delta(s) \sqcup \kappa(s)$ and its convergence condition as $o(s) = \bigsqcup_{o+k \in s} (o \sqcap k \propto obs) \sqcup \bigsqcup_{\underline{o+k} \in s} (o \sqcap k \propto obs)$.

We write $\mathcal{SS}$ for the set of sequences of constraints.

Definition 6. *The instantiation of a sequence with an observability constraint is defined as $o \cdot \langle\rangle = \langle\rangle$ and $o \cdot \langle v_1, \ldots, v_n \rangle = \langle o \cdot v_1, \ldots, o \cdot v_n \rangle$, where $o \cdot \widetilde{\langle o' + k' \rangle} = \widetilde{\langle o \sqcap o' + k' \rangle}$, $o \cdot \underline{\langle o' + k' \rangle} = \underline{\langle o \sqcap o' + k' \rangle}$, $o \cdot \overline{\langle o' + k' \rangle} = \overline{\langle o \sqcap o' + k' \rangle}$ and $o \cdot \langle o' + k' \rangle = \langle o \sqcap o' + k' \rangle$.*

Definition 7. *The instantiation of a sequence with a kernel constraint is defined as $k \circ \langle\rangle = \langle\rangle$ and $k \circ \langle v_1, \ldots, v_n \rangle = \langle k \circ v_1, \ldots, k \circ v_n \rangle$, where $k \circ \widetilde{\langle o' + k' \rangle} = \widetilde{\langle k \bullet o' + k \wedge k' \rangle}$, $k \circ \underline{\langle o' + k' \rangle} = \underline{\langle k \bullet o' + k \wedge k' \rangle}$, $k \circ \overline{\langle o' + k' \rangle} = \overline{\langle k \bullet o' + k \wedge k' \rangle}$ and $k \circ \langle o' + k' \rangle = \langle k \bullet o' + k \wedge k' \rangle$.*

The semantical operators are defined as follows:

$$[\![c]\!] = \langle \widetilde{true \propto obs} + c \rangle \ , \qquad s_1 \oplus s_2 = s_1 :: -\beta(s_1) \cdot s_2 \ .$$

$$\begin{aligned} !(\langle\rangle) &= \langle\rangle \\ !(\langle o+k \rangle) &= \langle \underline{o+k} \rangle & \qquad !(\langle \underline{o+k} \rangle) &= \langle \underline{o+k} \rangle \\ !(\langle \overline{o+k} \rangle) &= \langle \overline{o+k} \rangle & \qquad !(\langle \widetilde{o+k} \rangle) &= \langle \widetilde{o+k} \rangle \ . \end{aligned}$$

Moreover, if $length(s) \geq 2$ and therefore $s = s_1 :: s_2$, where s_1 and s_2 are non empty sequences, we define $!(s) = !(s_1) :: -o(s_1) \cdot !(s_2)$.

$$\begin{aligned} ¡(\langle\rangle) &= \langle\rangle \\ ¡(\langle o+k \rangle) &= \langle o+k \rangle & \qquad ¡(\langle \underline{o+k} \rangle) &= \langle o+k \rangle \\ ¡(\langle \overline{o+k} \rangle) &= \langle\rangle & \qquad ¡(\langle \widetilde{o+k} \rangle) &= \langle \widetilde{o+k} \rangle \ . \end{aligned}$$

Moreover, if $length(s) \geq 2$ and thus $s = s_1 :: s_2$, where s_1 and s_2 are non empty sequences, we define $¡(s) = ¡(s_1) :: ¡(s_2)$.

$$\begin{aligned} \mathbb{E}_x(\langle\rangle) &= \langle\rangle \\ \mathbb{E}_x(\langle o+k \rangle) &= \langle \exists_x o + \exists_x k \rangle & \qquad \mathbb{E}_x(\langle \underline{o+k} \rangle) &= \langle \underline{\exists_x o + \exists_x k} \rangle \\ \mathbb{E}_x(\langle \overline{o+k} \rangle) &= \langle \overline{\exists_x o + \exists_x k} \rangle & \qquad \mathbb{E}_x(\langle \widetilde{o+k} \rangle) &= \langle \widetilde{\exists_x o + \exists_x k} \rangle \ . \end{aligned}$$

Moreover, if $length(s) \geq 2$ and thus $s = s_1 :: s_2$, where s_1 and s_2 are non empty sequences, we define $\mathbb{E}_x(s) = \mathbb{E}_x(s_1) :: \mathbb{E}_x(s_2)$.

$$\begin{aligned} \langle \widetilde{o+k} \rangle \otimes s &= \langle \widetilde{o+k} \rangle & \qquad \langle \overline{o+k} \rangle \otimes s &= \langle \overline{o+k} \rangle \\ \langle o+k \rangle \otimes s &= o \cdot (k \circ s) & \qquad \langle \underline{o+k} \rangle \otimes s &= \langle \underline{o+k} \rangle :: o \cdot (k \circ s) \ . \end{aligned}$$

Moreover, if $length(s') \geq 2$ then there exist non empty sequences s_1 and s_2, such that $s' = s_1 :: s_2$. In this case we define: $s' \otimes s = s_1 \otimes s :: -\xi(s_1, s) \cdot (s_2 \otimes s)$, where: $\xi(s_1, s) = \bigsqcup_{o+k \in s_1} o \sqcap (k \bullet \delta(s)) \sqcup \bigsqcup_{\underline{o+k} \in s_1} o \sqcap (k \bullet \delta(s))$.

An interpretation is a map I from the set of predicate symbols Π to the set of sequences of constraints. Interpretations form a poset whose bottom element is I^0, such that $I^0[\![p]\!] = \langle \widetilde{true \propto obs} + true \rangle$ for every predicate p.

Note that our semantics is defined as $\mathcal{S}_P = \bigsqcup_{i \geq 0} T_P^i(I^0)$. Actually, while divergent or cut constraints are useful for a precise and compositional definition of the observability conditions, we are not interested in them when we turn to the use of the collected abstract information. In this second step, we only need to know the abstract information of every convergent constraint and the related observability constraint. Therefore it is sensible to discard all the redundant information (from this point of view), through an auxiliary function $\mathcal{O}$, defined as $\mathcal{O}(s) = \bigsqcup_{o+k \in s}(o \sqcap k \propto obs) \sqcup \bigsqcup_{\underline{o+k} \in s}(o \sqcap k \propto obs)$. $\mathcal{O}$ is extended to interpretations in the obvious way. Namely, $(\mathcal{O}(I))[\![p]\!] = \mathcal{O}(I[\![p]\!])$ for every predicate p. $\mathcal{O}$ is an observability condition which is satisfied by those constraint stores in which a computed answer is observed. Note that the unique difference with the case of the classical s-semantics is that observability conditions are taken into account in the computation of $\mathcal{O}$. Moreover, if s contains two constraints $o_1 + k_1$ and $o_2 + k_2$ and if o_1 entails o_2 and k_1 entails k_2, then $\mathcal{O}(s) = o_2 \sqcap k_2$. The pointwise partial order is defined on the range of $\mathcal{O}$.

In [8] it is shown that the set of consistent and observable kernel constraints in $\mathcal{O}(\mathcal{E}[\![G]\!]\mathcal{S}_P)$ is the set of Prolog computed answers for the goal G in the program P. Moreover, we showed that

Theorem 1. *A kernel constraint k is a Prolog computed answer constraint for a goal G executed in a program P if and only if $k \le \mathcal{O}(\mathcal{E}[\![G]\!](\mathcal{S}_P))$.*

3.1 A Call Pattern Semantics

In the section above we have shown a semantics able to characterize computed answers of a Prolog program. However, program analysis is more often concerned with call patterns, more precisely with the set of all possible call patterns for any predicate. This information is useful for an optimizing compiler, since it allows one to compile predicate calls, by specializing them for the specific call patterns which can actually arise at run-time.

In this section we sketch the definitions of the operators for a call pattern semantics for Prolog. We omit some details, since they can easily be reconstructed as an extension of the case of the computed answer semantics.

The main difference between call patterns and computed answers is that call patterns belong to the internal part of the SLD tree, and not only to its frontier. Moreover, a call pattern constraint is associated to a specific predicate. Hence we define the following kinds of constraint.

- $\widetilde{o+k; \mathrm{p}}$ (divergent constraints);
- $o+k$ (convergent constraints);
- $\overline{o+k; \mathrm{p}}$ (internal constraints);
- $\underline{o+k}$ (cut constraints);
- $\overline{\underline{o+k}}$ (cut internal constraints).

Note the introduction of a new kind of internal constraint. The definitions of the semantic operators is similar to the case of the computed answer semantics, with a few relevant differences: $[\![c]\!] = \langle \mathit{true} \propto \mathit{obs} + c\rangle$, $s_1 \oplus s_2 = s_1 :: -\beta(s_1) \cdot s_2$. The definition of $\exists_x$ is essentially unchanged while the basic cases of the definition of the ! operator are the following.

$$
\begin{aligned}
!(\langle\rangle) &= \langle\rangle & !(\langle o+k\rangle) &= \langle \underline{o+k}\rangle\\
!(\langle \overline{o+k;\mathrm{p}}\rangle) &= \langle \overline{o+k;\mathrm{p}}\rangle & !(\langle \underline{o+k}\rangle) &= \langle \underline{o+k}\rangle\\
!(\langle \overline{\underline{o+k}}\rangle) &= \langle \overline{\underline{o+k}}\rangle & !(\langle \widetilde{o+k;\mathrm{p}}\rangle) &= \langle \widetilde{o+k;\mathrm{p}}\rangle\ .
\end{aligned}
$$

The only relevant difference is that internal constraints are left untouched, because a constraint is allowed to cut only if it is a solution of the goal, i.e., only if it is a convergent constraint.

Even the basic cases of the definition of i are an extension of the old definition:

$$\begin{aligned} \mathsf{i}(\langle\rangle) &= \langle\rangle & \mathsf{i}(\langle o+k\rangle) &= \langle o+k\rangle \\ \mathsf{i}(\langle\overline{o+k;\mathrm{p}}\rangle) &= \langle\overline{o+k;\mathrm{p}}\rangle & \mathsf{i}(\langle\underline{o+k}\rangle) &= \langle o+k\rangle \\ \mathsf{i}(\langle\overline{\underline{o+k}}\rangle) &= \langle\rangle & \mathsf{i}(\langle\widetilde{o+k;\mathrm{p}}\rangle) &= \langle\widetilde{o+k;\mathrm{p}}\rangle\ . \end{aligned}$$

Finally, the basic cases of the definition of $\otimes$ are

$$\begin{aligned} \langle\widetilde{o+k;\mathrm{p}}\rangle \otimes s &= \langle\widetilde{o+k;\mathrm{p}}\rangle & \langle\overline{o+k;\mathrm{p}}\rangle \otimes s &= \langle\overline{o+k;\mathrm{p}}\rangle \\ \langle\overline{\underline{o+k}}\rangle \otimes s &= \langle\overline{\underline{o+k}}\rangle & \langle o+k\rangle \otimes s &= o\cdot(k\circ s) \\ \langle\underline{o+k}\rangle \otimes s &= \langle\overline{\underline{o+k}}\rangle :: o\cdot(k\circ s)\ . \end{aligned}$$

A difference is found in the definition of the T_P operator. Roughly speaking, in the denotation of $\mathrm{p}(x)$ there exists a call pattern which is precisely $\mathrm{p}(x)$, and then there are all the call patterns which arise in its execution. This leads to the following definition: $T_P(I)(\mathrm{p}) = \langle\overline{\mathit{true} \propto \mathit{obs} + \mathit{true};\mathrm{p}}\rangle :: \exists_x([\![\delta_{\alpha,x}]\!] \otimes \mathsf{i}(\mathcal{E}[\![B]\!]I))$. With this simple changes, the semantics scheme of definition 2 gives rise to a call pattern semantics for Prolog. It is computed as the least fixpoint of the T_P operator defined above, starting from the environment I^0 such that $I^0[\mathrm{p}] = \langle\widetilde{\mathit{true} \propto \mathit{obs} + \mathit{true};\mathrm{p}}\rangle$, for every predicate p.

We now show how to abstract the denotational semantics for computed answers. The following considerations can be easily extended to the case of call patterns.

4 Abstract Semantics

A concrete observability constraint tells us what we know of the constraint store in the case the conditional constraint is observable. It is the most precise information we could know in that case. Similarly, an abstract observability constraint should say what we know of the constraint store *w.r.t. the abstract domain we have at hand* in the case the conditional constraint is observable. This information will be called *success approximation* of the observability constraint. However, we compute even negations of observability constraints. The negation of a success approximation of an observability constraint o is *not* a success approximation of the observability constraint $-o$. Therefore, we need a *failure approximation* of a concrete observability constraint. It will become a success approximation of $-o$. Therefore, a failure approximation of o is nothing more than a success approximation of $-o$. We show now how these approximations can be computed. Note that we need these approximations even for kernel constraints, since they are mapped into observability constraints through the $\propto \mathit{obs}$ map.

Consider a concrete observability or kernel constraint c. If it is satisfiable in a constraint store, then we know that c is satisfiable in that constraint store. Therefore, the most concrete success condition is obviously c itself. Similarly, the most concrete failure condition for c is $-c$, that is "c is not satisfiable in the current store". So every concrete constraint is isomorphic to a success/failure pair $\langle c, -c\rangle$, which we will write as $\left[\begin{smallmatrix} c \\ -c \end{smallmatrix}\right]$. This is the most concrete success/failure approximation of c. Actually, it is a precise approximation of c. Assume now we are interested in a given abstract analysis, whose domain is $\mathcal{D}$ and whose abstraction map is $\rho : \mathcal{B} \mapsto \mathcal{D}$. It can easily be seen that a success condition for c is the constraint $s(c) = \bigvee_{\models S\wedge c} \rho(S)$. The previous formula should be read as follows. We look for the most precise condition which is satisfied by every constraint store S which is consistent with c. Dually, a failure condition for c is the constraint $f(c) = \bigvee_{\not\models S\wedge c} \rho(S)$. These two formulas can be generalized to observability constraints as follows:

$$s(o) = \bigvee\nolimits_{\models S\bullet o} \rho(S) \tag{1}$$

$$f(o) = \bigvee\nolimits_{\not\models S\bullet o} \rho(S) \,. \tag{2}$$

Hence we abstract an observability constraint into the pair $\chi(o) = \left[\begin{smallmatrix} s(o) \\ f(o) \end{smallmatrix}\right]$.

In the case of kernel constraints, we need also to know what happens (from an abstract point of view) if a kernel constraint k is satisfied. This information is obviously $\rho(k)$. Hence a kernel constraint will be abstracted into $\chi(k) = \left[\begin{smallmatrix} s(k) \\ \rho(k) \\ f(k) \end{smallmatrix}\right]$.

This leads to a straightforward definition for $(\propto obs)^\sharp$, i.e., the abstract counterpart of $\propto obs$: $\left[\begin{smallmatrix} s \\ a \\ f \end{smallmatrix}\right] \propto obs^\sharp = \left[\begin{smallmatrix} s \\ f \end{smallmatrix}\right]$.

The conjunction of kernel and observability constraints is defined as

$$\left[\begin{smallmatrix} s_1 \\ a_1 \\ f_1 \end{smallmatrix}\right] \sqcap^\sharp \left[\begin{smallmatrix} s_2 \\ a_2 \\ f_2 \end{smallmatrix}\right] = \left[\begin{smallmatrix} s_1\wedge s_2 \\ a_1\wedge a_2 \\ f_1\vee f_2 \end{smallmatrix}\right] \quad \text{and} \quad \left[\begin{smallmatrix} s_1 \\ f_1 \end{smallmatrix}\right] \sqcap^\sharp \left[\begin{smallmatrix} s_2 \\ f_2 \end{smallmatrix}\right] = \left[\begin{smallmatrix} s_1\wedge s_2 \\ f_1\vee f_2 \end{smallmatrix}\right] \,.$$

The disjunction of kernel and observability constraints is dually defined as

$$\left[\begin{smallmatrix} s_1 \\ a_1 \\ f_1 \end{smallmatrix}\right] \sqcup^\sharp \left[\begin{smallmatrix} s_2 \\ a_2 \\ f_2 \end{smallmatrix}\right] = \left[\begin{smallmatrix} s_1\vee s_2 \\ a_1\vee a_2 \\ f_1\wedge f_2 \end{smallmatrix}\right] \quad \text{and} \quad \left[\begin{smallmatrix} s_1 \\ f_1 \end{smallmatrix}\right] \sqcup^\sharp \left[\begin{smallmatrix} s_2 \\ f_2 \end{smallmatrix}\right] = \left[\begin{smallmatrix} s_1\vee s_2 \\ f_1\wedge f_2 \end{smallmatrix}\right] \,.$$

Abstract negation and cylindrification are defined as

$$-^\sharp\left[\begin{smallmatrix} s \\ f \end{smallmatrix}\right] = \left[\begin{smallmatrix} f \\ s \end{smallmatrix}\right], \quad \exists^\sharp_x \left[\begin{smallmatrix} s \\ f \end{smallmatrix}\right] = \left[\begin{smallmatrix} \exists_x s \\ \exists_x f \end{smallmatrix}\right] \quad \text{and} \quad \exists^\sharp_x \left[\begin{smallmatrix} s \\ a \\ f \end{smallmatrix}\right] = \left[\begin{smallmatrix} \exists_x s \\ \exists_x a \\ \exists_x f \end{smallmatrix}\right] \,.$$

With the above definitions, and assuming: $[\![c]\!]^\sharp = \left\langle \left[\begin{smallmatrix} \top \\ \bot \end{smallmatrix}\right] + \left[\begin{smallmatrix} s(c) \\ \rho(c) \\ f(c) \end{smallmatrix}\right] \right\rangle$, the semantic scheme of definition 2 can be directly translated into an abstract semantics definition.

Finally we define an abstract version of the $\mathcal{O}$ function as

$$\mathcal{O}^\sharp(l^\sharp) = \bigvee\nolimits_{\left[\begin{smallmatrix} s \\ f \end{smallmatrix}\right] + \left[\begin{smallmatrix} s' \\ d \\ f' \end{smallmatrix}\right] \in l^\sharp} (s \wedge s' \wedge d) \vee \bigvee\nolimits_{\underline{\left[\begin{smallmatrix} s \\ f \end{smallmatrix}\right] + \left[\begin{smallmatrix} s' \\ d \\ f' \end{smallmatrix}\right] \in l^\sharp}} (s \wedge s' \wedge d) \,.$$

For a call pattern version of the abstract semantics, one should only perform the same changes suggested for the concrete call pattern semantics. For an example of computation of this semantics, see later.

4.1 Correctness of the Abstraction

In this section we show that the abstract denotational semantics is correct with respect to the concrete one. The abstraction map χ can be extended to sequences of constraints by defining $\chi(o+k) = \chi(o) + \chi(k)$ and $\chi(\langle a_1, \ldots, a_n \rangle) = \langle \chi(a_1), \ldots, \chi(a_n) \rangle$. The approximation ordering $\leq$ on sequences is the elementwise extension of the approximation ordering on observability/kernel pairs: $o_1 + k_1 \leq o_2 + k_2$ if and only if $o_1 \leq o_2$ and $k_1 \leq k_2$. Similarly for the other possible pairs of the same type: $\left[\begin{smallmatrix} s_1 \\ f_1 \end{smallmatrix}\right] \leq \left[\begin{smallmatrix} s_2 \\ f_2 \end{smallmatrix}\right]$ if and only if $s_1 \leq s_2$ and $f_1 \leq f_2$. Similarly $\left[\begin{smallmatrix} s_1 \\ a_1 \\ f_1 \end{smallmatrix}\right] \leq \left[\begin{smallmatrix} s_2 \\ a_2 \\ f_2 \end{smallmatrix}\right]$ if and only if $s_1 \leq s_2$, $a_1 \leq a_2$ and $f_1 \leq f_2$. If we assume that the $\wedge$ and $\vee$ operators are correct w.r.t. the concrete ones, we conclude easily that all the operators defined in section 4 are correct with respect to the concrete ones. This in turn implies that the abstract semantical operators are correct with respect to the concrete ones, since the abstract operators are obtained from the concrete ones substituting the concrete operations with the abstract ones. In conclusion, it is a straightforward result that $\chi(T_P(I)) \leq T_P^\sharp(\chi(I))$. Note that it can be easily shown that all the semantical operators are monotonic with respect to the approximation ordering. Hence $T_P^\sharp$ is monotonic with respect to this previous ordering. This implies that

$$\chi(T_P^i(I^0)) \leq (T_P^\sharp)^i(\chi(I^0)) \tag{3}$$

for every $i \geq 0$. The case $i = 0$ is straightforward. By inductive hypothesis, we have

$$\begin{aligned} \chi(T_P^{i+1}(I^0)) &= \chi(T_P^i(T_P(I^0))) \leq (T_P^\sharp)^i(\chi(T_P(I^0))) \\ \text{(monotonicity)} &\leq (T_P^\sharp)^i(T_P^\sharp(\chi(I^0))) = (T_P^\sharp)^{i+1}(\chi(I^0)) \ . \end{aligned}$$

By monotonicity of $\mathcal{O}^\sharp$ with respect to the approximation ordering and using equation (3), we conclude that

$$\bigvee_{i \geq 0} \mathcal{O}^\sharp \left(\chi(T_P^i(I^0))\right) \leq \bigvee_{i \geq 0} \mathcal{O}^\sharp \left((T_P^\sharp)^i(\chi(I^0))\right) , \tag{4}$$

where $\leq$ is now the partial ordering on abstract constraints.

In can be shown that

Proposition 2. *$\mathcal{O}^\sharp\chi$ is a monotonic map with respect to the computational ordering $\sqsubseteq$ (see [8] for the definition of $\sqsubseteq$).*

By the proposition above, we conclude that the extension $\widetilde{\mathcal{O}^\sharp\chi}$ of $\mathcal{O}\chi$ is continuous and we have

$$\widetilde{\mathcal{O}^\sharp\chi}\left(\bigsqcup_{i \geq 0} T_P^i(I^0)\right) = \bigvee_{i \geq 0} \widetilde{\mathcal{O}^\sharp\chi}(T_P^i(I^0)) = \bigvee_{i \geq 0} \mathcal{O}^\sharp \left(\chi(T_P^i(I^0))\right) ,$$

which is the left hand side of equation (4). Hence

$$\widetilde{\mathcal{O}^\sharp\chi}(\mathcal{S}_P) \leq \bigvee_{i\geq 0} \mathcal{O}^\sharp\left((T_P^\sharp)^i(\chi(I^0))\right) . \tag{5}$$

Equation (5) will be taken as our correctness result. Since $\widetilde{\chi\mathcal{O}}(s) \leq \widetilde{\mathcal{O}\#\chi}(s)$ for every $s \in \mathcal{SS}$ (by the correctness of $\wedge$ and $\vee$), it says that $T_P^\sharp$ can be used to compute an approximation of the abstract behaviour of $\mathcal{S}_P$.

It can be similarly shown that

$$\widetilde{\mathcal{O}\#\chi}(\mathcal{E}[\![G]\!]\mathcal{S}_P) \leq \bigvee_{i\geq 0} \mathcal{O}^\#\left(\mathcal{E}^a[\![G]\!](T_P^\#)^i\chi(I^0)\right) ,$$

i.e., our abstract semantics is correct even for the denotation of a single goal.

We would like to be able now to compute the right hand side of equation (5) in a finite number of steps. Next section shows when and how this can be accomplished.

4.2 Towards a Finitely Computable Abstract Semantics

In the case of abstract analysis of pure logic programs, if the abstract domain is finite or noetherian (or at the least the set of abstract constraints on a finite set of variables is finite or noetherian), then the abstract fixpoint is finitely computable. This is because a semantics for pure logic programs uses as computational domain sets of constraints, rather than sequences, as we do in our approach. This problem was already tackled in [8]. As already noted, the semantics in [8] was quite complex, since cut and divergence conditions were "declared" rather than "applied" (as we do in this paper). This in turn led to complex and incomplete "reduction rules" on sequences. The present approach leads to a simple solution. We will show that it is safe to remove a constraint from a sequence if it entails another constraint of the same type and precedes it in the sequence. As a consequence it is not possible to have multiple copies of a constraint in a sequence and therefore the abstract domain of sequences becomes finite. Finally, the right hand side of equation (5) is finitely computable.

The reduction rule on sequences can be viewed as a further abstract interpretation process, such that every abstract sequence is abstracted into another abstract sequence where all the "entailing" constraints are removed. Formally, we define an abstraction map as $\lambda(\langle\rangle) = \langle\rangle$, $\lambda(\langle v\rangle) = v$ and $\lambda(s_1 :: s_2) = \lambda(s_1) :: \lambda'(s_2)$, where $\lambda'(s_2)$ is $\lambda(s_2)$ deprived of all constraints v such that there exists a constraint $w \in \lambda(s_1)$ of the same type as v such that $|v|_1 \sqcup^\sharp |w|_1 = |w|_1$ and $|v|_2 \sqcup^\sharp |w|_2 = |w|_2$, where the expression "of the same type of" means that the two (abstract) constraints should be both closed convergent, or both open convergent and so on. Intuitively, if a constraint is preceded by another constraint and entails it, it is not useful for computing the observable properties we are interested in. Moreover, it will not be useful, even if we combine the sequence in any possible compositional context, as it will be shown in the following paragraphs.

Definition 8. *Let $\mathcal{RSS}$ be the set of reduced sequences, that is the set of sequences such that no constraint precedes another constraint of the same type and which entails it. Obviously, we have $\lambda : \mathcal{SS} \mapsto \mathcal{RSS}$. The following abstract operators are defined on $\mathcal{RSS}$:*

$$\propto obs^a(c) = \propto obs^\sharp(c) \qquad [\![c]\!]^a = \left\langle [\,\top\,] + \begin{bmatrix} s(c) \\ \rho(c) \\ d(c) \end{bmatrix} \right\rangle$$

$$s_1^a \oplus^a s_2^a = \lambda(s_1^a \oplus^\sharp s_2^a) \qquad s_1^a \otimes^a s_2^a = \lambda(s_1^a \otimes^\sharp s_2^a)$$

$$\exists_x^a s^a = \lambda(\exists_x^\sharp s^a) \qquad !^a(s^a) = \lambda(!^\sharp(s^a))$$

$$¡^a(s^a) = \lambda(¡^\sharp(s^a)) \qquad \mathcal{O}^a(s^a) = \mathcal{O}^\sharp(s^a)\ .$$

The above definition induces a semantic scheme based on the one described in definition 2.

Lemma 1. *The following conditions hold*

i) $\lambda([\![c]\!]^\sharp) = [\![c]\!]^a$; *ii)* $\lambda(s_1 \oplus^\sharp s_2) = \lambda(s_1) \oplus^a \lambda(s_2)$;

iii) $\lambda(s_1 \otimes^\sharp s_2) = \lambda(s_1) \otimes^a \lambda(s_2)$; *iv)* $\lambda(\exists_x^\sharp s) = \exists_x^a \lambda(s)$;

v) $\lambda(!^\sharp(s)) = !^a(\lambda(s))$; *vi)* $\lambda(¡^\sharp(s)) = ¡^a(\lambda(s))$.

Proof.

i) Obvious.

ii) Assume $o \cdot^\sharp s = \lambda(o \cdot s)$. *We have*

$$\begin{aligned} \lambda(s_1 \oplus^\sharp s_2) &= \lambda(s_1 :: -\beta(s_1) \cdot^\sharp s_2) = \lambda(\lambda(s_1) :: -\beta(s_1) \cdot^\sharp s_2) \\ &= \lambda(\lambda(s_1) :: -\beta(\lambda(s_1)) \cdot^\sharp \lambda(s_2)) = \lambda(s_1) \oplus^a \lambda(s_2)\ . \end{aligned}$$

iii) $\lambda(s_1 \otimes^\sharp s_2) = \lambda(s_1 \otimes^\sharp \lambda(s_2)) = \lambda(\lambda(s_1) \otimes^\sharp \lambda(s_2)) = \lambda(s_1) \otimes^a \lambda(s_2)$.

iv) $\lambda(\exists_x^\sharp s) = \lambda(\exists_x^\sharp \lambda(s)) = \exists_x^a \lambda(s)$.

v) $\lambda(!^\sharp(s)) = \lambda(!^\sharp(\lambda(s))) = !^a(\lambda(s))$, *because if v entails and follows w in s, then, after the $!^\sharp$ operator has been applied, all the cut conditions added to w are added to v too, and therefore v will still entail w and will be removed by the λ abstraction present in $\lambda(!^\sharp(s))$. Hence it is equivalent to remove v before applying the $!^\sharp$ operator. Note that v does not contribute to the function o, because, if it is convergent, then w is convergent too and $v \sqcup^\sharp w = w$.*

vi) $\lambda(¡^\sharp(s)) = \lambda(¡^\sharp(\lambda(s))) = ¡^a\lambda(s)$.

Proposition 3. *Given an interpretation I, we have $\lambda(T_P^\sharp(I)) = T_P^a(\lambda(I))$ where, by definition, $\lambda(I)[\mathrm{p}] = \lambda(I[\mathrm{p}])$.*

Proof. Using lemma 1 it can easily be shown that for every goal G we have $\lambda(\mathcal{E}[\![G]\!]^\sharp I) = \mathcal{E}^a[\![G]\!]\lambda(I)$, by straightforward induction on the structure of G. Using lemma 1 again, we get the thesis.

Corollary 1. *For every $i \geq 0$ we have $\lambda((T^{\sharp}_P)^i(I^0)) = (T^a_P)^i(\lambda(I^0))$.*

Proof. *By induction on i.*

If v entails w then the success condition of v is lower than the success condition of w and similarly the abstract approximation of v is lower than the abstract approximation of w. Therefore $\mathcal{O}^a(\langle w \rangle) = \mathcal{O}^a(\langle v, w \rangle)$. This allows us to conclude that $\mathcal{O}^a(\lambda(s)) = \mathcal{O}^a(s)$. Hence we have

$$\begin{aligned} \bigvee_{i \geq 0} \mathcal{O}^a((T^{\sharp}_P)^i(I^0)) &= \bigvee_{i \geq 0} \mathcal{O}^a(\lambda((T^{\sharp}_P)^i(I^0))) \\ \text{(corollary 1)} &= \bigvee_{i \geq 0} \mathcal{O}^a\left((T^a_P)^i(\lambda(I^0))\right) . \end{aligned}$$

Note that, assuming the set of abstract constraints on a finite set of variables to be finite, we conclude that the right hand side of the above equality can be computed in a finite number of steps. This means that the right hand side of equation (5) can be computed in a finite number of steps using T^a_P.

A similar reduction can be applied to the call pattern analysis described in subsection 3.1 (see the following example).

5 An Example

In this section we show an example of the computation of the abstract call pattern semantics for a simple program. We use an abstract domain able to model groundness and non–freeness, without directionality. Even this "weak" domain is able to show the usefulness of our approach. Consider the Prolog program

```
p(X):-X=4,!.
p(X):-q(X).
q(X):-X=5.
q(X):-p(X).
```

which is translated into our abstract syntax as

$$\begin{aligned} &\mathtt{p}(x) : -\mathtt{cut}(x = 4) \text{ or } \mathtt{q}(x). \\ &\mathtt{q}(x) : -x = 5 \text{ or } \mathtt{p}(x). \end{aligned}$$

and then abstractly compiled into:

$$\begin{aligned} &\mathtt{p}(x) : -\mathtt{cut}\left(\begin{bmatrix}\top\\\bot\end{bmatrix} + \begin{bmatrix}\top\\ g(x)\\ nf(x)\end{bmatrix}\right) \text{ or } \mathtt{q}(x). \\ &\mathtt{q}(x) : -\begin{bmatrix}\top\\\bot\end{bmatrix} + \begin{bmatrix}\top\\ g(x)\\ nf(x)\end{bmatrix} \text{ or } \mathtt{p}(x). \end{aligned}$$

The computation of the abstract fixpoint (with sequence reductions) stops at the second iteration in such a way that $lfp(T^a_P)[\mathtt{p}]$ is the sequence

$$\left\langle \overline{\begin{bmatrix}\top\\\bot\end{bmatrix} + \begin{bmatrix}\top\\\top\\\bot\end{bmatrix}};\mathtt{p}, \begin{bmatrix}\top\\\bot\end{bmatrix} + \begin{bmatrix}\top\\ g(\alpha)\\ nf(\alpha)\end{bmatrix}, \overline{\begin{bmatrix}nf(\alpha)\\\top\end{bmatrix} + \begin{bmatrix}\top\\\top\\\bot\end{bmatrix}};\mathtt{q}, \begin{bmatrix}nf(\alpha)\\\top\end{bmatrix} + \widetilde{\begin{bmatrix}nf(\alpha)\\\top\\\bot\end{bmatrix}};\mathtt{p} \right\rangle .$$

From the third constraint of this denotation we conclude that the unique call pattern for q, which can arise from the execution of p, is observable only in a constraint store in which α (or, equivalently, x) is a non free variable. This information can be used to optimize the compilation of the unification for the clauses of the definition of q. Note that it would not have been possible to determine this information if we had discarded the cut operator in our analysis.

6 Conclusions and Future Work

We have shown a general framework for the abstract interpretation of Prolog programs both for computed answers and for call patterns. The main difference with classical abstract interpretation of logic programs is the use of sequences rather than sets of constraints and the use of success and failure conditions. While a general approach for defining success and failure conditions is available (equations (1) and (2)), one should not think that the problem of finding success and failure conditions is definitely solved. Actually, we only shifted the problem in the definition of the abstract domain, which must be devised in such a way that equations (1) and (2) do not give trivial approximations (like $\top$, for instance). Future work will be spent in devising such domains, and in showing the feasibility of the approach with its implementation.

Finally, we think that the problem of assuring termination in the presence of sequences rather than sets of constraints could be related with termination of abstract interpretation of concurrent logic programs. We will investigate this relation.

References

[1] R. Barbuti, M. Codish, R. Giacobazzi, and G. Levi. Modelling Prolog Control. *Journal of Logic and Computation*, 3:579–603, 1993.

[2] R. Barbuti, M. Codish, R. Giacobazzi, and M. Maher. Oracle Semantics for PROLOG. *Information and Computation*, 1995.

[3] P. Cousot and R. Cousot. Abstract Interpretation: A Unified Lattice Model for Static Analysis of Programs by Construction or Approximation of Fixpoints. In *Proc. Fourth ACM Symp. Principles of Programming Languages*, pages 238–252, 1977.

[4] P. Cousot and R. Cousot. Abstract Interpretation and Applications to Logic Programs. *Journal of Logic Programming*, 13(2 & 3):103–179, 1992.

[5] B. A. Davey and H. A. Priestley. *Introduction to Lattices and Order*. Cambridge University Press, 1990.

[6] B. Le Charlier, Rossi S., and P. Van Hentenryck. An Abstract Interpretation Framework which Accurately Handles PROLOG Search Rule and the Cut. In M. Bruynooghe, editor, *Proceedings of the 1994 Int'l Symposium on Logic Programming*, pages 157–171. The MIT Press, 1994.

[7] G. Levi and D. Micciancio. Analysis of pure PROLOG programs. In M.I. Sessa, editor, *Proceedings GULP-PRODE '95, 1995 Joint Conference on Declarative Programming*, 1995.
[8] F. Spoto and G. Levi. A denotational semantics for PROLOG. In M. Falaschi, M. Navarro, and Policriti A., editors, *Proceedings of the Joint Conference on Declarative Programming, APPIA-GULP-PRODE '97*, pages 201–212, 1997.

Factorizing Equivalent Variable Pairs in ROBDD-Based Implementations of *Pos*

Roberto Bagnara[1,*] and Peter Schachte[2]

[1] Dipartimento di Matematica, Università di Parma, Via M. D'Azeglio 85/A, Parma, Italy, bagnara@prmat.math.unipr.it

[2] Department of Computer Science, The University of Melbourne, Parkville, Victoria 3052, Australia, pets@cs.mu.oz.au

Abstract. The subject of groundness analysis for (constraint) logic programs has been widely studied, and interesting domains have been proposed. *Pos* has been recognized as the most suitable domain for capturing the kind of dependencies arising in groundness analysis, and *Reduced Ordered Binary Decision Diagrams* (ROBDDs) are generally accepted to be the most efficient representation for *Pos*. Unfortunately, the size of an ROBDDs is, in the worst case, exponential in the number of variables it depends upon. Earlier work [2] has shown that a hybrid representation that separates the definite information from the dependency information is considerably more efficient than keeping the two together. The aim of the present paper is to push this idea further, also separating out certain dependency information, in particular all pairs of variables that are always either both ground or neither ground. We find that this new hybrid representation is a significant improvement over previous work.

1 Introduction

The aim of *groundness analysis* (sometimes called *definiteness* analysis) is to derive statically, for all the program points of interest, which variables are bound to unique values (or *ground*). This kind of information is very important: it allows substantial optimizations to be performed at compile-time, and is also crucial to most semantics-based program manipulation tools. Moreover, many other analyses are made more precise by the availability of groundness information. For these reasons, the subject of groundness analysis for (constraint) logic programs has been widely studied. After the early attempts, some classes of Boolean functions have been recognized as constituting good abstract domains for groundness analysis [10, 13]. In particular, the set of *positive Boolean functions*, (namely, those functions that assume the *true* value under the valuation assigning *true* to all variables), which is denoted by *Pos*, allows to express Boolean properties of program variables where the property of one variable may depend on that property of other variables. For groundness analysis, since variables can be bound to terms containing other variables, the groundness of one variable may depend on

* Much of this work was supported by EPSRC grant GR/L19515.

A.M. Haeberer (Ed.): AMAST'98, LNCS 1548, pp. 471-485, 1998.

the groundness of other variables. *Pos* has been recognized as the most precise domain for capturing the kind of dependencies arising in groundness analysis.

This ability to express dependencies makes analysis based on *Pos* very precise, but also makes it relatively expensive, as many operations on Boolean formulae have exponential worst case complexity. Armstrong et al. [1] analyzed many representations of positive Boolean formulae for abstract interpretation, and found ***Reduced Ordered Binary Decision Diagrams*** (ROBDDs) [6] to give the best performance.

ROBDDs generated during program analysis often contain many variables that are definitely true. In the context of groundness analysis, this means that the corresponding program variable must be ground at that point in the program. It is shown in [2] that a hybrid representation for Boolean functions that keeps these definite variables separate is more efficient than ROBDDs alone. However, ROBDDs generated during program analysis also contain many pairs of variables that are equivalent. In terms of groundness, this means that either both variables are ground, or neither is. Such equivalent variables of course appear for a program goal of the form `X = Y`, but they also frequently appear naturally during the analysis process. For example, for a goal `X = [Y|Z]`, where it can be established that `Y` is ground, the analyzer will deduce that `X` and `Z` are equivalent. Such equivalent pairs can greatly increase the size of ROBDDs, which in turn makes ROBDD operations much more expensive. For example, the ROBDD for the Boolean function z comprises one node (not counting the **1** and **0** terminal nodes), while $(x \leftrightarrow y) \land z$ comprises 4 or 5 (usually 5). However, since $x \leftrightarrow y$ simply means that x and y are equivalent, we may remove y from the Boolean function altogether, leaving us again with a single node, and replace y by x in the formulae being analyzed. Since the time complexity of most ROBDD algorithms is at best quadratic in the sizes of the graphs involved, this can significantly speed up analysis.

There is another reason for our interest in equivalent variables. A recursive definition of the form

$$f(x_1, \ldots, x_n) = A \lor \bigl(B \land f(x_1, \ldots, x_n)\bigr),$$

always has least fixpoint A, as can be seen by Kleene iteration. This is a special instance of Søndergaard's immediate fixpoint theorem [16]. The key point here is that the formal parameters of the definition must be the same as the actual parameters in the recursive reference. We can establish this if we have a definition of the form

$$f(x_1, \ldots, x_n) = A \lor \bigl(B \land f(y_1, \ldots, y_n) \land (x_1 \leftrightarrow y_1) \land \cdots \land (x_n \leftrightarrow y_n)\bigr).$$

To show that our definition has this form, we need to find the equivalent variables in the recursive arm of the definition.

In this paper we present a hybrid representation for Boolean functions that uses a set to represent definite variables, a set of pairs of equivalent variables to represent equivalences, and an ROBDD to represent more complex dependencies

This hybrid representation proves to be significantly more efficient overall than that of [2].

Notice that Boolean functions are used in the more general context of *dependency analysis*, including *finiteness analysis* for deductive database languages [5] *suspension analysis* for concurrent (constraint) logic programming languages [11], and functional dependency (or determinacy) analysis [17]. The hybrid representation we propose might be useful also in these contexts, although we have not studied this yet.

The balance of this paper proceeds as follows. In Sect. 2 we briefly review the usage of Boolean functions for groundness analysis of (constraint) logic programs (even though we assume familiarity with this subject) and we discuss the representation we use for Boolean functions. Section 3 presents our hybrid representation, with the necessary algorithms appearing in Sect. 4. Experimental results are presented in Sect. 5, and Sect. 6 concludes with some final remarks.

2 Preliminaries

Let U be a set. The set of all subsets of U will be denoted by $\wp(U)$. The set of all *finite* subsets of U will be denoted by $\wp_{\mathrm{f}}(U)$. The notation $S \subseteq_{\mathrm{f}} T$ stands for $S \in \wp_{\mathrm{f}}(T)$.

2.1 Boolean Functions for Groundness Analysis

After the early approaches to groundness analysis [14, 12], which suffered from serious precision drawbacks, the use of Boolean functions [10, 13] has become customary in the field. The reason is that Boolean functions allow to capture in a very precise way the *groundness dependencies* that are implicit in unification constraints such as $z = f(g(x), y)$: the corresponding Boolean function is $(x \wedge y) \leftrightarrow z$, meaning that z is ground if and only if x and y are so. They also capture dependencies arising from other constraint domains: for instance, under CLP($\mathcal{R}$) $x + 2y + z = 4$ can be abstracted as $((x \wedge y) \to z) \wedge ((x \wedge z) \to y) \wedge ((y \wedge z) \to x)$, indicating that determining any two variables is sufficient to determine the third.

Vars is a fixed denumerable set of variable symbols. The variables are ordered by the total order relation $\prec$. For convenience we sometimes use $y \succ x$ as an alternative for $x \prec y$. We also use $x \preceq y$ and $y \succeq x$ to mean that either $x \prec y$ or $x = y$. We call the least variable α, that is, $\forall v \in \mathit{Vars} : \alpha \preceq v$. For a set of variables S we will denote by $\min^{\prec}(S)$ the minimum element of S with respect to $\prec$. We also define the succ (successor) function over *Vars* as follows:

Definition 1. (The function succ: *Vars* $\to$ *Vars*.**)**

$$\mathrm{succ}(v) \stackrel{\mathrm{def}}{=} x, \quad \text{if } v \prec x \text{ and } \neg\exists y \in \mathit{Vars} \,.\, v \prec y \prec x.$$

Note that x is unique.

We now introduce Boolean functions based on the notion of Boolean valuation.

Definition 2. (Boolean valuations.) *The set of* Boolean valuations over *Vars* *is* $\mathcal{A} \stackrel{\text{def}}{=} \mathit{Vars} \to \{0,1\}$. *For each* $a \in \mathcal{A}$, *each* $x \in \mathit{Vars}$, *and each* $c \in \{0,1\}$ *the valuation* $a[c/x] \in \mathcal{A}$ *is given, for each* $y \in \mathit{Vars}$, *by*

$$a[c/x](y) \stackrel{\text{def}}{=} \begin{cases} c, & \text{if } x = y; \\ a(y), & \text{otherwise.} \end{cases}$$

For $X = \{x_1, x_2, \dots\} \subseteq \mathit{Vars}$, *we write* $a[c/X]$ *for* $a[c/x_1][c/x_2]\cdots$.

Definition 3. (Boolean functions.) *The set of* Boolean functions over *Vars* *is* $\mathcal{F} \stackrel{\text{def}}{=} \mathcal{A} \to \{0,1\}$. *The distinguished elements* $\top, \bot \in \mathcal{F}$ *are the functions defined by* $\top \stackrel{\text{def}}{=} \lambda a \in \mathcal{A} \,.\, 1$ *and* $\bot \stackrel{\text{def}}{=} \lambda a \in \mathcal{A} \,.\, 0$. *For* $f \in \mathcal{F}$, $x \in \mathit{Vars}$, *and* $c \in \{0,1\}$, *the function* $f[c/x] \in \mathcal{F}$ *is given, for each* $a \in \mathcal{A}$, *by* $f[c/x](a) \stackrel{\text{def}}{=} f\bigl(a[c/x]\bigr)$. *When* $X \subseteq \mathit{Vars}$, $f[c/X]$ *is defined in the obvious way. If* $f \in \mathcal{F}$ *and* $x, y \in \mathit{Vars}$ *the function* $f[y/x] \in \mathcal{F}$ *is given, for each* $a \in \mathcal{A}$, *by*

$$f[y/x](a) \stackrel{\text{def}}{=} f\Bigl(a\bigl[a(y)/x\bigr]\Bigr).$$

Boolean functions are constructed from the elementary functions corresponding to variables, and by means of the usual logical connectives. Thus x *denotes the Boolean function* f *such that, for each* $a \in \mathcal{A}$, $f(a) = 1$ *if and only if* $a(x) = 1$. *For* $f_1, f_2 \in \mathcal{F}$, *we write* $f_1 \wedge f_2$ *to denote the function* g *such that, for each* $a \in \mathcal{A}$, $g(a) = 1$ *if and only if both* $f_1(a) = 1$ *and* $f_2(a) = 1$. *The other Boolean connectives and quantifiers are handled similarly.*

The question of whether a Boolean function f entails particular variable x (which is what, in the context of groundness analysis, we call *definite groundness information*) is equivalent to the question whether $f \to x$ is a tautology (namely, $f \to x = \top$). In what follows we will also need the notion of *dependent variables* of a function, as well as disentailed, or definitely false, variables.

Definition 4. (Dependent, true, false, and equivalent variables.) *For* $f \in \mathcal{F}$, *the set of* variables on which f depends, *the set of* variables necessarily true for f, *the set of* variables necessarily false for f, *and the set of* equivalent variables for f, *are given, respectively, by*

$$\begin{aligned}
\mathit{vars}(f) &\stackrel{\text{def}}{=} \bigl\{\, x \in \mathit{Vars} \bigm| \exists a \in \mathcal{A} \,.\, f\bigl(a[0/x]\bigr) \neq f\bigl(a[1/x]\bigr) \,\bigr\}, \\
\mathit{true}(f) &\stackrel{\text{def}}{=} \bigl\{\, x \in \mathit{Vars} \bigm| \forall a \in \mathcal{A} : f(a) = 1 \implies a(x) = 1 \,\bigr\}, \\
\mathit{false}(f) &\stackrel{\text{def}}{=} \bigl\{\, x \in \mathit{Vars} \bigm| \forall a \in \mathcal{A} : f(a) = 1 \implies a(x) = 0 \,\bigr\}, \\
\mathit{equiv}(f) &\stackrel{\text{def}}{=} \bigl\{\, (x,y) \in \mathit{Vars}^2 \bigm| x \neq y, \forall a \in \mathcal{A} : f(a) = 1 \implies a(x) = a(y) \,\bigr\}.
\end{aligned}$$

2.2 Binary Decision Diagrams

Binary Decision Diagrams (BDDs) are a well-known representations of Boolean functions [6, 7]. A BDD is a rooted directed acyclic graph where each interna node is labeled with a Boolean variable and has two out edges, leading to the node's *true* and *false successors*. External (leaf) nodes are either **1** or **0**. The Boolean function represented by an BDD can be evaluated for a given truth value assignment by traversing the graph from the root node, taking the *true* edge for nodes whose label is assigned 1 and the *false* edge when the label is assigned 0. The terminal node reached in this traversal is the function value for that assignment.

When a total ordering on the variables is available, we can define *Ordered Binary Decision Diagrams* (OBDDs) as BDDs with the restriction that the label of a node is always less than the label of any internal node in its successors. *Reduced Ordered Binary Decision Diagrams* (OBDDs) are OBDDs with the additional condition that they do not contain any two distinct nodes which represent the same Boolean function. This means that the two terminal nodes must be unique, no two distinct nodes may have the same label and true and false successors, and no node may have two identical successors (because then it would represent the same Boolean function as the successors).

We now define ROBDDs formally. Although an *ROBDD* is a particular kind of rooted, directed, and acyclic graph, we prefer not to use the standard notation for graphs. Thus an ROBDD is identified with the set of its *nodes*, one of which is designated as *the root*, the edges being formally part of the nodes themselves.

Definition 5. (ROBDD) *If N is the set of nodes of an ROBDD then N satisfies*

$$N \subseteq \{\mathbf{0}, \mathbf{1}\} \cup \mathit{Vars} \times N \times N.$$

The nodes **0** *and* **1** *are called* terminal nodes. *All the other nodes in N are called* non-terminal nodes. *For each non-terminal node $n \in N$, $n_{\mathrm{var}} \in \mathit{Vars}$ denotes the variable associated with n, $n_{\mathrm{false}} \in N$ denotes the false successor of n, and $n_{\mathrm{true}} \in N$ denotes the true successor of n. With this notation, N must also satisfy the irredundancy and the ordering conditions: for each non-terminal node $n \in N$ $n_{\mathrm{false}} \neq n_{\mathrm{true}}$ and $(m = n_{\mathrm{false}} \text{ or } m = n_{\mathrm{true}}) \implies (m \in \{\mathbf{0}, \mathbf{1}\} \text{ or } n_{\mathrm{var}} \prec m_{\mathrm{var}})$. Moreover, N is* rooted *and* connected, *that is, there exists $r \in N$ (the* root*) such that*

$$\forall n \in N \setminus \{r\} : \exists m \in N \,.\, (n = m_{\mathrm{false}} \text{ or } n = m_{\mathrm{true}}).$$

A ROBDDs is a pair (r, N) that satisfies the above conditions. The set of all ROBDDs is denoted by $\mathcal{D}$.

The meaning of an ROBDD is given as follows.

Definition 6. (Semantics of ROBDDs.) *The function $[\![\cdot]\!]_{\mathcal{D}} : \mathcal{D} \to \mathcal{F}$ is given, for each $(r, N) \in \mathcal{D}$, by $N' \stackrel{\mathrm{def}}{=} N \setminus \{r\}$ and*

$$[\![(r,N)]\!]_{\mathcal{D}} \stackrel{\mathrm{def}}{=} \begin{cases} \bot, & \text{if } r = \mathbf{0}; \\ \top, & \text{if } r = \mathbf{1}; \\ \Bigl(r_{\mathrm{var}} \wedge [\![(r_{\mathrm{true}}, N')]\!]_{\mathcal{D}}\Bigr) \vee \Bigl(\neg r_{\mathrm{var}} \wedge [\![(r_{\mathrm{false}}, N')]\!]_{\mathcal{D}}\Bigr), & \text{otherwise.} \end{cases}$$

For simplicity, we will identify an ROBDD with the ROBDD node that constitutes its root, since the set of all the nodes can be recovered by any traversal that starts from the root.

In the implementation, a new ROBDD node is created, given a label variable v and true and false successors n and m respectively, by the make_node(v, n, m) function. This is defined such that, if $n = m$, n will be returned. Furthermore, if an identical call to make_node has previously been made, the result of that call will be returned. This guarantees that if n and m are reduced, then so is the resulting node. Note that it is an error if $v \succeq n_{\mathrm{var}}$ or $v \succeq m_{\mathrm{var}}$.

ROBDDs have one very important property: they are *canonical.* This means that, for each fixed variable ordering, two ROBDDs represent the same function if and only if they are identical [6]. In fact, the definition of make_node is such that two ROBDDs are identical if and only if they are stored at the same memory address. This is important to the efficiency of many ROBDD operations.

We will often confuse ROBDDs with the Boolean functions they represent. For instance, for $n \in \mathcal{D}$, when we write $vars(n)$ or $true(n)$ what we really mean is $vars(\llbracket n \rrbracket_{\mathcal{D}})$ or $true(\llbracket n \rrbracket_{\mathcal{D}})$. This convention of referring to the semantics simplifies the presentation and should not cause problems.

3 A New Representation for *Pos*

We introduce a new representation for *Pos*. It is made up of three components: a set of *ground* variables, a set of *equivalent* variables, and an *ROBDD*, whence the name *GER representation.*[1] A set of ground variables is trivially an element of $\mathcal{V} \stackrel{\mathrm{def}}{=} \wp_{\mathrm{f}}(Vars)$. For $G \in \mathcal{V}$ we define $\llbracket G \rrbracket_{\mathcal{V}} \stackrel{\mathrm{def}}{=} \bigwedge(G)$, where $\bigwedge\{x_1, \ldots, x_n\} \stackrel{\mathrm{def}}{=} x_1 \wedge \cdots \wedge x_n$ and $\bigwedge\varnothing \stackrel{\mathrm{def}}{=} \top$.

The set of equivalent variables is simply given by a transitively closed set of ordered pairs of variables.

Definition 7. (A representation for equivalent variables.) *Sets of equivalent variables are represented by means of elements of $\mathcal{L} \subseteq \wp_{\mathrm{f}}(Vars \times Vars)$ such that*

1. $\forall L \in \mathcal{L} : \forall x, y \in Vars : (x, y) \in L \implies x \prec y;$
2. $\forall L \in \mathcal{L} : \forall x, y, z \in Vars : (x, y), (y, z) \in L \implies (x, z) \in L.$

For $L \in \mathcal{L}$ we use the following notation:

$$L|1 \stackrel{\mathrm{def}}{=} \{\, x \in Vars \mid (x, y) \in L \,\}, \qquad vars(L) \stackrel{\mathrm{def}}{=} L|1 \cup L|2,$$
$$L|2 \stackrel{\mathrm{def}}{=} \{\, y \in Vars \mid (x, y) \in L \,\}.$$

The family of functions $\lambda_L\colon Vars \to Vars$ is defined, for each $L \in \mathcal{L}$ and each $x \in Vars$, by $\lambda_L(x) \stackrel{\mathrm{def}}{=} \min^{\prec}\bigl(\{x\} \cup \{\, y \in Vars \mid (y, x) \in L \,\}\bigr)$. λ_L maps each variable to the least variable of its equivalence class, which we call its leader. $(\mathcal{L}, \supseteq)$ *is*

[1] In [2] we had only a set of ground variables and a ROBDD.

clearly a lattice. We will denote the glb *and the* lub *over* $(\mathcal{L}, \supseteq)$ *by* $\wedge_{\mathcal{L}}$ *(transitive closure of the union) and* $\vee_{\mathcal{L}}$ *(intersection), respectively. The semantics function* $[\![\cdot]\!]_{\mathcal{L}} : \mathcal{L} \to \mathcal{F}$ *is given by* $[\![L]\!]_{\mathcal{L}} \stackrel{\text{def}}{=} \bigwedge\{\, x \leftrightarrow y \mid (x, y) \in L \,\}$.

In the GER representation, an element of *Pos* is represented by an element of $\mathcal{V} \times \mathcal{L} \times \mathcal{D}$. There are elements of *Pos* that can be represented by several such triples and, in the GER representation, we need to make a choice among those. This choice must be *canonical* and *economical*. Economy can be explained as follows: true variables are most efficiently represented in the first component (a bit-vector at the implementation level) and should not occur anywhere else in the representation. Equivalent variables are best represented in the second component of the GER representation (implemented as a vector of integers). As equivalent variables partition the space of variables into equivalence classes, only one variable per equivalence class must occur in the ROBDD constituting the third component of the representation. If we choose, say, the least variable (with respect to the $\preceq$ ordering on *Vars*) of each equivalence class as the representative of the class, we have also ensured canonicity.

Definition 8. (GER representation.) *The* GER *representation for Pos is given by the set*

$$\mathcal{G} \stackrel{\text{def}}{=} \left\{ \langle G, L, n \rangle \;\middle|\; \begin{array}{l} G \in \mathcal{V}, L \in \mathcal{L}, n \in \mathcal{D}, \\ G \cap vars(L) = G \cap vars(n) = L|2 \cap vars(n) = \varnothing, \\ true(n) = equiv(n) = \varnothing \end{array} \right\}.$$

The meaning of $\mathcal{G}$*'s elements is given by the function* $[\![\cdot]\!]_{\mathcal{G}} : \mathcal{G} \to \mathcal{F}$*:*

$$[\![\langle G, L, n \rangle]\!]_{\mathcal{G}} \stackrel{\text{def}}{=} [\![G]\!]_{\mathcal{V}} \wedge [\![L]\!]_{\mathcal{L}} \wedge [\![n]\!]_{\mathcal{D}},$$

What is required now is a normalization function mapping each element of $\mathcal{V} \times \mathcal{L} \times \mathcal{D}$ into the right representative in $\mathcal{G}$.

Definition 9. (Normalization function η.) *The function* $\eta : \mathcal{V} \times \mathcal{L} \times \mathcal{D} \to \mathcal{V} \times \mathcal{L} \times \mathcal{D}$ *is given by*

$$\eta(\langle G, L, n \rangle) \stackrel{\text{def}}{=} \langle \hat{G}, \hat{L}, \hat{n} \rangle$$

where

$$\hat{G} \stackrel{\text{def}}{=} true\Big([\![\langle G, L, n \rangle]\!]_{\mathcal{G}}\Big),$$

$$\hat{L} \stackrel{\text{def}}{=} equiv\Big([\![\langle G, L, n \rangle]\!]_{\mathcal{G}}\Big) \setminus \{\, (x, y) \in \hat{G}^2 \mid x \prec y \,\},$$

$$\hat{n} \stackrel{\text{def}}{=} n[1/\hat{G}][\lambda_{\hat{L}}(x_1)/x_1] \cdots [\lambda_{\hat{L}}(x_n)/x_n], \quad \textit{if } vars(n) \setminus \hat{G} = \{x_1, \ldots, x_n\}.$$

A very basic implementation for η is given by the normalize function depicted in Alg. 1. The need for looping can be understood by means of the following examples. Forcing a variable to true in a ROBDD can result in new entailed

Require: an element $\langle G, L, n \rangle \in \mathcal{V} \times \mathcal{L} \times \mathcal{D}$

function normalize($\langle G, L, n \rangle$)

$G_{\text{new}} := G$; $L_{\text{new}} := L$; $n_{\text{new}} := n$;
repeat
  $G_{\text{old}} := G_{\text{new}}$; $L_{\text{old}} := L_{\text{new}}$; $n_{\text{old}} := n_{\text{new}}$;
  $G_{\text{new}} := G_{\text{new}} \cup \{ x, y \mid (x, y) \in L_{\text{new}}, \{x, y\} \cap G_{\text{new}} \neq \varnothing \}$
  $L_{\text{new}} := L_{\text{new}} \setminus \{ (x, y) \in G_{\text{new}}^2 \mid x \prec y \}$
  $n_{\text{new}} := n_{\text{new}}[1/G_{\text{new}}]$;
  $G_{\text{new}} := G_{\text{new}} \cup \mathit{true}(n_{\text{new}})$;
  $L_{\text{new}} := L_{\text{new}} \wedge_{\mathcal{L}} \mathit{equiv}(n_{\text{new}})$;
  $\{x_1, \ldots, x_k\} := \mathit{vars}(n_{\text{new}})$;
  $n_{\text{new}} := n_{\text{new}}[\lambda_{L_{\text{new}}}(x_1)/x_1] \cdots [\lambda_{L_{\text{new}}}(x_k)/x_k]$
until $G_{\text{new}} = G_{\text{old}}$ and $L_{\text{new}} = L_{\text{old}}$ and $n_{\text{new}} = n_{\text{old}}$;
return $\langle G_{\text{new}}, L_{\text{new}}, n_{\text{new}} \rangle$;

Algorithm 1: The normalize function.

variables: if n represents $x \to y$ then $n[1/x]$ represents y. Renaming a ROBDD node n by means of a set of equivalent variables L can also give rise to new entailed variables. Suppose that n represents the Boolean formula $x \vee y$ and that $L = \{(x, y)\}$. Then $n[\lambda_L(y)/y]$ represents x. Renaming can also result in new equivalent variables: take n representing $x \leftrightarrow (y \wedge z)$ and $L = \{(y, z)\}$ for an example.

Theorem 1. *We have that* $\eta \colon \mathcal{V} \times \mathcal{L} \times \mathcal{D} \to \mathcal{G}$. *Furthermore, for each triple* $\langle G, L, N \rangle \in \mathcal{V} \times \mathcal{L} \times \mathcal{D}$*, we have*

$$[\![\langle G, L, N \rangle]\!]_{\mathcal{G}} = [\![\eta(\langle G, L, N \rangle)]\!]_{\mathcal{G}}.$$

Finally, the normalize *function in* Alg. 1 *is a correct implementation of* η.

It is important to remark that in the actual implementation several specializations are used instead of Alg. 1. In other words, for every possible use of normalize, conditions can be granted so as to use a simpler algorithm instead. While space limitations do not allows us to be more precise, we just observe that roughly 50% of the times normalize would be called with the ROBDD **1**. This indicates that definitely ground variables and equivalent variables constitute a significant proportion of the dependencies that arise in practice.

3.1 Operations for the Analysis

Let us briefly review the operations we need over *Pos* for the purpose of groundness analysis. Modeling forward execution of (constraint) logic programs requires computing the logical conjunction of two functions, the merge over different computation paths amounts to logical disjunction, whereas projection onto a designated set of variables is handled through existential quantification. Conjunction with functions of the form $x \leftrightarrow (y_1 \wedge \cdots \wedge y_k)$, for $k \geq 0$, accommodate both abstract *mgus* and the *combination* operation in domains like $\mathtt{Pat}(Pos)$ [9].

Let Ω be an operation over *Pos*. The corresponding operation over $\mathcal{G}$ can be specified, roughly speaking, as $\eta \circ \Omega \circ [\![\cdot]\!]_{\mathcal{G}}$. However, this is simply a specification: the problem is how to compute $\eta \circ \Omega \circ [\![\cdot]\!]_{\mathcal{G}}$ more efficiently exploiting the fact that both definitely ground variables and pair of equivalent variables are kept separate in the GER representation. The intuitive recipe (which has been extensively validated through experimentation) for achieving efficiency can be synthesized in the *motto* "keep the ROBDD component as small as possible and touch it as little as possible". The specification above does the contrary: it pushes all the information into the ROBDD component, performs the operation on the ROBDD, and normalizes the result. Let us take the conjunction operation $\wedge_{\mathcal{G}} : \mathcal{G} \times \mathcal{G} \to \mathcal{G}$ and suppose we want to compute $\langle G_1, L_1, n_1 \rangle \wedge_{\mathcal{G}} \langle G_2, L_2, n_2 \rangle$. A first approximation is to compute

$$\eta\bigl(\langle G_1 \cup G_2, L_1 \wedge_{\mathcal{L}} L_2, n_1 \wedge_{\mathcal{D}} n_2 \rangle\bigr), \tag{1}$$

but we can do better if we reduce the ROBDDs n_1 and n_2 before computing the conjunction (whose complexity is $O(|n_1| \cdot |n_2|)$, where $|n|$ denotes the number of nodes in the ROBDD n). In order to apply the $\wedge_{\mathcal{D}}$ operator to the smallest possible ROBDD nodes we can use the alternative expression

$$\eta\Bigl(\langle G'_1 \cup G'_2, L'_1 \wedge_{\mathcal{L}} L'_2, n'_1 \wedge_{\mathcal{D}} n'_2 \rangle\Bigr), \tag{2}$$

where $\langle G'_i, L'_i, n'_i \rangle = \eta\bigl(\langle G_1 \cup G_2, L_1 \wedge_{\mathcal{L}} L_2, n_i\bigr)$, for $i = 1, 2$. For lack of space we cannot enter into details, but the current implementation uses an expression which is intermediate between (1) and (2). Indeed, the attentive reader will have noticed that there is a tradeoff in the above *motto*: keeping the ROBDDs as small as possible, as in (2), implies performing several (possibly fruitless) visits of the ROBDDs in order to collect entailed and equivalent variables.

Disjunction is computationally less complex than conjunction in that it does not require normalization through η. This, however, comes at the price of some extra complication in the definition.

$$\langle G_1, L_1, n_1 \rangle \vee_{\mathcal{G}} \langle G_2, L_2, n_2 \rangle \stackrel{\text{def}}{=} \langle G_1 \cap G_2, L', n'_1 \vee_{\mathcal{D}} n'_2 \rangle,$$

with $L' \stackrel{\text{def}}{=} L'_1 \vee_{\mathcal{L}} L'_2$ and

$$L'_1 \stackrel{\text{def}}{=} L_1 \wedge_{\mathcal{L}} \bigwedge\nolimits_{\mathcal{L}} {}_{\substack{(x,y) \in G_1 \setminus G_2 \\ x \prec y}} \{(x, y)\}, \qquad L'_2 \stackrel{\text{def}}{=} L_2 \wedge_{\mathcal{L}} \bigwedge\nolimits_{\mathcal{L}} {}_{\substack{(x,y) \in G_2 \setminus G_1 \\ x \prec y}} \{(x, y)\},$$

$$G'_1 \stackrel{\text{def}}{=} \{\, \lambda_{L'}(x) \mid x \in G_1 \setminus G_2 \,\}, \qquad G'_2 \stackrel{\text{def}}{=} \{\, \lambda_{L'}(x) \mid x \in G_2 \setminus G_1 \,\},$$

$$L''_1 \stackrel{\text{def}}{=} (L'_1 \setminus L'_2) \vee_{\mathcal{L}} L_1, \qquad L''_2 \stackrel{\text{def}}{=} (L'_2 \setminus L'_1) \vee_{\mathcal{L}} L_2,$$

$$n'_1 \stackrel{\text{def}}{=} n_1 \wedge_{\mathcal{D}} \bigwedge\nolimits_{\mathcal{D}} {}_{x \in G'_1} x \wedge_{\mathcal{D}} \bigwedge\nolimits_{\mathcal{D}} {}_{(x,y) \in L''_1} x \leftrightarrow y, \qquad n'_2 \stackrel{\text{def}}{=} n_2 \wedge_{\mathcal{D}} \bigwedge\nolimits_{\mathcal{D}} {}_{x \in G'_2} x \wedge_{\mathcal{D}} \bigwedge\nolimits_{\mathcal{D}} {}_{(x,y) \in L''_2} x \leftrightarrow y.$$

For $i = 1, 2$, L_i' contains the equivalent variables in L_i plus those implied by the groundness not shared by the two representations that are about to be disjoined (since $x \wedge y$ implies $x \leftrightarrow y$). Thus L' contains the common equivalent pairs. For $i = 1, 2$, G_i' contains the non-common ground variables to be restored into the ROBDD components, taking into account the common equivalences. Similarly, L_i'' contains the non-common equivalences to be restored into the respective ROBDD: notice that, for $x, y \in G_i \setminus G_{(i \bmod 2)+1}$, care is taken not to restore both $x \wedge y$ and $x \leftrightarrow y$.

For the projection operation over $\mathcal{G}$, which is indeed quite simple, we refer the reader to [4].

4 Some Specialized Algorithms

In order to implement the normalize function, its specializations, and the other operations for the analysis, we need efficient algorithms for several operations. Algorithms for finding all the variables entailed in an ROBDD have been presented in [2, 15], while the operation $n[1/V]$ (called *valuation* or *co-factoring*) can be easily implemented as described in [7].

4.1 Finding Equivalent Variables in ROBDDs

An algorithm for finding all the pairs of variables in an ROBDD that are equivalent is presented as Alg. 2. The algorithm follows directly from the following theorem.

Require: an ROBDD node n
  function equiv_vars(n)
  equiv_vars_aux(n, $\{\langle x, y\rangle : \alpha \preceq x \prec y \preceq \max \mathit{vars}(n)\}$)

  function equiv_vars_aux(n, U)
  if $n = \mathbf{1}$ **then**
    $\varnothing$
  else if $n = \mathbf{0}$ **then**
    U
  else
    $\Big\{ \langle n_{\mathrm{var}}, v\rangle \;\Big|\; v \in \big(\text{vars\_entailed}(n_{\mathrm{true}}) \cap \text{vars\_disentailed}(n_{\mathrm{false}})\big) \Big\}$
      $\cup\big(\text{equiv\_vars\_aux}(n_{\mathrm{true}}, U) \cap \text{equiv\_vars\_aux}(n_{\mathrm{false}}, U)\big)$

Algorithm 2: The equiv_vars function.

Theorem 2. *$[\![n]\!]_{\mathcal{D}}$ entails $x \leftrightarrow y$ where $x \prec y$ if and only if $n = \mathbf{0}$, or $n_{\mathrm{var}} = x$ and $[\![n_{\mathrm{true}}]\!]_{\mathcal{D}}$ entails y and $[\![n_{\mathrm{false}}]\!]_{\mathcal{D}}$ disentails y, or $n_{\mathrm{var}} \prec x$ and $[\![n_{\mathrm{true}}]\!]_{\mathcal{D}}$ and $[\![n_{\mathrm{false}}]\!]_{\mathcal{D}}$ both entail $x \leftrightarrow y$.*

We refer the reader to [2, 15] for the possible implementations of vars_entailed (and, by duality, of vars_disentailed). Observe that a crucial ingredient for the efficiency of the implementation is caching the results of the calls to equiv_vars, vars_entailed, and vars_disentailed.

4.2 Removing Equivalent Variables

Once we have identified which variables are equivalent to which others, we can significantly reduce the size of an ROBDD by removing all but one of each equivalence class of variables. Defining the leader function for an ROBDD node n as

$$\lambda_n \stackrel{\text{def}}{=} \lambda_{equiv(n)},$$

our aim is to restrict away all but the first variable in each equivalence class, that is, all variables v such that $\lambda_n(v) \neq v$. To motivate the algorithm, we begin with a simple theorem.

Theorem 3. *Given an ROBDD rooted at n, and its corresponding leader function λ_n, for every node $m \neq n$ appearing in the ROBDD such that $\lambda_n(m_{\text{var}}) = n_{\text{var}}$, either $m_{\text{true}} = \mathbf{0}$ or $m_{\text{false}} = \mathbf{0}$.*

We "remove" a variable from a Boolean function using existential quantification. For an ROBDD node m, removing m_{var} leaves disjoin$(m_{\text{true}}, m_{\text{false}})$. So Theorem 3 tells us that when $\lambda_n(m_{\text{var}}) \neq m_{\text{var}}$, either m_{true} or m_{false} will be $\mathbf{0}$, making the disjunction trivial.

This suggests the algorithm shown as Algorithm 3 for removing all the "unneeded" variables in an ROBDD n given its leader function λ_n. Two obvious optimizations of this algorithm immediately suggest themselves. Firstly, we may easily compute the last variable (in the ordering) z such that $\lambda_n(z) \neq z$; we may then add the case **else if** $n_{\text{var}} > z$ **then** n immediately after the initial **if**. The second and more important optimization is to avoid recomputing the squeeze_equiv function by the usual caching technique, returning the result of an earlier call with the same arguments. Since the λ_n function is the same in all recursive calls to squeeze_equiv, we may simplify this by clearing our table of previous results whenever squeeze_equiv is called non-recursively (from outside). This allows us to use only the n argument as a parameter to this cache.

When we conjoin two Boolean functions in their GER representation, we also have the opportunity to use the variable equivalences of each argument to reduce the size of the ROBDD component of the *other* argument. In order to do this, we need an algorithm to compute, given any ROBDD m and equivalent variable set L, the ROBDD n whose semantics is

$$[\![n]\!]_{\mathcal{D}} = \exists L|2 \,.\, [\![L]\!]_{\mathcal{L}} \wedge [\![m]\!]_{\mathcal{D}}.$$

Space limitations preclude a full exposition of this algorithm, but it may be found in [4].

Require: an ROBDD node n and a leader function λ
function squeeze_equiv(n, λ)
if is_terminal(n) **then**
  n
else if $\lambda(n_{\mathrm{var}}) = n_{\mathrm{var}}$ **then**
  make_node(n_{var}, squeeze_equiv($n_{\mathrm{true}}, \lambda$), squeeze_equiv($n_{\mathrm{false}}, \lambda$))
else if $n_{\mathrm{true}} = \mathbf{0}$ **then**
  squeeze_equiv($n_{\mathrm{false}}, \lambda$)
else
  squeeze_equiv($n_{\mathrm{true}}, \lambda$)

Algorithm 3: The squeeze_equiv function.

5 Experimental Evaluation

The ideas presented in this section have been experimentally validated in the context of the development of the CHINA analyzer [3]. CHINA is a data-flow analyzer for CLP($\mathcal{H}_\mathcal{N}$) languages (i.e., Prolog, CLP($\mathcal{R}$), `clp(FD)` and so forth) written in C++ and Prolog. It performs bottom-up analysis deriving information about success-patterns and, optionally, call-patterns by means of program transformations and optimized fixpoint computation techniques. We have performed the analysis of a suite comprising 170 programs on the domain Pattern(*Pos*) (similar to `Pat`(*Pos*) [3]), switching off all the other domains currently supported by CHINA[2], and switching off the widening operations normally used to throttle the complexity of the analysis.

A selection of the experimental results is reported in Tables 1 and 2. These tables give, for each program, the analysis times and the number of ROBDD nodes allocated for the standard implementation based on ROBDDs only, but making use of the optimized algorithms described in [15] (R), for the implementation where definitely ground variables are factored out from the ROBDDs as explained in [2] (GR), and for the implementation based on the ideas presented in this paper (GER). The analysis has been considered impractical (and thus stopped) as soon as the amount of memory used by CHINA exceeded 16 MB (for medium sized programs this corresponds to roughly 320.000 ROBDD nodes). This is indicated by ∞ in Table 1 and by ✠ in Table 2.

The computation times have been taken on a Pentium II machine clocked at 233MHz, with 64 MB of RAM, and running Linux 2.0.32.

As it can be seen from the tables, the proposed technique improves the state-of-the-art of groundness analysis with Pattern(*Pos*) considerably. Programs that were out of reach for previous implementations are now analyzable in reasonable time, while for most other programs the measured speedup is between a factor of 2 and an order of magnitude. As far as the the memory requirements of the analysis are concerned, the new representation allows for big savings, as indicated by Table 2. Comparing the results with those of [8, page 45], and

[2] Namely, numerical bounds and relations, aliasing, freeness, and polymorphic types.

	Goal independent			Goal dependent		
Program	R	GR	GER	R	GR	GER
action.pl	1.59	1.58	0.17	3.21	2.78	1.44
bp0-6.pl	0.18	0.09	0.04	0.18	0.06	0.07
bridge.clpr	0.3	0.33	0.11	0.1	0.02	0.02
chat_parser.pl	∞	∞	0.54	∞	∞	2.11
critical.clpr	0.18	0.17	0.03	∞	∞	0.14
cs2.pl	0.11	0.09	0.04	0.08	0.03	0.04
csg.clpr	0.11	0.11	0.01	0.06	0.04	0.02
ime_v2-2-1.pl	0.28	0.19	0.08	0.53	0.2	0.12
kalah.pl	0.23	0.1	0.05	0.24	0.09	0.12
log_interpreter.pl	0.51	0.43	0.17	2.95	2.56	0.6
peval.pl	0.87	0.73	0.31	1.97	1.58	0.55
read.pl	0.41	0.16	0.1	0.76	0.54	0.24
reducer.pl	0.11	0.1	0.07	0.9	0.83	0.25
rubik.pl	∞	∞	0.13	∞	∞	0.64
scc.pl	∞	∞	0.62	1.04	0.15	0.14
sdda.pl	0.11	0.09	0.03	1.94	1.47	0.16
sim_v5-2.pl	0.24	0.21	0.19	0.37	0.25	0.29
simple_analyzer.pl	∞	∞	0.16	∞	∞	4.2
unify.pl	1.39	0.66	0.14	∞	∞	0.78

Table 1. Results obtained with CHINA: analysis time in seconds.

	Goal independent			Goal dependent		
Program	R	GR	GER	R	GR	GER
action.pl	228913	228027	6301	186745	173167	20861
bp0-6.pl	33838	12694	2016	12162	1219	103
bridge.clpr	14765	14762	6324	4044	3243	2174
chat_parser.pl	✠	✠	17291	✠	✠	26634
critical.clpr	14824	14284	1893	✠	✠	7846
cs2.pl	16044	11359	1698	4425	214	64
csg.clpr	317	106	23	196	30	27
ime_v2-2-1.pl	42088	21210	3336	59203	20693	2634
kalah.pl	42008	10962	2253	8487	322	114
log_interpreter.pl	61249	50083	3070	213388	167080	9354
peval.pl	96883	75218	14256	190905	147545	20357
read.pl	49710	14883	2108	55804	32764	3095
reducer.pl	13534	11542	2435	92485	87306	7317
rubik.pl	✠	✠	3825	✠	✠	5261
scc.pl	✠	✠	16751	82788	5762	215
sdda.pl	19561	14360	786	201732	157191	2798
sim_v5-2.pl	18600	13073	4969	5958	319	120
simple_analyzer.pl	✠	✠	6772	✠	✠	65125
unify.pl	188476	94923	9569	✠	✠	25679

Table 2. Results obtained with CHINA: number of BDD nodes.

scaling the timings in order to account for the difference in performance between a Pentium-II at 233MHz and a Sun SparcStation 10/30, it can be seen that we have significantly pushed forward the practicality of *Pos*.

It is worth noticing that while the analyses based on Pattern(*Pos*) are computationally more complex than those simply based on *Pos* (Cortesi et al. measured a slowdown of around 20), they are also significantly more precise [8].

6 Conclusion

We have studied the problem of efficient dependency analysis, and in particular groundness analysis, of (constraint) logic programs, using the *Pos* domain. As others have concluded that ROBDDs are the most efficient representation for use in this sort of analysis, we have concentrated on improving the efficiency of the operations needed during program analysis for ROBDDs. However, since many ROBDD operations have super-linear time cost, we sought to reduce the size of the ROBDDs being manipulated by removing certain information from the ROBDDs and representing it in a way specialized to its nature. We remove definite variables as in [2], storing them in a bit vector. The main accomplishment of this work, however, has been to remove all pairs of equivalent variables, storing them as an array of variable numbers. We have shown how this new hybrid representation significantly decreases the size of the ROBDDs being manipulated. More importantly, analysis times are significantly improved beyond the significant speedup achieved in [2].

References

[1] T. Armstrong, K. Marriott, P. Schachte, and H. Søndergaard. Two classes of Boolean functions for dependency analysis. *Science of Computer Programming*, 31(1):3–45, 1998.

[2] R. Bagnara. A reactive implementation of *Pos* using ROBDDs. In H. Kuchen and S. D. Swierstra, editors, *Programming Languages: Implementations, Logics and Programs, Proceedings of the Eighth International Symposium*, volume 1140 of *Lecture Notes in Computer Science*, pages 107–121, Aachen, Germany, 1996. Springer-Verlag, Berlin.

[3] R. Bagnara. *Data-Flow Analysis for Constraint Logic-Based Languages.* PhD thesis, Dipartimento di Informatica, Università di Pisa, Corso Italia 40, I-56125 Pisa, Italy, March 1997. Printed as Report TD-1/97.

[4] R. Bagnara and P. Schachte. Efficient implementation of *Pos*. Technical Report 98/5, Department of Computer Science, The University of Melbourne, Australia, 1998.

[5] P. Bigot, S. K. Debray, and K. Marriott. Understanding finiteness analysis using abstract interpretation. In K. Apt, editor, *Logic Programming: Proceedings of the Joint International Conference and Symposium on Logic Programming*, MIT Press Series in Logic Programming, pages 735–749, Washington, USA, 1992. The MIT Press.

[6] R. E. Bryant. Graph-based algorithms for Boolean function manipulation. *IEEE Transactions on Computers*, C-35(8):677–691, August 1986.
[7] R. E. Bryant. Symbolic boolean manipulation with ordered binary-decision diagrams. *ACM Computing Surveys*, 24(3):293–318, September 1992.
[8] A. Cortesi, B. Le Charlier, and P. Van Hentenryck. Conceptual and software support for abstract domain design: Generic structural domain and open product. Technical Report CS-93-13, Brown University, Providence, RI, 1993.
[9] A. Cortesi, B. Le Charlier, and P. Van Hentenryck. Combinations of abstract domains for logic programming. In *Conference Record of POPL '94: 21st ACM SIGPLAN-SIGACT Symposium on Principles of Programming Languages*, pages 227–239, Portland, Oregon, 1994.
[10] P. W. Dart. *Dependency Analysis and Query Interfaces for Deductive Databases.* PhD thesis, The University of Melbourne, Department of Computer Science, 1988. Printed as Technical Report 88/35.
[11] M. Falaschi, M. Gabbrielli, K. Marriott, and C. Palamidessi. Confluence and concurrent constraint programming. In V. S. Alagar and M. Nivat, editors, *Proceedings of the Fourth International Conference on Algebraic Methodology and Software Technology (AMAST'95)*, volume 936 of *Lecture Notes in Computer Science*, pages 531–545. Springer-Verlag, Berlin, 1995.
[12] N. D. Jones and H. Søndergaard. A semantics-based framework for the abstract interpretation of Prolog. In S. Abramsky and C. Hankin, editors, *Abstract Interpretation of Declarative Languages*, chapter 6, pages 123–142. Ellis Horwood Ltd, West Sussex, England, 1987.
[13] K. Marriott and H. Søndergaard. Notes for a tutorial on abstract interpretation of logic programs. North American Conference on Logic Programming, Cleveland, Ohio, USA, 1989.
[14] C. S. Mellish. Some global optimizations for a Prolog compiler. *Journal of Logic Programming*, 2(1):43–66, 1985.
[15] P. Schachte. Efficient ROBDD operations for program analysis. In K. Ramamohanarao, editor, *ACSC'96: Proceedings of the 19th Australasian Computer Science Conference*, pages 347–356. Australian Computer Science Communications, 1996.
[16] H. Søndergaard. Immediate fixpoints and their use in groundness analysis. In V. Chandru and V. Vinay, editors, *Foundations of Software Technology and Theoretical Computer Science*, volume 1180 of *Lecture Notes in Computer Science*, pages 359–370. Springer-Verlag, Berlin, 1996.
[17] J. Zobel. *Analysis of Logic Programs.* PhD thesis, The University of Melbourne, 1990.

A Single Perspective on Arrows between Institutions

Alfio Martini* and Uwe Wolter

Technische Universität Berlin, FB Informatik, Sekr. 6-1, Franklinstr. 28/29
D-10587 Berlin, Germany
{alfio,wolter}@cs.tu-berlin.de

Abstract. The purpose of this paper is to bring the most important and influential concepts of arrows between institutions, i.e., institution morphisms, plain maps of institutions, simulations, and (simple) maps of institutions into a common perspective. Based on three simple constructions for institutions — reindexing, change of syntax, change of semantics — we show, firstly, that each of these arrows can be equivalently characterized by the existence of a correspond *intermediate institution* that is related to both involved institutions syntactically or semantically, respectively. Secondly, we show that taking into account reindexing and restriction of semantics, we can describe any of these arrows as an institution morphism (or dually as a plain map) between institutions of the same scheme. We also discuss the possible role of the intermediate institutions in applications.

1 Introduction

It is well-known that the (formal) software development process is usually carried out in a heterogeneous environment of methods, languages, formalisms, and platforms. More specifically, the need to use different logical formalisms in order to adequately specify different views of systems, brings with it the necessity of developing sophisticated tools in order to provide some sort of formal inter-operability in such a complex setting. Interesting and exciting work in this direction can be exemplified by the development of suitable notions of mappings between logics [5,8,2,12], of very flexible logical frameworks [9,8], of combination and synchronization of logics [11,13,14], and of multi-paradigm languages [3]. In this paper we take the concept of *institution* [5] as a formalization of the notion of logic. This concept focuses particularly on the model-theoretic aspects of a logical system.

Briefly, an *institution* $\mathcal{I} = (\mathsf{Sign}, Sen, Mod, \models)$ consists of

- a category Sign whose objects are called *signatures*;
- a functor $Sen : \mathsf{Sign} \to \mathsf{Set}$, giving for each signature a set whose elements are called *sentences* over that signature;

* Research supported in part by a CNPq-grant 200529/94-3

A.M. Haeberer (Ed.): AMAST'98, LNCS 1548, pp. 486–501, 1998.

- a functor $Mod : \mathsf{Sign}^{\mathrm{op}} \to \mathsf{Cat}$, giving for each signature Σ a category whose objects are called Σ-models, and whose arrows are called Σ-morphisms;
- and a function $\models$ associating to each signature Σ a relation $\models_\Sigma \subseteq |Mod(\Sigma)| \times Sen(\Sigma)$, called *$\Sigma$-satisfaction relation,*

such that for each arrow $\phi : \Sigma_1 \to \Sigma_2$ in Sign the *satisfaction condition*

$$M_2 \models_{\Sigma_2} Sen(\phi)(\varphi_1) \iff Mod(\phi)(M_2) \models_{\Sigma_1} \varphi_1,$$

holds for any $M_2 \in |Mod(\Sigma_2)|$ and any $\varphi_1 \in Sen(\Sigma_1)$.

Based on the satisfaction relation above, one is able to introduce the notion of *semantical consequence* between a set of sentences Γ and a sentence φ. More precisely, let $\Sigma \in |\mathsf{Sign}|$. Then a sentence $\varphi \in Sen(\Sigma)$ is said to be a semantical consequence in $\mathcal{I}$ of a set of sentences $\Gamma \subseteq Sen(\Sigma)$, written $\Gamma \models_\Sigma \varphi$, iff $\forall M \in |Mod(\Sigma)| : M \models_\Sigma \Gamma \Longrightarrow M \models_\Sigma \varphi$.

Two derived structures of an institution, which will be of particular interest in this paper, concern the notions of the category of theories and of the generalized model functor. The category Th of *theories(specifications)* has as objects pairs (Σ, Γ) with $\Sigma \in |\mathsf{Sign}|$ and $\Gamma \subseteq Sen(\Sigma)$, and as arrows $(\phi, \models) : (\Sigma_1, \Gamma_1) \to (\Sigma_2, \Gamma_2)$ signature morphisms $\phi : \Sigma_1 \to \Sigma_2$ such that $\Gamma_2 \models_{\Sigma_2} Sen(\phi)(\Gamma_1)$. For a theory (Σ, Γ) we consider $Mod_{\models}(\Sigma, \Gamma)$ to be the full subcategory of $Mod(\Sigma)$ with $M \in |Mod_{\models}(\Sigma, \Gamma)|$ iff $M \models_\Sigma \varphi$ for all $\varphi \in \Gamma$. The satisfaction condition ensures that for each arrow $(\phi, \models)$ in Th $Mod(\phi)(Mod_{\models}(\Sigma_2, \Gamma_2))$ is contained in $Mod_{\models}(\Sigma_1, \Gamma_1)$, i.e., we can restrict the functor $Mod(\phi) : Mod(\Sigma_2) \to Mod(\Sigma_1)$ to a functor $Mod_{\models}(\phi, \models) : Mod_{\models}(\Sigma_2, \Gamma_2) \to Mod_{\models}(\Sigma_1, \Gamma_1)$. Globally this means that we can extend the original model functor $Mod : \mathsf{Sign}^{\mathrm{op}} \to \mathsf{Cat}$ to a *generalized model functor* $Mod_{\models} : \mathsf{Th}^{op} \to \mathsf{Cat}$. The obvious *projection functor* from Th to Sign will be denoted by $sign : \mathsf{Th} \to \mathsf{Sign}$ and the embedding functor assigning Σ to $(\Sigma, \emptyset)$ by $inj : \mathsf{Sign} \to \mathsf{Th}$.

In the last decade, a number of different concepts of arrows between institutions were introduced. This variety is justified in principle by two reasons: firstly, the need to provide more flexible and general concepts in order to formalize properly relationships between logics. Secondly, to provide technical machinery suited to some particular application, as for instance, logical semantics of multi-paradigm languages [4], or specification development with constraints expressed in different logical systems [5].

Arrows between institutions, say $\mathcal{I}$ and $\mathcal{I}'$, are usually based on a translation Φ of signatures, a (natural) translation α of sentences, and a (natural) translation β of models. The essential point is that sentences and models are translated contravariantly. This contravariance means that, in general, only a part of $\mathcal{I}$ is related to a part of $\mathcal{I}'$. This relational nature of arrows can be made structurally explicit by a corresponding span characterization of each arrow. That is, each arrow between $\mathcal{I}$ and $\mathcal{I}'$ indicates the existence of an *intermediate institution* $\mathcal{I}^m$ that is related covariantly to both institutions $\mathcal{I}$ and $\mathcal{I}'$.

The purpose of this paper is to bring the most important and influential concepts of arrows between institutions, i.e., institution morphisms, plain maps of institutions, simulations, and (simple) maps of institutions into a common perspective. The technical basis for our results is provided by the observation that each of the above mentioned translations gives rise to a simple construction principle for institutions, i.e., Φ allows *reindexing*, α allows *change of syntax*, and β allows *change of semantics.* Taking into account reindexing and *restriction of semantics*, i.e., change of semantics with β a (natural) inclusion, we are able, moreover, to describe equivalently any of the considered concepts of arrows as an institution morphism (or dually as a plain map) between institutions of the same *scheme*, i.e., with the same category of signatures.

Each concept of an arrow between institution determines a corresponding category. Readers interested in a comprehensive presentation of all these categories, and in a systematic unifying study of all the functorial relations between these different categories, are referred to [6]

2 Institution Morphisms

We start our analysis with the original proposal for a formalization of arrows between institutions.

Definition 1 (Institution morphism,[5]). *Let* $\mathcal{I} = (\mathsf{Sign}, Sen, Mod, \models)$ *and* $\mathcal{I}' = (\mathsf{Sign}', Sen', Mod', \models')$ *be institutions. An* institution morphism $(\Phi, \alpha, \beta) : \mathcal{I} \to \mathcal{I}'$ *is given by*

- *a functor* $\Phi : \mathsf{Sign} \to \mathsf{Sign}'$,
- *a natural transformation* $\alpha : \Phi; Sen' \Rightarrow Sen : \mathsf{Sign} \to \mathsf{Set}$, *and*
- *a natural transformation* $\beta : Mod \Rightarrow \Phi^{op}; Mod' : \mathsf{Sign}^{op} \to \mathsf{Cat}$,

such that for each $\Sigma \in |\mathsf{Sign}|$ *the* institution morphism condition

$$M \models_{\Sigma} \alpha(\Sigma)(\varphi') \iff \beta(\Sigma)(M) \models'_{\Phi(\Sigma)} \varphi',$$

holds for any $M \in |Mod(\Sigma)|$ *and* $\varphi' \in Sen'(\Phi(\Sigma))$. □

Example 1. The running example presented in [5] concerns the relation between the institution of many-sorted first order logic with equality $MSFOL^{=}$ and the institution of many-sorted equational logic $MSEL$.

Firstly, forgetting predicate symbols, defines a functor $\Phi : \mathsf{Sign}_{MSFOL^{=}} \to \mathsf{Sign}_{MSEL}$ with $\Phi(S, OP, P) = (S, OP)$ for any first order signature (S, OP, P). Secondly, any equation in context $(X \vdash t = u) \in Sen_{MSEL}(S, OP)$ can be translated into a closed first order formula $(\forall X : t = u) \in Sen_{MSFOL^{=}}(S, OP, P)$, i.e., we actually have a natural transformation $\alpha : \Phi; Sen_{MSEL} \Rightarrow Sen_{MSFOL^{=}}$. Thirdly, any first order model $M \in |Mod_{MSFOL^{=}}(S, OP, P)|$ has an underlying total algebra $M\!\upharpoonright \in |Mod_{MSEL}(S, OP)|$, where the interpretation of the predicate symbols is forgotten, i.e., we actually have a natural transformation $\beta : Mod_{MSFOL^{=}} \Rightarrow \Phi^{op}; Mod_{MSEL}$. The institution morphism condition for $(\Phi, \alpha, \beta) : MSFOL^{=} \to MSEL$ holds obviously. □

One essential observation for logics with model theory concerns the contravariance of translating sentences or models, respectively, along a signature morphism. This observation is formally fixed in the concept of institution and it seems "natural" to require such a contravariance also for arrows between institutions. Goguen and Burstall use the category Trel of *twisted relations* to justify categorically the "naturality" of this choice.

A crucial point for applications of arrows between institutions, is how semantical consequence in both institutions is related.

Proposition 1. *Institution morphisms* $(\Phi, \alpha, \beta) : \mathcal{I} \to \mathcal{I}'$ *reflect semantical consequence, i.e., for every* $\Sigma \in |\mathsf{Sign}|$*, every* $\Gamma' \subseteq Sen'(\Phi(\Sigma))$*, and every* $\varphi' \in Sen'(\Phi(\Sigma))$*:* $\alpha(\Sigma)(\Gamma') \models_\Sigma \alpha(\Sigma)(\varphi') \Longleftarrow \Gamma' \models_{\Phi(\Sigma)} \varphi'$. □

Now, we take a step back and try to look more structurally on the definition of institution morphisms. At a second glance we see that the category Sign can be considered to be the "syntactic scheme" of an institution $\mathcal{I}$. Taking into account reindexing of institutions we can describe institution morphisms as arrows between institutions of the same syntactic scheme.

Proposition 2 (Reindexed institution). *Let* $\mathcal{I}' = (\mathsf{Sign}', Sen', Mod', \models')$ *be an institution and let* $\Phi : \mathsf{Sign} \to \mathsf{Sign}'$ *be a functor. Then there is a* reindexed institution $\mathcal{I}'_{\restriction\Phi}$ *, which can be defined as*

$$\mathcal{I}'_{\restriction\Phi} = (\mathsf{Sign}, \Phi; Sen', \Phi^{op}; Mod', \models'_\Phi),$$

where for each $\Sigma \in |\mathsf{Sign}|$: $\models'_{\Phi,\Sigma} \stackrel{def}{=} \models'_{\Phi(\Sigma)}$. □

Proposition 3. $(\Phi, \alpha, \beta) : \mathcal{I} \to \mathcal{I}'$ *is an institution morphism iff* $(id_{\mathsf{Sign}}, \alpha, \beta) : \mathcal{I} \to \mathcal{I}'_{\restriction\Phi}$ *is an institution morphism.* □

Proposition 3 suggests to look for further constructions of new institutions that may be helpful in understanding and applying the different notions of arrows between institutions.

Following this suggestion we will show next that the "syntax" of an institution can be changed via a natural transformation α mapping the new syntax to the old syntax and that, analogously, the "semantics" of an institution can be changed via a natural transformation β mapping the new semantics to the old one.

Lemma 1. *Let be given an institution* $\mathcal{I} = (\mathsf{Sign}, Sen, Mod, \models)$*, a functor* $Sen'' : \mathsf{Sign} \to \mathsf{Set}$ *with a natural transformation* $\alpha : Sen'' \Rightarrow Sen : \mathsf{Sign} \to \mathsf{Set}$*, and a functor* $Mod'' : \mathsf{Sign}^{\mathsf{op}} \to \mathsf{Cat}$ *with a natural transformation* $\beta : Mod'' \Rightarrow Mod : \mathsf{Sign}^{\mathsf{op}} \to \mathsf{Cat}$*. Then for any* $\phi : \Sigma_1 \to \Sigma_2$ *in* Sign *the following hold:*

1. $M \models_{\Sigma_2} \alpha(\Sigma_2)(Sen''(\phi)(\varphi'')) \iff Mod(\phi)(M) \models_{\Sigma_1} \alpha(\Sigma_1)(\varphi'')$ *for any* $M \in |Mod(\Sigma_2)|$ *and* $\varphi'' \in Sen''(\Sigma_1)$*;*

2. $\beta(\Sigma_2)(M'') \models_{\Sigma_2} Sen(\phi)(\varphi) \iff \beta(\Sigma_1)(Mod''(\phi)(M'')) \models_{\Sigma_1} \varphi$ *for any* $M'' \in |Mod''(\Sigma_2)|$ *and* $\varphi \in Sen(\Sigma_1)$.

Proof. "1": By the naturality of α and the satisfaction condition in $\mathcal{I}$, we have immediately that $M \models_{\Sigma_2} \alpha(\Sigma_2)(Sen''(\phi)(\varphi'')) \iff M \models_{\Sigma_2} Sen(\phi)(\alpha(\Sigma_1)(\varphi''))$ $\iff Mod(\phi)(M) \models_{\Sigma_1} \alpha(\Sigma_1)(\varphi'')$.
"2": Analogously. □

If we consider the patterns $\_ \models_\Sigma \alpha(\Sigma)(\_)$ and $\beta(\Sigma)(\_) \models_\Sigma \_$ as describing relations between $|Mod(\Sigma)|$ and $Sen''(\Sigma)$ or between $|Mod''(\Sigma)|$ and $Sen(\Sigma)$, respectively, it is not difficult to see that the equivalences in lemma 1 own the shape of satisfaction conditions. For what institutions these satisfaction conditions hold is made precise in the next

Proposition 4 (Change of syntax and semantics). *Let be given an institution* $\mathcal{I} = (\mathsf{Sign}, Sen, Mod, \models)$.

1. *For any natural transformation* $\alpha : Sen'' \Rightarrow Sen : \mathsf{Sign} \to \mathsf{Set}$ *there is an institution* $\mathcal{I}_{\restriction\alpha}$, *called* $\mathcal{I}$ prefixed by α, *which can be defined as* $\mathcal{I}_{\restriction\alpha} = (\mathsf{Sign}, Sen'', Mod, \models_\alpha)$, *where for each* $\Sigma \in |\mathsf{Sign}|$
$$\models_{\alpha,\Sigma} \stackrel{def}{=} (id_{Mod(\Sigma)} \times \alpha(\Sigma)^{-1})(\models_\Sigma) \subseteq |Mod(\Sigma)| \times Sen''(\Sigma),$$
i.e., $M \models_{\alpha,\Sigma} \varphi''$ *iff* $M \models_\Sigma \alpha(\Sigma)(\varphi'')$.
2. *For any natural transformation* $\beta : Mod'' \Rightarrow Mod : \mathsf{Sign}^{op} \to \mathsf{Cat}$ *there is an institution* $\mathcal{I}_{\restriction\beta}$, *called* $\mathcal{I}$ prefixed by β, *which can be defined as* $\mathcal{I}_{\restriction\beta} = (\mathsf{Sign}, Sen, Mod'', \models_\beta)$, *where for each* $\Sigma \in |\mathsf{Sign}|$
$$\models_{\beta,\Sigma} \stackrel{def}{=} (\beta(\Sigma)^{-1} \times id_{Sen(\Sigma)})(\models_\Sigma) \subseteq |Mod''(\Sigma)| \times Sen(\Sigma),$$
i.e., $M'' \models_{\beta,\Sigma} \varphi$ *iff* $\beta(\Sigma)(M'') \models_\Sigma \varphi$. □

According to proposition 2 and proposition 4 we can consider for any institution morphism $(\Phi, \alpha, \beta) : \mathcal{I} \to \mathcal{I}'$, firstly, the institution $\mathcal{I}$ prefixed by α
$$\mathcal{I}_{\restriction\alpha} = (\mathsf{Sign}, \Phi; Sen', Mod, \models_\alpha)$$
and, secondly, the reindexed institution $\mathcal{I}'_{\restriction\Phi}$ prefixed by β
$$\mathcal{I}'_{\restriction\Phi,\beta} = (\mathsf{Sign}, \Phi; Sen', Mod, \models'_{\Phi,\beta}).$$

Both institutions have the same category of signatures and the same sentence and model functor, respectively. Moreover, the institution morphism condition is equivalent to the requirement $\models_\alpha = \models'_{\Phi,\beta}$.

Proposition 5 (Institution morphism condition). $(\Phi, \alpha, \beta) : \mathcal{I} \to \mathcal{I}'$ *is an institution morphism* *iff* $\mathcal{I}_{\restriction\alpha} = \mathcal{I}'_{\restriction\Phi,\beta}$.

Proof. According to the definition of $\mathcal{I}_{\restriction\alpha}$ and $\mathcal{I}'_{\restriction\Phi,\beta}$ we have that for any $M \in |Mod(\Sigma)|$ and any $\varphi' \in Sen'(\Phi(\Sigma))$ the conditions ($M \models_\Sigma \alpha(\Sigma)(\varphi') \iff \beta(\Sigma)(M) \models'_{\Phi(\Sigma)} \varphi'$) and ($M \models_{\alpha,\Sigma} \varphi' \iff M \models'_{\Phi,\beta,\Sigma} \varphi'$) are equivalent. □

To be able to describe the relation between $\mathcal{I}_{\restriction\alpha}$ and $\mathcal{I}$ or $\mathcal{I}'_{\restriction\Phi,\beta}$ and $\mathcal{I}'$, respectively, we introduce the concept of institution transformation that can be understood to be a "pointwise" variant (for institutions) of the concept of pre-institution transformation in [12].

Definition 2 (Institution transformation,[7]). *Let $\mathcal{I}$ and $\mathcal{I}'$ be institutions. An* institution transformation $(\Phi, \alpha, \beta) : \mathcal{I} \to \mathcal{I}'$ *is given by*

- *a functor* $\Phi : \mathsf{Sign} \to \mathsf{Sign}'$,
- *a natural transformation* $\alpha : Sen \Rightarrow \Phi; Sen' : \mathsf{Sign} \to \mathsf{Set}$, *and*
- *a natural transformation* $\beta : Mod \Rightarrow \Phi^{op}; Mod' : \mathsf{Sign}^{op} \to \mathsf{Cat}$

such that the institution transformation condition

$$M \models_\Sigma \varphi \iff \beta(\Sigma)(M) \models'_{\Phi(\Sigma)} \alpha(\Sigma)(\varphi),$$

holds for any $\Sigma \in |\mathsf{Sign}|$, $M \in |Mod(\Sigma)|$, *and* $\varphi \in Sen(\Sigma)$. □

The above concept of arrow between institutions correlates to the usual concept of (strong) arrow between first order structures: Abstracting from indexing, we have structures with two carriers Mod, Sen and with a binary predicate $\models \subseteq |Mod| \times Sen$. A (strong) arrow between two structures $(Mod, Sen, \models)$ and $(Mod', Sen', \models')$ is given by two translations $\beta : Mod \to Mod'$ and $\alpha : Sen \to Sen'$, so that the truth of the binary predicate is not only preserved but also reflected. If we would like more intentional interpretation, we could also say that the satisfaction relation $\models'$ can be used to "simulate" the satisfaction relation $\models$ via α and β. That is, analogously to proposition 5 we can prove the institution transformation condition to be equivalent to the requirement $\models_\Sigma = \models_{\Phi,\alpha,\beta,\Sigma}$ where $\models_{\Phi,\alpha,\beta,\Sigma} = (\beta(\Sigma)^{-1} \times \alpha(\Sigma)^{-1})(\models_{\Phi(\Sigma)})$ for each $\Sigma \in |\mathsf{Sign}|$.
Now we can summarize our considerations so far by characterizing institution morphisms as spans of institution transformations.

Theorem 1 (Intermediate institution).

1. *Let $(\Phi, \alpha, \beta) : \mathcal{I} \to \mathcal{I}'$ be an institution morphism. Then there is an institution $\mathcal{I}^m = (Sign, \Phi; Sen', Mod, \models^m)$, such that the two triples*

   $$(id_{Sign}, \alpha, id_{Mod}) : \mathcal{I}^m \to \mathcal{I} \quad \text{and} \quad (\Phi, id_{\Phi;Sen'}, \beta) : \mathcal{I}^m \to \mathcal{I}',$$

   define institution transformations.
2. *Vice-versa, given an institution $\mathcal{I}^m = (Sign, \Phi; Sen', Mod, \models^m)$ and institutions transformations*

   $$(id_{Sign}, \alpha, id_{Mod}) : \mathcal{I}^m \to \mathcal{I} \quad \text{and} \quad (\Phi, id_{\Phi;Sen'}, \beta) : \mathcal{I}^m \to \mathcal{I}',$$

   the triple $(\Phi, \alpha, \beta) : \mathcal{I} \to \mathcal{I}'$ defines an institution morphism.

Proof. "1": We set $\mathcal{I}^m \stackrel{def}{=} \mathcal{I}_{\restriction\alpha} = \mathcal{I}'_{\restriction\Phi,\beta}$ according to proposition 5. Since $\models^m_\Sigma = \models_{\alpha,\Sigma}$ we have for any $M \in |Mod(\Sigma)|$, $\varphi' \in Sen'(\Phi(\Sigma))$: $M \models^m_\Sigma \varphi' \iff M \models_\Sigma \alpha(\Sigma)(\varphi')$. This shows that $(id_{Sign}, \alpha, id_{Mod}) : \mathcal{I}^m \to \mathcal{I}$ is an institution transformation. Further, we have $M \models^m_\Sigma \varphi' \iff \beta(\Sigma)(M) \models'_{\Phi(\Sigma)} \varphi'$, since $\models^m_\Sigma = \models'_{\Phi,\beta,\Sigma}$. This shows that $(\Phi, id_{\Phi;Sen'}, \beta) : \mathcal{I}^m \to \mathcal{I}'$ is also an institution transformation.
"2": The institution transformation conditions ensure $\models^m = \models_\alpha$ and $\models^m = \models'_{\Phi,\beta}$, respectively, thus we are done by proposition 5. □

Theorem 1 tells that the presence of an institution morphism $(\Phi, \alpha, \beta) : \mathcal{I} \to \mathcal{I}'$ indicates the existence of an *intermediate institution* that is related syntactically to $\mathcal{I}$ and semantically (up to reindexing) to $\mathcal{I}'$. The contravariance of α and β in $(\Phi, \alpha, \beta) : \mathcal{I} \to \mathcal{I}'$ means that, in general, only a part of $\mathcal{I}$ is related to a part of $\mathcal{I}'$. This relational nature of institution morphisms is now made structurally explicit by the span characterization.

Example 1 (Continued). The intermediate institution indicated by the institution morphism $(\Phi, \alpha, \beta) : MSFOL^= \to MSEL$ is $\mathcal{I}^m = (\mathsf{Sign}_{MSFOL^=}, \Phi; Sen_{MSEL}, Mod_{MSFOL^=}, \models^m)$. $\alpha : \Phi; Sen_{MSEL} \Rightarrow Sen_{MSFOL^=}$ is a natural injection thus $\mathcal{I}^m = MSFOL^=_{\restriction\alpha}$ can be seen as a syntactic subinstitution of $MSFOL^=$, i.e., as $MSFOL^=$ restricted to equational logic. Note, that there are still predicate symbols in $\mathcal{I}^m$ but the predicate symbols are not used in the sentences. $\Phi : \mathsf{Sign}_{MSFOL^=} \to \mathsf{Sign}_{MSEL}$ and $\beta : Mod_{MSFOL^=} \Rightarrow \Phi^{op}; Mod_{MSEL}$ are surjective thus $\mathcal{I}^m = MSEL_{\restriction\Phi,\beta}$ can be seen as a "semantical multiple" of $MSEL$. That is, for any algebra $A \in |Mod_{MSEL}(S, OP)|$ and any $(S, OP, P) \in |\mathsf{Sign}_{MSFOL^=}|$ there are (infinite) many representatives of A, i.e., first order structures $M \in |Mod_{MSFOL^=}(S, OP, P)|$ with $M\restriction = A$. Note, that as well the set P of predicate symbols as the interpretation of the predicate symbols is completely arbitrary. □

We are not intending to start generalizing arrows between institutions on the basis of theorem 1. We will even not carry out the straightforward definition of corresponding categories of spans. Besides adding to the general understanding of different concepts of arrows between institutions, we are mainly interested to draw attention to the involved construction principles for institutions and to clarify the possible role of the intermediate institutions in applications.

Note, e.g., that $\mathcal{I}$ and $\mathcal{I}_{\restriction\alpha} = \mathcal{I}'_{\restriction\Phi,\beta}$ have the same category of signatures and the same model functor such that the pattern $\alpha(\Sigma)(\_) \models_\Sigma \alpha(\Sigma)(\_)$ can be taken to denote semantical consequence in $\mathcal{I}_{\restriction\alpha} = \mathcal{I}'_{\restriction\Phi,\beta}$. In such a way proposition 1 could be interpreted also as stating that semantical consequences are reflected by the institution transformation $(\Phi, id_{\Phi;Sen'}, \beta) : \mathcal{I}_{\restriction\alpha} = \mathcal{I}'_{\restriction\Phi,\beta} \longrightarrow \mathcal{I}'$. In general, semantical consequences are reflected by any institution transformation.

One of the main applications of institution morphisms presented in [5] is the construction of so-called duplex (and multiplex) institutions. We will sketch this construction in view of our above results: let $\mathcal{I}'$ be an institution which admits some special constraints, say, *free constraints*, as for instance the institution of many-sorted equational logic. Thus, out of $\mathcal{I}'$, we might define a new institution $C(\mathcal{I}') = (\mathsf{Sign}', Sen_{free}, Mod, \models^{free})$, a new sentence functor $Sen_{free} : \mathsf{Sign} \to \mathsf{Set}$ and a new satisfaction relation $\models^{free}$ (see [5] for details about free constraints). Given an institution morphism $(\Phi, \alpha, \beta) : \mathcal{I} \to \mathcal{I}'$, we can relate $\mathcal{I}$ and $C(\mathcal{I}')$ with the pair $(\Phi, \beta) : \mathcal{I} \to C(\mathcal{I}')$ (without any condition) called an *institution semi-morphism* in [15,1]. Now we can construct the institution

$$C(\mathcal{I}')_{\restriction \Phi,\beta} = (\mathsf{Sign}, \Phi; Sen_{free}, Mod, \models^{free}_{\Phi,\beta})$$

with the same category of signatures and the same model functor as $\mathcal{I}$, thus we can finally make a disjoint union of both sentence functors with satisfaction defined separately.

Proposition 6 (Sum of sentence functors). *Given two institutions* $\mathcal{I}_1 = (\mathsf{Sign}, Sen_1, Mod, \models^1)$ *and* $\mathcal{I}_2 = (\mathsf{Sign}, Sen_2, Mod, \models^2)$ *we can define a new institution* $\mathcal{I}_1 + \mathcal{I}_2 = (\mathsf{Sign}, Sen_{1+2}, Mod, \models^{1+2})$ *with* $Sen_{1+2}(\Sigma) \stackrel{def}{=} Sen_1(\Sigma) \stackrel{+}{\cup} Sen_2(\Sigma)$ *and* $\models^{1+2}_{\Sigma} \stackrel{def}{=} \models^1_{\Sigma} \stackrel{+}{\cup} \models^2_{\Sigma}$ *for each* $\Sigma \in |\mathsf{Sign}|$. □

The institution $\mathcal{I} + C(\mathcal{I}')_{\restriction \Phi,\beta}$ is called the *duplex institution over* (Φ, α, β) in [5]. Note, however, that α and thus the institution morphism condition are not necessary for constructing $\mathcal{I} + C(\mathcal{I}')_{\restriction \Phi,\beta}$. Using the above mechanism we can extend the syntax of $\mathcal{I}$ by the syntax of as many institutions $\mathcal{I}''$ as we want as long as we find a way to map signatures and models from $\mathcal{I}$ to $\mathcal{I}''$.

3 Plain Maps

To account for relevant examples of relations between institutions a group of new concepts of arrows was introduced in [8]. The plain variant mapping signatures to signatures is presented in

Definition 3 (Plain map of institutions, [8]). *Let* $\mathcal{I} = (\mathsf{Sign}, Sen, Mod, \models)$ *and* $\mathcal{I}' = (\mathsf{Sign}', Sen', Mod', \models')$ *be institutions. A* plain map of institutions $(\Phi, \alpha, \beta) : \mathcal{I} \to \mathcal{I}'$ *is given by*

- *a functor* $\Phi : \mathsf{Sign} \to \mathsf{Sign}'$,
- *a natural transformation* $\alpha : Sen \Rightarrow \Phi; Sen' : \mathsf{Sign} \to \mathsf{Set}$, *and*
- *a natural transformation* $\beta : \Phi^{op}; Mod' \Rightarrow Mod : \mathsf{Sign}^{op} \to \mathsf{Cat}$

such that for each $\Sigma \in |\mathsf{Sign}|$ *the* plain map condition

$$\beta(\Sigma)(M') \models_\Sigma \varphi \iff M' \models'_{\Phi(\Sigma)} \alpha(\Sigma)(\varphi)$$

holds for any $M' \in |Mod'(\Phi(\Sigma))|$ *and* $\varphi \in Sen(\Sigma)$. □

Example 2 ([8]). Consider the institutions of many-sorted equational logic $MSEL$ and of unsorted equational logic EL, respectively.

Firstly, forgetting sort symbols defines a functor $\Phi : \mathsf{Sign}_{MSEL} \to \mathsf{Sign}_{EL}$ with $\Phi(S, OP) = OP$ for each many-sorted equational signature $(S, OP) \in |\mathsf{Sign}_{MSEL}|$. Secondly, any equation in context $(X \vdash t = u) \in Sen_{MSEL}(S, OP)$ can be considered as a sentence in $(t = u) \in Sen_{EL}(\Phi(S, OP))$ once we omit the sort declaration of variables. This gives a natural transformation $\alpha : Sen_{MSEL} \Rightarrow \Phi; Sen_{EL}$. Thirdly, any unsorted algebra $M \in |Mod_{EL}(\Phi(S, OP))|$ gives a many-sorted algebra $\beta(S, OP)(M) \in |Mod_{MSEL}(S, OP)|$, where $\beta(S, OP)(M)_s \stackrel{def}{=} M$ for every $s \in S$, and $op^{\beta(S,OP)(M)} \stackrel{def}{=} op^M$ for each $op \in OP$. This situation delivers a functor $\beta(S, OP) : Mod_{EL}(\Phi(S, OP)) \to Mod(S, OP)$, and globally a natural injection $\beta : \Phi^{op}; Mod_{EL} \Rightarrow Mod_{MSEL}$.

The plain map condition can be validated by observing that the assignments of X into $\beta(S, OP)(M)$ are in one-to-one correspondence to the assignments of the corresponding unsorted set $\bigcup X$ of variables into M since all components of $\beta(S, OP)(M)$ equal the carrier of M. □

Taking into account reindexing of institutions plain maps of institutions can be presented as the dual of institution morphisms.

Proposition 7. *$(\Phi, \alpha, \beta) : \mathcal{I} \to \mathcal{I}'$ is a plain map of institutions*

(a) iff $(id_{\mathsf{Sign}}, \alpha, \beta) : \mathcal{I} \to \mathcal{I}'_{\restriction\Phi}$ is a plain map of institutions

(b) iff $(id_{\mathsf{Sign}}, \alpha, \beta) : \mathcal{I}'_{\restriction\Phi} \to \mathcal{I}$ is an institution morphism. □

This duality allows to adopt immediately all results of the last section for plain maps: We can consider for any plain map of institutions $(\Phi, \alpha, \beta) : \mathcal{I} \to \mathcal{I}'$, firstly, the the reindexed institution $\mathcal{I}'_{\restriction\Phi}$ prefixed by α

$$\mathcal{I}'_{\restriction\Phi,\alpha} = (\mathsf{Sign}, Sen, \Phi^{op}; Mod', \models'_{\Phi,\alpha})$$

with $\models'_{\Phi,\alpha,\Sigma} = (id_{Mod'(\Phi(\Sigma))} \times \alpha(\Sigma)^{-1})(\models'_{\Phi(\Sigma)})$ for each $\Sigma \in |\mathsf{Sign}|$, and, secondly, the institution $\mathcal{I}$ prefixed by β

$$\mathcal{I}_{\restriction\beta} = (\mathsf{Sign}, Sen, \Phi^{op}; Mod', \models_\beta)$$

with $\models_{\beta,\Sigma} = (\beta(\Sigma)^{-1} \times id_{Sen(\Sigma)})(\models_\Sigma)$ for each $\Sigma \in |\mathsf{Sign}|$. The plain map condition is equivalent to the requirement $\models_\beta = \models'_{\Phi,\alpha}$. Finally a plain map $(\Phi, \alpha, \beta) : \mathcal{I} \to \mathcal{I}'$ can be characterized uniquely by the existence of an intermediate institution $\mathcal{I}^m \stackrel{def}{=} \mathcal{I}_{\restriction\beta} = \mathcal{I}'_{\restriction\Phi,\alpha}$ such that the two triples

$$(id_{\mathsf{Sign}}, id_{Sen}, \beta) : \mathcal{I}^m \to \mathcal{I} \quad \text{and} \quad (\Phi, \alpha, id_{\Phi^{op};Mod'}) : \mathcal{I}^m \to \mathcal{I}',$$

define institution transformations. This follows directly from proposition 7 and theorem 1 if we take into account that the institution transformation $(\Phi, \alpha$

$id_{\Phi^{op};Mod'}) : \mathcal{I}^m \to \mathcal{I}'$ can be decomposed into $(id_{Sign}, \alpha, id_{\Phi^{op};Mod'}) : \mathcal{I}^m \to \mathcal{I}'_{\upharpoonright\Phi}$ and $(\Phi, id_{\Phi;Sen'}, id_{\Phi^{op};Mod'}) : \mathcal{I}'_{\upharpoonright\Phi} \to \mathcal{I}'$.

Example 2 (Continued). The intermediate institution indicated by the plain map $(\Phi, \alpha, \beta) : MSEL \to EL$ is $\mathcal{I}^m = (\mathsf{Sign}_{MSEL}, Sen_{MSEL}, \Phi^{op}; Mod_{EL}, \models^m)$. It represents what both institutions $MSEL$ and EL have in common according to (Φ, α, β). $\mathcal{I}^m = EL_{\upharpoonright\Phi,\alpha}$ is as EL but with possible additional (sort) restrictions for building terms. Compared to $MSEL$ we lose in $\mathcal{I}^m = MSEL_{\upharpoonright\beta}$ the flexibility of choosing in models different components for different sorts. □

Typical applications of (plain) maps within system specifications are situations where Φ and α represent an extension or change of the syntax of institution $\mathcal{I}$. As long as the models of the new syntax can be related to the models of $\mathcal{I}$ we can (partly) save the work on specifications already done in $\mathcal{I}$. By $\mathcal{I}'_{\upharpoonright\Phi,\alpha} = \mathcal{I}_{\upharpoonright\beta}$ we can keep the old syntax but now equiped with new semantics. Note, that only in case β surjective consistency of specifications will be saved. The institution transformation $(\Phi, \alpha, id_{\Phi^{op};Mod'}) : \mathcal{I}_{\upharpoonright\beta} \to \mathcal{I}'$ provides further the basis for building structured specifications with mixed syntax (see the next section).

Example 1 (Continued). The functor $\Phi : \mathsf{Sign}_{MSFOL^=} \to \mathsf{Sign}_{MSEL}$ has a left-adjoint $\Psi : \mathsf{Sign}_{MSEL} \to \mathsf{Sign}_{MSFOL^=}$ with $\Psi(S, OP) \stackrel{def}{=} (S, OP, \emptyset)$. This means, by a general result [1,6], that the institution morphism $(\Phi, \alpha, \beta) : MSFOL^= \to MSEL$ can be equivalently represented by a plain map $(\Psi, \alpha', \beta') : MSEL \to MSFOL^=$ where in this example $\alpha' : Sen_{MSEL} \Rightarrow \Psi; Sen_{MSFOL^=}$ becomes a natural injection and $\beta' : \Psi; Mod_{MSFOL^=} \Rightarrow Mod_{MSEL}$ becomes even a natural isomorphism. That is, $MSEL_{\upharpoonright\beta'} = MSFOL^=_{\upharpoonright\Psi,\alpha'}$ is just as $MSEL$ but now considering algebras $M \in |Mod_{MSEL}(S, OP)|$ as first order structures $M \in |Mod_{MSFOL^=}(S, OP, \emptyset)|$. □

4 Simple Maps and Simulations

There are many situations where we can translate signatures and sentences of an institution $\mathcal{I}$ into signatures and sentences of another institution $\mathcal{I}'$, but, where only subclasses of the corresponding model classes in $\mathcal{I}'$ can be translated back into models of $\mathcal{I}$. Fortunately, we are able in most cases to axiomatize these subclasses within $\mathcal{I}'$.

Definition 4 (Simple maps of institutions, [8]). *Let $\mathcal{I}$ and $\mathcal{I}'$ be institutions. A* simple map of institutions *$(\Phi, \Psi, \alpha, \beta) : \mathcal{I} \to \mathcal{I}'$ is given by*

- *a functor $\Phi : \mathsf{Sign} \to \mathsf{Sign}'$,*
- *a functor $\Psi : \mathsf{Sign} \to \mathsf{Th}'$ with $\Psi; sign' = \Phi$,*
- *a natural transformation $\alpha : Sen \Rightarrow \Phi; Sen' : \mathsf{Sign} \to \mathsf{Set}$, and*
- *a natural transformation $\beta : \Psi^{op}; Mod'_{\models'} \Rightarrow Mod : \mathsf{Sign}^{op} \to \mathsf{Cat}$*

such that for each $\Sigma \in |\mathsf{Sign}|$ *the* simple map condition

$$\beta(\Sigma)(M') \models_\Sigma \varphi \iff M' \models'_{\Phi(\Sigma)} \alpha(\Sigma)(\varphi)$$

holds for any $M' \in |Mod'_{\models'}(\Psi(\Sigma))|$ *and* $\varphi \in Sen(\Sigma)$. □

Remark 1. We denote by $\emptyset'_\Sigma \subseteq Sen'(\Phi(\Sigma))$ the set of axioms induced by the translation of the signature Σ along the functor Ψ, i.e., $\Psi(\Sigma) = (\Phi(\Sigma), \emptyset'_\Sigma)$. □

Example 3 ([2]). Consider the institutions of many-sorted equational logic $MSEL$ and of (unsorted) first order predicate logic with equality $FOL^=$, respectively.

Firstly, the functor $\Phi : \mathsf{Sign}_{MSEL} \to \mathsf{Sign}_{FOL^=}$ is given by translating any many sorted signature $\Sigma = (S, OP)$ into the first order signature (OP, P_S) with unary typing predicates $P_S \stackrel{def}{=} \{\pi_s \mid s \in S\}$. Secondly, any equation in context $\varphi = (X \vdash t = u) \in Sen_{MSEL}(\Sigma)$, where X is an S-sorted family of variables, can be translated into a first order sentence $(\forall x_1, \ldots, x_n : \pi_{s_1}(x_1) \wedge \ldots \wedge \pi_{s_n}(x_n) \to t = u) \in Sen_{FOL^=}(OP, P_S)$, i.e., we actually have a natural transformation $\alpha : Sen_{MSEL} \Rightarrow \Phi; Sen_{FOL^=}$. Thirdly, the introduction of typing predicates allows to extract a Σ-algebra A out of any (OP, P_S)-structure M where $A_s \stackrel{def}{=} \{m \mid \pi_s^M(m)\}$ and $op^A : A_{s_1} \times \cdots \times A_{s_n} \to A_s$ is the corresponding restriction of $op^M : M \times \cdots \times M \to M$. In general, we obtain by this procedure partial operations op^A, so that we have to restrict the translation to those (OP, P_S)-structures M that represent total Σ-algebras, i.e., structures M satisfying the set of additional axioms $\emptyset'_\Sigma = \{\forall x_1, \ldots, x_n : \pi_{s_1}(x_1) \wedge \ldots \wedge \pi_{s_n}(x_n) \longrightarrow \pi_s(op(x_1, \ldots, x_n)) | op \in OP\}$. This gives rise to a functor $\Psi : \mathsf{Sign}_{MSEL} \to \mathsf{Th}_{FOL^=}$ and a natural transformation $\beta : \Psi^{op}; Mod_{FOL^= \models} \Rightarrow Mod_{MSEL}$. Note, finally, that the typing premise $\pi_{s_1}(x_1) \wedge \ldots \wedge \pi_{s_n}(x_n)$ in $\alpha(\Sigma)(\varphi)$ ensures that all first order representatives of an algebra A will satisfy $\alpha(\Sigma)(\varphi)$ if A satisfies φ. The implication into the other direction would be valid even if we omit the typing premise. □

According to the definition of generalized model functors and the assumption $\Psi; sign' = \Phi$ we have

Corollary 1. *Let be given* $\mathcal{I}$, Φ, *and* Ψ *as in definition 4. Then there is a natural inclusion* $in_\Psi : \Psi^{op}; Mod'_{\models'} \Rightarrow \Phi; Mod' : \mathsf{Sign}^{op} \to \mathsf{Cat}$. □

Using the functor $\Phi : \mathsf{Sign} \to \mathsf{Sign}'$ and the natural inclusion in_Ψ we obtain according to proposition 2 and proposition 4 the *(semantically) restricted reindexed institution*

$$\mathcal{I}'_{\restriction \Phi, in_\Psi} = (\mathsf{Sign}, \Phi; Sen', \Psi^{op}; Mod'_{\models'}, \models'_{\Phi, in_\Psi}),$$

where $\models'_{\Phi, in_\Psi, \Sigma} = \models'_{\Phi(\Sigma)} \cap |Mod'_{\models'}(\Psi(\Sigma))| \times Sen'(\Phi(\Sigma))$, since pre-images w.r.t inclusions are actually intersections. The institution $\mathcal{I}'_{\restriction \Phi, in_\Psi}$ fixes the result c cutting out those models of $\mathcal{I}'_{\restriction\Phi}$ that can be related to models in $\mathcal{I}$. As an immediate consequence of the definition of $\mathcal{I}'_{\restriction\Phi,in_\Psi}$ we have

Corollary 2. $(\Phi, id_{\Phi;Sen}, in_\Psi) : \mathcal{I}'_{\restriction\Phi,in_\Psi} \to \mathcal{I}'$ *is an institution transformation.*

The simple map condition includes universal quantification on $|Mod'_{\models'}(\Psi(\Sigma))|$, and thus it can be seen as relating satisfaction in $\mathcal{I}$ and $\mathcal{I}'_{\restriction\Phi,in_\Psi}$ in a plain way.

Proposition 8. $(\Phi, \Psi, \alpha, \beta) : \mathcal{I} \to \mathcal{I}'$ *is a simple map of institutions*

(a) iff $(id_{Sign}, \alpha, \beta) : \mathcal{I} \to \mathcal{I}'_{\restriction\Phi,in_\Psi}$ *is a plain map of institutions*
(b) iff $(id_{Sign}, \alpha, \beta) : \mathcal{I}'_{\restriction\Phi,in_\Psi} \to \mathcal{I}$ *is an institution morphism.* □

The span characterization of institution morphisms in theorem 1 provides now for any simple map $(\Phi, \Psi, \alpha, \beta) : \mathcal{I} \to \mathcal{I}'$ an intermediate institution

$$\mathcal{I}^m \overset{def}{=} \mathcal{I}_{\restriction\beta} = \mathcal{I}'_{\restriction\Phi,in_\Psi,\alpha} = (\mathsf{Sign}, Sen, \Psi^{op}; Mod'_{\models'}, \models^m)$$

and a span of institution transformations

$$(id_{Sign}, id_{Sen}, \beta) : \mathcal{I}^m \to \mathcal{I} \quad \text{and} \quad (\Phi, \alpha, in_\Psi) : \mathcal{I}^m \to \mathcal{I}'.$$

In contrast to plain maps we can not fully characterize simple maps by the corresponding span of institution transformations, since only the *intermediate model functor* $\Psi^{op}; Mod'_{\models'} : \mathsf{Sign}^{op} \to \mathsf{Cat}$ and not the *axiomatization* $\Psi : \mathsf{Sign} \to \mathsf{Th}'$ of this functor can be reconstructed from the span.

The application of simple maps within system specifications is the same as for plain maps. By $\mathcal{I}^m = \mathcal{I}_{\restriction\beta} = \mathcal{I}'_{\restriction\Phi,in_\Psi,\alpha}$ we have the old syntax of $\mathcal{I}$ with new models borrowed from $\mathcal{I}'$. The crucial difference is that the borrowing is not complete, i.e., the institution transformation $(\Phi, \alpha, in_\Psi) : \mathcal{I}^m \to \mathcal{I}'$ is not surjective on models. The good point, however, is that the codomains of in_Ψ can be axiomatized within $\mathcal{I}'$, thus we have still a proper basis for building structured specifications with mixed syntax: Structure specifications in an institution $\mathcal{I}$ are usually modeled as diagrams of theories (specifications) and theory morphism in Th. Using $(\Phi, \alpha, in_\Psi) : \mathcal{I}^m \to \mathcal{I}'$ we can now relate theories in Th^m and Th, where Th^m denotes the category of theories derived from the intermediate institution $\mathcal{I}^m = \mathcal{I}_{\restriction\beta} = \mathcal{I}'_{\restriction\Phi,in_\Psi,\alpha}$. Th is a subcategory of Th^m with $|\mathsf{Th}| = |\mathsf{Th}^m|$ where Th becomes a proper subcategory of Th^m if $\beta : \Psi^{op}; Mod'_{\models'} \Rightarrow Mod$ is non-surjective. Any natural transformation $\alpha : Sen \Rightarrow \Phi; Sen'$ allows to extend the functor $\Psi : \mathsf{Sign} \to \mathsf{Th}'$ to a functor $\Psi_\alpha : \mathsf{Th} \to \mathsf{Th}'$ by defining $\Psi_\alpha(\Sigma, \Gamma) \overset{def}{=} (\Phi(\Sigma), \alpha(\Sigma)(\Gamma) \cup \emptyset'_\Sigma)$. The simple map condition ensures that this also defines a functor $\Psi_\alpha : \mathsf{Th}^m \to \mathsf{Th}'$ where we have $(\Psi^{op}; Mod'_{\models'})_{\models^m} = \Psi_\alpha^{op}; Mod'_{\models'}$ for the generalized model functor $(\Psi^{op}; Mod'_{\models'})_{\models^m} : \mathsf{Th}^m \to \mathsf{Cat}$ of $\mathcal{I}^m$. Note, that the equation makes explicit that the semantical effects of the partial model translation β can be syntactically coded by Ψ and α.

Structured specifications with mixed syntax can be given, now, by theories in Th^m inTh' and morphisms in Th^m and Th', respectively, together with morphisms $(\phi, \models') : \Psi_\alpha(\Sigma, \Gamma) \to (\Sigma', \Gamma')$ relating specifications of different syntax (compare the *extra theory morphisms* in [3]).

Example 3 (Continued). The intermediate institution indicated by the simple map $(\Phi, \Psi, \alpha, \beta) : MSEL \to FOL^=$ is $\mathcal{I}^m = (\mathsf{Sign}_{MSEL}, Sen_{MSEL}, \Psi^{op}; Mod_{FOL^=\models}, \models^m)$. It can be proved that $\beta : \Psi^{op}; Mod_{FOL^=\models} \Rightarrow Mod_{MSEL}$ is a natural surjection such that $\mathcal{I}^m = MSEL_{\restriction\beta}$ can be seen as a semantical multiple of $MSEL$, i.e., any algebra in $MSEL$ is now represented by possibly infinite many first order structures. $\Phi : \mathsf{Sign}_{MSEL} \to \mathsf{Sign}_{FOL^=}$ is injective and $\alpha : Sen_{MSEL} \Rightarrow \Phi; Sen_{FOL^=}$ is a natural injection thus $\mathcal{I}^m = FOL^=_{\restriction\Phi, in_\Psi, \alpha}$ can be considered as a semantical *and* a syntactical subinstitution of $FOL^=$. □

Remark 2. Note, that functors $\Psi : \mathsf{Sign} \to \mathsf{Th}'$ and the corresponding construction of the institution $\mathcal{I}'_{\restriction\Phi, in_\Psi}$ appear in practice if we want to use signatures with "syntactic sugar" as, e.g., attributes for operation symbols like total, associative, and so on. Sign is the category of the intended signatures with "syntactic sugar", $\Psi : \mathsf{Sign} \to \mathsf{Th}'$ resolves the "syntactic sugar", and $\mathcal{I}'_{\restriction\Phi, in_\Psi}$ is the institution $\mathcal{I}'$ but now equiped with "sugared signatures". This technique is. e.g., used in [10] to define the institution underlying the specification language CASL, i.e., the institution of coherent order-sorted algebras enriched with sort contraints. In this case $\mathcal{I}'$ is the institution of many-sorted partial first order logic with equality, Sign is the category of subsorted signatures, and Ψ translates relations between sorts into extra operation symbols and conditional equations. □

In cases where the logic of $\mathcal{I}'$ is not strong enough to axiomatize the intermediate model functor we can use the notion of *simulation* [2] that is based directly on a partial translation of models. Note, that we could force such a situation in example 3 by chosing EL instead of $FOL^=$.

Definition 5 ((Weak) Simulation, [2]). *Let $\mathcal{I}$ and $\mathcal{I}'$ be institutions. A* weak simulation *$(\Phi, \alpha, \beta_p) : \mathcal{I} \to \mathcal{I}'$ is given by a functor $\Phi : \mathsf{Sign} \to \mathsf{Sign}'$, a natural transformation $\alpha : Sen \Rightarrow \Phi; Sen' : \mathsf{Sign} \to \mathsf{Set}$, and a* partial natural transformation *$\beta_p : \Phi^{op}; Mod' \Rightarrow Mod : \mathsf{Sign}^{op} \to \mathsf{Cat}$, i.e., a family of functors $\beta(\Sigma) : dom(\beta_p(\Sigma)) \to Mod(\Sigma)$ with $dom(\beta_p(\Sigma) \subseteq Mod'(\Phi(\Sigma))$ such that for any $\phi : \Sigma_1 \to \Sigma_2$ in Sign*

(a) $Mod'(\Phi(\phi))(dom(\beta_p(\Sigma_2))) \subseteq dom(\beta_p(\Sigma_1))$, and
(b) $dom(\beta_p(\phi)); \beta(\Sigma_1) = \beta(\Sigma_2); Mod(\phi)$

for the corresponding restriction $dom(\beta_p(\phi)) : dom(\beta_p(\Sigma_2)) \to dom(\beta_p(\Sigma_1))$ of the functor $Mod'(\Phi(\phi)) : Mod'(\Phi(\Sigma_2)) \to Mod'(\Phi(\Sigma_1))$ that is well-defined by condition (a). Moreover, for each $\Sigma \in |\mathsf{Sign}|$ the simulation condition,

$$\beta(\Sigma)(M') \models_\Sigma \varphi \iff M' \models'_{\Phi(\Sigma)} \alpha(\Sigma)(\varphi)$$

holds for any $M' \in |dom(\beta_p(\Sigma)|$ and $\varphi \in Sen(\Sigma)$.

(Φ, α, β_p) *is called a* simulation *if additionally* $\beta(\Sigma) : dom(\beta_p(\Sigma)) \to Mod(\Sigma)$ *is surjective on* $|Mod(\Sigma)|$ *for each* $\Sigma \in |\mathsf{Sign}|$. □

The information assumed about the partial natural transformation above can be presented in a more structured way.

Corollary 3 (Partial natural transformation). *A partial natural transformation* $\beta_p : \Phi^{op}; Mod' \Rightarrow Mod : \mathsf{Sign}^{op} \to \mathsf{Cat}$ *as in definition 5 can be equivalently described by*

- *a functor* $dom(\beta_p) : \mathsf{Sign}^{op} \to \mathsf{Cat}$,
- *a natural transformation* $\beta : dom(\beta_p) \Rightarrow Mod : \mathsf{Sign}^{op} \to \mathsf{Cat}$, *and*
- *a natural inclusion* $in : dom(\beta_p) \Rightarrow \Phi^{op}; Mod' : \mathsf{Sign}^{op} \to \mathsf{Cat}$. □

The absence of an axiomatization of the intermediate model functor $dom(\beta_p) : \mathsf{Sign}^{op} \to \mathsf{Cat}$ via a functor $\Psi : \mathsf{Sign} \to \mathsf{Th}'$ is the only difference between weak simulations and simple maps. Hence, we can directly assume the corresponding versions of corollary 2 and proposition 8 for simulations. The span characterization, however, applies to weak simulation right because there is no axiomatization to be taken into account.

Theorem 2 (Intermediate institution).

1. *Let* $(\Phi, \alpha, \beta_p) : \mathcal{I} \to \mathcal{I}'$ *be a weak simulation. Then there is an institution* $\mathcal{I}^m = (\mathsf{Sign}, Sen, dom(\beta_p), \models^m)$, *such that the two triples*
$$(id_{Sign}, id_{Sen}, \beta) : \mathcal{I}^m \to \mathcal{I} \quad \text{and} \quad (\Phi, \alpha, in) : \mathcal{I}^m \to \mathcal{I}',$$
define institution transformations.
2. *Vice-versa, given an institution* $\mathcal{I}^m = (\mathsf{Sign}, Sen, dom(\beta_p), \models^m)$ *and institution transformations*
$$(id_{Sign}, id_{Sen}, \beta) : \mathcal{I}^m \to \mathcal{I} \quad \text{and} \quad (\Phi, \alpha, in) : \mathcal{I}^m \to \mathcal{I}',$$
with $in : dom(\beta_p) \Rightarrow \Phi^{op}; Mod' : \mathsf{Sign}^{op} \to \mathsf{Cat}$ *a natural inclusion, we can define a partial natural transfromation* $\beta_p : \Phi^{op}; Mod' \Rightarrow Mod : \mathsf{Sign}^{op} \to \mathsf{Cat}$, *such that the triple* $(\Phi, \alpha, \beta_p) : \mathcal{I} \to \mathcal{I}'$ *defines a weak simulation.*

Proof. Follows directly from proposition 8 and theorem 1 where we have to take into account that the institution transformation $(\Phi, \alpha, in) : \mathcal{I}^m \to \mathcal{I}'$ can be decomposed into $(id_{Sign}, \alpha, id_{dom(\beta_p)}) : \mathcal{I}^m \to \mathcal{I}'_{\restriction \Phi, in}$ and $(\Phi, id_{\Phi;Sen'}, in)\mathcal{I}'_{\restriction \Phi, in} \to \mathcal{I}'$. □

To make the paper complete to a certain point we will finally state how general maps of institutions [8] could be described (see [6] for details).

Proposition 9. *Let be given institutions* $\mathcal{I}$, $\mathcal{I}'$ *and functors* $\Phi : \mathsf{Sign} \to \mathsf{Sign}'$, $\Upsilon : \mathsf{Th} \to \mathsf{Th}'$ *such that* $sign; \Phi = \Upsilon; sign'$. *Then* $(\Phi, \Upsilon, \alpha, \beta) : \mathcal{I} \to \mathcal{I}'$ *defines a map of institutions in the sense of [8] iff* $((inj; \Upsilon)_\alpha)^{op}; Mod'_{\models'} = \Upsilon^{op}; Mod'_{\models'}$ *and* $(\Phi, inj; \Upsilon, \alpha, \beta) : \mathcal{I} \to \mathcal{I}'$ *is a simple map of institutions in the sense of definition 4.* □

Note, that $(inj;\Upsilon)_\alpha$ and Υ are different. A map is just a simple map with an additional translation of theories. This additional translation, however, does not matter so much since its semantical effects, and thus also its logical effects, were already described by the simple map. In such a way there is nothing to say additionally concerning general maps in view of the considerations and results of this paper.

5 Concluding Remarks

The concept of institution is a very abstract formalization of the notion of a logical system. Nevertheless, it became very helpful during the last decade to present logical systems in a uniform way, to develop a general theory of structured and modular specifications, to describe and investigate translations of logics into universal frameworks, and so on.

Driven by different application needs, concepts of arrows between institutions are often formulated in a very involving technical way. By keeping the abstract level of consideration, we were able to give a well-structured presentation of these concepts and of their relations. The crucial observation was that any translation of signatures, sentences, or models, respectively, gives rise to the construction of a new institution. Based on these constructions we were able to describe any arrow as an arrow between institutions of the same scheme, i.e., with the same category of signatures. Thus at this abstract level, institution morphisms and plain maps appear as dual concepts. Further, we have seen that any arrow can be characterized by the existence of an intermediate institution.

We hope that the achieved clarification of the nature of the considered concepts of arrows will provide a better basis for a general understanding and a broader use of these important concepts. Moreover, we think that the concepts and results of this paper should provide an appropriate conceptual level for a concise and well-structured development and presentation of all the results concerning programming in the huge, multi-paradigm languages, combination of logics, and so on.

Acknowledgment: We are indebted to the anonymous referees of this paper for the very helpful criticisms. They gave us a chance to make a significant revision, and hopefully also to achieve a considerable improvement in the presentation as well.

References

1. M . Arrais and J. L. Fiadeiro. Unifying theories in different institutions. In *Recen[t] Trends in Type Specification*, pages 81–101. Springer, LNCS 1130, 1996.
2. Maura Cerioli. *Relationships between Logical Formalisms*. PhD thesis, Universit[à] di Pisa–Genova–Udine, 1993. TD-4/93.

3. R. Diaconescu. Extra-theory morphisms in institutions: logical semantics for multi-paradigm languages. *Journal Applied Categorical Structures*, 1998. To appear.
4. R. Diaconescu and K Futatsugi. Logical semantics for CafeOBJ. In *Precise Semantics for Software Modeling Techniques*, pages 31–54, 1998. TU Muenchen, Technical Report TUM-19803.
5. J. A. Goguen and R. M. Burstall. Institutions: Abstract Model Theory for Specification and Programming. *Journal of the ACM*, 39(1):95–146, January 1992.
6. A. Martini. *Unifying Arrows between Institutions*. PhD thesis, Technical University of Berlin, 1998. To appear.
7. A. Martini and U. Wolter. A systematic study of mappings between institutions. In *Recent Trends in Algebraic Development Techniques*, pages 300–315. Springer, LNCS 1376, 1998.
8. J. Meseguer. General logics. In H.-D. Ebbinghaus et. al., editor, *Logic colloquium '87*, pages 275–329. Elsevier Science Publishers B. V.,North Holland, 1989.
9. J. Meseguer and N. Martin-Olliet. From abstract data types to logical frameworks. In *Recent Trends in Data Type Specification*, pages 48–80. Springer, LNCS 906, 1995.
10. T. Mossakowski. Translating *OBJ3* into *CASL*. 1998. Submitted to WADT'98.
11. T. Mossakowski, A. Tarlecki, and W. Pawlowski. Combining and representing logical systems using model-theoretic parchments. In *Recent Trends in Algebraic Development Techniques*, pages 349–364. Springer, LNCS 1376, 1998.
12. S. Salibra and G. Scollo. Interpolation and compactness in categories of pre-institutions. *Mathematical Structures in Computer Science*, 6:261–286, 1996.
13. A. Sernadas, C. Sernadas, and C. Caleiro. Synchronization of logics. *Studia Logica*, 59(2):217–247, 1997.
14. A. Sernadas, C. Sernadas, and C. Caleiro. Synchronization of logics with mixed rules: Completeness preservation. In *Algebraic Methodology and Software Technology - AMAST*, pages 465–478. Springer, LNCS 1349, 1997.
15. A. Tarlecki. Moving between logical systems. In *Recent Trends in Data Type Specification*, pages 478–502. Springer, LNCS 1130, 1996.

On Oracles for Interpreting Test Results against Algebraic Specifications

Patrícia D. L. Machado*

LFCS, Division of Informatics, University of Edinburgh, JCMB, King's Buildings, Edinburgh EH9 3JZ, UK pdlm@dcs.ed.ac.uk

Abstract. This paper presents a solution to the so-called oracle problem for oracles derived from flat algebraic specifications expressed in first-order logic. The oracle problem in this context reduces to the general problem of comparing two values of a non-observable sort and also to the limitations imposed by quantifiers. The solution is based on constructing an oracle where equality between values of a non-observable sort is computed by one of two "approximate" equalities, according to the context in which the equation occurs. The interpretation given by the oracle does not make any assumptions on test sets and depends on how these equalities approximate the behavioural equality, which is based on behavioural theories, and also on the use of quantifiers.

Keywords. specification based-testing, algebraic specification, formal verification, oracle, correctness.

1 Introduction

A considerable number of works in the area of specification-based testing have pointed out that testing can be successfully used in the formal development of software. The aim is to derive test suites from a specification so that testing can be applied whenever formal proofs are not cost-effective. Moreover, the combination of formal proofs and testing can help to produce high integrity systems in a cost-effective way [7, 8]. However, a great effort is still needed in order to have testing as a standard activity in formal frameworks. For instance, the accurate interpretation of testing results seems to be a crucial point.

The process of testing programs consists not only in submitting various combinations of input values to them but also giving a coherent interpretation to the results produced by them when receiving these values. Along with selecting test data sets, specialised mechanisms to assist the testing process, widely called *oracles*, may also be carefully planned. However, the literature on specification based testing has mainly concentrated on refining test sets. Some exceptions are the works of Bernot [1] and Gaudel [6] in the context of positive conditional specifications and Le Gall and Arnould [5] which gives an interpretation by using institutions of observable results obtained from dynamic testing. Also, Dick

* Supported by CNPq - Brazilian Research Council (Proc. 200651/95-1) and on leave from DSC, Federal University of Paraíba, Brazil

A.M. Haeberer (Ed.): AMAST'98, LNCS 1548, pp. 502–518, 1998.

and Faivre [3] and Richardson et al [11] pointed out that formal specifications can be used as oracles for software testing. Although some theoretical issues about oracles have already been considered, a feasible method has not come out yet. Furthermore, the reasons for defining automated oracles are quite obvious: human oracles are usually imprecise and error-prone while automated oracles derived from consistent and unambiguous sources can lead to more efficiency, feasibility and reliability in the testing process. Also, any efforts to refine test sets can be useless without having an effective and accurate oracle, but we show in this paper that oracles can be defined independently from test sets.

Here we concentrate on generating oracles from algebraic specifications expressed in first-order logic rather than either refining test sets or defining levels of correctness, although some hints on these are given insofar as it is quite unreasonable to investigate one of these subjects without considering the others. We do not only stay in the field of pure specification-based testing, but also consider "white-box" techniques. More precisely, we are concerned with the *oracle problem*, that is, whether there is a decidable oracle for interpreting the results of a testing experiment. As Gaudel [6] pointed out, the oracle problem often reduces to the more general problem of comparing values of a non-observable sort which make oracles undecidable in general. However, the use of quantifiers in specifications can make the oracle problem even more difficult than in the context of positive conditional specifications which have been investigated so far [1, 6, 5], as infinite test sets may be required. We aim at giving a solution to this problem, that is, given a specification of programming task, how an oracle and, more precisely, an equality procedure for non-observable sorts may be constructed and what can be concluded about the correctness of the program being tested.

The paper is organised as follows. Section 2 presents a formal definition of oracle and test set along with a general correctness theorem. Section 3 approaches the oracle problem and presents some proposals on how to define equality for non-observable sorts. Section 4 introduces the idea of an approximate equality and oracle and the main important theorems of the paper which underlie the method of generating oracles proposed in section 5. Section 6 presents an application of this method to the unification algorithm. Finally, some concluding remarks along with pointers for further work is given in section 7. In the sequel, we assume the reader is familiar with general concepts of algebraic specifications [13]. Extended ML [9] (EML) is used in the examples, but the notation is mostly self-explanatory. Let $SP = \langle \Sigma, \Phi \rangle$ be a specification, where $\Sigma = (S, F)$ is a signature and Φ is a set of axioms, and let $T_\Sigma(X)$ be the Σ-term algebra, where X is an S-indexed set of countable infinite sets of variables. For any two Σ-terms t and t' of the same sort, $t = t'$ is a Σ-equation and first-order Σ-formulas are built from Σ-equations, logical connectives ($\neg$,$\wedge$,$\vee$,$\Rightarrow$,$\Leftrightarrow$) and quantifiers ($\forall$,$\exists$). A Σ-formula without free variables is called a Σ-sentence. Also, let $Var(t)$ denote the set of variables occurring in a Σ-term t. For any Σ-algebra A and valuation $v : X \to |A|$, there exists a unique Σ-homomorphism $v^{\#} : T_\Sigma(X) \to A$ which extends v. The value of $t \in |T_\Sigma(X)|_s$ in A under v is $v^{\#}(t) \in |A|_s$, where $s \in S$.

If $t \in T_\Sigma$, that is, t is a ground Σ-term, the value of t in A is $^\#(t)$, where $\# : T_\Sigma \to A$ is unique.

2 Basic Concepts

When verifying a program against a specification, oracles should check the results obtained from tests against exactly what was specified for each function. The best way of verifying a function, which is the procedure adopted by formal proofs, is to check if it satisfies its axioms rather than comparing the values it produces with pre-defined targets. So, testing a function may not be different if we would like to reach similar conclusions regarding correctness. Definitions 1 and 3 follow this idea and give our view of what a test set and an oracle for a specification are. These definitions differ from the standard ones in the sense that test sets are defined as a subset of the operations' domain in the standard theory of testing and as a subset of the carrier sets of an algebra in definition 1, that is, test sets are defined for each operation in the former and for a whole algebra in the latter. Moreover, the standard definition of oracle is based on the standard practice of testing where each operation is individually executed for all values of a test set and the results are compared with the target ones, while definition 3 is a specific and automated way of interpreting the general notion of behavioural satisfaction of Σ-formulas [2, 12] and the results are verified rather than compared with targets. The reason for this is that in order to formally verify an implementation, all axioms are proved without being sorted out rather than treating individual functions separately. As we are considering behavioural satisfaction, test sets should only have values which can be expressed by a ground term in T_Σ.

Definition 1 (*A*-test set). *Let A be a Σ-algebra. $T = \{T_s\}_{s \in S}$ is an A-test set if $T \subseteq |^\#(T_\Sigma)|$, i.e., $T_s \subseteq |^\#(T_\Sigma)|_s$ for all $s \in S$.*

Let a *behavioural equality* be a partial congruence relation $\approx_A = (\approx_{A,s})_{s \in S}$, that is, a symmetric and transitive relation on A which is compatible[1] with Σ [2]. A behavioural equality is *total* if it is reflexive, that is, for all a in A, $a \approx_A a$ (the subscript s is omitted for the sake of simplicity). Also, for any set Obs of observable sorts, $\approx_{Obs,A}$ will denote an observational equality with respect to Obs, a special case of the behavioural equality where related elements are those that cannot be distinguished by observable computations.[2] In the sequel, the family $\approx = (\approx_A)_{A \in Alg(\Sigma)}$ will denote a Σ-behavioural equality.

[1] $\forall f : s_1 \ldots s_n \to s; a_i, b_i : A_{s_i} \cdot a_i \approx_{A,s_i} b_i \Rightarrow f^A(a_1 \ldots a_n) \approx_{A,s} f^A(b_1 \ldots b_n)$

[2] Let C_{Obs} be the set of all Σ-contexts $T_\Sigma(X \cup z_s)$ of observable sorts where z_s is a context variable. Values a and b of a non-observable sort s are observationally equal $a \approx_{Obs,A} b$ iff they cannot be distinguished by observable contexts, i.e., $\forall C \in C_{Obs} \forall \alpha : X \to |A| \cdot \alpha_a^\#(C) = \alpha_b^\#(C)$, where $\alpha^\# : T_\Sigma(X) \to |A|$, and $\alpha_a, \alpha_b : X \cup z_s \to |A|$ are the unique extensions of α defined by $\alpha_a(z_s) = a$ and $\alpha_b(z_s) = b$.

Definition 2. *Let A be a Σ-algebra, let T be an A-test set, and let $\approx$ be a Σ-behavioural equality. The interpretation of a first-order Σ-formula under a valuation $\alpha : X \to |A|$ is given as follows.*

1. $[\![t = t']\!]^{\approx,T}_{\alpha} \stackrel{def}{=} [\![t]\!]_{\alpha} \approx_A [\![t']\!]_{\alpha}$
2. $[\![\neg\psi]\!]^{\approx,T}_{\alpha} \stackrel{def}{=} \neg[\![\psi]\!]^{\approx,T}_{\alpha}$
3. $[\![\psi_1 \odot \psi_2]\!]^{\approx,T}_{\alpha} \stackrel{def}{=} [\![\psi_1]\!]^{\approx,T}_{\alpha} \odot [\![\psi_2]\!]^{\approx,T}_{\alpha}$, *where* $\odot = \{\wedge, \vee\}$
4. $[\![\psi_1 \Rightarrow \psi_2]\!]^{\approx,T}_{\alpha} \stackrel{def}{=} [\![\psi_1]\!]^{\approx,T}_{\alpha} \Rightarrow [\![\psi_2]\!]^{\approx,T}_{\alpha}$
5. $[\![\psi_1 \Leftrightarrow \psi_2]\!]^{\approx,T}_{\alpha} \stackrel{def}{=} [\![\psi_1 \Rightarrow \psi_2]\!]^{\approx,T}_{\alpha} \wedge [\![\psi_2 \Rightarrow \psi_1]\!]^{\approx,T}_{\alpha}$
6. $[\![\forall x : s \cdot \psi]\!]^{\approx,T}_{\alpha} \stackrel{def}{=} \bigwedge_{v \in T_s} [\![\psi]\!]^{\approx,T}_{\alpha[x \mapsto v]}$
7. $[\![\exists x : s \cdot \psi]\!]^{\approx,T}_{\alpha} \stackrel{def}{=} \bigvee_{v \in T_s} [\![\psi]\!]^{\approx,T}_{\alpha[x \mapsto v]}$

where $\alpha[x \mapsto v]$ denotes the valuation α superceded at x by v.

When Ψ is a Σ-sentence, then $[\![\Psi]\!]^{\approx,T}_{\alpha} = [\![\Psi]\!]^{\approx,T}_{\beta}$, for any valuations α and β. So, we write $[\![\Psi]\!]^{\approx,T}$ without the subscript.

Definition 3 (SP-Oracle). *Let $\Phi = \{\phi_1, \phi_2, \ldots, \phi_n\}$ be the set of axioms of a flat specification SP, where each ϕ_i is a first-order Σ-sentence, let $\approx$ be a Σ-behavioural equality, let A be a Σ-algebra, and let T be an A-test set. A SP-*oracle *O for verifying whether A conforms to SP on T is as follows.*

$$O(T) = [\![\phi_1 \wedge \phi_2 \wedge \ldots \phi_n]\!]^{\approx,T}$$

Example 1 (SP-oracle). A SP-oracle for a stack specification S with the usual operations and axioms is given by the *o_S* function below where axioms are grouped to test all functions at the same time. The test set consists of a list of stacks and a list of integers. The *o_forall* function which implements the $\forall$ quantifier is also given below.

```
fun o_forall [] pred = true |
    o_forall (x::xs) pred = (pred x) andalso o_forall xs pred;

fun o_S (ls,ln) = o_forall ls (fn s => o_forall ln (fn n =>
    (top(push(n,s)) = n) andalso (pop(push(n,s)) == s) andalso
    (is_empty(push(n,s)) = false) andalso (is_empty(empty) = true)))
```

where "==" is a specificational equality construction in EML, which makes the *o_S* function non-executable, unless an explicity equality on *stacks* is defined. Section 3 addresses this problem. □

Exhaustive test sets which exercise a program with all possible combinations of values are usually infinite and some refinements must be applied to make them finite. Test sets may be also valid and/or unbiased [1, 6] in the following sense.

Definition 4 (Exhaustive, Valid and Unbiased A-test sets). *Let O be a SP-oracle, T be an A-test set and $D = |^{\#}(T_{\Sigma})|$.*

- D *is an* exhaustive A-test set;
- T *is a* valid A-test set *iff* $O(T) = true$ implies $O(D) = true$;
- T *is an* unbiased A-test set *iff* $O(D) = true$ implies $O(T) = true$.

Any $T \subseteq D$ is always unbiased in the context of positive conditional specifications investigated in [6]. Nevertheless, T is not always unbiased in general. For example, if Φ has only existential quantifiers then $O(T) = true \Rightarrow O(D) = true$ for any T but the converse may not hold.

Theorem 1 (Correctness). *Let $\approx$ be a Σ-behavioural equality, such that $\forall v \cdot v \approx v \Rightarrow v \in {}^{\#}T_\Sigma$, let T be an A-test set and let O be an SP-oracle where the standard equality is interpreted as $\approx$. If T is valid and unbiased, then $A \models_\approx \Phi$ iff $O(T) = true$, where $\models_\approx$ is the behavioural satisfaction relation w.r.t. $\approx$.*

Proof. Follows from definitions 3 and 4 and the fact that $\approx$ is reflexive on values of ground terms by the compatibility condition. □

3 The Oracle Problem

In this section, we introduce some approaches to tackle the oracle problem. Inspired by work on behavioural theories where equality between two elements of an algebra is interpreted by the behavioural equality [2], we focus on comparing two values of a non-observable sort, along with the limitations imposed by quantifiers which are also central to the oracle problem.

Revisiting example 1, it can be noticed that o_S is not computable because the evaluation of "==" is undefined. Also, the arguments ls and ln are likely to be infinite due to the $\forall$ quantifier. One way of verifying "==" is to interpret it up to observational equivalence, that is, "==" can be interpreted by the observational equality $\approx_{Obs}$ for an appropriate choice of Obs. However, this does not solve the problem raised by the quantifiers and also an oracle which interprets equality up to observational equivalence should check all observable contexts which is likely to lead to an infinite process. Gaudel [6] argued that "oracle hypotheses" may be applied in order to reduce the number of observable contexts and restricted the use of observable contexts as oracle only to those positive conditional specifications where all equations in preconditions are of observable sorts. The reason for this restriction was that when considering only a subset of observable contexts to interpret equality, an equality can be erroneously found to be satisfied and if this equality is in the premises of a conditional axiom then the axiom can be erroneously found not to be satisfied, leading to the rejection of correct programs. However, even if this restriction is applied, incorrect programs are likely to be accepted, which is not a crucial mistake since testing has been regarded as not suitable for proving correctness, that is, the complete absence of errors, but for showing the presence of errors [4].

The problem that arises here is how $\approx$ can be derived from a finite set of observable contexts. As this will not always be possible [6, 5], we suggest that an equality $\sim$ which approximates $\approx$ may be defined instead. For this, following [2] we define a "lifted" signature $\mathcal{L}(\Sigma) \stackrel{\text{def}}{=} \Sigma \cup \{\sim_s: s\ s\}_{s \in S}$ where $\sim_s$ is a predicate defining equality between two terms of sort s, introducing an explicit denotation for $\approx_{A,s}$. From a Σ-algebra A we define a lifted $\mathcal{L}(\Sigma)$-algebra $\mathcal{L}(A)$ satisfying the property that it is the unique extension of A defined by $\mathcal{L}(A)|_\Sigma \stackrel{\text{def}}{=} A$ and for any $s \in S$, $\sim_s^{\mathcal{L}(A)} \stackrel{\text{def}}{=} \approx_{A,s}$. If ϕ is a Σ-sentence, $\mathcal{L}(\phi)$ is its lifted version generated by substituting "$\sim$" for "$=$" in ϕ. Then, $\mathcal{L}(SP) = \langle \mathcal{L}(\Sigma), \mathcal{L}(\Phi) \rangle$ is the lifted version of SP, where $\mathcal{L}(\Phi) = \{\mathcal{L}(\phi_1), \ldots \mathcal{L}(\phi_n)\}$. A lifted version of a SP-oracle O is a $\mathcal{L}(SP)$-oracle named $\mathcal{L}(O)$. The idea is to test $\mathcal{L}(A)$ against $\mathcal{L}(SP)$ using $\mathcal{L}(O)$ instead of A against SP insofar as it might not be possible to test A against SP because O is not executable. From [2], $A \models_\approx \phi$ if and only if $\mathcal{L}(A) \models \mathcal{L}(\phi)$, where $\models_\approx$ is the behavioural satisfaction relation.

One may consider three possible ways of defining $\sim_s$ for non-observable sorts. For instance, the result of comparing two values $v = v'$ of sort s in a Σ-algebra A may depend on the internal and concrete implementation of s. Then, one way of defining $\sim_s$ which is named the *white-box approach* relies on defining a congruence relation by using details about the concrete representation of s. This approach does not consist in explicitly specifying and implementing equality for all new sorts, because the equality computation may only be necessary at testing time. The idea is that the implementation under test remain unchanged while the lifted version incorporates the explicit definitions of equality which avoids introducing bugs in the programs.

Another way of defining $\sim_s$ which is called the *black-box approach* consists in using observable contexts provided by the abstract formal specification. This approach was first investigated by Bernot [1] and Gaudel [6] in the context of positive conditional specifications and by Le Gall and Arnould [5]. We propose that a finite axiomatisation of the observational equality may be derived from a method which is based on the guidelines and results obtained by Bidoit and Hennicker [2] in order to prove the behavioural validity of first-order formulas. This method can be applied in the context of first-order specifications including quantifiers. Basically, an appropriate finite subset of observable contexts, namely the so-called *crucial* ones, is chosen and if this does not lead to an adequate axiomatisation of the equality, than the initial specification is incremented with a hidden part. For lack of space, the full method is not presented here. This approach would be appropriate in most of the general cases where the chosen set of observable contexts is descriptive enough to avoid defining a hidden part. However, when a hidden part is necessary, it may be difficult to find an adequate axiomatization of $\approx$, because defining a hidden part usually relies on intuition and clever choices.

Whenever neither the white-box approach nor the black-box approach can be successfully applied, a *grey-box approach* which is the result of combining them can be used instead. Basically, the grey-box approach consists in using two approximate equalities – one derived by using the black-box approach and the other derived by the white-box approach – which are not necessarily congruence relations but, together, are enough to solve the equality problem in oracle definitions under certain restrictions. Depending on the contexts where these equalities are used, either the oracle interpretation may imply that the program being tested is correct or whenever the program is correct the oracle interpretation may also say it is correct. In other words, this means that the oracle may either reject correct programs or accept incorrect programs respectively. However, despite these restrictions, approximate equalities are more likely to work in practice, since giving a decidable definition of a particular congruence relation may be either impossible or too complicated and require a great amount of effort. The grey-box approach is presented in section 5. However, before going into details about this approach, section 4 describes more precisely what an approximate equality is, how it can be used to define an "approximate oracle" and how quantifiers can influence the choice of the appropriate equality.

4 The Approximate Oracle

This section presents some important definitions and results which underlie the grey box-approach and introduces the definition of an approximate oracle. Initially, we define what an approximate equality is. In the sequel, let any Σ-algebra A be reachable and any $\approx$ be a *total* Σ-behavioural equality.

Definition 5 (Approximate equality). *Let A be a Σ-algebra. A binary relation $\doteq_A$ on A is called an* approximate equality.

It would have been reasonable to request $\doteq_A$ to be reflexive in definition 5. However, this assumption is not necessary to the results presented in this paper. In the sequel, $\doteq$ will denote a Σ-approximate equality. When compared with the behavioural equality, an approximate equality may be classified as sound if all values that it identifies are indeed equal, or complete if all equal values are identified. In general, the white-box approach will produce sound equalities while the black-box will produce complete equalities.

Definition 6 (Sound Equality). *Let $\doteq_A$ be an approximate equality. Then, $\doteq_A$ is a* sound equality *if and only if $\forall a, a' \cdot a \doteq_A a' \Rightarrow a \approx_A a'$*

Definition 7 (Complete Equality). *Let $\doteq_A$ be an approximate equality. Then, $\doteq_A$ is a* complete equality *if and only if $\forall a, a' \cdot a \approx_A a' \Rightarrow a \doteq_A a'$*

Sound or complete equalities can be successfully applied instead of the behavioural equality in specific contexts. These contexts are regarded as equality occurrences in a Σ-formula which might be either positive or negative.

Definition 8 (Equality occurrences). *The set of occurrences of Σ-equations in a first-order Σ-formula is defined as follows.*

1. $Occ[t = t'] \stackrel{def}{=} \{\epsilon\}$
2. $Occ[\neg\psi] \stackrel{def}{=} \{-1 \cdot \omega \mid \omega \in Occ(\psi)\}$

3. $Occ[\psi_1 \odot \psi_2] \stackrel{def}{=} \{1 \cdot \omega \mid \omega \in Occ(\psi_1)\} \cup \{2 \cdot \omega \mid \omega \in Occ(\psi_2)\}$, *where* $\odot = \{\wedge, \vee\}$
4. $Occ[\psi_1 \Rightarrow \psi_2] \stackrel{def}{=} \{-1 \cdot \omega \mid \omega \in Occ(\psi_1)\} \cup \{2 \cdot \omega \mid \omega \in Occ(\psi_2)\}$
5. $Occ[\psi_1 \Leftrightarrow \psi_2] \stackrel{def}{=} Occ(\psi_1 \Rightarrow \psi_2) \cup Occ(\psi_2 \Rightarrow \psi_1)$
6. $Occ[\mathcal{Q}x\psi] \stackrel{def}{=} \{1 \cdot \omega \mid \omega \in Occ(\psi)\}$, *where* $\mathcal{Q} = \{\forall, \exists\}$

where "-" expresses that the formula is in a negative position. An occurrence in $Occ[\Psi]$ *is* positive *if it has an even number of* -1, *otherwise it is* negative.

Both left and right equality occurrences in $\Leftrightarrow$ can be positive and negative depending on which direction of implication we are looking at.

Definition 9 (Approximate Interpretation). *Let A be a Σ-algebra, T be an A-test set, $\hat{=}$ and $\bar{\bar{=}}$ be two Σ-approximate equalities. The approximate interpretation of a first-order Σ-formula under a valuation $\alpha : X \rightarrow |A|$ is given as follows.*

1. $[\![t = t']\!]_\alpha^{\hat{=},\bar{\bar{=}},T} \stackrel{def}{=} [\![t]\!]_\alpha \hat{=}_A [\![t']\!]_\alpha$
2. $[\![\neg\psi]\!]_\alpha^{\hat{=},\bar{\bar{=}},T} \stackrel{def}{=} \neg[\![\psi]\!]_\alpha^{\bar{\bar{=}},\hat{=},T}$
3. $[\![\psi_1 \odot \psi_2]\!]_\alpha^{\hat{=},\bar{\bar{=}},T} \stackrel{def}{=} [\![\psi_1]\!]_\alpha^{\hat{=},\bar{\bar{=}},T} \odot [\![\psi_2]\!]_\alpha^{\hat{=},\bar{\bar{=}},T}$, *where* $\odot = \{\wedge, \vee\}$
4. $[\![\psi_1 \Rightarrow \psi_2]\!]_\alpha^{\hat{=},\bar{\bar{=}},T} \stackrel{def}{=} [\![\psi_1]\!]_\alpha^{\bar{\bar{=}},\hat{=},T} \Rightarrow [\![\psi_2]\!]_\alpha^{\hat{=},\bar{\bar{=}},T}$
5. $[\![\psi_1 \Leftrightarrow \psi_2]\!]_\alpha^{\hat{=},\bar{\bar{=}},T} \stackrel{def}{=} [\![\psi_1 \Rightarrow \psi_2]\!]_\alpha^{\hat{=},\bar{\bar{=}},T} \wedge [\![\psi_2 \Rightarrow \psi_1]\!]_\alpha^{\hat{=},\bar{\bar{=}},T}$
6. $[\![\forall x : s \cdot \psi]\!]_\alpha^{\hat{=},\bar{\bar{=}},T} \stackrel{def}{=} \bigwedge_{v \in T_s} [\![\psi]\!]_{\alpha[x \mapsto v]}^{\hat{=},\bar{\bar{=}},T}$
7. $[\![\exists x : s \cdot \psi]\!]_\alpha^{\hat{=},\bar{\bar{=}},T} \stackrel{def}{=} \bigvee_{v \in T_s} [\![\psi]\!]_{\alpha[x \mapsto v]}^{\hat{=},\bar{\bar{=}},T}$

where $\alpha[x \mapsto v]$ denotes the valuation α superceded at x by v.

Again, when Ψ is a Σ-sentence, then $[\![\Psi]\!]_\alpha^{\hat{=},\bar{\bar{=}},T} = [\![\Psi]\!]_\beta^{\hat{=},\bar{\bar{=}},T}$, for any valuations α and β. So, we write $[\![\Psi]\!]^{\hat{=},\bar{\bar{=}},T}$ without the subscript.

The approximate interpretation is similar to the *behavioural interpretation* from [2] which we refer as $[\![]\!]_\alpha$. The main differences are that $[\![t = t']\!]_\alpha \stackrel{def}{=} [\![t]\!]_\alpha \approx_A [\![t']\!]_\alpha$ is replaced by $[\![t = t']\!]_\alpha^{\hat{=},\bar{\bar{=}},T} \stackrel{def}{=} [\![t]\!]_\alpha \hat{=}_A [\![t']\!]_\alpha$ and the roles of $\hat{=}$ and $\bar{\bar{=}}$ are reversed when interpreting a formula in a negative position. Also, bound variables have their domains restricted to T.

Lemma 1. *Let Ψ be a Σ-formula. The interpretation $[\![\Psi]\!]_\alpha^{\hat{=},\bar{\bar{=}},T}$ applies $\hat{=}$ only to positive occurrences of equations and $\bar{\bar{=}}$ only to negative occurrences of equations.*

Proof. Follows directly from definitions 8 and 9. □

Definition 10 (Approximate SP-Oracle). *Let $\Phi = \{\phi_1, \phi_2, \ldots, \phi_n\}$ be the set of axioms of a flat specification SP, where each ϕ_i is a first-order Σ-sentence, let $\hat{=}$ and $\bar{\bar{=}}$ be two Σ-approximate equalities, let A be a Σ-algebra and let T be an A-test set. An approximate oracle O is given as follows.*

$$O(T) = [\![\phi_1 \wedge \ldots \wedge \phi_n]\!]_\alpha^{\hat{=},\bar{\bar{=}},T}$$

The main results of this section are presented by the following theorems.

Theorem 2. *Let $\eqsim$ be a complete equality, $\simeq$ be a sound equality and Φ be a Σ-formula. If Φ contains only positive occurrences of $\forall$ and negative occurrences of $\exists$ then $[\![\Phi]\!]_\alpha \Rightarrow [\![\Phi]\!]_\alpha^{\eqsim,\simeq,T}$.*

Proof. Let Φ be normalised to a Σ-formula Ψ by applying the following laws:

- $\neg(P \wedge Q)$ is equivalent to $\neg P \vee \neg Q$ and $\neg(P \vee Q)$ is equivalent to $\neg P \wedge \neg Q$
- $P \Rightarrow Q$ is equivalent to $\neg P \vee Q$
- $\neg\exists x \cdot P(x)$ is equivalent to $\forall x \cdot \neg P(x)$ and $\neg\forall x \cdot P(x)$ is equivalent to $\exists x \cdot \neg P(x)$
- $\neg\neg P$ is equivalent to P
- $P \Leftrightarrow Q$ is equivalent to $(P \Rightarrow Q) \wedge (Q \Rightarrow P)$

such that for every sub-formula $\neg\psi$ of Ψ, ψ is an equation. Any positive occurrence of an equation ψ in Φ remains positive in Ψ and each negative occurrence remains negative in Ψ. Also, regarding quantifiers, Ψ has only positive $\forall$. The proof is conducted by structural induction on Ψ.

(*i*) $\Psi \equiv t = t'$. In this case, $[\![t = t']\!]_\alpha \stackrel{\text{def}}{=} [\![t]\!]_\alpha \approx_A [\![t']\!]_\alpha$ and $[\![t = t']\!]_\alpha^{\eqsim,\simeq,T} \stackrel{\text{def}}{=} [\![t]\!]_\alpha \eqsim_A [\![t']\!]_\alpha$. As $\eqsim$ is complete then $[\![t]\!]_\alpha \approx_A [\![t']\!]_\alpha \Rightarrow [\![t]\!]_\alpha \eqsim_A [\![t']\!]_\alpha$.

(*ii*) $\Psi \equiv \neg\psi$. In this case, $[\![\neg\psi]\!]_\alpha \stackrel{\text{def}}{=} \neg[\![\psi]\!]_\alpha$ and $[\![\neg\psi]\!]_\alpha^{\eqsim,\simeq,T} \stackrel{\text{def}}{=} \neg[\![\psi]\!]_\alpha^{\simeq,\eqsim,T}$. By the contrapositive law, $\neg[\![\psi]\!]_\alpha \Rightarrow \neg[\![\psi]\!]_\alpha^{\simeq,\eqsim,T}$ iff $[\![\psi]\!]_\alpha^{\simeq,\eqsim,T} \Rightarrow [\![\psi]\!]_\alpha$. As ψ is an equation $t = t'$ and $\simeq$ is a sound equality then $[\![t]\!]_\alpha \simeq_A [\![t']\!]_\alpha \Rightarrow [\![t]\!]_\alpha \approx_A [\![t']\!]_\alpha$. Hence, $[\![\psi]\!]_\alpha^{\simeq,\eqsim,T} \Rightarrow [\![\psi]\!]_\alpha$.

(*iii*) $\Psi \equiv \psi_1 \odot \psi_2$. In this case, $[\![\psi_1 \odot \psi_2]\!]_\alpha^{\eqsim,\simeq,T} \stackrel{\text{def}}{=} [\![\psi_1]\!]_\alpha^{\eqsim,\simeq,T} \odot [\![\psi_2]\!]_\alpha^{\eqsim,\simeq,T}$ and $[\![\psi_1 \odot \psi_2]\!]_\alpha \stackrel{\text{def}}{=} [\![\psi_1]\!]_\alpha \odot [\![\psi_2]\!]_\alpha$, where $\odot = \{\wedge, \vee\}$. By induction hypothesis, $[\![\psi_1]\!]_\alpha \Rightarrow [\![\psi_1]\!]_\alpha^{\eqsim,\simeq,T}$ and $[\![\psi_2]\!]_\alpha \Rightarrow [\![\psi_2]\!]_\alpha^{\eqsim,\simeq,T}$. Thus, because $\odot$ has the same interpretation in both, then $[\![\psi_1 \odot \psi_2]\!]_\alpha \Rightarrow [\![\psi_1 \odot \psi_2]\!]_\alpha^{\eqsim,\simeq,T}$

(*iv*) $\Psi \equiv \psi_1 \Rightarrow \psi_2$. This case need not be considered as all occurrences of $\Rightarrow$ in Φ are replaced by using the conditional law.

(*v*) $\Psi \equiv \psi_1 \Leftrightarrow \psi_2$. This case need not be considered as all occurrences of $\Leftrightarrow$ in Φ are replaced by using the biconditional law.

(*vi*) $\Psi \equiv \forall x : s \cdot \psi$. In this case, $[\![\forall x : s \cdot \psi]\!]_\alpha^{\eqsim,\simeq,T} \stackrel{\text{def}}{=} \bigwedge_{v \in T_s} [\![\psi]\!]_{\alpha[x \mapsto v]}^{\eqsim,\simeq,T}$ and $[\![\forall x : s \cdot \psi]\!]_\alpha \stackrel{\text{def}}{=} \bigwedge_{v \in |A_s|} [\![\psi]\!]_{\alpha[x \mapsto v]}$. By induction hypothesis, $[\![\psi]\!]_{\alpha[x \mapsto v]} \Rightarrow [\![\psi]\!]_{\alpha[x \mapsto v]}^{\eqsim,\simeq,T}$ and as $T \subseteq |A|$ by definitions 1 and 9 then $[\![\forall x : s \cdot \psi]\!]_\alpha \Rightarrow [\![\forall x : s \cdot \psi]\!]_\alpha^{\eqsim,\simeq,T}$.

(*vii*) $\Psi \equiv \exists x : s \cdot \psi$. This case need not be considered as only negative occurrences of $\exists$ are possible and these are converted by quantifier laws. □

Theorem 2 is a generalisation of the results obtained by Gaudel [6] for positive conditional specifications with conditions of observable sorts. There, the only negative occurrences of equations are in the conditions and the restriction to observable equations is so that the computational equality, which is sound and complete, can be used. The black-box equality, which is complete but need not be sound, is used in the conclusion, which is a positive occurrence. Finally, the only quantifiers in conditional equations are universal quantifiers outermost, that is in positive position. As in the context investigated by Gaudel, theorem 2, which covers a prevalent use of ∀ and ∃, implies that incorrect programs can be accepted by an approximate oracle. On the other hand, theorem 3 given below which is the dual of 2 implies that correct programs can be rejected.

Theorem 3. *Let* $\simeq$ *be a sound equality,* $\eqsim$ *be a complete equality and* Φ *be a* Σ*-formula. If* Φ *contains only negative occurrences of* ∀ *and positive occurrences of* ∃ *then* $[\![\Phi]\!]_\alpha^{\simeq,\eqsim,T} \Rightarrow [\![\Phi]\!]_\alpha$.

If only finitary algebras are considered, no restrictions on quantifiers are needed.

Theorem 4. *Let* $\eqsim$ *be a complete equality,* $\simeq$ *be a sound equality and* Φ *be a* Σ*-formula. If* A *is a finitary* Σ*-algebra then* $[\![\Phi]\!]_\alpha \Rightarrow [\![\Phi]\!]^{\eqsim,\simeq,|A|}$.

Proof. Follows the same line of the proof of theorem 2, having that ∀ and ∃ has the same interpretation in $[\![\Phi]\!]_\alpha$ and $[\![\Phi]\!]_\alpha^{\eqsim,\simeq,|A|}$. □

Theorem 5. *Let* $\simeq$ *be a sound equality,* $\eqsim$ *be a complete equality and* Φ *be a* Σ*-formula. If* A *is a finitary* Σ*-algebra then* $[\![\Phi]\!]^{\simeq,\eqsim,|A|} \Rightarrow [\![\Phi]\!]_\alpha$.

If T is finite, valid and unbiased, no restrictions on quantifiers are necessary as well. Nevertheless, these may be quite strong assumptions, not straightforward to be fulfilled in practice. The fact that theorems 2, 3, 4 and 5 do not have assumptions on test sets brings out a way of interpreting testing results without relying on whether the test set is valid, unbiased, or whatever. These assumptions on test sets would mainly replace the assumptions on quantifiers in theorems 2 and 3, but the conclusions of these theorems would remain the same.

5 The Grey-Box Approach

This approach is based on combining the white-box and black-box approaches in order to define an approximate oracle, where the former may be used to produce a sound equality $\simeq$ and the second may be used to produce a complete equality $\eqsim$. It can be noticed that equalities defined from a subset of the set of all observable contexts is always complete. Thus, the black-box approach can be applied to find an axiomatization over the crucial contexts[3] as defined by Bidoit and Hennicker [2]. The equality induced by these contexts either coincides with the observational equality or is a complete approximate equality.

Sound and complete equalities approximate the behavioural equality from opposite extremes ($\simeq \subseteq \approx \subseteq \eqsim$). Whenever the black-box equality is only complete, it is reasonable to define a sound equality by using the white-box approach. Structural equalities based on the behavioural equality of the values of the concrete representation of a sort s, even though not always complete, are always sound w.r.t. a Σ-algebra A.

The grey-box approach can be applied with one of the following objectives.

[3] $f(x_1 \ldots z_{s_k} \ldots x_n)$, with $f : s_1 \ldots s_k \ldots s_n \to s \in F$, $s \in Obs$ and $s_k \in S \setminus Obs$

(i) Avoid rejecting correct programs. Based on theorems 2 and 4, this approach consists in applying a sound equality in negative occurrences of Σ-equations and a complete equality in positive occurrences.

(ii) Avoid accepting incorrect programs. Based on theorems 3 and 5, this approach consists in applying the complete equality in negative occurrences of a Σ-equations and the sound equality in positive occurrences.

Depending on the approach used, different conclusions about correctness and incorrectness can be achieved. Let $\models_{\approx}$ be the behavioural satisfaction relation.

Theorem 6 (Incorrectness). *Let O be an approximate oracle. If $\forall\alpha \cdot [\![\phi_1 \wedge \ldots \wedge \phi_n]\!]_\alpha \Rightarrow [\![\phi_1 \wedge \ldots \wedge \phi_n]\!]_\alpha^{\eqsim,\simeq,T}$, then $O(T) = false$ implies $A \not\models_{\approx} \Phi$.*

Proof. As $O(T) = false$, there exists α such that $[\![\phi_1 \wedge \ldots \wedge \phi_n]\!]_\alpha^{\eqsim,\simeq,T}$ does not hold. Thus, because by assumption $[\![\phi_1 \wedge \ldots \wedge \phi_n]\!]_\alpha \Rightarrow [\![\phi_1 \wedge \ldots \wedge \phi_n]\!]_\alpha^{\eqsim,\simeq,T}$, $[\![\phi_1 \wedge \ldots \wedge \phi_n]\!]_\alpha$ does not hold as well. □

Theorem 6 states that whenever testing is not successful in the approach (i) we can conclude that the program is incorrect. However, it is easy to check that the converse does not hold: if testing is successful we cannot conclude that the program is correct. On the other hand, theorem 7 states that whenever testing is successful in the approach (ii) we can conclude that the program is correct, but once again the converse does not hold.

Theorem 7 (Correctness). *Let O be an approximate oracle. If $\forall\alpha \cdot [\![\phi_1 \wedge \ldots \wedge \phi_n]\!]_\alpha^{\simeq,\eqsim,T} \Rightarrow [\![\phi_1 \wedge \ldots \wedge \phi_n]\!]_\alpha$, then $O(T) = true$ implies $A \models_{\approx} \Phi$.*

Proof. As $O(T) = true$, then $\forall\alpha : X \to |A| \cdot [\![\phi_1 \wedge \ldots \wedge \phi_n]\!]_\alpha^{\simeq,\eqsim,T}$ holds. Thus, because by assumption $[\![\phi_1 \wedge \ldots \wedge \phi_n]\!]_\alpha^{\simeq,\eqsim,T} \Rightarrow [\![\phi_1 \wedge \ldots \wedge \phi_n]\!]_\alpha$, $[\![\phi_1 \wedge \ldots \wedge \phi_n]\!]_\alpha$ holds as well. Hence, $A \models_{\approx} \Phi$. □

Approach (i) is the classical one to testing whereas approach (ii) comes up with an innovation in the testing theory and practice. If the test succeed in approach (i) the program might be incorrect which is expected as an intrinsic feature of testing. Also, even if the test does not succeed in approach (ii), the program might be correct. Whenever the premises of theorems 2 or 4 and of their respective duals are satisfied, these approaches can be combined in order to achieve a higher degree of confidence in the interpretation given by the oracle. For example, if the test fails in approach (i) and (ii), the program is incorrect while if it succeeds in approach(i) and (ii), then the program is correct.

6 Example

This section presents an example in EML which applies the grey-box approach to the unification problem. Unification, which plays a central role in theorem proving, is the process of finding a common instance of two expressions, and if such an instance exists, the algorithm produces the most general substitution which yields it. The specification presented here is inspired by the one given in [10]. Initially, a specification of expressions is given as follows. An expression can be a constant, a variable or a function application to a list of expressions.

```
signature Func = sig eqtype function; val arity: function → int; ... end

signature Expression =
 sig
  structure S : Set;
  structure F : Func;
  eqtype constant;
  eqtype variable;
  type expression;
  val const : constant → expression;
  val var : variable → expression;
  val func : F.function * expression list → expression;
  val is_valid : expression → bool;
   axiom (forall c ⇒ is_valid(const(c))) andalso (forall v ⇒ is_valid(var(v)))
   axiom forall (f,l) ⇒ is_valid(func(f,l)) = (F.arity(f) = length(l))
  val vars : expression → variable S.set;
  val varsl : expression list → variable S.set;
   axiom forall c ⇒ vars(const(c)) == S.emptyset
   axiom forall v ⇒ vars(var(v)) == S.add(v,S.emptyset)
   axiom forall (f,l) ⇒ vars(func(f,l)) == varsl(l)
   axiom varsl([]) == S.emptyset
   axiom forall (a,x) ⇒ varsl((a::x)) == S.union(vars(a),varsl(x));
end;
```

where *length* returns the size of a list and *Set* specifies sets with usual operations. *Substitution* is an operation that replaces specific variables of an expression by other expressions.

```
signature Substitution =
 sig
  include Expression;
  type substitution;
  val empty : substitution;
  val dom : substitution → variable S.set;
  val rng : substitution → variable S.set;
  val add : variable * expression * substitution → substitution;
   axiom (dom(empty) == S.emptyset) andalso (rng(empty) == S.emptyset)
   axiom forall (s,x,e) ⇒
      ((S.member(x,dom(s)) = false) andalso ((e == var(x)) = false)) implies
         ( (dom(add(x,e,s)) == S.add(x,dom(s))) andalso
           (rng(add(x,e,s)) == S.union(vars(e),rng(s))) )
  val apply : expression * substitution → expression;
  val applyl : expression list * substitution → expression list;
   axiom forall e ⇒ apply(e,empty) == e
```

```
  axiom forall (c,s) ⇒ apply(const(c),s) == const(c)
  axiom forall (x,e,s) ⇒ apply(var(x),add(x,e,s)) == e
  axiom forall (x,y,e,s) ⇒
     (x<>y) implies (apply(var(y),add(x,e,s)) == apply(var(y),s))
  axiom forall (f,l,s) ⇒ apply(func(f,l),s) == func(f,applyl(l,s))
  axiom forall s ⇒ applyl([],s) == []
  axiom forall (e,l,s) ⇒ applyl((e::l),s) == apply(e,s)::applyl(l,s)
 val compose : substitution * substitution → substitution;
  axiom forall (s,s',e)⇒ apply(e,compose(s,s')) == apply(apply(e,s),s')
  axiom forall s ⇒
     (compose(s,empty) == s) andalso (compose(empty,s) == s)
  axiom forall (x,y,z)⇒
     compose(compose(x,y),z) == compose(x,compose(y,z))
 val is_moregeneral : substitution * substitution → bool;
  axiom forall (s,s') ⇒
     is_moregeneral(s,s') iff (exists r ⇒ s' == compose(s,r))
 val is_idempotent : substitution → bool;
  axiom forall s ⇒
     is_idempotent(s) iff S.intersect(dom(s),rng(s)) == S.emptyset;
end;
```

The unification specification is given as follows in the *Unification* signature.

```
signature Unification =
 sig
  include Substitution;
  val unify : expression * expression → substitution;
   axiom forall (e,e') ⇒ (exists s ⇒ apply(e,s) == apply(e',s)) implies
        ( (apply(e,unify(e,e')) == apply(e',unify(e,e'))) andalso
          (forall s' ⇒ apply(e,s') == apply(e',s')
                implies is_moregeneral(unify(e,e'),s')) andalso
          (is_idempotent(unify(e,e'))) )
end;
```

Let *bool* and *constant* be observable sorts. An oracle for the *Unification* signature, and more precisely for the *unify* function needs to compute equality on *expression*, which is taken, at first instance, as a non-observable sort. The black-box approach is not appropriate to directly define an observational equality on *expression*, because its crucial contexts ($isvalid(z_{exp})$,$vars(z_{exp})$) are not descriptive enough and then hidden functions must be added. Thus, the grey-box approach is a better choice. It can be noticed that the unify axiom has only positive *forall* and negative *exists*. Then, to apply theorem 2, we define a complete equality on expressions by using the black-box approach and a sound equality by using the white-box approach, and this implies from theorem 6 that we will only be able to detect incorrectness of the implementation. A lifted version of an implementation of the signature *Expression* is given as follows, where the equality on expressions is implemented by *eq_s* which is clearly a structural sound equality.

```
functor LExpression
   (eqtype Constant eqtype Variable structure set: Set structure f : Func) :
 sig include Expression;
    val eq_s : expression * expression → bool;
    sharing S = set and F = function and
           type constant = Constant and type variable = Variable end =
 struct
   ...
  datatype expression =
     const of constant | var of variable | func of F.function * (expression list)
  ...
  fun eq_s (const(c),const(c')) = c = c' |
      eq_s (var(v),var(v')) = v = v' |
      eq_s (func(f,l),func(f',l')) = (f = f') andalso eq_sl(l,l')
  and eq_sl ([],[]) = true |
      eq_sl ((e::l),(e'::l')) = eq_s(e,e') andalso eq_sl(l,l') |
      eq_sl (l,l') = false
end;
```

For sake of simplicity, *is_valid*, *vars* and *varsl* are omitted. Obviously, different *eq_s* may be defined from other implementations of *Expression*. The complete equality which is defined from the finite set of crucial contexts of *expression* is as follows.

$$\forall e, e' \cdot eq_c(e, e') = (is\_valid(e) = is\_valid(e')) \wedge (vars(e) = vars(e'))$$

Finally, the oracle function for the unification algorithm can be as follows.

```
fun o_unify (le,ls) = o_forall le (fn e ⇒ o_forall le (fn e' ⇒
  o_exists ls (fn s ⇒ eq_s(apply(e,s),apply(e',s))) implies
   ( (eq_c(apply(e,unify(e,e')),apply(e',unify(e,e')))) andalso
     (o_forall ls (fn s' ⇒ eq_s(apply(e,s'),apply(e',s')) implies
         is_moregeneral(unify(e,e'),s'))) andalso
    is_idempotent(unify(e,e')) ) ))
```

Now, consider how an oracle for *Substitution* can be defined. Let *substitution* and *expression* be non-observable sorts. The crucial contexts on *substitution* are $dom(z_{subst})$, $rng(z_{subst})$, $is\_moregeneral(z_{subst}, s)$ and $idempotent(z_{subst})$. However, it is clear that substitutions are equal if they produce the same result when applied to all expressions. As *expression* is a non-observable sort, *apply* cannot be used as a crucial context. Thus, as the set of crucial contexts is not enough to construct a sound and complete equality, the grey-box approach seems to be the best choice again. Consider the axiom which specifies the *is_moregeneral* operation. This axiom can be converted to the following by expanding iff.

axiom forall $(s,s') \Rightarrow$
(is_moregeneral(s,s') implies (exists $r \Rightarrow s'$ == compose(s,r))) (ϕ)
andalso
((exists $r \Rightarrow s'$ == compose(s,r)) implies is_moregeneral(s,s')) (ρ)

Here we have a positive occurrence of $\forall$ and a positive and a negative occurrence of $\exists$ and then neither theorem 2 nor 3 is applicable for the whole axiom. Whenever quantifiers occur as one side of *iff* predicates these theorems cannot be applied in both directions of the biconditional. The direction which does not meet the premises of the theorem being considered must be discarded. It is easy to check that from the standard interpretation of $\wedge$ and theorem 2, $[\![\phi \wedge \rho]\!]_\alpha \Rightarrow [\![\rho]\!]_\alpha \Rightarrow [\![\rho]\!]_\alpha^{\widetilde{\approx},\simeq,T}$, but $[\![\phi]\!]_\alpha \Rightarrow [\![\phi]\!]_\alpha^{\widetilde{\approx},\simeq,T}$ may not hold, unless either the algebra is finitary (theorem 4) or T is unbiased (due to the positive $\exists$). Also, when considering theorem 3 dual conclusions can be reached. If (ϕ) is added to the *Substitution* oracle, then correct programs can be rejected and incorrect programs can be accepted at the same time. The reason for this is that if $is\_moregeneral(s, s')$ happens to be true, then the test set must have a witness r in order to avoid rejecting a correct program which requires either a finite exhaustive or an unbiased test set. In case $is\_moregeneral(s, s') = false$ and a witness r can be found in the test set, then incorrect programs can be accepted. Hence, only $\forall s, s' \cdot \rho$ is added to the *Substitution* oracle, implying that incorrect programs can be accepted w.r.t. a simplified *Substitution* specification without (ϕ). This reflects the limitations of testing which is aimed at detecting the presence of errors, but often not all of them can be detected. From theorem 2, a sound equality can be used to compute (s' == compose(s,r)).

Finally, suppose *expression* is an observable sort. Then, the black-box approach seems to be a good choice as $apply(e, z_{subst})$ can be added to the set of crucial contexts. After applying the black-box approach, a sound and complete equality on *substitution* can be defined as $\forall s, s' \cdot eq(s, s') = (\forall e(apply(e, s) = apply(e, s')))$. However, it is necessary to check whether this equality is a congruence relation [2]. Otherwise, theorem 2 can help interpreting test results as it does not require *eq* to be a congruence relation.

7 Concluding Remarks

Having defined an SP-oracle in section 2 as a boolean function which is constructed from the conjunction of the axioms of SP, the oracle problem in the context of algebraic specifications expressed in first-order logic reduces to the general problem of finding an equality procedure for non-observable sorts and the limitations imposed by quantifiers. This paper introduces an approach to tackle this problem, the grey-box approach. In this approach, a sound and a complete approximate equality are constructed as close as possible or even equal to the behavioural equality and taking the presence of quantifiers into account. An approximate equality might be neither an equivalence nor a congruence relation and no restriction is made on test sets. So, it is not necessary to define valid and unbiased test sets nor to check whether both equalities are congruence relations. If approximate equalities are defined in a systematic way, they can be guaranteed to be either complete or sound. For instance, it is always possible and simple to define complete equalities from a set of crucial contexts (black-box) and structural equalities are always sound (white-box). Furthermore, the grey-box approach leads to similar levels of confidence when compared to the one presented in [6], but it can be applied in a wider context.

As further work, we aim to define a method for applying the grey-box approach, extend this approach for generating oracles to structured specifications, and to provide a basis for integration testing. The EML framework [9] seems to be an appropriate basis. Also, an investigation of how testing using grey-box oracles would work when used together with refinement and proofs as part of a conventional formal development process seems to be quite important.

8 Acknowledgments

This research has been partially supported by EPSRC grant GR/K63795. The author wishes to thank specially Don Sannella for important discussions and ideas and also the referees for comments which helped to improve the final version.

References

[1] G. Bernot. Testing against formal specifications: a theoretical view. In S. Abramsky and T.S.E. Maibaum, editors, *Proceedings of Theory and Practice of Software Development - TAPSOFT'91*, volume 494 of *Lecture Notes in Computer Science*. Springer, 1991.

[2] M. Bidoit and R. Hennicker. Behavioural theories and the proof of behavioural properties. *Theoretical Computer Science*, 165(1):3–55, 1996.

[3] J. Dick and A. Faivre. Automating the generation and sequencing of test cases from model-based specifications. In *Formal Methods Europe- FME'93*, volume 670 of *Lecture Notes in Computer Science*, pages 268–284. Springer, 1993.

[4] E. W. Dijkstra. Why correctness must be a mathematical concern. In R. S. Boyer and J. S. Moore, editors, *The Correctness Problem in Computer Science*. Academic Press, 1981.

[5] P. Le Gall and A. Arnould. Formal specification and test: Correctness and oracle. In *Recent Trends in Data Type Specification*, volume 1130 of *Lecture Notes in Computer Science*. Springer, 1996.

[6] M. Gaudel. Testing can be formal, too. In Peter D. Mosses, Mogens Nielsen, and Michael I. Schwartzbach, editors, *Proceedings of Theory and Practice of Software Development - TAPSOFT'95*, volume 915 of *Lecture Notes in Computer Science*. Springer, 1995.

[7] J. B. Goodenough and S. L. Gerhart. Towards a theory of test data selection. *IEEE Transactions on Software Engineering*, SE-1(2):156–173, June 1975.

[8] C. A. R. Hoare. How did software get so reliable without proof ? In *Formal Methods Europe - FME'96*, volume 1051 of *Lecture Notes in Computer Science*, pages 1–17. Springer, 1996.

[9] S. Kahrs, D. T. Sannella, and A. Tarlecki. The definition of Extended ML: a gentle introduction. *Theoretical Computer Science*, 173:445–484, 1997.

[10] Z. Manna and R. Waldinger. Deductive synthesis of the unification algorithm. *Science of Computer Programming*, 1:5–48, 1981.

[11] D. J. Richardson, S. L. Aha, and T. O. O'Malley. Specification-based test oracles for reactive systems. In *Proceedings of the Fourteenth International Conference on Software Engineering*, pages 105–118, Melbourne, Australia, May 1992.

[12] D. Sannella and A. Tarlecki. Essential concepts of algebraic specification and program development. *Formal Aspects of Computing*, 9:229–269, 1997.

[13] M. Wirsing. Algebraic specification. In J. van Leeuwen, editor, *Handbook of Theoretical Computer Science*, volume B, chapter 13, pages 675–788. Elsevier Science Publishers, 1990.

RECOPLA: An Extendible Graphic Meta-Editor*

Zoltan Gassmann[1], Luis Mandel[2], and Roshan Sembacuttiaratchy[2]

[1] Ludwig-Maximilians-Universität München.
Oettingenstr. 67, D-80538 Munich, Germany
[2] Forschungsinstitut für Angewandte Software Technologie (FAST e. V.)
Arabellastr. 17, D-81925 Munich, Germany
recopla@fast.de

System Description

Diagrams in general, and graphs in particular, are commonplace for researchers of science. Up to 70% of the papers presented in every Computer Science conference use specialised diagrams or graphs for clarification of ideas and also for demonstrations. Graph theory is a discipline in itself, constantly growing inside Discrete Mathematics. New specialised graph editors are a must for different applications. Examples of such applications in computer science are Petri nets, Pert diagrams, binary trees, Entity Relationship Diagrams, Class diagrams for OO-design, State charts, finite automaton, etc. However, diagrams are used not only in computer science; they are also used in other broad areas such as electrical engineering, chemistry and architecture, of which examples are electrical circuits, chemical reaction chains, molecular models and pipes. Such diagrams or graphs have their own grammar, i.e. their own lexica and their own editing rules. A grammar for binary trees is trivial, but the inner structure of the grammar of a molecular chain could be rather complicated.

The objective of the RECOPLA Meta-Editor is to generate specialised editors for the applications mentioned above. At the Meta-level, it provides graphic editors for nodes and arcs. Both nodes and Arcs can have an unlimited number of typed attributes with or without a default value. Tables are also allowed as attributes. Attributes have names and they can be defined to be visible or non-visible. In the first case they will appear on the canvas showing its value whereas in the latter they will be visible only through the inspector. For visible attributes, the font-family, font-colour, size, alignment, padding and background can be dynamically defined. RECOPLA has a number of distinguished attributes, like URL and File. Each node/arc can have an associated URL, which is used later when generating an HTML of the created diagram. While saving a diagram, a `gif` file will be generated with a picture of the diagram; and also an `html` file will be generated with hyperlinks pointing to the URLs in the nodes and arcs. A Java serialised object (a persistent object) of the edited diagram will also be saved. I.e. any other Java application can read that serialised object and cast it to a RECOPLA diagram and make proper diagram analysis and refinements.

* This project was partially supported by the LMF-DI PUC-RIO, FAST e.V. and by the Institut für Informatik, Ludwig-Maximilians-Universität München.

J.M. Haeberer (Ed.): AMAST'98, LNCS 1548, pp. 519–521, 1998.

A node or an arc could also be be associated with a file, and for each type of file (for each extension) a helper application can be defined using the "Preferences Browser". Double clicking on a node or arc will call the associated application of the File attribute using the value of this attribute as a parameter. That is, if the File value is a Postscript file and the helper for ps extensions is Ghostview, then it will be opened with the corresponding Postscript file.

Arcs can be coloured; different colours for the left and right arrows can be specified and so too for the body. Typical arrows have already been predefined and are modifiable. The body of an arrow can be defined to be drawn with hyphens and the space between hyphens is also adjustable. Nodes can contain any graphic element such as circles, rectangles, lines, images and text. Colours can also be given for any of these elements. Images in `gif` or `jpg` format can be bound to any graphic element contained in a node. The shape of arcs can be defined to be polygonal or curved using a spline interpolator.

In addition to nodes and arcs, some other diagram elements can be specified for a given editor in RECOPLA. They are graphic elements which do not belong to the grammar of this editor. These elements are text for titles, images for logos, and normal graphic elements such as lines, circles and rectangles.

The grammar of a given editor is specified by edition rules using prolog syntax. The edition rules drive the syntax of the editor using an event based model. The predicates to be evaluated are the actions to be performed on objects on the canvas. Examples of these actions are **`canInsertNode`**, **`canConnect`**, **`canDeleteNode`**, **`canDeleteArc`**, **`canDeleteSelection`** and **`canMove`**. Using these actions not only ensures a syntactically correct diagram but also adds control to the layout, i.e. by combining the predicates **`canInsert`** and **`canMove`** one can draw non-overlapping graphs. In the case of an editor for electric circuits the rule **`canConnect`** will ensure a shortcircuit-free diagram.

At user-level RECOPLA has been provided with a toolbar with buttons for the most common operations such as copy, paste, undo, abstract -for the abstraction of a given selection, explode -for the explosion of an abstracted node, a button for the inspector -which shows all the hidden and visible attributes and properties of the selected object-, a tool-box button -linked to the geometry browser which allows the user to change the geometry of the selected node, that is, to change its shape, its position, to rotate it, to flip it, to mirror it and to lock it. RECOPLA provides a grid for the background which is resizable via the grid browser and could be hidden. The editor can be set to snap objects to the grid.

RECOPLA allows editing of hypergraphs, that is, the arcs can not only connect nodes-to-nodes but also arcs-to-nodes and arcs-to-arcs. This feature is very useful for UML editors. RECOPLA also allows the edition of hierarchical graphs of which the nodes also contain graphs. The user has to select the part of the graph to be abstracted and click the correspondent option. A nodes browser will be opened showing the nodes available to be selected for the abstraction. By clicking in one of them, the selected part of the diagram will be replaced by the selected node in the nodes browser. Abstracted nodes can be exploded. By exploding a node a new editor will be opened with the contents of that node for further edition

Deleting an abstracted node will delete only the abstraction and not the contents of the node. That is, the abstracted node will be replaced by its content.

Programs can be bound to RECOPLA generated editors. In this way, graph animation can be achieved, as well as pretty printing, routing, minimal paths.

When generating an editor by giving the nodes, arcs, graphic elements and rules, RECOPLA will automatically generates an `html` page with the documentation of the current editor with hyperlinks to `html` pages containing descriptions of the nodes and arcs included in the editor.

The architecture of RECOPLA is done in three layers: the graphical interface - where the user can draw a diagram using a drag-and-drop paradigm, the underlying semantics of the graphic editor - given by the edition rules, and finally the persistence of the saved diagrams. RECOPLA diagrams will be saved as serialized objects and can then be saved in a repository using a CORBA interface.

RECOPLA is written in Java which means it can run under any platform. Generated editors can be copied to other machines and other platforms (along with all other required files), and will be able to run without having to be recompiled, due to the cross-platform nature of Java.

A new approach for diagram grammars is under development. The idea is to add to RECOPLA the possibility of specifying the grammar of the graph in an abstract way using relational grammars (see [BMHZ]), instead of the event based model -which is more appropriated for context-free diagrams. This new approach is thought to allow the edition of context-dependent diagrams allowing the specification on a high level the rules for inserting, deleting, moving etc. of nodes and arcs. The grammar should be confluent, but also non confluent grammar specifications are supported.

As an ongoing project, RECOPLA generated editors are being ported to run as signed applets inside web browsers for Intranet applications. It is planned to incorporate scaling and zooming capabilities into RECOPLA. Semantic rules will be added in order to generate code direct from diagrams. If the diagram represents a finite automata then its equivalent regular expression can be generated. If the graph is a class diagram, different semantic rules could be written in order to generate java code, C++ code, etc. Multiple windows will be added, with the ability to copy and paste among these.

RECOPLA generated editors were successfully used for animated petri nets, class and time diagrams among others.

References

[BMHZ] Bernd Meyer and Hubert Zweckstetter. Interpretation of Visual Notations in the RECOPLA Editor Generator. University of Munich. Proceedings of the AAAI 1998 Fall Symposium Formalizing Reasoning with Visual and Diagrammatic Representations Orlando, Florida.

[ZGA] Zoltan Gassmann. A Graphic Meta-Editor for the Generation of Syntax-Oriented Graph Editors. Diplomarbeit. University of Munich. 1998.

The State of PEP*

Bernd Grahlmann

Universität Oldenburg, FB 10, Postfach 2503, D-26111 Oldenburg
grahlmann@informatik.uni-oldenburg.de,
http://theoretica.informatik.uni-oldenburg.de/~bernd

Abstract. The PEP tool can be considered to be one of the most widely distributed Petri net based tools. A continuously increasing functionality and an adequate graphical user interface may have been good reasons for its acceptance. Currently the tool contains approximately 500,000 lines of source code, and supports (to the best of our knowledge) the widest variety of input formalisms and verification methods of all verification tools. We briefly review the most recent developments.
Ftp-able versions of the tool and PEP related papers are available via http://theoretica.informatik.uni-oldenburg.de/~pep.
Keywords: 3D visualisation, C code generation, Parallel programs, PEP, Petri nets, SDL, Simulation, Verification.

1 New Petri Net Generators

One of the key features of the PEP tool [3] is that simulation, analysis and verification of different input formalisms

- SDL (Specification Description Language [6]) systems,
- parallel programs written in $B(PN)^2$ [4],
- parallel finite automata (PFA) [11],
- process algebra terms expressed in the PBC [1],
- high-level (HL) Petri nets in the M-nets algebra [2], and
- low-level (LL) Petri nets [1]

are based on Petri net theory. PEP's existing transformers between different formalisms (PFA $\Rightarrow$ $B(PN)^2$, $B(PN)^2$ $\Rightarrow$ PBC, PBC $\Rightarrow$ LL net, $B(PN)^2$ $\Rightarrow$ HL net, and HL net $\Rightarrow$ LL net) were accomplished with two new compilers.

An HL Petri net semantics of SDL [10] was implemented. In particular, dynamic creation as well as termination of processes and (also recursive) procedures are covered. Its integration in PEP supports almost the same simulation, analysis and verification methods as there are available for $B(PN)^2$ programs:

- An SDL simulation may be triggered by a Petri net simulation.
- Standard analyses, such as deadlock detection are possible.

* PEP has been a joint project between the 'Universität Hildesheim', the 'Carl-von-Ossietzky Universität Oldenburg' and the 'Humboldt-Universität zu Berlin' which has been financed by the DFG (German Research Foundation).

A.M. Haeberer (Ed.): AMAST'98, LNCS 1548, pp. 522–526, 1998.

- CTL and LTL properties of SDL specifications may be checked. The user may, e.g., ask whether there exists an instance of a certain process Sender, such that whenever this instance is in state wait while its input queue is non-empty, then it may reach a state send while its variable X has value 3: EXISTS *id* IDSET(Sender):

$$\mathsf{AG}\Big(\big(\mathsf{Sender}[id].\mathtt{state} = \mathsf{wait} \wedge \mathsf{Sender}[id].\mathsf{noqueue} > 0\big) \Rightarrow \big(\mathsf{EF}(\mathsf{Sender}[id].\mathtt{state} = \mathsf{send} \wedge \mathsf{Sender}[id].\mathsf{X} = 3)\big)\Big).$$

  Such a formula is transparently tranformed into a Petri net formula, which is then checked against the Petri net semantics. A resulting counter example may finally be simulated in the SDL editor.

In order to improve the efficiency and to support hybrid modelling of parallel systems we added a compiler from PFA into M-nets [14]. A collection of finite automata together with a number of variable declarations is compiled in a $B(PN)^2$ specific or SDL specific way into an HL Petri net.

2 New HL Petri Net Editing Facilities

In addition to the usual net editing facilities, the HL Petri net editor of the PEP tool now supports a 'programming with nets' approach.

1. The user may compose nets in parallel, in sequence, or in choice; transitions may be substituted by nets; and nets may be synchronised.
2. Moreover, parameterised macro nets (e.g., variable, channel, or procedure nets) may be loaded from libraries and be instantiated (e.g., with the name or type of a variable). Likewise, program semantics may be constructed.

3 New HL Petri Net Simulation Improvements

The performance of the HL Petri net simulator was increased:

- More elaborated analyses of the inscriptions as well as different caching mechanisms give a speed up.
- Petri net simulation may now be distributed over a network of computers (with possibly different operating systems).

4 New 3D Visualisation Component

Simulation plays a major role within the PEP tool. Based on a Petri net simulation engine, not only Petri nets but also, e.g., parallel programs can be simulated using a dedicated reference scheme [13,15].

In particular for non Petri net experts, an interactive 3D visualisation of the modelled parallel system (e.g., an elevator) is more appropriate than a Petri net simulation. Thus, we included such an additional Java based feature [19].

We have chosen to base this visualisation on VRML (Virtual Reality Modelling Language). Thus, the user models, e.g., an elevator as a 3D VRML world using an arbitrary VRML editor. After this, the PEP tool offers the possibility to define certain graphical actions (e.g., that the door of the elevator is closed, the elevator moves up one floor, and the door is opened) on top of these worlds, and to bind such actions either directly to transition occurrences of the corresponding Petri net or to action occurrences of the corresponding parallel program. This is done in such an abstract way, that (in general) these bindings are automatically updated upon changes of the corresponding program (and thus resulting net).

The new component may be used in three complementary ways:

1. A random Petri net simulation which triggers a 3D visualisation may exhibit errors (even to non-experts) if, e.g., an elevator moves with an open door.
2. Test scenarios may be created and checked easily interacting with the 3D visualisation (e.g, mouse clicks on elevator buttons may guide the simulation).
3. Transition sequences resulting from a verification may be used to trigger the simulation of a 3D elevator visualising specification errors to non-experts.

5 New Code Generation

Jäger [17] extended the C code generator for the parallel programming language $B(PN)^2$, which uses the parallelism of the operating system, to the handling of procedures. POSIX conformity and thus operating system and compiler independency was improved. Thus, PEP now supports generation of executable code.

6 New Verification Facilities

PEP's verification component (a partial order based model checker [8,9,16], an interface to the BDD based CTL model checker SMV [7,23], and an interface to the structural analysis component of INA [22]) was extended and improved:

- We speeded up (by a factor of up to more than 100) the program verification with the SMV verifier using information provided by the different net generators. Compared with the original approach, we are, e.g., able to replace different variables which correspond to control flow points by one single variable using invariant information calculated in almost zero time [15].
- INA was extended by a CTL model checker which performs state graph analysis supporting reductions based on stubborn sets and symmetries.
- Spin [18] was integrated [12] including editing facilities for LTL formulae and a graphical interface to the options of the LTL model checker which is almost identical to the XSPIN interface. Based on an automatic translation of Petri nets into PROMELA, the user may now use the SPIN verifier in a transparent way for the verification of all kinds of input formalisms of PEP

- We integrated more algorithms based on linear programming:
  - a deadlock checker using net unfoldings [21] and
  - an extended reachability analyser which inputs the LL net [20].
- Moreover, we implemented a compiler from Petri nets into networks of communicating automata in the FC2 format. Based on this translation we are integrating parts of the FC2tool set [5]. As a first step we provided deadlock detection based on BDD/implicit representations of states.

7 Conclusion

We briefly presented some of the new features of the PEP tool. For a more detailed overview we refer to [15] and the various papers which are available together with the tool at http://theoretica.informatik.uni-oldenburg.de/~pep.
Acknowledgement: A lot of people (theoreticians and implementors) contributed to the development of the PEP system. Thanks!

References

1. E. Best, R. Devillers, and J.G. Hall. The Box Calculus: a New Causal Algebra with Multi-Label Communication. In *APN'92*, *LNCS* 609, 21–69. Springer, 1992.
2. E. Best, H. Fleischhack, W. Frączak, R.P. Hopkins, H. Klaudel, and E. Pelz. A Class of Composable High Level Petri Nets. In *Proc. of ATPN'95*, *LNCS* 935, 103–118. Springer, 1995.
3. E. Best and B. Grahlmann. *PEP: Documentation and User Guide.* Universität Hildesheim. Available together with the tool via: http://theoretica.informatik.uni-oldenburg.de/~pep.
4. E. Best and R.P. Hopkins. $B(PN)^2$ – a Basic Petri Net Programming Notation. In *Proc. of PARLE*, *LNCS* 694, 379–390, Springer, 1993.
5. A. Bouali, A. Ressouche, V. Roy, and R. de Simone. The FC2tools Set. In *Proc. of CAV'96*, *LNCS* 1102, 441–444, Springer, 1996.
6. CCITT. *Specification and Description Language*, CCITT Z.100, Geneva, 1992.
7. E. Clarke, K. McMillan, S. Campos, and V. Hartonas-Garmhausen. Symbolic Model Checking. In *Proc. of CAV'96*, *LNCS* 1102, 419–422, Springer, 1996.
8. J. Esparza, S. Römer, and W. Vogler. An Improvement of McMillan's Unfolding Algorithm. In *Proc. of TACAS'96*, *LNCS* 1055, 87–106, Springer, 1996.
9. J. Esparza. *Model Checking Using Net Unfoldings.* In Number 23 in *Science of Computer Programming*, 151–195, Elsevier, 1994.
10. H. Fleischhack and B. Grahlmann. A Compositional Petri Net Semantics for SDL. In *Proc. of ATPN'98*, *LNCS 1420*, 144–164, Springer, 1998.
11. B. Grahlmann, M. Moeller, and U. Anhalt. A New Interface for the PEP Tool – Parallel Finite Automata. In *Proc. of AWPN'95*, AIS 22, 21–26. FB 10 Universität Oldenburg, 1995.
12. B. Grahlmann and C. Pohl. Profiting from Spin in PEP. In *Proc. of the SPIN'98 Workshop*, 1998.
13. B. Grahlmann. The Reference Component of PEP. In *Proc. of TACAS'97*, *LNCS 1217*, 65–80, Springer, 1997.

14. B. Grahlmann. Combining Finite Automata, Parallel Programs and SDL using Petri Nets. In *Proc. of TACAS'98, LNCS 1384*, 102–117, Springer, 1998.
15. B. Grahlmann. *Parallel Programs as Petri Nets.* Ph.D. thesis, Universität Hildesheim, 1998.
16. B. Graves. Computing Reachability Properties Hidden in Finite Net Unfoldings. In *Proc. of FST&TCS'97, LNCS 1346*, 327–341, Springer. 1997.
17. J. Jäger. *Portable Codegenerierung für eine parallele Programmiersprache.* Diploma thesis, Universität Hildesheim, 1997.
18. G. J. Holzmann and D. Peled. The State of SPIN. In *Proc. of CAV'96, LNCS 1102*, 385–389. Springer, 1996.
19. M. Kater. SimPEP: 3D-Visualisierung und Animation paralleler Prozesse. Diploma thesis, Universität Hildesheim, 1998.
20. S. Melzer and J. Esparza. Checking System Properties via Integer Programming. In *Proc. of ESOP'96, LNCS 1058*, 250–264, Springer, 1996.
21. S. Melzer and S. Römer. Deadlock Checking using Net Unfoldings. In *Proc. of CAV'97, LNCS 1254*, 352–363, Springer, 1997.
22. S. Roch and P. H. Starke. INA: Integrated Net Analyzer Version 2.1 Handbuch. Humboldt-Universität zu Berlin, 1998. Available together with the tool via: http://www.informatik.hu-berlin.de/lehrstuehle/automaten/ina.
23. G. Wimmel. A BDD-based Model Checker for the PEP Tool. *Technical Report, University of Newcastle upon Tyne*, 1997. Available at PEP's Web page.

The ABACO System – An Algebraic Based Action COmpiler

Hermano Perrelli de Moura and Luis Carlos de Sousa Menezes

Federal University of Pernambuco - Department of Informatics
hermano@di.ufpe.br, lcsm@di.ufpe.br

Abstract. We show the main features of the ABACO system, a semantic directed compiler generator, that produces object-oriented compilers from action semantics descriptions of programming languages.

1 Introduction

Action semantics [2] is a useful formalism to describe the semantics of programming languages. Its modular structure and formal notation, based on terms of the English language, eases the writing of real programming languages' descriptions. Some examples of real programming languages described in action semantics are: PASCAL [3] and STANDARD ML [5].

Action semantics formal notation is based on an algebraic specification model known as unified algebras [1]. This model defines entities called sorts and operations that manipulate them.

An action semantic description for a programming language is divided in the following modules:

- Abstract Syntax: describes the abstract syntax for the programming language.
- Semantic Functions: describes a mapping from the abstract syntax tree (AST) of a program to its meaning. The meaning of a program uses action notation, the formal notation used in action semantics, to describe the program's semantics.
- Semantic Entities: defines the data types used by the language, and auxiliary sorts and operators used in the description.

2 The ABACO System

We built an automatic compiler generation tool named ABACO (Algebraic Based Action COmpiler). The system combines object-orientation and action semantics to produce implementations of programming languages from their action semantic descriptions. The system is composed by the following tools:

- A generic *unified algebras translator*: this tool accepts unified algebras specifications and produces an object-oriented library that implements the given

A.M. Haeberer (Ed.): AMAST'98, LNCS 1548, pp. 527–529, 1998.

specification. The choice of object-oriented languages as target language is based on the similarities between the existing concepts in both models, for example the concepts of sort inclusion (in unified algebras) and class inheritance (in object-orientation).
- A *parser generator*: this tool accepts an action semantic description of a programming language and builds a description that can be used with parsers generators like Lex and Yacc to produce the parser for the specified language.
- An *action compiler*: generates a C++ code which implements an action that represents the meaning of a program in the described language (program action).

The process of building a compiler for a described programming language has the following steps, showed graphicaly in Figure 1:

1. The definition of the abstract syntax of the programming language is processed by the *parser generator* that will generate the parser. The generated parser recognizes the source programs and produces their abstract syntax trees (AST).
2. The programming language description is processed by the *unified algebras translator* that will produce a library, named dynamic library, that will be able to give the meaning of a program, represented as an AST and produced by the parser, according the action semantics description language's.
3. The programs writen in the specified language will need a library that defines the data types used by the language to be correctly compiled. This library, named static library, is obtained processing the semantics entities of the programming language description with the unified algebras translator.
4. The dynamic library is linked with the *action compiler* to produce the code generator for the specified language. This program is able to produce C++ programs from AST produced by the parser generated in step 1. The C++ programs generated by the code generator can be compiled using a generic C++ compiler.

The generated compiler is formed by the parser, the code generator, the static library and a generic C++ compiler. Its architeture and compilation process is showed in Figure 2.

3 Conclusions

The ABACO system is useful to prototype semantic descriptions of programming languages. His main characteristics is to produce a dynamic implementation for the datatypes of the described language. It enables the generated compiler to represent more accurately the peculiarities of the specified programming language.

A prototipe of this system was implemented using the C++ language and was tested using the GNU C++ compiler (GCC) in the Solaris operating system but it could work in every system supporting the C++ language. The source code of this implementation can be obtained at the Recife Action Tools home page [4].

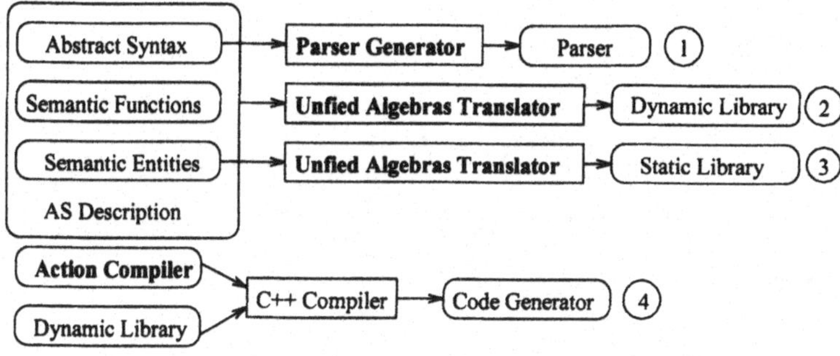

Fig. 1. Compiler Generation Using the ABACO System

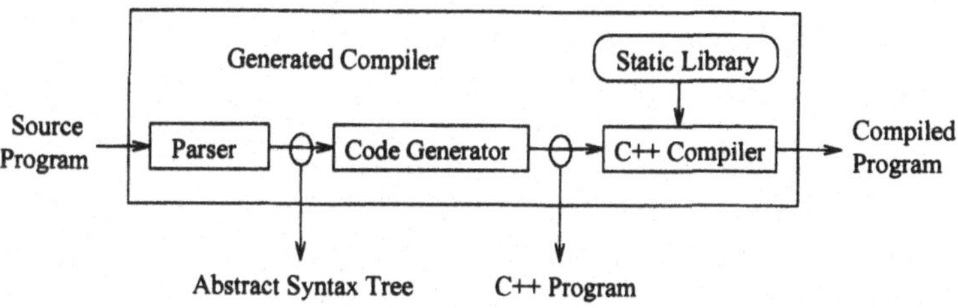

Fig. 2. Architeture of the Generated Compiler

References

[1] P. D. Mosses. Unified algebras and modules. Departamental Report DAIMI PB–266, Aarhus University, Computer Science Department, Denmark, 1988.

[2] Peter D. Mosses. *Action Semantics*. Number 26 in Cambridge Tracts in Theoretical Computer Science. Cambridge University Press, 1992.

[3] Peter D. Mosses and David A. Watt. Pascal action semantics, version 0.6. Available by FTP as ftp.brics.dk/pub/BRICS/Projects/AS/Papers/-MossesWatt93DRAFT/pas-0.6.ps.Z, March 1993.

[4] Homepage of the RAT Project. http://www.di.ufpe.br/~rat.

[5] David A. Watt. An action semantics of Standard ML. In *Proc. Third Workshop on Math. Foundations of Programming Language Semantics, New Orleans*, volume 298 of *Lecture Notes in Computer Science*, pages 572–598. Springer-Verlag, 1988.

Author Index

Lecture Notes in Computer Science

For information about Vols. 1–1463
please contact your bookseller or Springer-Verlag

Vol. 1464: H.H.S. Ip, A.W.M. Smeulders (Eds.), Multimedia Information Analysis and Retrieval. Proceedings, 1998. VIII, 264 pages. 1998.

Vol. 1465: R. Hirschfeld (Ed.), Financial Cryptography. Proceedings, 1998. VIII, 311 pages. 1998.

Vol. 1466: D. Sangiorgi, R. de Simone (Eds.), CONCUR'98: Concurrency Theory. Proceedings, 1998. XI, 657 pages. 1998.

Vol. 1467: C. Clack, K. Hammond, T. Davie (Eds.), Implementation of Functional Languages. Proceedings, 1997. X, 375 pages. 1998.

Vol. 1468: P. Husbands, J.-A. Meyer (Eds.), Evolutionary Robotics. Proceedings, 1998. VIII, 247 pages. 1998.

Vol. 1469: R. Puigjaner, N.N. Savino, B. Serra (Eds.), Computer Performance Evaluation. Proceedings, 1998. XIII, 376 pages. 1998.

Vol. 1470: D. Pritchard, J. Reeve (Eds.), Euro-Par'98: Parallel Processing. Proceedings, 1998. XXII, 1157 pages. 1998.

Vol. 1471: J. Dix, L. Moniz Pereira, T.C. Przymusinski (Eds.), Logic Programming and Knowledge Representation. Proceedings, 1997. IX, 246 pages. 1998. (Subseries LNAI).

Vol. 1472: B. Freitag, H. Decker, M. Kifer, A. Voronkov (Eds.), Transactions and Change in Logic Databases. Proceedings, 1996, 1997. X, 396 pages. 1998.

Vol. 1473: X. Leroy, A. Ohori (Eds.), Types in Compilation. Proceedings, 1998. VIII, 299 pages. 1998.

Vol. 1474: F. Mueller, A. Bestavros (Eds.), Languages, Compilers, and Tools for Embedded Systems. Proceedings, 1998. XIV, 261 pages. 1998.

Vol. 1475: W. Litwin, T. Morzy, G. Vossen (Eds.), Advances in Databases and Information Systems. Proceedings, 1998. XIV, 369 pages. 1998.

Vol. 1476: J. Calmet, J. Plaza (Eds.), Artificial Intelligence and Symbolic Computation. Proceedings, 1998. XI, 309 pages. 1998. (Subseries LNAI).

Vol. 1477: K. Rothermel, F. Hohl (Eds.), Mobile Agents. Proceedings, 1998. VIII, 285 pages. 1998.

Vol. 1478: M. Sipper, D. Mange, A. Pérez-Uribe (Eds.), Evolvable Systems: From Biology to Hardware. Proceedings, 1998. IX, 382 pages. 1998.

Vol. 1479: J. Grundy, M. Newey (Eds.), Theorem Proving in Higher Order Logics. Proceedings, 1998. VIII, 497 pages. 1998.

Vol. 1480: F. Giunchiglia (Ed.), Artificial Intelligence: Methodology, Systems, and Applications. Proceedings, 1998. IX, 502 pages. 1998. (Subseries LNAI).

Vol. 1481: E.V. Munson, C. Nicholas, D. Wood (Eds.), Principles of Digital Document Processing. Proceedings, 1998. VII, 152 pages. 1998.

Vol. 1482: R.W. Hartenstein, A. Keevallik (Eds.), Field-Programmable Logic and Applications. Proceedings, 1998. XI, 533 pages. 1998.

Vol. 1483: T. Plagemann, V. Goebel (Eds.), Interactive Distributed Multimedia Systems and Telecommunication Services. Proceedings, 1998. XV, 326 pages. 1998.

Vol. 1484: H. Coelho (Ed.), Progress in Artificial Intelligence – IBERAMIA 98. Proceedings, 1998. XIII, 421 pages. 1998. (Subseries LNAI).

Vol. 1485: J.-J. Quisquater, Y. Deswarte, C. Meadows, D. Gollmann (Eds.), Computer Security – ESORICS 98. Proceedings, 1998. X, 377 pages. 1998.

Vol. 1486: A.P. Ravn, H. Rischel (Eds.), Formal Techniques in Real-Time and Fault-Tolerant Systems. Proceedings, 1998. VIII, 339 pages. 1998.

Vol. 1487: V. Gruhn (Ed.), Software Process Technology. Proceedings, 1998. VIII, 157 pages. 1998.

Vol. 1488: B. Smyth, P. Cunningham (Eds.), Advances in Case-Based Reasoning. Proceedings, 1998. XI, 482 pages. 1998. (Subseries LNAI).

Vol. 1489: J. Dix, L. Fariñas del Cerro, U. Furbach (Eds.), Logics in Artificial Intelligence. Proceedings, 1998. X, 391 pages. 1998. (Subseries LNAI).

Vol. 1490: C. Palamidessi, H. Glaser, K. Meinke (Eds.), Principles of Declarative Programming. Proceedings, 1998. XI, 497 pages. 1998.

Vol. 1491: W. Reisig, G. Rozenberg (Eds.), Lectures on Petri Nets I: Basic Models. XII, 683 pages. 1998.

Vol. 1492: W. Reisig, G. Rozenberg (Eds.), Lectures on Petri Nets II: Applications. XII, 479 pages. 1998.

Vol. 1493: J.P. Bowen, A. Fett, M.G. Hinchey (Eds.), ZUM '98: The Z Formal Specification Notation. Proceedings, 1998. XV, 417 pages. 1998.

Vol. 1494: G. Rozenberg, F. Vaandrager (Eds.), Lectures on Embedded Systems. Proceedings, 1996. VIII, 423 pages. 1998.

Vol. 1495: T. Andreasen, H. Christiansen, H.L. Larsen (Eds.), Flexible Query Answering Systems. IX, 393 pages. 1998. (Subseries LNAI).

Vol. 1496: W.M. Wells, A. Colchester, S. Delp (Eds.), Medical Image Computing and Computer-Assisted Intervention – MICCAI'98. Proceedings, 1998. XXII, 1256 pages. 1998.

Vol. 1497: V. Alexandrov, J. Dongarra (Eds.), Recent Advances in Parallel Virtual Machine and Message Passing Interface. Proceedings, 1998. XII, 412 pages. 1998.

Vol. 1498: A.E. Eiben, T. Bäck, M. Schoenauer, H.-P. Schwefel (Eds.), Parallel Problem Solving from Nature – PPSN V. Proceedings, 1998. XXIII, 1041 pages. 1998.

Vol. 1499: S. Kutten (Ed.), Distributed Computing. Proceedings, 1998. XII, 419 pages. 1998.

Vol. 1501: M.M. Richter, C.H. Smith, R. Wiehagen, T. Zeugmann (Eds.), Algorithmic Learning Theory. Proceedings, 1998. XI, 439 pages. 1998. (Subseries LNAI).

Vol. 1502: G. Antoniou, J. Slaney (Eds.), Advanced Topics in Artificial Intelligence. Proceedings, 1998. XI, 333 pages. 1998. (Subseries LNAI).

Vol. 1503: G. Levi (Ed.), Static Analysis. Proceedings, 1998. IX, 383 pages. 1998.

Vol. 1504: O. Herzog, A. Günter (Eds.), KI-98: Advances in Artificial Intelligence. Proceedings, 1998. XI, 355 pages. 1998. (Subseries LNAI).

Vol. 1505: D. Caromel, R.R. Oldehoeft, M. Tholburn (Eds.), Computing in Object-Oriented Parallel Environments. Proceedings, 1998. XI, 243 pages. 1998.

Vol. 1506: R. Koch, L. Van Gool (Eds.), 3D Structure from Multiple Images of Large-Scale Environments. Proceedings, 1998. VIII, 347 pages. 1998.

Vol. 1507: T.W. Ling, S. Ram, M.L. Lee (Eds.), Conceptual Modeling – ER '98. Proceedings, 1998. XVI, 482 pages. 1998.

Vol. 1508: S. Jajodia, M.T. Özsu, A. Dogac (Eds.), Advances in Multimedia Information Systems. Proceedings, 1998. VIII, 207 pages. 1998.

Vol. 1510: J.M. Zytkow, M. Quafafou (Eds.), Principles of Data Mining and Knowledge Discovery. Proceedings, 1998. XI, 482 pages. 1998. (Subseries LNAI).

Vol. 1511: D. O'Hallaron (Ed.), Languages, Compilers, and Run-Time Systems for Scalable Computers. Proceedings, 1998. IX, 412 pages. 1998.

Vol. 1512: E. Giménez, C. Paulin-Mohring (Eds.), Types for Proofs and Programs. Proceedings, 1996. VIII, 373 pages. 1998.

Vol. 1513: C. Nikolaou, C. Stephanidis (Eds.), Research and Advanced Technology for Digital Libraries. Proceedings, 1998. XV, 912 pages. 1998.

Vol. 1514: K. Ohta, D. Pei (Eds.), Advances in Cryptology – ASIACRYPT'98. Proceedings, 1998. XII, 436 pages. 1998.

Vol. 1515: F. Moreira de Oliveira (Ed.), Advances in Artificial Intelligence. Proceedings, 1998. X, 259 pages. 1998. (Subseries LNAI).

Vol. 1516: W. Ehrenberger (Ed.), Computer Safety, Reliability and Security. Proceedings, 1998. XVI, 392 pages. 1998.

Vol. 1517: J. Hromkovič, O. Sýkora (Eds.), Graph-Theoretic Concepts in Computer Science. Proceedings, 1998. X, 385 pages. 1998.

Vol. 1518: M. Luby, J. Rolim, M. Serna (Eds.), Randomization and Approximation Techniques in Computer Science. Proceedings, 1998. IX, 385 pages. 1998.

1519: T. Ishida (Ed.), Community Computing and Support Systems. VIII, 393 pages. 1998.

Vol. 1520: M. Maher, J.-F. Puget (Eds.), Principles and Practice of Constraint Programming - CP98. Proceedings, 1998. XI, 482 pages. 1998.

Vol. 1521: B. Rovan (Ed.), SOFSEM'98: Theory and Practice of Informatics. Proceedings, 1998. XI, 453 pages. 1998.

Vol. 1522: G. Gopalakrishnan, P. Windley (Eds.), Formal Methods in Computer-Aided Design. Proceedings, 1998. IX, 529 pages. 1998.

Vol. 1524: G.B. Orr, K.-R. Müller (Eds.), Neural Networks: Tricks of the Trade. VI, 432 pages. 1998.

Vol. 1525: D. Aucsmith (Ed.), Information Hiding. Proceedings, 1998. IX, 369 pages. 1998.

Vol. 1526: M. Broy, B. Rumpe (Eds.), Requirements Targeting Software and Systems Engineering. Proceedings, 1997. VIII, 357 pages. 1998.

Vol. 1528: B. Preneel, V. Rijmen (Eds.), State of the Art in Applied Cryptography. Revised Lectures, 1997. VIII, 395 pages. 1998.

Vol. 1529: D. Farwell, L. Gerber, E. Hovy (Eds.), Machine Translation and the Information Soup. Proceedings, 1998. XIX, 532 pages. 1998. (Subseries LNAI).

Vol. 1530: V. Arvind, R. Ramanujam (Eds.), Foundations of Software Technology and Theoretical Computer Science. XII, 369 pages. 1998.

Vol. 1531: H.-Y. Lee, H. Motoda (Eds.), PRICAI'98: Topics in Artificial Intelligence. XIX, 646 pages. 1998. (Subseries LNAI).

Vol. 1096: T. Schael, Workflow Management Systems for Process Organisations. Second Edition. XII, 229 pages 1998.

Vol. 1532: S. Arikawa, H. Motoda (Eds.), Discovery Sci ence. Proceedings, 1998. XI, 456 pages. 1998. (Subserie LNAI).

Vol. 1533: K.-Y. Chwa, O.H. Ibarra (Eds.), Algorithm and Computation. Proceedings, 1998. XIII, 478 pages 1998.

Vol. 1538: J. Hsiang, A. Ohori (Eds.), Advances in Com puting Science – ASIAN'98. Proceedings, 1998. X, 30 pages. 1998.

Vol. 1540: C. Beeri, P. Buneman (Eds.), Database Theor – ICDT'99. Proceedings, 1999. XI, 489 pages. 1999.

Vol. 1541: B. Kågström, J. Dongarra, E. Elmroth, Waśniewski (Eds.), Applied Parallel Computing. Procee ings, 1998. XIV, 586 pages. 1998.

Vol. 1542: H.I. Christensen (Ed.), Computer Vision Sy tems. Proceedings, 1999. XI, 554 pages. 1999.

Vol. 1543: S. Demeyer, J. Bosch (Eds.), Object-Orient Technology ECOOP'98 Workshop Reader. 1998. XX 573 pages. 1998.

Vol. 1544: C. Zhang, D. Lukose (Eds.), Multi-Agent Sy tems. Proceedings, 1998. VII, 195 pages. 1998. (Subser LNAI).

Vol. 1546: B. Möller, J.V. Tucker (Eds.), Prospects Hardware Foundations. Survey Chapters, 1998. X, pages. 1998.

Vol. 1548: A.M. Haeberer (Ed.), Algebraic Methodol and Software Technology. Proceedings, 1999. XI, pages. 1999.